Fundamentals of Music Processing

Meinard Müller

Fundamentals of Music Processing

Using Python and Jupyter Notebooks

Second Edition

Meinard Müller
International Audio Laboratories Erlangen
Erlangen, Germany

ISBN 978-3-030-69810-2 ISBN 978-3-030-69808-9 (eBook)
https://doi.org/10.1007/978-3-030-69808-9

This Springer imprint is published by the registered company Springer Nature Switzerland AG
The registered company address is: Gewerbestrasse 11, 6330 Cham, Switzerland

To Michael Clausen and Hans-Peter Seidel

Preface to the Second Edition

When writing the first edition of this book, my motivation was to provide a textbook on the emerging fields of music processing and music information retrieval (MIR) with a focus on audio signal processing. Using well-established music analysis and retrieval topics as motivating application scenarios, the book introduces fundamental techniques and algorithms relevant for general courses in various fields, including computer science, multimedia engineering, information science, and digital humanities. The book is intended for Master and advanced Bachelor students in these fields as well as any reader interested in delving into the field of music processing (and not being frightened by some mathematics). While providing profound technological knowledge as well as a comprehensive treatment of music processing applications, the book also includes numerous examples and illustrations to convey the main ideas in an intuitive fashion. In recent years, suitably designed software packages and freely accessible web-based frameworks have made education in computer science and signal processing more interactive. Such novel technology allows for designing courses that aid students in moving from recalling and reciting theoretical concepts towards comprehension and application.

These new developments are precisely the motivation for the second edition of this book. It extends the first edition by providing additional material (called **FMP Notebooks**), yielding an interactive foundation for teaching and learning fundamentals of music processing (FMP). The FMP notebooks are built upon the Jupyter notebook framework, which has become a standard in educational settings. This open-source web application allows users to create documents that contain executable code, text-based information, mathematical formulas, plots, images, sound examples, and videos. By leveraging the Jupyter framework, the FMP notebooks bridge the gap between theory and practice by interleaving technical concepts, mathematical details, code examples, illustrations, and sound examples within a unifying setting. The FMP notebooks closely follow the eight chapters of the textbook and, as such, provide an explicit link between structured educational environments and current professional practices, in line with current curricular recommendations for computer science.

One primary purpose of the FMP notebooks is to provide audio-visual material and Python code examples that implement the computational approaches step by step. Additionally, the FMP notebooks yield an interactive framework that allows students to experiment with their music examples, explore the effect of parameter settings, and understand the computed results by suitable visualizations and sonifications. When teaching and learning music processing, it is essential to have a

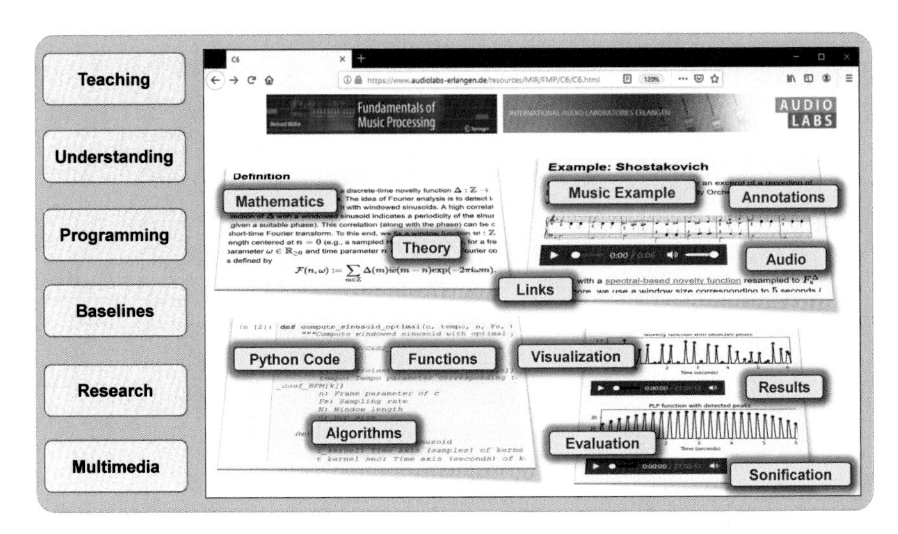

Overview of various components and didactical aspects of the FMP notebooks.

holistic view of the MIR task at hand, the algorithmic approach, and its practical implementation. Looking at all the steps of the processing pipeline sheds light on the input data and its biases, possible violations of model assumptions, and the shortcomings of quantitative evaluation measures. Only by an interactive examination of all these aspects will students acquire a deeper understanding of the concepts, transitioning from merely understanding concepts to applying their music processing approaches both conceptually and in code.

The main body of the FMP notebooks consists of eight parts, structured along with the eight chapters of this textbook. In the book's second edition, we provide at the end of each chapter an additional section titled **FMP Notebooks**. These sections serve two purposes. First, we give a comprehensive guide by systematically describing the content and purpose of all the notebooks related to the corresponding chapter. As a second objective, we make concrete suggestions on using the FMP notebooks to create an enriching, interactive, and interdisciplinary supplement in the form of experiments and advanced studies in a music processing curriculum. The textbook's guide can be best appreciated and understood when the FMP notebooks run in a browser simultaneously while reading.

The FMP notebooks are publicly available under a Creative Commons license at `https://www.audiolabs-erlangen.de/FMP` in the form of Jupyter notebooks as well as HTML exports, which can be accessed through a conventional web browser. Using the static HTML version, all multimedia material, including the music examples, audio files, video files, and images, can be directly accessed without any specific technical requirements beyond a standard web browser. To run the FMP notebooks' code, one needs to install Python, Jupyter, and additional Python packages. All necessary steps for installing, running, and updating the required software packages are described in a separate part (called **Part B**) of the FMP notebooks. This part also contains short introductions to Python programming, Jupyter notebooks, multimedia integration, as well as data annotation, visualization, and sonification. Rather than being comprehensive, **Part B** gives instructive code examples that become relevant in the other parts and documents how the FMP notebooks were created.

Besides its substantial extensions through the FMP notebooks, another major change in the second edition is the thorough revision of the sections called **Summary and Further Readings** (previously called **Further Notes** in the first edition). These sections have been streamlined, now containing more compact and focused summaries. Furthermore, the references and the links to literature for further readings have been revised and updated. Rather than providing an extensive literature review, we have deliberately limited ourselves to citing only selected core literature and overview articles, where one can find further pointers to relevant and more advanced work. As in general multimedia processing, many recent advances in music processing have been driven by techniques based on deep learning (DL). For example, DL-based techniques have led to significant improvements for many tasks such as music source separation, music transcription, chord recognition, melody estimation, beat tracking, tempo estimation, and lyrics alignment, to name a few. In particular, major improvements could be achieved for music scenarios where sufficient training data is available. A particular strength of DL-based approaches is their ability to extract complex features directly from raw audio data, which can then be used to make predictions based on hidden structures and relations. Furthermore, powerful software packages allow for easily designing, implementing, and experimenting with machine learning algorithms based on deep neural networks (DNNs). Covering the fast-growing and dynamic field of deep learning goes beyond the scope of this textbook. Instead, we focus on classical signal and music processing techniques, yielding fundamental insights into the problem at hand and providing explicit baseline approaches one may (and should) compare against when exploring more powerful yet often difficult-to-interpret DNN-based learning approaches. For further readings, we provided links to selected references that apply recent DNN-based techniques to music processing. We hope that these references help students and researchers transition from model-based approaches as introduced in this textbook to the world of deep learning applied to specific music processing tasks. Our literature choice is undoubtedly subjective, and we would like to apologize to all those whose work we have not mentioned or adequately appreciated.

I want to thank Springer-Verlag for the opportunity afforded by the preparation of this second edition. The initial idea of extending and complementing the textbook by the FMP notebooks arose in 2018 during my visit to the Center for Computer Research in Music and Acoustics (CCRMA) at Stanford University, where I was invited to give a summer workshop on MIR jointly with Steve Tjoa and Brian McFee. In particular, I started learning Python myself using the Python package `librosa`, which offers advanced music and audio processing pipelines. Many thanks, Brian, for developing this fantastic software package, which is not only extensively used in the FMP notebooks but has also been a source of inspiration. Furthermore, I would like to express my gratitude to Frank Zalkow, who has helped me in the last two years with all the technical aspects related to the FMP notebooks. Frank, you have been my patient teacher and companion when struggling through the pitfalls of Python programming. Many more people have helped me in creating the FMP notebooks, and I will confine myself to only mentioning their names in alphabetical order: Vlora Arifi-Müller, Stefan Balke, Rachel Bittner, Eran Egozy, Katherine Kinnaird, Michael Krause, Patricio López-Serrano, Brian McFee, Sebastian Rosenzweig, Bob Sturm, Steve Tjoa, Angel Villar-Corrales, Christof Weiß, Frank Zalkow, and Tim Zunner. Of course, in the second edition of this book, I also corrected the errors that have come to my attention, and I express my thanks to those colleagues and students who pointed out these errors in the first edition. I hope that this book will continue to serve as a basis for those interested in learning music processing and undertaking research in this, to my view, beautiful and challenging field.

Erlangen, *Meinard Müller*
January 2021

Preface to the First Edition

Music is a ubiquitous and vital part of the lives of billions of people worldwide. Musical creations and performances are amongst the most complex and intricate of our cultural artifacts, and the emotional power of music can touch us in surprising and profound ways. Music spans an enormous range of forms and styles, from simple, unaccompanied folk songs, to popular and jazz music, to symphonies for full orchestras. The digital revolution in music distribution and storage has simultaneously fueled tremendous interest in and attention to the ways that information technology can be applied to this kind of content. From browsing personal collections, to discovering new artists, to managing and protecting the rights of music creators, computers are now deeply involved in almost every aspect of music consumption, not to mention their vital role in much of today's music production.

Despite the importance of music, *music processing* is still a relatively young discipline compared with speech processing, a research field with a long tradition. A research community represented by the International Society for Music Information Retrieval (ISMIR), which systematically deals with a wide range of computer-based music analysis, processing, and retrieval topics, was formed in the year 2000. Traditionally, computer-based music research has mostly been conducted on the basis of symbolic representations using music notation or MIDI representations. Because of the increasing availability of digitized audio material and an explosion of computing power, automated processing of waveform-based audio signals is now increasingly in the focus of research efforts.

Many of these research efforts are directed towards the development of technologies that allow users to access and explore music in all its different facets. For example, audio fingerprinting techniques are nowadays integrated into commercial products that help users automatically identify songs they hear. Music processing techniques are used in extended audio players that highlight the current measures within sheet music while playing back a recording of a symphony. On demand, additional information about melodic and harmonic progressions or rhythm and tempo is automatically presented to the listener. Interactive music interfaces display structural parts of the current piece of music and allow users to directly jump to any section such as the chorus, the main musical theme, or a solo section without tedious fast-forwarding and rewinding. Furthermore, listeners are equipped with Google-like search engines that enable them to explore large music collections in various ways. For example, the user may create a query by specifying a certain note constellation, or some harmonic or rhythmic pattern by whistling a melody or tapping a rhythm, or simply by selecting a short passage from an audio recording; the system then provides the user with a ranked list of available music excerpts from the collection that are musically related to the query. In music processing, one main objective is to contribute concepts, models, algorithms, implementations, and evaluations for tackling such types of analysis and retrieval problems.

This textbook is devoted to the emerging fields of music processing and music information retrieval (MIR)—interdisciplinary research areas which are related to various disciplines including signal processing, information retrieval, machine learning, multimedia engineering, library science, musicology, and digital humanities. The main goal of this book is to give an introduction to this vibrant and exciting new research area for a wide readership. Well-established topics in music analysis and retrieval have been selected to serve as motivating application scenarios. Within these scenarios, fundamental techniques and algorithms that are applicable to a wide range of analysis and retrieval problems are presented in depth.

This book is meant to be a *textbook* that is suitable for courses at the advanced undergraduate and beginning master level. By mixing theory and practice, the book provides both deep technological knowledge as well as a comprehensive treatment of music processing applications. Furthermore, by including numerous examples, illustrations (the book contains more than 300 figures), and exercises, I hope that the book provides interesting material for courses in various fields such as computer science, multimedia engineering, information science, and digital humanities.

The subsequent sections of this preface contain further information on the overall structure of the book, the interconnections between the various topics and techniques, and suggestions on how this book may be used as a basis for different courses. We first give an overview of the book's content by quickly going through the individual chapters. Then, we explain various ways of reading and using the book, each time focusing on a different aspect. We start with the view of a lecturer who wants to use this textbook as a basis for an introductory course in music processing or music information retrieval. Then, we show how the book may be used for an introductory course on Fourier analysis and its applications. Finally, we assume the view of a computer scientist who wants to teach fundamental issues on data representations and algorithms, where music may serve as an underlying application domain. Describing these different views, we try to work out the dependencies between the chapters as well as the conceptual relationships between the various music processing tasks.

Content

This textbook consists of eight chapters. The first two chapters cover fundamental material on music representations and the Fourier transform—concepts that are required throughout the book. These two chapters make the book self-contained to a great extent. In the subsequent chapters, concrete music processing tasks serve as starting points for our investigations. Each of these chapters is organized in a similar fashion. A chapter starts with a general description of the music processing scenario

Chapter		Music Processing Scenario	Notions, Techniques & Algorithms
1		**Music Representations**	Music notation, MIDI, audio signal, waveform, pitch, loudness, timbre
2		**Fourier Analysis of Signals**	Discrete/analog signal, sinusoid, exponential, Fourier transform, Fourier representation, DFT, FFT, STFT
3		**Music Synchronization**	Chroma feature, dyamic programming, dyamic time warping (DTW), alignment, user interface
4		**Music Structure Analysis**	Similarity matrix, repetition, thumbnail, homogeneity, novelty, evaluation, precision, recall, F-measure, visualization, scape plot
5		**Chord Recognition**	Harmony, music theory, chords, scales, templates, hidden Markov model (HMM), evaluation
6		**Tempo and Beat Tracking**	Onset, novelty, tempo, tempogram, beat, periodicity, Fourier analysis, autocorrelation
7		**Content-Based Audio Retrieval**	Identification, fingerprint, indexing, inverted list, matching, version, cover song
8		**Musically Informed Audio Decomposition**	Harmonic/percussive component, signal reconstruction, instanteneous frequency, fundamental frequency (F0), trajectory, nonnegative matrix factorization (NMF)

at hand and integrates the topic into a wider context. Motivated by the scenario at hand, each chapter discusses important techniques and algorithms that are generally applicable to a wide range of analysis, classification, and retrieval problems. All these techniques are treated in a mathematically rigorous way. At the same time, the techniques are immediately applied to a concrete music processing task. By mixing theory and practice, the book's goal is to convey both profound technological knowledge as well as a solid understanding of music processing applications. Each of the chapters ends with a section that includes links to the research literature, hints for further reading, a list of references, and exercises. Before we discuss how this textbook may be employed in a course or used for self-study, we first give an overview of the individual chapters and the main topics.

Musical information can be represented in many different ways. In **Chapter 1**, we consider three widely used music representations: sheet music, symbolic, and audio representations. This first chapter also introduces basic terminology that is used throughout the book. In particular, we discuss musical and acoustic properties of audio signals including aspects such as frequency, pitch, dynamics, and timbre.

Important technical terminology is covered in **Chapter 2**. In particular, we approach the Fourier transform—which is perhaps the most fundamental tool in signal processing—from various perspectives. For the reader who is more interested in the musical aspects of the book, Section 2.1 provides a summary of the most important facts on the Fourier transform. In particular, the notion of a spectrogram, which yields a time–frequency representation of an audio signal, is introduced. The remainder of the chapter treats the Fourier transform in greater mathematical depth and also includes the fast Fourier transform (FFT)—an algorithm of great beauty and high practical relevance.

As a first music processing task, we study in **Chapter 3** the problem of music synchronization. The objective is to temporally align compatible representations of the same piece of music. Considering this scenario, we explain the need for musically informed audio features. In particular, we introduce the concept of chroma-based music features, which capture properties that are related to harmony and melody. Furthermore, we study an alignment technique known as dynamic time warping (DTW), a concept that is applicable for the analysis of general time series. For its efficient computation, we discuss an algorithm based on dynamic programming—a widely used method for solving a complex problem by breaking it down into a collection of simpler subproblems.

In **Chapter 4**, we address a central and well-researched area within MIR known as music structure analysis. Given a music recording, the objective is to identify important structural elements and to temporally segment the recording according to these elements. Within this scenario, we discuss fundamental segmentation principles based on repetitions, homogeneity, and novelty—principles that also apply to other types of multimedia beyond music. As an important technical tool, we study in detail the concept of self-similarity matrices and discuss their structural properties. Finally, we briefly touch the topic of evaluation, introducing the notions of precision, recall, and F-measure. These measures are used to compare the computed results that are obtained by an automated procedure with so-called ground truth annotations that are typically generated manually by some domain expert.

In **Chapter 5**, we consider the problem of analyzing harmonic properties of a piece of music by determining a descriptive progression of chords from a given audio recording. We take this opportunity to first discuss some basic theory of harmony including concepts such as intervals, chords, and scales. Then, motivated by the automated chord recognition scenario, we introduce template-based matching procedures and hidden Markov models—a concept of central importance for the analysis of temporal patterns in time-dependent data streams including speech, gestures, and music.

Tempo and beat are further fundamental properties of music. In **Chapter 6**, we introduce the basic ideas on how to extract tempo-related information from audio recordings. In this scenario, a first challenge is to locate note onset information—a task that requires methods for detecting changes in energy and spectral content. To derive tempo and beat information, note onset candidates are then analyzed with regard to quasiperiodic patterns. This leads us to the study of general methods for local periodicity analysis of time series.

One important topic in information retrieval is concerned with the development of search engines that enable users to explore music collections in a flexible and intuitive way. In **Chapter 7**, we discuss audio retrieval strategies that follow the query-by-example paradigm: given an audio query, the task is to retrieve all documents that are somehow similar or related to the query. Starting with audio identification, a technique used in many commercial applications such as *Shazam*, we study various retrieval strategies to handle different degrees of similarity. Furthermore, considering efficiency issues, we discuss fundamental indexing techniques based on inverted lists—a concept originally used in text retrieval.

In the final **Chapter 8** on audio decomposition, we present a challenging research direction that is closely related to source separation. Within this wide research area, we consider three subproblems: harmonic–percussive separation, main melody extraction, and score-informed audio decomposition. Within these scenarios, we discuss a number of key techniques including instantaneous frequency estimation, fundamental frequency (F0) estimation, spectrogram inversion, and nonnegative matrix factorization (NMF). Furthermore, we encounter a number of acoustic and musical properties of audio recordings that have been introduced and discussed in previous chapters, which rounds off the book.

Target Readership

In the last fifteen years, music processing and music information retrieval (MIR) have developed into a vibrant and multidisciplinary area of research. Because of the diversity and richness of music, this area brings together researchers and students from a multitude of fields including information science, audio engineering, computer science, and musicology. This book's intention is to offer interesting material for courses in these fields. The main target groups of this book are Master and advanced Bachelor students. Furthermore, we also hope that researchers who are interested in delving into the field of music processing will benefit from this textbook. The eight chapters are organized in a modular fashion, thus offering lecturers and readers many ways to choose, rearrange, or supplement the material. In this way, it should be possible to easily integrate selected chapters or individual sections into courses that are related to general multimedia, information science, signal processing, music informatics, or digital humanities.

Of course, writing a textbook requires making some choices. The topics selected for this textbook play an important role in music processing and MIR, but they also reflect the research areas of the author—I want to apologize to my colleagues for having ignored many other important topics. The focus of this textbook is not to give a comprehensive overview of music processing, but to provide a solid understanding of the concepts introduced within a small number of important application scenarios. The layout, the tempo of presentation, and the pattern of figures have been kept consistent throughout the textbook. We hope that this helps lecturers and students to quickly get comfortable with the style of presentation and to flexibly use the material. In particular, great care has been taken with the illustrations. One way to approach a new topic is to first go through all figures of a section or chapter. Not only should this hone one's intuition, but also yield a first visual overview of the concepts to be studied.

In the following, we describe the dependencies between the chapters and sections by assuming different views on the book. Each view focuses on different aspects and may serve as a basis for designing a one-semester or even two-semester course (with two to four hours weekly per semester plus exercises). Even though the views are presented from the perspective of a lecturer, we hope that they are also helpful for a student or reader to gain a comprehensive overview and a better understanding of the crosslinks between sections and chapters. A more abstract goal of describing the different views is to highlight the general applicability of the presented techniques and the conceptual relationships between the various music processing tasks.

View: A First Course in Music Processing

We start with the view of a lecturer who wants to use this textbook as a basis for an introductory course in music processing or music information retrieval. To lay the foundation for such a course and to fix important notions, we recommend to begin with Chapter 1 on music representations. By going through Section 1.1, the student should get an intuitive idea on the various attributes of music such as notes, pitch, chroma, note length, dynamics, or time signature. We also hope that students who are not familiar with Western music notation will benefit from this section by gaining some intuitive understanding—the intricacies of music notation are not required for the subsequent chapters. Section 1.2 contains background information on symbolic representations. As with the sheet music section, an understanding of all details, e.g., concerning the MIDI format or optical music recognition, is not required. These details, however, become important when working with this kind of data in practice. For most tasks and techniques presented in this book, the piano-roll representation (Section 1.2.1) may serve as an intuitive substitute for sheet music or symbolic representations.

The material on audio representations (Section 1.3) is fundamental for a music processing course based on this book. Many notions such as waveform, sinusoid, frequency, phase, pitch, harmonic, partial, decibel, timbre, transient, or spectrogram are introduced in a more informal way—concepts that will be revisited in the subsequent chapters in more detail.

To make this textbook self-contained and accessible to a wide audience, the required tools from signal processing have been confined to a small number of key

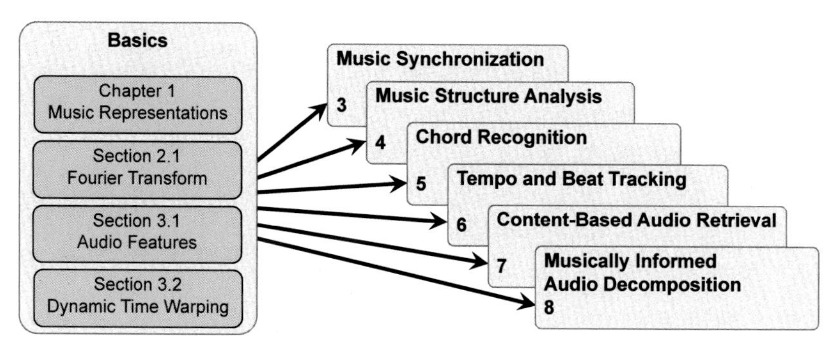

View: A First Course in Music Processing

techniques. Basically all audio processing steps as presented in this book are derived from standard Fourier analysis. The Fourier transform becomes our main signal processing tool, and a good understanding of this transform is indispensable. In Section 2.1, the most important facts on Fourier analysis are introduced in a mathematically rigorous, yet compact fashion. Omitting the proofs, this section aims to convey the main ideas (using many illustrations and examples), while introducing the required technical notions. This section contains all material that is required to understand the subsequent chapters. For a course with a focus on music processing, we recommend to skip the remaining sections of Chapter 2 (and to come back to them at a later stage if required). However, Section 2.1 should be covered in detail.

Motivated by the music synchronization application, Chapter 3 introduces further basic concepts that run like a thread through this book. To make music data comparable and algorithmically accessible, the first step in most music processing tasks is to convert the data into suitable feature representations that capture the relevant aspects while suppressing irrelevant details. In Section 3.1, we address the issue of converting an audio signal into musically informed feature representations. As our main example, we discuss the construction of time–chroma representations, which are based on the equal-tempered scale. Besides music synchronization, these features play an important role in many other applications including music structure analysis (Chapter 4), chord recognition (Chapter 5), and content-based audio retrieval (Chapter 7).

The second important concept introduced in Chapter 3 is known as sequence alignment—a general technique for arranging two time-dependent sequences to identify regions of similarity. To compute an optimal alignment, there are efficient algorithms that are based on dynamic programming—a general paradigm for solving a complex problem by breaking it down into a collection of simpler subproblems. In Section 3.2, we study an alignment technique referred to as dynamic time warping (DTW) as well as an efficient algorithm. In later chapters, we encounter similar alignment techniques, e.g., in the context of audio thumbnailing (Section 4.3), chord recognition (Section 5.3), beat tracking (Section 6.3), audio matching (Section 7.2), and version identification (Section 7.3).

While we recommend covering the fundamental material presented in Chapter 1, Section 2.1, Section 3.1, and Section 3.2 in a course on music processing, there is a lot of freedom on how to proceed afterwards. The remaining chapters are kept mostly independent, excluding a few exceptions that are suitably referenced. One possible continuation of a course is to cover the applications of music synchronization (Section 3.3) and then to proceed with Chapter 4 on music structure analysis. As opposed to music synchronization, where one compares two given sequences, in music structure analysis a single sequence is compared with itself. This leads to the notion of self-similarity matrices—a concept that is related to recurrence plots as used for the analysis of general time series. The study of self-similarity matrices yields deep insights into structural properties of music representations as well as into the properties of the underlying feature representations. By suitably visualizing self-similarity matrices, these aspects can be conveyed in a nontechnical and intuitive fashion. On the other hand, the automated extraction of musically relevant structures from self-similarity matrices—even if they seem obvious for humans—is anything but a trivial problem. In Chapter 4, various challenges as well as algorithmic approaches are presented.

As an alternative, after having introduced chroma-based audio features (Section 3.1), one may directly jump to Chapter 5. The task of automated chord recognition yields a natural motivation for this type of feature. The reason is that chroma features capture a signal's short-time tonal content, which is closely correlated to the harmonic progression of the underlying piece. For a more musically oriented course, Section 5.1 provides some background material on harmony theory including concepts such as intervals, chords, and scales. In a more technically oriented course, most of this material may be skipped. One can then directly proceed with the classification approaches based on templates (Section 5.2) and hidden Markov models (Section 5.3). In view of their great importance, Section 5.3 provides a detailed technical account on Markov chains and hidden Markov models using chord recognition as a motivating application. In particular, the Viterbi algorithm (Section 5.3.3.2) and its close relation to the DTW algorithm (Section 3.2) can be elaborated in a lecture and in homework problems.

Being of high practical relevance and widely known by smartphone users, the topic of audio identification (Section 7.1) is well suited to delve into the topic of content-based audio retrieval. Only requiring the spectrogram representation as prerequisite, this section may be covered directly after Section 2.1. Furthermore, the audio identification application provides a good opportunity for raising efficiency and indexing issues—a topic that is often neglected in music processing and MIR. The next two sections on audio matching (Section 7.2) and version identification (Section 7.3) deal with retrieval scenarios of lower specificity, where the query and the documents to be retrieved may reveal only a low degree of similarity. Requiring chroma-based audio features and alignment techniques, Section 7.2 and Section 7.3 form a nice continuation of Chapter 3 and Chapter 4.

Along with Section 7.1, Chapter 6 and Chapter 8 focus more on technical aspects. Requiring Fourier analysis of audio signals, this material may be used after covering Section 1.3 and Section 2.1. In Chapter 6, which deals with tempo and beat tracking, the Fourier transform is used on two different levels. On the first level, it is used to convert an audio signal into a novelty representation that indicates note onset candidates (Section 6.1). On the second level, Fourier analysis is applied as a means to detect locally periodic patterns in the novelty function. This type of periodicity analysis not only yields a tempogram representation (Section 6.2.2), but also reveals locally periodic pulse trains that can be used for beat tracking applications (Section 6.3.1). Having a close personal relation to rhythm and dance, many students are immediately receptive to the topic of beat and tempo tracking. Therefore, also in my experience as a lecturer, this topic generates a lot of interest and inspiration.

As said before, Chapter 8 is also quite independent from previous chapters and can be studied after Section 1.3 and Section 2.1. The topic of harmonic–percussive separation (Section 8.1) is a direct application of the spectrogram representation. Applying some simple median filtering and binary masking techniques allows for decomposing a music signal into a percussive component and a harmonic component. In this context, we also cover the issue of reconstructing time-domain signals from modified spectral representations—a topic that is fraught with unanticipated pitfalls (Section 8.1.2). Using melody extraction as a motivating music processing application, Section 8.2 details further important topics including fundamental and instantaneous frequency estimation. This scenario provides the opportunity to have a closer look at the phase information supplied by Fourier analysis—a rather technical yet important topic that is not easy to understand when studied for the first time (Section 8.2.1).

In Section 8.3, we touch on another central research field related to source separation. Within this area, a general concept known as nonnegative matrix factorization (NMF) has turned out to be a key technique. Among its many variants, we discuss the most basic NMF version in Section 8.3.1. This technique is then employed for decomposing a music signal into more elementary sound events. Doing so, one can highlight another general strategy that is widely applied in music processing to cope with the complexity of music signals. In order to make certain problems tractable, current approaches often exploit musical knowledge in one way or another. In this chapter, we study several score-informed approaches that make use of the availability of score representations in order to support an audio processing task. This strategy, in turn, requires note information aligned to the audio signal to be processed, which brings us back to Chapter 3 on music synchronization.

View: Introduction to Fourier Analysis and Applications

As said before, the Fourier transform is one of the most important tools for a wide range of applications in engineering and computer science. Due to a large number of variants and the complex-valued formulation, students often have difficulties in understanding the Fourier transform when encountering this concept for the first time. The music domain offers a natural access to the main ideas of Fourier analy-

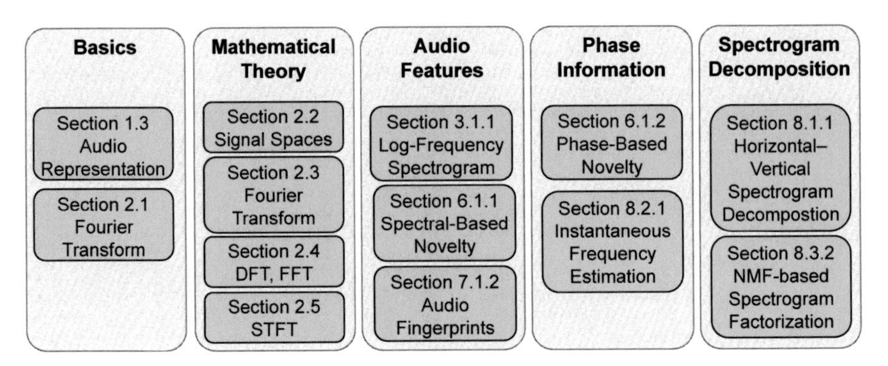

View: Introduction to Fourier Analysis and Applications

sis thanks to intuitive relations between abstract concepts and musical counterparts such as sinusoids and musical tones, frequency and pitch, magnitude and tone intensity, and so on. This textbook can be used as a basis for an introductory course on Fourier analysis. Starting with some basics on audio representations and their properties (Section 1.3), one can continue with Section 2.1 to introduce the most important facts on Fourier analysis. This section contains all material that is actually needed to understand the subsequent chapters. For an in-depth treatment of signals, signal spaces, and Fourier analysis—including many of the mathematical proofs—one may proceed with the remaining sections of Chapter 2. One algorithmic highlight is definitely the fast Fourier transform (FFT), which is treated in Section 2.4.3.

As example applications of the Fourier transform and its short-time versions (STFT, spectrogram), one can then discuss log-frequency spectrograms and their relation to musical pitch (Section 3.1.1), spectrum-based novelty detection as used in note onset detection (Section 6.1.2), and spectral peak fingerprints applied to audio identification (Section 7.1). Using the many concrete examples and illustrations provided by the book, these applications can be treated in a nontechnical fashion without needing to go through all the material of the respective chapter.

Considering only the magnitude information, the phases of the complex-valued Fourier coefficients are often neglected in many applications. With Section 6.1.3 and Section 8.2.1, the book offers material to illustrate the importance of the phase and to approach this difficult topic. Using phase-based novelty detection and instantaneous frequency estimation as motivating applications, the meaning of phase becomes evident when considering possible phase inconsistencies over subsequent frames. These applications also put the STFT and its properties in a different light.

To round off an introductory course on Fourier analysis, one may look into how to decompose time–frequency representations with applications to source separation. In particular, the decomposition of audio signals into harmonic and percussive components by considering horizontal and vertical time–frequency patterns is a simple and very instructive application (Section 8.1.1). This scenario also offers a nice motivation for discussing important topics such as binary and soft spectral masking (Section 8.1.1.2), as well as Fourier inversion and signal reconstruction (Section 8.1.2). Finally, as another more advanced application, one may consider Section 8.3 on audio decomposition using a technique known as nonnegative matrix factorization (NMF). In this application, a music signal is decomposed into a set of notewise audio events, where each audio event is directly associated with a note of a given musical score.

View: Data Representations and Algorithms

We finally want to assume the view of a computer scientist who may be interested in making his or her basic course on data representations and algorithms a bit more "musical." As a multimedia domain, music offers a wide range of data types and formats including text, symbolic data, audio, image, and video. For example, as discussed in Chapter 1, music can be represented as printed sheet music (image domain), encoded as MIDI or MusicXML files (symbolic domain), and played back as audio recordings (acoustic domain). Using music as an example, one can discuss fundamental issues of data representations including bitmap and vector graphic encodings for images, XML-like markup languages for symbolic music, communication protocols for electronic musical instruments such as MIDI, or audio file formats including WAV or MP3. The immediate relationships between different music representations yield a natural motivation for data conversion issues including image rendering, optical character/music recognition, sound synthesis, and so on (see Figure 1.24).

The first step in most computer-based analysis and classification applications consists in transforming the input data into suitable feature representations, which capture relevant information while suppressing redundancies. The spectrogram representation (Section 2.1) and the derived audio features (Section 3.1) can be seen as typical examples for such a transformation process. In many cases, feature extraction can be seen as a kind of dimensionality reduction. A prominent example are the twelve-dimensional chroma features, which capture tonal information of a music signal (Section 3.1.2).

After introducing data representations, a computer science course may continue with the discussion of algorithms. This textbook offers a number of interesting algorithms that are relevant for a wide range of applications going far beyond the music processing scenarios considered. Many of these algorithms are based on dynamic programming, which is a fundamental algorithmic paradigm for solving optimization problems. This method appears—in one form or another—in the curriculum of basically any computer science student. The idea of dynamic programming is to break down a complex problem into smaller "overlapping" subproblems in some recursive manner. An optimal solution of the global problem is obtained by efficiently assembling optimal solutions for the subproblems. Dynamic programming is widely used for alignment tasks as occurring in bioinformatics (e.g., to determine the similarity of DNA sequences) or in text processing (e.g., to compute the distance

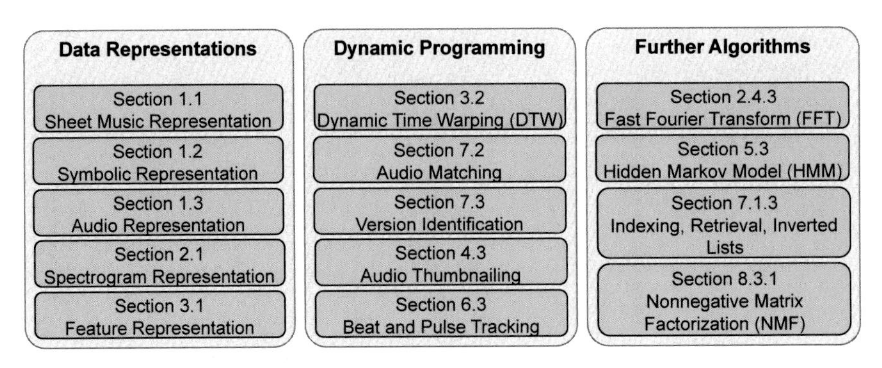

View: Data Representations and Algorithms

between text strings). In this book, we consider a variant of this technique referred to as dynamic time warping (DTW), which allows us to temporally align feature sequences extracted from music representations. Motivated by a music synchronization application, Section 3.2 covers DTW in detail including careful mathematical modeling of the optimization problem, the algorithm based on dynamic programming, and the mathematical proofs. Furthermore, numerous illustrations, examples, and exercises are provided.

Besides DTW, further algorithms based on dynamic programming are presented throughout the book. For example, subsequence variants of DTW are discussed in the context of audio matching (Section 7.2) and version identification (Section 7.3). In our audio thumbnailing application (Section 4.3), dynamic programming is used to efficiently compute a fitness measure for audio segments. Furthermore, the well-known Viterbi algorithm for finding an optimizing state sequence is based on dynamic programming—a concept that is applied in this book for estimating chord sequences (Section 5.3). Finally, a dynamic programming approach is introduced to derive an optimal beat sequence (Section 6.3). In all these problems, which are motivated by concrete applications, the objective is to find a sequence or an alignment between two sequences that is optimal in one or another way. By considering various scenarios, the student should acquire a solid understanding of the underlying principles of dynamic programming.

There are a number of other important algorithms treated in this book, which may be integrated into a basic computer science curriculum. First of all, Section 2.4.3 covers the classic fast Fourier transform (FFT), which goes back to Carl Friedrich Gauß (1805, published posthumously in 1866). Being a typical example for a divide-and-conquer strategy, the basic idea of the FFT algorithm is to divide the discrete Fourier transform (DFT) into two pieces of half the size. The FFT algorithm can also be interpreted as a factorization of the DFT matrix into a product of sparse matrices.

In Section 8.3, we study another matrix factorization technique known as non-negative matrix factorization (NMF). This technique is studied within an audio decomposition scenario. The general objective of NMF is to factorize a given real-valued matrix with no negative elements into a product of two other matrices that also have no negative elements. Usually, the two matrices in the product have a much lower rank than the original matrix. In this case, the product can be thought of as a compressed and more structured version of the original matrix. As a typical example for how to approach nonconvex optimization problems in machine learning, we discuss an iterative procedure for learning an NMF decomposition (Section 8.3.1).

Originally applied for speech recognition, hidden Markov models (HMMs) are now a standard tool for applications in temporal pattern recognition. Motivated by a chord recognition application, we introduce this mathematical concept in Section 5.3 as a typical example for a statistical data model. A rigorous treatment of statistical data analysis goes beyond the scope of this book. With Section 5.3.2 we provide, at least, a glimpse into this important area. Furthermore, by considering HMMs, one can also show how alignment concepts such as DTW can be extended using a probabilistic framework.

As a final fundamental topic that may be covered in an introductory course in computer science, we address the issue of data indexing, where the objective is to speed up a retrieval process. The basic procedure is similar to what we do when using a traditional book index. When looking for a specific passage in a book, an index allows us to directly access the page numbers where certain key words occur. In Section 7.1, we study such techniques in the context of an audio identification application. Here, the key words correspond to audio fingerprints (e.g., spectral peaks or combinations thereof), while the page numbers correspond to the time positions where these fingerprints appear.

With these comments, we hope to have convinced lecturers that music processing may serve as a beautiful and instructive application scenario for teaching basic concepts on data representations and algorithms. In my experience as a lecturer in computer science and engineering, starting a lecture with music processing applications, in particular playing music to students, opens them up and raises their interest. This makes it much easier to get the students engaged with the mathematical theory and technical details. Mixing theory and practice by immediately applying algorithms to concrete music processing tasks helps to develop the necessary intuition behind the abstract concepts and awakens the student's fascination and enthusiasm for the topic.

Acknowledgements

This textbook reflects my experience as a researcher and lecturer over the last twelve years. During these years, I have closely collaborated on, discussed, struggled with, learned from, and enjoyed research with many different people. I would like to take the opportunity to express my gratitude to all these people, without whom I would never have been able to write this book.

I want to dedicate this book to Michael Clausen and Hans-Peter Seidel—two people who have played a very special role in my academic career. It was Michael Clausen who first got me interested in the research areas of music processing and computer algebra. Doing my PhD as well as my Habilitation in his group (at the Computer Science Department, University of Bonn), Michael Clausen gave me all the freedom and support to pursue my own research goals. His analytic thinking, open feedback, enthusiasm, and integrity have had a huge influence on me as a scientist, teacher, and human being. Thank you, Michael, for all your mental, intellectual, and financial support—I will try to pass down your spirit to future student generations.

A second key person in my academic career is Hans-Peter Seidel—the head of the Computer Graphics Department of the Max-Planck-Institut für Informatik. I was very fortunate to join his group, working as a senior researcher within the Cluster of Excellence on Multimodal Computing and Interaction from 2007 to 2012. Within an open and inspiring atmosphere, I was able to independently conduct research having my own PhD students while enjoying all the academic freedom one can dream of. It was at this time that the idea of writing this book originated—even though it took another two years to actually start with this endeavor. Thank you so much, Hans-Peter, for your support, guidance, and trust over all these years.

In September 2012, I joined the International Audio Laboratories Erlangen (AudioLabs), a joint institution of the Friedrich-Alexander-Universität Erlangen-Nürnberg (FAU) and the Fraunhofer Institut für Intergrierte Schaltungen IIS. Leading the research group on Semantic Audio Processing, I am proud to be part of a team that shapes the future of audio and multimedia technologies in research areas such as audio coding, audio signal analysis, and spatial audio processing. Being in the vicinity of both the university and the Audio & Multimedia division of Fraunhofer IIS, the AudioLabs offer an excellent infrastructure that enables close scientific collaborations in an ideal setting. I want to thank Heinz Gerhäuser as well as Albert Heuberger, Bernhard Grill, and Jürgen Herre as representatives for all those who have established this fantastic research environment. At this point, I also want to thank all my colleagues from the AudioLabs and the university for the very pleasant and productive daily cooperation: Tom Bäckström, Sascha Disch, Bernd Edler, Emanuël Habets, Tracy Harris, Jürgen Herre, Walter Kellermann, Frederik Nagel, Rudolf Rabenstein, Stefan Turowski, Christian Uhle, Elke Weiland, and many more.

As said, this textbook is based on results, material, and insights that have been obtained in close collaboration with different people. I would like to express my gratitude to my former and current PhD students, collaborators, and colleagues who have influenced and supported me in writing this textbook. Many of these people have also helped me with numerous discussions on the book's content, constructive suggestions for improvements, and various rounds of proofreading. I will confine myself to only mentioning their names in alphabetical order: Andreas Baak, Stefan Balke, Juan Bello, Rachel Bittner, David Damm, Christian Dittmar, Jonathan Driedger, Zhiyao Duan and his students, Dan Ellis, Sebastian Ewert, Derry Fitzgerald, Christian Fremerey, Emilia Gómez, Masataka Goto, Harald Grohganz, Peter Grosche, Thomas Helten, Alex Hollenbeck, Nanzhu Jiang, Anssi Klapuri, Verena Konz, Verena Kriesel, Frank Kurth, Lukas Lamprecht, Cynthia Liem, Patricio López-Serrano, Oriol Nieto, Bryan Pardo, Jouni Paulus, Thomas Prätzlich, Sanu Pulimootil Achankunju, Gaël Richard, Tido Röder, Shigeki Sagayama, Justin Salamon, Mark Sandler, Hendrik Schreiber, Joan Serrà, Jordan Smith, Timothy J. Tsai, Avery Wang, Christof Weiß, Gordon Wichern, Frans Wiering, Geraint Wiggins, Aaron Wishnick, Frank Wu, and Udo Zölzer. Thank you so much for your help, support, stimulation, and encouragement.

Before and during the process of writing this textbook, I had the opportunity to teach most of the material as graduate courses at the Department of Computer Science, Rheinische Friedrich-Wilhelms-Universität Bonn; at the Department of Computer Science, Universität des Saarlandes; and at the Department Elektrotechnik-Elektronik-Informationstechnik, Friedrich-Alexander-Universität Erlangen-Nürnberg (FAU). I want to thank the students for their comments and valuable feedback. I also want to thank Ralf Gerstner and Viktoria Meyer from Springer-Verlag for helping me in organizing, editing, and publishing this book. Many research results that have entered this textbook were achieved within projects funded by the German Research Foundation (Deutsche Forschungsgemeinschaft, DFG). I want to thank the DFG for their financial support and the unbureaucratic help when shifting the projects from one location to another.

Last but not least, I am grateful to my family and friends for all the support and encouragement I have received in my life. First and foremost, I want to thank my wife Vlora from the bottom of my heart for being extremely supportive and for standing beside me throughout my career. I also thank my wonderful children, Hana and Zanfina, for reminding me of the really important things in life—you are the best kids a dad could hope for. Finally, I am grateful to my parents, Irmin and Hans-Georg Müller, for always supporting my ambitions throughout my life.

Erlangen,
June 2015

Meinard Müller

Contents

Basic Symbols and Notions

The following basic symbols and notions are used throughout this book:

$\mathbb{N} = \{1,2,3\ldots\}$	natural numbers
$\mathbb{N}_0 = \mathbb{N} \cup \{0\}$	whole numbers
$\mathbb{Z} = \{\ldots,-2,-1,0,1,2,\ldots\}$	integers
$[a:b] := \{a,a+1,\ldots,b\} \subset \mathbb{Z}$	integers from a to b for $a,b \in \mathbb{Z}$
$\mathbb{Q}$	rational numbers
$\mathbb{R}$	real numbers
$\mathbb{R}_{>0} = \{a \in \mathbb{R} \mid a > 0\}$	positive real numbers
$\mathbb{R}_{\geq 0} = \{a \in \mathbb{R} \mid a \geq 0\}$	nonnegative real numbers
$[a,b] := \{r \in \mathbb{R} \mid a \leq r \leq b\} \subset \mathbb{R}$	interval of real numbers from a to b for $a,b \in \mathbb{R}$
$\mathbb{C}$	complex numbers
$i := \sqrt{-1}$	imaginary unit
$\lvert a \rvert$	absolute value of a number $a \in \mathbb{R}$ (or $a \in \mathbb{C}$)
$\mathbb{R}^N$	real coordinate space of dimension $N \in \mathbb{N}$
$\mathbb{C}^N$	complex coordinate space of dimension $N \in \mathbb{N}$
$\lVert x \rVert$	norm of a vector $x \in \mathbb{R}^N$ (or $x \in \mathbb{C}^N$)
$\langle x \mid y \rangle$	inner product of two vectors $x,y \in \mathbb{R}^N$ (or $x,y \in \mathbb{C}^N$)
$x^\top$	transpose of a vector x
$A^\top$	transpose of a matrix A

Chapter 1
Music Representations

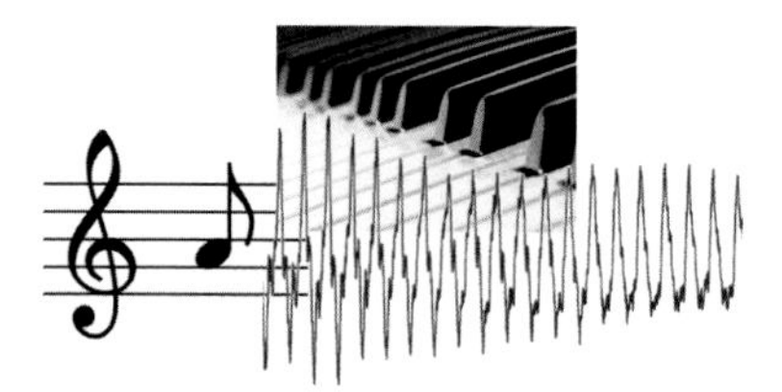

Music can be represented in many different ways and formats. For example, a composer may write down a composition in the form of a musical score. In a score, musical symbols are used to visually encode notes and how these notes are to be played by a musician. The printed form of a musical score is also referred to as **sheet music**. The original medium of this representation is paper, although it is now also accessible on computer screens through digital images. For electronic instruments and computers, music may be communicated by means of standard protocols such as the widely used Musical Instrument Digital Interface (MIDI) protocol, where event messages specify pitches, velocities, and other parameters to generate the intended sounds. In this book, we use the term **symbolic** to refer to any machine-readable data format that explicitly represents musical entities. These musical entities may range from timed note events, as is the case of MIDI files, to graphical shapes with attached musical meaning, as is the case of music engraving systems. Unlike symbolic representations, audio representations such as WAV or MP3 files do not explicitly specify musical events. These files encode acoustic waves, which are generated when a source (e.g., an instrument) creates a sound that travels to the human ear as air pressure oscillations.

In this book, we distinguish between three main classes of music representations: sheet music, symbolic, and audio. To put it simply, the term **sheet music** stands for visual representations of a score given in printed form or in the form of digitized images. The term **symbolic** comprises any kind of score representation with an explicit encoding of notes or other musical events. Finally, the term **audio** refers to representations of acoustic sound waves. Each of these representations reflects certain aspects of a musical object, but no single representation encompasses all its properties. In this sense, each representation can be considered a projection or a realization of what we generally refer to as a piece of music. In this introductory chapter, we discuss some basic properties of music by means of these different music representations. We start by describing basic elements of Western music notation as used in

M. Müller, *Fundamentals of Music Processing*, https://doi.org/10.1007/978-3-030-69808-9_1

Fig. 1.1 Sheet music representation of the first five measures of Symphony No. 5 by Ludwig van Beethoven in a piano reduced version.

sheet music representations (Section 1.1). Even though the exact specifications of music notation are not essential in this book, we require basic notions of the pitch, duration, and onset time of musical notes. Then, we summarize basic properties of symbolic representations with a specific focus on MIDI, which is the prevailing standard for controlling music synthesizers (Section 1.2). Finally, we discuss audio representations, which are at the heart of this book. In particular, we deal with aspects concerning the properties of sound waves including frequency, dynamics, and timbre (Section 1.3).

1.1 Sheet Music Representations

Sheet music, also referred to as **musical score**, provides a visual representation of what we commonly refer to—in particular for Western classical music—as the "piece of music." Sheet music describes a musical work using a formal language based on musical symbols and letters, which are depicted in a graphical–textual form. Reading sheet music, a musician can create a performance by following the given instructions. Performing a piece from sheet music, however, not only requires a special form of literacy, i.e., the ability to understand the music notation, but also involves a creative process. A musical score is rarely played mechanically. Musicians may shape the flow of the music by varying the tempo, dynamics, and articulation, thus resulting in a personal interpretation of the given musical score. In this sense, rather than giving rigid specifications, sheet music can be considered as a guide for performing a piece of music leaving room for different interpretations.

As a first example, let us consider Symphony No. 5 in C minor by Ludwig van Beethoven, which is one of the most popular and best-known compositions in classical music. It begins with a short musical idea, the famous "short-short-short-long" **motif**, which is commonly referred to as the "fate motif" of Beethoven's Fifth. Figure 1.1 shows a sheet music representation of the first five measures in a piano reduced version. In the following sections, we explain the meaning of the musical symbols in more detail while introducing some music notations used throughout this book. The Beethoven piece will serve as our running example.

1.1.1 Musical Notes and Pitches

In music, the term **note** is often used in a rather loose way and may refer to both a musical symbol (when talking about score representations) as well as a pitched sound (when talking about audio representations). In this section, we employ the term to refer to musical symbols used in Western music notation. Each note has several attributes that determine the relative duration and the pitch of a sound to be performed by a musician. For example, in the case of a piano, the pitch of a note tells a musician which key is to be pressed on the keyboard, and the duration of the note determines how long this key is to be held. The notion of **pitch** is not strict and refers to a perceptual property that allows a listener to order a sound on a frequency-related scale. As we will discuss in Section 1.3 in more detail, playing a note on an instrument results in a (more or less) periodic sound of a certain fundamental frequency. This fundamental frequency is closely related to what is meant by the pitch of a note. In the following discussion, we use the term "pitch" in an intuitive way. It allows us to order pitched sounds from "lower" to "higher"—similarly to the keys of a piano keyboard ordered from left to right.

Two notes with fundamental frequencies in a ratio equal to any power of two (e.g., half, twice, or four times) are perceived as very similar. Because of that, all notes with this kind of relation can be grouped under the same **pitch class**. This observation also leads to the fundamental notion of an **octave**, which is defined to be the interval between one musical note and another with half or double its fundamental frequency. Using this definition, a pitch class is a set of all pitches or notes that are an integer number of octaves apart.

In order to describe music using a finite number of symbols, one needs to discretize the space of all possible pitches. This leads to the notion of a **musical scale**, which can be thought of as a finite set of representative pitches. Because of the close octave relationship of pitches, scales are generally considered to span a single octave, with higher or lower octaves simply repeating the pattern. A musical scale can then be specified by a division of the octave space into a certain number of scale steps. The elements of a scale are often simply referred to as the **notes** of the scale and are ordered according to their respective pitches.

In music history, many different scales have been suggested and used, and there have been fierce discussions about the suitability of specific scales. The appropriateness of a scale very much depends on the kind of music to be described, the instruments used, the musical genre, or the cultural background. A scale that is suited for representing Western piano music may not be suited for representing Indian sitar music. A scale used for Gregorian chant of the 10th century may not be a good choice for describing experimental music of the 20th century. There is no universally valid musical scale, and the choice of a musical scale necessarily goes along with simplifications typically imposed by practical considerations.

In this book, we only consider the case of the **twelve-tone equal-tempered scale**, where an octave is subdivided into twelve scale steps. The fundamental frequencies of these steps are equally spaced on a logarithmic frequency axis (see Section 1.3.2).

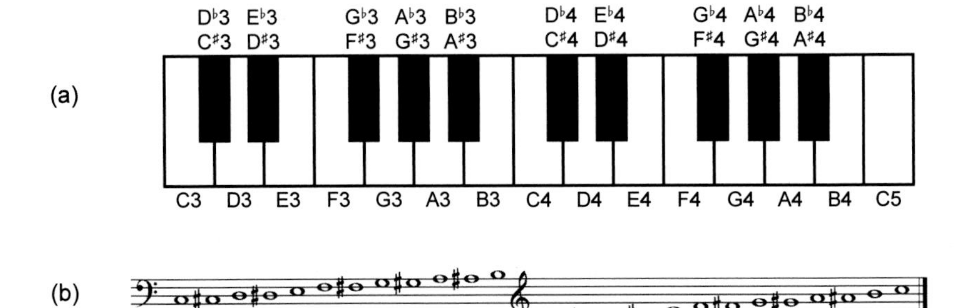

Fig. 1.2 **(a)** Section of piano keyboard with keys ranging from C3 to C5. **(b)** Corresponding notes using Western music notation.

The difference between the fundamental frequencies of two subsequent scale steps is also called a **semitone**, which is the smallest possible interval in this scale.

In the twelve-tone equal-tempered scale, there are twelve pitch classes. In Western music notation, these pitch classes are denoted by combining a letter-name and accidentals. Seven of the pitch classes (corresponding to C major) are denoted by the letters C, D, E, F, G, A, B. These pitch classes correspond to the white keys of a piano keyboard (see Figure 1.2). The remaining five pitch classes correspond to the black keys of a piano keyboard and are denoted by a combination of a letter and an **accidental** (♯, ♭). A sharp (♯) raises a note by a semitone, and a flat (♭) lowers it by a semitone. The accidentals are written after the note name. For example, $\mathrm{D}^\sharp$ represents D-sharp and $\mathrm{D}^\flat$ represents D-flat. In the equal-tempered scale, the remaining five pitches can be either denoted by $\mathrm{C}^\sharp$, $\mathrm{D}^\sharp$, $\mathrm{F}^\sharp$, $\mathrm{G}^\sharp$, $\mathrm{A}^\sharp$ or by $\mathrm{D}^\flat$, $\mathrm{E}^\flat$, $\mathrm{G}^\flat$, $\mathrm{A}^\flat$, $\mathrm{B}^\flat$. For example, $\mathrm{C}^\sharp$ and $\mathrm{D}^\flat$ represent the same pitch class,[1] even though from a musical point of view one distinguishes between these two concepts.

To name the notes of the twelve-tone equal-tempered scale, in addition to indicating the pitch class, one needs to provide an identifier for the octave. Following the **Scientific Pitch Notation**, each note is specified by the pitch class name, followed by a number that indicates the octave. The note A4 is determined to have a fundamental frequency of 440 Hz and serves as a reference. The **octave number** increases by one upon an ascension from a note with pitch class B to one with pitch class C. For example, the note B4 is followed by the note C5. Similarly, the octave number decreases by 1 upon a descent from a C to a B. The lowest note C0 in this notation has a fundamental frequency in the region of 16 Hz, which is already below what a human can acoustically perceive. Figure 1.2 shows the notes from C3 to C5 along with the corresponding keys of a piano keyboard.

Ordering all notes of the equal-tempered scale according to their pitches, one obtains an equal-tempered **chromatic scale**, where all notes of the scale are equally spaced. The term **chromatic** is derived from the Greek word **chroma**, meaning color. In the music context, the term "chroma" closely relates to the twelve different

[1] This phenomenon is also known as **enharmonic equivalence**.

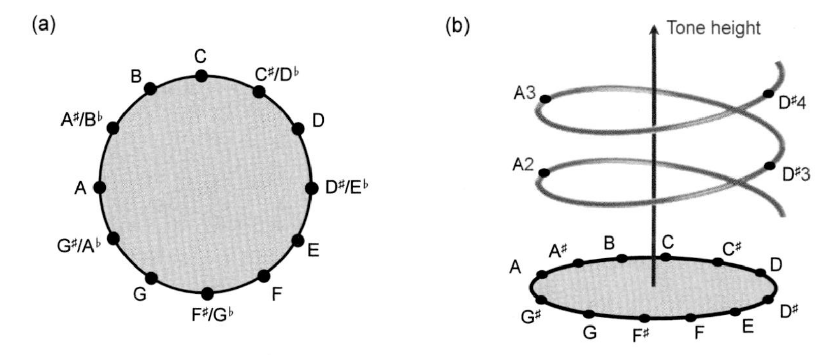

Fig. 1.3 **(a)** Chromatic circle. **(b)** Shepard's helix of pitch [23].

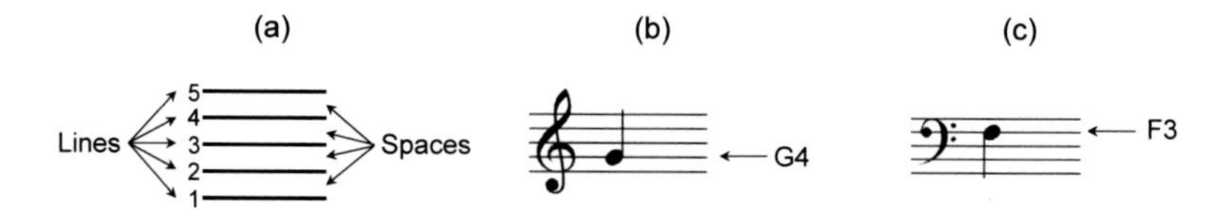

Fig. 1.4 **(a)** Staff. **(b)** Staff with G-clef. **(c)** Staff with F-clef.

pitch classes. For example, the notes C2 and C5 both have the same chroma value C. In other words, all notes that have the same chroma value belong to the same pitch class. Recall that notes that belong to the same pitch class (or have the same chroma value) are perceived as similar in a certain way. In contrast, notes that belong to different pitch classes (or have different chroma values) are perceived as dissimilar. This justifies the usage of the term "chroma" in the sense that notes with different chroma values have a different "sound color." The cyclic nature of chroma values is illustrated by the **chromatic circle** as shown in Figure 1.3a. Extending this notion, **Shepard's helix of pitch** represents the linear pitch space as a helix wrapped around a cylinder so that octave-related pitches lie along a single vertical line [23]. The projection of the cylinder onto the horizontal plane yields the chromatic circle. The factorization of a pitch into a chroma value and an octave number will play an important role in this book. The chroma components of pitches can be used to yield mid-level representations, which turn out to be a powerful tool for various music analysis and retrieval applications.

1.1.2 Western Music Notation

Generally speaking, **music notation** refers to a system for graphically representing music through symbols. The standard Western music notation is based on a **staff**, which is a set of five horizontal lines and four spaces each representing a different

Fig. 1.5 **(a)** Musical score of a C-major scale starting with C4 and ending with C5. **(b)** Key signature consisting of three flats converting the notes into a C-minor scale.

musical pitch (see Figure 1.4a). Appropriate music symbols, depending upon the intended effect, are placed on the staff according to their corresponding pitch or function. Pitch is shown by placement of note symbols on the staff—sometimes modified by accidentals. The higher the placement within a given staff, the higher the pitch of the corresponding note. Furthermore, the duration is indicated by the shapes of the note symbols as well as additional symbols such as dots and ties.

Notation is read from left to right. A staff generally begins with a **clef** symbol, which indicates the position of one particular note on the staff. For example, by convention, the **treble clef** (𝄞), also known as the G-clef, indicates that the second line is the pitch G4 (see Figure 1.4b). Similarly, the **bass clef** (𝄢), also known as the F-clef, indicates that the fourth line is the pitch F3 (see Figure 1.4c). There are also further clef symbols and clef positions. The details are not important in this book. However, one should keep in mind that the clef symbol, along with its position, serves as a reference in relation to which the meaning of the notes positioned on any line or space of the staff can be determined. Notes representing a pitch outside the scope of the five-line staff can be described using **ledger lines**, which provide a single note with additional lines and spaces (see, e.g., the C4 in Figure 1.5).

Following the clef, the **key signature** on a staff indicates the key of the piece by specifying that certain notes are flat or sharp throughout the piece, unless otherwise specified. For example, the notes shown in Figure 1.5a are C4, D4, E4, F4, G4, A4, B4, C5 thus forming a C-major scale. Using the key signature consisting of three flats as shown in Figure 1.5b, the notes become C4, D4, E$^\flat$4, F4, G4, A$^\flat$4, B$^\flat$4, C5 thus forming a (natural) C-minor scale.

Music is typically organized into temporal units, referred to as **beats**. Repeating sequences of stressed and unstressed beats, in turn, form higher temporal patterns, which are related to what is called the **rhythm** of music and is expressed in terms of the musical **meter**. A **measure** (or **bar**) is a segment of time defined by a given number of beats. Dividing music into measures not only reflects it rhythmic nature, but also provides regular reference points within it. In music notation, the temporal structure of a piece is indicated by the **time signature**, which appears in a staff after the key signature. Typically, a time signature consists of two numerals, one stacked above the other. The lower numeral indicates the note duration that represents one beat (given as a fraction with regard to a whole note), while the upper numeral indicates how many such beats are in a measure. For example, the time signature $\frac{6}{8}$ shown in Figure 1.6b indicates that a measure consists of six beats, where a beat has the duration of an eighth note. In sheet music, two subsequent measures are

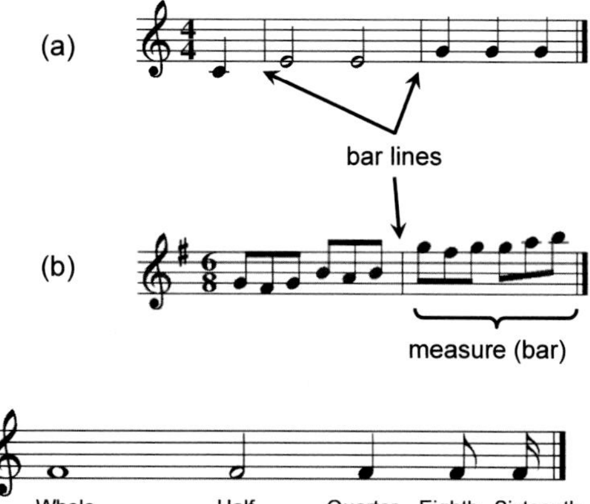

Fig. 1.6 Notation of time signature. **(a)** Four quarter notes per measure with upbeat. **(b)** Six eighth notes per measure.

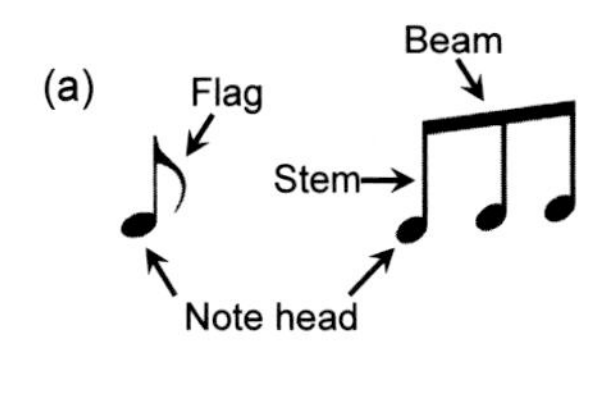

Fig. 1.7 **(a)** Parts of a note. **(b)** Notation for different durations of notes. **(c)** Notation for different durations of rests.

separated by a vertical line drawn through the staff, which are referred to as **bar lines**.

After specifying the clef, the key signature, and the time signature, which all reflect global characteristics of the piece and hold for the entire staff (if not redefined explicitly), the actual notes are specified. As illustrated by Figure 1.7a, each note is represented by a symbol that consists of a **note head** and possibly a **stem** and a **flag**. Sometimes several notes are combined by a **beam**. As discussed above, a note's pitch is indicated by its placement on the staff and possibly by an accidental, where the clef symbol serves as a reference pitch. The duration of a note is defined in a relative fashion by means of its **note value**, which is indicated by the color or shape of the note head, the presence or absence of a stem, and the presence or absence of flags (see Figure 1.7b). The whole note is the reference value, and the other notes are named in accordance. For example, a **half note** has half the length of a whole note, a **quarter note** has a quarter the length of a whole note, and so on. For each note value, there also exists a **rest symbol** of equivalent duration, which expresses an interval of silence in a piece of music (see Figure 1.7c).

The musical onset times of the notes are specified in a relative fashion and follow from the horizontal formation of the note symbols. Notes that are to be played at the same time are given by vertically aligned musical symbols. In this case, different notes may share the same stem and flag as illustrated by Figure 1.1. Once the physical duration of a beat is known, the physical onset times of the notes can be derived from the relative timing. The duration of a beat is given by the **tempo** indication specified in beats per minute (BPM). For example, a specification of 120 BPM

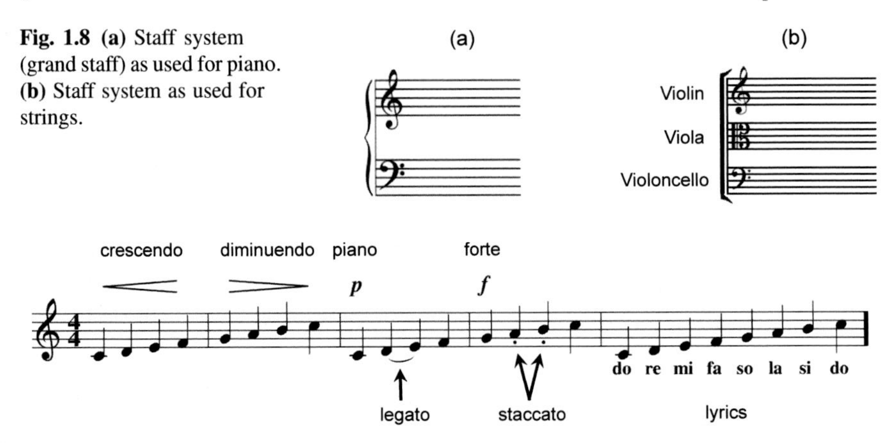

Fig. 1.8 **(a)** Staff system (grand staff) as used for piano. **(b)** Staff system as used for strings.

Fig. 1.9 Musical score with various symbols used for indicating dynamics and articulation.

means that 120 beats are to be played within one minute. In the case that a beat corresponds to a quarter note, 120 BPM implies that the duration of a quarter note is half a second. Composers often suggest a tempo notated above the first staff line of the piece. For example, in Figure 1.1, the suggested tempo is 108 BPM with a beat being a half note (𝅗𝅥 = 108). However, when performing a piece, musicians often significantly deviate from the suggested tempo.

To notate music that is played on a piano or is played by different musicians on various instruments, one often uses several staves to notate the various musical voices. A single vertical line drawn to the left of multiple staves creates a **staff system**, which indicates that the music on all the staves is to be played simultaneously. A bracket is an additional vertically aligned symbol joining staves. This symbol shows groupings of instruments that function as a unit, such as the string section of an orchestra (see Figure 1.8b). When music notated across different staves is intended to be played at once by a single performer (usually a keyboard instrument or the harp), a **grand staff** is created by joining the two staves by a brace. For example, in the case of piano music, one has two staves, where the upper staff uses a treble clef and the lower staff uses a bass clef (see Figure 1.8a). When playing the piano, the upper staff is normally played with the right hand and the lower staff with the left hand. This is the case with our Beethoven example shown in Figure 1.1.

Besides the aforementioned attributes, music notation may contain many more instructions to the musician regarding matters such as tempo, dynamics, and expression. For example, the overall **tempo** and **style** of the piece may be specified by textual notations such as *Allegro con brio* (fast with vigor and spirit) or *Andante con moto* (moderate tempo with motion). Other directions such as **accelerando** (gradually becoming faster) or **ritardando** (gradually becoming slower) refer to local tempo deviations. Similarly, **dynamics**, which refers to the volume of a sound or note, may be described by terms such as **forte** (loud), **piano** (soft), **crescendo** (gradually becoming louder), or **diminuendo** (gradually becoming softer). For vo-

Fig. 1.10 Sheet music representation of the full orchestral score of the beginning of Beethoven's Fifth Symphony (from Breitkopf & Härtel, Leipzig, 1862).

cal music, **lyrics** may be written above or below staff lines. Other symbols such as **articulation** marks are used to indicate how certain notes are to be played. For example, a **staccato** mark (a dot placed above or below a note) signifies that a note is to be played with shortened duration detached from the subsequent note, whereas a **legato** mark (a curved line placed above or below a group of notes) indicates that musical notes are played smoothly and connected (see Figure 1.9).

We close this section by coming back to our Beethoven example. In the piano reduced version shown in Figure 1.1, the score shows a system with two staves, where the upper staff for the right hand starts with a G-clef 𝄞 and the lower staff for the left hand with an F-clef 𝄢. Both staves are equipped with a key signature (three flats ♭) and a time signature (two quarter-note beats per measure $\frac{2}{4}$). The score reveals that the first five measures consist of two "short-short-short-long" patterns, where the second fate motif is played lower than the first fate motif. Further instructions are given in the form of additional symbols and textual notations. For example, a fermata sign (𝄐) indicates that the respective note duration should be prolonged. The pedaling information (𝆮 ✻) tells the musician to hold the sustain pedal, which can have a significant impact on the sound. The overall tempo is indicated by "Allegro con brio" and a metronome specification (108 half notes per minute) (see Exercise 1.1). Finally, the symbol *ff* stands for "fortissimo" or "very loud" and indicates the dynamics.

Let us now have a look at Figure 1.10, which shows the full orchestral score of the beginning of Beethoven's Fifth as used by conductors to direct rehearsals and performances. The shown excerpt is the scanned version of an edition published by Breitkop & Härtel in the year 1862. The various staves of the system specify the music according to the different instruments lined up in a fixed order. From top to bottom, the voices for the woodwinds (flute, oboe, clarinet, bassoon), the brass (French horn, trumpet) and percussion (timpani), and the strings (violin, viola, cello, double bass) are listed. For certain instruments such as the violin, there may be more than one musical voice to be played, each specified by a separate staff (e.g., violin I and violin II).

In this section, we have only scratched the surface of Western music notation. Rather than giving a comprehensive overview, our goal was to build up some intuition while introducing some basic terminology. Furthermore, we wanted to indicate that music notation is far from being comprehensive. Many of the symbols only give a vague description of how the notes should be played leaving room for artistic freedom and creativity. Furthermore, as indicated by the full score and the piano transcript of Beethoven's Fifth, there may exist different score versions of the same piece of music. For most parts of this book, it suffices to have a rough understanding of musical concepts examined in this chapter. The aspects of pitch and timing will be picked up again when discussing various kinds of derived music representations.

1.2 Symbolic Representations

As discussed at the beginning of this chapter, symbolic representations describe music by means of entities that have an explicit musical meaning and, given in some digital format, can be parsed by a computer. Any kind of digital data format may be regarded as "symbolic" since it is based on a finite alphabet of letters or symbols. For example, the pixels in a digital image file or the samples in a digital audio file may be regarded as symbols or basic entities. However, considering these entities individually, no musical meaning can be inferred. Therefore, neither scanned images nor digitized music recordings are regarded as being symbolic music formats. Similarly, graphical shapes in vector graphics representations are not considered to be musical entities as long as no musically meaningful specification of the shapes is given. Still, there is a wide range of what may be considered as symbolic music. In this section, we discuss some examples including piano-roll representations, MIDI representations, and other symbolic formats that encode sheet music. Furthermore, we touch on optical music recognition (OMR), which is the process of converting digital scans of printed sheet music into symbolic representations.

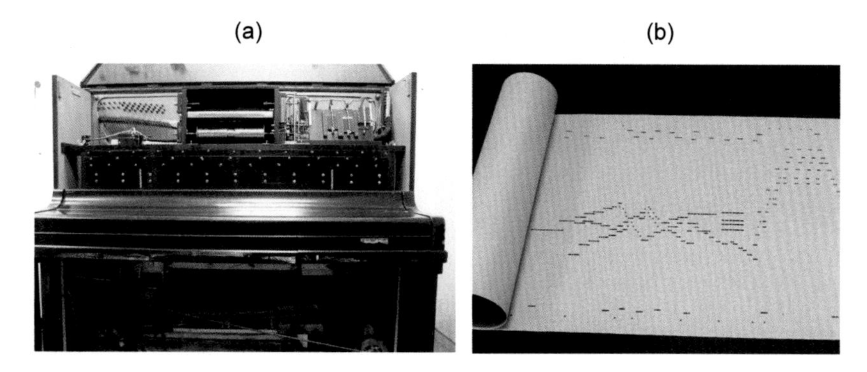

Fig. 1.11 (**a**) Player piano. (**b**) Piano roll. (Reprinted by kind permission of the Institut für Musikwissenschaft der Goethe-Universität Frankfurt)

1.2.1 Piano-Roll Representations

We start with a symbolic representation having a history of more than one hundred years. In the late 19th and early 20th century, self-playing pianos, so-called **player pianos** (see Figure 1.11a), became quite popular with a peak in 1924, before being replaced by phonograph recordings. Player pianos contained pneumatic mechanisms to automatically operate the key and pedal movements according to the instructions specified by a prestored piano-roll medium. A **piano roll** is a continuous roll of paper with perforations (holes) punched into it. The perforations represent note control data (see Figure 1.11b). The roll moves over a reading system known as a tracker bar, and the playing cycle for each musical note is triggered when a perforation crosses the bar and is read. Rolls for player pianos were generally made from recorded performances of musicians. This way, the playing of many famous pianists and composers including Gustav Mahler, Edvard Grieg, Scott Joplin, or George Gershwin is preserved on piano rolls. Typically, a pianist would sit at a specially designed player piano, and the pitch and duration of any notes played would be perforated into a blank roll, together with the duration of the sustain and soft pedal. Player pianos can also recreate the dynamics of a pianist's performance by means of specially encoded control perforations placed towards the edges of a music roll.

In the following, a **piano-roll representation** is understood to be a geometric visualization of the note information as specified by a piano roll. The horizontal axis of this two-dimensional representation encodes time, whereas the vertical axis encodes pitch. Every note is described by an axis-parallel rectangle coding three parameters. The first parameter is the onset time, given by the leftmost horizontal coordinate of the rectangle, and the second is the pitch, given by the lower vertical coordinate of the rectangle. Finally, the third parameter is the duration of the note, encoded by the width of the rectangle.

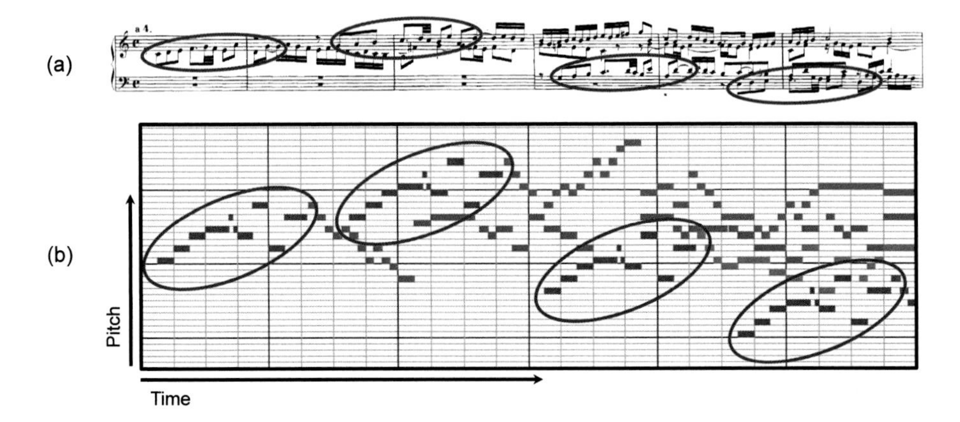

Fig. 1.12 **(a)** Sheet music representation and **(b)** piano-roll representation of the beginning of Fugue BWV 846 in C major by Johann Sebastian Bach. The four occurrences of the theme are highlighted.

Figure 1.12b shows a piano-roll representation of the beginning of Fugue BWV 846 in C major by Johann Sebastian Bach. For comparison, Figure 1.12a shows the corresponding part in a sheet music representation. Generally, a **fugue** is a compositional technique using two or more musical voices, built on a musical theme (or subject) that is introduced at the beginning by one voice and then repeated at different pitches in the other voices. Fugue BWV 846 consists of four voices. Although played on a keyboard instrument, the four voices are referred to as soprano (highest voice), alto (second highest voice), tenor (third highest voice), and bass (lowest voice). The fugue starts with the main theme in the alto, which is then repeated in the soprano, the tenor, and finally in the bass. As shown by Figure 1.12, the four occurrences of the theme are hard to detect in the sheet music representation, but can be easily seen in the piano-roll representation, where each one corresponds to a pattern shifted in the time–pitch plane.

While they are a considerable simplification of what is notated in sheet music, piano-roll representations visually describe the most important attributes of musical notes in an easy-to-understand way. Therefore, we will often use piano-roll representations when describing and talking about symbolic music. Furthermore, as we will see in later chapters, one can also derive similar representations from other music encodings including MIDI and audio. In this sense, piano rolls can be seen as a kind of **mid-level representation** on the basis of which semantic relations can be established across various manifestations of music.

1.2.2 MIDI Representations

The next symbolic representation we want to discuss is based on the **MIDI** standard, which stands for **Musical Instrument Digital Interface**. Although MIDI was not originally developed to be used as a symbolic music format and imposes many limitations on what can actually be represented, the importance of MIDI is due to its widespread usage over the last three decades, and the abundance of MIDI data freely available on the web. From a music encoding point of view, one needs to keep in mind that the quality of available MIDI data is sometimes questionable.

MIDI was originally developed as an industry standard to get digital electronic musical instruments from different manufacturers to work and play together. It was the advent of MIDI in 1981–1983 that caused a rapid growth of the electronic musical instrument market. MIDI allows a musician to remotely and automatically control an electronic instrument or a digital synthesizer in real time. As an example, let us consider a digital piano, where a musician pushes down a key of the piano keyboard to start a sound. The intensity of the sound is controlled by the velocity of the keystroke. Releasing the key stops the sound. Instead of physically pushing and releasing the piano key, the musician may also trigger the instrument to produce the same sound by transmitting suitable MIDI messages, which encode the note-on, the velocity, the note-off, and other information. These MIDI messages may be automatically generated by some other electronic instrument or may be provided by a computer. It is an important fact that MIDI does not represent musical sound directly, but only represents performance information encoding the instructions about how an instrument has been played or how music is to be produced.

The original MIDI standard was later augmented to include the **Standard MIDI File** (SMF) specification, which describes how MIDI data should be stored on a computer. In the following, we denote SMF files simply as **MIDI files** or **MIDI representations**. The SMF file format allows users to exchange MIDI data regardless of the computer operating system and has provided a basis for an efficient internet-wide distribution of music data, including numerous websites devoted to the sale and exchange of music. A MIDI file contains a list of MIDI messages together with timestamps, which are required to determine the timing of the messages. Further information (called meta messages) is relevant to software that processes MIDI files.

For our purposes, the most important MIDI messages are the note-on and the note-off commands, which correspond to the start and the end of a note, respectively. Each note-on and note-off message is, among others, equipped with a MIDI note number, a value for the key velocity, a channel specification, as well as a timestamp. The **MIDI note number** is an integer between 0 and 127 and encodes a note's pitch. Here, MIDI pitches are based on the equal-tempered scale as discussed in Section 1.1.1. Similarly to an acoustic piano, where the 88 keys of the keyboard correspond to the musical pitches A0 to C8, the MIDI note numbers encode, in increasing order, the musical pitches C0 to G$^\sharp$9. For example, note C4 has the MIDI note number 60, whereas the concert pitch A4 has the MIDI note number 69.

The **key velocity** is again an integer between 0 and 127, which controls the intensity of the sound—in the case of a note-on event it determines the volume, whereas

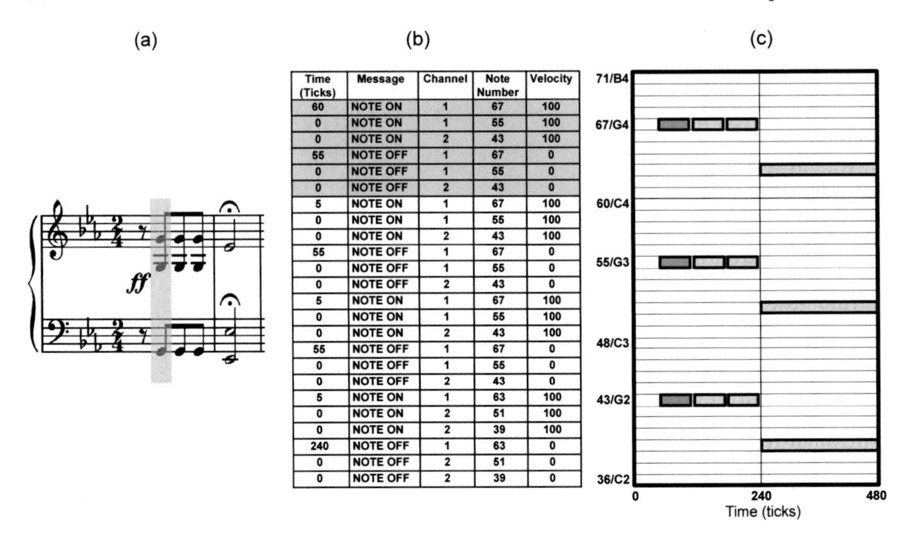

Time (Ticks)	Message	Channel	Note Number	Velocity
60	NOTE ON	1	67	100
0	NOTE ON	1	55	100
0	NOTE ON	2	43	100
55	NOTE OFF	1	67	0
0	NOTE OFF	1	55	0
0	NOTE OFF	2	43	0
5	NOTE ON	1	67	100
0	NOTE ON	1	55	100
0	NOTE ON	2	43	100
55	NOTE OFF	1	67	0
0	NOTE OFF	1	55	0
0	NOTE OFF	2	43	0
5	NOTE ON	1	67	100
0	NOTE ON	1	55	100
0	NOTE ON	2	43	100
55	NOTE OFF	1	67	0
0	NOTE OFF	1	55	0
0	NOTE OFF	2	43	0
5	NOTE ON	1	63	100
0	NOTE ON	2	51	100
0	NOTE ON	2	39	100
240	NOTE OFF	1	63	0
0	NOTE OFF	2	51	0
0	NOTE OFF	2	39	0

Fig. 1.13 Various symbolic music representations of the first twelve notes of Beethoven's Fifth. **(a)** Sheet music representation. **(b)** MIDI representation (in a simplified, tabular form). **(c)** Piano-roll representation.

in the case of a note-off event it controls the decay during the release phase of the tone. The exact interpretation of the key velocity, however, depends on the respective instrument or synthesizer. The **MIDI channel** is an integer between 0 and 15. Intuitively speaking, this number prompts the synthesizer to use the instrument that has been previously assigned to the respective channel number. Note that each channel, in turn, supports polyphony, i.e., multiple simultaneous notes. Finally, the **time stamp** is an integer value that represents how many clock pulses or **ticks** to wait before the respective note-on or note-off command is executed. Before we comment in more detail on the timing concept employed by MIDI, we illustrate the MIDI representation by means of our Beethoven example. Figure 1.13b shows a (simplified and tabular) MIDI encoding of the first fate motif corresponding to the twelve notes of the score in Figure 1.13a. In this example, the notes of the right hand are assigned to channel 1 and the notes of the left hand to channel 2. The notes specified by corresponding note-on and note-off events in the MIDI file can also be visualized by a piano-roll representation (see Figure 1.13c). In case we are only interested in the note events (and not the channel and velocity information), this is how we represent MIDI information.

An important feature of the MIDI format is that it can handle musical as well as physical onset times and note durations. Similarly to sheet music representations, MIDI can express timing information in terms of musical entities rather than using absolute time units such as microseconds. To this end, MIDI subdivides a quarter note into basic time units referred to as **clock pulses** or **ticks**. The number of pulses per quarter note (PPQN) is to be specified at the beginning, in the so-called **header** of a MIDI file, and refers to all subsequent MIDI messages. A common value is

120 PPQN, which determines the resolution of the time stamps associated to note events. As mentioned above, a time stamp indicates how many ticks to wait before a certain MIDI message is executed, relative to the previous MIDI message. For example, the first note-on message with MIDI note number 67 is executed after 60 ticks, corresponding to the eighth rest at the beginning of Beethoven's Fifth. The second and third note-on messages are executed at the same time as the first one, encoded by the tick value zero. Then, after 55 ticks, MIDI note 67 is switched off by the note-off message and so on.

Like the sheet music representation, MIDI also allows for encoding and storing absolute timing information, however, at a much finer resolution level and in a more flexible way. To this end, one can include additional tempo messages that specify the number of microseconds per quarter note. From the tempo message, one can compute the absolute duration of a tick. For example, having 600000 μs per quarter note and 120 PPQN, each tick corresponds to 5000 μs. Furthermore, one can derive from the tempo message the number of quarter notes played in a minute, which yields the tempo measured in **beats per minute** (BPM). For example, the 600000 μs per quarter note correspond to 100 BPM. While the number of pulses per quarter note is fixed throughout a MIDI file, the absolute tempo information may be changed by inserting a tempo message between any two note-on or other MIDI messages. This makes it possible to account not only for global tempo information but also for local tempo changes such as accelerandi, ritardandi, or fermate.

In this section, we have briefly touched on MIDI and its functionality. As noted above, MIDI was originally designed to solve problems in electronic music performance and is limited in terms of the musical aspects it represents. For example, MIDI is not capable of distinguishing between a D$^\sharp$4 and an E$^\flat$4, both of which have the MIDI note number 63. Also, information on the representation of beams, stem directions, or clefs is not encoded by MIDI. Furthermore, MIDI does not define a note element explicitly; rather, notes are bounded by note-on and note-off events (or note-on events with velocity 0). Rests are not represented at all and must be inferred from the absence of notes.

1.2.3 Score Representations

Within the class of symbolic music representations, we want to distinguish one subclass we refer to as **score representations**. A representation from this subclass is defined to yield explicit information about musical symbols such as the staff system, clefs, time signatures, notes, rests, accidentals, and dynamics. In this sense, score representations are, compared with MIDI representations, much closer to what is actually shown in sheet music. For example, in a score representation, the notes D$^\sharp$4 and E$^\flat$4 would be distinguishable, and the musical onset times are specified. However, a score representation may not contain a description of the final layout and the particular shape of the musical symbols. The process of generating or rendering visually pleasing sheet music representations from score representations is an

Fig. 1.14 Different sheet music representations corresponding to the same score representation of the beginning of Prelude BWV 846 (C major) by Johann Sebastian Bach. From top left to bottom right, a computer-generated, a handwritten, and two traditionally engraved representations are shown.

art in itself. In former days, the art of drawing high-quality music notation for mechanical reproduction was called **music engraving**. Nowadays, computer software or **scorewriters** have been designed for the purpose of writing, editing, and printing music, though only a few produce results comparable to high-quality traditional engraving. Figure 1.14 illustrates this by showing different sheet music representations corresponding to the same score.

In this book, we do not give an overview of existing symbolic score formats. Instead, as an example, we discuss some aspects of **MusicXML**, which has been developed to serve as a universal format for storing music files and sharing them between different music notation applications. Following the general XML (Extensible Markup Language) paradigm, MusicXML is a textual data format that defines a set of rules for encoding documents in a way that is both human and machine readable. For example, Figure 1.15 shows how a note $\mathrm{E}^\flat 4$ is encoded. In the MusicXML encoding of the half note $\mathrm{E}^\flat 4$, the tags `<note>` and `</note>` mark the beginning and the end of a MusicXML note element. The pitch element, delimited by the tags `<pitch>` and `</pitch>`, consists of a pitch class element `E` (denoting the letter name of the pitch), the alter element `-1` (changing E to E flat), and the octave element `4` (fixing the octave). Thus, the resulting note is an $\mathrm{E}^\flat 4$. The element `<duration>2</duration>` encodes the duration of the note measured in quarter notes. Finally, the element `<type>half</type>` tells us how this note is actually depicted in the rendered sheet music.

There are various ways to generate digital score representations. For example, one could manually input the score information in a format such as MusicXML. This, however, is a tedious and error-prone procedure. Music notation software or scorewriters support users in the task of writing and editing digitized sheet music. Such software allows a user to conveniently input and modify note objects by standard computer input devices or electronic keyboards. In the next section, we discuss another way for generating score representations from scanned images of printed sheet music, which is, in a sense, the inverse of a rendering process.

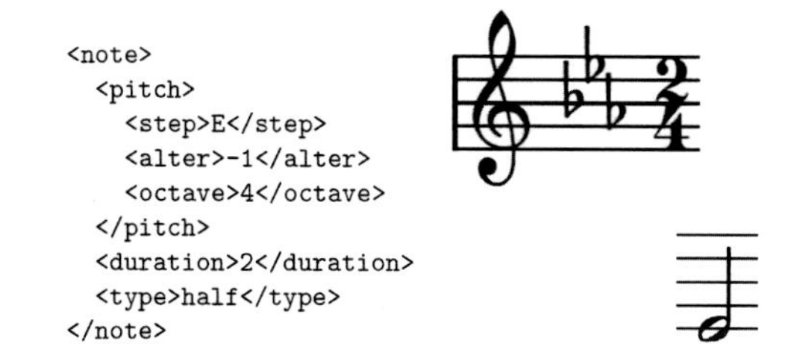

Fig. 1.15 Textual description in the MusicXML format of a half note E$^\flat$4. The clef, key signature, and time signature are defined at the beginning of the MusicXML file.

1.2.4 Optical Music Recognition

Sheet music is widely available, and many people are trained to use music notation for studying and playing music. For centuries, music has been documented, transmitted, and distributed in the form of printed sheet music. Music libraries and archives possess huge collections comprising millions of sheet music books, which are now successively being transferred into the digital domain using scanning devices. A digital image resulting from such a scanning process consists of a number of rows and columns of pixels, each pixel encoding the color at a specific point of the scanned page. In other words, a digital image of a sheet music page is by itself a mere accumulation of colored (often black and white) pixels without expressing any deeper musical meaning.

The process of converting digital images of sheet music into symbolic music representations such as MIDI or MusicXML is commonly referred to as **optical music recognition** (OMR).[2] During this process, the image pixels have to be suitably grouped and interpreted in terms of musical symbols. This process is not easy, because of the many ways musical symbols may be engraved into sheet music. As discussed in the last section and illustrated by Figure 1.14, there may be substantial variations in the layout of the symbols and the staff system. Symbols do not always look exactly the same across different editions and may also be degraded in quality by artifacts of the printing or scanning process. Furthermore, musical symbols often intersect with staff lines, and several symbols may be stacked and combined (e.g., several notes sharing the same stem or combined with a beam). As a result, musical scores and the interrelations between musical symbols can become quite complex.

Correctly recognizing and interpreting the meaning of all the musical symbols is easy for a trained human, but hard for a computer. Figure 1.16a shows some examples of typical errors produced by automated OMR procedures. Some of these errors such as missing notes, flags or beams are of local nature, while other errors, such as an incorrectly detected key signature, affect all notes of a staff line. Even worse is the presence of a **transposing instrument**, whose music is notated at a pitch different from the pitch that is actually played (see Figure 1.16c). For example, a clarinet in B$^\flat$ is a transposed instrument, where a C in a score sounds like a B$^\flat$. Missing this information, which is encoded in textual form in front of a staff line, leads to a mis-

[2] The equivalent in the text domain is known as **optical character recognition** (OCR) with the goal of converting scanned images of printed text into machine-encoded text.

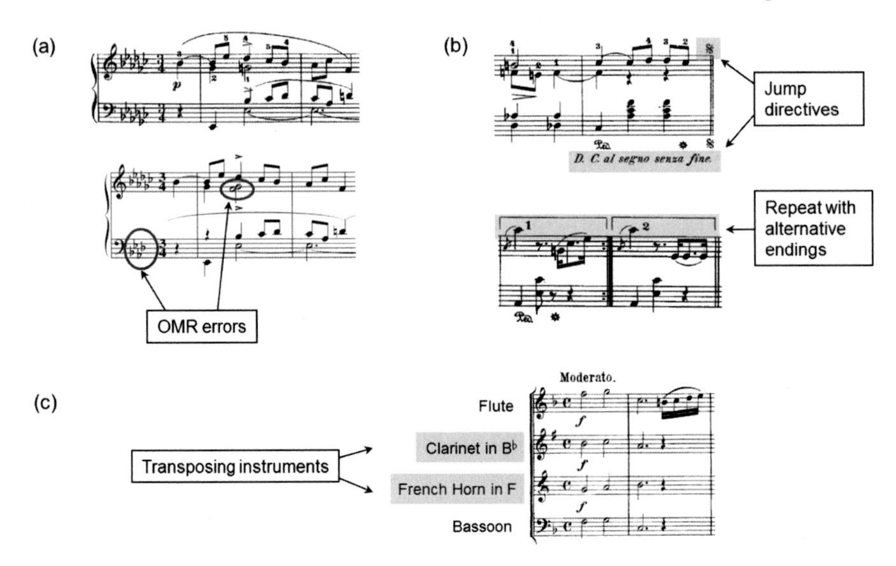

Fig. 1.16 **(a)** Examples of typical OMR errors (top: original score; bottom: OMR result). **(b)** Jump directives and repeats often not detected by OMR. **(c)** Transposed instruments often not interpreted correctly by OMR.

representation of all the notes' pitches. A score may also contain repeat signs with alternative endings or textual jump directives as shown in Figure 1.16b. This information is required to derive the correct sequence of measures to be performed by a musician. Consequently, an error in detecting jump directives may lead to structural misinterpretations of the score. Another problem is that even small artifacts in the scan may lead to confusion with musical symbols, e.g., a small dot being mixed up with a staccato mark. Even though current OMR software is reported to yield highly accurate results, manual postprocessing still seems necessary to obtain high-quality symbolic representation.

1.3 Audio Representation

Music is much more than a symbolic description of the notes to be played. Music is about making, creating, and shaping sounds. When musicians start delving into the music, the playing instructions recede into the background. The musical meter turns into a rhythmic flow, the different note objects melt into harmonic sounds and smooth melody lines, and the instruments communicate with each other. Musicians get emotionally involved with their music and react to it by continuously adapting tempo, dynamics, and articulation. Instead of playing mechanically, they speed up at some points and slow down at others in order to shape a piece of music. Similarly, they continuously change the sound intensity and stress certain notes. All of this results in a unique performance or an interpretation of the piece of music.

From a physical point of view, performing music results in **sounds** or **acoustic waves**, which are transmitted through the air as pressure oscillations. The term **audio** is used to refer to the transmission, reception, or reproduction of sounds that lie within the limits of human hearing. An **audio signal** is a representation of sound. As opposed to sheet music and symbolic representations, an audio representation encodes all information needed to reproduce an acoustic realization of a piece of music. This includes the temporal, dynamic, and tonal microdeviations that make up the specific performance style of a musician. However, in an audio representation, note parameters such as onset times, pitches or note durations are not given explicitly. This makes the analysis and comparison of music signals a difficult task, in particular with regard to polyphonic music, where different instruments and voices are superimposed upon each other. Furthermore, the perception of sounds does not only depend on objective properties of the acoustic wave, but also on subjective criteria as a result of the complex processing a sound undergoes by both the human ear and the brain. The study of subjective human sound perception is called **psychoacoustics**—for further details see [5, 17]. In this section, after having a look at waves and waveforms, we summarize the most important properties of audio representations: frequency and pitch, dynamics, intensity and loudness, as well as timbre.

1.3.1 Waves and Waveforms

A **sound** is generated by a vibrating object such as the vocal cords of a singer, the string and soundboard of a violin, the diaphragm of a kettledrum, or the prongs of a tuning fork. These vibrations cause displacements and oscillations of air molecules, resulting in local regions of compression and rarefaction. The alternating pressure travels through the air as a **wave**, from its source to a listener or a microphone. At its destination, it can then be perceived as sound by the human or converted into an electrical signal by a microphone (see Figure 1.17). In the case of a listener, the outer part of the ear captures the sound wave and passes it to the eardrum, which in turn starts vibrating according to the pressure oscillations. After further processing in the middle and inner ear, the sound wave is transformed into nerve impulses, which are finally sent to and interpreted by the brain. Graphically, the change in air pressure at a certain location can be represented by a **pressure–time plot**, also referred to as the **waveform** of the sound. The waveform shows the deviation of the air pressure from the average air pressure. Figure 1.18 shows a waveform representation of a recording of Beethoven's Fifth Symphony.

In general terms, a (mechanical) **wave** can be described as an oscillation that travels through space, where energy is transferred from one point to another. When a wave travels through some medium, the substance of this medium is temporarily deformed. As described above, sound waves propagate via air molecules colliding with their neighbors. After air molecules collide, they bounce away from each other

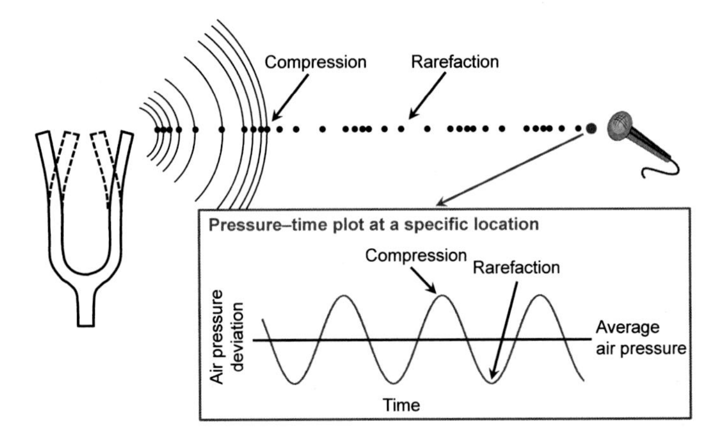

Fig. 1.17 Vibrating tuning fork resulting in a back and forth vibration of the surrounding air particles. The pressure oscillation propagates as a longitudinal wave through the air. The waveform shows the deviation over time of the air pressure from the average air pressure at a specific location (as indicated by the microphone).

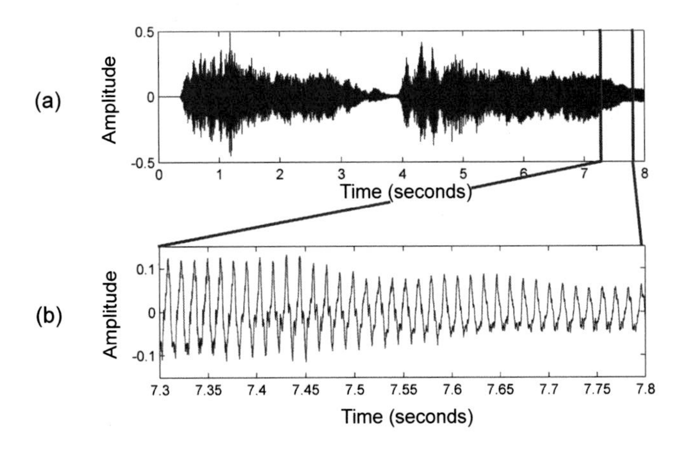

Fig. 1.18 **(a)** Waveform of the first eight seconds of a recording of the first five measures of Beethoven's Fifth as indicated by Figure 1.1. **(b)** Enlargement of the section between 7.3 and 7.8 seconds.

(a restoring force). This keeps the molecules from continuing to travel in the direction of the wave. Instead, they oscillate around almost fixed locations. A general wave can be **transverse** or **longitudinal**, depending on the direction of its oscillation. Transverse waves occur when a disturbance creates oscillations perpendicular (at right angles) to the propagation (the direction of energy transfer). Longitudinal waves occur when the oscillations are parallel to the direction of propagation. According to this definition, a vibration in a string is an example of a transverse wave, whereas a sound wave has the form of a longitudinal wave. A transverse wave can

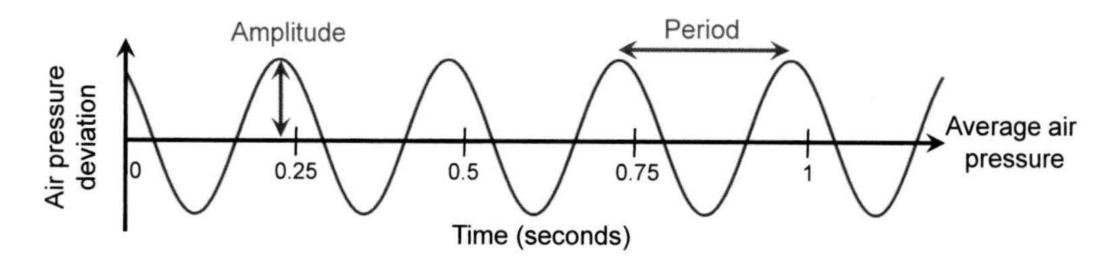

Fig. 1.19 Waveform of a sinusoid with a frequency of 4 Hz.

in fact generate a longitudinal wave and visa versa. An instrument's vibrating string, which oscillates between the two fixed end points, gradually emits its energy to the air, generating a longitudinal sound wave. If this wave, in turn, hits an eardrum, again a transverse wave is generated.

1.3.2 Frequency and Pitch

We have seen that a sound wave can be visually represented by a waveform. If the points of high and low air pressure repeat in an alternating and regular fashion, the resulting waveform is called **periodic**. In this case, the **period** of the wave is defined as the time required to complete a cycle. The **frequency**, measured in **Hertz** (Hz), is the reciprocal of the period. Figure 1.19 shows a **sinusoid**, which is the simplest type of periodic waveform. In this example, the waveform has a period of a quarter second and hence a frequency of 4 Hz. A sinusoid is completely specified by its frequency, its **amplitude** (the peak deviation of the sinusoid from its mean), and its **phase** (determining where in its cycle the sinusoid is at time zero). These three attributes of a sinusoid will become important when analyzing general audio signals (see Section 2.3).

The higher the frequency of a sinusoidal wave, the higher it sounds. The audible frequency range for humans is between about 20 Hz and 20,000 Hz (20 kHz). Other species have different hearing ranges. For example, the top end of a dog's hearing range is about 45 kHz, a cat's is 64 kHz, while bats can even detect frequencies beyond 100 kHz. This is why one can use a dog whistle, which emits **ultrasonic sound** beyond the human hearing capability, to train and to command animals without disturbing nearby people.

The sinusoid can be considered the prototype of an acoustic realization of a musical note. Sometimes the sound resulting from a sinusoid is called a **harmonic sound** or **pure tone**. As indicated in Section 1.1.1, the notion of frequency is closely related to what determines the **pitch** of a sound. In general, pitch is a subjective attribute of sound. In the case of complex sound mixtures its relation to frequency can be especially ambiguous. In the case of pure tones, however, the relation between frequency and pitch is clear. For example, a sinusoid having a frequency of 440 Hz corresponds to the pitch A4. This particular pitch is known as **concert pitch**, and it is used as the reference pitch to which a group of musical instruments are tuned for a performance. Since a slight change in frequency does not necessarily lead to a perceived change, one usually associates an entire range of frequencies with a single pitch.

As mentioned in Section 1.1.1, two frequencies are perceived as similar if they differ by a power of two, which has motivated the notion of an octave. For example, the pitches A3 (220 Hz), A4 (440 Hz), and A5 (880 Hz) sound similar. Furthermore, the perceived distance between the pitches A3 and A4 is the same as the perceived distance between the pitches A4 and A5. In other words, the human perception of pitch is logarithmic in nature. This perceptual property has already been used in Section 1.1.1 when defining the equal-tempered scale that subdivides an octave into twelve semitones based on a logarithmic frequency axis. More formally, using the MIDI note numbers introduced in Section 1.2.2, we can associate to each pitch $p \in [0:127]$ a **center frequency** $F_{\mathrm{pitch}}(p)$ (measured in Hz) by

$$F_{\mathrm{pitch}}(p) = 2^{(p-69)/12} \cdot 440. \tag{1.1}$$

Indeed, this formula yields the frequency $F_{\mathrm{pitch}}(p) = 440$ for the reference pitch $p = 69$ (A4). Increasing the pitch number by 12 (an octave) leads to an increase by a factor of two, i.e., $F_{\mathrm{pitch}}(p+12) = 2 \cdot F_{\mathrm{pitch}}(p)$. Similarly, it is easy to show that the frequency ratio

$$F_{\mathrm{pitch}}(p+1)/F_{\mathrm{pitch}}(p) = 2^{1/12} \approx 1.059463 \tag{1.2}$$

of two subsequent pitches $p+1$ and p is constant (see Exercise 1.6). In other words, multiplying the center frequency of an arbitrary pitch by this constant, the pitch is raised by a semitone. Generalizing the notion of semitones, the **cent** denotes a logarithmic unit of measure used for musical intervals. By definition, an octave is divided into 1200 cents, so that each semitone corresponds to 100 cents. Again the ratio of frequencies one cent apart is constant, yielding the value

$$2^{1/1200} \approx 1.0005777895. \tag{1.3}$$

The difference in cents between two frequencies, say ω_1 and ω_2, is given by

$$\log_2\left(\frac{\omega_1}{\omega_2}\right) \cdot 1200. \tag{1.4}$$

The interval of one cent is much too small to be heard between successive notes. The threshold of what is perceptible, also called the **just noticeable difference**, varies from person to person and depends on other aspects such as the timbre (Section 1.3.4) and the musical context. As a rule of thumb, normal adults are able to recognize pitch differences as small as 25 cents very reliably, with differences of 10 cents being recognizable only by trained listeners.

Real-world sounds are far from being a simple pure tone with a well-defined frequency. Playing a single note on an instrument may result in a complex sound that contains a mixture of different frequencies changing over time. Intuitively, such a **musical tone** can be described as a superposition of pure tones or sinusoids, each with its own frequency of vibration, amplitude, and phase. A **partial** is any of the sinusoids by which a musical tone is described. The frequency of the lowest partial present is called the **fundamental frequency** of the sound. The pitch of a musical tone is usually determined by the fundamental frequency, which is the one created by vibration over the full length of a string or air column of an instrument. A **harmonic** (or a **harmonic partial**) is a partial that is an integer multiple of the fundamental frequency. Partials, as well as harmonics, are counted upwards along the frequency axis. This convention implies that the fundamental frequency is the first partial, as well as the first harmonic of a musical tone. The term **inharmonicity** is used to denote a measure of the deviation of a partial from the closest ideal harmonic, typically measured in cents for each partial. Finally, another term that is often used in music theory is the **overtone**, which is any partial except the lowest. This can lead to numbering confusion when comparing overtones with partials, since the first overtone is the second partial.

Most pitched instruments are designed to have partials that are close to being harmonics, with very low inharmonicity. Thus, for simplicity, one often speaks of the partials in those instruments' sounds as harmonics, even if they have some inharmonicity. Other pitched instruments, especially certain percussion instruments, such as the marimba, vibraphone, bells, and kettledrums (timpani), contain nonharmonic partials, yet give the ear a good sense of pitch. Nonpitched, or indefinite-pitched, instruments, such as cymbals, gongs, or tam-tams, make sounds rich in inharmonic partials. As an example of a harmonic sound, Figure 1.18 shows in its lower part an enlargement of the waveform of the section between 7.3 and 7.8 seconds, which reveals the almost periodic nature of the sound signal. The waveform within these 500 ms corresponds to the sound of a decaying D, which is played by the orchestra in unison in the fourth and fifth measure (see Figure 1.1). Indeed, one counts 37 periods within this section, corresponding to a frequency of 74 Hz—the fundamental frequency of D2.

We close this section on frequency and pitch by looking at harmonics in terms of musical pitches. Let ω denote the center frequency of a musical note, e.g., $\omega = 65.4$ Hz for C2 (having MIDI note number $p = 36$). The harmonic series is an arithmetic series $\omega, 2\omega, 3\omega, 4\omega, \ldots$, where the difference between consecutive harmonics is constant and equal to the fundamental. Since our perception of pitch is logarithmic in frequency, we perceive higher harmonics as "closer together" than lower ones. On the other hand, the octave series is a geometric progression ω, 2ω, 4ω, $8\omega, \ldots$, and we hear these distances as "the same" in the sense of musical interval. Consequently, in terms of what we hear, each octave in the harmonic series is divided into increasingly "smaller" and more numerous intervals. In our example, the second harmonic (2ω) sounds like a C3 (one octave higher), the third harmonic (3ω) like a G3 (a so-called **perfect fifth** above C3), and the fourth harmonic (4ω) like a C4 (two octaves higher). Starting with a C2, Figure 1.20 shows for each of the first 16 harmonics the musical note that is closest in terms of the difference between the harmonic's frequency and the center frequency of the note as specified

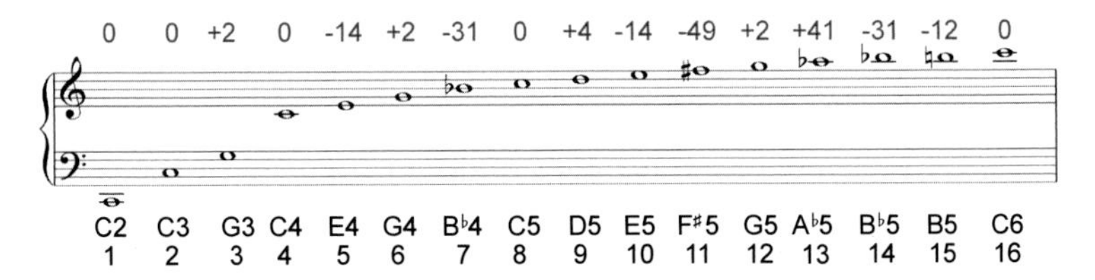

Fig. 1.20 Illustration of the harmonic series in music notation. Starting with the note C2, for each of the first 16 harmonics the closest musical note is shown. On top, the difference (in cents) between a harmonic's frequency and the center frequency of the closest note is shown.

in (1.1) (see also Exercise 1.9). For example, the frequency of the third harmonic is just 2 cents above the center frequency of G3, which is much smaller than the just noticeable difference. In contrast, the frequency of the 11th harmonic is 49 cents below the center frequency of the note F♯5, which is nearly half a semitone and clearly audible. If the harmonics are transposed into the span of one octave (by suitably multiplying or dividing the frequencies by a power of two), they approximate certain notes of the twelve-tone equal-tempered scale. Some of the twelve scale steps are approximated well such as the ones for C (1st harmonic), G (3rd harmonic), or D (9th harmonic), whereas others are problematic such as F♯ (11th harmonic), A♭ (13th harmonic), or B♭ (7th harmonic).

1.3.3 Dynamics, Intensity, and Loudness

As mentioned in Section 1.1.2, a further important property of music concerns the **dynamics**, a general term that is used to refer to the volume of a sound as well as to the musical symbols that indicate the volume. For example, a piano (notated as ***p***) indicates that notes are to be played softly, whereas a forte (notated as ***f***) indicates that notes are to be played loudly. There are many more indicators for describing the dynamics of notes in sheet music. On the audio side, dynamics correlate with a perceptual property called **loudness**, by which sounds can be ordered on a scale extending from quiet to loud. Similarly to the relation between pitch and frequency, loudness is a subjective measure which correlates to objective measures of **sound intensity** and **sound power**. However, loudness also depends on other sound characteristics such as duration or frequency. We will come back to some of these subjective phenomena after having a closer look at the objective measures.

From a physical point of view, it is not easy to strictly define the intensity or power of a sound. In the following, we only give some intuitive explanations. In general, **power** is the rate at which energy is transferred, used, or transformed. Power is measured in units of **watt** (W), which is defined as one joule per second. For example, the rate at which a light bulb transforms electrical energy into heat and light is measured in watts—the more wattage, the more power, or equivalently the more electrical energy is used per unit time. Similarly, **sound power** expresses how much

Table 1.1 Typical intensity values given in $\mathrm{W/m^2}$ (intensity), in decibels (intensity level), and by a factor compared with the TOH.

Source	Intensity	Intensity level	× TOH
Threshold of hearing (TOH)	10^{-12}	0 dB	1
Whisper	10^{-10}	20 dB	10^{2}
Pianissimo	10^{-8}	40 dB	10^{4}
Normal conversation	10^{-6}	60 dB	10^{6}
Fortissimo	10^{-2}	100 dB	10^{10}
Threshold of pain	10	130 dB	10^{13}
Jet take-off	10^{2}	140 dB	10^{14}
Instant perforation of eardrum	10^{4}	160 dB	10^{16}

energy per unit time is emitted by a sound source passing in all directions through the air. The term **sound intensity** is then used to denote the sound power per unit area.

In practice, sound power and sound intensity can show extremely small values that are still relevant for human listeners. For example, the **threshold of hearing** (TOH), which is the minimum sound intensity of a pure tone a human can hear, is as small as

$$I_{\mathrm{TOH}} := 10^{-12}\ \mathrm{W/m^2}. \tag{1.5}$$

Furthermore, the range of intensities a human can perceive is extremely large with $I_{\mathrm{TOP}} := 10\ \mathrm{W/m^2}$ being the **threshold of pain** (TOP). For practical reasons, one switches to a logarithmic scale to express power and intensity. More precisely, one uses a **decibel** (dB) scale, which is a logarithmic unit expressing the ratio between two values. Typically, one of the values serves as a reference, such as I_{TOH} in the case of sound intensity. Then the intensity measured in dB is defined as

$$\mathrm{dB}(I) := 10 \cdot \log_{10}\left(\frac{I}{I_{\mathrm{TOH}}}\right). \tag{1.6}$$

From this definition, one obtains $\mathrm{dB}(I_{\mathrm{TOH}}) = 0$, and a doubling of the intensity results in an increase of roughly 3 dB:

$$\mathrm{dB}(2 \cdot I) = 10 \cdot \log_{10}(2) + \mathrm{dB}(I) \approx 3 + \mathrm{dB}(I). \tag{1.7}$$

When specifying intensity values in terms of decibels, one also speaks of **intensity levels**. Table 1.1 shows some typical intensity values given in $\mathrm{W/m^2}$ as well as in decibels for some sound sources and dynamics indicators. For example, notes being played pianissimo ("very softly") typically result in intensity levels around 40 dB, whereas notes being played fortissimo ("very loudly") can reach levels up to 100 dB.

We now come back to the concept of **loudness**, which is the perceptual correlate to sound intensity [6, 17]. As said before, the loudness is affected by a number of factors. First of all, the same sound may be perceived to have different loudness depending on the individual. In particular, age is one factor that affects the human ear's response to a sound. Also, the duration of the sound influences perception, since the human auditory system averages the effect of sound intensity over an interval up to a second. Therefore, a human has the feeling that a sound lasting for 200 ms

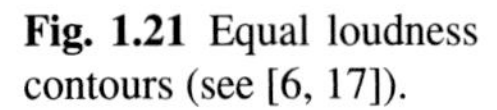

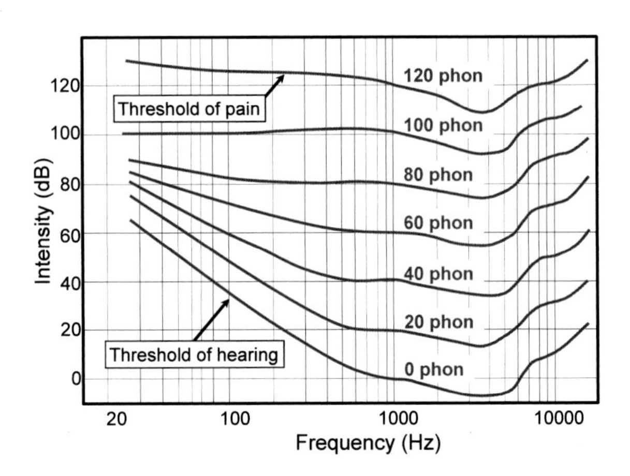

Fig. 1.21 Equal loudness contours (see [6, 17]).

is louder than a similar sound only lasting 50 ms. Furthermore, two sounds with the same intensity but different frequencies are generally not perceived to have the same loudness. Humans with normal hearing are most sensitive to sounds around 2 to 4 kHz, with sensitivity declining for lower as well as higher frequencies. Based on psychoacoustic experiments, the perceived loudness of pure tones depending on the frequency has been determined and expressed by the unit **phon**. Figure 1.21 shows **equal loudness contours**. Each contour line specifies for a fixed loudness given in phons the sound intensities over a (logarithmically spaced) frequency axis. The unit of a phon is normalized with respect to the frequency of 1,000 Hz, where a phon value equals the intensity level in dB. The contour for 0 phon shows how the threshold of hearing depends on frequency.

1.3.4 Timbre

Besides pitch, loudness, and duration, there is another fundamental aspect of sound referred to as **timbre** or **tone color**. Timbre allows a listener to distinguish the musical tone of a violin, an oboe, or a trumpet even if the tone is played at the same pitch and with the same loudness. As with pitch and loudness, timbre is a perceptual property of sound [22]. However, timbre is very hard to grasp, and because of its vagueness, it is often described in an indirect way: timbre is the attribute whereby a listener can judge two sounds as dissimilar using any criterion other than pitch, loudness, and duration. For example, timbre information allows us to tell apart the sounds produced by the oboe and the violin, even when the pitch and loudness of the sounds are identical [19]. The sound of a musical instrument may be described with such words as bright, dark, warm, harsh, and other terms. Researchers have tried to approach timbre by looking at correlations to more objective sound characteristics such as the temporal and spectral evolution, the absence or presence of tonal and

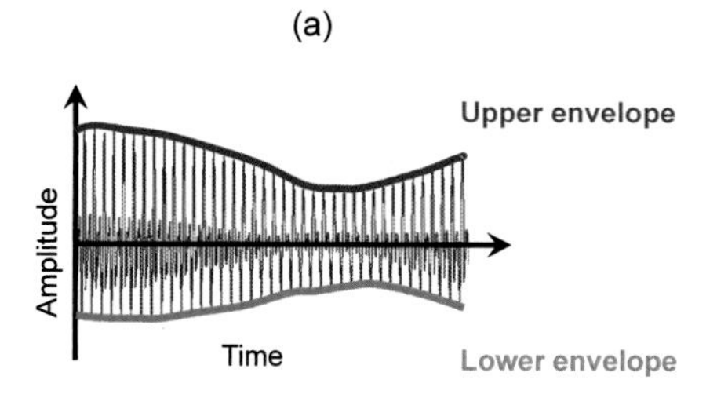

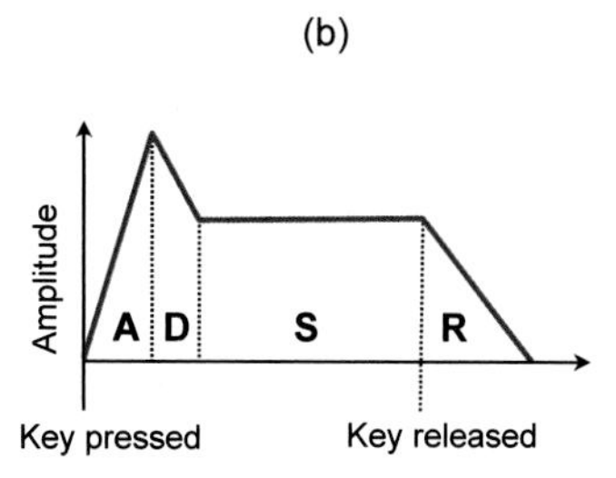

Fig. 1.22 **(a)** Envelope of a signal. **(b)** Schematic view of an ADSR envelope.

noise-like components, or the energy distribution across the partials of a tone. In the following, we take a closer look at some of these characteristics.

When striking a piano key, the resulting sound is much more than a superposition of pure sinusoids that correspond to the fundamental frequency and its overtones. Playing a single note already produces a complex sound mixture with characteristics that may constantly change over time, containing periodic as well as nonperiodic components. At the beginning of a musical tone, there is often a sudden increase of energy. In this short phase, the **attack phase** of the tone, the sound builds up. It contains a high degree of nonperiodic components that are spread over the entire range of frequencies, a property that is also inherent to **noise**. In acoustics, such a noise-like short-duration sound of high amplitude occurring at the beginning of a waveform is also called a **transient**. In the case of a piano, striking a key triggers an entire chain of mechanical actions before a hammer hits one or several strings. All these actions, starting with the finger touching the key and ending with the hammer hitting the strings, produce mechanical noise that merges with the acoustic effects of the strings' excitation. After the attack phase, the sound of a musical tone stabilizes (**decay phase**) and reaches a steady phase with a (more or less) periodic pattern. This third phase, which is also called the **sustain phase**, makes up most of the duration of a musical tone, where the energy remains more or less constant or slightly decreases as is the case with a piano sound. In the final phase of a musical tone, also called the **release phase**, the musical tone fades away. For a piano, this phase starts as soon as the finger leaves the key and the damper stops the strings' vibrations.

Intuitively, the **envelope** of a waveform can be regarded to be a smooth curve outlining its extremes in amplitude (see Figure 1.22a). The different phases as described above have a strong influence on the shape of the envelope of a musical tone. In sound synthesis, the envelope of a signal to be generated is often described by a model called **ADSR**, which consists of an attack (A), decay (D), sustain (S), and release (R) phase (see Figure 1.22b). The relative durations and the amplitudes of the four phases have a significant impact on how the synthesized tone will sound.

The ADSR model is a strong simplification and only yields a meaningful approximation for amplitude envelopes of tones that are generated by certain instruments. For example, the musical tone shown in Figure 1.23a, which is the note C4 played on a piano, has an envelope that is similar to the one suggested by the ADSR model.

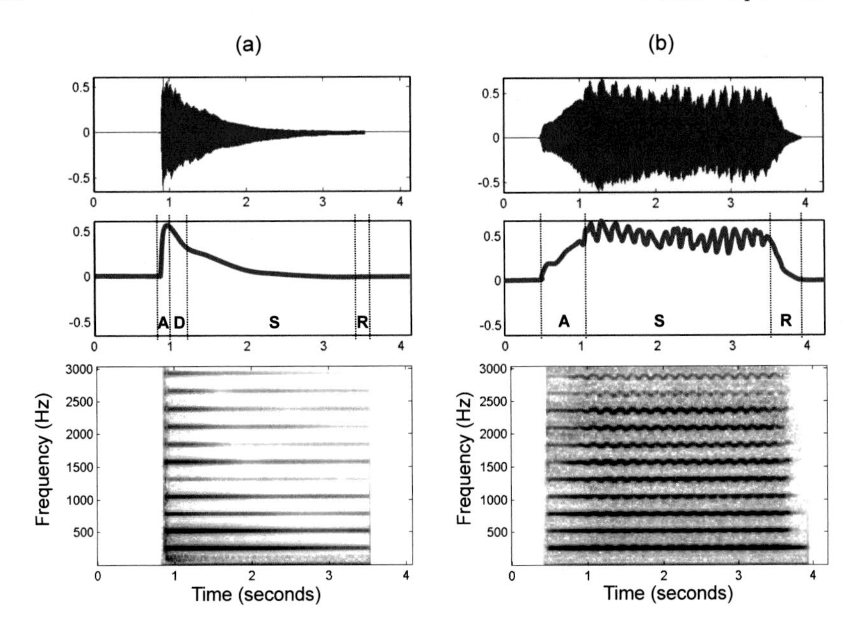

Fig. 1.23 Waveform, amplitude envelope, and spectrogram representation for different instruments playing the same note C4 (261.6 Hz). **(a)** Piano. **(b)** Violin.

After a sharp attack (when the hammer hits the string) and a stabilizing decay, the tone continuously fades out. In the case of a piano sound, the decrease in sound intensity is very slow as long as the damper does not touch the string. Therefore, one can regard this phase as a kind of sustain phase. When the piano key is released and the damper stops the string's vibration, the sound quickly comes to an end. For other instruments, however, the amplitude may evolve in a completely different fashion. This is illustrated by Figure 1.23b, which shows an envelope for the note C4 played on a violin. First of all, since the tone is played softly with a gradual increase in volume, the attack phase is spread out in time. Furthermore, there does not seem to be any decay phase and the subsequent sustain phase is not steady; instead, the amplitude envelope oscillates in a regular fashion. The release phase starts when the violinist stops exciting the string with the bow. The sound then quickly fades out.

For our violin example, one can observe periodic variations in amplitude. This phenomenon, known as **tremolo**, is generated by certain playing styles used for string or wind instruments. The effect of tremolo often goes along with **vibrato**, which is a musical effect consisting of a regular, pulsating change of frequency. Besides string music, vibrato is mainly used by human singers to add expression. In technical terms, tremolo corresponds to an **amplitude modulation**, whereas vibrato corresponds to a **frequency modulation**. Both tremolo and vibrato depend on two parameters: the extent of the variation and the rate at which the amplitude or frequency is varied. Even though tremolo and vibrato are simply local changes in intensity and frequency, they do not necessarily evoke a perceived change in loudness or pitch of the overall musical tone. Rather, they are features that influence the timbre of a musical tone.

Perhaps the most important and well-known property for characterizing timbre is the existence of certain partials and their relative strengths [19]. Recall from Section 1.3.2 that partials are the dominant frequencies of a musical tone with the lowest partial being the fundamental frequency. The inharmonicity expresses the extent to which a partial deviates from the closest ideal harmonic. For harmonic sounds such as a musical tone with a clearly perceivable pitch, most of the partials are close to being harmonics. However, not all partials need to occur with the same strength, as we will see in a moment.

The composition of a sound in terms of its partials can be visualized by a so-called **spectrogram**, which shows the intensity of the occurring frequencies over time. For a detailed introduction on such time–frequency representations refer to Section 2.5. Figure 1.23a shows at the bottom a spectrogram for the note C4 played on a piano, where the intensity is reflected by the shade of gray (the darker the more intense). Both the fundamental frequency of the note (261.6 Hz) as well as its harmonics (integer multiples of 261.6 Hz) are visible as horizontal lines. The decay of the musical tone is reflected by a corresponding decay in each of the partials. Most of the tone's energy is contained in the lower partials, and the energy tends to be lower for the higher partials. Such a distribution is typical for many instruments.

For string instruments, sounds tend to have a rich spectrum of partials, where lots of energy may also be contained in the upper harmonics (see Figure 1.23b). This figure also reveals the vibrato as a regular oscillation in the time–frequency plane. Certain classes of wind instruments including the clarinet (so-called **closed-pipe** wind instruments) produce a very characteristic spectrum of partials. For a cylindrical wind instrument that is open at one end, but closed at the other (at the mouthpiece), one can show that the even harmonics do not show up. In other words, most energy is contained in the odd harmonics $\omega_0, 3\omega_0, 5\omega_0, \ldots$, with ω_0 denoting the fundamental frequency. For a musical tone played on a bassoon, the fundamental frequency often contains much less energy compared with the higher partials. In contrast, for a tuning fork, most energy is contained in the fundamental frequency, resulting in a sound that is close to a synthesized sinusoid. Instruments such as bells have a very complex spectrum with lots of inharmonicities, which often evokes in the listener the feeling of a bell being out-of-tune. For stringed instruments, one can often measure substantial deviations between higher partials and the theoretical harmonics. The less elastic a string is (that is, the shorter, thicker, higher tension or stiffer it is), the more inharmonicity it may exhibit. This particularly holds for the piano, where such inharmonicities have a crucial influence on the timbre.

With this discussion, we want to indicate that timbre is a multidimensional phenomenon that is hard to measure. It is the irregularities and variations that make a musical tone sound interesting and that give it a particular and natural quality.

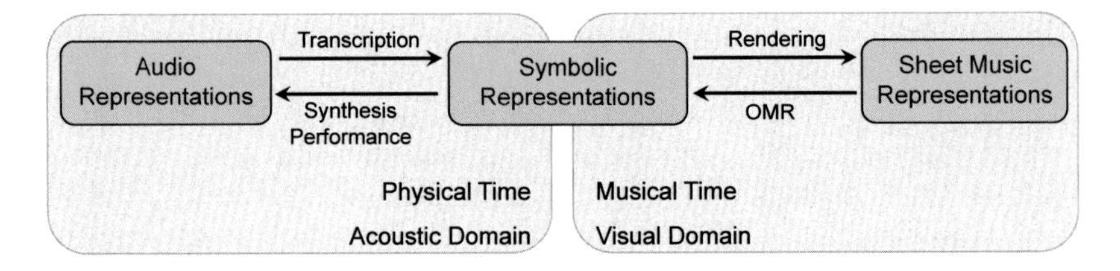

Fig. 1.24 Illustration of three classes of music representation and their relations.

1.4 Summary and Further Readings

In this chapter, we looked at three different classes of music representations while introducing some musical and technical terminology that is used throughout this book. We use the term **sheet music** to refer to visual representations of a musical score either given in printed form or encoded digitally in some image format. The term **symbolic** stands for any kind of symbolic representation where the entities have an explicit musical meaning. Finally, the term **audio** is used to denote music recordings given in the form of acoustic waveforms. The boundaries between these classes are not clear. In particular, as illustrated by Figure 1.24, symbolic representations may be close to both sheet music as well as audio representations [24]. On the one hand, symbolic representations such as MusicXML are used for **rendering** sheet music, where the shape of the note objects and their arrangement on a page are determined. As we have seen, optical music recognition (OMR) is the inverse process with the goal of transforming sheet music into a symbolic representation. On the other hand, symbolic representations such as MIDI are used for **synthesizing** audio, where the note objects are transformed into musical tones and real sounds. The inverse process is known as **music transcription**, where the objective is to extract note events, key signature, time signature, instrumentation, and other score parameters from a given music recording [2, 13].

In a sense, symbolic representations can be regarded as the link between the visual (or graphical) domain accommodating sheet music representations and the acoustic (or physical) domain accommodating audio representations [24]. In the first case, timing is specified in terms of the shape and the relative arrangement of the musical symbols and is typically given in musical units such as measures or beats. In the latter case, timing is specified in physical units such as seconds. For music recordings, there are often no sharp note onsets or offsets (think of a soft onset for a note played on a violin or a gradual fade-out) and the specification of the beginning and the end of musical events becomes an ill-defined problem. For a general discussion of alignment procedures to bridge the gap between sheet music and audio representations, we refer to [24].

Of course, any kind of categorization of music representations goes along with an oversimplification. Our categorization is far from being comprehensive. We have seen that, when describing musical attributes such as pitch, loudness, and timbre, human perception is a crucial factor. Therefore, besides the acoustic and visual domain, Babbitt [1] considers an additional **auditory** domain. In his taxonomy, a graphemic note (the blob on the page) corresponds in meaning with the (auditory) percept of the note. From a philosophical point of view, as argued by Wiggins et al. [25], music is actually something abstract and intangible which does not have real existence in itself. In this sense, all of the domain-specific representations are *aspects* of music, but none of them *is* music, individually. Mazzola [15] considers music to be the universe of all different perspectives one may assume. For a psychologically based approach to music along with the expectations and emotions it evokes, we refer to the book by Huron [12]. Running the risk of oversimplification, we adopt in this book a more technically oriented view of music processing and leave out perhaps the most important aspect of music: the human mind.

Sheet music has a history of hundreds of years, and the basic concepts we have presented can be found in introductory textbooks on music notation [9], and we also refer to Wikipedia as a rich source of useful information on this topic. Because of significant digitization efforts, sheet music is now widely available in digital formats. In particular for Western classical music, scanned versions of musical editions out of copyright are now freely accessible on the world wide web. One prominent example is the *Petrucci Music Library*, which is a virtual library of public-domain music scores organized and created by the *International Music Score Library Project* (IMSLP).[3] For symbolic music, many formats have been suggested in the literature to represent sheet music in a digital, machine-readable form. A comprehensive account on MIDI[4] and its use with electronic instruments and sequencers can be found in [11]. Extensions and challenges of the MIDI format are summarized in [14]. In the book edited by Selfridge-Field [21], one not only finds an introduction to the MIDI format but also a detailed overview and description of symbolic formats up to the year 1997. Since then, many new formats have been proposed and developed, including both open and well-documented formats, as well as proprietary formats that are bound to specific software packages. The MusicXML[5] format [8] and the MEI[6] format developed by the community-driven Music Encoding Initiative [10], are only two prominent examples. Similarly, a multitude of commercial and noncommercial OMR software systems have been developed. While many of these systems only work for printed sheet music, others also address the much harder problem of recognizing handwritten scores. In recent decades, significant research efforts have been directed towards improving, comparing, and evaluating OMR systems [3]. Even though substantial improvements could be achieved, also thanks to recent data-driven techniques based on deep learning, OMR can still not be regarded as a solved problem. For a comprehensive overview of OMR literature, we refer to the *Bibliography on Optical Music Recognition*.[7]

---

[3] `http://imslp.org`

[4] `http://www.midi.org`

[5] `http://www.musicxml.com`

[6] `https://music-encoding.org/`

[7] `https://omr-research.github.io/`

There are many excellent books on the foundations of the acoustical properties of music and audio signals. For example, the classic book by Fletcher and Rossing [7] gives a detailed account on musical sound waves and the physics behind their generation by musical instruments. The book by Fastl and Zwicker [5] as well as the one by Moore [17] give deeper insights into the field of auditory perception and psychoacoustics for general audio signals. A source of inspiration for this chapter has been the book by Sethares [22] on tuning, timbre, spectrum, and scale, which provides interesting insights (along with sound examples) on how these concepts are related. A signal-processing-oriented approach to the concepts of timbre and instrumentation can be found in [19].

We finally want to remark that deep learning techniques have opened up new avenues for various tasks related to processing, converting, and linking music representations. For example, this holds for the task of OMR when a sufficient amount of well-annotated training data is available [3]. Similarly, major progress could be achieved in music transcription using deep learning techniques [2]. Data-driven techniques are also increasingly used for cross-modal retrieval and alignment tasks [4, 18]. However, music turns out to be a hard domain due to the complexity and diversity of music, which would require vast amounts of data to efficiently cover all these aspects. For example, OMR is still a hard problem for handwritten music or sheet music with a dense and complex layout. Similarly, while automated methods for music transcription work well for piano recordings of high acoustic quality (where one has a lot of training data), the automatic conversion of complex orchestral or choir performances into score notation—a task Mozart was capable of after listening to a polyphonic choral piece only once—is still a largely open problem despite decades of research.

1.5 FMP Notebooks

In this chapter, we have seen that musical information can be represented in many different ways, including sheet music, symbolic, and audio representations. In **Part 1** of the FMP notebooks [20], which is closely associated with this first chapter, we offer visual and acoustic material as well as Python code examples to study musical and acoustic properties of music. We now briefly go through the FMP notebooks of **Part 1** one by one while indicating how these can be used for possible experiments and exercises.

We start with the **FMP Notebook *Sheet Music Representations***, where we take up the example of Beethoven's Fifth Symphony. Besides the piano reduced version (see Figure 1.1) and a full orchestral score (see Figure 1.10), we also show a computer-generated sheet music representation. The comparison of these versions is instructive, since it demonstrates the huge differences one may have between different layouts, also indicating that the generation of visually pleasing sheet music representations from score representations is an art in itself. Besides the visual data, the notebook also provides different recordings of this passage, including a synthesized orchestral version created from a full score and a recording by the Vienna Philharmonic orchestra conducted by Herbert von Karajan (1946). The comparison between the mechanical and performed versions shows that one requires additional knowledge not directly specified in the sheet music to make the music come alive. In the data folder of **Part 1** (`data/C1`), one finds additional representations of our Beethoven example including Sibelius files[8] of a piano, orchestral, and string quartet version. These files, in turn, have been exported in symbolic formats (`.mid`, `.sib`, `.xml`), image formats (`.png`), and audio formats (`.wav`, `.mp3`). Such systematically generated data is well suited for hands-on exercises that allow teachers and students to experiment within a controlled setting. This is also one reason why we will take up the Beethoven (and other) examples again and again throughout the FMP notebooks.

In the **FMP Notebook *Musical Notes and Pitches***, we deepen the concepts as introduced in Section 1.1.1. We show how to generate musical sounds using a simple sinusoidal model, which can then be used to obtain acoustic representations of concepts such as octaves, pitch classes, and musical scales. In the **FMP Notebook *Chroma and Shepard Tones***, we generate Shepard tones, which are weighted superpositions of sine waves separated by octaves. These tones can be used to sonify the chromatic circle and Shepard's helix of pitch (see Figure 1.3). Extending the notion of the twelve-tone discrete chromatic circle, one can generate a pitch-continuous version, where the Shepard tones ascend (or descend) continuously. Originally created by the French composer Jean-Claude Risset, this continuous version is also known as the **Shepard–Risset glissando**. To implement such a glissando, one requires a chirp function with an exponential (rather than a linear) frequency increase. Experimenting with Shepard tones and glissandi not only leads to interesting sound effects that may be used even for musical compositions, but also deepens the understanding of concepts such as frequency, pitch, and the role of overtones. The concept of Shepard tones can also be used to obtain a sonification of chroma features as introduced in Section 3.1.2 (see also the **FMP Notebook *Sonification*** of **Part B**).

In the subsequent FMP notebooks, we discuss Python code for parsing, converting, and visualizing various symbolic music formats. In particular, for students who are not familiar with Western music notation, the piano-roll representation yields an easy-to-understand geometric encoding of symbolic music. Motivated by traditional piano rolls, the horizontal axis of this two-dimensional representation encodes time, whereas the vertical axis encodes pitch. The notes are visualized as axis-parallel rectangles, where the color of the rectangles can be used to encode additional note parameters such as velocity or instrumentation. A piano-roll representation can be easily stored in a comma-separated values (`.csv`) file, where each line encodes a note event specified by parameters such as `start`, `duration`, `pitch`, `velocity`, and an additional `label` (e.g., encoding the instrumentation). This slim and explicit format, even though representing symbolic music in a simplified way, is used throughout most parts of the FMP notebooks, where the focus is on

[8] These files have a `.sib` extension and are generated by the `Sibelius` music notation software application.

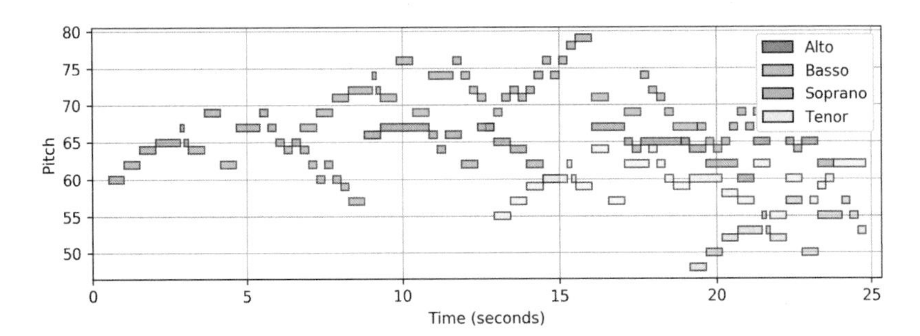

Fig. 1.25 Visualization of a piano-roll representation generated by the **FMP Notebook *Symbolic Format: CSV***. The figure shows the beginning of the four-voice Fugue BWV 846 in C major by Johann Sebastian Bach.

the processing of waveform-based audio signals. In the **FMP Notebook *Symbolic Format: CSV***, we introduce the Python library `pandas`, which provides easy-to-use data structures and data analysis tools for parsing and modifying text files [16]. Furthermore, we introduce a function for visualizing a piano-roll representation as shown in Figure 1.25. The implementation of such visualization functions is an instructive exercise for students to get familiar with fundamental musical concepts as well as to gain experience in standard concepts of Python programming.

As discussed in Section 1.2 , there are numerous formats for encoding symbolic music. Describing and handling these formats goes beyond this textbook. The good news is that there are various Python software tools for parsing, manipulating, synthesizing, and storing music files. In the **FMP Notebook *Symbolic Format: MIDI***, we introduce the Python package `PrettyMIDI` for handling MIDI files. This package allows for transforming the (often cryptic) MIDI messages into a list of easy-to-understand note events, which may then be stored using simple CSV files. Similarly, in the **FMP Notebook *Symbolic Format: MusicXML***, we indicate how the Python package `music21` can be used for parsing and handling symbolic music given as a MusicXML file. This package is a toolkit for computer-aided musicology which allows users to study large datasets of symbolically encoded music, to generate musical examples, to teach fundamentals of music theory, to edit musical notation, to study music and the brain, and to compose music. Finally, in the **FMP Notebook *Symbolic Format: Rendering***, we discuss some software tools for rendering sheet music from a given symbolic music representation. By mentioning a few open-source tools, our FMP notebooks only scratch the surface on symbolic music processing and are intended to yield entry points to this area.

The next FMP notebooks cover aspects of audio representations and their properties (Section 1.3). In the **FMP Notebook *Waves and Waveforms***, we provide functions for simulating transverse and longitudinal waves as well as combinations thereof. Furthermore, one finds Python code for generating videos of these simulations, thus indicating how the FMP notebooks can be used for generating educational material (see Figure 1.26). In the **FMP Notebook *Frequency and Pitch***,

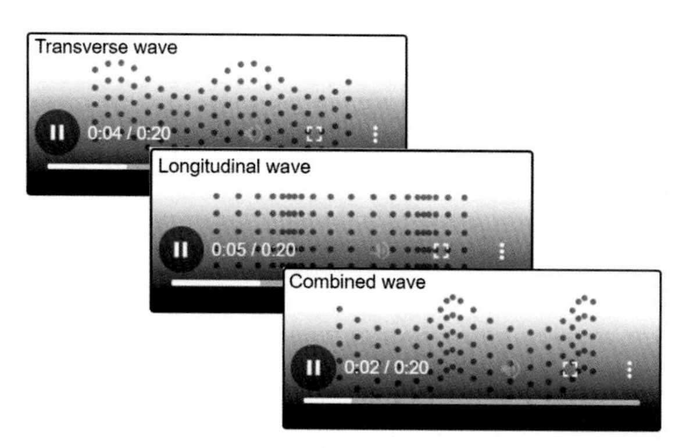

Fig. 1.26 Videos generated by the **FMP Notebook *Waves and Waveforms*** to illustrate the concepts of transverse, longitudinal, and combined waves.

we discuss some experiments on the audible frequency range and the just-noticeable difference in pitch perception. In the **FMP Notebook *Harmonic Series***, one finds an acoustic comparison of the musical scale based on harmonics with the twelve-tone equal-tempered scale (see Figure 1.20). Similarly, the **FMP Notebook *Pythagorean Tuning*** considers the Pythagorean scale (see Exercise 1.10). In both of these notebooks, we again use simple sinusoidal models for the sonification. The **FMP Notebook *Dynamics, Intensity, and Loudness*** yields an implementation for visualizing the sound power level over time for our Beethoven example. Furthermore, we present an experiment using a chirp signal to illustrate the relation between signal power and perceived loudness (see Figure 1.21). In the **FMP Notebook *Timbre***, we introduce simple yet instructive experiments that are also suitable as programming exercises. First, we give an example on how one may compute an envelope of a waveform by applying a windowed maximum filter (see Figure 1.22a). Then, we provide some implementations for generating synthetic sinusoidal signals with vibrato (frequency modulations) and tremolo (amplitude modulations). Finally, we demonstrate that the perception of the perceived pitch depends not only on the fundamental frequency but also on its higher harmonics and their relationships. In particular, we show that a human may perceive the pitch of a tone even if the fundamental frequency associated to this pitch is completely missing.

In summary, in the FMP notebooks of **Part 1**, we provide basic Python code examples for parsing and visualizing various music representations. Furthermore, we consider tangible music examples and suggest various experiments for deepening understanding of musical and acoustic properties of audio signals including aspects such as frequency, pitch, dynamics, and timbre. At the same time, the material is also intended for developing Python programming skills as required in subsequent FMP notebooks.

References

1. M. BABBITT, *The use of computers in musicological research*, Perspectives of New Music, 3 (1965), pp. 74–83.
2. E. BENETOS, S. DIXON, Z. DUAN, AND S. EWERT, *Automatic music transcription: An overview*, IEEE Signal Processing Magazine, 36 (2019), pp. 20–30.
3. J. CALVO-ZARAGOZA, J. HAJIČ JR., AND A. PACHA, *Understanding optical music recognition*, ACM Computing Surveys, 53 (2020).
4. M. DORFER, J. HAJIČ JR., A. ARZT, H. FROSTEL, AND G. WIDMER, *Learning audio-sheet music correspondences for cross-modal retrieval and piece identification*, Transactions of the International Society for Music Information (TISMIR), 1 (2018), pp. 22–31.
5. H. FASTL AND E. ZWICKER, *Psychoacoustics: Facts and Models*, Springer, 3rd ed., 2007.
6. H. FLETCHER AND W. A. MUNSON, *Loudness, its definition, measurement and calculation*, Journal of the Acoustic Society of America, 5 (1933), pp. 82–108.
7. N. H. FLETCHER AND T. D. ROSSING, *The Physics of Musical Instruments*, Springer, 1991.
8. M. GOOD, *Lessons from the adoption of MusicXML as an interchange standard*, in Proceedings of XML, Boston, Massachusetts, USA, 2006.
9. E. GOULD, *Behind Bars: The Definitive Guide to Music Notation*, Alfred Music, 2011.
10. A. HANKINSON, P. ROLAND, AND I. FUJINAGA, *The music encoding initiative as a document-encoding framework*, in Proceedings of the International Society for Music Information Retrieval Conference (ISMIR), Miami, Florida, USA, October 2011, pp. 293–298.
11. D. M. HUBER, *The MIDI Manual*, Focal Press, 3rd ed., 2006.
12. D. B. HURON, *Sweet Anticipation: Music and the Psychology of Expectation*, The MIT Press, 2006.
13. A. P. KLAPURI AND M. DAVY, eds., *Signal Processing Methods for Music Transcription*, Springer, New York, 2006.
14. P. D. LEHRMAN, *MIDI 2.0: Promises and challenges*, in Proceedings of Music Encoding Conference (MEC), Tufts University, USA, 2020.
15. G. MAZZOLA, *The Topos of Music*, Birkhäuser, 2002.
16. W. MCKINNEY, *Data structures for statistical computing in python*, in Proceedings Python in Science Conference, S. van der Walt and J. Millman, eds., 2010, pp. 56–61.
17. B. C. MOORE, *An Introduction to the Psychology of Hearing*, Brill Academic Publisher, 6th ed., 2013.
18. M. MÜLLER, A. ARZT, S. BALKE, M. DORFER, AND G. WIDMER, *Cross-modal music retrieval and applications: An overview of key methodologies*, IEEE Signal Processing Magazine, 36 (2019), pp. 52–62.
19. M. MÜLLER AND A. KLAPURI, *Music signal processing*, in Image, Video Processing and Analysis, Hardware, Audio, Acoustic and Speech Processing, vol. 4 of Library in Signal Processing, Academic Press, 2013, ch. 27, pp. 713–756.
20. M. MÜLLER AND F. ZALKOW, *FMP Notebooks: Educational material for teaching and learning fundamentals of music processing*, in Proceedings of the International Society for Music Information Retrieval Conference (ISMIR), Delft, The Netherlands, 2019, pp. 573–580.
21. E. SELFRIDGE-FIELD, ed., *Beyond MIDI: The Handbook of Musical Codes*, MIT Press, Cambridge, Massachusetts, USA, 1997.
22. W. A. SETHARES, *Tuning, Timbre, Spectrum, Scale*, Springer, London, 1998.
23. R. N. SHEPARD, *Circularity in judgments of relative pitch*, Journal of the Acoustic Society of America, 36 (1964), pp. 2346–2353.
24. V. THOMAS, C. FREMEREY, M. MÜLLER, AND M. CLAUSEN, *Linking sheet music and audio – challenges and new approaches*, in Multimodal Music Processing, M. Müller, M. Goto, and M. Schedl, eds., vol. 3 of Dagstuhl Follow-Ups, Schloss Dagstuhl–Leibniz-Zentrum für Informatik, Dagstuhl, Germany, 2012, pp. 1–22.
25. G. WIGGINS, D. MÜLLENSIEFEN, AND M. PEARCE, *On the non-existence of music: Why music theory is a figment of the imagination*, Musicae Scientiae, 5 (2010), pp. 231–255.

Exercises

Exercise 1.1. Assume that a pianist exactly follows the specifications given in the Beethoven example from Figure 1.1. Determine the duration (in milliseconds) of a quarter note and a measure, respectively.

Exercise 1.2. Specify the MIDI representation (in tabular form) and sketch the piano-roll representation (similar to Figure 1.13) of the following sheet music representations. Assume that a quarter note corresponds to 120 ticks. Set the velocity to a value of 100 for all active note events. Furthermore, assign the notes of the G-clef to channel 1 and the notes of the F-clef to channel 2.

[**Hint:** In this exercise, we assume that the reader has some basic knowledge of Western music notation.]

Exercise 1.3. In this exercise, a **melody** is regarded as a linear succession of musical notes. A **transposition** of a given melody moves all notes up or down in pitch by a constant interval. Furthermore, an **inversion** of a melody turns all the intervals upside-down. For instance, if the original melody rises by three semitones, the inverted melody falls by three semitones. Finally, the **retrograde** of a melody is the reverse, where the notes are played from back to front. Let us consider the following two melodies given in piano-roll representation:

Specify for each of the two melodies the piano-roll representation of the transposition by two semitones upwards, the inversion (keeping the first note fixed), the retrograde, and the retrograde of the inversion. Furthermore, regarding melodies only up to pitch classes (by ignoring octave information), determine the number of different melodies that can be generated by successively applying an arbitrary number of transpositions, inversions, and retrogrades.

Exercise 1.4. The **speed of sound** is the distance traveled per unit of time by a sound wave propagating through an elastic medium. Look up the speed of sound in air. Assume that a concert hall has a length of 50 meters. How long does it take for a sound wave to travel from the front to the back of the hall?

Exercise 1.5. Using (1.1), compute the center frequencies for all notes of the C-major scale C4, D4, E4, F4, G4, A4, B4, C5 and for all notes of the C-minor scale C4, D4, E$^\flat$4, F4, G4, A$^\flat$4, B$^\flat$4, C5 (see also Figure 1.5).

Exercise 1.6. Using (1.1), compute the frequency ratio $F_{\mathrm{pitch}}(p+1)/F_{\mathrm{pitch}}(p)$ of two subsequent pitches $p+1$ and p (see (1.2)). How does the frequency $F_{\mathrm{pitch}}(p+k)$ for some $k \in \mathbb{Z}$ relate to $F_{\mathrm{pitch}}(p)$? Furthermore, derive a formula for the distance (in semitones) for two arbitrary frequencies ω_1 and ω_2.

Exercise 1.7. Let us have a look at Figure 1.18b, which shows a waveform obtained from a recording of Beethoven's Fifth. Estimate the fundamental frequency of the sound played by counting the number of oscillation cycles in the section between 7.3 and 7.8 seconds. Furthermore, determine the musical note that has a center frequency closest to the estimated fundamental frequency. Compare the result with the sheet music representation of Figure 1.1.

Exercise 1.8. Assume an equal-tempered scale that consists of 17 tones per octave and a reference pitch $p = 100$ having a center frequency of 1000 Hz. Specify a formula similar to (1.1), which yields the center frequencies for the pitches $p \in [0:255]$. In particular, determine the center frequency for the pitches $p = 83$, $p = 66$, and $p = 49$ in this scale. What is the difference (in cents) between two subsequent pitches in this scale?

Exercise 1.9. Write a small computer program to calculate the differences (in cents) between the first 16 harmonics of the note C2 and the center frequencies of the closest notes of the twelve-tone equal-tempered scale (see Figure 1.20). What are the corresponding differences when considering the harmonics of another note such as $B^{\flat}4$?

Exercise 1.10. Pythagorean tuning (named after the ancient Greek mathematician and philosopher Pythagoras) is a system of musical tuning in which the frequency ratios of all intervals are based on the ratio 3 : 2 as found in the harmonic series. This ratio is also known as the **perfect fifth**. A **Pythagorean scale** is a scale constructed from only pure perfect fifths (3 : 2) and octaves (2 : 1). To obtain such a scale, start with the center frequency of the note C2, successively multiply the frequency value by a factor of $3/2$, and if necessary, divide it by two such that all frequency values lie between C2 and C3. Repeat this procedure to produce 13 frequency values (including the one for C2). As in Exercise 1.9, determine for each such frequency value the closest note of the equal-tempered scale (along with the difference in cents). The last of the produced frequency values is closest to the fundamental frequency of the note C3. The difference between the produced frequency and the center frequency of C3 is known as the **Pythagorean comma**, which indicates the degree of inconsistency when trying to define a twelve-tone scale using only perfect fifths.

Exercise 1.11. Investigate the typical frequency range as well as pitch range of musical instruments (including the human voice) and graphically display this information as indicated by the following figure. For example, consider the ranges of standard instruments as used in Western orchestras including the piano, human voice (bass, tenor, alto, soprano), double bass, cello, viola, violin, bass guitar, guitar, trumpet. Similarly, consider instruments you are familiar with.

20 30 44 70 100 150 200 300 440 700 1000 1500 2000 3000 4400 Hz
C0 C1 C2 C3 C4 C5 C6 C7 C8
Piano

Exercise 1.12. Suppose that the intensity of a sound has been increased by 17 dB as defined in (1.6). Determine the factor by which the sound intensity has been increased.

Chapter 2
Fourier Analysis of Signals

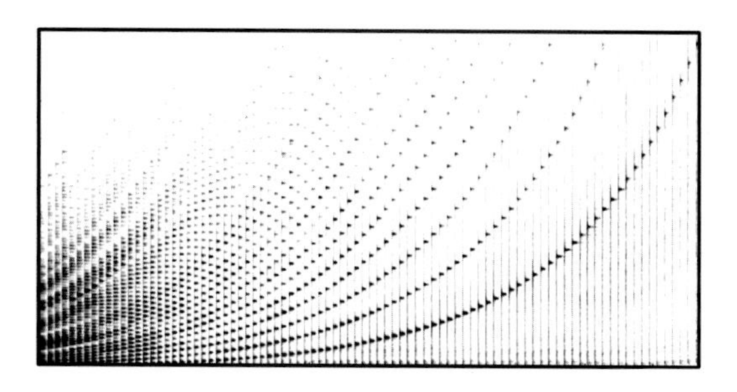

As we have seen in the last chapter, music signals are generally complex sound mixtures that consist of a multitude of different sound components. Because of this complexity, the extraction of musically relevant information from a waveform constitutes a difficult problem. A first step in better understanding a given signal is to decompose it into building blocks that are more accessible for the subsequent processing steps. In the case that these building blocks consist of sinusoidal functions, such a process is also called Fourier analysis. Sinusoidal functions are special in the sense that they possess an explicit physical meaning in terms of frequency. As a consequence, the resulting decomposition unfolds the frequency spectrum of the signal—similar to a prism that can be used to break light up into its constituent spectral colors. The Fourier transform converts a signal that depends on time into a representation that depends on frequency. Being one of the most important tools in signal processing, we will encounter the Fourier transform in a variety of music processing tasks.

In Section 2.1, we introduce the main ideas of the Fourier transform and summarize the most important facts that are needed for understanding the subsequent chapters of the book. Furthermore, we introduce the required mathematical notions. A good understanding of Section 2.1 is essential for the various music processing tasks to be discussed. In Section 2.2 to Section 2.5, we cover the Fourier transform in greater mathematical depth. The reader who is mainly interested in the music processing applications may skip these more technical sections on a first reading.

In Section 2.2, we take a closer look at signals and discuss their properties from a more abstract perspective. In particular, we consider two classes of signals: analog signals that give us the right physical interpretation and digital signals that are needed for actual digital processing by computers. The different signal classes lead to different versions of the Fourier transform, which we introduce with math-

M. Müller, *Fundamentals of Music Processing*, https://doi.org/10.1007/978-3-030-69808-9_2

ematical rigor along with intuitive explanations and numerous illustrating examples (Section 2.3). In particular, we explain how the different versions are interrelated and how they can be approximated by means of the discrete Fourier transform (DFT). The DFT can be computed efficiently by means of the fast Fourier transform (FFT), which will be discussed in Section 2.4. Finally, we introduce the short-time Fourier transform (STFT), which is a local variant of the Fourier transform yielding a time–frequency representation of a signal (Section 2.5). By presenting this material from a different perspective as typically encountered in an engineering course, we hope to refine and sharpen the understanding of these important and beautiful concepts.

2.1 The Fourier Transform in a Nutshell

Let us start with an audio signal that represents the sound of some music. For example, let us analyze the sound of a single note played on a piano (see Figure 2.1a). How can we find out which note has actually been played? Recall from Section 1.3.2 that the pitch of a musical tone is closely related to its fundamental frequency, the frequency of the lowest partial of the sound. Therefore, we need to determine the frequency content, the main periodic oscillations of the signal. Let us zoom into the signal considering only a 10-ms section (see Figure 2.1b). The figure shows that the signal behaves in a nearly periodic way within this section. In particular, one can observe three main crests of a sinusoidal-like oscillation (see also Figure 2.1c). Having approximately three oscillation cycles within a 10-ms section means that the signal contains a frequency component of roughly 300 Hz.

The main idea of **Fourier analysis** is to compare the signal with sinusoids of various[1] frequencies $\omega \in \mathbb{R}$ (measured in Hz). Each such sinusoid or pure tone may be thought of as a prototype oscillation. As a result, we obtain for each considered frequency parameter $\omega \in \mathbb{R}$ a magnitude coefficient $d_\omega \in \mathbb{R}_{\geq 0}$ (along with a phase coefficient $\varphi_\omega \in \mathbb{R}$, the role of which is explained later). In the case that the coefficient d_ω is large, there is a high similarity between the signal and the sinusoid of frequency ω, and the signal contains a periodic oscillation at that frequency (see Figure 2.1c). In the case that d_ω is small, the signal does not contain a periodic component at that frequency (see Figure 2.1d).

Let us plot the coefficients d_ω over the various frequency parameters $\omega \in \mathbb{R}$. This yields a graph as shown in Figure 2.1f. In this graph, the highest value is assumed for the frequency parameter $\omega = 262$ Hz. By (1.1), this is roughly the center frequency of the pitch $p = 60$ or the note C4. Indeed, this is exactly the note played in our piano example. Furthermore, as illustrated by Figure 2.1e, one can also observe a

[1] In the following, we also consider *negative frequencies* for mathematical reasons without explaining this concept in more detail. In our musical context, negative frequencies are redundant (having the same interpretation as positive frequencies), but simplify the mathematical formulation of the Fourier transform.

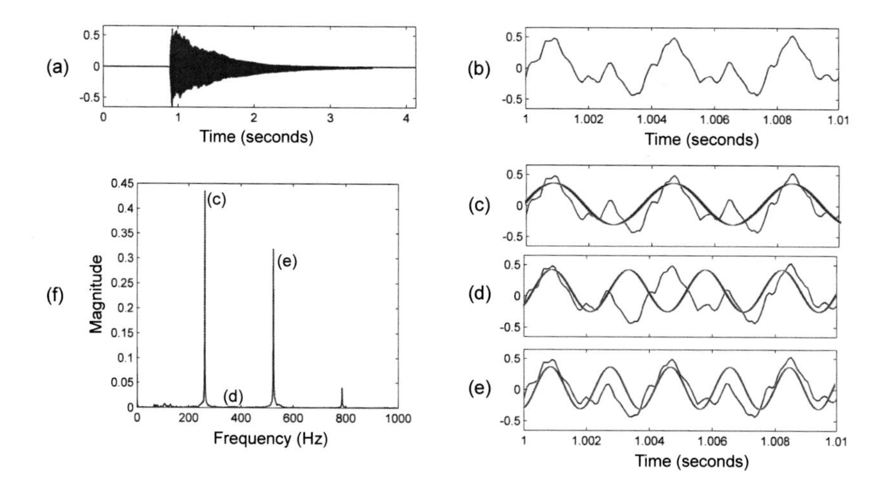

Fig. 2.1 **(a)** Waveform of a note C4 (261.6 Hz) played on a piano. **(b)** Zoom into a 10-ms section starting at time position $t = 1$ sec. **(c–e)** Comparison of the waveform with sinusoids of various frequencies ω. **(f)** Magnitude coefficients d_ω in dependence on the frequency ω.

high similarity between the signal and the sinusoid of frequency $\omega = 523$ Hz. This is roughly the frequency for the second partial of the tone C4.

With this example, we have already seen the main idea behind the **Fourier transform**. The Fourier transform breaks up a signal into its frequency components. For each frequency $\omega \in \mathbb{R}$, the Fourier transforms yields a coefficient d_ω (and a phase φ_ω) that tells us to which extent the given signal matches a sinusoidal prototype oscillation of that frequency.

One important property of the Fourier transform is that the original signal can be reconstructed from the coefficients d_ω (along with the coefficients φ_ω). To this end, one basically superimposes the sinusoids of all possible frequencies, each weighted by the respective coefficient d_ω (and shifted by φ_ω). This weighted superposition is also called the **Fourier representation** of the original signal. The original signal and the Fourier transform contain the same amount of information. This information, however, is represented in different ways. While the signal displays the information across time, the Fourier transform displays the information across frequency. As put by Hubbard [9], the signal tells us when certain notes are played in time, but hides the information about frequencies. In contrast, the Fourier transform of music displays which notes (frequencies) are played, but hides the information about when the notes are played.

In the following sections, we take a more detailed look at the Fourier transform and some of its main properties.

2.1.1 Fourier Transform for Analog Signals

In Section 1.3.1, we saw that a signal or sound wave yields a function that assigns to each point in time the deviation of the air pressure from the average air pressure at a specific location. Let us consider the case of an **analog** signal, where both the time as well as the amplitude (or deviation) are continuous, real-valued parameters. In this case, a signal can be modeled as a function $f\colon \mathbb{R} \to \mathbb{R}$, which assigns to each time point $t \in \mathbb{R}$ an amplitude value $f(t) \in \mathbb{R}$. Plotting the amplitude over time, one obtains a graph of this function that corresponds to the waveform of the signal (see Figure 1.17).

The term **function** may need some explanation. In mathematics, a function yields a relation between a set of input elements and a set of output elements, where each input element is related to exactly one output element. For example, a function can be a polynomial $f\colon \mathbb{R} \to \mathbb{R}$ that assigns for each input element $t \in \mathbb{R}$ an output element $f(t) = t^2 \in \mathbb{R}$. At this point, we want to emphasize that one needs to differentiate between a function f and its output element $f(t)$ (also referred to as the **value**) at a particular input element t (also referred to as the **argument**). In other words, mathematicians think of a function f in an abstract way, where the symbol or physical meaning of the argument does not matter. As opposed to this, engineers often like to emphasize the meaning of the input argument and loosely speak of a function $f(t)$, even though this is strictly speaking an output value. In this book, we assume the viewpoint of a mathematician.

2.1.1.1 The Role of the Phase

After this side note, let us turn towards the spectral analysis of a given analog signal $f\colon \mathbb{R} \to \mathbb{R}$. As explained in our introductory example, we compare the signal f with prototype oscillations that are given in the form of sinusoids. In Section 1.3.2 and Figure 1.19, we have already encountered such sinusoidal signals. Mathematically, a **sinusoid** is a function $g\colon \mathbb{R} \to \mathbb{R}$ defined by

$$g(t) := A\sin(2\pi(\omega t - \varphi)) \tag{2.1}$$

for $t \in \mathbb{R}$. The parameter A corresponds to the **amplitude**, the parameter ω to the **frequency** (measured in Hz), and the parameter φ to the **phase** (measured in normalized radians with 1 corresponding to an angle of 360°). In Fourier analysis, we consider prototype oscillations that are normalized with regard to their power (average energy) by setting $A = \sqrt{2}$. Thus for each frequency parameter ω and phase parameter φ we obtain a sinusoid $\mathbf{cos}_{\omega,\varphi} : \mathbb{R} \to \mathbb{R}$ given by

$$\mathbf{cos}_{\omega,\varphi}(t) := \sqrt{2}\cos(2\pi(\omega t - \varphi)) \tag{2.2}$$

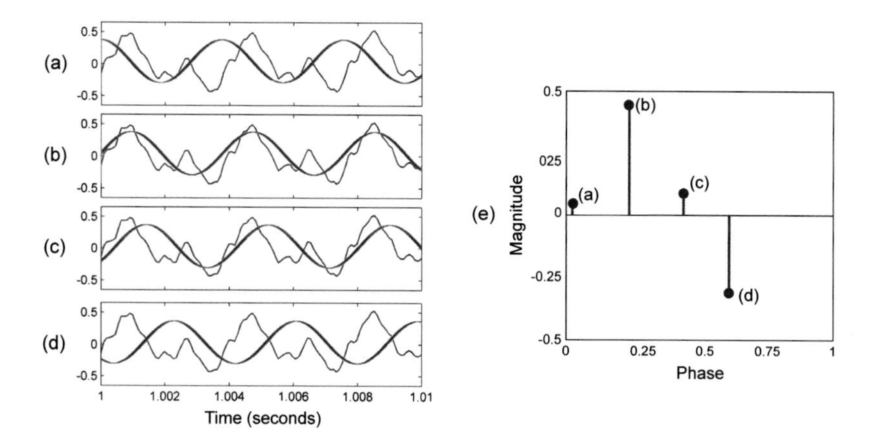

Fig. 2.2 **(a–d)** Waveform and different sinusoids of a fixed frequency $\omega = 262$ Hz but different phases $\varphi \in \{0.05, 0.24, 0.45, 0.6\}$. **(e)** Values that express the degree of similarity between the waveform and the four different sinusoids.

for $t \in \mathbb{R}$. Since the cosine function is periodic, the parameters φ and $\varphi + k$ for integers $k \in \mathbb{Z}$ yield the same function. Therefore, the phase parameter only needs to be considered for $\varphi \in [0, 1)$.

When measuring how well the given signal coincides with a sinusoid of frequency ω, we have the freedom of shifting the sinusoid in time. This degree of freedom is expressed by the phase parameter φ. As illustrated by Figure 2.2, the degree of similarity between the signal and the sinusoid of fixed frequency crucially depends on the phase. What have we done with the phase when computing the coefficients d_ω as illustrated by Figure 2.1? The procedure outlined in the introduction was only half the story. When comparing the signal f with a sinusoid $\mathbf{cos}_{\omega,\varphi}$ of frequency ω, we have implicitly used the phase φ_ω that yields the maximal possible similarity. To understand this better, we first need to explain how we actually compare the signal and a sinusoid or, more generally, how we compare two given functions.

2.1.1.2 Computing Similarity with Integrals

Let us assume that we are given two functions of time $f : \mathbb{R} \to \mathbb{R}$ and $g : \mathbb{R} \to \mathbb{R}$. What does it mean for f and g to be similar? Intuitively, one may agree that f and g are similar if they show a similar behavior over time: if f assumes positive values, then so should g, and if f becomes negative, the same should happen to g. The joint behavior of these functions can be captured by forming the integral of the product of the two functions:

$$\int_{t\in\mathbb{R}} f(t) \cdot g(t)dt. \tag{2.3}$$

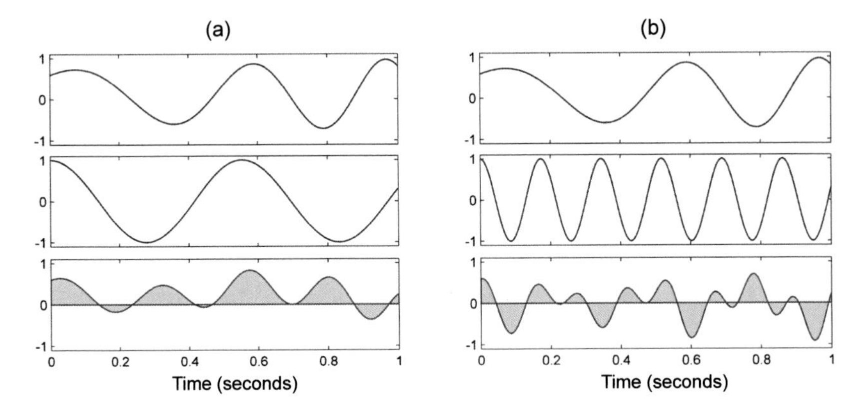

Fig. 2.3 Measuring the similarity of two functions f (top) and g (middle) by computing the integral of the product (bottom). **(a)** Two functions having high similarity. **(b)** Two functions having low similarity.

The integral measures the area delimited by the graph of the product $f \cdot g$, where the negative area (below the horizontal axis) is subtracted from the positive area (above the horizontal axis) (see Figure 2.3). In the case that f and g are either both positive or both negative at most time instances, the product is positive for most of the time and the integral becomes large (see Figure 2.3a). However, if the two functions are dissimilar, then the overall positive and the overall negative areas cancel out, yielding a small overall integral (see Figure 2.3b). Further examples are discussed in Exercise 2.1.

There are many more ways for comparing two given signals. For example, the integral of the absolute difference between the functions also yields a notion of how similar the signals are. In the formulation of the Fourier transform, however, one encounters the measure as considered in (2.3), which generalizes the **inner product** known from linear algebra (see 2.37). We continue this discussion in Section 2.2.3.

2.1.1.3 First Definition of the Fourier Transform

Based on the similarity measure (2.3), we compare the original signal f with sinusoids $g = \mathbf{cos}_{\omega,\varphi}$ as defined in (2.2). For a fixed frequency $\omega \in \mathbb{R}$, we define

$$d_\omega := \max_{\varphi \in [0,1)} \left(\int_{t \in \mathbb{R}} f(t) \mathbf{cos}_{\omega,\varphi}(t) dt \right), \tag{2.4}$$

$$\varphi_\omega := \operatorname*{argmax}_{\varphi \in [0,1)} \left(\int_{t \in \mathbb{R}} f(t) \mathbf{cos}_{\omega,\varphi}(t) dt \right). \tag{2.5}$$

As previously discussed, the magnitude coefficient d_ω expresses the intensity of frequency ω within the signal f. Additionally, the phase coefficient $\varphi_\omega \in [0,1)$ tells

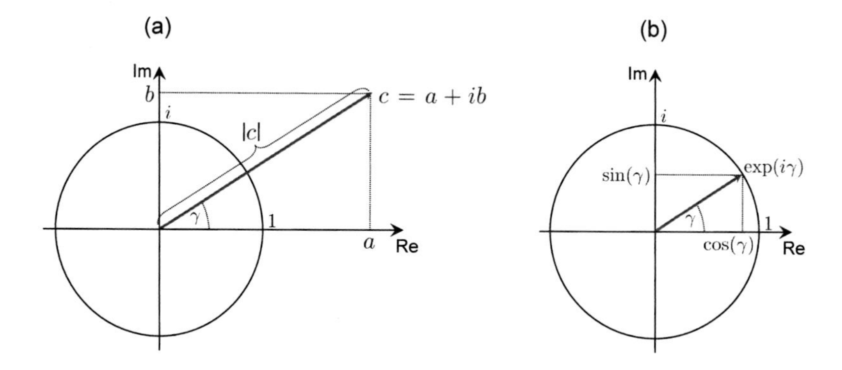

Fig. 2.4 **(a)** Polar coordinate representation of a complex number $c = a + ib$. **(b)** Definition of the exponential function.

us how the sinusoid of frequency ω needs to be displaced in time to best fit the signal f. The **Fourier transform** of a function $f : \mathbb{R} \to \mathbb{R}$ is defined to be the "collection" of all coefficients d_ω and φ_ω for $\omega \in \mathbb{R}$. Shortly, we will state this definition in a more formal way.

The computation of d_ω and φ_ω feels a bit awkward, since it involves an optimization step. The good news is that there is a simple solution to this optimization problem, which results from the existence of certain trigonometric identities that relate phases and amplitudes of certain sinusoidal functions. Using the concept of complex numbers, these trigonometric identities become simple and lead to an elegant formulation of the Fourier transform. We discuss such issues in more detail in Section 2.3. In the following, we introduce the standard complex-valued formulation of the Fourier transform without giving any proofs.

2.1.1.4 Complex Numbers

Let us first review the concept of complex numbers. The complex numbers extend the real numbers by introducing the imaginary number $i := \sqrt{-1}$ with the property $i^2 = -1$. Each complex number can be written as $c = a + ib$, where $a \in \mathbb{R}$ is the real part and $b \in \mathbb{R}$ the imaginary part of c. The set of all complex numbers is written as $\mathbb{C}$, which can be thought of as a two-dimensional plane: the horizontal dimension corresponds to the real part, and the vertical dimension to the imaginary part. In this plane, the number $c = a + ib$ is specified by the Cartesian coordinates (a, b). As illustrated by Figure 2.4a, there is another way of representing a complex number, which is known as the **polar coordinate** representation. In this case, a complex number c is described by its absolute value $|c|$ (distance from the origin) and the angle γ between the positive horizontal axis and the line from the origin and c. The polar coordinates $|c| \in \mathbb{R}_{\geq 0}$ and $\gamma \in [0, 2\pi)$ (given in radians) can be derived from the coordinates (a, b) via the following formulas:

$$|c| := \sqrt{a^2+b^2}, \tag{2.6}$$
$$\gamma := \mathrm{atan2}(b,a). \tag{2.7}$$

Further details on polar coordinates and the function atan2, which is a variant of the inverse of the tangent function, are explained in Section 2.3.2.2. To regain the complex number c from its polar coordinates, one uses the **exponential function**, which maps an angle $\gamma \in \mathbb{R}$ (given in radians) to a complex number defined by

$$\exp(i\gamma) := \cos(\gamma) + i\sin(\gamma) \tag{2.8}$$

(see also Figure 2.4b). The values of this function turn around the unit circle of the complex plane with a period of 2π (see Section 2.3.2.1). From this, we obtain the following **polar coordinate representation** for a complex number c:

$$c = |c| \cdot \exp(i\gamma). \tag{2.9}$$

2.1.1.5 Complex Definition of the Fourier Transform

What have we gained by bringing complex numbers into play? Recall that we have obtained a positive coefficient $d_\omega \in \mathbb{R}_{\geq 0}$ from (2.4) and a phase coefficient $\varphi_\omega \in [0,1)$ from (2.5). The basic idea is to use these coefficients as polar coordinates and to encode both coefficients by a single complex number. Because of some technical reasons (a normalization issue that becomes clearer when discussing the mathematical details), one introduces some additional factors and a sign in the phase to yield the complex coefficient

$$c_\omega := \frac{d_\omega}{\sqrt{2}} \cdot \exp(2\pi i(-\varphi_\omega)). \tag{2.10}$$

This complex formulation directly leads us to the Fourier transform of a real-valued function $f : \mathbb{R} \to \mathbb{R}$. For each frequency $\omega \in \mathbb{R}$, we obtain a complex-valued coefficient $c_\omega \in \mathbb{C}$ as defined by (2.4), (2.5), and (2.10). This collection of coefficients can be encoded by a complex-valued function $\hat{f} : \mathbb{R} \to \mathbb{C}$ (called "f hat"), which assigns to each frequency parameter the coefficient c_ω:

$$\hat{f}(\omega) := c_\omega. \tag{2.11}$$

The function $\hat{f}$ is referred to as the **Fourier transform** of f, and its values $\hat{f}(\omega) = c_\omega$ are called the **Fourier coefficients**. One main result in Fourier analysis is that the Fourier transform can be computed via the following compact formula:

$$\hat{f}(\omega) = \int_{t\in\mathbb{R}} f(t)\exp(-2\pi i\omega t)dt \tag{2.12}$$
$$= \int_{t\in\mathbb{R}} f(t)\cos(-2\pi\omega t)dt + i\int_{t\in\mathbb{R}} f(t)\sin(-2\pi\omega t)dt. \tag{2.13}$$

In other words, the real part of the complex coefficient $\hat{f}(\omega)$ is obtained by comparing the original signal f with a cosine function of frequency ω, and the imaginary part is obtained by comparing with a sine function of frequency ω. The absolute value $|\hat{f}(\omega)|$ is also called the **magnitude** of the Fourier coefficient. Similarly, the real-valued function $|\hat{f}| : \mathbb{R} \to \mathbb{R}$, which assigns to each frequency parameter ω the magnitude $|\hat{f}(\omega)|$, is called the **magnitude Fourier transform** of f.

In the standard literature on signal processing, the formula (2.12) is often used to define the Fourier transform $\hat{f}$ and, then, the physical interpretation of the Fourier coefficients is discussed. In particular, the real-valued coefficients d_ω in (2.4) and φ_ω in (2.5) can be derived from $\hat{f}(\omega)$. Using (2.10), one obtains

$$d_\omega = \sqrt{2}|\hat{f}(\omega)|, \tag{2.14}$$

$$\varphi_\omega = -\frac{\gamma_\omega}{2\pi}, \tag{2.15}$$

where $|\hat{f}(\omega)|$ and γ_ω are the polar coordinates of $\hat{f}(\omega)$.

2.1.1.6 Fourier Representation

As mentioned above, the original signal f can be reconstructed from its Fourier transform. In principle, the reconstruction is straightforward: one superimposes the sinusoids of all possible frequency parameters $\omega \in \mathbb{R}$, each weighted by the respective coefficient d_ω and shifted by φ_ω. Both kinds of information are encoded in the complex Fourier coefficient c_ω. In the analog case considered so far, we are dealing with a continuum of frequency parameters, where the superposition becomes an integration over the parameter space. The reconstruction is given by the formulas

$$f(t) = \int_{\omega\in\mathbb{R}_{\geq 0}} d_\omega \sqrt{2}\cos(2\pi(\omega t - \varphi_\omega))d\omega \tag{2.16}$$

$$= \int_{\omega\in\mathbb{R}} c_\omega \exp(2\pi i\omega t)d\omega, \tag{2.17}$$

first given in the real-valued formulation, and then given in the complex-valued formulation with $c_\omega = \hat{f}(\omega)$. As said before, the representation of a signal in terms of a weighted superposition of sinusoidal prototype oscillations is also called the **Fourier representation** of the signal. Notice that the formula (2.12) for the Fourier transform and the formula (2.17) for the Fourier representation are nearly identical. The main difference is that the roles of the time parameter t and frequency parameter ω are interchanged. The beautiful relationship between these two formulas will be further discussed in later sections of this chapter.

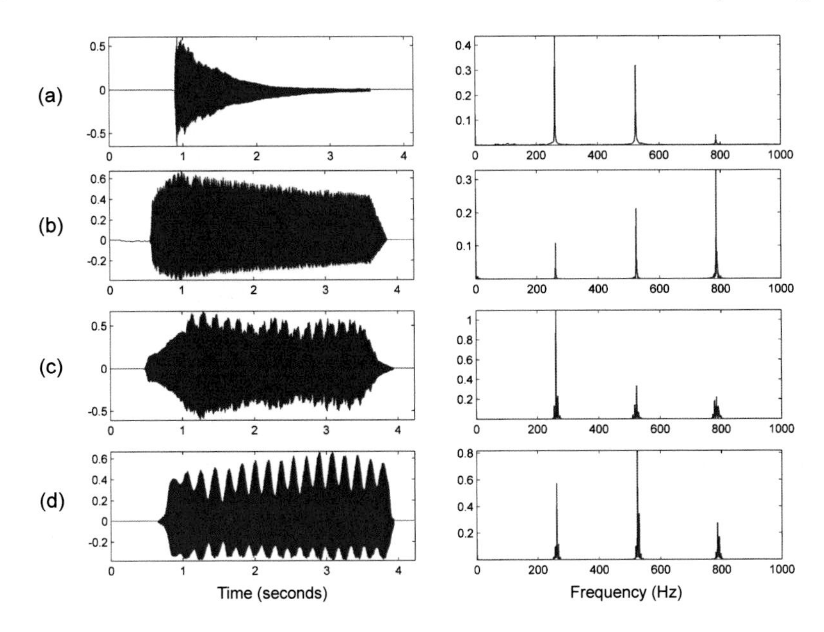

Fig. 2.5 Waveform and magnitude Fourier transform of a tone C4 (261.6 Hz) played by different instruments (see also Figure 1.23). **(a)** Piano. **(b)** Trumpet. **(c)** Violin. **(d)** Flute.

2.1.2 Examples

Let us consider some examples including the one introduced in Figure 2.1. Figure 2.5 shows the waveform and the magnitude Fourier transform for some audio signals, where a single note C4 is played on different instruments: a piano, a trumpet, a violin, and a flute. We have already encountered this example in Figure 1.23 of Section 1.3.4, where we discussed the aspect of timbre. Recall that the existence of certain partials and their relative strengths have a crucial influence on the timbre of a musical tone. In the case of the piano tone (Figure 2.5a), the Fourier transform has a sharp peak at 262 Hz, which reveals that most of the signal's energy is contained in the first partial or the fundamental frequency of the note C4. Further peaks (also beyond the shown frequency range from 0 to 1000 Hz) can be found at integer multiples of the fundamental frequency corresponding to the higher partials.

Figure 2.5b shows that the same note played on a trumpet results in a similar frequency spectrum, where the peaks appear again at integer multiples of the fundamental frequency. However, most of the energy is now contained in the third partial, and the relative heights of the peaks are different compared with the piano. This is one reason why a trumpet sounds different from a piano. For a violin, as shown by Figure 2.5c, most energy is again contained in the first partial. Observe that the peaks are blurred in frequency, which is the result of the vibrato (see also Figure 1.23b). The time-dependent frequency modulations of the vibrato are aver-

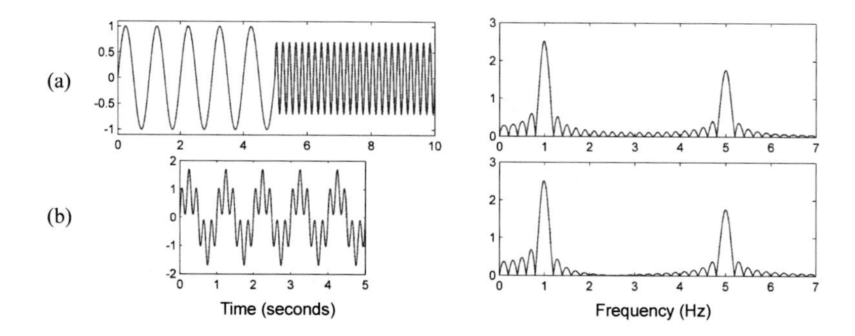

Fig. 2.6 Missing time information of the Fourier transform illustrated by two different signals and their magnitude Fourier transforms. **(a)** Two subsequent sinusoids of frequency 1 Hz and 5 Hz. **(b)** Superposition of the same sinusoids.

aged by the Fourier transform. This yields a single coefficient for each frequency independent of spectro-temporal fluctuations. A similar explanation holds for the flute tone shown in Figure 2.5d.

We have seen that the magnitude of the Fourier transform tells us about the signal's overall frequency content, but it does not tell us at which time the frequency content occurs. Figure 2.6 illustrates this fact, showing the waveform and the magnitude Fourier transform for two signals. The first signal consists of two parts with a sinusoid of $\omega = 1$ Hz and amplitude $A = 1$ in the first part and a sinusoid of $\omega = 5$ Hz and amplitude $A = 0.7$ in the second part. Furthermore, the signal is zero outside the interval $[0, 10]$. In contrast, the second signal is a superposition of these two sinusoids, being zero outside the interval $[0, 5]$. Even though the two signals are different in nature, the resulting magnitude Fourier transforms are more or less the same. This demonstrates the drawbacks of the Fourier transform when analyzing signals with changing characteristics over time. In Section 2.1.4 and Section 2.5 we discuss a short-time version of the Fourier transform, where time information is recovered at least to some degree. Besides the two peaks, one can observe in Figure 2.6 a large number of small "ripples." Such phenomena as well as further properties of the Fourier transform are discussed in Section 2.3.3.

2.1.3 Discrete Fourier Transform

When using digital technology, only a finite number of parameters can be stored and processed. To this end, analog signals need to be converted into finite representations—a process commonly referred to as **digitization**. One step that is often applied in an analog-to-digital conversion is known as **equidistant sampling**. Given an analog signal $f : \mathbb{R} \to \mathbb{R}$ and a positive real number $T > 0$, one defines a function $x : \mathbb{Z} \to \mathbb{R}$ by setting

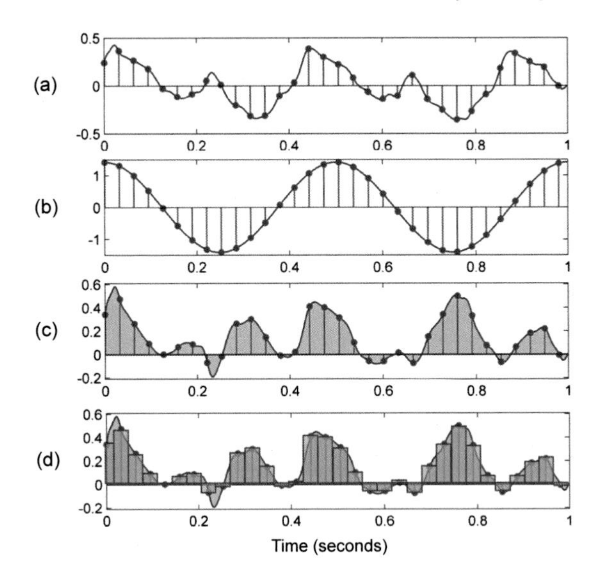

Fig. 2.7 Illustration of the sampling process using a sampling rate of $F_\mathrm{s} = 32$. The waveforms of the analog signals are shown as curves and the sampled versions as stem plots. **(a)** Signal f. **(b)** Sinusoid $\mathbf{cos}_{\omega,\varphi}$ with $\omega = 2$ and $\varphi = 0$. **(c)** Product $f \cdot \mathbf{cos}_{\omega,\varphi}$ and its area. **(d)** Approximation of the integral by a Riemann sum obtained from the sampled version.

$$x(n) := f(n \cdot T). \tag{2.18}$$

Since x is only defined on a discrete set of time points, it is also referred to as a **discrete-time** (DT) signal (see Section 2.2.2.1). The value $x(n)$ is called a **sample** taken at time $t = n \cdot T$ of the original analog signal f. This procedure is also known as T**-sampling**, where the number T is referred to as the **sampling period**. The inverse

$$F_\mathrm{s} := 1/T \tag{2.19}$$

of the sampling period is also called the **sampling rate** of the process. It specifies the number of samples per second and is measured in Hertz (Hz). Figure 2.7a shows an example of sampling an analog signal using $F_\mathrm{s} = 32$ Hz.

In general, one loses information in the sampling process. The famous **sampling theorem** says that the original analog signal f can be reconstructed perfectly from its sampled version x, if f does not contain any frequencies higher than

$$\Omega := F_\mathrm{s}/2 = 1/(2T) \text{ Hz}. \tag{2.20}$$

In this case, we also say that f is an Ω**-bandlimited** signal, where the frequency Ω is known as the **Nyquist frequency**. In the case that f contains higher frequencies, sampling may cause artifacts referred to as **aliasing** (see Section 2.2.2 for details). The sampling theorem will be further discussed in Exercise 2.28.

In the following, we assume that the analog signal f satisfies suitable requirements so that the sampled signal x does not contain major artifacts. Now, having a discrete number of samples to represent our signal, how do we calculate the Fourier transform? Recall that the idea of the Fourier transform is to compare the signal with a sinusoidal prototype oscillation by computing the integral over the pointwise product (see (2.12)). Therefore, in the digital domain, it seems reasonable to sample the sinusoidal prototype oscillation in the same fashion as the signal (see Figure 2.7b). By multiplying the two sampled functions in a pointwise fashion, we obtain a sampled product (see Figure 2.7c). Finally, integration in the analog case becomes summation in the discrete case, where the summands need to be weighted by the sampling period T. As a result, one obtains the following approximation:

$$\sum_{n\in\mathbb{Z}} Tf(nT)\exp(-2\pi i\omega nT) \approx \hat{f}(\omega). \tag{2.21}$$

In mathematical terms, the sum can be interpreted as the overall area of rectangular shapes that approximates the area corresponding to the integral (see Figure 2.7d). Such an approximation is also known as a **Riemann sum**. As we will show in Section 2.3.4, the quality of the approximation is good for "well-behaved" signals f and "small" frequency parameters ω.

One defines a discrete version of the Fourier transform for a given DT-signal $x:\mathbb{Z}\to\mathbb{R}$ by setting

$$\hat{x}(\omega) := \sum_{n\in\mathbb{Z}} x(n)\exp(-2\pi i\omega n). \tag{2.22}$$

In this definition, where a simple 1-sampling (i.e., T-sampling with $T=1$) of the exponential function is used, one does not assume that one knows the relation between x and the original signal f. If one is interested in recovering the relation to the Fourier transform $\hat{f}$, one needs to know the sampling period T. Based on (2.21), an easy calculation shows that

$$\hat{x}(\omega) \approx \frac{1}{T}\hat{f}\left(\frac{\omega}{T}\right). \tag{2.23}$$

In this approximation, the frequency parameter ω used for $\hat{x}$ corresponds to the frequency ω/T for $\hat{f}$. In particular, $\omega=1/2$ for $\hat{x}$ corresponds to the Nyquist frequency $\Omega=1/(2T)$ of the sampling process. Therefore, assuming that f is bandlimited by $\Omega=1/(2T)$, one needs to consider only the frequencies with $0\leq\omega\leq 1/2$ for $\hat{x}$. In the digital case, all other frequency parameters are redundant and yield meaningless approximations.

For doing computations on digital machines, we still have some problems. One problem is that the sum in (2.22) involves an infinite number of summands. Another problem is that the frequency parameter ω is a continuous parameter. For both problems, there are some pragmatic solutions. Regarding the first problem, we assume that most of the relevant information of f is limited to a certain duration in time.[2] For example, a music recording of a song hardly lasts for more than ten minutes. Having a finite duration means that the analog signal f is assumed to be zero outside a compact interval. By possibly shifting the signal, we may assume that this interval starts at time $t=0$. This means that we only need to consider a finite number of

[2] Strictly speaking, this assumption is problematic since it conflicts with the requirement of f being bandlimited. A mathematical fact states that there are no functions that are both limited in frequency (bandlimited) and limited in time (having finite duration).

samples $x(0), x(1), \ldots, x(N-1)$ for some suitable number $N \in \mathbb{N}$. As a result, the sum in (2.22) becomes finite.

Regarding the second problem, one computes the Fourier transform only for a finite number of frequencies. Similar to the sampling of the time axis, one typically samples the frequency axis by considering the frequencies $\omega = k/M$ for some suitable $M \in \mathbb{N}$ and $k \in [0:M-1]$. In practice, one often couples the number N of samples and the number M that determines the frequency resolution by setting $N = M$. Note that the two numbers N and M refer to different aspects. However, the coupling is convenient. It not only makes the resulting transform invertible, but also leads to a computationally efficient algorithm, as we will see in Section 2.4.3. Setting $X(k) := \hat{x}(k/N)$ and assuming that $x(0), x(1), \ldots, x(N-1)$ are the relevant samples (all others being zero), we obtain from (2.22) the formula

$$X(k) = \hat{x}(k/N) = \sum_{n=0}^{N-1} x(n) \exp(-2\pi i k n/N) \tag{2.24}$$

for integers $k \in [0:M-1] = [0:N-1]$. This transform is also known as the **discrete Fourier transform** (DFT), which is covered in Section 2.4.

Next, let us have a look at the frequency information supplied by the Fourier coefficient $X(k)$. By (2.23) the frequency ω of $\hat{x}$ corresponds to ω/T of $\hat{f}$. Therefore, the index k of $X(k)$ corresponds to the physical frequency

$$F_{\mathrm{coef}}(k) := \frac{k}{N \cdot T} = \frac{k \cdot F_{\mathrm{s}}}{N} \tag{2.25}$$

given in Hertz. As we will discuss in Section 2.4.4, the coefficients $X(k)$ need to be taken with care. First, the approximation quality in (2.23) may be rather poor, in particular for frequencies close to the Nyquist frequency. Second, for a real-valued signal x, the Fourier transform fulfills certain symmetry properties (see Exercise 2.24). As a result, the upper half of the Fourier coefficients are redundant, and one only needs to consider the coefficients $X(k)$ for $k \in [0:\lfloor N/2 \rfloor]$. Note that, in the case of an even number N, the index $k = N/2$ corresponds to $F_{\mathrm{coef}}(k) = F_{\mathrm{s}}/2$, which is the Nyquist frequency of the sampling process.

Finally, we consider some efficiency issues when computing the DFT. To compute a single Fourier coefficient $X(k)$, one requires a number of multiplications and additions linear in N. Therefore, to compute all coefficients $X(k)$ for $k \in [0:N/2]$ one after another, one requires a number of operations on the order of N^2. Despite being a finite number of operations, such a computational approach is too slow for many practical applications, in particular when N is large.

The number of operations can be reduced drastically by using an efficient algorithm known as the **fast Fourier transform** (FFT). The FFT algorithm, which was discovered by Gauss and Fourier two hundred years ago, has changed whole industries and is now being used in billions of telecommunication and other devices. The FFT exploits redundancies across sinusoids of different frequencies to jointly compute all Fourier coefficients by a recursion. This recursion works particularly well in the case that N is a power of two. As a result, the FFT reduces the overall number of operations from the order of N^2 to the order of $N\log_2 N$. The savings are enormous. For example, using $N = 2^{10} = 1024$, the FFT requires roughly $N\log_2 N = 10240$ instead of $N^2 = 1048576$ operations in the naive approach—a savings factor of about 100. In the case of $N = 2^{20}$, the savings amount to a factor of about 50000 (see Exercise 2.6). In Section 2.4.3, we discuss the algorithmic details of the FFT.

2.1.4 Short-Time Fourier Transform

The Fourier transform yields frequency information that is averaged over the entire time domain. However, the information on *when* these frequencies occur is hidden in the transform. We have already seen this phenomenon in Figure 2.6a, where the change in frequency is not revealed when looking at the magnitude of the Fourier transform. To recover the hidden time information, Dennis Gabor introduced in the year 1946 the **short-time Fourier transform** (STFT). Instead of considering the entire signal, the main idea of the STFT is to consider only a small section of the signal. To this end, one fixes a so-called **window function**, which is a function that is nonzero for only a short period of time (defining the considered section). The original signal is then multiplied with the window function to yield a **windowed signal**. To obtain frequency information at different time instances, one shifts the window function across time and computes a Fourier transform for each of the resulting windowed signals.

This idea is illustrated by Figure 2.8, which continues our example from Figure 2.6a. To obtain local sections of the original signal, one multiplies the signal with suitably shifted rectangular window functions. In Figure 2.8b, the resulting local section only contains frequency content at 1 Hz, which leads to a single main peak in the Fourier transform at $\omega = 1$. Further shifting the time window to the right, the resulting section contains 1 Hz as well as 5 Hz components (see Figure 2.8c). These components are reflected by the two peaks at $\omega = 1$ and $\omega = 5$. Finally, the section shown in Figure 2.8d only contains frequency content at 5 Hz.

Already at this point, we want to emphasize that the STFT reflects not only the properties of the original signal but also those of the window function. First of all, the STFT depends on the length of the window, which determines the size of the section. Then, the STFT is influenced by the shape of the window function. For example, the sharp edges of the rectangular window typically introduce "ripple" artifacts. In Section 2.5.1, we discuss such issues in more detail. In particular, we introduce more suitable, bell-shaped window functions, which typically reduce such artifacts.

In Section 2.5, one finds a detailed treatment of the analog and discrete versions of the STFT and their relationship. In the following, we only consider the discrete case and specify the most important mathematical formulas as needed in practical applications. Let $x : \mathbb{Z} \to \mathbb{R}$ be a real-valued DT-signal obtained by equidistant sampling with respect to a fixed sampling rate F_s given in Hertz. Furthermore, let $w : [0 : N-1] \to \mathbb{R}$ be a sampled window function of length $N \in \mathbb{N}$. For example,

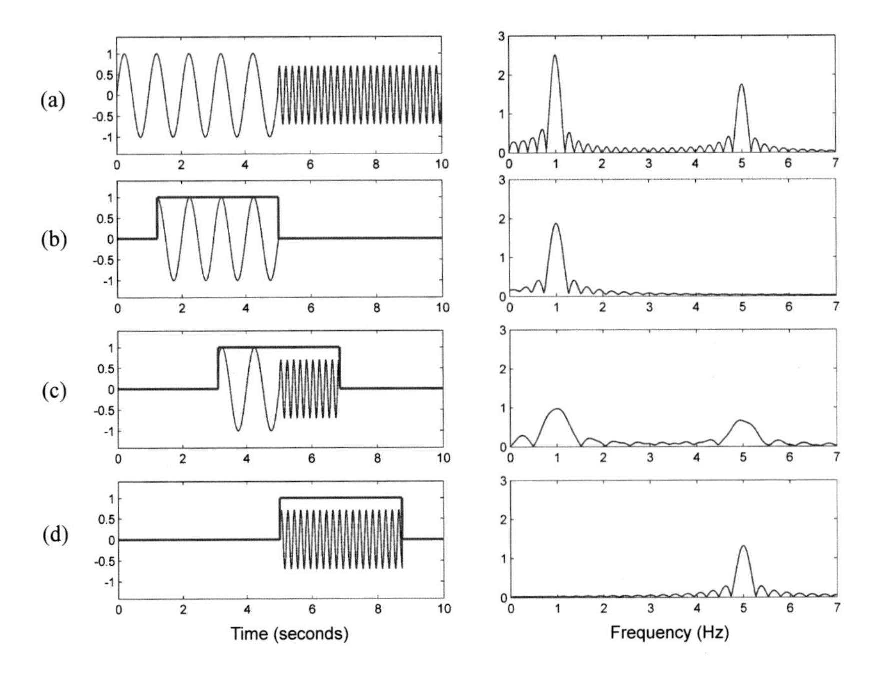

Fig. 2.8 Signal and Fourier transform consisting of two subsequent sinusoids of frequency 1 Hz and 5 Hz (see Figure 2.6a). **(a)** Original signal. **(b)** Windowed signal centered at $t = 3$. **(c)** Windowed signal centered at $t = 5$. **(d)** Windowed signal centered at $t = 7$.

in the case of a rectangular window one has $w(n) = 1$ for $n \in [0 : N-1]$. Implicitly, one assumes that $w(n) = 0$ for all other time parameters $n \in \mathbb{Z} \setminus [0 : N-1]$ outside this window. The length parameter N determines the duration of the considered sections, which amounts to N/F_s seconds. One also introduces an additional parameter $H \in \mathbb{N}$, which is referred to as the **hop size**. The hop size parameter is specified in samples and determines the step size in which the window is to be shifted across the signal.

With regard to these parameters, the **discrete STFT** $\mathcal{X}$ of the signal x is given by

$$\mathcal{X}(m,k) := \sum_{n=0}^{N-1} x(n+mH)w(n)\exp(-2\pi ikn/N) \tag{2.26}$$

with $m \in \mathbb{Z}$ and $k \in [0 : K]$. The number $K = N/2$ (assuming that N is even) is the frequency index corresponding to the Nyquist frequency. The complex number $\mathcal{X}(m,k)$ denotes the kth Fourier coefficient for the mth time frame. Note that for each fixed time frame m, one obtains a **spectral vector** of size $K+1$ given by the coefficients $\mathcal{X}(m,k)$ for $k \in [0 : K]$. The computation of each such spectral vector amounts to a DFT of size N as in (2.24), which can be done efficiently using the FFT.

What have we actually computed in (2.26) in relation to the original analog signal f? As for the temporal dimension, each Fourier coefficient $\mathcal{X}(m,k)$ is associated with the physical time position

$$T_{\mathrm{coef}}(m) := \frac{m \cdot H}{F_{\mathrm{s}}} \tag{2.27}$$

given in seconds. For example, for the smallest possible hop size $H = 1$, one obtains $T_{\mathrm{coef}}(m) = m/F_{\mathrm{s}} = m \cdot T$ sec. In this case, one obtains a spectral vector for each sample of the DT-signal x, which results in a huge increase in data volume. Furthermore, considering sections that are only shifted by one sample generally yields very similar spectral vectors. To reduce this type of redundancy, one typically relates the hop size to the length N of the window. For example, one often chooses $H = N/2$, which constitutes a good trade-off between a reasonable temporal resolution and the data volume comprising all generated spectral coefficients. As for the frequency dimension, we have seen in (2.25) that the index k of $\mathcal{X}(m,k)$ corresponds to the physical frequency

$$F_{\mathrm{coef}}(k) := \frac{k \cdot F_{\mathrm{s}}}{N} \tag{2.28}$$

given in Hertz.

Before we look at some concrete examples, we first introduce the concept of a **spectrogram**, which we denote by $\mathcal{Y}$. The spectrogram is a two-dimensional representation of the squared magnitude of the STFT:

$$\mathcal{Y}(m,k) := |\mathcal{X}(m,k)|^2. \tag{2.29}$$

It can be visualized by means of a two-dimensional image, where the horizontal axis represents time and the vertical axis represents frequency. In this image, the spectrogram value $\mathcal{Y}(m,k)$ is represented by the intensity or color in the image at the coordinate (m,k). Note that in the discrete case, the time axis is indexed by the frame indices m and the frequency axis is indexed by the frequency indices k.

Continuing our running example from Figure 2.8, we now consider a sampled version of the analog signal using a sampling rate of $F_{\mathrm{s}} = 32$ Hz. Having a physical duration of 10 sec, this results in 320 samples (see Figure 2.9a). Using a window length of $N = 64$ samples and a hop size of $H = 8$ samples, we obtain the spectrogram as shown in Figure 2.9b. In the image, the shade of gray encodes the magnitude of a spectral coefficient, where darker colors correspond to larger values. By (2.27), the mth frame corresponds to the physical time $T_{\mathrm{coef}}(m) = m/4$ sec. In other words, the STFT has a time resolution of four frames per second. Furthermore, by (2.28), the kth Fourier coefficient corresponds to the physical frequency $F_{\mathrm{coef}}(k) := k/2$ Hz. In other words, one obtains a frequency resolution of two coefficients per Hertz. The plots of the waveform and the spectrogram with the physically correct time and frequency axes are shown in Figure 2.9c and Figure 2.9d, respectively.

Let us consider some typical settings as encountered when processing music signals. For example, in the case of CD recordings one has a sampling rate of $F_{\mathrm{s}} = 44100$ Hz. Using a window length of $N = 4096$ and a hop size of $H = N/2$,

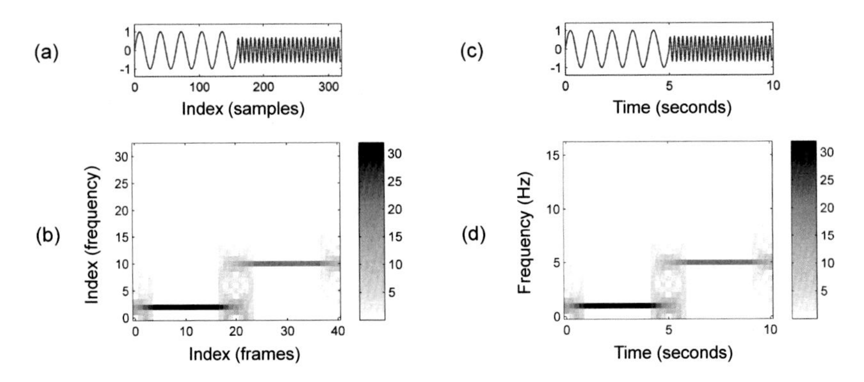

Fig. 2.9 DT-signal sampled with $F_s = 32$ Hz and spectrogram using a window length of $N = 64$ and a hop size of $H = 8$. **(a)** DT-signal with time axis given in samples. **(b)** Spectrogram with time axis given in frames and frequency axis given in indices. **(c)** DT-signal with time axis given in seconds. **(d)** Spectrogram with time axis given in seconds and frequency axis given in Hertz.

this results in a time resolution of $H/F_s \approx 46.4$ ms by (2.27) and a frequency resolution of $F_s/N \approx 10.8$ Hz by (2.28). To obtain a better frequency resolution, one may increase the window length N. This, however, leads to a poorer localization in time so that the resulting STFT loses its capability of capturing local phenomena in the signal. This kind of trade-off is further discussed in Section 2.5.2 and in the exercises.

We close this section with a further example shown in Figure 2.10, which is a recording of a C-major scale played on a piano. The first note of this scale is C4, which we have already considered in Figure 2.1. In Figure 2.10c, the spectrogram representation of the recording is shown, where the time and frequency axes are labeled in a physically meaningful way. The spectrogram reveals the frequency information of the played notes over time. For each note, one can observe horizontal lines that are stacked on top of each other. As discussed in Section 1.3.4, these equally spaced lines correspond to the partials, the integer multiples of the fundamental frequency of a note. Obviously, the higher partials contain less and less of the signal's energy. Furthermore, the decay of each note over time is reflected by the fading out of the horizontal lines. To enhance small sound components that may still be perceptually relevant, one often uses a logarithmic dB scale (see Section 1.3.3). Figure 2.10d illustrates the effect when applying the dB scale to the values of the spectrogram. Besides an enhancement of the higher partials, one can now observe vertical structures at the notes' onset positions. These structures correspond to the noise-like transients that occur in the attack phase of the piano sound (see Section 1.3.4).

This concludes our "nutshell section" covering the most important definitions and properties of the Fourier transform as needed for the subsequent chapters of this book. In particular, the formula (2.26) of the discrete STFT as well as the physical interpretation of the time parameter (2.27) and the frequency parameter (2.28) are

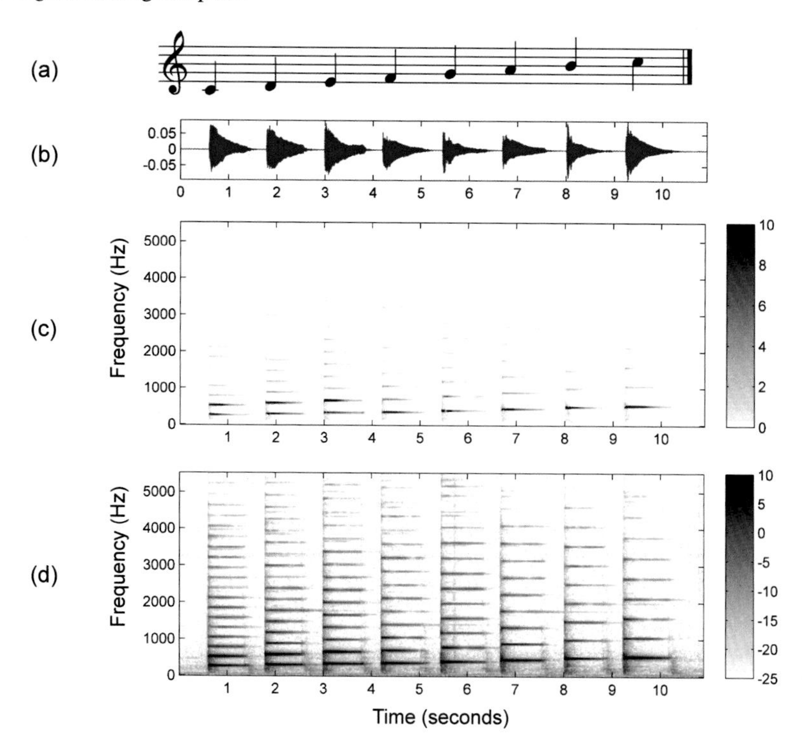

Fig. 2.10 Waveform and spectrogram of a music recording of a C-major scale played on a piano. **(a)** The recording's underlying musical score. **(b)** Waveform. **(c)** Spectrogram. **(d)** Spectrogram with the magnitudes given in dB.

of central importance for most music processing applications to be discussed. As said in the introduction, we provide in the subsequent sections of this chapter some deeper insights into the mathematics underlying the Fourier transform. In particular, we explain in more detail the connection between the various kinds of signals and associated Fourier transforms.

2.2 Signals and Signal Spaces

In technical fields such as engineering or computer science, a **signal** is a function that conveys information about the state or behavior of a physical system. For example, a signal may describe the time-varying sound pressure at some place, the motion of a particle through some space, the distribution of light on a screen representing an image, or the sequence of images as in the case of a video signal. In the following, we consider the case of audio signals as discussed in Section 1.3. We have seen that such a signal can be graphically represented by its waveform, which

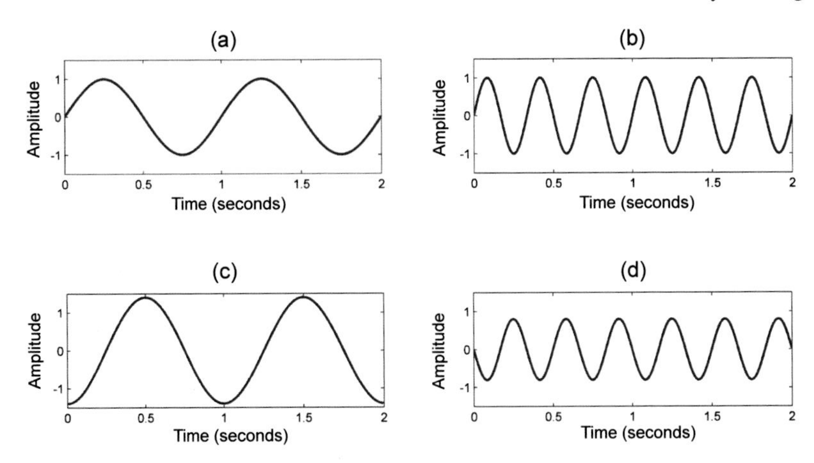

Fig. 2.11 The sinusoid $f(t) = A\sin(2\pi(\omega t - \varphi))$ displayed for $t \in [0,2]$ and for various values of A, ω, and φ. **(a)** $A = 1$, $\omega = 1, \varphi = 0$. **(b)** $A = 1$, $\omega = 3$, $\varphi = 0$. **(c)** $A = 1.4$, $\omega = 1$, $\varphi = 0.25$. **(d)** $A = 0.8$, $\omega = 3$, $\varphi = 0.5$.

depicts the amplitude of the air pressure over time. In the following, we introduce the mathematical notation that is necessary to formally model such a signal. Doing so, we distinguish between two different types of signals: **analog signals** as occur around us in the real world and **digital signals** as are processed by computers. We show how signals can be modified and combined to yield new signals by applying mathematical operations. Some operations can be applied only if the involved signals satisfy certain properties. This leads us to the concept of **signal spaces**, a kind of universe that comprises signals that share a certain property.

2.2.1 Analog Signals

As already defined in Section 2.1.1, an **analog** signal is a function $f\colon \mathbb{R} \to \mathbb{R}$, which assigns an amplitude value $f(t) \in \mathbb{R}$ to each time point $t \in \mathbb{R}$. In the analog case, both the time domain as well as the range of the amplitude values are represented by the set $\mathbb{R}$ of real numbers, which is a continuous range of values. This makes it possible to model infinitesimally small changes in both time and amplitude. In the case of having a continuous time axis (given by $\mathbb{R}$), one also speaks of **continuous-time** (CT) signals. A signal f is called **periodic** with **period** $\lambda \in \mathbb{R}_{>0}$ if $f(t) = f(t+\lambda)$ holds for all $t \in \mathbb{R}$. If there exists a least positive constant with this property, it is called the **prime period** of the signal (see Exercise 2.7 and Exercise 2.8).

In Section 1.3.2 and Section 2.1.1.1, we have already encountered an entire class of analog signals: the **sinusoids**. Recall from (2.1) that a sinusoid is a periodic function f defined by $f(t) := A\sin(2\pi(\omega t - \varphi))$, $t \in \mathbb{R}$. The parameter A describes the **amplitude**, the parameter ω the **frequency**, and the parameter φ the **phase**. The

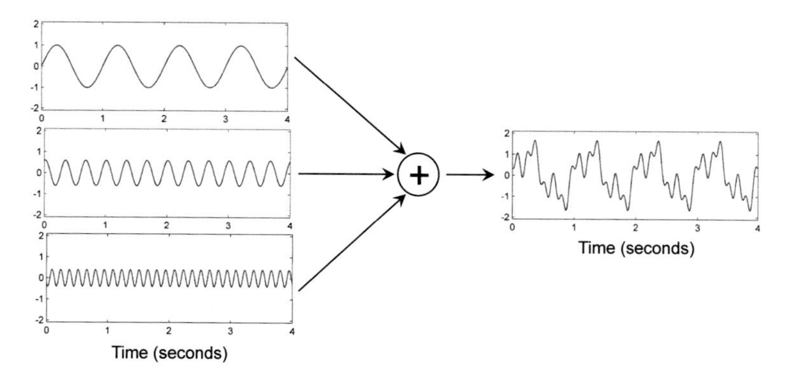

Fig. 2.12 Superposition of three analog signals.

frequency parameter ω determines the period of the sinusoid, which is $\lambda = 1/\omega$. In other words, a sinusoid of frequency ω repeats every $\lambda = 1/\omega$ unit times. In the following, we use seconds as the units of time if not specified otherwise. Figure 2.11 shows various sinusoids resulting from different parameter settings.

Besides having a compact description, sinusoids also have an explicit physical meaning with a perceptual correspondence: the amplitude A corresponds to the loudness and the frequency ω to the pitch of a sinusoidal sound. Only the phase φ, which indicates the relative position of an oscillation within its cycle, does not have a direct perceptual correspondence. Note that, because of the periodicity of a sinusoid, a phase shift by $\varphi + 1$ has the same effect as a phase shift by φ. In other words, integer shifts leave a sinusoid unaltered and the parameter φ needs to be considered only in the interval $[0, 1)$.

Regarding a signal as a mathematical function is convenient, since this allows us to express modifications of signals in terms of mathematical operations. For example, the **superposition** of two signals f and g can be expressed by the sum $f + g$ defined as pointwise addition

$$(f+g)(t) := f(t) + g(t) \tag{2.30}$$

for $t \in \mathbb{R}$. Similarly, the **scaling** of a signal f by a real factor a is the scalar multiple af, which is also defined pointwise by

$$(af)(t) := a \cdot f(t). \tag{2.31}$$

Figure 2.12 shows an example of a superposition of three signals. We have seen in Section 2.1 that the Fourier transform can be regarded as a kind of inverse operation, where a given signal is decomposed into a weighted superposition of elementary signals.

2.2.2 Digital Signals

Analog signals have a continuous range of values in both time and amplitude, which generally leads to an infinite number of values. Since a computer can only store and process a finite number of values, one has to convert the waveform into some **discrete** representation—a process commonly referred to as **digitization**. Some analog signals such as sinusoids are already characterized by a small number of parameters, which can be used to represent the signal, but for general analog signals one needs other ways for deriving a model that can be described by a finite number of parameters. Furthermore, it should be possible to perform signal manipulations directly in the parameter domain such that computations become feasible and efficient. The most common approach for digitizing audio signals consists of two steps called **sampling** and **quantization** (see Figure 2.13 for an illustration). We now explain these two steps in more detail.

2.2.2.1 Sampling

In signal processing, the term **sampling** refers to the process of reducing a continuous-time (CT) signal to a **discrete-time** (DT) signal, which is defined only on a discrete subset of the time axis. By means of a suitable encoding, one often assumes that this discrete set is a subset I of the set $\mathbb{Z}$ of integers. Then a DT-signal is defined to be a function $x\colon I \to \mathbb{R}$, where the domain I corresponds to points in time. Since one can extend any DT-signal from the domain I to the domain $\mathbb{Z}$ simply by setting all values to zeros for points in $\mathbb{Z} \setminus I$, we may assume $I = \mathbb{Z}$. The most common sampling procedure to transform a CT-signal $f\colon \mathbb{R} \to \mathbb{R}$ into a DT-signal $x\colon \mathbb{Z} \to \mathbb{R}$ is known as **equidistant sampling**. For convenience, we repeat the definitions from Section 2.1.3. Fixing a positive real number $T > 0$, the DT-signal x is obtained by setting

$$x(n) := f(n \cdot T) \tag{2.32}$$

for $n \in \mathbb{Z}$. The value $x(n)$ is called the **sample** taken at time $t = n \cdot T$ of the original analog signal f. In short, this procedure is also called **T-sampling**. The number T is referred to as the **sampling period** and the inverse $F_\mathrm{s} := 1/T$ as the **sampling rate**. The sampling rate specifies the number of samples per second and is measured in Hertz (Hz).

Figure 2.13 shows an illustrative example, where the DT-signal x is represented by the red stem plot. In this example, one has 13 samples in the first two seconds. Thus, the sampling rate is roughly 6.5 Hz and the sampling period 0.154 seconds. In practical applications, typical sampling rates are 8 kHz (8,000 Hz) for telephony, 32 kHz for digital radio, 44.1 kHz for CD recordings, and 48 kHz up to 96 kHz for professional studio technology.

In general, sampling is a **lossy** operation in the sense that information is lost in this process and that the original analog signal cannot be recovered from its sampled version. Only if the analog signal has additional properties in terms of its frequency

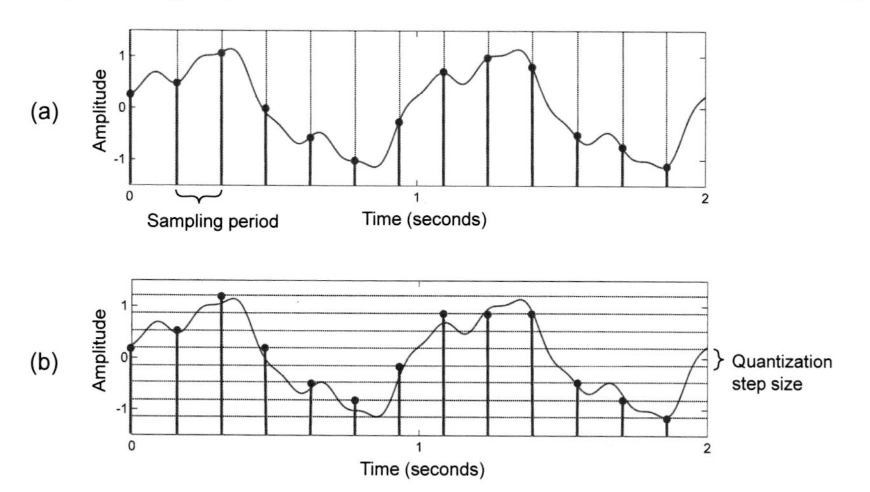

Fig. 2.13 Two steps of a digitization process to transform an analog signal (solid curve) into a digital signal (stem plot). **(a)** Sampling. **(b)** Quantization.

spectrum is a perfect reconstruction possible. This is the assertion of the famous **sampling theorem**, which we discuss in Exercise 2.28 in more detail. Without such additional properties, sampling may cause an effect known as **aliasing**, where certain frequency components of the signal become indistinguishable. This effect is illustrated by Figure 2.14, which shows an analog signal that is the superposition of two sinusoids. Using a high sampling rate as in Figure 2.14a, the analog signal can be reconstructed with high accuracy. However, when decreasing the sampling rate, the higher-frequency component is not captured well and only a coarse approximation of the original signal remains (see Figure 2.14c).

2.2.2.2 Quantization

We have seen how sampling transforms a continuous time axis (encoded by $\mathbb{R}$) into a discrete time axis (encoded by $\mathbb{Z}$). This is only the first step in an analog-to-digital conversion of a signal. In the second step, one needs to replace the continuous range of possible amplitudes (again encoded by $\mathbb{R}$) by a discrete range of possible values (encoded by a discrete set $\Gamma \subset \mathbb{R}$). This process is commonly known as **quantization**. Such a quantization can be modeled by a function $Q : \mathbb{R} \to \Gamma$, referred to as the **quantizer**, which assigns to each amplitude value $a \in \mathbb{R}$ a value $Q(a) \in \Gamma$. Many of the quantizers used simply round off or truncate the analog value to some units of precision. For example, a typical uniform quantizer with a **quantization step size** equal to some value Δ can be defined by

$$Q(a) := \mathrm{sgn}(a) \cdot \Delta \cdot \left\lfloor \frac{|a|}{\Delta} + \frac{1}{2} \right\rfloor \tag{2.33}$$

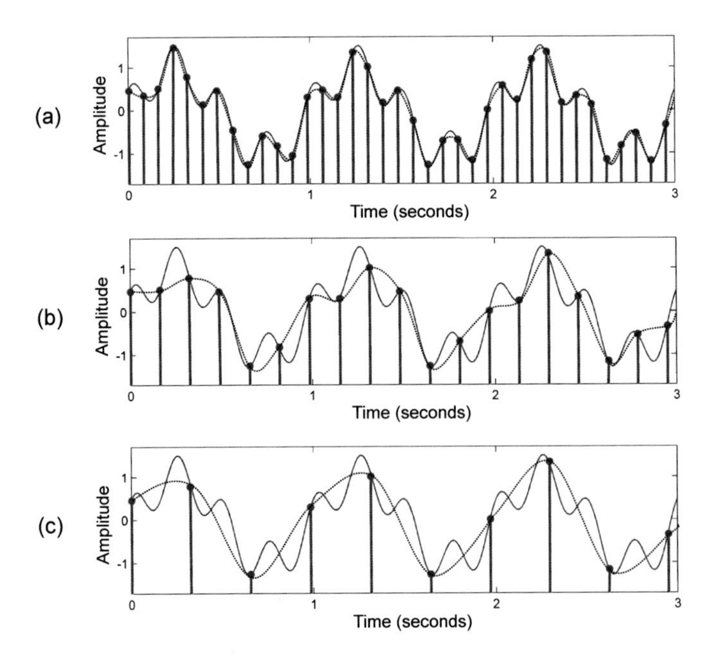

Fig. 2.14 Illustration of the aliasing effect when reducing the sampling rate. The figures show the original analog signal (solid curve), the sampled version (stem plot), and the reconstructed analog signal (dotted curve) for sampling rates of **(a)** 12 Hz, **(b)** 6 Hz, and **(b)** 3 Hz.

for $a \in \mathbb{R}$, were $\operatorname{sgn}(\cdot)$ is the signum function that yields the sign of a real number and the brackets $\lfloor \cdot \rfloor$ truncate a real number to yield the largest integer below this number. Note that, in the case of $\Delta = 1$, the quantizer Q is simple rounding to the nearest integer. Like sampling, quantization is generally a lossy operation, because different analog values may be mapped to the same digital value. The difference between the actual analog value and the quantized value is called the **quantization error** (see Exercise 2.9). Reducing the quantization step size Δ typically leads to smaller quantization errors. However, at the same time, the number of quantized values (and therefore also the number of bits needed to encode these values) increases. Figure 2.13b shows the result after sampling and quantizing an analog signal. In this example, the quantization step size $\Delta = 1/3$ is used, resulting in 8 different quantization values for the given signal. Hence, a 3-bit coding scheme may be used to represent the quantized values. For CD recordings, a 16-bit coding scheme is used, which allows representation of 65536 possible values.

In summary, after using an analog-to-digital conversion based on sampling and quantization, it is generally not possible to reconstruct the original waveform from the digital representation. Aliasing and quantization may introduce audible sound artifacts such as harsh buzzing sounds or noise. For digital representations as used for CDs, however, the sampling rate as well as the quantization resolution are chosen in such ways that the degradation of the waveform is not noticeable by the human ear.

2.2.3 Signal Spaces

In the previous sections, we considered analog and digital signals, which were modeled as CT-signals $f\colon \mathbb{R} \to \mathbb{R}$ and as DT-signals $x\colon \mathbb{Z} \to \mathbb{R}$, respectively. In the following discussion, we use the symbols f and g to denote CT-signals and the symbols x and y to denote DT-signals. For the time parameter, we typically use the parameter t in the CT case and the parameter n in the DT case.

2.2.3.1 Complex Numbers

In view of the complex-valued formulation of the Fourier transform one needs to extend the range $\mathbb{R}$ of real numbers to the range $\mathbb{C}$ of **complex numbers**. Recall from Section 2.1.1.4 that each complex number $c \in \mathbb{C}$ can be regarded as a pair $(a,b) \in \mathbb{R}^2$ of real numbers, where $a = \mathrm{Re}(c)$ denotes the real part and $b = \mathrm{Im}(c)$ the imaginary part of c. One also often writes $c = a + ib$, where i is the imaginary unit. The complex number field $\mathbb{C}$ possesses a multiplication that extends the multiplication of the real number field $\mathbb{R}$. Given two complex numbers $c_1 = a_1 + ib_1, c_2 = a_2 + ib_2 \in \mathbb{C}$, the product is defined by

$$c_1 \cdot c_2 = a_1 a_2 - b_1 b_2 + i(a_1 b_2 + a_2 b_1). \tag{2.34}$$

Furthermore, the **complex conjugate** $\overline{c}$ of a complex number $c = a + ib \in \mathbb{C}$ is defined as

$$\overline{c} = a - ib. \tag{2.35}$$

Various computation rules for complex numbers are discussed in Exercise 2.12. Extending the notion of real-valued signals, a complex-valued CT-signal is a function $f\colon \mathbb{R} \to \mathbb{C}$ and a complex-valued DT-signal a function $x\colon \mathbb{Z} \to \mathbb{C}$. As is the case with complex numbers, each complex-valued signal can be considered as a pair of two real-valued signals. Furthermore, each real-valued signal can be regarded as a complex-valued signal simply by defining the imaginary part to be zero. In the following, we therefore only consider the more general complex-valued case, which includes the real-valued case.

2.2.3.2 Vector Spaces

A general principle in mathematics is to form suitable spaces that comprise all objects under consideration. These spaces can then be equipped with additional structures that can be used to manipulate and organize the objects. For example, for a given natural number $N \in \mathbb{N}$, one may consider the space $\mathbb{R}^N$ consisting of all real-valued N-tuples. This space can be equipped with an addition and a scalar multiplication such that $\mathbb{R}^N$ becomes a **vector space** over $\mathbb{R}$. Similarly, one can define the space $\mathbb{C}^N$, which consists of all complex-valued N-tuples. In our case, the objects under consideration are complex-valued CT- and DT-signals. The resulting signal spaces are defined as

$$\mathbb{C}^{\mathbb{R}} := \{f | f \colon \mathbb{R} \to \mathbb{C}\} \quad \text{and} \quad \mathbb{C}^{\mathbb{Z}} := \{x | x \colon \mathbb{Z} \to \mathbb{C}\}, \tag{2.36}$$

for the CT and DT case, respectively. We have already seen in (2.30) and (2.31) how one can define an addition of two signals and a scalar multiplication of a real factor and a signal. These definitions directly carry over to the case of complex-valued signals using complex summation and multiplication, which makes $\mathbb{C}^{\mathbb{R}}$ a vector space over $\mathbb{C}$. Similarly, one can define addition and scalar multiplication in the DT case, making $\mathbb{C}^{\mathbb{Z}}$ a vector space over $\mathbb{C}$.

One may need to get used to the fact that elements (the "points") of a space such as $\mathbb{C}^{\mathbb{R}}$ or $\mathbb{C}^{\mathbb{Z}}$ can be entire signals. As opposed to the case $\mathbb{C}^N$, which defines a vector space of (complex) dimension N, the vector spaces $\mathbb{C}^{\mathbb{R}}$ and $\mathbb{C}^{\mathbb{Z}}$ have infinite dimension. Still, many of the geometric structures known for the finite-dimensional space $\mathbb{C}^N$ can be transferred to suitably defined infinite-dimensional subspaces of $\mathbb{C}^{\mathbb{R}}$ and $\mathbb{C}^{\mathbb{Z}}$. This is what we show next.

2.2.3.3 Inner Products

We start by reviewing some concepts from linear algebra. Usually, an element $x \in \mathbb{C}^N$ is thought of as a column vector of size N. The **transposed** vector, which we denote by $x^\top$, is then the corresponding row vector. The vector space $\mathbb{C}^N$ can be equipped with an additional structure called an **inner product**. This additional structure associates to each pair of vectors a scalar quantity which is called the inner product of the two vectors. Mathematically, the inner product of $\mathbb{C}^N$ is a mapping $\langle \cdot | \cdot \rangle \colon \mathbb{C}^N \times \mathbb{C}^N \to \mathbb{C}$ defined by

$$\langle x | y \rangle := \sum_{n=0}^{N-1} x(n)\overline{y(n)} \tag{2.37}$$

for $x = (x(0), x(1), \ldots, x(N-1))^\top \in \mathbb{C}^N$ and $y = (y(0), y(1), \ldots, y(N-1))^\top \in \mathbb{C}^N$. The inner product satisfies three mathematical properties, which are also used for an axiomatic definition of general inner products. First, it is **positive definite**; i.e., $\langle x|x \rangle \geq 0$ and $\langle x|x \rangle = 0$ if and only if x is the all-zero vector. Second, it is **conjugate symmetric**; i.e., $\langle x|y \rangle = \overline{\langle y|x \rangle}$. And third, it is **$\mathbb{C}$-linear** in the first argument; i.e., $\langle x_1 + x_2|y \rangle = \langle x_1|y \rangle + \langle x_2|y \rangle$ and $\langle cx|y \rangle = c\langle x|y \rangle$ for any $x_1, x_2, x, y \in \mathbb{C}^N$ and $c \in \mathbb{C}$.

The importance of inner products is that they allow the introduction of intuitive geometrical notions such as the length of a vector, the angle between two vectors, and orthogonality between vectors (see Figure 2.15 for an illustration). More pre-

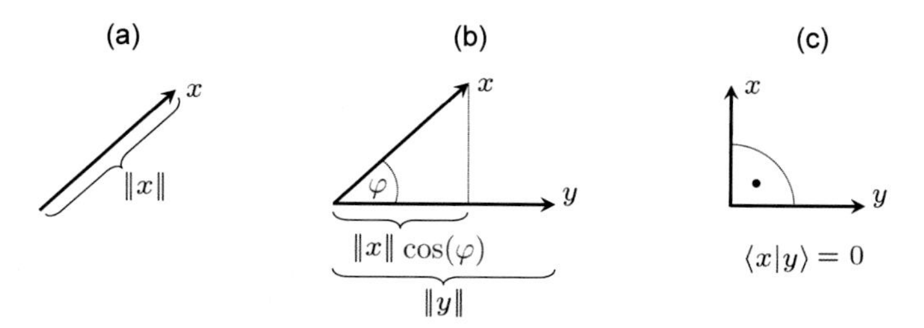

Fig. 2.15 Geometrical notions defined in terms of the inner product. **(a)** Length of a vector. **(b)** Angle between two vectors. **(b)** Orthogonality of two vectors.

cisely, the inner product induces a norm on $\mathbb{C}^N$ via

$$\|x\| := \sqrt{\langle x|x\rangle}. \tag{2.38}$$

In general, a **norm** satisfies $\|x\| = 0$ if and only if $x = 0$, $\|ax\| = |a|\|x\|$ for any $a \in \mathbb{C}$ (positive scalability), and $\|x+y\| \leq \|x\| + \|y\|$ for any vectors x and y (**triangle inequality**). The positive number $\|x-y\|$ is also called the **distance** between the vectors x and y. The relation between the inner product and the angle φ between two vectors x and y is given by

$$\cos(\varphi) = \frac{|\langle x|y\rangle|}{\|x\| \cdot \|y\|}. \tag{2.39}$$

In other words, the angle φ is determined by the inner product: it is given by taking the inverse of the cosine of the absolute value of the inner product of the normalized vectors. The basis for this relation is the **Cauchy–Schwarz inequality**

$$|\langle x|y\rangle| \leq \|x\| \cdot \|y\|, \tag{2.40}$$

which is an indispensable mathematical tool for many estimations. Finally, two vectors $x, y \in \mathbb{C}^N$ are said to be **orthogonal** if $\langle x|y\rangle = 0$ (see Figure 2.15c). This concept can then be used to define orthogonal subspaces, orthogonal complements, projection operators, and so on.

2.2.3.4 The Space $\ell^2(\mathbb{Z})$

Given an arbitrary vector space, one can introduce the same geometric concepts once one has an inner product. It turns out that the signal spaces $\mathbb{C}^\mathbb{R}$ or $\mathbb{C}^\mathbb{Z}$ are too general. One strategy is to only consider signals with certain properties by passing over to suitable signal subspaces. We make this point clearer by first considering the space $\mathbb{C}^\mathbb{Z}$ of DT-signals. One idea for defining an inner product on this space is to simply extend the definition of (2.37) for $\mathbb{C}^N$. However, in contrast to $\mathbb{C}^N$, there may be an infinite number of nonzero summands in the case of $\mathbb{C}^\mathbb{Z}$, with the consequence that the sum may be infinite. This leads to the following definitions: First, we define the **energy** $\mathrm{E}(x)$ of a signal $x \in \mathbb{C}^{\mathbb{Z}}$ to be

$$\mathrm{E}(x) := \sum_{n\in\mathbb{Z}} |x(n)|^2. \tag{2.41}$$

Then the space $\ell^2(\mathbb{Z}) \subset \mathbb{C}^{\mathbb{Z}}$ is defined to be the set of all signals having finite energy:

$$\ell^2(\mathbb{Z}) := \{x\colon \mathbb{Z} \to \mathbb{C} \mid \mathrm{E}(x) < \infty\}. \tag{2.42}$$

In mathematical terms, $\ell^2(\mathbb{Z})$ is also referred to as the space of **square-summable** sequences. Obviously, there are many DT-signals that do not have finite energy. For example, the sampled sinusoid x given by $x(n) = \sin(\pi n/16)$ is not square-summable since it assumes the value 1 for infinitely many n. On the other hand, any DT-signal with a finite number of nonzero entries obviously has finite energy. The space $\mathbb{C}^N$ for arbitrary $N \in \mathbb{N}$ can be regarded as a subspace of $\ell^2(\mathbb{Z})$ by extending a vector $x = (x(0), x(1), \ldots, x(N-1))^\top \in \mathbb{C}^N$ to a sequence by setting $x(n) = 0$ for all $n < 0$ and $n \geq N$. Furthermore, it is not hard to show that $\ell^2(\mathbb{Z})$ is a vector space (see Exercise 2.13). For the restricted space $\ell^2(\mathbb{Z}) \subset \mathbb{C}^{\mathbb{Z}}$, it is now possible to introduce an inner product that extends the one for $\mathbb{C}^N$. Indeed, one can show that

$$\langle x|y\rangle := \sum_{n\in\mathbb{Z}} x(n)\overline{y(n)} \tag{2.43}$$

is finite and hence well defined for any two signals $x, y \in \ell^2(\mathbb{Z})$ (see again Exercise 2.13). From this point on, everything works as in the finite-dimensional case $\mathbb{C}^N$. The inner product satisfies the Cauchy–Schwarz inequality (2.40), one can define an angle as in (2.39), one can talk about signals being orthogonal, and so on.

2.2.3.5 The Space $L^2(\mathbb{R})$

For the space $\mathbb{C}^{\mathbb{R}}$ of CT-signals, an inner product is defined in a similar fashion. However, technically, the definitions become more sophisticated in the continuous case, where summation becomes integration. In order to define an integral for a signal $f \in \mathbb{C}^{\mathbb{R}}$, it needs to fulfill certain integrability conditions, which in turn depend on the notion of integration to be used. For example, the notion of the well-known **Riemann integral** turns out to be too weak for many mathematical constructions. The technical deficiencies in Riemann integration can be remedied with the **Lebesgue integral**, which can be defined for a class of signals called **measurable**. At this point, since we may assume that basically all signals that we encounter are measurable, we do not want to go further into this issue. Similarly to the case of DT-signals, the **energy** $\mathrm{E}(f)$ of a measurable signal $f \in \mathbb{C}^{\mathbb{R}}$ is defined by

$$\mathrm{E}(f) := \int_{t\in\mathbb{R}} |f(t)|^2 dt. \tag{2.44}$$

Furthermore, the space $L^2(\mathbb{R}) \subset \mathbb{C}^{\mathbb{R}}$ is defined to be the set of all signals of finite energy:

$$L^2(\mathbb{R}) := \{f\colon \mathbb{R} \to \mathbb{C} \mid f \text{ measurable and } \mathrm{E}(f) < \infty\}. \tag{2.45}$$

In mathematical terms, $L^2(\mathbb{R})$ is also referred to as the Lebesgue space[3] of square-integrable functions. Again, there are many CT-signals that do not have finite energy. For example, any nonzero sinusoid has infinite energy. As with the DT case, it is not hard to show that $L^2(\mathbb{R})$ is a vector space. In the CT case, the inner product is defined by

$$\langle f|g\rangle := \int_{t\in\mathbb{R}} f(t)\overline{g(t)}dt \tag{2.46}$$

for any $f,g \in L^2(\mathbb{R})$. Again this makes it possible to introduce the geometric concepts known from linear algebra.

2.2.3.6 The Space $L^2([0,1))$

Finally, we want to consider another class of CT-signals of fundamental importance: the class of **periodic signals**. As already mentioned above, nonzero periodic functions[4] are not contained in $L^2(\mathbb{R})$. However, also for periodic functions one can define a suitable signal subspace of $\mathbb{C}^{\mathbb{R}}$ that possesses an inner product. Recall from Section 2.2.1 that a signal $f\colon \mathbb{R} \to \mathbb{C}$ is periodic with period $\lambda \in \mathbb{R}_{>0}$ if $f(t) = f(t+\lambda)$ holds for all $t \in \mathbb{R}$. A λ-periodic signal f can be transformed into a 1-periodic signal $t \mapsto f(\lambda \cdot t)$ by applying the linear transform $t \mapsto \lambda \cdot t$. Hence, in the following discussion, we only consider the case $\lambda = 1$. Obviously, any 1-periodic function f is already known when restricted to the interval $[0,1)$. In contrast, any function $g\colon [0,1) \to \mathbb{C}$ can be extended in an obvious fashion to a 1-periodic function $f\colon \mathbb{R} \to \mathbb{C}$. In other words, there is a one-to-one correspondence between the 1-periodic functions in $\mathbb{C}^{\mathbb{R}}$ and the signal space $\mathbb{C}^{[0,1)} := \{f\colon [0,1) \to \mathbb{C}\}$. Similar to the nonperiodic case, one can define the **energy** $\mathrm{E}_{[0,1)}(f)$ by

$$\mathrm{E}_{[0,1)}(f) := \int_{t\in[0,1)} |f(t)|^2 dt \tag{2.47}$$

and the space $L^2([0,1)) \subset \mathbb{C}^{[0,1)}$ by

$$L^2([0,1)) := \{f\colon [0,1) \to \mathbb{C} \mid f \text{ measurable and } \mathrm{E}_{[0,1)}(f) < \infty\}. \tag{2.48}$$

Furthermore, one can show that the inner product

$$\langle f|g\rangle := \int_{t\in[0,1)} f(t)\overline{g(t)}dt \tag{2.49}$$

---

[3] From a strict technical point of view, $L^2(\mathbb{R})$ is defined as a quotient space, where all functions that are zero almost everywhere are identified.

[4] Strictly speaking, we mean here periodic functions that are not zero almost everywhere.

is well defined for any $f,g \in L^2([0,1))$. Generalizing these definitions, one can introduce a space $L^2([a,b))$ with an inner product for any $a,b \in \mathbb{R}$, $a < b$, which consists of λ-periodic signals with $\lambda = b - a$.

2.2.3.7 Hilbert Spaces

In summary, we have introduced the signal spaces $\ell^2(\mathbb{Z})$, $L^2(\mathbb{R})$, and $L^2([0,1))$, which all possess an inner product similar to the one of the finite-dimensional vector space $\mathbb{C}^N$. All of these spaces are special cases of what is known as **Hilbert space**. By definition, a Hilbert space is a vector space $\mathcal{H}$ equipped with an inner product $\langle\cdot|\cdot\rangle\colon \mathcal{H} \times \mathcal{H} \to \mathbb{C}$ satisfying the three axiomatic conditions mentioned in Section 2.2.3. Furthermore, one requires that a Hilbert space is **complete** in the sense that every Cauchy sequence[5] in $\mathcal{H}$ converges in $\mathcal{H}$. Intuitively, a space is complete if no points are missing from it. For example, the set of rational numbers is not complete, because there are numbers such as $\sqrt{2}$ missing from it, even though one can construct Cauchy sequences of rational numbers that converge to such irrational numbers. As one can show, this nontrivial completeness condition is satisfied for the signal spaces $\ell^2(\mathbb{Z})$, $L^2(\mathbb{R})$, and $L^2([0,1))$. As we will see in the next sections, the geometric concepts provided by the inner product help to develop our intuition and to simplify the formulation of the Fourier transform.

A particularly important concept that generalizes from the finite-dimensional space $\mathbb{C}^N$ to arbitrary Hilbert spaces is the existence of orthonormal bases. Let I be a countable set, then a subset $(x_i)_{i\in I}$ of $\mathcal{H}$ is called an **orthonormal basis** (ON-basis) if the following three conditions hold:

$$\langle x_i|x_j\rangle = 0 \quad \text{for} \quad i,j \in I, i \neq j, \tag{2.50}$$

$$\|x_i\|^2 = 1 \quad \text{for} \quad i \in I, \tag{2.51}$$

$$x = \sum_{i\in I} \langle x|x_i\rangle x_i \quad \text{for} \quad x \in X. \tag{2.52}$$

The first condition means that any two distinct elements x_i and x_j are orthogonal, and the second one that each of the elements x_i has unit energy. The third condition, also referred to as the **completeness condition**, requires that any element of $x \in \mathcal{H}$ can be represented as a weighted superposition of the **basis vectors** x_i, $i \in I$. Intuitively, completeness means that everything in $\mathcal{H}$ can be captured by the basis vectors. Furthermore, the weights are given by the inner products $\langle x|x_i\rangle$. One can show that for a Hilbert space there always exists an ON-basis and, in general, even a very large number of different ON-bases. As we will see, the Fourier transforms for DT-signals and periodic CT-signals are based on very specific choices of such ON-bases.

[5] A Cauchy sequence is a sequence whose elements become arbitrarily close to each other as the sequence progresses. More precisely, given any small positive distance, all but a finite number of elements of the sequence are less than that given distance from each other.

2.3 Fourier Transform

The Fourier transform is the most important mathematical tool in audio signal processing. As discussed in Section 2.1, the Fourier transform converts a time-dependent signal into a frequency-dependent function. The inverse process is realized by the Fourier representation, which represents a signal as a weighted superposition of independent elementary functions. Each of the weights expresses the extent to which the corresponding elementary function contributes to the original signal, thus revealing a certain aspect of the signal. Because of their explicit physical interpretation in terms of frequency, sinusoids are particularly suited to serve as elementary functions. Each of the weights is then associated to a frequency value and expresses the degree to which the signal contains a periodic oscillation of that frequency. The Fourier transform can be regarded as a way to compute the frequency-dependent weights.

In the following, depending on the underlying signal space, we introduce several variants of the Fourier transform and its inverse, the Fourier representation. We start with the signal space $L^2([0,1))$ consisting of 1-periodic finite-energy CT-signals (Section 2.3.1). We continue by showing how the formulation of the Fourier transform in terms of complex-valued exponential functions (instead of real-valued sinusoids) makes the mathematical handling much more convenient (Section 2.3.2). We then discuss the Fourier transform for the signal space $L^2(\mathbb{R})$ (Section 2.3.3) as well as for the signal space $\ell^2(\mathbb{Z})$ (Section 2.3.4). It is important to note that each of these signal spaces possesses its own Fourier transform and the mathematical concepts needed to prove the existence and properties of the respective Fourier transform are different for the variants. While giving mathematically rigorous definitions of the various Fourier transforms, we do not provide the proofs. In particular for the analog case, the proofs require results from measure and integration theory, which are outside the scope of this book. Instead, we will try to give some intuitive explanations while highlighting the meaning and the interrelations of the various variants.

2.3.1 *Fourier Transform for Periodic CT-Signals*

We start our discussion by considering the case of all **real-valued** signals in $L^2([0,1))$. Let us denote this subspace by $L^2_{\mathbb{R}}([0,1)) \subset L^2([0,1))$. Note that any constant as well as any $(1/k)$-periodic function for an integer $k \in \mathbb{N}$ is 1-periodic too. The sinusoid $t \mapsto \sqrt{2}\cos(2\pi kt)$ may be regarded as the archetype of a $(1/k)$-periodic function, which represents a pure tone of k Hz. The factor $\sqrt{2}$ is introduced to normalize the sinusoid to have unit energy or, equivalently, to have norm one (see Exercise 2.14). Of course, also the sinusoid $t \mapsto \sqrt{2}\sin(2\pi kt)$ or all phase-shifted versions $t \mapsto \sqrt{2}\cos(2\pi(kt-\varphi))$ have the same interpretation. One important theorem in Fourier analysis is that any real-valued signal $f \in L^2_{\mathbb{R}}([0,1))$ can be written as a superposition

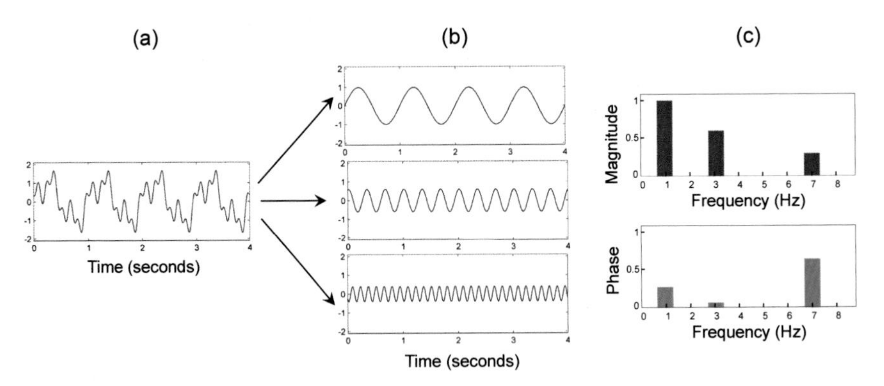

Fig. 2.16 **(a)** Analog 1-periodic signal. **(b)** Decomposition of the signal into three sinusoids. **(c)** Magnitude and phase coefficients of the Fourier transform.

$$f(t) = d_0 + \sum_{k\in\mathbb{N}} d_k\sqrt{2}\cos(2\pi(kt - \varphi_k)) \tag{2.53}$$

of 1-periodic sinusoids with suitable amplitudes $d_k \in \mathbb{R}_{\geq 0}$ and phases $\varphi_k \in [0,1)$. The superposition exhibits the frequency content of f as follows: the coefficient d_k, also referred to as the **magnitude**, reflects the contribution of the sinusoid of k Hz, whereas the coefficient φ_k, also referred to as the **phase**, shows how the sinusoid has to be shifted to best "explain" or "match" the original signal. Note that the phase coefficients are determined only up to an integer and can therefore be assumed to lie in the interval $[0,1)$. Figure 2.16 shows an example of a 1-periodic signal and the resulting magnitude and phase coefficients. The superposition in (2.53) is the **Fourier representation** of the signal f, whereas the magnitude and phase coefficients are called the **Fourier coefficients**.

In our first reformulation, we exploit the fact that any sinusoid with arbitrary phase can be represented as a weighted sum of two specific sinusoids of the same frequency having fixed phases. Indeed, using the trigonometric identity $\cos(\alpha - \beta) = \cos(\alpha)\cos(\beta) + \sin(\alpha)\sin(\beta)$ for arbitrary angles α and β, one obtains

$$\cos(2\pi(kt - \varphi)) = \cos(2\pi kt)\cos(2\pi\varphi) + \sin(2\pi kt)\sin(2\pi\varphi) \tag{2.54}$$

when setting $\alpha = 2\pi kt$ and $\beta = 2\pi\varphi$. Let $\mathbf{cos}_k, \mathbf{sin}_k \in L^2_{\mathbb{R}}([0,1))$ be the two specific sinusoids defined by

$$\mathbf{cos}_k(t) := \sqrt{2}\cos(2\pi kt), \tag{2.55}$$
$$\mathbf{sin}_k(t) := \sqrt{2}\sin(2\pi kt), \tag{2.56}$$

for $k \in \mathbb{N}$. Then plugging (2.54) into (2.53), one obtains the following Fourier representation, which is also known as the **Fourier series**:

$$f(t) = a_0 + \sum_{k\in\mathbb{N}} a_k\mathbf{cos}_k(t) + \sum_{k\in\mathbb{N}} b_k\mathbf{sin}_k(t). \tag{2.57}$$

It readily follows that the **Fourier coefficients** a_0, a_k, and b_k are given by

$$a_0 = d_0, \tag{2.58}$$
$$a_k = \cos(2\pi\varphi_k)d_k, \tag{2.59}$$
$$b_k = \sin(2\pi\varphi_k)d_k \tag{2.60}$$

for $k \in \mathbb{N}$. Vice versa, the magnitudes and phases can be computed from the a_k and b_k via

$$d_k = \sqrt{a_k^2 + b_k^2}, \tag{2.61}$$
$$\varphi_k = \frac{1}{2\pi}\,\mathrm{atan2}(b_k, a_k). \tag{2.62}$$

The atan2 function, which is a variant of the inverse of the tangent function, will be explained in Section 2.3.2.2. A nice property of the Fourier representation in (2.57) is that its Fourier coefficients can be easily computed using Hilbert space theory. To this end, one needs to show that the set

$$\{\mathbf{1}, \mathbf{cos}_k, \mathbf{sin}_k | k \in \mathbb{N}\}, \tag{2.63}$$

is an ON-basis of the Hilbert space $L^2_{\mathbb{R}}([0,1))$, where $\mathbf{1}$ denotes the all-one signal (i.e., $\mathbf{1}(t) = 1$ for $t \in [0,1)$). The two conditions specified in (2.50) and (2.51) follow from trigonometric identities (see Exercise 2.14). Only the completeness condition specified in (2.52) is harder to show and requires some more involved mathematical tools that are outside the scope of this book. From (2.52), one not only recovers the Fourier series in (2.57), but also a formula for how to compute the Fourier coefficients as inner products of the signal f with the basis functions of the ON-basis:

$$a_0 = \langle f|\mathbf{1}\rangle = \int_{t\in[0,1)} f(t)dt, \tag{2.64}$$
$$a_k = \langle f|\mathbf{cos}_k\rangle = \sqrt{2}\int_{t\in[0,1)} f(t)\cos(2\pi kt)dt, \tag{2.65}$$
$$b_k = \langle f|\mathbf{sin}_k\rangle = \sqrt{2}\int_{t\in[0,1)} f(t)\sin(2\pi kt)dt. \tag{2.66}$$

2.3.2 Complex Formulation of the Fourier Transform

As often in mathematics, the transfer of a problem from the real into the complex world can lead to significant simplifications. A famous example is the problem of finding solutions of polynomial equations. The equation $z^2 - 1 = 0$ has the two solutions $z = +1$ and $z = -1$, however the equation $z^2 + 1 = 0$ does not have any solution when only considering real numbers. Extending $\mathbb{R}$ to $\mathbb{C}$, however, one also finds for the second equation two solutions given by $z = +i$ and $z = -i$, where

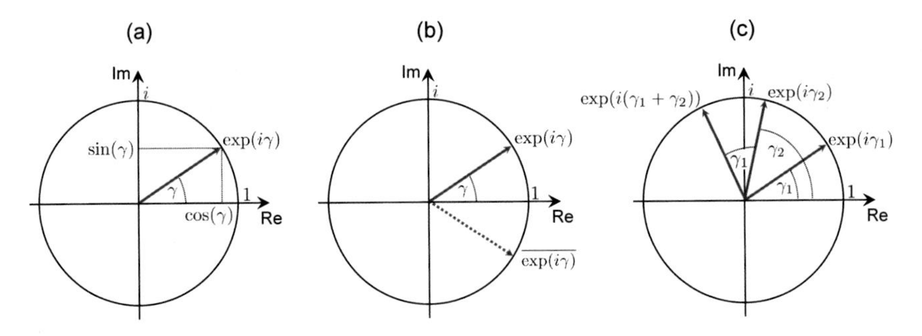

Fig. 2.17 Illustration of the complex exponential function.

i denotes the complex unit. Considering polynomial equations over $\mathbb{C}$ makes the problem much easier to understand. In general, an extension of the real numbers to the complex numbers not only gives a broader view but also provides additional tools and structures. For example, the complex multiplication as defined by (2.34), which extends the usual multiplication of real numbers, yields such a powerful tool. Also, the trigonometric identities are considerably simplified when using a complex formulation.

2.3.2.1 Exponential Function

Converting the Fourier transform from the real into the complex domain has several advantages. First, the concept of Fourier series can be naturally generalized from real-valued to complex-valued signals. Second, one obtains compact and elegant formulas, where the magnitude and phase are naturally expressed by a single complex Fourier coefficient. Recall from Section 2.1.1.4 that the **exponential function** combines the two real-valued sinusoids given by the cosine and sine into a single complex-valued function:

$$\exp(i\gamma) = \cos(\gamma) + i\sin(\gamma). \tag{2.67}$$

This equation, which can be used as a defining relation, is also known as **Euler's formula**. However, there are many other ways in which the exponential function may be characterized, e.g., in terms of a power series expansion or by means of a differential equation. The exponential function has some important properties, which are also illustrated by Figure 2.17:

$$\exp(i\gamma) = \exp(i(\gamma+2\pi)), \tag{2.68}$$
$$|\exp(i\gamma)| = 1, \tag{2.69}$$
$$\overline{\exp(i\gamma)} = \exp(-i\gamma), \tag{2.70}$$
$$\exp(i(\gamma_1+\gamma_2)) = \exp(i\gamma_1)\exp(i\gamma_2) \tag{2.71}$$

for $\gamma, \gamma_1, \gamma_2 \in \mathbb{R}$. For a proof of these properties, we refer to Exercise 2.15. The property (2.68) means that the exponential function is 2π-periodic. The property (2.69) implies that all values of this function live on the unit circle of $\mathbb{C}$. By successively increasing the angle γ starting with $\gamma = 0$ and ending with $\gamma = 2\pi$, one travels exactly once along the unit circle in a counterclockwise fashion. The property (2.70) shows that complex conjugation results in changing the direction of this travel. Finally, the property (2.71) is the complex formulation of the real-valued trigonometric identities that hold for the cosine and sine functions (see also Exercise 2.15).

2.3.2.2 Polar Coordinates

A complex number $c = a + ib \in \mathbb{C}$ is specified by its Cartesian coordinates $(a, b) \in \mathbb{R}^2$ in the two-dimensional plane. The complex exponential function makes it possible to represent a complex number in the form of **polar coordinates**, which we discussed in Section 2.1.1.4. In the polar coordinate system, the point $c = a + ib$ is determined by the distance $|c|$ from the origin and the angle γ (in radians) between the positive horizontal axis and the point given by the coordinates (a, b) (see Figure 2.4). Repeating the formulas from (2.6) and (2.7), we obtain the following relations between Cartesian and polar coordinates:

$$|c| = \sqrt{a^2 + b^2}, \tag{2.72}$$

$$\gamma = \mathrm{atan2}(b, a), \tag{2.73}$$

$$a = |c| \,\mathrm{Re}(\exp(i\gamma)) = |c| \cos(\gamma), \tag{2.74}$$

$$b = |c| \,\mathrm{Im}(\exp(i\gamma)) = |c| \sin(\gamma). \tag{2.75}$$

The atan2 function is a generalization of the arctangent function (denoted as arctan), which is the inverse of the principal branch of the tangent function (see Figure 2.18b). The arctan function requires a real-valued argument $v \in \mathbb{R}$ and computes an angle $\arctan(v) \in (-\pi/2, \pi/2)$ (given in radians), which is called the principal value. As opposed to the arctan function, the atan2 function has two real-valued arguments. This makes it possible to capture the quadrant of the computed angle, which is not possible for the single-argument arctan function. In terms of the standard arctan function, the atan2 function is given by

$$\mathrm{atan2}(b, a) := \begin{cases} \arctan(b/a), & a > 0, \\ \arctan(b/a) + \pi, & b \geq 0,\ a < 0, \\ \arctan(b/a) - \pi, & b < 0,\ a < 0, \\ +\pi/2, & b > 0,\ a = 0, \\ -\pi/2, & b < 0,\ a = 0, \\ \text{undefined} & b = 0,\ a = 0 \end{cases} \tag{2.76}$$

for $(a, b) \in \mathbb{R}^2$ (see Figure 2.18c). The angle computed by the atan2 function is positive for complex numbers $c = a + ib$ with positive imaginary part $b > 0$ (upper half-plane) and negative for those with negative imaginary part $b < 0$ (lower half-plane).

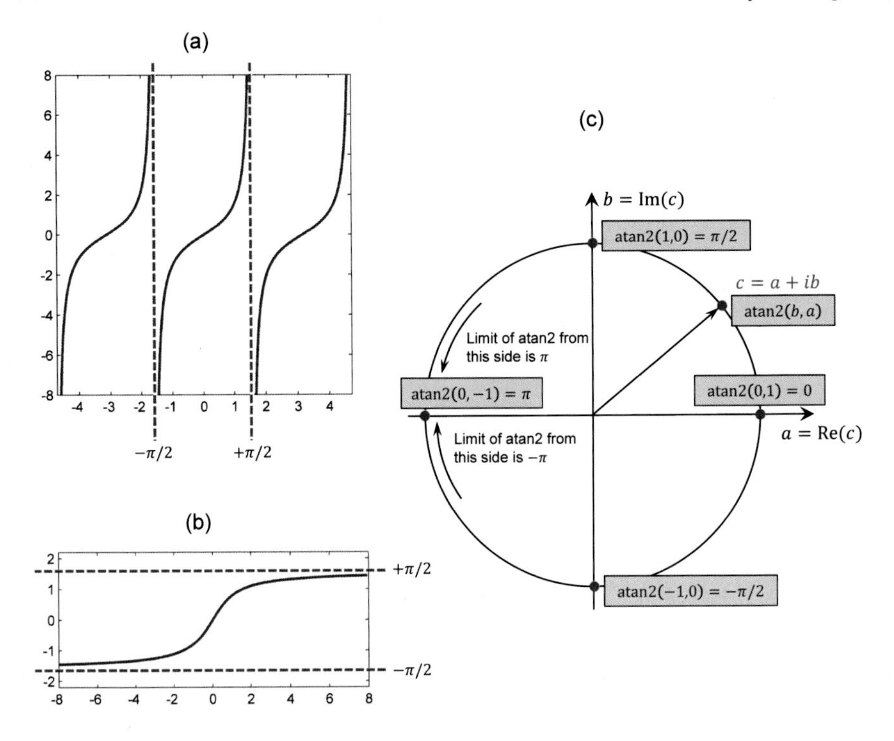

Fig. 2.18 **(a)** Tangent function with different branches. **(b)** Arctangent function inverting the principal branch of the tangent function. **(c)** Illustration of the values assumed by the atan2 function.

The range $(-\pi,\pi]$ of angles can be mapped to $[0,2\pi)$ by adding 2π to negative values. Further properties of the atan2 function are discussed in Exercise 2.17.

2.3.2.3 Complex Fourier Series

We are now ready for the complex formulation of the Fourier series. To this end, we replace in (2.57) the real-valued sinusoids $\mathbf{cos}_k$ and $\mathbf{sin}_k$ defined for $k \in \mathbb{N}$ by the complex-valued exponential functions $\mathbf{exp}_k : [0,1) \to \mathbb{C}$ defined by

$$\mathbf{exp}_k(t) := \exp(2\pi ikt). \tag{2.77}$$

Obviously, $\mathbf{exp}_k$ is a $(1/k)$-periodic signal for $k \neq 0$ and $\mathbf{exp}_k$ is the all-one signal $\mathbf{1}$ for $k = 0$. Furthermore, as in (2.63), it can be shown that the set

$$\{\mathbf{exp}_k \mid k \in \mathbb{Z}\} \tag{2.78}$$

is an ON-basis of the (complex) Hilbert space $L^2[0,1)$. The properties $\|\mathbf{exp}_k\| = 1$ for $k \in \mathbb{Z}$ and $\langle \mathbf{exp}_k | \mathbf{exp}_\ell \rangle = 0$ for $k \neq \ell$, $k, \ell \in \mathbb{Z}$, are shown in Exercise 2.16. Again, as in the real-valued case, the completeness property is more difficult to prove and is not discussed in this book. The resulting expansion of a signal $f \in L^2([0,1))$ with respect to this ON-basis leads to the equality[6]

$$f(t) = \sum_{k\in\mathbb{Z}} c_k \mathbf{exp}_k(t) = \sum_{k\in\mathbb{Z}} c_k \exp(2\pi i k t), \tag{2.79}$$

which is also referred to as the (complex) **Fourier series**. The corresponding (complex) **Fourier coefficients** $c_k \in \mathbb{C}$ are given by

$$c_k = \langle f|\mathbf{exp}_k\rangle = \int_{t\in[0,1)} f(t)\overline{\exp(2\pi i k t)}dt = \int_{t\in[0,1)} f(t)\exp(-2\pi i k t)dt, \tag{2.80}$$

where we used (2.70) in the last equation. As in (2.11), the function

$$\hat{f}:\mathbb{Z}\to\mathbb{C}, \quad \hat{f}(k) := c_k \tag{2.81}$$

is called the **Fourier transform** of $f \in L^2([0,1))$. Note that, in this case, a 1-periodic *continuous-time* signal f is mapped to a *discrete-time* signal $\hat{f}$. Furthermore, one can show that the Fourier transform is **energy preserving** in the sense that the energy of $\hat{f}$ is the same as the energy of f:

$$\|f\|_{L^2([0,1))} = \|\hat{f}\|_{\ell^2(\mathbb{Z})}. \tag{2.82}$$

At this point, using the signal spaces as subscripts of the norms, we want to emphasize that the energy of $\hat{f}$ is measured in the space $\ell^2(\mathbb{Z})$ and the energy of f is measured in $L^2([0,1))$. Mathematically, such an energy-preserving map between Hilbert spaces is also called an **isometry**. As a consequence, the inverse mapping $\hat{f}\mapsto f$ given by the Fourier series (2.79) is again an isometry. We will see that the Fourier transforms for the other finite-energy signal spaces have similar properties.

2.3.2.4 Relation Between Complex and Real Fourier Series

Note that the complex Fourier series can be used to represent **complex-valued** signals, thus extending the Fourier series of (2.57) for **real-valued** signals. Being a special case of a complex-valued function, a real-valued signal $f \in L^2_{\mathbb{R}}([0,1)) \subset L^2([0,1))$ can also be represented using a complex Fourier series. In this case, each signal value $f(t)$ coincides with its complex conjugate $\overline{f(t)}$. Using the computation rules for complex numbers (see Exercise 2.12) and (2.70), one obtains

$$\sum_{k\in\mathbb{Z}} c_k \mathbf{exp}_k(t) = f(t) = \overline{f(t)} = \overline{\sum_{k\in\mathbb{Z}} c_k \mathbf{exp}_k(t)} = \sum_{k\in\mathbb{Z}} \overline{c_k}\mathbf{exp}_{-k}(t). \tag{2.83}$$

[6] Strictly speaking, this equality only holds for almost all $t \in [0,1)$. In the following, even though a bit sloppy in a strict mathematical sense, we do not further mention such issues.

This implies $c_{-k} = \overline{c_k}$ for $k \in \mathbb{Z}$. In other words, for real-valued signals, the coefficients with negative indices are redundant. Furthermore, the complex coefficients c_k of a real-valued signal relate to the real coefficients a_k and b_k of the Fourier series in (2.57) in the following way:

$$a_0 = c_0, \tag{2.84}$$
$$a_k = \sqrt{2}\,\mathrm{Re}(c_k), \tag{2.85}$$
$$b_k = -\sqrt{2}\,\mathrm{Im}(c_k) \tag{2.86}$$

for $k \in \mathbb{N}$. To see this, one needs to use $c_{-k} = \overline{c_k}$ and the definitions (2.77) of $\mathbf{exp}_k$, (2.55) of $\mathbf{cos}_k$, and (2.56) of $\mathbf{sin}_k$. Since the proof is an instructive example of how to compute with complex numbers, we conduct the calculation in detail:

$$\begin{aligned}
f(t) &= \sum_{k\in\mathbb{Z}} c_k \mathbf{exp}_k(t) \\
&= c_0 + \sum_{k=1}^{\infty} c_k \mathbf{exp}_k(t) + \sum_{k=1}^{\infty} c_{-k} \mathbf{exp}_{-k}(t) \\
&= c_0 + \sum_{k=1}^{\infty} \big(c_k \mathbf{exp}_k(t) + \overline{c_k \mathbf{exp}_k(t)}\big) \\
&= c_0 + \sum_{k=1}^{\infty} 2\mathrm{Re}\big(c_k \mathbf{exp}_k(t)\big) \\
&= c_0 + \sum_{k=1}^{\infty} \big(2\mathrm{Re}(c_k)\cos(2\pi kt) - 2\mathrm{Im}(c_k)\sin(2\pi kt)\big) \\
&= c_0 + \sum_{k=1}^{\infty} \sqrt{2}\mathrm{Re}(c_k)\mathbf{cos}_k(t) + \sum_{k=1}^{\infty} \big(-\sqrt{2}\mathrm{Im}(c_k)\big)\mathbf{sin}_k(t).
\end{aligned} \tag{2.87}$$

Comparing coefficients with (2.57) yields the assertion.

Finally, let us come back to our first version of the Fourier series in (2.53), where we introduced the magnitude coefficients d_k and phase coefficients φ_k. How are these coefficients related to the complex Fourier coefficients c_k in the case of real-valued signals? This question can be easily answered when using (2.61) and (2.62) in combination with the polar coordinate representation $c_k = |c_k|\exp(i\gamma_k)$ and the above identities:

$$d_k = \sqrt{a_k^2 + b_k^2} = \sqrt{2\,\mathrm{Re}(c_k)^2 + 2\,\mathrm{Im}(c_k)^2} = \sqrt{2}\,|c_k|, \tag{2.88}$$

$$\begin{aligned}
\varphi_k &= \frac{1}{2\pi}\,\mathrm{atan2}(b_k, a_k) = \frac{1}{2\pi}\,\mathrm{atan2}(-\sqrt{2}\,\mathrm{Im}(c_k), \sqrt{2}\,\mathrm{Re}(c_k)) \\
&= \frac{1}{2\pi}\,\mathrm{atan2}(-\mathrm{Im}(c_k), \mathrm{Re}(c_k)) = -\frac{\gamma_k}{2\pi}.
\end{aligned} \tag{2.89}$$

In the last equations, we used the fact that atan2 is invariant under scaling with a nonzero constant and assumes the negative angle for the conjugate of a complex number (see Exercise 2.17). These identities correspond to (2.14) and (2.15).

2.3.3 Fourier Transform for CT-Signals

The general idea of the Fourier transform carries over from the case of periodic to the case of nonperiodic signals in $L^2(\mathbb{R})$. In the nonperiodic case, however, the exponential functions $\mathbf{exp}_k$ of integer frequency $k \in \mathbb{Z}$ do not suffice to "describe" a signal. Instead, one needs exponential functions

$$\mathbf{exp}_\omega : \mathbb{R} \to \mathbb{C}, \quad \mathbf{exp}_\omega(t) := \exp(2\pi i \omega t) \tag{2.90}$$

for all frequencies $\omega \in \mathbb{R}$. Then, replacing summation by integration one obtains the following nonperiodic analog of the Fourier representation:

$$f(t) = \int_{\omega \in \mathbb{R}} c_\omega \mathbf{exp}_\omega(t) d\omega = \int_{\omega \in \mathbb{R}} c_\omega \exp(2\pi i \omega t) d\omega \tag{2.91}$$

for $t \in \mathbb{R}$. The coefficients c_ω have the same interpretation as the Fourier coefficients c_k. The frequency-dependent function $\hat{f} : \mathbb{R} \to \mathbb{C}$ defined by

$$\hat{f}(\omega) := c_\omega = \int_{t \in \mathbb{R}} f(t) \overline{\mathbf{exp}_\omega(t)} dt = \int_{t \in \mathbb{R}} f(t) \exp(-2\pi i \omega t) dt \tag{2.92}$$

is called the **Fourier transform** of f. Again, it can be shown that the Fourier transform is energy preserving. In other words, if $f \in L^2(\mathbb{R})$, then $\hat{f} \in L^2(\mathbb{R})$ and $\|f\|_{L^2(\mathbb{R})} = \|\hat{f}\|_{L^2(\mathbb{R})}$.

Strictly speaking, there are some mathematical issues that need to be considered for the nonperiodic case. Recall that, in the periodic case, the elementary functions $\mathbf{exp}_k$ have finite energy over the interval $[0,1)$ and are therefore elements of $L^2([0,1))$. This is the reason why the Fourier transform and the Fourier representation can be expressed by means of inner products. Unfortunately, this is no longer the case for the nonperiodic case, since the elementary functions $\mathbf{exp}_\omega$ do not have finite energy over the real time axis $\mathbb{R}$ and are therefore *not* elements in the space $L^2(\mathbb{R})$. As a consequence, the inner product is not defined between a signal $f \in L^2(\mathbb{R})$ and $\mathbf{exp}_\omega$. Furthermore, the integrals in (2.91) and (2.92) need to be defined as limits over increasing finite integration domains. For example,

$$\hat{f}(\omega) := \lim_{N \to \infty} \int_{t \in [-N,N]} f(t) \exp(-2\pi i \omega t) dt. \tag{2.93}$$

Similarly, one has to define the Fourier representation. However, these technical issues will not play any further role in this book. Furthermore, most of the signals we consider in this book have **compact support**; i.e., they are zero outside an interval of finite length. For such signals, no problems occur in the integrals even from a strict mathematical point of view.

The Fourier representation in (2.91) yields a quite surprising result. It states that every **nonperiodic** function of finite energy can be represented as a weighted (infinitesimal) superposition of **periodic** elementary frequency functions $\mathbf{exp}_\omega$ that continue out to infinity without decaying. For example, even noise-like short-

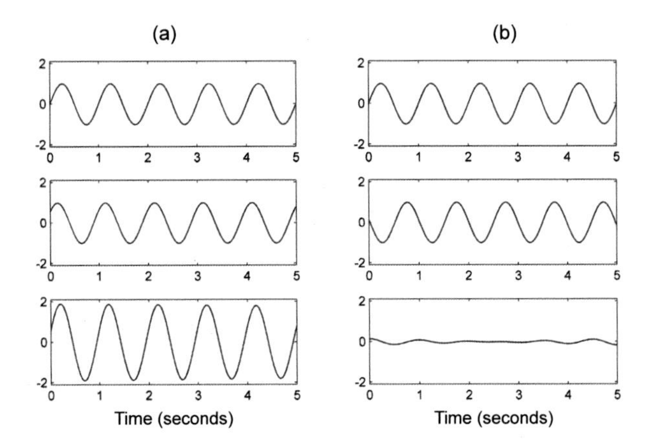

Fig. 2.19 Interference of two sinusoids of similar frequency. **(a)** Constructive interference. **(b)** Destructive interference.

duration sounds such as transients, which often occur in the attack phase of a tone, can be represented by ceaselessly oscillating sinusoids.

2.3.3.1 Interference

In Section 2.1.2, we have already discussed some real as well as synthetic signals to illustrate important properties of the Fourier transform. In the following, we take a closer look at some of the encountered phenomena. Let us start with the example from Figure 2.6b. Besides the two peaks, we could observe in the magnitude Fourier transform $|\hat{f}|$ a number of "ripples" of decreasing amplitude. Where do these ripples come from? In the figure, the analog signal f is shown only for the time interval $[0,5]$ and is (implicitly) assumed to be zero outside this compact interval. The ripples in the spectrum come from a phenomenon known as destructive interference, where many different frequency components are involved for generating the compact support of f.

In general, **interference** occurs when a wave is superimposed with another wave of similar frequency. When a crest of one wave meets a crest of the other wave at some point, then the individual magnitudes add up for a certain period of time, which is known as **constructive interference** (see Figure 2.19a). Vice versa, when a crest of one wave meets a trough of the other wave, then the magnitudes cancel out for a certain period of time, which is known as **destructive interference** (see Figure 2.19b).

Coming back to Figure 2.6b, one needs the sinusoids of frequency $\omega = 1$ Hz and $\omega = 5$ Hz to generate the main components of the signal f within the interval $[0,5]$. Note that these two sinusoids also oscillate outside the visualized interval $[0,5]$, where the signal is assumed to be zero. Therefore, to cancel out these oscillations

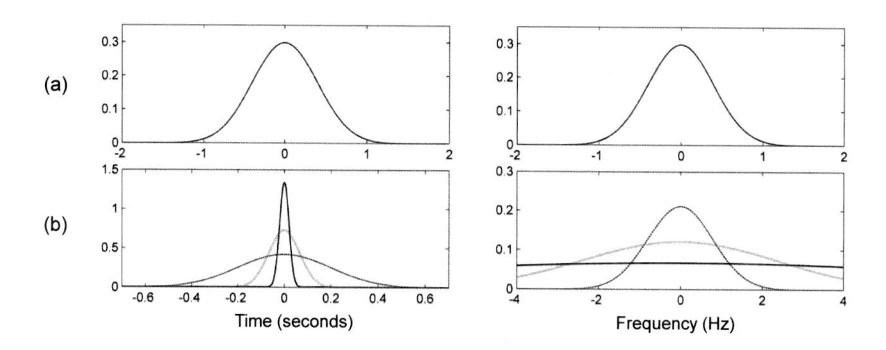

Fig. 2.20 **(a)** Gaussian function (left) and its Fourier transform (right). **(b)** Dirac sequence (left) with corresponding Fourier transforms (right).

outside $[0,5]$ by destructive interference, one needs to add many more sinusoids of different frequencies and weights. These additional sinusoidal components are reflected by the ripples. Interference effects are further discussed in Exercise 2.19 and in the subsequent examples.

2.3.3.2 Fourier Transform for Impulses

The synthetic signals shown in Figure 2.20 illustrate further properties of the Fourier transform. First of all, the **Gaussian function** defined by the formula

$$f(t) = (2\pi)^{-\frac{1}{2}}\pi^{-\frac{1}{4}}\exp(-\pi t^2) \tag{2.94}$$

has the remarkable property that it coincides with its Fourier transform (see Figure 2.20a). In particular, its Fourier transform is real-valued and positive. Therefore, it agrees with its magnitude Fourier transform. The Fourier representation (2.91) tells us that the Gaussian function is obtained as an (infinitesimal) weighted superposition of periodic sine waves, where the weights are again given by the Gaussian function. The next question we consider is how the Fourier transform behaves, if we start to make the Gaussian function somewhat narrower (see Figure 2.20b). This leads to the notion of a **Dirac sequence**, which is a sequence of functions $(f_n)_{n\in\mathbb{N}}$ of norm $\|f_n\| = 1$ such that for increasing n the functions f_n "concentrate" more and more around the point $t = 0$. The limit of this sequence is the **Dirac delta function** or **impulse function** (often denoted by the symbol δ), which can be thought of as a function that is zero everywhere except for $t = 0$. At $t = 0$, it has an infinitely narrow spike of infinite height, which integrates to a value of one. Strictly speaking, this impulse is not a function, but a so-called **distribution**. As illustrated by Figure 2.20b, the magnitude Fourier transform of a Dirac sequence becomes broader and broader. This scaling property of the Fourier transform is shown in Exercise 2.20. In the limit case, the Fourier transform approaches a con-

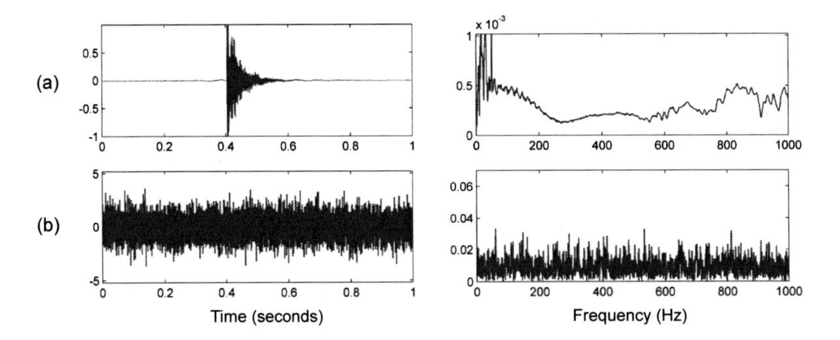

Fig. 2.21 Waveform and its magnitude Fourier transform for **(a)** a clapping sound and **(b)** white noise.

stant function, where the magnitudes of all frequency components have the same, yet infinitesimally small value.

The interpretation of this property is important in view of practical applications. It says that impulse-like sounds such as a drum hit or a transient as occurring in the attack phase of a musical tone (see Section 1.3.4) lead to a flat magnitude Fourier transform with many small, yet nonzero Fourier coefficients. In other words, for a sudden sharp sound, the signal's energy is spread across the entire spectrum of frequencies. This is also illustrated by Figure 2.21a, which shows the waveform and its magnitude Fourier transform for a real clapping sound. Another type of sound that results in an energy spread across the entire frequency spectrum are noise-like signals. Generally speaking, random signals such as white noise also remain random when transformed into the Fourier domain. For example, Figure 2.21b shows white Gaussian noise and its magnitude Fourier transform, which also looks like noise that is equally spread over the entire frequency range.

2.3.3.3 Translation and Modulation

As a final example, which is shown in Figure 2.22, we consider the **rectangular function**

$$f(t) := \begin{cases} 1, & \text{if } -0.5 \leq t \leq 0.5, \\ 0, & \text{otherwise.} \end{cases} \tag{2.95}$$

Its Fourier transform is the **sinc function**, which is defined by

$$\operatorname{sinc}(t) := \begin{cases} \frac{\sin \pi t}{\pi t}, & \text{if } t \neq 0, \\ 1, & \text{if } t = 0. \end{cases} \tag{2.96}$$

For the proof of this fact, we refer to Exercise 2.21. The rectangular and the sinc function play an important role in the sampling theorem (see Exercise 2.28). In the case that the rectangular function is centered around $t = 0$, its Fourier transform is

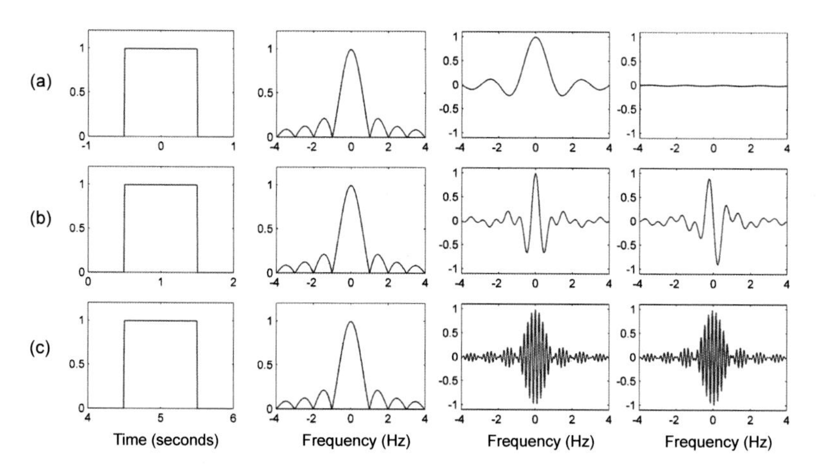

Fig. 2.22 Behavior of Fourier transform under translations. From left to right, the signal as well as the magnitude, real part, and imaginary part of the Fourier transform are shown. **(a)** Rectangular function. **(b)** Translation by one second. **(c)** Translation by five seconds.

a real-valued function (see Figure 2.22a). However, this is no longer the case if we start to shift the rectangle in time. For example, translating the rectangular function one second to the right, as illustrated by Figure 2.22b, leaves the magnitude of the Fourier transform unchanged. However, the translation has a significant impact on the phase as well as on the real and imaginary parts of the Fourier transform. This again demonstrates that time information is not revealed by the magnitude, but that it is encoded in the phase of the Fourier transform. Let us have a more general look at this phenomenon. Let $f \in L^2(\mathbb{R})$ be a signal, then the function f_{t_0} defined by

$$f_{t_0}(t) := f(t - t_0) \tag{2.97}$$

is called the **translation** of f by $t_0 \in \mathbb{R}$, and the function f^{ω_0} defined by

$$f^{\omega_0}(t) := \exp(2\pi i \omega_0 t) f(t) \tag{2.98}$$

is called the **modulation** of f by $\omega_0 \in \mathbb{R}$. It is not hard to show (Exercise 2.22) that for the Fourier transform one obtains

$$\widehat{f_{t_0}}(\omega) = \exp(-2\pi i \omega t_0) \hat{f}(\omega) \tag{2.99}$$

and

$$\widehat{f^{\omega_0}}(\omega) = \hat{f}(\omega + \omega_0). \tag{2.100}$$

In other words, a translation of the signal in the time domain leads to a modulation in the Fourier domain, and vice versa.

2.3.4 Fourier Transform for DT-Signals

We finally introduce the Fourier transform for the signal space $\ell^2(\mathbb{Z})$, which consists of the finite-energy DT-signals. Recall from (2.32) that the most common discretization procedure to transform a CT-signal $f\colon \mathbb{R} \to \mathbb{R}$ into a DT-signal $x\colon \mathbb{Z} \to \mathbb{R}$ is equidistant sampling, where the samples are defined by $x(n) = f(n \cdot T)$, $n \in \mathbb{Z}$, for a given sampling rate $F_s = 1/T$ and sampling period $T > 0$.

Let $x \in \ell^2(\mathbb{Z})$ be an arbitrary DT-signal of finite energy, then the **Fourier representation** of x is

$$x(n) = \int_{\omega\in[0,1)} c_\omega \mathbf{exp}_\omega(n) d\omega = \int_{\omega\in[0,1)} c_\omega \exp(2\pi i \omega n) d\omega \tag{2.101}$$

for $n \in \mathbb{Z}$. Furthermore, the coefficients c_ω are given by the frequency-dependent function $\hat{x}\colon [0,1) \to \mathbb{C}$ defined by

$$c_\omega = \hat{x}(\omega) := \sum_{n\in\mathbb{Z}} x(n)\overline{\mathbf{exp}_\omega(n)} = \sum_{n\in\mathbb{Z}} x(n) \exp(-2\pi i \omega n), \tag{2.102}$$

which is called the **Fourier transform** of x. Both the Fourier representation as well as the Fourier transform are nontrivial facts that require mathematical proofs. Although similar in nature, the Fourier transform for DT-signals cannot be directly derived from the Fourier transform for CT-signals. However, as we will see, the case of DT-signals can be regarded to be dual to the case of periodic CT-signals. Also, the Fourier transform of a sampled analog signal can be regarded as a kind of approximation of the Fourier transform of the analog signal.

2.3.4.1 Periodicity and Aliasing

The Fourier representation (2.101) says that the signal x can be represented as an infinitesimal superposition of the elementary frequency functions $\mathbf{exp}_\omega$ sampled with $T = 1$ (see (2.32)). In this case, only the frequencies $\omega \in [0,1)$ are needed. Intuitively, the restriction of the frequency parameters to the set $[0,1)$ can be explained as follows: For an integer frequency parameter $k \in \mathbb{Z}$ and sampling points $n \in \mathbb{Z}$ one has $\exp(2\pi i k n) = 1$. Therefore,

$$\mathbf{exp}_{\omega+k}(n) = \exp(2\pi i(\omega + k)n) = \exp(2\pi i \omega n)\exp(2\pi i k n) = \mathbf{exp}_\omega(n). \tag{2.103}$$

In other words, two exponential functions with an integer difference in their frequency parameter coincide on the set of sampling points $n \in \mathbb{Z}$. Consequently, they cannot be distinguished when considered as 1-sampled DT-signals. We have encountered this **aliasing** phenomenon already in Figure 2.14 of Section 2.2.2. Using a sampling rate of 1 Hz, the **Nyquist frequency** is $\omega = 0.5$ Hz. All oscillations with a frequency above this rate are not captured by 1-sampling and lead to the

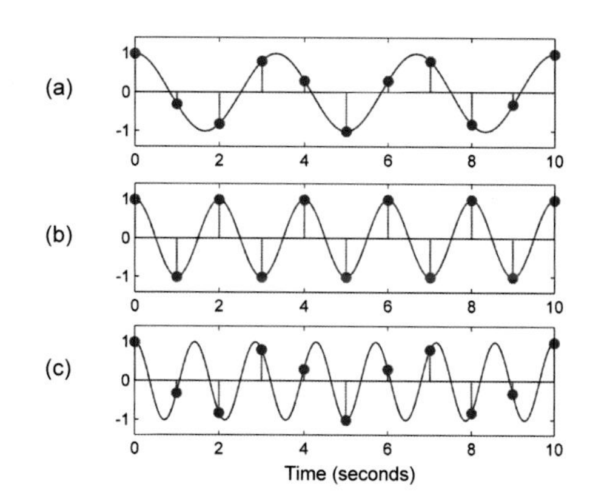

Fig. 2.23 Sinusoids of different frequencies ω sampled at a rate of $F_s = 1$ Hz. **(a)** $\omega = 0.3$ Hz. **(b)** $\omega = 0.5$ Hz. **(c)** $\omega = 0.7$ Hz.

same samples as oscillations of lower frequencies. This fact is also illustrated by Figure 2.23.

Next, let us have a closer look at the Fourier transform (2.102). Note that (2.103) implies that the function $\omega \mapsto \exp(-2\pi i\omega n)$ is 1-periodic for all $n \in \mathbb{Z}$. Being a superposition of 1-periodic functions, also the Fourier transform $\hat{x}$ is 1-periodic. Furthermore, one can show that the Fourier transform is energy preserving, i.e., $\|x\|_{\ell^2(\mathbb{Z})} = \|\hat{x}\|_{L^2([0,1))}$. Note that this is exactly the reverse of the situation we have seen for 1-periodic signals $f \in L^2([0,1))$, where the Fourier transform was a DT-signal $\hat{f} \in \ell^2(\mathbb{Z})$. Replacing the frequency parameter ω by the time parameter t, the formula (2.102) for the Fourier transform of $\ell^2(\mathbb{Z})$ becomes (up to a sign in the exponential function) the formula (2.79) for the Fourier representation of $L^2([0,1))$. A similar relation holds between the Fourier representation (2.101) for $\ell^2(\mathbb{Z})$ and the Fourier transform (2.80) for $L^2([0,1))$. From this it also follows that the Fourier transform for $\ell^2(\mathbb{Z})$ applied to the Fourier transform $\hat{f}$ of a signal $f \in L^2([0,1))$ gives back the 1-periodic signal f up to a sign, i.e., $\hat{\hat{f}}(t) = f(-t)$. In mathematics, the close relation between the spaces $\ell^2(\mathbb{Z})$ and $L^2([0,1))$ and their Fourier transforms is also referred to as **duality**.

2.3.4.2 Riemann Approximation

Let us now investigate the relation between the Fourier transform of $L^2([0,1))$ and the one of $\ell^2(\mathbb{Z})$. Starting with a CT-signal $f \in L^2(\mathbb{R})$, let x be its T-sampled version. Then one obtains

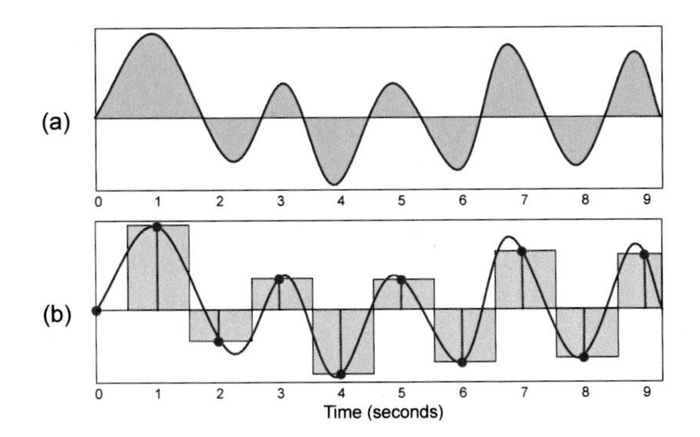

Fig. 2.24 Approximation of the integral of an analog signal by a Riemann sum obtained from a 1-sampling. **(a)** Integral. **(b)** Riemann sum.

$$\begin{aligned}\hat{x}(\omega) &= \sum_{n\in\mathbb{Z}} x(n)\exp(-2\pi i\omega n) \\ &= \sum_{n\in\mathbb{Z}} f(nT)\exp(-2\pi i\omega n) \\ &\approx \int_{t\in\mathbb{R}} f(tT)\exp(-2\pi i\omega t)dt \\ &= \frac{1}{T}\int_{t\in\mathbb{R}} f(t)\exp\left(\frac{-2\pi i\omega t}{T}\right)dt \\ &= \frac{1}{T}\hat{f}\left(\frac{\omega}{T}\right),\end{aligned} \tag{2.104}$$

where we have used the substitution rule for indefinite integrals to replace tT by t. The approximation sign expresses that the value $\hat{x}(\omega)$ obtained by a sum has roughly the same size as the value $\hat{f}(\omega/T)/T$ obtained by an integral. This is a special case of the Riemann sum approximation, which we explain next.

Recall that the integral of a function is the (weighted) area determined by the function's graph and the time axis. In case of a complex-valued function, the complex-valued integral is defined by the integral of the real part and of the imaginary part of the function. For many functions, the integral can be approximated by partitioning the time axis into small intervals, picking the function value at the mid-point of each interval, and then summing up the interval lengths weighted by the respective value (see Figure 2.24). The resulting sum is also called the **Riemann sum** for the integral. The accuracy of the approximation very much depends on the resolution of the partition (the finer, the better the approximation) and the properties of the integrand (the slower it oscillates, the better the approximation).

In our case, the intervals of the partitioning have length one. Furthermore, the integrand is the function $h:\mathbb{R}\to\mathbb{C}$ defined by $h(t) := f(tT)\exp(-2\pi i\omega t)$, which basically is the product of the signal and an exponential function. Because of aliasing effects, in particular arising from the factor $\exp(-2\pi i\omega t)$, the Riemann sum does not yield a meaningful approximation for $\omega\in\mathbb{R}\setminus\left[-\frac{1}{2},\frac{1}{2}\right]$. In particular, while $\hat{x}$ is 1-periodic, the function $\omega\mapsto\hat{f}(\omega/T)/T$ is nonperiodic and approaches zero for $\omega\to\pm\infty$. Within the interval $\left[-\frac{1}{2},\frac{1}{2}\right]$, however, in particular when approaching the

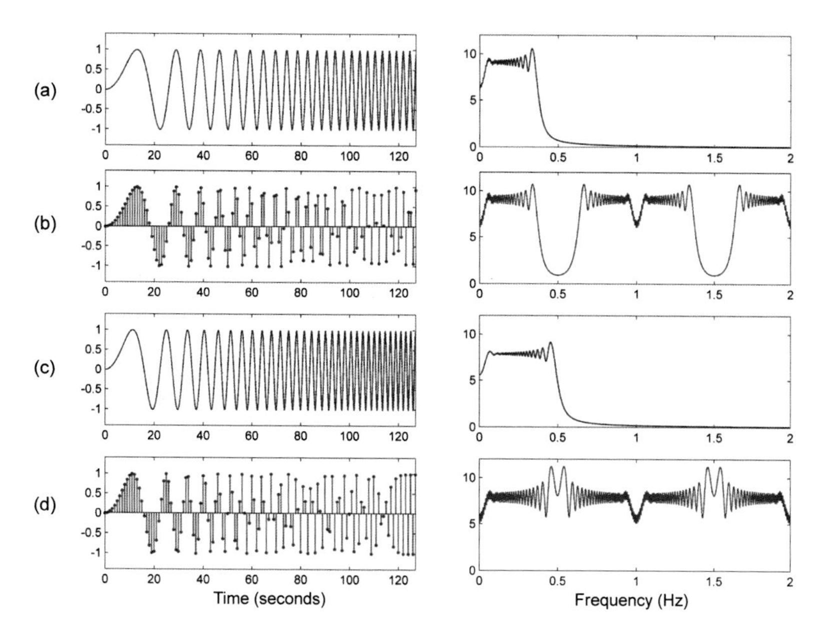

Fig. 2.25 Relation between the Fourier transform of a CT-signal and that of the DT-signal obtained by 1-sampling. Each row shows a signal (left) and its magnitude Fourier transform (right). **(a)** Analog chirp signal with $\lambda = 0.003$ and **(b)** its 1-sampled version. **(c)** Analog chirp signal with $\lambda = 0.004$ and **(d)** its 1-sampled version showing strong aliasing artifacts around the Nyquist frequency.

frequency $\omega = 0$, the Riemann sum $\hat{x}(\omega)$ approximates the value $\hat{f}(\omega/T)/T$ with increasing accuracy.

2.3.4.3 Chirp Signal Example

To further illustrate the relation between CT- and DT-signals and their Fourier transforms, we consider a signal in which the frequency increases with time. Such a signal is also called a **chirp signal** or **sweep signal**. In particular, for a given positive constant $\lambda > 0$, the function

$$f(t) := \begin{cases} \sin(\lambda \cdot \pi t^2), & \text{for } t \geq 0, \\ 0, & \text{for } t < 0, \end{cases} \tag{2.105}$$

defines a **linear chirp**, which is a sinusoidal wave that increases in frequency linearly over time. It can be shown that the **instantaneous frequency** at time $t = t_0$ is $\omega_0 = \lambda t_0$, which is the derivative of the phase divided by 2π. Figure 2.25 shows two chirp signals for different values of λ. In the first case (Figure 2.25a), the main frequencies are below $\omega \leq 0.4$, which is also shown by the magnitude Fourier transform. As a result, there is little aliasing when 1-sampling the signal (Figure 2.25b).

The Fourier transform $\hat{x}$ of the resulting DT-signal x yields a good approximation of the Fourier transform $\hat{f}$ in the range $[0,0.5)$. Note that $\hat{x}$ is 1-periodic whereas $\hat{f}$ is not. Now, increasing the constant λ results in a chirp signal with frequency components above the Nyquist frequency of 0.5 (Figure 2.25c). Therefore, when 1-sampling the signal, there are aliasing artifacts where frequencies $0.5+\omega$ are identified with frequencies $0.5-\omega$ (see Figure 2.25d). In this case, the Riemann sum (2.104) yields a poor approximation of the actual integral.

2.4 Discrete Fourier Transform (DFT)

Computing the Fourier transform of signals involves the evaluation of integrals or infinite sums, which is, in general, computationally infeasible. In practice, as we have already discussed in Section 2.1.3, one typically approximates the Fourier transform by finite sums. Furthermore, the Fourier transform is evaluated only for a finite number of frequencies. In this section, we show how the finite sums and the Fourier coefficients must be chosen to obtain a linear transform known as the **discrete Fourier transform** (DFT). The important point is that the DFT can be computed efficiently by means of an algorithm, the famous **fast Fourier transform** (FFT). The FFT is considered one of the most important algorithms, being widely used for many applications in engineering and mathematics. In the following, we introduce the case of finite-length signals and their Fourier transform, which can then be formulated in terms of the DFT. We then describe in detail the FFT algorithm and discuss its computational complexity.

2.4.1 Signals of Finite Length

To derive the DFT, we start to reinvestigate the Fourier transform for a DT-signal $x\in\ell^2(\mathbb{Z})$. We assume that the energy of x is concentrated in the interval $[0:N-1]$, i.e., $x(n)\approx 0$ for $n\in\mathbb{Z}\setminus[0:N-1]$. Then we obtain from (2.102)

$$\hat{x}(\omega)=\sum_{n\in\mathbb{Z}}x(n)\overline{\mathbf{exp}_\omega(n)}\approx\sum_{n=0}^{N-1}x(n)\overline{\mathbf{exp}_\omega(n)} \tag{2.106}$$

for a frequency parameter ω. Recall that since $\hat{x}$ is 1-periodic only the frequencies $\omega\in[0,1)$ need to be considered. In practice, one often computes the Fourier transform only for a finite subset of frequencies. In particular, fixing a number $K\in\mathbb{N}$, one considers the frequencies $\omega=k/K$ for $k\in[0:K-1]$, which corresponds to a $1/K$-sampling of the frequency space $[0,1)$. Even though the number N of points in time and the number K of frequencies are not related at all, it is convenient to assume $N=K$. This assumption, as we will see, leads to a compact matrix-theoretic formulation of the Fourier transform along with an efficient algorithm for computing the transform.

In the following, we assume $N = K$. Furthermore, let $x \in \ell^2(\mathbb{Z})$ be a signal that is zero outside the interval $[0:N-1]$ so that one obtains equality in (2.106). Such DT-signals are also referred to as **finite-length** signals, where N is the **length** of the signal. Each such signal x can be identified with a vector $\mathbf{x} := (x(0), x(1), \ldots, x(N-1))^\top \in \mathbb{C}^N$. This way, we can regard $\mathbb{C}^N$ as a subspace of $\ell^2(\mathbb{Z})$, where the inner product (2.43) of $\ell^2(\mathbb{Z})$ reduces to the inner product (2.37) of $\mathbb{C}^N$. Not all frequencies $\omega \in [0,1)$ are needed to characterize a signal of length N. Indeed, only the frequencies k/N for $k \in [0:N-1]$ suffice to represent such signals. To see this, we define a vector $\mathbf{u}_k \in \mathbb{C}^N$ for each $k \in [0:N-1]$ by setting

$$\mathbf{u}_k(n) := \mathbf{exp}_{k/N}(n) = \exp(2\pi i k n/N), \tag{2.107}$$

$n \in [0:N-1]$. In other words, the vector $\mathbf{u}_k$ consists of the first N samples of the exponential function $\mathbf{exp}_{k/N}$. Then (2.106) can be expressed as

$$\hat{x}(k/N) = \sum_{n=0}^{N-1} x(n)\overline{\mathbf{exp}_{k/N}(n)} = \mathbf{x}^\top \overline{\mathbf{u}_k} = \langle \mathbf{x} | \mathbf{u}_k \rangle. \tag{2.108}$$

Thus, the Fourier transform of a signal of length N can be obtained by inner products with the sampled and truncated exponential functions $\mathbf{u}_k$. We now show that these exponential functions (after rescaling) form an ON-basis of the Hilbert space $\mathbb{C}^N$. First, we define the number $\rho_N := \exp(2\pi i/N)$. Obviously, $\rho_N^N = 1$ and $\rho_N^k \neq 1$ for $k \in [1:N-1]$. Such a number is also called a **primitive** N^{th} root of unity (see also Exercise 2.23). Using the properties (2.70) and (2.71) of the exponential function, one obtains

$$\langle \mathbf{u}_k | \mathbf{u}_\ell \rangle = \sum_{n=0}^{N-1} \exp(2\pi i k n/N)\overline{\exp(2\pi i \ell n/N)} \tag{2.109}$$

$$= \sum_{n=0}^{N-1} \exp(2\pi i (k-\ell)n/N) = \sum_{n=0}^{N-1} \rho_N^{(k-\ell)n}. \tag{2.110}$$

for $k, \ell \in [0:N-1]$. In the case $k = \ell$, this implies $\|\mathbf{u}_k\|^2 = \langle \mathbf{u}_k | \mathbf{u}_k \rangle = N$. In the case $k \neq \ell$, one has $\rho_N^{(k-\ell)} \neq 1$. Therefore, one can apply the sum formula

$$\sum_{n=0}^{N-1} a^n = (1-a^N)/(1-a) \tag{2.111}$$

for geometric series, which holds for any complex number $a \neq 1$ (see Exercise 2.18). Setting $a = \rho_N^{(k-\ell)}$, one obtains

$$\langle \mathbf{u}_k | \mathbf{u}_\ell \rangle = \frac{1 - \rho_N^{N(k-\ell)}}{1 - \rho_N^{(k-\ell)}} = 0. \tag{2.112}$$

This shows that

$$\{\mathbf{u}_k/\sqrt{N} | k \in [0:N-1]\} \tag{2.113}$$

is an ON-basis of the complex Hilbert space $\mathbb{C}^N$. In particular, from (2.52), one obtains the Fourier representation

$$\mathbf{x} = \frac{1}{N}\sum_{k=0}^{N-1} \langle \mathbf{x}|\mathbf{u}_k\rangle \mathbf{u}_k. \tag{2.114}$$

In other words, a finite-length signal can be represented as a weighted superposition of sampled and truncated exponential functions $\mathbf{u}_k$, where the weights are the Fourier coefficients given by (2.108). Next, we show how the Fourier transform and Fourier representation for finite-length signals relate to the discrete Fourier transform (DFT).

2.4.2 Definition of the DFT

Recall from (2.108) that the Fourier coefficients of a signal x of finite length N are given by

$$X(k) := \langle \mathbf{x}|\mathbf{u}_k\rangle = \sum_{n=0}^{N-1} x(n)\exp(-2\pi ikn/N) \tag{2.115}$$

for $k \in [0:N-1]$. Let $\mathbf{X} := (X(0), X(1), \ldots, X(N-1))^\top \in \mathbb{C}^N$ denote the vector of Fourier coefficients. By definition, the **discrete Fourier transform** (DFT) is the mapping $\mathbb{C}^N \to \mathbb{C}^N$ that maps the input vector $\mathbf{x}$ to the output vector $\mathbf{X}$. From (2.115) it is clear that this is a linear mapping, which can be described by the $(N \times N)$ matrix DFT_N given by

$$\mathrm{DFT}_N(n,k) = \exp(-2\pi ikn/N). \tag{2.116}$$

One crucial observation is that there are many relations between the numbers $\exp(2\pi ikn/N)$ for $k,n \in [0, N-1]$. Using the primitive N^{th} root of unity $\rho_N = \exp(2\pi i/N)$ as well as the relations $\rho_N^{kn} = \exp(2\pi ikn/N)$ and $\sigma_N := \overline{\rho_N} = \exp(-2\pi i/N)$, one obtains $\mathrm{DFT}_N(n,k) = \sigma_N^{kn}$. This yields the famous matrix

$$\mathrm{DFT}_N = \begin{pmatrix} 1 & 1 & 1 & \cdots & 1 \\ 1 & \sigma_N & \sigma_N^2 & \cdots & \sigma_N^{N-1} \\ 1 & \sigma_N^2 & \sigma_N^4 & \cdots & \sigma_N^{2(N-1)} \\ \vdots & \vdots & \vdots & \ddots & \vdots \\ 1 & \sigma_N^{N-1} & \sigma_N^{2(N-1)} & \cdots & \sigma_N^{(N-1)(N-1)} \end{pmatrix}. \tag{2.117}$$

Obviously, DFT_N is a symmetric matrix. Its columns are given by $\overline{\mathbf{u}_k}$ and its rows by $\overline{\mathbf{u}_k}^\top$. In summary, we have seen that the Fourier transform $\hat{x}$ of a DT-signal x of finite length N can be computed for frequencies $\omega = k/N$, $k \in [0:N-1]$ by a single matrix–vector product $\mathbf{X} = \mathrm{DFT}_N \cdot \mathbf{x}$.

The Fourier representation given by (2.114) is the inverse of the Fourier transform. For a spectral vector $\mathbf{X}$, it outputs the original signal $\mathbf{x}$. Again, being a linear mapping $\mathbb{C}^N \to \mathbb{C}^N$, the Fourier representation is given by a matrix, the inverse of the matrix DFT_N. From (2.113) it directly follows that

$$\mathrm{DFT}_N^{-1} = \frac{1}{N}\overline{\mathrm{DFT}_N}^\top = \frac{1}{N}\begin{pmatrix} 1 & 1 & 1 & \cdots & 1 \\ 1 & \rho_N & \rho_N^2 & \cdots & \rho_N^{N-1} \\ 1 & \rho_N^2 & \rho_N^4 & \cdots & \rho_N^{2(N-1)} \\ \vdots & \vdots & \vdots & \ddots & \vdots \\ 1 & \rho_N^{N-1} & \rho_N^{2(N-1)} & \cdots & \rho_N^{(N-1)(N-1)} \end{pmatrix}. \tag{2.118}$$

In other words, the inverse essentially coincides with the DFT matrix up to some normalizing factor and complex conjugation.

2.4.3 Fast Fourier Transform (FFT)

Note that the usual computation of the matrix–vector product $\mathbf{X} = \mathrm{DFT}_N \cdot \mathbf{x}$ requires $O(N^2)$ multiplications and additions, which is too many for most applications. For example, having a signal with one thousand samples ($N = 10^3$) would require already a number of operations on the order of a million ($N^2 = 10^6$). In many cases one has to deal with much larger $N \gg 10^5$, which makes a naive computation of a DFT infeasible. The good news is that the DFT matrix is highly structured, which can be exploited when computing a matrix–vector product. The main idea lies in a factorization of the DFT matrix into a product of $O(\log N)$ sparse matrices, each of which can be evaluated with $O(N)$ operations. This leads to an efficient algorithm, the so-called **fast Fourier transform** (FFT), which only requires $O(N \log N)$ multiplications and additions. The FFT algorithm was originally found by Gauss in about 1805 and then rediscovered by Cooley and Tukey in 1965.

The FFT algorithm is based on the observation that applying a DFT of even size $N = 2M$ can be expressed in terms of applying two DFTs of half the size M. Let $\sigma_N = \exp(-2\pi i/N)$ be the primitive root of unity used in DFT_N so that $\mathrm{DFT}_N(n,k) = \sigma_N^{kn}$ for $n,k \in [0:N-1]$. Similarly, we define $\sigma_M = \exp(-2\pi i/M)$ so that $\mathrm{DFT}_M(n,k) = \sigma_M^{kn}$ for $n,k \in [0:M-1]$. Obviously, $\sigma_M = \sigma_N^2$. Let $\mathbf{x} \in \mathbb{C}^N$ be an input vector and $\mathbf{X} = \mathrm{DFT}_N \cdot \mathbf{x}$ as before. Then for the first M entries $X(k)$, $k \in [0:M-1]$ one has

$$X(k) = \sum_{n=0}^{N-1} x(n)\sigma_N^{kn} \tag{2.119}$$
$$= \sum_{n=0}^{M-1} x(2n)\sigma_N^{k2n} + \sum_{n=0}^{M-1} x(2n+1)\sigma_N^{k(2n+1)} \tag{2.120}$$
$$= \sum_{n=0}^{M-1} x(2n)\sigma_M^{kn} + \sigma_N^k \sum_{n=0}^{M-1} x(2n+1)\sigma_M^{kn}. \tag{2.121}$$

In other words, the first M entries of $\mathbf{X}$ are obtained by first applying a DFT_M on the even-indexed entries of $\mathbf{x}$ as well as a DFT_M on the odd-indexed entries of $\mathbf{x}$. The final result is then obtained by adding up the two output vectors, where the second one is adjusted by the factors σ_N^k, which are also known as **twiddle factors**. Similarly, for the last M entries $X(M+k)$, $k \in [M-1]$ one has

$$X(M+k) = \sum_{n=0}^{N-1} x(n)\sigma_N^{(M+k)n} \tag{2.122}$$
$$= \sum_{n=0}^{M-1} x(2n)\sigma_N^{(M+k)2n} + \sum_{n=0}^{M-1} x(2n+1)\sigma_N^{(M+k)(2n+1)} \tag{2.123}$$
$$= \sum_{n=0}^{M-1} x(2n)\sigma_M^{kn} - \sigma_N^k \sum_{n=0}^{M-1} x(2n+1)\sigma_M^{kn}, \tag{2.124}$$

where we have used $\sigma_N^{M(2n+1)} = -1$. This shows that the last M entries of $\mathbf{X}$ are obtained by the same computation scheme as the first M ones, except for using the twiddle factors $-\sigma_N^k$ instead of σ_N^k. The following matrix factorization summarizes this result:

$$\mathrm{DFT}_N \cdot \begin{pmatrix} x(0) \\ x(1) \\ \vdots \\ x(N-1) \end{pmatrix} = \left(\begin{array}{c|c} \mathrm{id}_M & \Delta_M \\ \hline \mathrm{id}_M & -\Delta_M \end{array}\right) \left(\begin{array}{c|c} \mathrm{DFT}_M & 0 \\ \hline 0 & \mathrm{DFT}_M \end{array}\right) \begin{pmatrix} x(0) \\ x(2) \\ \vdots \\ x(N-2) \\ \hline x(1) \\ x(3) \\ \vdots \\ x(N-1) \end{pmatrix}. \tag{2.125}$$

The matrix $\mathrm{id}_M = \mathrm{diag}(1,1,\ldots,1)$ denotes the $(M \times M)$ identity matrix and $\Delta_M = \mathrm{diag}(1,\sigma_N,\ldots,\sigma_N^{M-1})$ the $(M \times M)$ diagonal matrix containing the twiddle factors. The rearrangement of the input vector into components with an even and components with an odd index can be expressed by an additional permutation matrix. Altogether, this leads to a factorization of the DFT_N matrix into a product of sparse matrices (having only few nonzero coefficients) and DFT_M matrices of half the size.

Algorithm: FFT

Input: The length $N = 2^L$ with N being a power of two
The vector $(x(0),\ldots,x(N-1))^\top \in \mathbb{C}^N$
Output: The vector $(X(0),\ldots,X(N-1))^\top = \mathrm{DFT}_N \cdot (x(0),\ldots,x(N-1))^\top$

Procedure: Let $(X(0),\ldots,X(N-1)) = \mathrm{FFT}(N,x(0),\ldots,x(N-1))$ denote the general form of the FFT algorithm.
If $N = 1$ then
$X(0) = x(0)$.
Otherwise compute recursively:
$(A(0),\ldots,A(N/2-1)) = \mathrm{FFT}(N/2,x(0),x(2),x(4)\ldots,x(N-2))$,
$(B(0),\ldots,B(N/2-1)) = \mathrm{FFT}(N/2,x(1),x(3),x(5),\ldots,x(N-1))$,
$C(k) = \sigma_N^k \cdot B(k)$ for $k \in [0:N/2-1]$ with $\sigma_N = \exp(-2\pi i/N)$,
$X(k) = A(k) + C(k)$ for $k \in [0:N/2-1]$,
$X(N/2+k) = A(k) - C(k)$ for $k \in [0:N/2-1]$.

Table 2.1 Recursive version of the FFT algorithm.

The FFT algorithm is again summarized by the compact recursive version shown in Table 2.1.

What have we gained when evaluating the DFT_N by means of this procedure? Let $\mu(N)$ be the number of multiplications and additions[7] needed to compute the matrix–vector product $\mathrm{DFT}_N \cdot \mathbf{x}$. By (2.125), one needs to evaluate two DFT_M, which takes $2\mu(M)$ operations. Furthermore, at first sight, one seems to require $2M = N$ multiplications for the twiddle factors and $2M = N$ additions to sum up the output vectors from the DFT_M step. A closer look shows that one can do even better. First note that the first twiddle factor ($k = 0$) is $\sigma_N^k = 1$, thus causing no multiplication cost. Furthermore, multiplication with the other twiddle factors ($k \in [1:M-1]$) needs to be done only once, but can be used twice (see $C(k)$ in Table 2.1, where it is used once in $X(k) = A(k) + C(k)$ and once in $X(N/2+k) = A(k) - C(k)$). As a result, one requires only $M-1$ multiplications for the twiddle factors (instead of $2M = N$). Altogether, one obtains the estimate

$$\mu(N) \leq 2\mu(N/2) + 1.5N. \tag{2.126}$$

Now, this procedure unfolds its full effect when applied recursively. To this end, one assumes that $N = 2^L$ is a power of two. Obviously $\mu(1) = 0$, since in the case $N = 1$ nothing has to be done. This leads to the following overall estimate:

[7] In the following, subtractions are counted as additions.

$$\mu(N) \leq 2\cdot\mu(N/2)+1.5N \tag{2.127}$$
$$\leq 4\cdot\mu(N/4)+1.5N+1.5N \tag{2.128}$$
$$\leq \ldots \tag{2.129}$$
$$\leq 2^L\cdot\mu(1)+\underbrace{1.5N+1.5N+\ldots+1.5N}_{L=\log_2(N)\text{ times}} \tag{2.130}$$
$$= 1.5N\log_2(N). \tag{2.131}$$

This equation can also be formally shown by a simple induction (see Exercise 2.26). The savings obtained from the FFT algorithm are huge, in particular for large N. For example, in the case $N = 10^3$, the FFT algorithm requires $2\cdot 10^4$ operations instead of 10^6 as needed for the naive approach, which is a reduction of operations by a factor of 50. For $N = 10^5$, this factor is already $3,000$, and for $N = 10^6$, it reaches $25,000$. In this case, if the FFT requires a second of computing time, the naive approach would require 7 hours.

2.4.4 Interpretation of the DFT

Let us summarize the results obtained so far. We started with a CT-signal $f \in L^2(\mathbb{R})$ and derived a DT-signal x by T-sampling. Fixing a number $N \in \mathbb{N}$ of samples, we computed $\mathbf{X} = \mathrm{DFT}_N\cdot\mathbf{x}$ for $\mathbf{x} = (x(0),\ldots,x(N-1))^\top$. What is the meaning of the Fourier coefficients $\mathbf{X} = (X(0),\ldots,X(N-1))^\top$ in relation to the original analog signal f? To answer this question, we need to combine the results induced by the DFT approximation (2.106) and the Riemann approximation (2.104):

$$X(k) \approx \hat{x}\left(\frac{k}{N}\right) \approx \frac{1}{T}\hat{f}\left(\frac{k}{N}\cdot\frac{1}{T}\right). \tag{2.132}$$

In other words, to obtain the "correct" physical interpretation of the coefficient $X(k)$ one needs to know the window size N and the sampling rate $1/T$. First, $X(k)$ needs to be scaled by the factor T. Second, the index k corresponds to the frequency $\omega = k/(NT)$. In other words, the DFT computes the frequencies only on a linear grid of frequencies with a resolution of $1/(NT)$ Hz.

However, the approximations in (2.132) need to be taken with care. The first approximation is only good if the samples of $x(n)$ are close to zero outside the interval $[0:N-1]$. Obviously, this is the case if the analog signal f is close to zero outside the interval $[0,(N-1)/T]$. Furthermore, recall that the second approximation is only good if f does not contain frequency components above the Nyquist frequency $1/(2T)$ Hz. Also, the approximation becomes poor for large k corresponding to high frequencies of the exponential functions. Assuming that f is real-valued, one can easily check that $\hat{f}(\omega) = \overline{\hat{f}(-\omega)}$, $\hat{x}(\omega) = \overline{\hat{x}(-\omega)}$, and $X(k) = \overline{X(N-k)}$ (see (2.83) and Exercise 2.24). Therefore, the coefficients $X(k)$ are redundant for

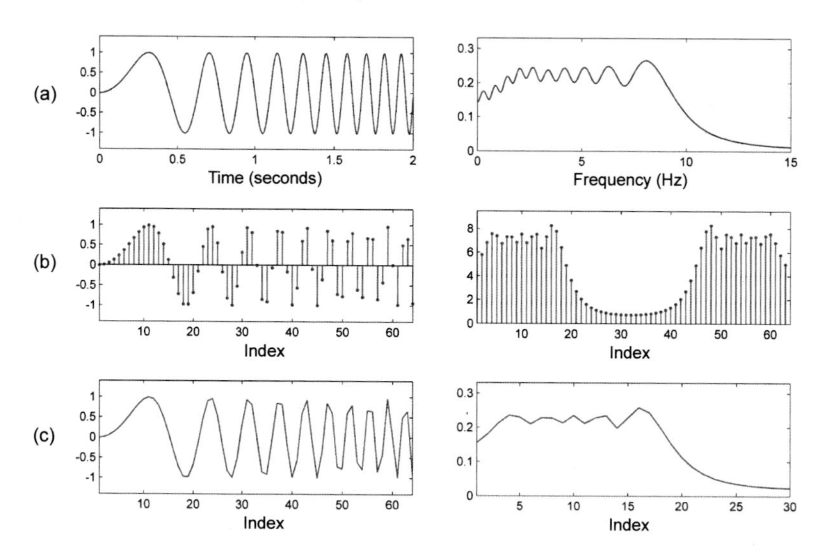

Fig. 2.26 DFT approximation of the Fourier transform. **(a)** Analog chirp signal and its Fourier transform. **(b)** Sampled signal using $T = 1/32$ and DFT coefficients using $N = 64$. **(c)** Interpolation of sampled signal and of DFT coefficients.

$k = \lfloor \frac{N}{2} \rfloor + 1, \ldots, N-1$, and one only needs to consider the coefficients $X(k)$ for $k = 0, 1, \ldots, \lfloor \frac{N}{2} \rfloor$.

As an example, let us consider the analog chirp signal shown in Figure 2.26a, where we assume that the signal is zero outside the shown interval $[0,2]$. The Fourier transform is shown for frequencies $\omega \in [0,15]$. Next, we sample the chirp signal using a sampling rate of $F_s = 32$ Hz and obtain a finite-length signal $\mathbf{x}$ of length $N = 64$. Applying a DFT_N results in a complex-valued vector $\mathbf{X} = \mathrm{DFT}_N \cdot \mathbf{x}$, the magnitude values of which are shown in Figure 2.26b. By (2.132), we obtain $X(k)/32 \approx \hat{f}(k/2)$. For example, the index $k = 30$ corresponds to the frequency $\omega = 15$ (see Figure 2.26c). The resulting frequency resolution is 0.5 Hz.

2.5 Short-Time Fourier Transform (STFT)

The Fourier transform $\hat{f}$ of a signal $f \in L^2(\mathbb{R})$ describes the frequency content of the signal. Comparing the signal with a periodic exponential function $t \mapsto \exp(2\pi i \omega t)$ results in a coefficient $\hat{f}(\omega)$ that exhibits the overall intensity of oscillations at ω Hz occurring in the signal. However, because of the nonlocal nature of the analysis function, the frequency information is always averaged over the entire time domain. Sudden changes and local variations of the signal such as the beginning and the end of events cannot be detected well by the Fourier transform. Local phenomena of the signal become global phenomena in the Fourier transform. In contrast, small changes in the phase of the Fourier transform can have considerable effects in the time domain.

To remedy the drawbacks of the Fourier transform, as we have already discussed in Section 2.1.4, Dennis Gabor introduced in the year 1946 the modified Fourier transform, now known as the **short-time Fourier transform** (STFT). This transform is a compromise between a time- and a frequency-based representation, determining the sinusoidal frequency and phase content of local sections of a signal as it changes over time. In this way, the STFT does not only tell which frequencies are "contained" in the signal but also at which points of times or, to be more precise, in which time intervals these frequencies appear. In the following, we start by introducing the STFT for the case of analog signals. From the STFT one can derive a spectrogram, which visually represents the time–frequency content of a signal. Finally, we introduce a discrete version of the STFT as it is typically used in practice. This is the version of the STFT we have already encountered in Section 2.1.4.

2.5.1 Definition of the STFT

For a given signal, we want to find a transform that exhibits the frequency content of f in a neighborhood of each point in time t. The basic idea is to consider only a small section of the signal around a point t, where the influence of a point within the section decreases with increasing distance from t. Mathematically, this weighting is modeled by multiplying the signal with a **window function**, which can be thought of as a weighting (often bell-shaped) function that localizes around t. Instead of using a different window function for each point t, one uses a single window function that localizes around the point $t = 0$. This function is then shifted across time. If $f \in L^2(\mathbb{R})$ is a signal and $g : \mathbb{R} \to \mathbb{R}$ is such a window function, then the function $f_{g,t}$ localized at point t is defined by

$$f_{g,t}(u) := f(u)g(u-t). \tag{2.133}$$

Figure 2.27 shows a chirp signal f as well as the resulting localized signals $f_{g,t}$ when using a bell-like window function g centered at zero for the shift parameters $t = 0.5$, $t = 1$, and $t = 1.5$, respectively.

In view of a general mathematical formulation, one often admits complex-valued window functions $g : \mathbb{R} \to \mathbb{C}$ and requires $g \in L^2(\mathbb{R})$ as well as $\|g\|_2 \neq 0$. Extending (2.133), the function $f_{g,t}$ is defined by

$$f_{g,t}(u) := f(u)\overline{g}(u-t). \tag{2.134}$$

Note that the complex conjugate does not play any role in case of a real-valued window g, which will always be the case in this book. Also, note that from a technical point of view, g does not need to have a particular shape.

Given a signal $f \in L^2(\mathbb{R})$ as well as a window function $g \in L^2(\mathbb{R})$, the (continuous-time) **short-time Fourier transform** (STFT) is a function $\widetilde{f}_g : \mathbb{R} \times \mathbb{R} \to$

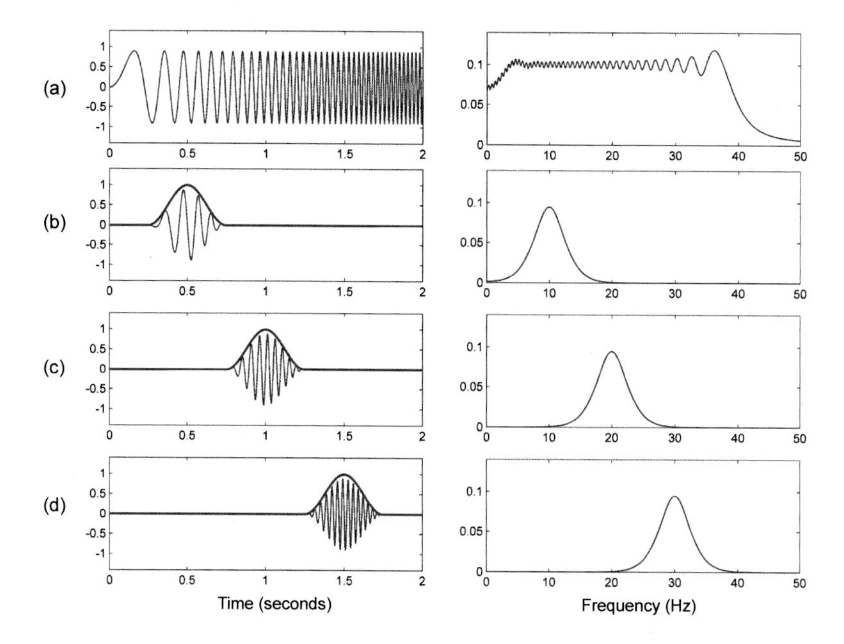

Fig. 2.27 Chirp signal and windowed versions along with their magnitude Fourier transforms. **(a)** Original signal. **(b)** Window centered at $t = 0.5$. **(c)** Window centered at $t = 1.0$. **(d)** Window centered at $t = 1.5$.

$\mathbb{C}$ defined by

$$\widetilde{f}_g(t,\omega) := \widehat{f_{g,t}}(\omega) = \int_{u\in\mathbb{R}} f(u)\overline{g}(u-t)\exp(-2\pi i\omega u)du. \tag{2.135}$$

In other words, $\widetilde{f}_g(t,\cdot)$ coincides with the Fourier transform of the localized signal $f_{g,t}$ for a fixed time instance $t \in \mathbb{R}$.

As an illustration, let us continue with the example of Figure 2.27, which shows the chirp signal $f(t) = \sin(20 \cdot \pi t^2)$ for $t \in [0,2]$. As we mentioned after (2.105), the instantaneous frequency at time t is $\omega = 20t$. Therefore, when considering the localized signal $f_{g,t}$ one may expect frequencies around $\omega = 20t$ Hz. Indeed, the Fourier transform $\widehat{f_{g,t}}$ reveals a peak at 10 Hz for $t = 0.5$ (Figure 2.27b), a peak at 20 Hz for $t = 1$ (Figure 2.27c), and a peak at 30 Hz for $t = 1.5$ (Figure 2.27d).

2.5.1.1 Alternative Definition of the STFT

When considering the short-time Fourier transform, one can assume a different viewpoint, which leads to a sightly different definition. In the above definition, we first windowed the original signal f with the time-shifted window g_t to obtain the localized signal $f_{g,t}$, which was then compared against the exponential func-

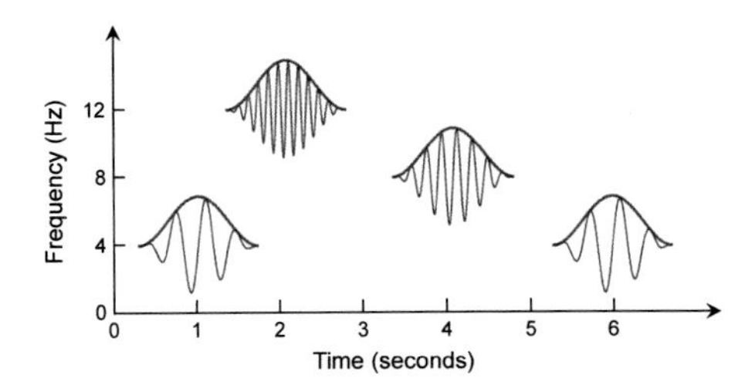

Fig. 2.28 Illustration of four different "musical notes" $g_{t,\omega}$ located in the time–frequency plane: $(t,\omega)=(1,4)$, $(t,\omega)=(2,12)$, $(t,\omega)=(4,8)$, and $(t,\omega)=(6,4)$.

tions $\mathbf{exp}_\omega$. A different viewpoint is to construct localized elementary functions $g_{t,\omega}:\mathbb{R}\to\mathbb{C}$ by defining

$$g_{t,\omega}(u) := \exp(2\pi i\omega(u-t))g(u-t), \tag{2.136}$$

$u\in\mathbb{R}$. In other words, $g_{t,\omega}$ is obtained by first modulating the window g by ω Hz, which is a frequency shift in the Fourier domain (see (2.100)). The resulting modulated window is then shifted in time by t sec (see (2.97)). Intuitively, $g_{t,\omega}$ may be thought of as a "musical note" of frequency ω that is active in a neighborhood of t. The parameters t and ω allow for shifting the musical note in the time–frequency plane (see Figure 2.28).

It is not hard to see that $\|g_{t,\omega}\| = \|g\|$ for a window function $g\in L^2(\mathbb{R})$ (see Exercise 2.22). Therefore, as opposed to the exponential functions $\mathbf{exp}_\omega$, which do not have finite energy, one has $g_{t,\omega}\in L^2(\mathbb{R})$. Therefore, we can define a function $\widetilde{f}^g:\mathbb{R}\times\mathbb{R}\to\mathbb{C}$ by setting

$$\widetilde{f}^g(t,\omega) = \langle f|g_{t,\omega}\rangle = \int_{u\in\mathbb{R}} f(u)\overline{g}(u-t)\exp(-2\pi i\omega(u-t))du. \tag{2.137}$$

The inner product $\langle f|g_{t,\omega}\rangle$ measures the similarity between the signal f and the musical note $g_{t,\omega}$. If f and $g_{t,\omega}$ oscillate with the same frequency within the window, the inner product $\langle f|g_{t,\omega}\rangle$ has a large absolute value. Vice versa, if f has no frequency components around ω, the inner product is close to zero and f and $g_{t,\omega}$ are more or less orthogonal. The signal

$$u\mapsto \langle f|g_{t,\omega}\rangle g_{t,\omega}(u) \tag{2.138}$$

can be considered as the "projection" of the signal f in the direction of the musical note $g_{t,\omega}$ (see Figure 2.15).

The original STFT $\widetilde{f}_g$ defined by (2.135) and the version $\widetilde{f}^g$ defined by (2.137) coincide up to some time-dependent modulation factor:

$$\widetilde{f}_g(t,\omega) = \widetilde{f}^g(t,\omega)\exp(2\pi i\omega t). \tag{2.139}$$

In the first version only the window is shifted, whereas in the second version also the exponential function is shifted along with the window. Often $\widetilde{f}_g$ is used for the ana-

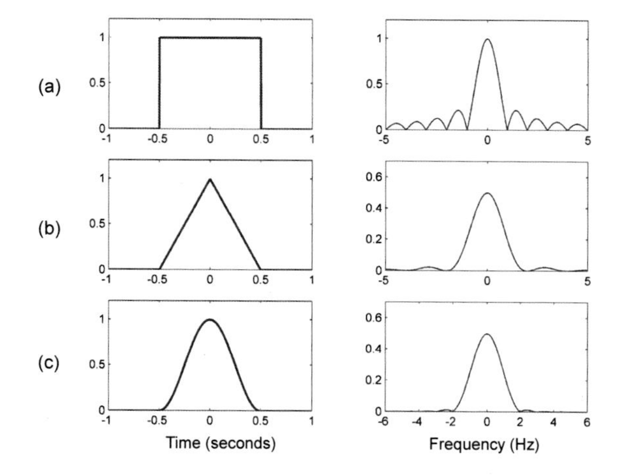

Fig. 2.29 Window functions and their Fourier transforms. **(a)** Rectangular window. **(b)** Triangular window. **(c)** Hann window.

log case, whereas $\widetilde{f^g}$ corresponds to what is used for the discrete Fourier transform (see for example (2.26)). We will come back to this issues in Section 2.5.3.

2.5.1.2 Role of the Window Function

We now discuss the role of the window function g, which plays an important role from a signal processing point of view. Typically, a window function is chosen to be zero-valued outside of some chosen section, so that when a signal is multiplied by the window function, the product is also zero-valued outside the section. The finite-length signal that is left can be regarded as a "view through the window." The definition (2.135) shows that the STFT depends on both the signal as well as the window function, although one is typically interested only in the signal's properties. The design of suitable window functions and their influence is a science by itself, which is outside the scope of this book. In the following, we discuss some examples that illustrate how the window may affect the spectral estimate computed by the STFT.

The seemingly simplest way to obtain a local view on the signal f is to leave it unaltered within the desired section and to set all values to zero outside the section. Such a localization is realized by a **rectangular window** as defined in (2.95) and again shown in Figure 2.29a. However, using the rectangular window has major drawbacks, since it generally leads to discontinuities at the section's boundaries in the localized signal $f_{g,t}$. As we have discussed before, such abrupt changes lead to artifacts due to interferences which are spread over the entire frequency spectrum. Rather than being part of the original signal f, these frequency components come from the properties of the rectangular window (see Figure 2.29a). Recall that the Fourier transform of the rectangular window is the sinc function defined in (2.96), which shows slowly decaying ripples across the entire spectrum. These ripples also become visible in the STFT of a chirp signal as demonstrated by Figure 2.30a.

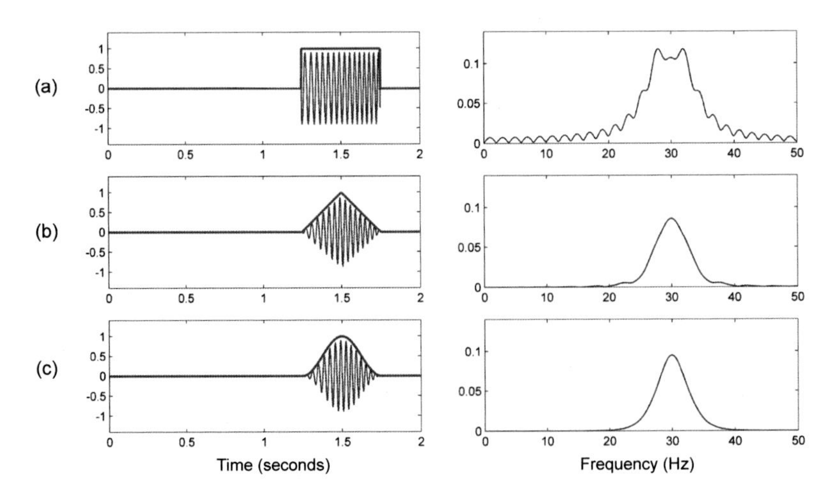

Fig. 2.30 Windowed chirp signal and its magnitude Fourier transform using different window functions. **(a)** Rectangular window. **(b)** Triangular window. **(c)** Hann window.

To attenuate the boundary effects, one often uses windows that are nonnegative within the desired section and continuously fall to zero towards the section's boundaries. One such example is the **triangular window** (Figure 2.29b), which leads to much smaller ripple artifacts (Figure 2.30b). A window often used in signal processing is the **Hann window** (also known as the **Hanning window**) named after Julius von Hann. The Hann window g is a raised cosine window defined by

$$g(u) := \begin{cases} (1+\cos(\pi u))/2 & \text{if } -0.5 \leq u \leq 0.5 \\ 0 & \text{otherwise} \end{cases} \tag{2.140}$$

(see Figure 2.29c). Dropping smoothly to zero at the section boundaries, the above-mentioned artifacts in the Fourier transform of the windowed signal are softened. This is also illustrated by Figure 2.30c. However, on the downside, the Hann window introduces some smearing of frequencies. As a result, the Fourier transform of a signal's windowed section may look smoother than the signal's properties suggest. In other words, the reduction of ripple artifacts introduced by the window is achieved at the expense of a poorer spectral localization. Similarly, as we will see in the next section, the size of the window crucially affects the STFT.

2.5.2 Spectrogram Representation

The STFT of a signal f yields for each point in time t and frequency ω a complex number $\widetilde{f}_g(t,\omega)$. This information is often visualized by means of a **spectrogram**, which is a two-dimensional representation of the squared magnitude:

$$\mathrm{Spec}(t,\omega) = |\widetilde{f}_g(t,\omega)|^2 = |\widetilde{f}^g(t,\omega)|^2. \tag{2.141}$$

For the definition of the spectrogram, the version of the STFT in (2.139) does not matter, since the modulation factor has a magnitude of one. When generating an image of a spectrogram, the horizontal axis represents time, the vertical axis is frequency, and the dimension indicating the spectrogram value of a particular frequency at a particular time is represented by the intensity or color in the image. There are many variations in visualizing a spectrogram. Sometimes the vertical and horizontal axes are switched, so time runs up and down. Sometimes the amplitude is represented as the height of a 3D surface instead of color or intensity. To emphasize musical or tonal relationships, the frequency axis is often plotted in a logarithmic fashion, which yields a **log-frequency representation** as we will encounter in the subsequent chapters. A logarithmic frequency axis also accounts for the fact that human perception of pitch is logarithmic in nature (see Section 1.3.2). Finally, in the case of audio signals, the amplitude values are also often visualized using a logarithmic scale, for example, by using a decibel scale. In this way, small intensity values of perceptual relevance become visible in the image. In the following, if not specified otherwise, we use in our visualizations a linear frequency axis and a logarithmic scale to represent amplitudes. The specific scale is not of importance, but only serves the purpose of enhancing the qualitative properties of the visualization.

In our first example, we again consider a chirp signal f defined by $f(t) = \sin(400\pi t^2)$ for $t \in [0,1]$, which is smoothly faded out towards $t = 1$ (see Figure 2.31a). For this chirp, the instantaneous frequency linearly raises from $\omega = 0$ Hz at $t = 0$ to $\omega = 400$ Hz at $t = 1$. For computing the STFT, we use a Hann window having a size of 62.5 ms. The resulting spectrogram is shown in Figure 2.31b. The logarithmic amplitude values are encoded by different gray levels, which are lighter for small values and darker for large values. Note that each column of the spectrogram corresponds to a plot of a Fourier transform as, for example, shown in Figure 2.30c.

The image of the spectrogram shows a strong diagonal stripe starting at the time–frequency point $(t,\omega) = (0,0)$ and ending at $(t,\omega) = (1,400)$, which reveals the linear frequency increase of the chirp signal. This diagonal stripe has a substantial width (roughly 40 Hz), which can be explained as follows: First, recall that at a given point t the STFT exhibits the frequency content of an entire neighborhood (a windowed section of the signal) around t, and the STFT averages the frequency information across this section. Second, as discussed in Section 2.5.1, the window introduces some additional smearing of frequencies in the Fourier domain. The artifacts introduced by the window function also explain the weaker diagonal stripes that run below and above the strong diagonal stripe. These weaker stripes correspond to the ripples occurring in the Fourier transform of the window function. As opposed to Figure 2.30c, where no such ripples can be seen for the Hann window, the ripples become visible in the visualization of the spectrogram only because we have used a logarithmic magnitude scale. We have already seen in Figure 2.30a that the ripple artifacts become much stronger when using a rectangular window instead of a Hann window. This phenomenon is illustrated by Figure 2.31c, which shows

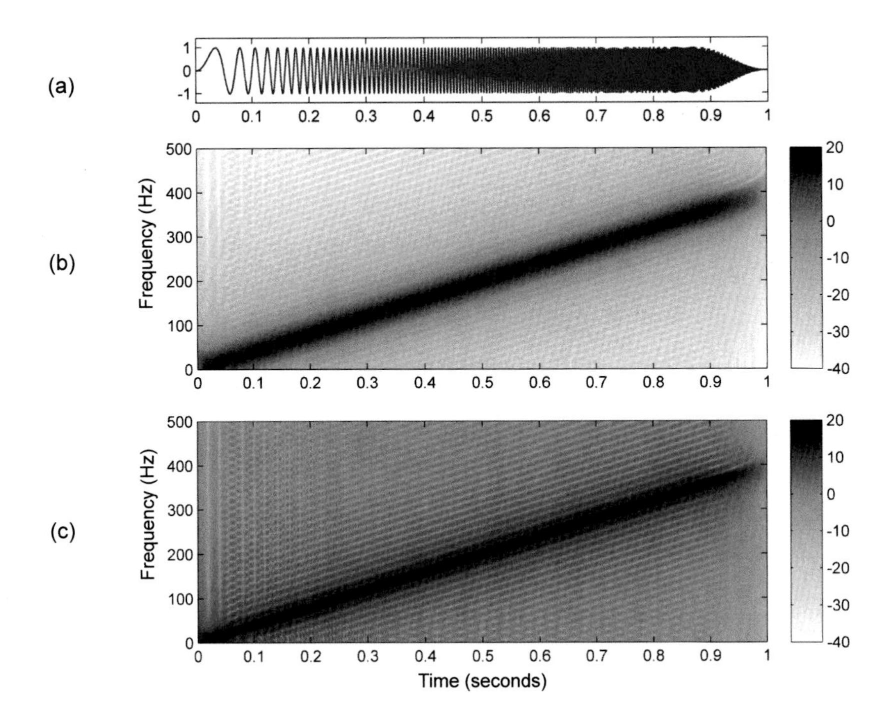

Fig. 2.31 Spectrogram of a chirp signal using two different window types. **(a)** Signal. **(b)** Spectrogram with Hann window of size 62.5 ms. **(c)** Spectrogram with rectangular window of size 62.5 ms.

a corresponding spectrogram. This visualization demonstrates the importance of choosing a suitable window function. In general, it is not easy to distinguish the characteristics of the signal and the effects introduced by the window function.

With the next example, we discuss the role of the size of the window function g. To this end, we consider the signal f shown in Figure 2.32a, which is defined by

$$f(t) = \sin(800\pi t) + \sin(900\pi t) + \delta(t-0.45) + \delta(t-0.5) \tag{2.142}$$

for $t \in [0,1]$. In this interval, f is a superposition of two sinusoids of frequency 400 and 450 Hz, respectively. Furthermore, two impulses are added at the points $t = 0.45$ and $t = 0.5$ sec. Again we assume that f is zero outside the shown interval $[0,1]$. This signal is interesting since it contains two components that are close in time (the two impulses that are 50 ms apart) and two components that are close in frequency (the two sinusoids that are 50 Hz apart). Figure 2.32b shows the spectrogram when using a Hann window of size 32 ms. The image contains a horizontal stripe in the region between 375 and 475 Hz, which corresponds to the sinusoids, as well as two vertical stripes at $t = 0.45$ and $t = 0.5$ sec, which correspond to the impulses. As illustrated by Figure 2.20b, each of the impulses results in many nonzero Fourier

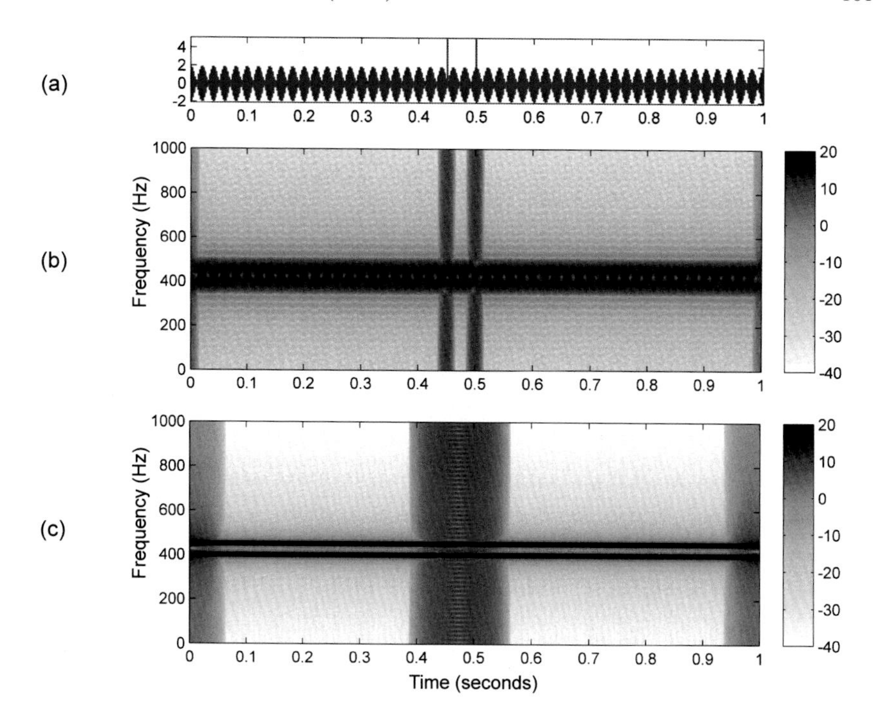

Fig. 2.32 Spectrogram using different window sizes. **(a)** Signal. **(b)** Spectrogram with short Hann window (32 ms). **(c)** Spectrogram with long Hann window (128 ms).

coefficients spread across the entire spectrum, which explains the vertical stripes. Since the window size of 32 ms implies that in each window there is at most one of the impulses, the two impulses can be clearly separated by the STFT. However, the STFT is not able to separate the two frequency components at $\omega = 400$ Hz and $\omega = 450$ Hz. The reason is that the chosen window introduces frequency smearing. The scaling property of the Fourier transform (Exercise 2.20) says that reducing the size by temporally compressing the window leads to a broadening of its Fourier transform. This, in turn, implies that the frequency smearing becomes more severe. Therefore, to separate the two frequency components, one strategy is to increase the window size, thus reducing the frequency smearing. Indeed, using a Hann window of size 128 ms results in a clear separation as shown by the two horizontal stripes (see Figure 2.32c). However, increasing the window size goes along with an increased smearing in the time domain. As a result, the two impulses are not separated any longer. As a side remark, we want to point to the two vertical stripes showing up at $t = 0$ and $t = 1$. An explanation is to be given in Exercise 2.27.

In summary, using a large window size results in a good localization in frequency, but a poor localization in time, whereas using a small window size has the opposite effect. Increasing the window size leads to an STFT which averages the frequencies of the signal over a greater time interval, resulting in a loss of time information. In the limit case of an "infinite window size" one ends up with the usual Fourier transform, which averages the frequencies over the entire time domain $\mathbb{R}$. Vice versa, successively decreasing the window size results in a Dirac sequence, where, in the limit case of g being an impulse, the STFT gives back the original signal: perfect time localization, no frequency localization.

The time localization property of the STFT depends on the temporal spread of the window function g, whereas the frequency localization property of the STFT depends on the spectral spread of the Fourier transform $\hat{g}$. We want to mention that one cannot have both properties at the same time. A variant of the **Heisenberg uncertainty principle** says that there is no window function that simultaneously localizes in time and frequency with arbitrary precision.

2.5.3 Discrete Version of the STFT

So far, we have discussed the STFT and spectrogram in the case of analog signals. In practice, one uses sampled signals and computes the STFT only on a finite time–frequency grid. Because of efficiency issues, one typically employs DFTs which can be computed by means of the FFT algorithm. As before, let x be a DT-signal obtained from a CT-signal f by T-sampling. Furthermore, let w be a sampled version of an analog window function g. In the discrete case, the window can be shifted only in a sample-wise fashion. Because of efficiency issues, one often shifts the window in even larger steps, which are specified by some **hop size** parameter $H \in \mathbb{N}$ (given in samples). Following the alternative definition (2.137) in the analog case, we define the (discrete-time) STFT $\widetilde{x^w}$ of the DT-signal x with respect to the window function w by

$$\widetilde{x^w}(m,\omega) := \sum_{n\in\mathbb{Z}} x(n)\overline{w}(n-mH)\exp(-2\pi i\omega(n-mH)) \tag{2.143}$$

$$= \sum_{n\in\mathbb{Z}} x(n+mH)\overline{w}(n)\exp(-2\pi i\omega n) \tag{2.144}$$

for $m \in \mathbb{Z}$ and $\omega \in [0,1)$. Now, if the sampled window function w is a finite signal, the sum in (2.144) becomes finite, and we can apply the DFT to compute the discrete STFT for certain frequencies.

In the analog case, we assumed that the window function g was centered at time zero. To simplify the formulas in the discrete case, we assume that the support of the window function is contained only in the positive part of the time axis centered at half the window length (i.e., the window is shifted by half a window length to the right compared with the zero-centered case). The zero-centered case can be easily restored by also shifting the original signal by half a window length.

Having said this, we assume that the nonzero samples of the discrete window w are $w(n)$ for $n \in [0:N-1]$. For each frame index $m \in \mathbb{Z}$, we define the vector $\mathbf{x}_m = (x_m(0),\ldots,x_m(N-1))^\top \in \mathbb{C}^N$ with

$$x_m(n) = x(n+mH)\overline{w}(n) \tag{2.145}$$

for $n \in [0:N-1]$ and compute the vector $\mathbf{X}_m = (X_m(0),\ldots,X_m(N-1))^\top \in \mathbb{C}^N$ via a DFT of size N:

$$\mathbf{X}_m = \mathrm{DFT}_N \cdot \mathbf{x}_m. \tag{2.146}$$

Then one obtains

$$\begin{aligned}\widetilde{x^w}(m,k/N) &= \sum_{n=0}^{N-1} x(n+mH)\overline{w}(n)\exp(-2\pi ikn/N)\\ &= \sum_{n=0}^{N-1} x_m(n)\exp(-2\pi ikn/N)\\ &= X_m(k)\end{aligned} \tag{2.147}$$

for $k \in [0:N-1]$. Thus, we have shown that, for each time frame $m \in \mathbb{Z}$, one can compute the discrete STFT at frequencies $\omega = k/N$ for $k \in [0:N-1]$ by means of a DFT_N. In the case that N is a power of two, this can be done efficiently using the FFT.

2.5.3.1 Summary

Altogether, we have reached exactly the version of the discrete STFT already introduced in Section 2.1.4. Let us again summarize the main results. Let x be a DT-signal obtained by T-sampling. Furthermore, let w be a discrete window of finite length N with coefficients $w(n)$ for $n \in [0:N-1]$. Then

$$\mathcal{X}(m,k) = \widetilde{x^w}(m,k/N) = \sum_{n=0}^{N-1} x(n+mH)\overline{w}(n)\exp(-2\pi ikn/N) \tag{2.148}$$

is the **discrete STFT** or simply the STFT of x (see also (2.26)). Each spectral vector for some time frame $m \in \mathbb{Z}$ can be computed by using a DFT_N, which can be evaluated efficiently by using an FFT if N is a power of two. The coefficients $\mathcal{X}(m,k)$ have a similar interpretation as discussed in Section 2.4.4. First recall that the upper half of the frequency coefficients are redundant if x and w are real-valued. In this case, one only considers the coefficients $k \in [0:N/2]$. By (2.132), the index k corresponds to the frequency

$$F_{\mathrm{coef}}(k) := \frac{k \cdot F_{\mathrm{s}}}{N} \tag{2.149}$$

(see also (2.28)). In particular, the index $k = N/2$ corresponds to the Nyquist frequency $\omega = 1/(2T)$.

Next, we discuss how the index m is to be interpreted. The interpretation is not straightforward since m refers to an entire windowed section of the signal rather than a specific point in time. In signal processing, such a windowed section is also called

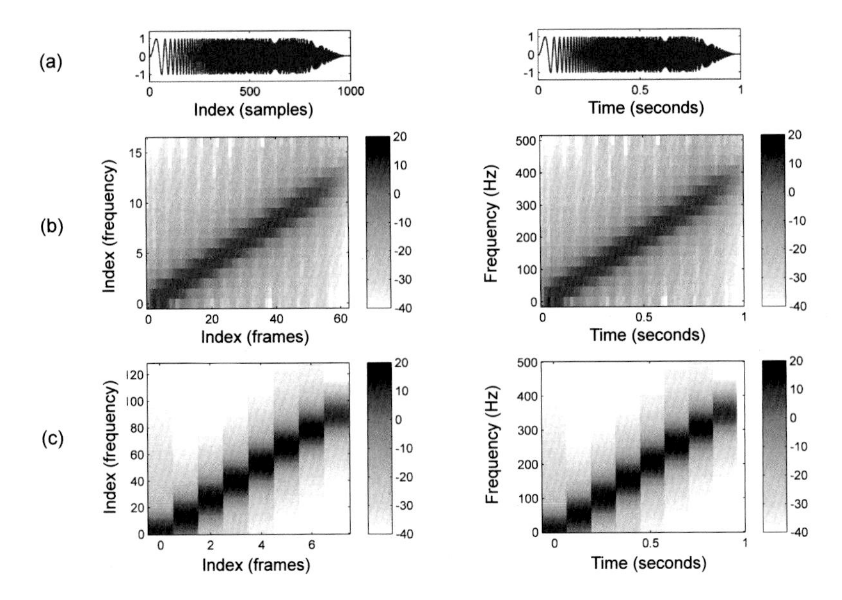

Fig. 2.33 Spectrogram representation of Discrete STFT. Shown are the original indices (left) and their physical interpretation (right). **(a)** Signal using $1/T = 1000$ Hz. **(b)** Spectrogram using $N = 32$ and $H = 16$. **(c)** Spectrogram using $N = 256$ and $H = 128$.

a **frame** and m is called the **frame index**. The physical duration of a frame is NT seconds. There are no strict conventions for associating a physical time position to a frame. When assuming that the window is centered at zero, as we did in the case of the continuous-time version of the STFT, one should take the center of the frame as a physical reference point. When assuming that the window starts at time position zero centered at half the window length, one may take the start of the frame as a physical reference point. As said before, the second convention can be transferred into the first one by shifting the original signal by half a window length. In the following, we want to adapt the second convention such that the frame index m is associated to the physical time position

$$T_{\mathrm{coef}}(m) := \frac{m \cdot H}{F_{\mathrm{s}}} \tag{2.150}$$

(see also (2.27)). Using this convention, the index $m = 0$ is associated with $t = 0$.

2.5.3.2 Examples

In Figure 2.9 we have already seen an example of how to interpret the frame and frequency indices in terms of physical units such as seconds and Hertz. Let us consider a second example to illustrate the effect of different parameter settings.

Figure 2.33a shows a DT-signal based on a sampling rate of $F_s = 1/T = 1000$ Hz. To compute the spectrogram of Figure 2.33b, a window length of $N = 32$ and a hop size of $H = 16$ were used. This yields a frame size of $NT = 32$ ms and the frame index m corresponds to time $T_{\mathrm{coef}}(m) = mTH = m \cdot 16$ ms, which is also the time resolution of the STFT. In particular, frame index $m = 62$ corresponds to $T_{\mathrm{coef}}(m) = 0.992 \approx 1$ sec. Furthermore, the frequency index k corresponds to frequency $F_{\mathrm{coef}}(k) = k/(NT) = k \cdot 31.25$ Hz. In particular $k = 16$ yields the Nyquist frequency $F_{\mathrm{coef}}(16) = 500$ Hz. A second parameter setting using $N = 256$ and $H = 128$ is shown in Figure 2.33c.

2.6 Summary and Further Readings

In this chapter, we studied fundamental techniques for analyzing signals by means of elementary sinusoidal functions, which possess an explicit physical meaning in terms of frequency. We considered various types of signals including analog or CT-signals as well as DT-signals or more general digital signals, which were obtained by sampling and quantization. Generally speaking, the CT-domain gives the "right" interpretation of physical phenomena, whereas the DT-domain is used to do the actual computations. Being the most important tool for processing audio signals, we introduced different variants of the Fourier transform for the CT- as well as for the DT-domain. The **Fourier transform** converts a time-dependent signal into frequency-dependent coefficients, each of which indicates the strength (and a strength-optimizing phase) of the respective elementary exponential function. The process of decomposing a signal into frequency components is also called **Fourier analysis**. In contrast, we have seen that the **Fourier representation** rebuilds a signal from the elementary functions, a process also called **Fourier synthesis**.

The Fourier transform and the Fourier representation are closely related, leading to very similar formulas (see Table 2.2 for an overview). Many of these formulas can be expressed by inner products—notions that make it possible to use the same geometric language one knows from finite-dimensional Euclidean spaces [15, 16]. The basic definitions and main properties of the Fourier transform are covered in most introductory books on signal processing. As example references, we want to mention the classical textbook on *Signals and Systems* by Oppenheim et al. [12] or the book on *Digital Signal Processsing* by Proakis and Manolakis [13].

In this chapter, we used clear mathematical modeling, which is necessary when one wants to understand the relation between the CT- and DT-domain. While we used the notion of Lebesgue spaces, we did not introduce them with rigor. In particular in the case of CT-signals, the definition of Lebesgue spaces becomes a bit tricky, since one needs the notion of measurability of the functions in order for the integrals to be defined. For a mathematically rigorous treatment of measure and Lebesgue theory, we refer to the book *Real Analysis* by Folland [5]. As we have already indicated before, the spaces $L^2(\mathbb{R})$ and $L^2([0,1))$ are actually quotient spaces where two functions f and g are considered to coincide if $\|f-g\|_2 = 0$ (i.e., if they

Signal space	$L^2(\mathbb{R})$	$L^2([0,1))$	$\ell^2(\mathbb{Z})$
Inner product	$\langle f\|g\rangle = \int_{t\in\mathbb{R}} f(t)\overline{g(t)}dt$	$\langle f\|g\rangle = \int_{t\in[0,1)} f(t)\overline{g(t)}dt$	$\langle x\|y\rangle = \sum_{n\in\mathbb{Z}} x(n)\overline{y(n)}$
Norm	$\\|f\\|_2 = \sqrt{\langle f\|f\rangle}$	$\\|f\\|_2 = \sqrt{\langle f\|f\rangle}$	$\\|x\\|_2 = \sqrt{\langle x\|x\rangle}$
Definition	$L^2(\mathbb{R}) :=$ $\{f:\mathbb{R}\to\mathbb{C} \mid \\|f\\|_2<\infty\}$	$L^2([0,1)) :=$ $\{f:[0,1)\to\mathbb{C} \mid \\|f\\|_2<\infty\}$	$\ell^2(\mathbb{Z}) :=$ $\{f:\mathbb{Z}\to\mathbb{C} \mid \\|x\\|_2<\infty\}$
Elementary frequency function	$\mathbb{R}\to\mathbb{C}$ $t\mapsto\exp(2\pi i\omega t)$	$[0,1)\to\mathbb{C}$ $t\mapsto\exp(2\pi ikt)$	$\mathbb{Z}\to\mathbb{C}$ $n\mapsto\exp(2\pi i\omega n)$
Frequency parameter	$\omega\in\mathbb{R}$	$k\in\mathbb{Z}$	$\omega\in[0,1)$
Fourier representation	$f(t) =$ $\int_{\omega\in\mathbb{R}} c_\omega\exp(2\pi i\omega t)d\omega$	$f(t) =$ $\sum_{k\in\mathbb{Z}} c_k\exp(2\pi ikt)$	$x(n) =$ $\int_{\omega\in[0,1)} c_\omega\exp(2\pi i\omega n)d\omega$
Fourier transform	$\hat{f}:\mathbb{R}\to\mathbb{C}$ $\hat{f}(\omega)=c_\omega=$ $\int_{t\in\mathbb{R}} f(t)\exp(-2\pi i\omega t)dt$	$\hat{f}:\mathbb{Z}\to\mathbb{C}$ $\hat{f}(k)=c_k=$ $\int_{t\in[0,1)} f(t)\exp(-2\pi ikt)dt$	$\hat{x}:[0,1)\to\mathbb{C}$ $\hat{x}(\omega)=c_\omega=$ $\sum_{n\in\mathbb{Z}} x(n)\exp(-2\pi i\omega n)$

Table 2.2 Overview of the signal spaces $L^2(\mathbb{R})$, $L^2([0,1))$, and $\ell^2(\mathbb{Z})$ and their respective Fourier representation and Fourier transform.

differ only up to a null set). The equality in the Fourier representation and in the Fourier transform is just an equality in the L^2-sense, which is a weaker notion than pointwise equality. We have also mentioned before that the integral in the definition (2.92) of the Fourier transform of a signal $f\in L^2(\mathbb{R})$ does not exist in general. Instead, one needs to define the integral by some limit process (2.93). The existence of the limit is based on the so-called *Hahn–Banach theorem* [5]. One main problem in the CT case is that the exponential functions $\mathbf{exp}_\omega : \mathbb{R}\to\mathbb{C}$ are not contained in $L^2(\mathbb{R})$. Therefore, the integral in (2.92) cannot be written as an inner product as is possible for the Fourier coefficients (2.80). The proofs for the existence and correctness of the considered Fourier transforms and Fourier representations, which are outside the scope of this book, can be found in the book by Folland [5].

Furthermore, we have only scratched the topics of sampling and aliasing, which are of crucial importance for digital signal processing. The famous **sampling theorem** says that an Ω-**bandlimited** signal $f\in L^2(\mathbb{R})$ (i.e., a signal where the Fourier transform $\hat{f}$ vanishes for $|\omega|>\Omega$ for a real number $\Omega>0$) can be reconstructed perfectly from the T-sampling of f with $T:=1/(2\Omega)$ (see [12, 13]). In Exercise 2.28, we cover this important result in more detail. The sampling theorem is often associated with the names Harry Nyquist and Claude Shannon. It is interesting to note that the theorem was also discovered independently by Edmund Taylor Whittaker, Vladimir Kotelnikov, and others (see [1, 8] for an overview and historical notes).

There exists a vast literature on the discrete Fourier transform (DFT) and associated fast Fourier transform (FFT) algorithms. In the original article by Cooley and Tukey [3], the authors describe an FFT algorithm that works in case that the length N of the DFT is an integral power of two. By applying several tricky modifications of the FFT, this result can be extended to an algorithm for evaluating a DFT of arbitrary length $N \in \mathbb{N}$ with time complexity of $O(N \log N)$. A detailed description of this result can be found in the book *Fast Fourier Transforms* by Clausen and Baum [2], which treats this topic from an algebraic point of view.

The short-time Fourier transform (STFT), which is also often referred to as the **windowed Fourier transform**, was pioneered in the year 1946 by Dennis Gabor for use in communication theory [6]. We have seen that the STFT is a compromise between a time- and a frequency-based representation of the signal. For a detailed discussion of the role of the window function used in the STFT calculation, we refer to [7]. One main drawback of the STFT is that the window function g implies a kind of rigid time–frequency resolution. As a result, properties of a signal that are much shorter than the window size are "synthesized" in the frequency domain, whereas properties of the signal that are much longer than the window size are "synthesized" in the time domain. In both cases, many of the "notes" $g_{\omega,t}$ (see Figure 2.28) are needed to represent the phenomena of the signal. To remedy this problem, numerous alternatives have been suggested, including time–frequency representations based on **wavelets**. For further readings and links on this topic, we refer to [4, 10, 14, 15]. Parts of Section 2.5, including the notation and the association of $g_{\omega,t}$ to "musical notes," were inspired by [10, Chapter 2]. An entertaining and nontechnical introduction to the main ideas of time–frequency analysis can be found in the book *The World According to Wavelets* by Hubbard [9].

2.7 FMP Notebooks

The Fourier transform serves as the main signal processing tool throughout this textbook. In **Part 2** of the FMP notebooks [11], we approach Fourier analysis from a practical perspective with a focus on the discrete Fourier transform (DFT). In particular, we cover the entire computational pipeline in a bottom-up fashion by providing Python code examples for deepening the understanding of complex numbers, exponential functions, the DFT, the FFT, and the STFT. In this context, we address practical issues such as digitization, padding, and axis conventions—issues that are often neglected in theory. Assuming that the reader has opened the FMP notebooks of **Part 2** (e.g., available as a static HTML version), we now briefly comment on the FMP notebooks in the order in which they appear.

We start with the **FMP Notebook *Complex Numbers***, where we review basic properties of complex numbers. In particular, we provide Python code for visualizing complex numbers using either Cartesian coordinates or polar coordinates. Such visualizations help students gain a geometric understanding of complex numbers and the effect of their algebraic operations. Subsequently, we consider in the **FMP**

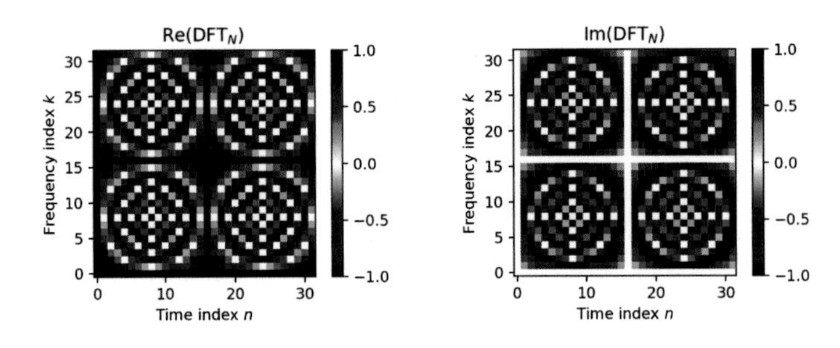

Fig. 2.34 The matrix DFT_N and a visualization of its real and imaginary parts for the case $N = 32$.

Notebook *Exponential Function* the complex version of the exponential function. Many students are familiar with the real version of this function, which is often introduced by its power series $\exp(a) = \sum_{n=0}^{\infty} a^n/n!$ for $a \in \mathbb{R}$. This definition can be extended by replacing the real variable $a \in \mathbb{R}$ by a complex one $c \in \mathbb{C}$. Studying the approximation quality of the power series (and other limit definitions of the exponential function) is instructive and can be combined well with small programming exercises. One important property of the complex exponential function, which is also central for the Fourier transform, is expressed by Euler's formula $\exp(i\gamma) = \cos(\gamma) + i\sin(\gamma)$ for $\gamma \in \mathbb{R}$. We provide a visualization that illustrates how the exponential function restricted to the unit circle relates to the real sine and cosine functions. Furthermore, we discuss the notion of roots of unity, which are the central building blocks that relate the exponential function to the DFT matrix. The study of these roots can be supported by small programming exercises, which may also cover mathematical concepts such as complex polynomials and the fundamental theorem of algebra.

In the **FMP Notebook *Discrete Fourier Transform***, we approach the DFT in various ways. Recall from (2.24) that, given $x = (x(0), x(1), \ldots, x(N-1))^\top \in \mathbb{R}^N$, the DFT is defined by $X(k) := \sum_{n=0}^{N-1} x(n) \exp(-2\pi ikn/N)$ for $k \in [0 : N-1]$. The vector $X \in \mathbb{C}^N$ can be interpreted as a frequency representation of the time-domain signal x. The real (imaginary) part of a Fourier coefficient $X(k)$ can be interpreted as the inner product of the input signal x and a sampled version of the cosine (sine) function of frequency $F_{\mathrm{coef}}(k)$ (see (2.28)). In the notebook, we provide a concrete example that illustrates how this inner product can be interpreted as the correlation between a signal x and the cosine (sine) function. We recommend that students experiment with different signals and frequency parameters k to deepen the intuition of these correlations. From a computational view, the vector X can be expressed by the product of the matrix DFT_N with the vector x. Defining the complex number $\sigma_N := \exp(-2\pi i/N)$ (which is a specific root of unity), one can express the DFT matrix in a very compact form given by $\mathrm{DFT}_N(n,k) = \sigma_N^{nk}$ for $n, k \in [0 : N-1]$. Visualizing the real and imaginary parts of the DFT matrix reveals its structural properties (see Figure 2.34). In particular, one can observe that the rows of the DFT matrix correspond to sampled cosine (real part) and sine (imaginary part) functions.

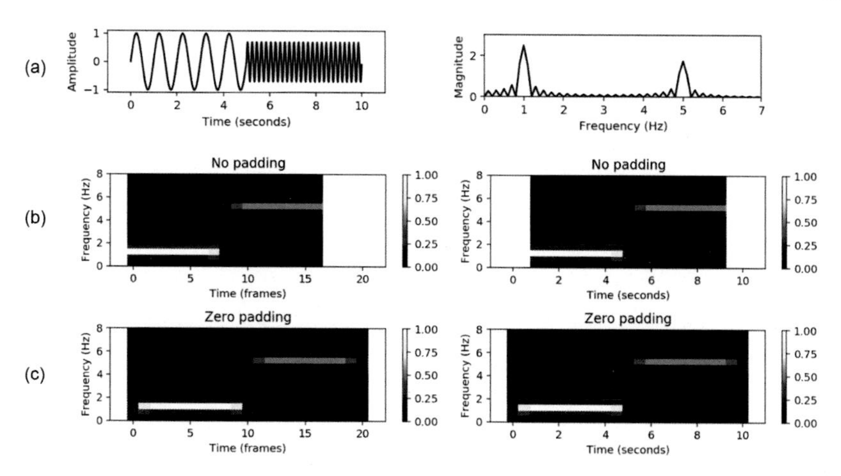

Fig. 2.35 **(a)** Time-domain signal ($F_s = 256$) and magnitude Fourier transform. **(b)** STFT ($N = 512, H = 128$) without padding. **(c)** STFT ($N = 512, H = 128$) with zero-padding.

The specific structure of the matrix DFT_N (with its relation to DFT_M for $M = N/2$) can be exploited in a recursive fashion, yielding the famous fast Fourier transform (FFT). In our notebook, we provide an explicit implementation of the FFT algorithm and present some experiments, where we compare the running time of a naive implementation with the FFT-based one. We think that implementing and experimenting with the FFT—an algorithm of great beauty and high practical relevance—is a computational eye opener and a must in every signal processing curriculum. Computing a DFT results in complex-valued Fourier coefficients, where each such coefficient can be represented by a magnitude and a phase component. In the **FMP Notebook *DFT: Phase***, we provide a Python code example that highlights the optimality property of the phase (similar to Figure 2.2). Studying this property is the core of understanding the role of the phase—a concept that is often difficult to access for students new to the field.

As another central topic of this chapter, we discussed in Section 2.5 various versions of the short-time Fourier transform (STFT). In the **FMP Notebook *Discrete Short-Time Fourier Transform (STFT)***, we implement a discrete version of the STFT from scratch and discuss various options for visualizing the resulting spectrogram. While the main idea of the STFT (i.e., applying a sliding window technique and computing for each windowed section a DFT) seems simple, computing the discrete STFT in practice can be tricky. In an applied signal processing course, it is essential to make students aware of the different parameters and conventions when applying windowing. Our notebooks provide Python implementations (applied to synthetic signals and real music recordings) that allow students to experiment with the STFT and to gain an understanding on how to physically interpret discrete objects such as samples, frames, and spectral coefficients. In the **FMP Notebook *STFT: Influence of Window Function***, we explore the role of the window type and window size. Furthermore, in the **FMP Notebook *STFT: Padding***, we discuss various padding conventions that become crucial to correctly interpret and visualize feature representations. This important topic, which is a typical source of inaccuracies and errors in music processing pipelines, is illustrated by simple examples as a basis for further exploration (see also Figure 2.35).

One main limitation of the discrete STFT is the linear frequency grid whose resolution is determined by the signal's sampling rate and the STFT window size. In the **FMP Notebook *STFT: Frequency Grid Density***, we deepen the understanding on the connection between the different parameters involved. In particular, we discuss how to make the frequency grid denser by suitably padding the windowed sections in the STFT computation. Often, one loosely says that this procedure increases the frequency resolution. This, however, is not true in a qualitative sense, as is explained in the notebook. As an alternative, we discuss in the **FMP Notebook *STFT: Frequency Interpolation*** another common procedure to adjust the frequency resolution. On the way, we give a quick introduction to interpolation and show how the Python package `scipy` can be applied for this task. Beside refining the frequency grid, we then show how interpolation techniques can be used for a nonlinear deformation of the frequency grid, resulting in a log-frequency spectrogram. This topic goes beyond the scope of the current chapter, but plays an important role in Section 3.1.1.

We have seen in (2.118) that the matrix DFT_N is invertible, and that its inverse DFT_N^{-1} coincides with the DFT matrix up to some normalizing factor and complex conjugation. This algebraic property can be proven using the properties of the roots of unity. In the **FMP Notebook *STFT: Inverse***, we show that the two matrices are indeed inverse to each other—up to some numerical issues due to rounding in floating-point arithmetic. While inverting the DFT is straightforward, the inversion of the discrete STFT is less obvious, since one needs to compensate for effects introduced by the sliding window technique. While we cover this important topic in Section 8.1.2.1, we provide some basic Python implementation of the inverse STFT in this notebook, since it sheds another light on the sliding window concept and its effects. Furthermore, we discuss numerical issues as well as typical errors that may creep into one's source code when losing sight of windowing and padding conventions. At this point, we want to emphasize again that the STFT is one of the most important tools in music and audio processing. Common software packages for audio processing offer STFT implementations, which include convenient presets and provide information on how to physically interpret time, frequency, and magnitude parameters. From a teaching perspective, we find it crucial to exactly understand the role of the STFT parameters and the conventions made implicitly in black-box implementations. In the **FMP Notebook *STFT: Conventions and Implementations***, we summarize various variants for computing and interpreting a discrete STFT, while fixing the conventions used throughout the FMP notebooks (if not specified otherwise explicitly).

The FMP notebooks of **Part 2** close with some experiments related to the digitization of waveforms and its effects (see Section 2.2.2). In the **FMP Notebook *Digital Signals: Sampling***, we implement the concept of equidistant sampling and

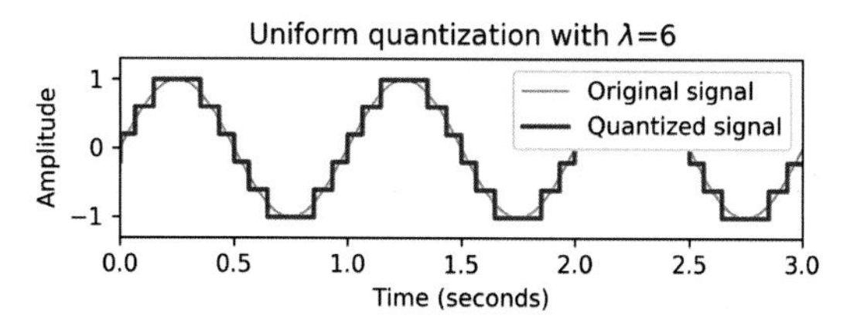

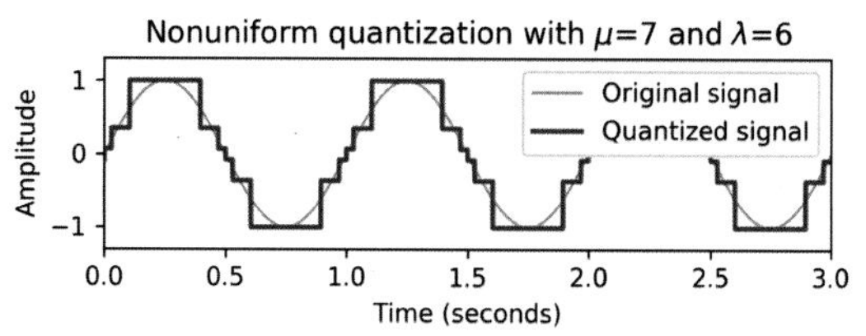

Fig. 2.36 Uniform and nonuniform quantization using $\lambda = 6$ quantization levels.

apply it to a synthetic example. We then reconstruct the signal from its samples (using the sinc interpolation of the sampling theorem as discussed in Exercise 2.28) and compare the result with the original signal. Based on the provided functions, one simple yet instructive experiment is to successively decrease the sampling rate and to look at the properties of the reconstructed signal (similar to Figure 2.14). Similarly, starting with a real music recording (e.g., we use a C-major scale played on a piano similar to the one shown in Figure 2.10), students may acoustically explore and understand aliasing effects. We continue with the **FMP Notebook *Digital Signals: Quantization***, where we have a closer look at the effects resulting from quantization (see Section 2.2.2.2). We provide a function for uniform quantization, which is then applied to a synthetic example and visually explored using different quantization parameters. Furthermore, using again the C-major scale recording, we reconstruct an analog signal from the quantized version, which allows for understanding the distortions introduced by quantization (also referred to as **quantization noise**). Going beyond the scope of this textbook, we finally introduce an approach for **nonuniform quantization**, where quantization levels are spaced in a logarithmic fashion. Besides theoretical explanations, we provide Python code that allows students to experiment, compare, and explore the various quantization strategies and their properties (see Figure 2.36). In the subsequent **FMP Notebook *Interference and Beating***, we pick up the topic of interference, which occurs when a wave is superimposed with another wave of similar frequency (see Section 2.3.3.1). We provide code that allows for reproducing the effects as shown in Figure 2.19. Furthermore, we present several experiments using sinusoidal as well as chirp functions to visually and acoustically study the related effect of **beating**.

References

1. P. L. BUTZER, W. SPLETTSTÖSSER, AND R. L. STENS, *The sampling theorem and linear prediction in signal analysis*, Jahresbericht der Deutschen Mathematiker-Vereinigung, 90 (1988), pp. 1–70.
2. M. CLAUSEN AND U. BAUM, *Fast Fourier Transforms*, BI Wissenschaftsverlag, 1993.
3. J. W. COOLEY AND J. W. TUKEY, *An algorithm for the machine calculation of complex Fourier series*, Mathematics of Computation, 19 (1965), pp. 297–301.
4. I. DAUBECHIES, *Ten Lectures on Wavelets*, Society for Industrial and Applied Mathematics (SIAM), 1992.
5. G. B. FOLLAND, *Real Analysis*, John Wiley & Sons, 1984.
6. D. GABOR, *Theory of communication*, Journal of the Institution of Electrical Engineers (IEE), 93 (1946), pp. 429–457.
7. F. J. HARRIS, *On the use of windows for harmonic analysis with the discrete Fourier transform*, Proceedings of the IEEE, 66 (1978), pp. 51–83.
8. J. R. HIGGINS, *Five short stories about the cardinal series*, Bulletin of the American Mathematical Society, 12 (1985), pp. 45–89.
9. B. B. HUBBARD, *The World According to Wavelets*, AK Peters, Wellesley, Massachusetts, 1996.
10. G. KAISER, *A Friendly Guide to Wavelets*, Modern Birkhäuser Classics, 2011.
11. M. MÜLLER AND F. ZALKOW, *FMP Notebooks: Educational material for teaching and learning fundamentals of music processing*, in Proceedings of the International Society for Music Information Retrieval Conference (ISMIR), Delft, The Netherlands, 2019, pp. 573–580.
12. A. V. OPPENHEIM, A. S. WILLSKY, AND H. NAWAB, *Signals and Systems*, Prentice Hall, 1996.
13. J. G. PROAKIS AND D. G. MANOLAKIS, *Digital Signal Processing*, Prentice Hall, 1996.
14. G. STRANG AND T. NGUYEN, *Wavelets and Filter Banks*, Wellesley-Cambridge Press, 2nd ed., 1996.
15. M. VETTERLI, J. KOVACEVIC, AND V. K. GOYAL, *Fourier and Wavelet Signal Processing*, Cambridge University Press, `http://fourierandwavelets.org/`, 1st ed., 2013.
16. ———, *Foundations of Signal Processing*, Cambridge University Press, `http://fourierandwavelets.org/`, 3rd ed., 2014.

Exercises

Exercise 2.1. Let $\langle f|g\rangle := \int_{t\in\mathbb{R}} f(t)\cdot g(t)dt$ be the similarity measure for two functions $f:\mathbb{R}\to\mathbb{R}$ and $g:\mathbb{R}\to\mathbb{R}$ as defined in (2.3). Consider the following six functions $f_n:\mathbb{R}\to\mathbb{R}$ for $n\in[1:6]$, which are defined to be zero outside the shown interval:

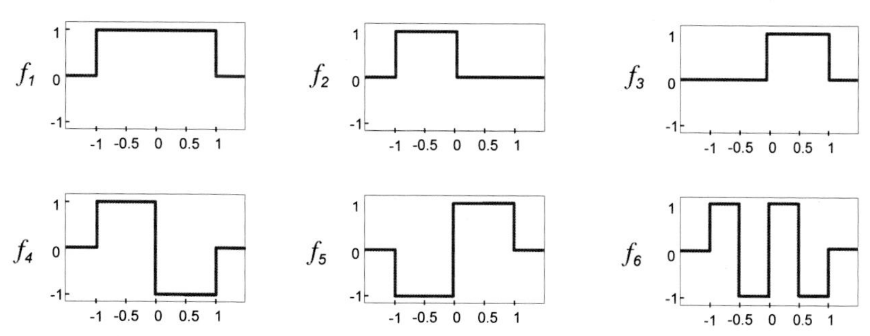

Determine the similarity values $\langle f_n|f_m\rangle$ for all pairs $(n,m)\in[1:6]\times[1:6]$.

Exercise 2.2. Sketch the magnitude Fourier transform of the following signals assuming that the signals are zero outside the shown intervals (see Figure 2.6 for similar examples):

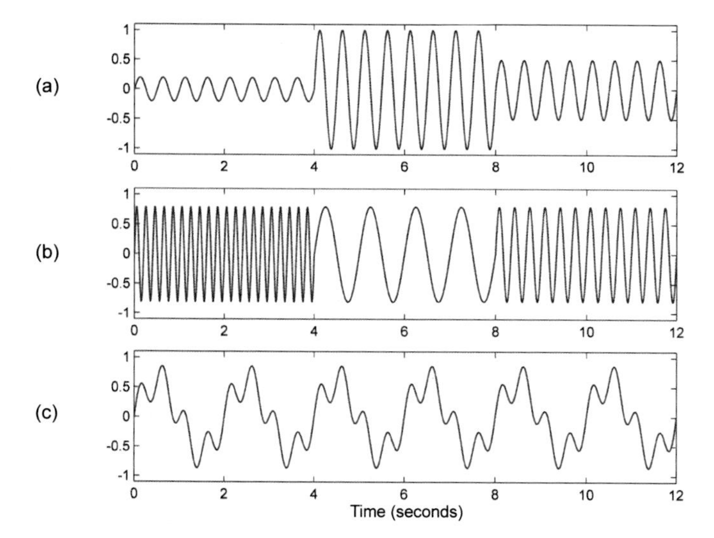

Exercise 2.3. Based on (2.27) and (2.28), compute the time resolution (in ms) and frequency resolution (in Hz) of a discrete STFT based on the following parameter settings:

(a) $F_s = 22050, N = 1024, H = 512$
(b) $F_s = 48000, N = 1024, H = 256$
(c) $F_s = 4000, N = 4096, H = 1024$

What are the respective Nyquist frequencies?

Exercise 2.4. Let $F_s = 44100$, $N = 2048$, and $H = 1024$ be the parameter settings of a discrete STFT $\mathcal{X}$ as defined in (2.26). What is the physical meaning of the Fourier coefficients $\mathcal{X}(1000,1000)$, $\mathcal{X}(17,0)$, and $\mathcal{X}(56,1024)$, respectively? Why is the coefficient $\mathcal{X}(56,1024)$ problematic?

Exercise 2.5. Sketch the spectrogram (as in Figure 2.9) for each of the three signals shown in Exercise 2.2. Assume a window length that corresponds to a physical duration of about one second.

Exercise 2.6. The naive approach for computing a DFT requires about N^2 operations, while the FFT requires about $N\log_2 N$ operations. Compute the factor for the savings when using the FFT for various N. In particular, consider $N = 2^n$ for $n = 5, 10, 15, 20, 25, 30$.

Exercise 2.7. Let f_1 and f_2 be two periodic analog signals with integer periods $\lambda_1 \in \mathbb{N}$ and $\lambda_2 \in \mathbb{N}$, respectively. Show that $g = f_1 + f_2$ is periodic with periods that are integer multiples of λ_1 as well as λ_2. In general, g may have additional periods not necessarily being integer multiples of λ_1 and λ_2. As an example, specify two signals f_1 and f_2 with prime period $\lambda_1 = \lambda_2 = 2$ such that $g = f_1 + f_2$ is periodic with prime period $\lambda = 1$.

Exercise 2.8. In this exercise, we show that there are periodic functions that do not have a prime period (i.e., that do not have a least positive constant being a period). The easiest example of such a function is a constant function. Show that the function $f : \mathbb{R} \to \mathbb{R}$ defined by

$$f(t) := \begin{cases} 1, & \text{for } t \in \mathbb{Q}, \\ 0, & \text{for } t \in \mathbb{R} \setminus \mathbb{Q} \end{cases}$$

is also periodic without having a prime period.
[**Hint:** In this exercise, we assume that the reader is familiar with the properties of rational numbers ($\mathbb{Q}$) and irrational numbers ($\mathbb{R} \setminus \mathbb{Q}$).]

Exercise 2.9. Sketch the graph of the quantization function $Q : \mathbb{R} \to \mathbb{R}$ defined by

$$Q(a) := \operatorname{sgn}(a) \cdot \Delta \cdot \left\lfloor \frac{|a|}{\Delta} + \frac{1}{2} \right\rfloor$$

for $a \in \mathbb{R}$ and some fixed quantization step size $\Delta > 0$ (see (2.33)). Furthermore, sketch the graph of the absolute quantization error.

Exercise 2.10. In mathematics, the term "operator" is used to denote a mapping from one vector space to another. Let V and W be two vector spaces over $\mathbb{R}$. An operator $M : V \to W$ is called **linear** if $M[a_1v_1 + a_2v_2] = a_1M[v_1] + a_2M[v_2]$ for any $v_1, v_2 \in V$ and $a_1, a_2 \in \mathbb{R}$. Show that $V := \{f \mid f\colon \mathbb{R} \to \mathbb{R}\}$ and $W := \{x \mid x\colon \mathbb{Z} \to \mathbb{R}\}$ are vector spaces. Fixing a sampling period $T > 0$, consider the operator M that maps a CT-signal $f \in V$ to the DT-signal $M[f] := x \in W$ obtained by T-sampling as defined in (2.32). Show that this defines a linear operator.

Exercise 2.11. Show that the quantization operator $Q : \mathbb{R} \to \mathbb{R}$ as defined in Exercise 2.9 and (2.33) is *not* a linear operator.

Exercise 2.12. In this exercise we discuss various computation rules for complex numbers and their conjugates. The complex multiplication is defined by $c_1 \cdot c_2 = a_1a_2 - b_1b_2 + i(a_1b_2 + a_2b_1)$ for two complex numbers $c_1 = a_1 + ib_1, c_2 = a_2 + ib_2 \in \mathbb{C}$ (see (2.34)). Furthermore, complex conjugation is defined by $\overline{c} = a - ib$ for a complex number $c = a + ib \in \mathbb{C}$ (see (2.35)). Finally, the absolute value of a complex number c is defined by $|c| = \sqrt{a^2 + b^2}$. Prove the following identities:

(a) $\mathrm{Re}(c) = (c + \overline{c})/2$
(b) $\mathrm{Im}(c) = (c - \overline{c})/(2i)$
(c) $\overline{c_1 + c_2} = \overline{c_1} + \overline{c_2}$
(d) $\overline{c_1 \cdot c_2} = \overline{c_1} \cdot \overline{c_2}$
(e) $c\overline{c} = a^2 + b^2 = |c|^2$
(f) $1/c = \overline{c}/(c\overline{c}) = \overline{c}/(a^2 + b^2) = \overline{c}/(|c|^2)$

Exercise 2.13. We have seen in Section 2.2.3.2 that the set $\mathbb{C}^{\mathbb{Z}} = \{x | x\colon \mathbb{Z} \to \mathbb{C}\}$ of complex-valued DT-signals defines a vector space. Show that the subset $\ell^2(\mathbb{Z}) \subset \mathbb{C}^{\mathbb{Z}}$ of DT-signals of finite energy is a linear subspace. To this end, you need to show that $x + y \in \ell^2(\mathbb{Z})$ and $ax \in \ell^2(\mathbb{Z})$ for any $x, y \in \ell^2(\mathbb{Z})$ and $a \in \mathbb{C}$.

Exercise 2.14. In Section 2.3.1, we defined the set $\{\mathbf{1}, \mathbf{sin}_k, \mathbf{cos}_k \mid k \in \mathbb{N}\} \subset L^2_{\mathbb{R}}([0,1))$. Prove that this set is an orthonormal set in $L^2_{\mathbb{R}}([0,1))$, i.e., that it satisfies (2.50) and (2.51).
[**Hint:** Use the following trigonometric identities:

(a) $\cos(\alpha)^2 + \sin(\alpha)^2 = 1$
(b) $\cos(\alpha)\cos(\beta) = (\cos(\alpha+\beta) + \cos(\alpha-\beta))/2$
(c) $\sin(\alpha)\sin(\beta) = (\cos(\alpha-\beta) - \cos(\alpha+\beta))/2$
(d) $\sin(\alpha)\cos(\beta) = (\sin(\alpha+\beta) + \sin(\alpha-\beta))/2$

To show (2.51), use (a) and the fact that $\mathbf{cos}_k^2$ and $\mathbf{sin}_k^2$ have the same area over a full period. The proof of (2.50) is a bit cumbersome, but not difficult when using (b), (c), and (d).]

Exercise 2.15. Let $\exp(i\gamma) := \cos(\gamma) + i\sin(\gamma)$, $\gamma \in \mathbb{R}$, be the complex exponential function as defined in (2.67). Prove the following properties (see (2.68) to (2.71)):

(a) $\exp(i\gamma) = \exp(i(\gamma + 2\pi))$
(b) $|\exp(i\gamma)| = 1$
(c) $\overline{\exp(i\gamma)} = \exp(-i\gamma)$
(d) $\exp(i(\gamma_1 + \gamma_2)) = \exp(i\gamma_1)\exp(i\gamma_2)$
(e) $\dfrac{d\exp(i\gamma)}{d\gamma} = i\exp(i\gamma)$

[**Hint:** To prove (d), you need the trigonometric identities $\cos(\alpha+\beta) = \cos(\alpha)\cos(\beta) - \sin(\alpha)\sin(\beta)$ and $\sin(\alpha+\beta) = \cos(\alpha)\sin(\beta) + \sin(\alpha)\cos(\beta)$. In (e), note that the real (imaginary) part of a derivative of a complex-valued function is obtained by computing the derivative of the real (imaginary) part of the function.]

Exercise 2.16. In (2.77), we defined for each $k \in \mathbb{Z}$ the complex-valued exponential function $\mathbf{exp}_k : [0,1) \to \mathbb{C}$ by $\mathbf{exp}_k(t) := \cos(2\pi kt) + i\sin(2\pi kt)$, $t \in \mathbb{R}$. As in Exercise 2.14, show that the set $\{\mathbf{exp}_k \mid k \in \mathbb{Z}\} \subset L^2([0,1))$ is an orthonormal set, i.e., $\|\mathbf{exp}_k\|^2 = 1$ for $k \in \mathbb{Z}$ (see (2.51)) and $\langle \mathbf{exp}_k | \mathbf{exp}_\ell \rangle = 0$ for $k \neq \ell$, $k, \ell \in \mathbb{Z}$ (see (2.50)).
[**Hint:** Use the properties of the exponential function introduced in Exercise 2.15. Furthermore, note that the real (imaginary) part of an integral of a complex-valued function is obtained by integrating the real (imaginary) part of the function.]

Exercise 2.17. Let atan2 be the function as defined in (2.76). For a complex number $c = a + ib \in \mathbb{C}$, we set $\mathrm{atan2}(c) := \mathrm{atan2}(b, a)$. Show that $\mathrm{atan2}(\lambda \cdot c) = \mathrm{atan2}(c)$ for any positive constant $\lambda \in \mathbb{R}_{>0}$. Furthermore, show that $\mathrm{atan2}(\overline{c}) = -\mathrm{atan2}(c)$.
[**Hint:** Use the fact that the arctan function is an odd function, i.e., $\arctan(-v) = -\arctan(v)$ for $v \in \mathbb{R}$.]

Exercise 2.18. In this exercise, we consider the geometric series for compex numbers, which is needed in (2.112). Prove that $\sum_{n=0}^{N-1} a^n = (1 - a^N)/(1 - a)$ for any complex number $a \neq 1$.
[**Hint:** For the proof, use mathematical induction on N.]

Exercise 2.19. We have seen that two sinusoids of similar frequency may add up (constructive interference) or cancel out (destructive interference); see Figure 2.19. Let $f_1(t) = \sin(2\pi\omega_1 t)$ and $f_2(t) = \sin(2\pi\omega_2 t)$ be two such sinusoids with distinct but nearby frequencies $\omega_1 \approx \omega_2$. In the following figure, for example, $\omega_1 = 1$ and $\omega_2 = 1.1$ is used.

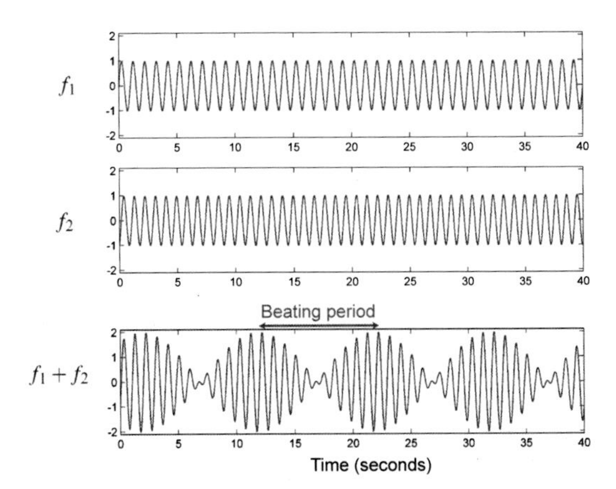

The figure also shows that the superposition $f_1 + f_2$ of these two sinusoids results in a function that looks like a single sine wave with a slowly varying amplitude, a phenomenon also known as *beating*. Determine the rate (reciprocal of the period) of the beating in dependency on ω_1 and ω_2. Compare this result with the plot of $f_1 + f_2$ in the figure.
[**Hint:** Use the trigonometric identity $\sin(\alpha)+\sin(\beta) = 2\cos\left(\frac{\alpha-\beta}{2}\right)\sin\left(\frac{\alpha+\beta}{2}\right)$ for $\alpha, \beta \in \mathbb{R}$.]

Exercise 2.20. Let $f \in L^2(\mathbb{R})$ be a signal of unit energy $\|f\|^2 = 1$. Show that the scaled signal g defined by $g(t) := s^{1/2} f(s \cdot t)$ also has unit energy for a positive real scaling factor $s > 0$. Furthermore show that $\hat{g}(\omega) = s^{-1/2}\hat{f}(\omega/s)$ for $\omega \in \mathbb{R}$. Discuss this result. Describe how one can obtain a Dirac sequence by changing the parameter s (see Section 2.3.3.2).

Exercise 2.21. Show that the Fourier transform of the rectangular function in (2.95) is the sinc function in (2.96). Also prove that the sinc function is continuous at $t = 0$.
[**Hint:** Use the fact that the derivative of $t \mapsto \exp(-2\pi i\omega t)$ is given by $t \mapsto -2\pi i\omega\exp(-2\pi i\omega t)$; see Exercise 2.15. From this, one can derive the indefinite integral of the exponential function. To prove the continuity at $t = 0$, look at the first terms of the Taylor series of the sine function.]

Exercise 2.22. For a signal $f \in L^2(\mathbb{R})$, consider the translation f_{t_0} defined by $f_{t_0}(t) := f(t - t_0)$ for $t \in \mathbb{R}$ (see (2.97)) and the modulation f^{ω_0} defined by $f^{\omega_0}(t) := \exp(2\pi i\omega_0 t)f(t)$ for $t \in \mathbb{R}$ (see (2.98)). Show that $\|f\| = \|f_{t_0}\| = \|f^{\omega_0}\|$. Furthermore, prove the properties (2.99) and (2.100):

$$\widehat{f_{t_0}}(\omega) = \exp(-2\pi i\omega t_0)\hat{f}(\omega) \qquad \text{and} \qquad \widehat{f^{\omega_0}}(\omega) = \hat{f}(\omega + \omega_0)$$

for $\omega \in \mathbb{R}$.

Exercise 2.23. Any complex number $c \in \mathbb{C}$ with $c^N = 1$ for a given $N \in \mathbb{N}$ is called an Nth **root of unity**. If in addition $c^k \neq 1$ for $1 < k < N$, the root c is called **primitive**. Show that $\sigma_N := \exp(-2\pi i/N)$ defines a primitive N^{th} root of unity. Furthermore, describe *all* N^{th} roots of unity. Which of these roots are primitive? Determine for $N \in \{4, 7, 12\}$ all primitive N^{th} roots of unity.
[**Hint:** In this exercise, one needs to know that a (nonzero) polynomial of degree N has at most N different roots, where a **root** of a function is an input value that produces an output of zero.]

Exercise 2.24. Let $\mathbf{x} = (x(0), \ldots, x(N-1))^\top$ be a real-valued vector consisting of samples $x(n) \in \mathbb{R}$ for $n \in [0 : N-1]$. Show that

$$\mathbf{X} = \mathrm{DFT}_N \cdot \mathbf{x}$$

with $\mathbf{X} = (X(0), \ldots, X(N-1))^\top$ fulfills the symmetry property $X(k) = \overline{X(N-k)}$ for all $k \in [1 : N-1]$ and $X(0) \in \mathbb{R}$. This shows that the upper half of the frequency coefficients are redundant if $\mathbf{x}$ is real-valued. Furthermore, show the converse. Given a spectral vector $\mathbf{X}$ with $X(0) \in \mathbb{R}$ and $X(k) = \overline{X(N-k)}$ for all $k \in [1 : N-1]$, then

$$\mathbf{x} = \mathrm{DFT}_N^{-1} \cdot \mathbf{X}$$

is a real-valued vector (see (2.118)).
[**Hint:** Use the computation rules for complex numbers from Exercise 2.12.]

Exercise 2.25. Specify the DFT_N matrix explicitly for $N \in \{1,2,4\}$. Count the number of multiplications and additions when performing the usual matrix–vector product $\mathrm{DFT}_4 \cdot \mathbf{x}$ for a vector $\mathbf{x} = (x_1, x_2, x_3, x_4)^\top$. Then conduct all steps of the FFT algorithm (two recursions are needed) and again count the overall number of multiplications and additions needed to compute $\mathrm{DFT}_4 \cdot \mathbf{x}$.

Exercise 2.26. Let $N = 2^n$ be a power of two. In (2.127), we derived the estimate $\mu(N) \leq 2\mu(N/2) + 1.5N$ for the number of multiplications and additions needed to compute the matrix–vector product $\mathrm{DFT}_N \cdot \mathbf{x}$. Using $\mu(1) = 0$ (the case $n = 0$), show by a mathematical induction on n that this implies $\mu(N) \leq 1.5N \log_2(N)$.

Exercise 2.27. In the spectrograms shown in Figure 2.32 one can notice vertical stripes at $t = 0$ and $t = 1$. Why?

Exercise 2.28. In this exercise, we prove the **sampling theorem**. A CT-signal $f \in L^2(\mathbb{R})$ is called **Ω-bandlimited** if the Fourier transform $\hat{f}$ vanishes for $|\omega| > \Omega$, i.e., $\hat{f}(\omega) = 0$ for $|\omega| > \Omega$. Let $f \in L^2(\mathbb{R})$ be an Ω-bandlimited function and let x be the T-sampled version of f with $T := 1/(2\Omega)$, i.e., $x(n) = f(nT)$, $n \in \mathbb{Z}$. Then f can be reconstructed from x by

$$f(t) = \sum_{n \in \mathbb{Z}} x(n) \mathrm{sinc}\left(\frac{t - nT}{T}\right) = \sum_{n \in \mathbb{Z}} f\left(\frac{n}{2\Omega}\right) \mathrm{sinc}\left(2\Omega t - n\right),$$

where the sinc function is defined in (2.96). In other words, the CT-signal f can be perfectly reconstructed from the DT-signal obtained by equidistant sampling if the bandlimit is no greater than half the sampling rate.
[**Hint:** Note that one may assume $\Omega = 1/2$ (and $T = 1$) by considering the scaled function $t \mapsto f(t/\Omega)$. In this case, f is $1/2$-bandlimited and can be extended to a 1-periodic function g. Represent g by its Fourier series (2.79) and compute the Fourier coefficients $c_n = \langle g | \mathbf{exp}_n \rangle$, $n \in \mathbb{Z}$. Compare these coefficients with the Fourier representation (2.91) of f evaluated at $t = n$ for $n \in \mathbb{Z}$ (again using the fact that f is $1/2$-bandlimited). As a result, one obtains $c_n = f(-n)$. Finally, reconstruct f from the Fourier series of g. To this end, you need the result of Exercise 2.21.]

Chapter 3
Music Synchronization

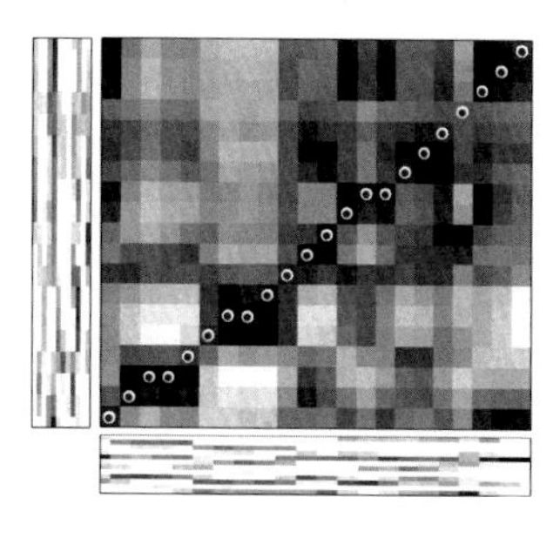

Music can be described and represented in many different ways including sheet music, symbolic representations, and audio recordings. For each of these representations, there may exist different versions that correspond to the same musical work. For example, for Beethoven's Fifth Symphony one can find a large number of music recordings performed by different orchestras and conductors. The general goal of **music synchronization** is to automatically link the various data streams, thus interrelating the multiple information sets related to a given musical work. More precisely, **synchronization** is taken to mean a procedure which, for a given position in one representation of a piece of music, determines the corresponding position within another representation.

As a motivating example, let us consider Figure 3.1, which shows a sheet music, an audio, and a piano-roll representation of the beginning of a piano piece. Recall that the sheet music representation gives an explicit specification of notes in terms of pitch and duration. However, sheet music typically leaves room for interpreting aspects such as tempo or dynamics, thus resulting in different performances of the same piece of music. These performances may be recorded and stored in the form of audio or MIDI files. In case of audio or symbolic representations there exists a time line, which may be specified in physical units such as seconds or in musical units such as measures or beats. Synchronization of such time-dependent data streams is the process of assigning each time position in one version to a musically corresponding time position in the other version. In Figure 3.1, each such assignment is visualized by a red bidirectional arrow. For other representations such as digitized images of sheet music, one has spatial positions rather than time positions, where note events may be located by bounding boxes given in pixel coordinates. Then the synchronization between, e.g., a sheet music and an audio representation is an

M. Müller, *Fundamentals of Music Processing*, https://doi.org/10.1007/978-3-030-69808-9_3

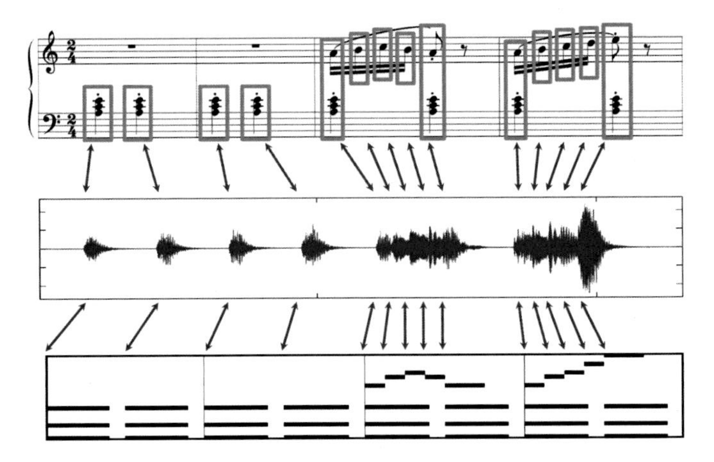

Fig. 3.1 First four measures of Op. 100, No. 2 by Friedrich Burgmüller in a sheet music, audio, and piano-roll representation. The red bidirectional arrows indicate the aligned time positions of corresponding note events in the different representations.

assignment of spatial positions (in 2D pixel coordinates) in the image file to time positions in the audio file.

Such synchronization results form the basis for novel interfaces that allow users to access, search, and browse musical content in a convenient way. For example, Figure 3.22 shows an interface for a time-synchronous display of the score position in sheet music during the playback of an audio recording of the same piece of music. A simple click on a measure within the sheet music representation allows a user to directly jump to the corresponding position in a music recording. The Interpretation Switcher shown in Figure 3.21 makes use of synchronization techniques for affording efficient and convenient access to several recordings of the same piece of music. This interface allows a user to listen to a specific recording and then, at any time during playback, seamlessly switch to any of the other versions. In Section 3.3, we will discuss a number of applications and interfaces that are built on music synchronization techniques.

In this chapter we study fundamental techniques for analyzing, comparing, and synchronizing different music representations that belong to the same piece of music. Given two different music representations, typical synchronization approaches proceed in two steps (see Figure 3.2). In the first step, the two representations are transformed into sequences of suitable features. In general, such feature representations need to find a compromise between two conflicting goals. On the one hand, features should show a large degree of robustness to variations that are to be left unconsidered for the task at hand. On the other hand, features should capture enough characteristic information to accomplish the given task. In Section 3.1, we introduce **chroma-based features**, which have turned out to be a powerful tool for analyzing music whose pitches can be meaningfully categorized (often into 12 categories) and whose tuning approximates to the equal-tempered scale. As we will see, such fea-

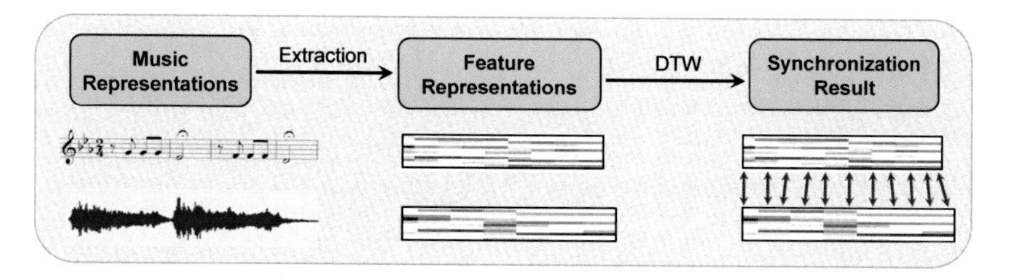

Fig. 3.2 Overview of the processing pipeline of a typical music synchronization procedure.

tures capture harmonic and melodic characteristics of music, while being robust to changes in timbre and instrumentation. In the second step, the derived feature sequences have to be brought into temporal correspondence. In Section 3.2, we discuss a technique known as **dynamic time warping** (DTW), which computes an optimal alignment between two given feature sequences. This alignment constitutes our synchronization result. Intuitively, the alignment can be thought of as a linking structure as indicated by the red bidirectional arrows shown in Figure 3.1. In Section 3.3, we indicate how music synchronization techniques can be applied for supporting users in experiencing and exploring music.

We want to emphasize that the task of music synchronization mainly serves as motivation for the techniques treated in this chapter. Both chroma-based music features as well as DTW-like alignment techniques play an important role also in other music processing tasks as demonstrated in subsequent chapters.

3.1 Audio Features

To make music data comparable and algorithmically accessible, the first step in basically all music processing tasks is to extract suitable **features** that capture relevant key aspects while suppressing irrelevant details. Obviously, the notion of **similarity** is of crucial importance in the design of music features. In some tasks such as music synchronization, one may be interested in characterizing an audio recording irrespective of certain details concerning the interpretation or instrumentation. Conversely, other applications may be concerned with measuring just the subtleties that relate to a musician's individual articulation or emotional expressiveness.

In this section, we introduce audio features and mid-level representations that are particularly useful in the music synchronization context and related tasks (see Figure 3.3 for a first impression). Assuming that we are dealing with music whose pitches can be meaningfully categorized according to the equal-tempered scale, we show how an audio recording can be transformed into a feature representation that reveals the distribution of the signal's energy across the different pitches (Section 3.1.1). Technically, these features are obtained from a spectrogram by converting the linear frequency axis (measured in Hertz) into a logarithmic axis (measured in pitches). From this log-frequency spectrogram, we then derive a time–chroma representation by suitably combining pitch bands that correspond to the same pitch class or chroma (Section 3.1.2). The resulting chroma features show a high degree of robustness to variations in timbre and instrumentation. Finally, we discuss various pre- and postprocessing steps modifying spectral, temporal, and dynamical aspects. This leads to a number of chroma feature variants, which may show a quite different behavior in the context of a specific music analysis scenario.

3.1.1 Log-Frequency Spectrogram

The Fourier transform and, in particular, the discrete STFT serve as a **front-end transform**, the first computing step, for deriving a large number of musically relevant audio features. We quickly recall the definition of the discrete STFT from Section 2.5.3 to fix the notation. Let x be a real-valued discrete signal obtained by equidistant sampling with respect to a fixed sampling rate F_s given in Hertz. Furthermore, let w be a real-valued discrete-time window such that $w(\ell)$ for $\ell \in [0:N-1]$ are the nonzero samples of w for some $N \in \mathbb{N}$, and let $H \in \mathbb{N}$ be the hop size parameter. Then, by (2.26), the STFT $\mathcal{X}$ of x is given by

$$\mathcal{X}(n,k) := \sum_{\ell=0}^{N-1} x(\ell+nH)w(\ell)\exp(-2\pi ik\ell/N) \tag{3.1}$$

with $n \in \mathbb{Z}$ and $k \in [0:K]$, where $K = N/2$ is the frequency index corresponding to the Nyquist frequency. The complex number $\mathcal{X}(n,k)$ denotes the kth Fourier coefficient for the nth time frame. Recall from Section 2.5.3 that each Fourier coefficient $\mathcal{X}(n,k)$ is associated with the physical time position $T_{\mathrm{coef}}(n) = nH/F_\mathrm{s}$ given in seconds (see (2.27)) and with the physical frequency $F_{\mathrm{coef}}(k) = kF_\mathrm{s}/N$ given in Hertz (see (2.28)).

In Section 1.3.2 we discussed the equal-tempered scale, where each octave is split up into twelve logarithmically spaced units. Recall that, in MIDI notation, the pitches of the equal-tempered scale are serially numbered, where the pitch A4 corresponds to the MIDI pitch $p = 69$. In the following, we do not distinguish between the different notations and often simply speak of a note while meaning a pitch. We have seen that the notes of the equal-tempered scale depend on their center frequencies in a logarithmic fashion. By (1.1), the center frequency $F_{\mathrm{pitch}}(p)$ of a pitch $p \in [0:127]$ is

$$F_{\mathrm{pitch}}(p) = 2^{(p-69)/12} \cdot 440 \tag{3.2}$$

(see also Table 3.1). As an illustration, we consider a chromatic scale played on a piano starting with the note A0 ($p = 21$) and ending with C8 ($p = 108$). The resulting spectrogram, as shown in Figure 3.3b, reveals the exponential dependency of the fundamental frequency on the pitches of the played notes. Also, as already discussed in the example of Figure 1.23a, the harmonics and the notes' onset positions

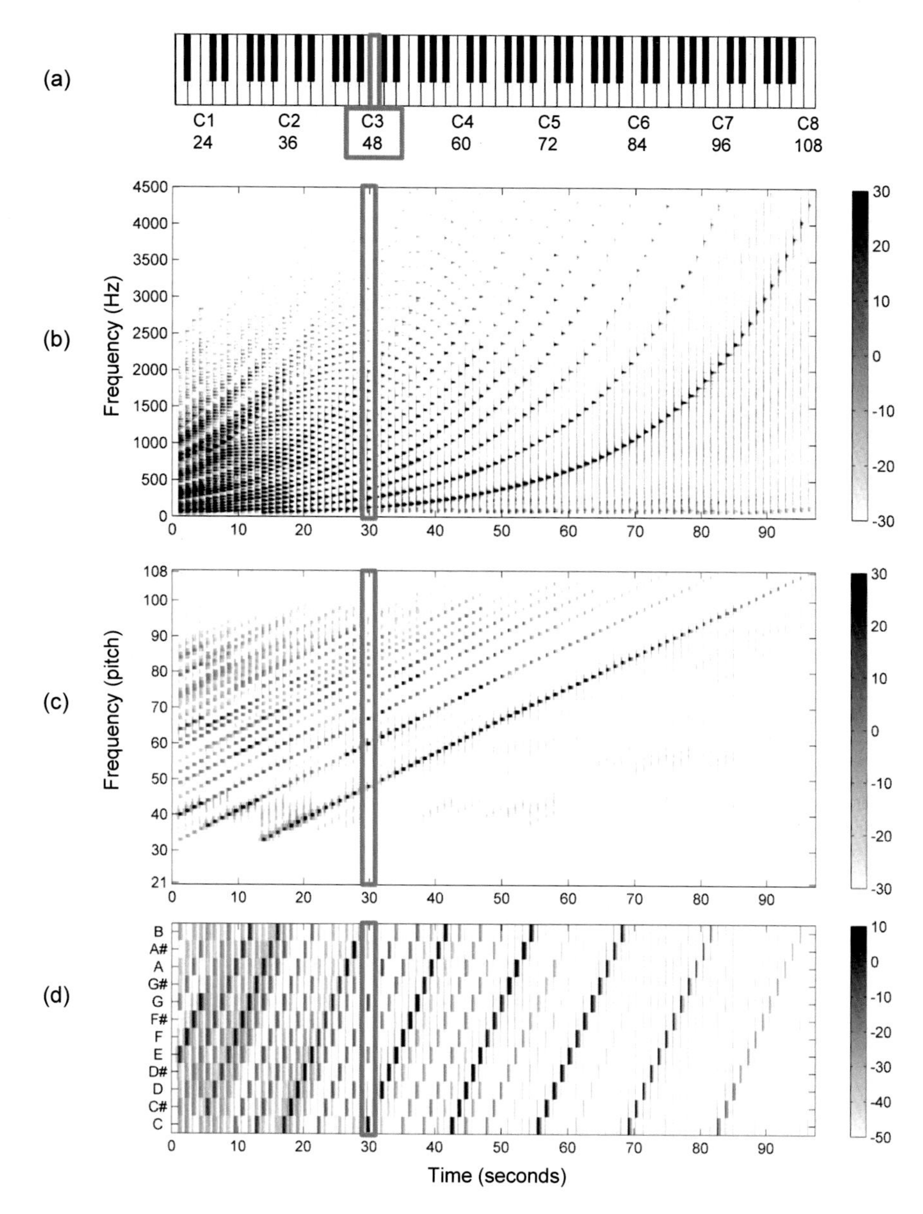

Fig. 3.3 Various representations for a recording of the chromatic scale played on a real piano. The scale ranges from A0 ($p = 21$) to C8 ($p = 108$). **(a)** Piano keys representing the chromatic scale. **(b)** Magnitude spectrogram. **(c)** Pitch-based log-frequency spectrogram. **(d)** Chromagram. For visualization purposes the values are encoded by shades of gray using a logarithmic scale. The C3 ($p = 48$) played at time $t = 30$ sec is highlighted by rectangular frames.

Note	p	$F_{\mathrm{pitch}}(p)$	$F_{\mathrm{pitch}}(p-0.5)$	$F_{\mathrm{pitch}}(p+0.5)$	$\mathrm{BW}(p)$
C4	60	261.63	254.18	269.29	15.11
C♯4	61	277.18	269.29	285.30	16.01
D4	62	293.66	285.30	302.27	16.97
D♯4	63	311.13	302.27	320.24	17.97
E4	64	329.63	320.24	339.29	19.04
F4	65	349.23	339.29	359.46	20.18
F♯4	66	369.99	359.46	380.84	21.37
G4	67	392.00	380.84	403.48	22.65
G♯4	68	415.30	403.48	427.47	23.99
A4	69	440.00	427.47	452.89	25.41
A♯4	70	466.16	452.89	479.82	26.93
B4	71	493.88	479.82	508.36	28.53
C5	72	523.25	508.36	538.58	30.23

Table 3.1 Various notes and their MIDI note number p, center frequency $F_{\mathrm{pitch}}(p)$, cutoff frequencies $F_{\mathrm{pitch}}(p-0.5)$ and $F_{\mathrm{pitch}}(p+0.5)$, and bandwidth $\mathrm{BW}(p)$.

(vertical structures) are clearly visible. In the following, the chromatic scale will serve as one of our running examples.

The logarithmic perception of frequency motivates the use of a time–frequency representation with a logarithmic frequency axis labeled by the pitches of the equal-tempered scale. To derive such a representation from a given spectrogram representation, the basic idea is to assign each spectral coefficient $\mathcal{X}(n,k)$ to the pitch with center frequency that is closest to the frequency $F_{\mathrm{coef}}(k)$. More precisely, we define for each pitch $p \in [0:127]$ the set

$$P(p) := \{k : F_{\mathrm{pitch}}(p-0.5) \leq F_{\mathrm{coef}}(k) < F_{\mathrm{pitch}}(p+0.5)\}. \tag{3.3}$$

From this, we obtain a log-frequency spectrogram $\mathcal{Y}_{\mathrm{LF}} : \mathbb{Z} \times [0:127]$ defined by

$$\mathcal{Y}_{\mathrm{LF}}(n,p) := \sum_{k \in P(p)} |\mathcal{X}(n,k)|^2. \tag{3.4}$$

Let us have a look at this pooling procedure by means of the small example shown in Figure 3.4. As before, we assume a sampling rate of $F_s = 44100$ Hz and a window length of $N = 4096$. By (2.28), we obtain the frequencies associated to the Fourier coefficients of the STFT. For example, one obtains $F_{\mathrm{coef}}(40) \approx 430.7$ Hz, $F_{\mathrm{coef}}(41) \approx 441.4$ Hz, and so on. To compute a coefficient $\mathcal{Y}_{\mathrm{LF}}(n,p)$ for some pitch p, we consider the **cutoff frequencies** $F_{\mathrm{pitch}}(p-0.5)$ and $F_{\mathrm{pitch}}(p+0.5)$ (see Table 3.1). The set $P(p)$ consists of all Fourier coefficients whose frequencies lie within these bounds. For example, we obtain $P(p) = \{40,41,42\}$ for $p = 69$ (see Figure 3.4). The pitch coefficient $\mathcal{Y}_{\mathrm{LF}}(n,69)$ for some time frame n is obtained by summing up the squared magnitudes of the three coefficients $\mathcal{X}(n,40)$, $\mathcal{X}(n,41)$, and $\mathcal{X}(n,42)$. Similarly, one obtains $\mathcal{Y}_{\mathrm{LF}}(n,68)$ by considering the coefficients for $k = 38$ and $k = 39$, and so on.

By this procedure, the frequency axis is partitioned logarithmically and labeled linearly according to MIDI pitches. Figure 3.3c shows the resulting log-frequency

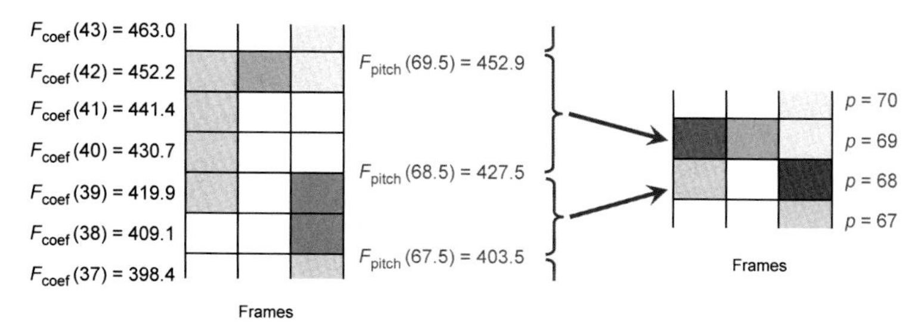

Fig. 3.4 Illustration of the pooling procedure for deriving a log-frequency spectrogram. In this example, a sampling rate of $F_s = 44100$ Hz and a window length of $N = 4096$ are assumed.

spectrogram, where the played notes of the chromatic scale now appear in a linearly increasing fashion. This figure again illustrates that playing a single note on a real instrument may already result in a complex mixture of different frequencies. As a general trend, the sounds for higher notes possess a cleaner harmonic spectrum than the ones for lower notes. For lower notes, the signal's energy is often contained in the higher harmonics, while the listener may still have the perception of a low-pitched sound. Furthermore, the frequency content of a sound depends on the microphone's frequency response. For example, the microphone may capture only frequencies above a certain threshold as in the case of the audio example used for Figure 3.3. This also explains why there is virtually no energy visible in the fundamental frequencies for the notes A0 ($p = 21$) to B0 ($p = 32$).

Besides acoustic properties, there is another reason for the rather poor representation of low pitches when using the pooling strategy based on a discrete STFT as described in (3.4). First note that the discrete STFT introduces a **linear** sampling of the frequency axis, as shown by (2.27). In contrast, the frequency range in (3.3) considered for each pitch p depends on the frequency in a **logarithmic** fashion. More precisely, we define the **bandwidth** $\mathrm{BW}(p)$ of pitch p by

$$\mathrm{BW}(p) := F_{\mathrm{pitch}}(p+0.5) - F_{\mathrm{pitch}}(p-0.5), \tag{3.5}$$

which becomes smaller for decreasing pitches (see Table 3.1). For example, for MIDI pitch $p = 66$ one has a bandwidth of roughly 21.4 Hz, whereas for $p = 54$ the bandwidth falls below 10.7 Hz. Now, using for example a sampling rate of $F_s = 44100$ Hz and a window length of $N = 4096$, the resulting STFT has a frequency resolution of 10.8 Hz by (2.27). In this case, the frequency resolution of the spectrogram does not suffice to separate the center frequencies of adjacent MIDI pitches below $p = 54$. As a result, the set $P(p)$ as defined in (3.3) may contain only very few spectral coefficients or may even be empty (see Exercise 3.2). This is one main reason for having a poor log-frequency representation in the lower pitch range.

As we discussed in Section 2.5.3 (see Figure 2.32 and Figure 2.33), the frequency resolution can be increased by enlarging the window length N, which, however, results in a decreased temporal resolution. As a consequence one may lose important short-time information such as note onsets. One alternative to using a single spectrogram is to use several spectrograms based on different sampling rates and window sizes. One crucial observation is that sudden sound events such as note onsets become prominent in particular in the high-frequency range of the spectrum, which necessitates a good time resolution in this range. Therefore, for high pitches, one often uses a large sampling rate and a short window (short with respect to the physical duration given in seconds) to obtain a good temporal resolution. For high pitches, the resulting poor frequency resolution is tolerable because of the exponential spread of the fundamental frequencies. Vice versa, for the analysis of low pitches, frequency resolution becomes more important. Therefore, in this case, one often uses a smaller sampling rate and a (physically) longer window at the cost of losing temporal resolution. In practice, one may process the highest octave to be considered with an STFT-based method based on a large sampling rate. For the next lowest octave, the sampling rate can be reduced by a factor of two and then processed by the same STFT-based method (which leads to an increase in physical duration of the window by a factor of two). This multiresolution approach may be repeated for as many octaves as desired.

As an illustration, we continue with the Burgmüller example from Figure 3.1. The log-frequency spectrogram shown in Figure 3.5 has been computed using a multiresolution approach as outlined above based on a window size that corresponds to roughly 10 ms. The time axis has been converted from frame indices to seconds, and only the frequency information corresponding to the pitches $p = 55$ to $p = 92$ is shown. For visualization purposes, as in Section 2.5.2, the values are encoded by shades of gray using a logarithmic scale (otherwise the low-energy values would not been visible in the plot). In the lower staff of the score (left hand), one can see that the chord consisting of the three notes A3 ($p = 57$), C4 ($p = 60$), and E4 ($p = 64$) is played every quarter beat, altogether eight times over the first four measures. These chords are also clearly visible in the time–pitch representation of the log-frequency spectrogram. In particular, one can observe some energy in the subbands corresponding to the pitches $p = 57$, $p = 60$, and $p = 64$. Further energy can be seen in the upper pitch bands that correspond to harmonics. For example, even though the note A4 ($p = 69$) only appears in the right hand of the score in the third and fourth measure, there is significant energy in this subband at the eight time positions corresponding to the quarter beat. This is due to the fact that the note A4 corresponds to the second harmonic of the note A3 ($p = 57$) played by the left hand. Interestingly, for this note there is more energy in the second harmonic than in the fundamental frequency, a phenomenon that often occurs for low-pitched notes. Furthermore, the two sixteenth-note phrases shown in the upper staff of the score (right hand) can be seen in the time–pitch representation. For example, the note B4 ($p = 71$) is played three times by the right hand, which is clearly revealed in the log-frequency spectrogram. It also becomes evident that some of the signal's energy is spread over large parts of the spectrum. The main reason for the energy spread is due to the inharmonicities of the piano sound caused by the keystroke (mechanical noise) as well as transient and resonance effects (see Section 1.3.4). Another general reason is the imperfection of the Fourier analysis also known as **spectral leakage**,

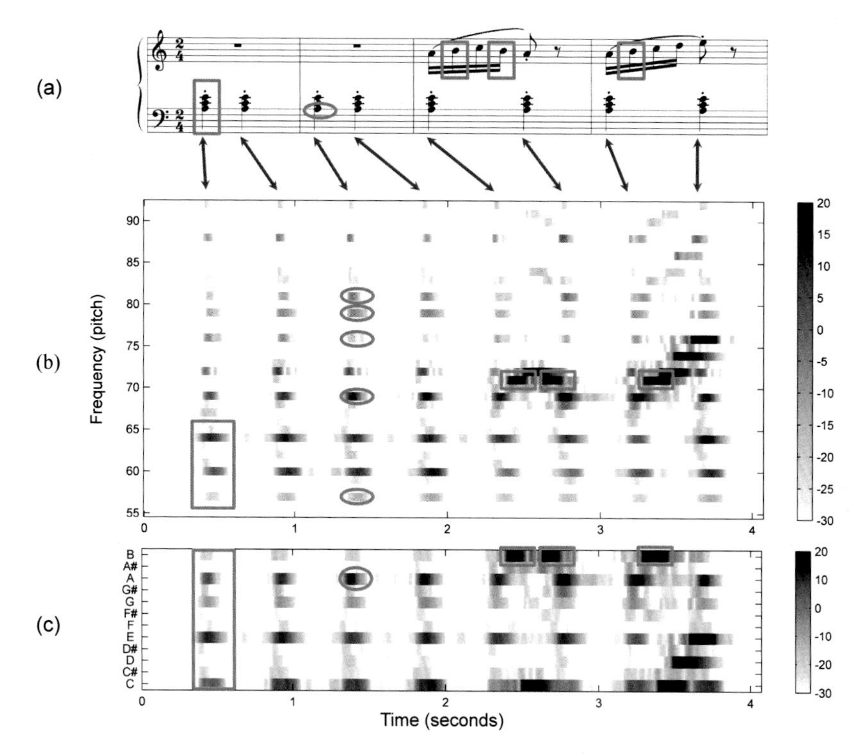

Fig. 3.5 Various representations for the Burgmüller example from Figure 3.1. **(a)** Sheet music representation. **(b)** Pitch-based log-frequency spectrogram. **(c)** Chromagram. Correspondences between selected notes and parts of the spectrogram are highlighted by rectangular and oval frames. For visualization purposes the values are encoded by shades of gray using a logarithmic scale.

which is the result of the frequency smearing introduced by the window function (see Section 2.5.2). However, in the present example, the leakage effect only causes marginal artifacts compared with those introduced by the inharmonicities.

3.1.2 Chroma Features

Recall from Section 1.1.1 that human perception of pitch is periodic in the sense that two pitches are perceived as similar in "color" (playing a similar harmonic role) if they differ by an octave. Based on this observation, a pitch can be separated into two components, which are referred to as **tone height** and **chroma**. The tone height refers to the octave number and the chroma to the respective pitch spelling attribute contained in the set $\{\mathrm{C},\mathrm{C}^\sharp,\mathrm{D},\ldots,\mathrm{B}\}$. Enumerating the chroma values, we identify this set with $[0:11]$ where 0 refers to chroma C, 1 to $\mathrm{C}^\sharp$, and so on. A **pitch class** is defined as the set of all pitches that share the same chroma. For example, the pitch class corresponding to the chroma C is the set $\{\ldots, \mathrm{C0}, \mathrm{C1}, \mathrm{C2}, \mathrm{C3}, \ldots\}$ consisting of all pitches separated by an integer number of octaves. For simplicity, we use the terms chroma and pitch class interchangeably.

The main idea of **chroma features** is to aggregate all spectral information that relates to a given pitch class into a single coefficient. Given a pitch-based log-frequency spectrogram $\mathcal{Y}_{\mathrm{LF}} : \mathbb{Z} \times [0:127] \to \mathbb{R}_{\geq 0}$ as defined in (3.4), a chroma representation or **chromagram** $\mathbb{Z} \times [0:11] \to \mathbb{R}_{\geq 0}$ can be derived by summing up all pitch coefficients that belong to the same chroma:

$$\mathcal{C}(n,c) := \sum_{\{p \in [0:127]\,:\, p \bmod 12 = c\}} \mathcal{Y}_{\mathrm{LF}}(n,p) \tag{3.6}$$

for $c \in [0:11]$. As a first example, Figure 3.3d shows the chromagram of the chromatic scale, where the cyclic nature of chroma features becomes evident. Because of the octave equivalence, the increasing notes of the chromatic scale are "wrapped around" the chroma axis. As with the log-frequency spectrogram, the resulting chromagram of the considered audio example is rather noisy, in particular for the lower notes. Furthermore, because of the presence of higher harmonics, the energy is typically spread across various chroma bands even when playing a single note at a time. For example, playing the note C3, the third harmonic corresponds to G4 and the fifth harmonic to E5. Therefore, when playing the note C3 on the piano, not only the chroma band C, but also the chroma bands G and E contain a substantial portion of the signal's energy.

Next, let us have a look at the chromagram for our Burgmüller example (see Figure 3.5c). The chords consisting of the three notes A3 ($p = 57$), C4 ($p = 60$), and E4 ($p = 64$) of the lower staff are clearly visible in the chromagram, where most of the signal's energy is contained in the chroma bands A, C, and E. The smaller amount of energy seen in band G comes from G5, which is the third harmonic of C4. Similarly, there is some energy in the band B coming from B5, which is the third harmonic of E4. The information contained in the 128 pitch bands of the log-frequency spectrogram is collapsed into the twelve chroma bands. As a result, a lot of information is irrecoverably lost. For example, in the chromagram, the notes A5 of the upper staff cannot be distinguished from the notes A4 in the lower staff. However, for certain applications, this loss in information is desired since it introduces a high degree of robustness to variations in timbre. To see this, recall from Section 1.3.4 that the timbre of a sound strongly relates to the energy distribution in the harmonics. Now, the 1st, 2nd, 4th, 8th, ... harmonics differ by octaves and their energy is gathered in the same chroma band. Similarly, the 3rd, 6th, 12th, ... harmonics share the same chroma. As a result, a lot of information that is relevant to distinguish the timbre of different instruments falls into the same chroma bands.

To get a better feeling for chroma features, let us consider some further examples. We continue with the example from Figure 1.23, where the note C4 is played by different instruments. The resulting chroma representations are shown in Figure 3.6a. In the case of the piano, most energy is contained in the chroma band C with a strong decay in amplitude. We have already seen in the spectrogram representations

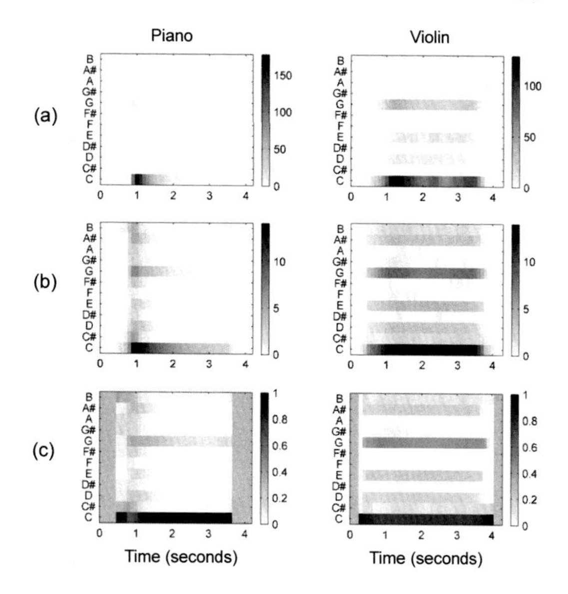

Fig. 3.6 Various chroma representations for the note C4 played by different instruments (see Figure 1.23). **(a)** Chromagram. **(b)** Chromagram after logarithmic compression. **(c)** Normalized chromagram obtained from (b).

of Figure 1.23 that the violin has a much richer spectrum of partials. As a result, the energy is spread more evenly over the harmonics, which explains the energy contained in the chroma bands G and E. This effect becomes more clearly visible when using a logarithmic scale, which we also have employed already in previous figures. The motivation for using such a scale is that the perception of a sound's intensity is logarithmic in nature (see Section 1.3.3). In other words, even sound components of very low energy may still bear perceptually important information.

3.1.2.1 Logarithmic Compression

In audio processing, one often applies a step referred to as **logarithmic compression**. This step, which can be seen as an alternative to using a decibel scale, works as follows: Let $\gamma \in \mathbb{R}_{>0}$ be a positive constant and $\Gamma_\gamma : \mathbb{R}_{>0} \rightarrow \mathbb{R}_{>0}$ the function defined by

$$\Gamma_\gamma(v) := \log(1+\gamma \cdot v) \tag{3.7}$$

for some positive value $v \in \mathbb{R}_{>0}$ (see Figure 3.7a). Hence, as opposed to the dB function in (1.6), the function Γ_γ yields a positive value $\Gamma_\gamma(v)$ for any positive value $v \in \mathbb{R}_{>0}$. Now, for a representation with positive values such as a spectrogram, log-frequency spectrogram or chromagram, one obtains a compressed version by applying the function Γ_γ to each of the values. For example, for the chromagram $\mathcal{C}$, the compressed version is the concatenation $\Gamma_\gamma \circ \mathcal{C}$ defined by

$$(\Gamma_\gamma \circ \mathcal{C})(n,c) := \log(1+\gamma \cdot \mathcal{C}(n,c)). \tag{3.8}$$

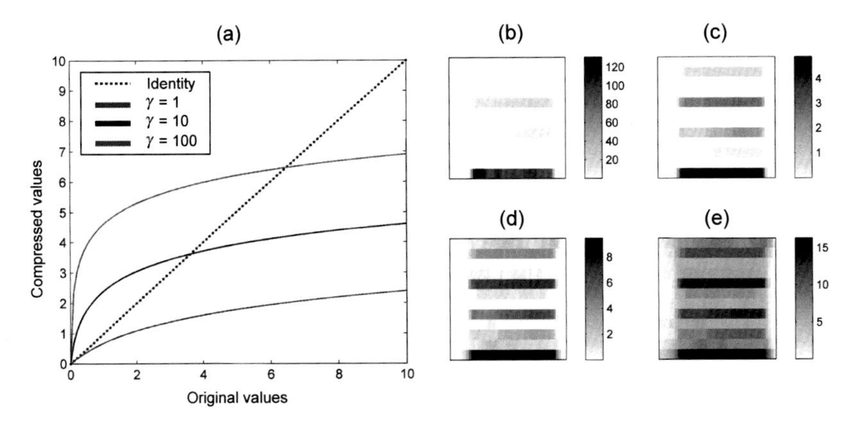

Fig. 3.7 **(a)** Compression function Γ_γ for different constants γ. **(b)** Original image showing a chromagram for the note C4 (similar to Figure 3.6). **(c)** Compressed image using $\gamma = 1$. **(d)** Compressed image using $\gamma = 100$. **(e)** Compressed image using $\gamma = 100000$.

Why is this operation called "compression," and what is the role of the constant $\gamma \in \mathbb{R}_{>0}$? The problem with representations such as a spectrogram is that its values possess a large dynamic range. As a result, small, but still relevant values may be dominated or obscured by large values. Therefore, the idea of compression is to balance out this discrepancy by reducing the difference between large and small values with the effect of enhancing the small values. This exactly is done by the function Γ_γ, where the degree of compression can be adjusted by the constant γ. The larger γ, the larger the resulting compression (see Figure 3.7). A suitable choice of γ very much depends on the data characteristics and the application in mind. In particular, in the presence of noise one needs to find a good balance between enhancing the weak but relevant signal components while not amplifying the undesired noise components too much.

Let us come back to our example with the note C4. Figure 3.6b shows the compressed versions of the chromagrams using the constant $\gamma = 100$. As can be seen, the two chromagrams of the piano and the violin look more or less the same, even though they sound quite different. This illustrates that chroma features are invariant to differences in timbre—at least to a certain degree. To make the chromagrams also invariant to changes in dynamics, one can normalize the features with respect to some suitable norm $\|\cdot\|$ (see Section 2.2.3). To this end, each of the twelve-dimensional chroma vectors $x = \mathcal{C}(n,\cdot) \in \mathbb{R}^{12}$ for a frame index $n \in \mathbb{Z}$ (n^{th} column of $\mathcal{C}$) is replaced by the vector $x/\|x\|$. In the following, we only consider the Euclidean norm, which we have already encoutered in (2.38). Recall that this norm is defined by

$$\|x\| := \left(\sum_{i=0}^{11} |x(i)|^2\right)^{1/2} \tag{3.9}$$

for a given chroma vector $x = (x(0), x(1), \ldots, x(11))^\top$. The normalization procedure is only possible if $\|x\| > 0$. Also for very small values $\|x\|$ that may occur in passages of silence before the actual start of the recording or during long pauses, normalization would lead to more or less random and therefore meaningless chroma value distributions. Therefore, if $\|x\|$ falls below a certain threshold, the vector x may be replaced by a uniform vector of norm one instead of dividing by $\|x\|$.

Mathematically, this normalization procedure can be described as follows: Let $S^{11} \subset \mathbb{R}^{12}$ be the unit sphere containing all 12-dimensional vectors of norm one. Then, for a given threshold $\varepsilon > 0$, we define a projection operator $\pi^\varepsilon : \mathbb{R}^{12} \to S^{11}$ by

$$\pi^\varepsilon(x) := \begin{cases} x/\|x\| & \text{if } \|x\| > \varepsilon, \\ (1,1,\ldots,1)^\top/\sqrt{12} & \text{if } \|x\| \leq \varepsilon. \end{cases} \tag{3.10}$$

Based on this operator, each chroma vector x is replaced by $\pi^\varepsilon(x)$. The threshold ε is a parameter that needs to be chosen with care. A suitable choice will depend on the requirements of the application in mind. Recall from Section 1.3.3 that the human ear is very sensitive so that sound components of even very low energy may still be relevant. Therefore, the threshold ε is generally chosen several orders of magnitude smaller (often a factor between 10^4 to 10^6) than the average sound level.

Figure 3.6c illustrates the effect when applying a frame-wise normalization to the chroma representations. In particular for the piano example, the strong decay in amplitude in the chroma band C has been compensated by the normalization. As another interesting fact, this example also shows that the relative energy decay in the chroma band G is stronger than in the chroma band C. In other words, the energy decay may not be proportional over the harmonics. This property is one of the characteristics that influence the timbre of a specific sound (see Section 1.3.4).

The effect of logarithmic compression and normalization also becomes evident in our next example shown in Figure 3.8, which is based on a C-major scale played on a piano. With this example, we also want to discuss some further issues that are related to using the twelve-tone equal-tempered scale as an underlying model for the chroma computation. Often, when playing a note on a real instrument, the fundamental frequency of the sound may deviate substantially from the note's theoretical center frequency. These deviations may be due to effects such as vibrato (frequency modulations) or to **portamento**, a musical term that describes continuous pitch sliding to smoothly connect subsequent notes. Furthermore, global frequency deviations may be the result of instruments that are tuned lower or higher than the expected reference pitch A4 with center frequency 440 Hz (see Section 1.3.2). For example, many modern orchestras are using a tuning frequency slightly above 440 Hz, whereas ensembles that play Baroque music are often tuned lower than concert pitch. As an extreme example, Figure 3.8d shows a chromagram in the case that the C-major scale is played on a piano tuned 40 cents upwards. As a result, the energy of the respective notes is no longer captured by individual chroma bands, but smeared across neighboring bands. To compensate for tuning effects, one therefore needs to perform an additional tuning estimation step to adjust the center frequencies of the MIDI pitches in (3.2) as well as the logarithmic partitioning in (3.3) for

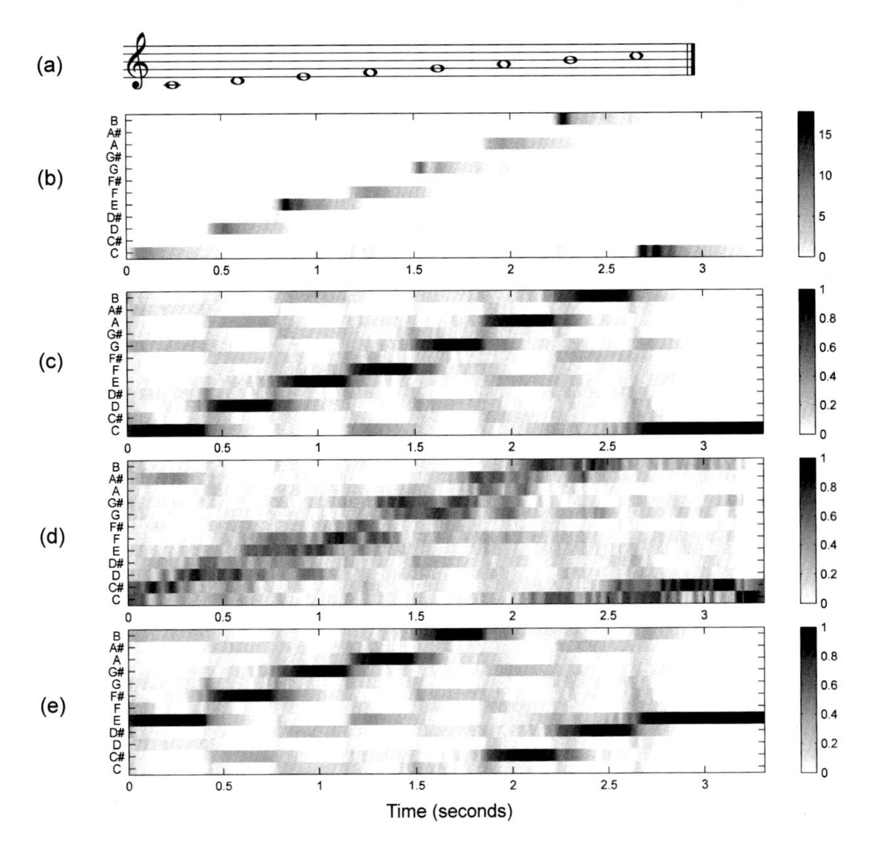

Fig. 3.8 Various chroma representations for a C-major scale played on a piano. **(a)** Sheet music representation. **(b)** Chromagram. **(c)** Chromagram after logarithmic compression and normalization. **(d)** Chromagram based on a piano tuned 40 cents upwards. **(e)** Chromagram after applying a cyclic shift of four semitones upwards.

computing the pitch-based log-frequency spectrogram. These issues will be further discussed in the exercises (see Exercise 3.5 and Exercise 3.6).

3.1.2.2 Transpositions

Next, we want to discuss a property of chroma features by which one can simulate a **transposition**, a concept used in music to shift a melody or an entire piece of music to another key. Such modifications are often applied to adapt the pitch range of a given piece to a different instrument or singer. For example, a song originally written for a soprano voice (highest female voice type) may be transposed seven semitones downwards to better fit a contralto voice (lowest female voice type). Technically speaking, a transposition refers to the process of moving a collection of notes up or down in pitch by a constant interval (see Exercise 1.3). To understand the effect on the chroma level, recall that the twelve chroma values, which we identified with the set $[0:11]$, are used to index a chroma vector $x=(x(0),x(1),\ldots,x(10),x(11))^\top \in \mathbb{R}^{12}$. As illustrated by Figure 1.3, the chroma values are cyclically ordered. This motivates the definition of the **cyclic shift operator** $\rho:\mathbb{R}^{12}\to\mathbb{R}^{12}$ given by

$$\rho(x) := (x(11),x(0),x(1),\ldots,x(10))^\top. \tag{3.11}$$

In other words, the chroma band C in x becomes the chroma band C$^\sharp$ in $\rho(x)$, the band C$^\sharp$ becomes D, and so on, and the last band B becomes C. The cyclic shift operator can be applied successively obtaining $\rho^i := \rho\circ\rho^{i-1}$ for $i\in\mathbb{N}$, which defines a cyclic shift of i semitones upwards. Obviously, $\rho^{12}(x)=x$, which means that, by cyclically shifting a chroma vector twelve semitones (one octave) upwards, one recovers the original vector. Applying the cyclic shift operator to all frames of a chromagram simultaneously leads to a cyclical shift of the entire chromagram in the vertical direction. This is illustrated by Figure 3.8e, where the original chromagram of the C-major scale has been shifted four semitones upwards. This results in a chromagram that looks like the one of an E-major scale, a transposition of four semitones.

3.1.2.3 Concluding Example

There are many ways of converting music recordings into chroma-based representations. In this section, we have only discussed the most basic version based on a spectrogram. By applying suitable pre- and postprocessing steps, the properties of chroma features can be changed significantly. For example, we have seen that one can increase the robustness to variations in timbre or sound intensity by performing additional compression and normalization steps. To illustrate the effect of other enhancement strategies, we consider two different performances of Beethoven's Fifth Symphony: an orchestral version conducted by Karajan and a piano version played by Glenn Gould (see Figure 3.9). Even though these two interpretations exhibit significant variations in articulation and instrumentation, the two compressed chromagrams shown in Figure 3.9b and Figure 3.9c, respectively, already show a similar progression of the chroma distribution over time. For certain applications such as the music synchronization task, these two chromagrams may be too detailed, and it is desirable to further increase the similarity between the two chromagrams. This can be achieved, for example, by applying additional quantization and smoothing procedures. Such a procedure will be discussed in more detail in Section 7.2.1. Further reducing the feature rate, one obtains the two enhanced chromagrams shown in Figure 3.9d and Figure 3.9e. As the figures demonstrate, the additional processing steps have not only reduced the noise level significantly, but also have further worked out the common characteristic chroma patterns of the two recordings. In the next section, we show how to account for the differences between the chromagrams that result from different tempi in the two recordings.

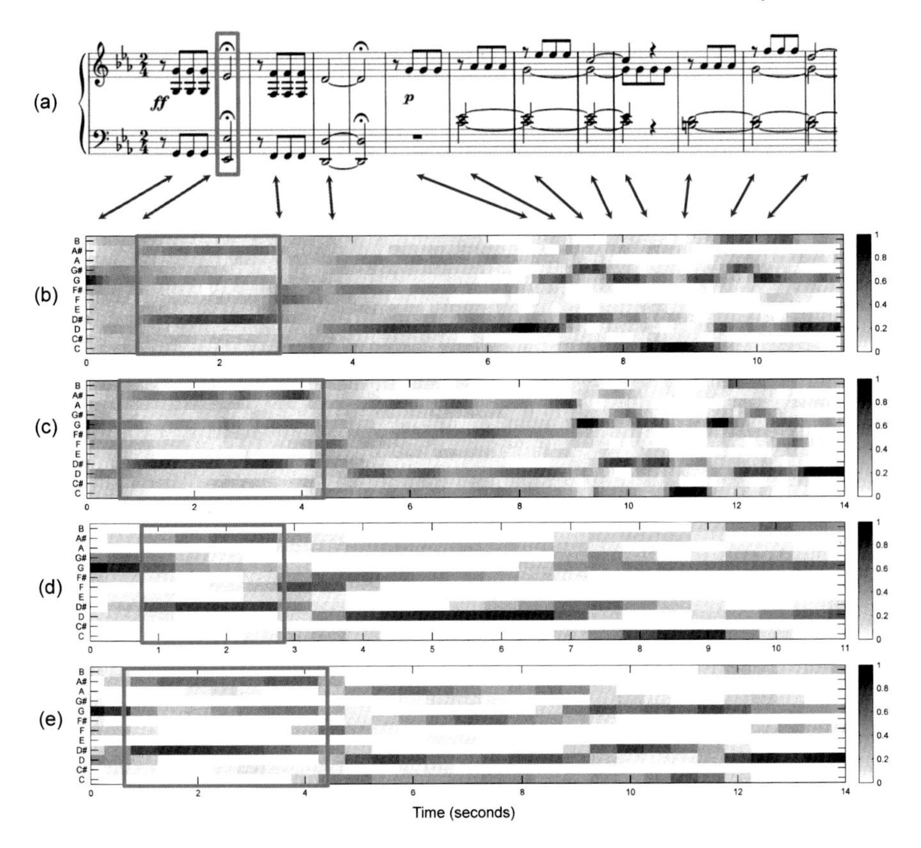

Fig. 3.9 Chroma representations for two different performances of Beethoven's Fifth Symphony. **(a)** Sheet music representation (in a piano reduced version). **(b)** Compressed and normalized chromagram for an orchestra performance. **(c)** Compressed and normalized chromagram for a piano performance. **(d)** Enhanced chromagram for an orchestra performance. **(e)** Enhanced chromagram for a piano performance.

In conclusion, we have seen that chroma features are a suitable representation for harmonic music with a broad class of pitch and tuning characteristics. Chroma representations closely correlate to the aspect of harmony, while showing a high degree of robustness to changes in timbre and dynamics. One important message of this section is that there are many ways of computing and enhancing chroma features resulting in a large number of chroma variants with different properties. We will see that there is no "best" chroma variant that performs equally well for all applications in mind. Therefore, in order to be successful, one needs to have a good understanding of both the feature design step as well as the requirements of the given application scenario.

3.2 Dynamic Time Warping

We have seen how different music representations can be made comparable by converting them into suitable feature representations. Next, we study how these feature representations can be aligned or synchronized to bring them into temporal correspondence. To this end, we introduce an important technique that is known as **dynamic time warping** (DTW).

The objective of DTW is to compare two given sequences. In our music synchronization scenario, these two sequences are, for example, chroma representations of two different versions of the same piece of music. Let us denote the first sequence by $X = (x_1, x_2, \ldots, x_N)$ with $N \in \mathbb{N}$ being the length of the sequence and the second sequence by $Y = (y_1, y_2, \ldots, y_M)$ having length $M \in \mathbb{N}$, where the elements x_n and y_m of the two sequences are chroma vectors. For example, X may be a chroma vector sequence obtained from an orchestral version of Beethoven's Fifth (as in Figure 3.9d) and Y a sequence obtained from a piano version (Figure 3.9e). Since the orchestral performance underlying the sequence X is played faster than the piano performance underlying the sequence Y, the two sequences do not have the same length even though they musically correspond to each other. In our example, as illustrated by Figure 3.10, the sequence X has length $N = 12$, whereas the sequence Y has length $M = 15$. The goal of DTW is to compensate for differences in tempo by finding a possibly nonlinear alignment between the elements of the two sequences. Intuitively speaking, this can be achieved by either skipping certain elements of a sequence or by using certain elements more than once. For example, in the alignment shown in Figure 3.10, the element x_3 is assigned to the two elements y_3 and y_4. Similarly, the elements x_5 and x_{12} are used twice in the overall alignment.

The general goal of DTW is to find an optimal alignment between two given (time-dependent) sequences under certain restrictions. Based on this alignment, the sequences can be warped in a nonlinear fashion to match each other. Originally, DTW was used to compare different speech patterns in automatic speech recognition. Closely related to concepts such as the edit distance or longest common subsequence, DTW-like procedures are now widely used in various fields such as data mining, information retrieval, and bioinformatics.

In this section, we introduce the main ideas of classical DTW as well as an efficient algorithm based on dynamic programming to compute an optimal alignment (Section 3.2.1). Then, we discuss several modifications to DTW which make it possible to influence certain local and global properties of the alignment and to further speed up the DTW computation (Section 3.2.2). A number of related algorithms and DTW variants are also discussed in the exercises and in subsequent chapters.

3.2.1 Basic Approach

As said above, the objective of DTW is to compare two sequences $X := (x_1, x_2, \ldots, x_N)$ of length $N \in \mathbb{N}$ and $Y := (y_1, y_2, \ldots, y_M)$ of length $M \in \mathbb{N}$. Going

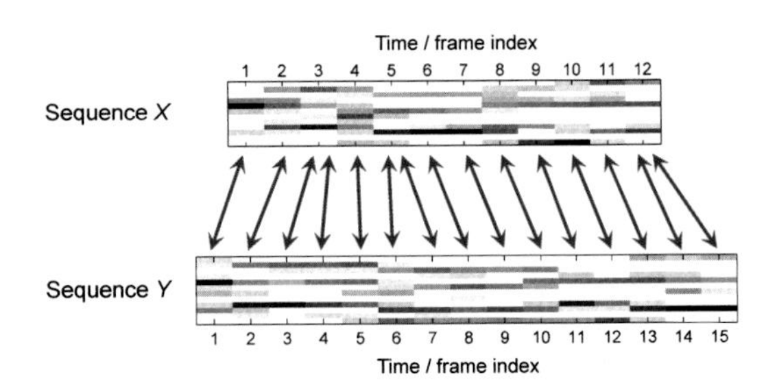

Fig. 3.10 Time alignment of two time-dependent or indexed sequences of feature vectors. Aligned points or frames are indicated by the arrows.

beyond the music synchronization scenario, these sequences may be discrete signals, feature sequences, sequences of characters, or any kind of time series. Often the indices of the sequences correspond to successive points in time that are spaced at uniform time intervals. In the following, we fix a **feature space** denoted by $\mathcal{F}$ and assume $x_n, y_m \in \mathcal{F}$ for $n \in [1:N]$ and $m \in [1:M]$. To compare two different features $x, y \in \mathcal{F}$, one needs a **local cost measure**, sometimes also referred to as a **local distance measure**, which is defined to be a function

$$c : \mathcal{F} \times \mathcal{F} \to \mathbb{R}. \tag{3.12}$$

Typically, $c(x,y)$ is small (low cost) if x and y are similar to each other, and otherwise $c(x,y)$ is large (high cost). Evaluating the local cost measure for each pair of elements of the sequences X and Y, one obtains a **cost matrix** $\mathbf{C} \in \mathbb{R}^{N \times M}$ defined by

$$\mathbf{C}(n,m) := c(x_n, y_m) \tag{3.13}$$

for $n \in [1:N]$ and $m \in [1:M]$. In the following, a tuple (n,m) representing an entry of the matrix $\mathbf{C}$ will be referred to as a **cell** of the matrix.

Let us come back to our example from Figure 3.10. Using a sequence of twelve-dimensional chroma vectors, the feature space is $\mathcal{F} = \mathbb{R}^{12}$. There are many ways to define a distance between two elements $x, y \in \mathcal{F}$. Maybe the most well-known distance is the Euclidean distance defined by $\|x - y\|$ using the Euclidean norm of $\mathbb{R}^{12}$. Another distance is referred to as the **cosine distance**, which we use as the local cost measure c in the subsequent examples. The cosine distance between two nonzero vectors $x, y \in \mathcal{F}$ is defined by

$$c(x,y) := 1 - \frac{\langle x|y \rangle}{\|x\| \cdot \|y\|}, \tag{3.14}$$

and $c(x,y) := 0$ if either x or y is zero. Recall from (2.39) that the quotient of the inner product and the product of the norms is simply the cosine of the angle between the two vectors x and y. Therefore, $c(x,y) \in [0,1]$, $c(x,y) = 0$ in the case that x and y point in the same direction, and $c(x,y) = 1$ in the case that x and y are orthogonal. Note that, as opposed to the Euclidean distance, the cosine distance does not depend on the actual length of the vectors. Therefore, when comparing chroma vectors, the measure only considers the energy distributions across the twelve chroma bands and

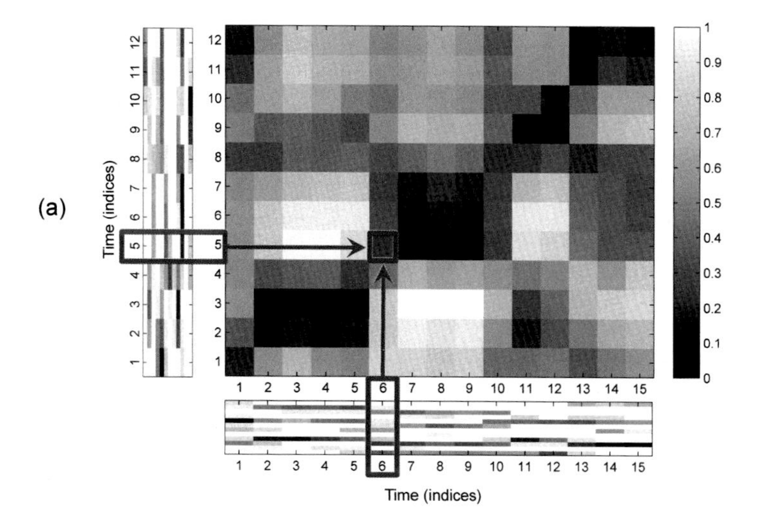

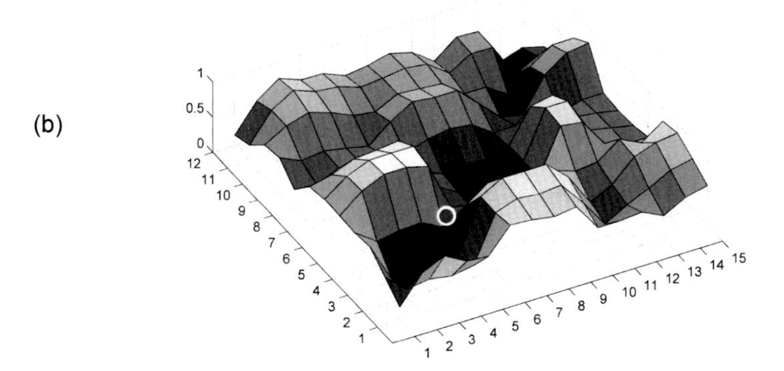

Fig. 3.11 **(a)** Cost matrix of the two chroma sequences X (vertical axis) and Y (horizontal axis) using the cosine distance (3.14) as the local cost measure c. Regions of low cost are indicated by dark colors, and regions of high cost are indicated by light colors. **(b)** 3D surface plot of the cost matrix.

disregards the actual local energy. In our music synchronization application, this property is desirable when the two versions to be compared may differ significantly in dynamics (which, for example, may be the case when comparing an orchestral and piano version of a piece of music). Furthermore, there is another practical reason why the cosine distance is beneficial: the computation of an entire cost matrix based on the cosine distance can be done efficiently using a simple matrix multiplication (see Exercise 3.16).

The cost matrix $\mathbf{C}$ obtained from our example is shown in Figure 3.11a. In this visualization, cells of low cost are depicted in dark colors and cells of high cost in light colors. Since the two sequences show a similar overall progression, except for

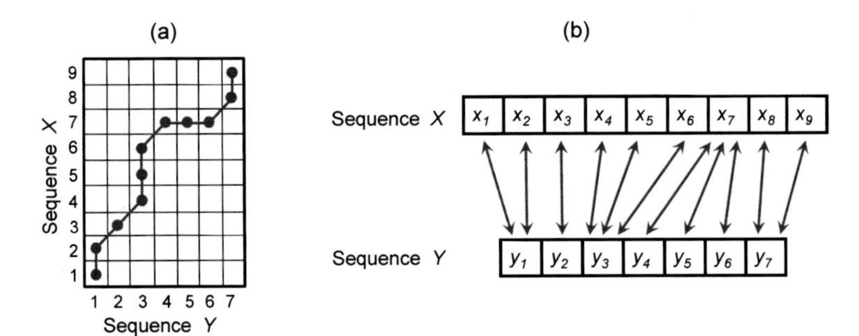

Fig. 3.12 **(a)** Illustration of a warping path and **(b)** its interpretation for some sequence X of length $N = 9$ and some sequence Y of length $M = 7$. Each cell (n,m) belonging to the warping path is indicated by a red dot and corresponds to an alignment between the elements x_n and y_m indicated by a red bidirectional arrow.

a global difference in tempo, the cost matrix has low values along the diagonal of the matrix. For example, the cell $(n,m) = (5,6)$, which indicates the distance between the vectors x_5 and y_6, has a small cost value. Now, the goal is to find an alignment between X and Y having minimal overall cost. Intuitively, such an optimal alignment runs along a "valley" of low cost within the cost matrix $\mathbf{C}$ (see Figure 3.11b for an illustration). Next, we formalize the notion of an alignment.

3.2.1.1 Warping Path

Given two sequences $X = (x_1, x_2, \ldots, x_N)$ and $Y = (y_1, y_2, \ldots, y_M)$, we have seen that a correspondence between two elements x_n and y_m can be modeled by the index pair or cell (n,m). In Figure 3.10, such a correspondence was indicated by a red bidirectional arrow. Therefore, to model a global alignment between the elements of the sequences X and Y, the idea is to consider a sequence of index pairs that fulfills certain constraints. This leads to the following definition: An (N,M)**-warping path** of length $L \in \mathbb{N}$ is a sequence

$$P = (p_1, \ldots, p_L) \tag{3.15}$$

with $p_\ell = (n_\ell, m_\ell) \in [1:N] \times [1:M]$ for $\ell \in [1:L]$ satisfying the following three conditions:

$$\text{Boundary condition: } p_1 = (1,1) \text{ and } p_L = (N,M). \tag{3.16}$$

$$\text{Monotonicity condition: } n_1 \leq n_2 \leq \ldots \leq n_L \text{ and } m_1 \leq m_2 \leq \ldots \leq m_L. \tag{3.17}$$

$$\text{Step size condition: } p_{\ell+1} - p_\ell \in \{(1,0),(0,1),(1,1)\} \text{ for } \ell \in [1:L-1]. \tag{3.18}$$

An (N,M)-warping path $P = (p_1, \ldots, p_L)$ defines an alignment between two sequences $X = (x_1, x_2, \ldots, x_N)$ and $Y = (y_1, y_2, \ldots, y_M)$ by assigning the element x_{n_ℓ}

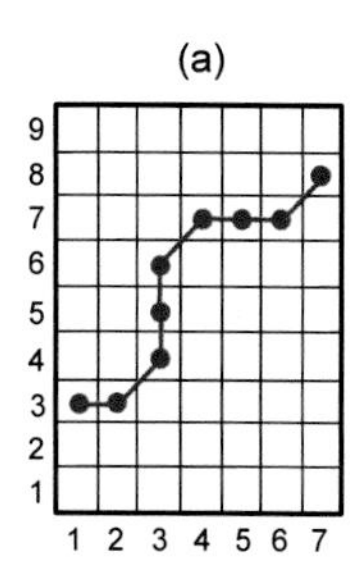

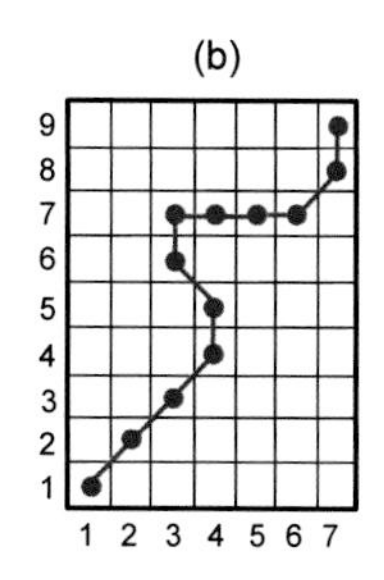

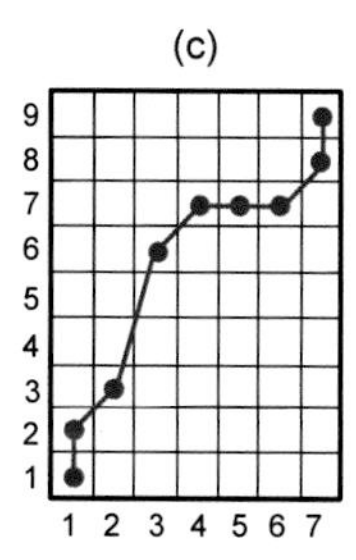

Fig. 3.13 Illustration of violations of the warping path conditions. **(a)** Violation of boundary condition. **(b)** Violation of monotonicity condition. **(c)** Violation of step size condition.

of X to the element y_{m_ℓ} of Y (see Figure 3.12 for an illustration). The boundary condition enforces that the first elements of X and Y as well as the last elements of X and Y are aligned to each other. In other words, the alignment refers to the entire sequences X and Y. The monotonicity condition reflects the requirement of faithful timing: if an element in X precedes a second element in X, then this should also hold for the corresponding elements in Y, and vice versa. Finally, the step size condition with respect to the set

$$\Sigma := \{(1,0),(0,1),(1,1)\} \tag{3.19}$$

expresses a kind of continuity condition: no element in X and Y can be omitted, and there are no replications in the alignment (in the sense that all index pairs contained in a warping path P are pairwise distinct). Note that the step size condition (3.18) implies the monotonicity condition (3.17), which nevertheless has been quoted explicitly for the sake of clarity. Figure 3.13 illustrates the conditions by some examples where the conditions are violated. In the following, if N and M are clear from the context, we simply speak of a warping path instead of an (N,M)-warping path.

3.2.1.2 Optimal Warping Path and DTW Distance

So far, in the definition of a warping path, the cost matrix does not play any role. A warping path simply encodes how to run through certain cells of a matrix with N rows and M columns starting with cell $(1,1)$ and ending with cell (N,M), while satisfying some monotonicity and step size conditions. Next, we introduce a notion that tells us something about the **quality** of a warping path. The **total cost** $c_P(X,Y)$ of a warping path P between two sequences X and Y with respect to the local cost measure c is defined as

$$c_P(X,Y) := \sum_{\ell=1}^{L} c(x_{n_\ell}, y_{m_\ell}) = \sum_{\ell=1}^{L} \mathbf{C}(n_\ell, m_\ell). \tag{3.20}$$

The intuition of this definition is that the warping path accumulates the cost of all cells it runs through. A warping path is "good" if its total cost is low, and it is "bad"

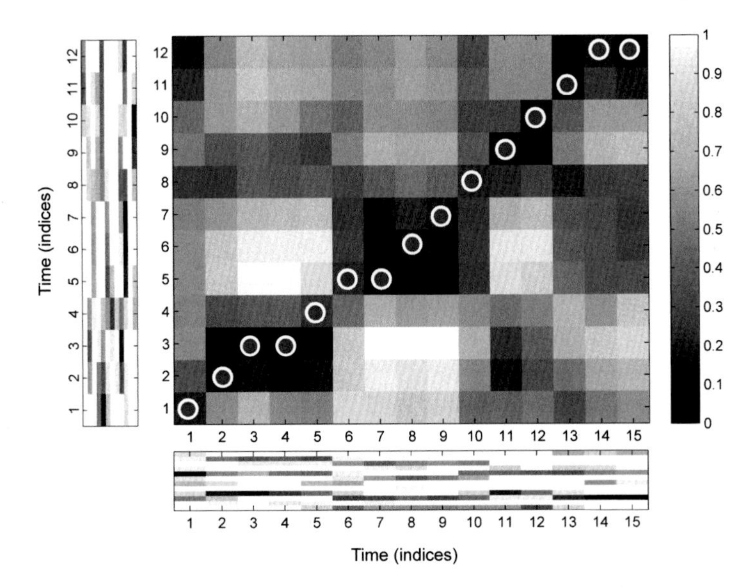

Fig. 3.14 Cost matrix from Figure 3.11 with an optimal warping path.

if its total cost is high. Now, we are interested in an **optimal warping path** between X and Y, which is defined to be a warping path P^* that has minimal total cost among all possible warping paths. Such an optimal warping path is shown in Figure 3.14 continuing the example of Figure 3.11. The cells of this warping path encode an overall optimal alignment between the chroma vectors of the two sequences, where the warping path conditions ensure that each element of sequence X is assigned to at least one element of Y and vice versa.

This leads us to the definition of the **DTW distance** denoted as $\mathrm{DTW}(X,Y)$ between the two sequences X of length N and Y of length M, which is defined as the total cost of an optimal (N,M)-warping path P^*:

$$\begin{aligned}\mathrm{DTW}(X,Y) &:= c_{P^*}(X,Y) \\ &= \min\{c_P(X,Y) \mid P \text{ is an } (N,M)\text{-warping path}\}.\end{aligned} \tag{3.21}$$

Note that in general there may exist more than one optimal warping path. For example, in the case that the cost matrix $\mathbf{C}$ is an all-zero matrix, every warping path is optimal, having a total cost of zero. Nevertheless, the DTW distance is well defined since all optimal warping paths obviously have the same total cost.

The number $\mathrm{DTW}(X,Y)$ defined in (3.21) is commonly referred to as the "DTW distance" between the sequences X and Y. However, from a mathematical point of view, the term "distance" is misused in this case. In mathematics, a **distance** is required to satisfy certain conditions, being symmetric, positive definite, and satisfying the triangle inequality. It is not hard to see that the DTW distance is symmetric, i.e., $\mathrm{DTW}(X,Y) = \mathrm{DTW}(Y,X)$, in case that the local cost measure c is symmetric (see Exercise 3.7). However, the DTW distance is in general not positive definite, where one requires that the distance between two elements is zero if and only if the elements are the same. For example, one obtains $\mathrm{DTW}(X,Y) = 0$ for the two different sequences $X := (x_1,x_2)$ and $Y := (x_1,x_1,x_2,x_2,x_2)$ in the case that $c(x_1,x_1) = c(x_2,x_2) = 0$. Intuitively, this property means that warping can be done without causing any cost. Even more surprising is the fact that the DTW distance generally does not satisfy the triangle inequality even if this holds for c. This fact will be illustrated by an example in Exercise 3.11.

3.2.1.3 Dynamic Programming Algorithm

To determine an optimal warping path P^* for two sequences X and Y, one could compute the total cost of all possible (N,M)-warping paths and then take the minimal cost. However, the number of different (N,M)-warping paths is exponential in N and M (see Exercise 3.9). Therefore, such a naive approach is computationally infeasible for large N and M. We now introduce an $O(NM)$ algorithm that is based on **dynamic programming**. The general idea behind dynamic programming is to break down a given problem into simpler subproblems and then to combine the solutions of the subproblems to reach an overall solution. In the case of DTW, the idea is to derive an optimal warping path for the original sequences from optimal warping paths for truncated subsequences. This idea can then be applied recursively. To formalize this idea, we define the prefix sequences $X(1\!:\!n) := (x_1,\ldots,x_n)$ for $n \in [1:N]$ and $Y(1\!:\!m) := (y_1,\ldots,y_m)$ for $m \in [1:M]$ and set

$$\mathbf{D}(n,m) := \mathrm{DTW}(X(1\!:\!n),Y(1\!:\!m)). \tag{3.22}$$

The values $\mathbf{D}(n,m)$ define an $(N \times M)$ matrix $\mathbf{D}$, which is also referred to as the **accumulated cost matrix**. Each value $\mathbf{D}(n,m)$ specifies the total (or accumulated) cost of an optimal warping path starting at cell $(1,1)$ and ending at cell (n,m). Obviously, one has $\mathbf{D}(N,M) = \mathrm{DTW}(X,Y)$. The next equations show how the accumulated cost matrix $\mathbf{D}$ can be computed recursively (see Figure 3.15 for an illustration):

$$\mathbf{D}(n,1) = \sum_{k=1}^{n} \mathbf{C}(k,1) \quad \text{for} \quad n \in [1:N], \tag{3.23}$$

$$\mathbf{D}(1,m) = \sum_{k=1}^{m} \mathbf{C}(1,k) \quad \text{for} \quad m \in [1:M], \tag{3.24}$$

$$\mathbf{D}(n,m) = \mathbf{C}(n,m) + \min \begin{cases} \mathbf{D}(n-1,m-1) \\ \mathbf{D}(n-1,m) \\ \mathbf{D}(n,m-1) \end{cases} \tag{3.25}$$

for $n \in [2:N]$ and $m \in [2:M]$.

We now give a formal proof of these equations. First, let $m = 1$ and $n \in [1:N]$. In this case, there is only one possible warping path, which assigns the single element

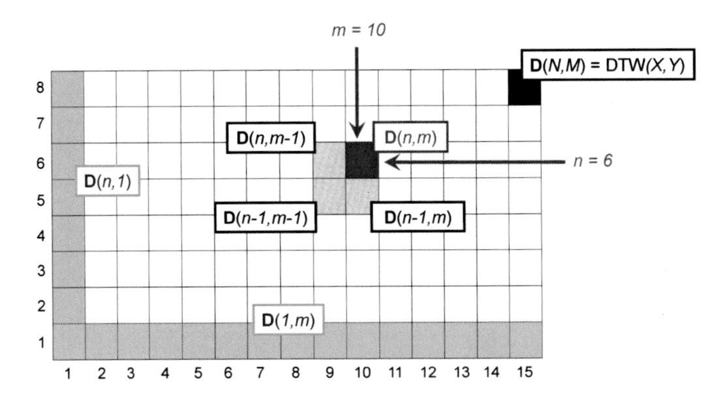

Fig. 3.15 Illustration of the recursive computation of the DTW distance. The blue cells indicate the entries $\mathbf{D}(n,1)$ and $\mathbf{D}(1,m)$ initialized by (3.23) and (3.24), respectively. The black cell indicates the final result $\mathbf{D}(N,M)$. The red cell indicates the current entry $\mathbf{D}(n,m)$ being computed by (3.25).

y_1 of $Y(1:1)$ to all elements of $X(1:n)$. Being the only possible warping path, this path is optimal. The cost of the path is obviously $\sum_{k=1}^{n} \mathbf{C}(k,1)$, which proves (3.23). Similarly, one obtains (3.24) for the case $n = 1$ and $m \in [1:M]$. Now, let $n > 1$ and $m > 1$ and let $q = (q_1,\ldots,q_{L-1},q_L)$ denote an optimal warping path for $X(1:n)$ and $Y(1:m)$. We now show how this path can be obtained by extending a previously constructed optimal warping path. First note that the boundary condition (3.16) implies $q_L = (n,m)$. Setting $(a,b) := q_{L-1}$, the step size condition (3.18) implies $(a,b) \in \{(n-1,m-1),(n-1,m),(n,m-1)\}$. Furthermore, it follows that $(q_1,\ldots,q_{L-1})$ must be an optimal warping path for $X(1:a)$ and $Y(1:b)$ (otherwise, q would not be optimal for $X(1:n)$ and $Y(1:m)$). Since

$$\mathbf{D}(n,m) = c_{(q_1,\ldots,q_{L-1})}(X(1:a),Y(1:b)) + \mathbf{C}(n,m), \tag{3.26}$$

the optimality of q implies the assertion of (3.25). This finishes the proof.

The equations (3.23) and (3.24) yield the initialization of a recursive procedure for computing $\mathbf{D}$. The values $\mathbf{D}(n,m)$ for $n > 1$ and $m > 1$ can then be computed via (3.25). The computation of $\mathbf{D}$ needs to be done by successively increasing n or m starting with the bottom left corner $(1,1)$ and ending with the upper right corner (N,M). This final cell (N,M) yields the DTW distance $\mathrm{DTW}(X,Y) = \mathbf{D}(N,M)$ (see Figure 3.15). The entries $\mathbf{D}(n,m)$ can be computed in different orders as long as n and m are increased monotonically. For example, one may proceed in a columnwise fashion, where the computation of the m-th column requires the values of the $(m-1)$-th column. This implies that, if one is only interested in the value $\mathrm{DTW}(X,Y) = \mathbf{D}(N,M)$, the storage requirement is $O(N)$. Similarly, one can proceed in a rowwise fashion, leading to a storage requirement of $O(M)$. However, in any case, the recursive step (3.25) is called $(N-1)\cdot(M-1)$ times. Since each step requires the minimization over three numbers as well as an addition, the overall complexity for computing $\mathbf{D}$ from a given cost matrix $\mathbf{C}$ is $O(NM)$.

Algorithm: DTW

Input: Cost matrix $\mathbf{C}$ of size $N \times M$
Output: Accumulated cost matrix $\mathbf{D}$
Optimal warping path P^*

Procedure: Initialize $(N \times M)$ matrix $\mathbf{D}$ by $\mathbf{D}(n,1) = \sum_{k=1}^{n} \mathbf{C}(k,1)$ for $n \in [1:N]$ and $\mathbf{D}(1,m) = \sum_{k=1}^{m} \mathbf{C}(1,k)$ for $m \in [1:M]$. Then compute in a nested loop for $n = 2,\ldots,N$ and $m = 2,\ldots,M$:

$$\mathbf{D}(n,m) = \mathbf{C}(n,m) + \min\{\mathbf{D}(n-1,m-1), \mathbf{D}(n-1,m), \mathbf{D}(n,m-1)\}.$$

Set $\ell = 1$ and $q_\ell = (N,M)$. Then repeat the following steps until $q_\ell = (1,1)$:

Increase ℓ by one and let $(n,m) = q_{\ell-1}$.
If $n = 1$, then $q_\ell = (1,m-1)$,
else if $m = 1$, then $q_\ell = (n-1,m)$,
else $q_\ell = \mathrm{argmin}\{\mathbf{D}(n-1,m-1), \mathbf{D}(n-1,m), \mathbf{D}(n,m-1)\}$.
(If 'argmin' is not unique, take lexicographically smallest cell.)

Set $L = \ell$ and return $P^* = (q_L, q_{L-1}, \ldots, q_1)$ as well as $\mathbf{D}$.

Table 3.2 DTW algorithm based on dynamic programming.

So far, we have computed the DTW distance, but we do not know how an optimal warping path looks. To determine such a path, one needs to recover the information about the minimizing cells in the recursion (3.25). Applying a **backtracking** procedure, the optimal warping path can be constructed incrementally in reverse order starting with the cell $q_1 = (N,M)$. Suppose $q_\ell = (n,m)$ has been computed. In case $(n,m) = (1,1)$, we are done and set $L = \ell$. The path $P^* = (q_L, q_{L-1}, \ldots, q_1)$ then defines an optimal warping path. Otherwise,

$$q_{\ell+1} = (1,m-1) \quad \text{if } n = 1, \tag{3.27}$$

$$q_{\ell+1} = (n-1,m) \quad \text{if } m = 1, \tag{3.28}$$

$$q_{\ell+1} = \mathrm{argmin} \begin{cases} \mathbf{D}(n-1,m-1), \\ \mathbf{D}(n-1,m), \\ \mathbf{D}(n,m-1) \end{cases} \tag{3.29}$$

if $n \in [2:N]$ and $m \in [2:M]$, where 'argmin' yields the cell leading to the minimum of the three values. Note that 'argmin' does not need to be unique, thus opening up the possibility of having more than one optimal warping path. To obtain a uniquely determined path, one may take, for example, the lexicographically smallest cell in case 'argmin' is not unique. Table 3.2 summarizes the entire procedure for computing the DTW distance as well as an optimal warping path.

We now look at a small example to illustrate how the algorithm in Table 3.2 is applied. To this end, we consider the feature space $\mathcal{F} = \mathbb{R}$ and the local measure $c : \mathcal{F} \times \mathcal{F} \to \mathbb{R}_{\geq 0}$ defined by $c(x,y) = |x-y|$, $x,y \in \mathbb{R}$. Furthermore, let $X = (1,3,3,8,1)$ and $Y = (2,0,0,8,7,2)$. Figure 3.16a shows the resulting cost

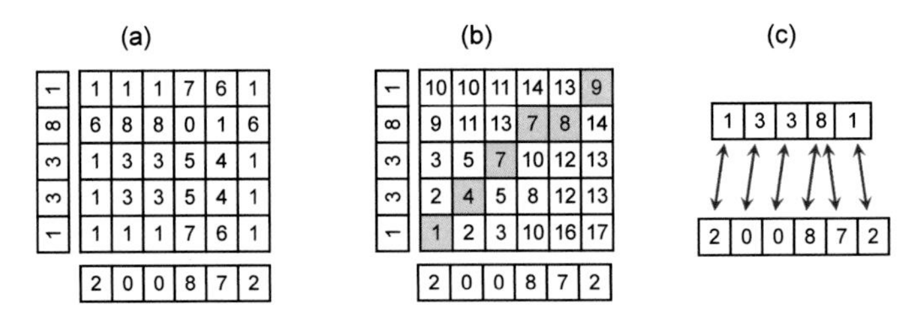

Fig. 3.16 **(a)** Cost matrix **C** for the two sequences $X = (1,3,3,8,1)$ and $Y = (2,0,0,8,7,2)$ over $\mathcal{F} = \mathbb{R}$ using the absolute differences as the local cost measure. **(b)** Accumulated cost matrix **D** with an optimal warping path. **(c)** Resulting alignment.

matrix **C**, where we have indexed the rows from bottom to top and the columns from left to right (as opposed to the usual convention according to which matrices are visualized, since we think of the two dimensions being two time axes). Figure 3.16b shows the accumulated cost matrix **D** along with the optimal warping path $P^* = ((1,1),(2,2),(3,3),(4,4),(4,5),(5,6))$. This path is obtained by starting with the cell $q_1 = (N,M) = (5,6)$ of the upper right corner. Applying backtracking, one looks at all cells from which (N,M) can be reached by applying a valid step from Σ, which are the cells $(N-1,M)$, $(N,M-1)$, and $(N-1,M-1)$. From these cells, the predecessor cell q_2 is obtained by looking at the cell with minimal accumulated cost. In our example, this is the cell $q_{L-1} = (N-1,M-1) = (4,5)$, which has an accumulated cost of $\mathbf{D}(4,5) = 8$. The procedure is repeated until one reaches the cell $q_L = (1,1)$. Since in each backtracking step of our example there is only one choice for 'argmin' in (3.29), there is only one optimal warping path. The induced alignment of this path is shown in Figure 3.16c. At this point we want to emphasize that the backtracking in **D** is essential to find an optimal warping path. Such a path cannot be found by, for example, starting with the cell $(1,1)$ and then proceeding in a greedy fashion in the forward direction. One can check that such a forward approach would yield the path $((1,1),(1,2),(1,3),(2,3),(3,3),(4,4),(4,5),(5,6))$, which is not optimal.

Note that for the backtracking the entire accumulated cost matrix **D** may be needed. Therefore, as opposed to the case where one is only interested in computing $\mathrm{DTW}(X,Y)$, the storage requirement is $O(NM)$ when an optimal warping path is to be computed.

Finally, we introduce a small trick for simplifying the initialization (3.23) and (3.24). To this end, we extend the matrix **D** with an additional row and column (indexed by 0) by formally setting $\mathbf{D}(n,0) := \infty$ for $n \in [1:N]$, $\mathbf{D}(0,m) := \infty$ for $m \in [1:M]$, and $\mathbf{D}(0,0) := 0$. Then the recursion of (3.25) can be applied for $n \in [1:N]$ and $m \in [1:M]$, yielding exactly the same values for **D** as before (see Exercise 3.13). This trick will be helpful when considering modifications and variants of dynamic time warping as discussed next.

3.2.2 DTW Variants

Various modifications have been proposed in order to speed up DTW computations as well as to better control the overall course of the warping paths. In the following, we discuss some of these DTW variants.

3.2.2.1 Step Size Condition

Recall that the step size condition (3.18) expressed by the set $\Sigma = \{(1,0),(0,1),(1,1)\}$ is a kind of local continuity condition, ensuring that a warping path aligns each element of the sequence $X = (x_1,x_2,\ldots,x_N)$ to an element of $Y = (y_1,y_2,\ldots,y_M)$ and vice versa. One drawback of this condition is that a single element of one sequence may be assigned to many consecutive elements of the other sequence, which leads to vertical and horizontal sections in the warping path (see Figure 3.17a). Intuitively, in such cases the warping path is stuck at some position in one of the sequences, while moving on in the other sequence. In terms of physical time, this situation corresponds to a strong temporal deformation in the alignment of the two time series. To avoid such degenerations, one can modify the step size condition by constraining the slope of the admissible warping paths, which can be done by replacing the set Σ. For example, instead of using the original set $\Sigma = \{(1,0),(0,1),(1,1)\}$, one can use the set

$$\Sigma = \{(2,1),(1,2),(1,1)\}. \tag{3.30}$$

This leads to warping paths having a local slope within the bounds $1/2$ and 2 (see Figure 3.17b). The resulting accumulated cost matrix $\mathbf{D}$ can then be computed by the recursion

$$\mathbf{D}(n,m) = \mathbf{C}(n,m) + \min \begin{cases} \mathbf{D}(n-1,m-1), \\ \mathbf{D}(n-2,m-1), \\ \mathbf{D}(n-1,m-2) \end{cases} \tag{3.31}$$

for $n \in [1:N]$ and $m \in [1:N]$ with $(n,m) \neq (1,1)$. For the initialization, we again use the trick of extending $\mathbf{D}$, this time by two additional rows and columns (indexed by -1 and 0) and set $\mathbf{D}(1,1) := \mathbf{C}(1,1)$, $\mathbf{D}(n,-1) := \mathbf{D}(n,0) := \infty$ for $n \in [-1:N]$, and $\mathbf{D}(-1,m) := \mathbf{D}(0,m) := \infty$ for $m \in [-1:M]$. Note that, with respect to the modified step size condition, there is a warping path of finite total cost between two sequences X and Y if and only if the lengths N and M differ by less than a factor of two (see Exercise 3.14). Furthermore, note that not all elements of X need to be assigned to some element of Y and vice versa. This is illustrated by Figure 3.17b, where x_1 is assigned to y_1, x_3 is assigned to y_2, but x_2 is not assigned to any element of Y. In other words, x_2 is omitted and does not cause any cost at all.

Figure 3.17c gives a second example of a step size condition which avoids such omissions while imposing constraints on the slope of the warping path. The recursion of the resulting accumulated cost matrix $\mathbf{D}$ is given by

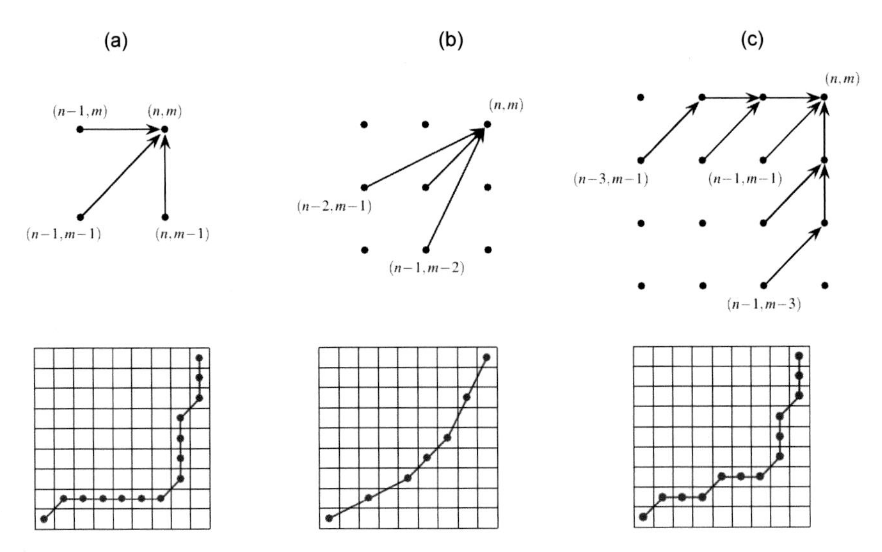

Fig. 3.17 Illustration of three different step size conditions (top), which express different local constraints on the admissible warping paths, along with some typical examples (bottom). The original step size condition based on the set $\Sigma = \{(1,0),(0,1),(1,1)\}$ is shown in (a).

$$\mathbf{D}(n,m) = \min \begin{cases} \mathbf{D}(n-1,m-1) + \mathbf{C}(n,m) \\ \mathbf{D}(n-2,m-1) + \mathbf{C}(n-1,m) + \mathbf{C}(n,m) \\ \mathbf{D}(n-1,m-2) + \mathbf{C}(n,m-1) + \mathbf{C}(n,m) \\ \mathbf{D}(n-3,m-1) + \mathbf{C}(n-2,m) + \mathbf{C}(n-1,m) + \mathbf{C}(n,m) \\ \mathbf{D}(n-1,m-3) + \mathbf{C}(n,m-2) + \mathbf{C}(n,m-1) + \mathbf{C}(n,m) \end{cases} \tag{3.32}$$

for $(n,m) \in [1:N] \times [1:M] \setminus \{(1,1)\}$. For the initialization, we extend the matrix by three additional rows and columns indexed by -2, -1, and 0. The initial values are set to $\mathbf{D}(1,1) := \mathbf{C}(1,1)$, $\mathbf{D}(n,-2) := \mathbf{D}(n,-1) := \mathbf{D}(n,0) := \infty$ for $n \in [-2:N]$, and $\mathbf{D}(-2,m) := \mathbf{D}(-1,m) := \mathbf{D}(0,m) := \infty$ for $m \in [-2:M]$. The global average slope of a resulting warping path lies between the values $1/3$ and 3. Note that this step size condition enforces that all elements of X are aligned to some element of Y and vice versa. In other words, in the recursion (3.32) all elements of X and Y generate some cost in the accumulated cost matrix $\mathbf{D}$—as opposed to the recursion (3.31). The examples in Figure 3.17 illustrate the differences of the resulting optimal warping paths computed with respect to different step size conditions.

3.2.2.2 Local Weights

To favor the vertical, horizontal, or diagonal direction in the alignment, one can introduce additional **local weights** $w_{\mathrm{d}}, w_{\mathrm{h}}, w_{\mathrm{v}} \in \mathbb{R}$. To compute the accumulated cost matrix $\mathbf{D}$, one uses the following initialization and recursion:

$$\mathbf{D}(1,1) := \mathbf{C}(1,1) \tag{3.33}$$

$$\mathbf{D}(n,1) = \sum_{k=1}^{n} w_{\mathrm{h}} \cdot \mathbf{C}(k,1) \text{ for } n \in [2:N] \tag{3.34}$$

$$\mathbf{D}(1,m) = \sum_{k=1}^{m} w_{\mathrm{v}} \cdot \mathbf{C}(1,k) \text{ for } m \in [2:M] \tag{3.35}$$

$$\mathbf{D}(n,m) = \min \begin{cases} \mathbf{D}(n-1,m-1) + w_{\mathrm{d}} \cdot \mathbf{C}(n,m) \\ \mathbf{D}(n-1,m) + w_{\mathrm{v}} \cdot \mathbf{C}(n,m) \\ \mathbf{D}(n,m-1) + w_{\mathrm{h}} \cdot \mathbf{C}(n,m) \end{cases} \tag{3.36}$$

for $n \in [2:N]$ and $m \in [2:M]$. The case $w_{\mathrm{d}} = w_{\mathrm{h}} = w_{\mathrm{v}} = 1$ reduces to classical DTW. Note that in the classical case one has a preference for the diagonal alignment direction, since one diagonal step (cost of one cell) corresponds to the combination of one horizontal and one vertical step (cost of two cells). To balance out this preference, one often chooses $w_{\mathrm{d}} = 2$ and $w_{\mathrm{h}} = w_{\mathrm{v}} = 1$. Similarly, one can introduce weights for other step size conditions.

3.2.2.3 Global Constraints

One common DTW variant is to impose **global constraints** on the admissible warping paths. Such constraints not only speed up DTW computations but also prevent "pathological" alignments by globally controlling the overall course of a warping path. More precisely, let $R \subseteq [1:N] \times [1:M]$ be a subset referred to as a global **constraint region**. Then a **warping path relative to** R is a warping path that entirely runs within the region R. The **optimal warping path relative to** R, denoted by P^*_R, is the cost-minimizing warping path among all warping paths relative to R.

Two well-known global constraint regions are the **Sakoe–Chiba band** and the **Itakura parallelogram**, as indicated by Figure 3.18. Alignments of cells can be selected only from the respective shaded region. The **Sakoe–Chiba band** runs along the main diagonal and has a fixed width (see Figure 3.18a). The **Itakura parallelogram** describes a region that constrains the slope of a warping path. More precisely, for a fixed $S \in \mathbb{R}_{>1}$, the Itakura parallelogram consists of all cells that lie within a **global** slope between the values $1/S$ and S (see Figure 3.18b). Note that a local step size condition may induce some global constraints. For example, using $\Sigma = \{(2,1),(1,2),(1,1)\}$ actually leads to warping paths that are contained in an Itakura parallelogram with $S = 2$. However, local step size conditions are stronger than global constraints, which do not enforce any **local** slope conditions on the warping paths.

For a general constraint region R, the path P^*_R can be computed similarly to the unconstrained case by formally setting $\mathbf{C}(n,m) := \infty$ for all $(n,m) \in [1:N] \times [1:M] \setminus R$. Therefore, in the computation of P^*_R only the cells that lie in R need to be evaluated. This may significantly speed up the DTW computation. For example, in

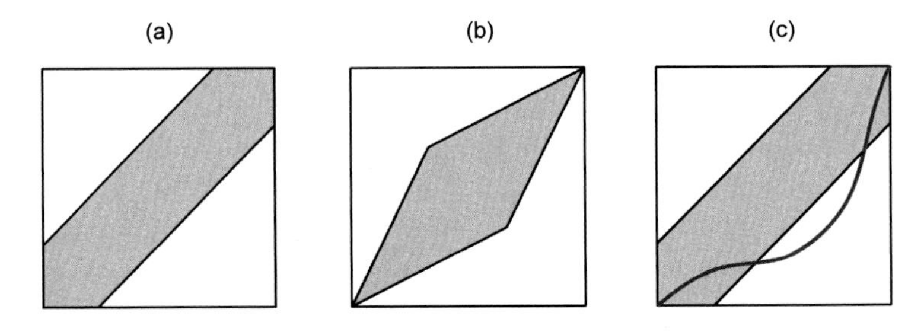

Fig. 3.18 - **(a)** Sakoe–Chiba band. **(b)** Itakura parallelogram . **(c)** Unconstraint optimal warping path P^* (red line) which does not run within the given constraint region.

case of a Sakoe–Chiba band of fixed width Δ, only $O(\Delta \cdot \max(N,M))$ computations need to be performed instead of $O(NM)$ as required in classical DTW. The savings can be substantial in particular in the case that $\Delta \ll M$ and $\Delta \ll N$. However, the usage of global constraint regions is also problematic, since the unconstrained optimal warping path P^* may traverse cells outside the specified constraint region. In this case, the constrained optimal warping path P^*_R does not coincide with P^* (see Figure 3.18c). Therefore, using a constraint region that is too strict may lead to undesirable or even completely useless alignment results. Only if the optimal unconstrained warping path P^* lies within R (which is of course not known in advance) does one obtain $P^* = P^*_R$ (see Exercise 3.17).

3.2.2.4 Multiscale DTW

We have seen that, when using the concept of global constraint regions, one needs to make sure that the optimal warping path to be computed actually lies within this region. Since this path is not known a priori, it is often difficult to find a good trade-off between choosing the constraint region as small as possible (to speed up computations) but large enough to contain the desired path. One possible strategy to increase the probability of finding the "right" path is to use data-dependent constraint regions instead of a data-independent, fixed constraint region. This idea can be realized by a multiscale approach to DTW, where the general strategy is to recursively project an optimal warping path computed at a coarse resolution level to the next highest level and then to refine the projected path. In the following, we summarize the main steps of such an approach (see Figure 3.19 for an overview).

Let $X_1 := X$ and $Y_1 := Y$ be two sequences having length $N_1 := N$ and $M_1 := M$, respectively. These two sequences represent the data at the highest resolution level, which we also refer to as level 1. The objective is to compute an optimal warping path P^* between X_1 and Y_1. The first step of multiscale DTW is to reduce the lengths of the sequences. This can be done, for example, by suitably coarsening X_1 and Y_1 and then reducing the feature sampling rate. Let us assume that we have a coarsening and downsampling procedure at hand that reduces the lengths by a factor $f_2 \in \mathbb{N}$.

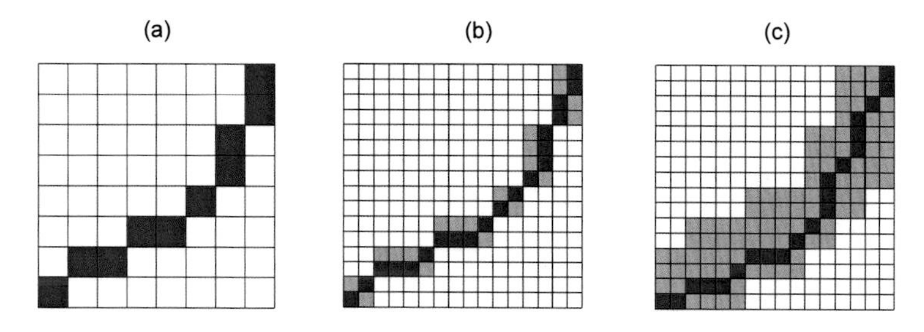

Fig. 3.19 **(a)** Optimal warping path P_2^* at level 2. **(b)** Optimal warping path P_R^* with respect to the constraint region R obtained by projecting path P_2^* to level 1. In this example, P_R^* does not coincide with the (unconstrained) optimal warping path P^*. **(c)** Optimal warping path $P_{R^\delta}^*$ using an increased constraint region $R^\delta \supset R$ with $\delta = 2$.

Furthermore, let us assume that f_2 divides N_1 and M_1, which can be achieved by suitably padding X_1 and Y_1. Let X_2 and Y_2 by the resulting feature sequences of length $N_2 := N_1/f_2$ and $M_2 := M_1/f_2$, respectively. Next, one computes an optimal warping path P_2^* of length L_2 between X_2 and Y_2 at the resulting resolution level, which we also call level 2. This path is projected onto level 1 to define a constraint region R, which consists of $L_2 \times (f_2)^2$ cells. Finally, an optimal warping path P_R^* relative to R is computed. We say that this procedure is **successful** if $P^* = P_R^*$. The overall number of cells to be computed in this procedure is $N_2M_2 + L_2(f_2)^2$, which is generally much smaller than the total number N_1M_1 of cells at level 1. This procedure can be recursively applied by introducing further levels of decreasing resolution. For a complexity analysis, we refer to Exercise 3.18.

One important issue with the multiscale approach is that the coarsened features at the different resolution levels need to be specified with great care. If the features are too coarse and the resolution too low, relevant information may be smoothed out or even lost. This may result in a poor warping path which does not lead to a meaningful constraint region for the next level. Note that a violation of the assumption $P_R^* = P^*$ at any level of the multiscale approach leads to irrecoverable errors of the overall procedure. Therefore, in practice, the recursion has to be stopped at a certain resolution level, where a full DTW needs to be computed. Also, to alleviate the problem that P_R^* may not coincide with P^*, one should increase the constraint region R—at the expense of efficiency—by a suitable neighborhood. This can be done, for example, by extending R to a constraint region R^δ, where, in addition to all cells in R, also the δ cells to the left, right, top, and bottom of all cells in R are included (see Figure 3.19c).

3.3 Applications

In the previous sections, we discussed an overall pipeline and the necessary techniques for synchronizing different music representations of the same underlying musical work. Assuming the equal-tempered scale, we converted the music representations into chroma sequences. These features show a high degree of robustness to variations in timbre and dynamics—aspects that are to be left unconsidered in the alignment. On the other hand, chroma features capture information related to the pitch distribution of sounds and are therefore well suited to characterize the melodic and harmonic progression of music. In a second step, we then applied dynamic time warping (DTW) to find optimal temporal correspondences between the elements of two given chroma sequences. These correspondences establish a musically meaningful linking structure between the given music representations.

In this section, we discuss different application scenarios, where automated synchronization methods play an important role for supporting the user in experiencing and exploring music. In a first scenario (Section 3.3.1), we describe user interfaces for multimodal (audiovisual) music presentation and navigation. In particular, such interfaces allow a user to listen to an audio recording while displaying the corresponding measures within a sheet music representation. In a second scenario (Section 3.3.2), we show how synchronization techniques can be used for extracting temporal information from music recordings in a fully automated fashion. This information is given in the form of tempo curves that are derived from measuring relative tempo differences between actual performances and reference representations of the given musical work.

3.3.1 Multimodal Music Navigation

Significant digitization efforts have resulted in large music collections, which comprise music-related documents of various types and formats including text, symbolic data, audio, image, and video expressing musical content at different semantic levels. Modern digital music libraries contain textual data including lyrics and libretti, symbolic data, visual data such as scanned sheet music or CD album covers, as well as music and video recordings of performances (see Figure 3.20). Therefore, beyond the mere recording and digitization of musical data, a key challenge in a real-life library application scenario is to integrate techniques and interfaces to organize, understand, and search musical content in a robust, efficient, and intelligent manner. In this context, music synchronization techniques are one way to automate the generation of cross-links, which can then be used for making musical data better accessible to the user.

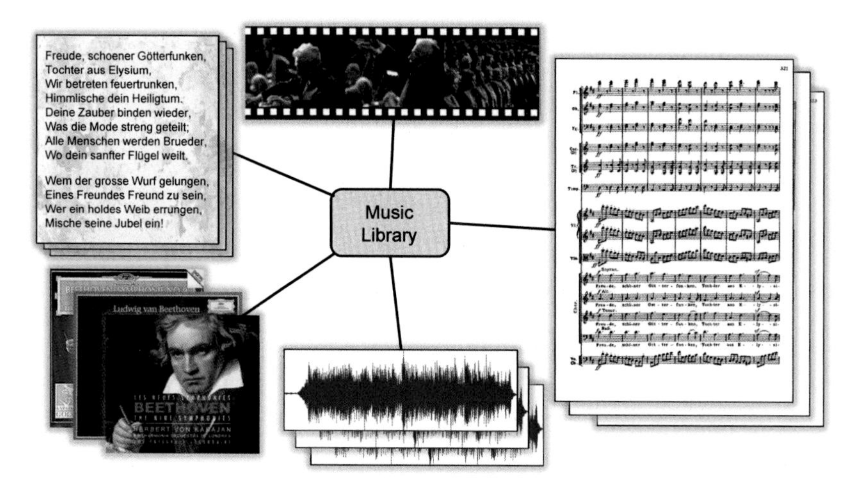

Fig. 3.20 Different document types typically available in a music library.

3.3.1.1 Interpretation Switcher Interface

As a first scenario, let us consider the case of having many different audio recordings for the same musical work. For example, for Beethoven's Fifth Symphony, a digital music library may contain interpretations by Karajan and Bernstein, historical recordings by Furtwängler and Toscanini, Liszt's piano transcription of Beethoven's Fifth played by Scherbakov and Glenn Gould, and some synthesized version generated from a MIDI file. We have seen how one can automatically link the various audio recordings by aligning musically corresponding time positions. We now describe an interface referred to as the **Interpretation Switcher**, which makes use of these alignments to enable efficient and convenient audio browsing (see Figure 3.21). This interface allows a user to select several recordings of the same piece of music, which have previously been synchronized using the techniques presented in the previous sections. Each of the recordings is represented by a slider bar indicating the current playback position with respect to the recording's particular timeline. Each timeline encodes absolute timing, where the length of a particular slider bar is proportional to the duration of the respective version. The user may listen to a specific recording by activating a slider bar and then, at any time during playback, seamlessly switch to any of the other versions. This kind of navigation between different documents is also referred to as **interdocument navigation**. As an example, in Figure 3.21, the user has selected four of the nine available recordings. As can be seen by the length of the respective slider bars, the Bernstein version has the slowest and the MIDI version the fastest tempo. At the current playback position indicated by the marker, the Bernstein recording is at time position 60 seconds, whereas the MIDI version is at position 44, the Sawallisch at 58, and the Scherbakov version at 52 seconds.

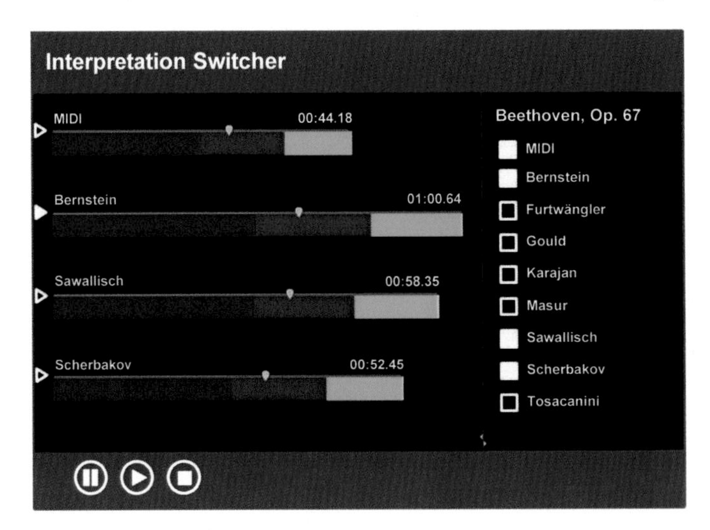

Fig. 3.21 User interface that facilitates navigation across different performances. The four sliders correspond to four different recordings of the exposition (first part) of Beethoven's Fifth Symphony. The color-coded blocks correspond to the first theme (blue), the second theme (red), and the end section (green) of the exposition.

In addition to the switching functionality, the Interpretation Switcher can be extended to also indicate available version-dependent annotations below each individual slider bar, where labeled segments are represented by color-coded blocks. Such annotations may encode chord labels generated manually or obtained by some automated chord recognition procedure (Chapter 5) or may correspond to the repetitive structure or the musical structure, which may have been extracted from the respective recording using automated structure analysis procedures (Chapter 4). Based on these annotations, the Interpretation Switcher may also facilitate navigation within a given document, where the user can directly jump to the beginning of any structural element simply by clicking on the corresponding block. This kind of navigation is also called **intradocument navigation**. For example, in Figure 3.21, the four slider bars represent the first part (the exposition) of Beethoven's Fifth Symphony. This part is further partitioned into three sections that correspond to the first theme (blue), the second theme (red), and the end section (green) of the exposition. In combination, inter- and intradocument navigation allow a user to conveniently browse through the different performances of a given musical work and to easily locate, playback, and compare musically interesting passages. As a further application, as we will discuss in Chapter 5, the Interpretation Switcher also opens up new possibilities for viewing, comparing, interacting with, and evaluating music analysis results, thus bridging the gap between signal processing and music sciences.

3.3.1.2 Score Viewer Interface

As a second scenario, let us consider a more multimodal setting dealing with audio-visual music data. We have seen that sheet music and audio recordings represent and describe music at different levels of abstraction. Sheet music specifies high-level parameters such as notes, keys, measures, or repetitions in a visual form. Because of its explicitness and compactness, Western music is often discussed and analyzed on the basis of sheet music. In contrast, most people enjoy music by listening to audio recordings, which represent music in an acoustic form. In particular, the nuances and subtleties of musical performances, which are generally not written down in the score, make the music come alive.

Let us assume that we are given scanned images of sheet music as well as an audio recording of a musical work. Then, as discussed in the introduction of this chapter, the synchronization task is to link regions (given as pixel coordinates) within the scanned images to semantically corresponding physical time positions within an audio recording. To accomplish this task, one may proceed as follows: First, using optical music recognition (OMR), the scanned images are converted into a piano-roll-like symbolic representation (see Section 1.2.4). Along with actual note information, also layout information is required, which may be specified in the form of a mapping between the musical objects given by the symbolic representation and the 2D coordinates of their depicted counterparts in the image representation. Note that a piano-roll representation has basically the same format as a log-frequency spectrogram, where the vertical axis corresponds to the 128 pitches of the equal-tempered scale (see Section 3.1.1). Therefore, the symbolic representation can be directly converted into a chroma representation as in (3.6). Furthermore, converting the audio recording into a chroma representation, the two resulting chroma feature sequences can be aligned as before. Using the synchronization result and the above mapping between the musical objects and 2D pixel coordinates, a correspondence between spatial regions in the sheet music and temporal regions in the audio recording can be derived. The quality of the resulting synchronization depends on several factors. In particular, differences between the audio and score representation may have a crucial impact on the final synchronization result. Such differences may be due to extraction errors in the OMR step, or the actual interpretation may deviate from the notated score.

Having computed the links between the images and the audio, one can realize novel ways of browsing through and experiencing music. As an example, let us discuss a **Score Viewer** interface for presenting sheet music while playing back associated audio recordings (see Figure 3.22). The main visualization mode is illustrated for two scanned pages of Beethoven's Piano Sonata Op. 13 (Pathétique). When starting audio playback, corresponding measures within the sheet music are synchronously highlighted based on the linking information generated by the synchronization procedure. In Figure 3.22a, a region in the center of the right page, corresponding to the eighth measure of the 3rd movement (Rondo), is currently highlighted by a surrounding rectangular frame. When reaching the end of an odd-numbered page during playback, pages are turned over automatically. Additional

Fig. 3.22 Score Viewer interface for multimodal music presentation and navigation (from [22]). **(a)** Synchronously to audio playback, corresponding musical measures within the sheet music are highlighted. **(b)** A page view allows the user to conveniently navigate through the currently selected musical work in a page-wise fashion. **(c)** Interpretation Switcher for seamless crossfade from one interpretation to another.

control elements allow the user to switch between measures of the currently selected musical work. By clicking on a measure, the playback position is changed and the audio recording is resumed at the appropriate time position. As indicated by Figure 3.22b, additional views such as a page view may be made available. Displaying all pages around the current playback position, the user can conveniently navigate through the entire piece of music in a page-wise fashion. In addition, if more than one music recording is available for the currently selected musical work, the Score Viewer interface may be combined with an Interpretation Switcher, as shown in Figure 3.22c. In summary, the described Score Viewer interface is a convenient way to enjoy a musical work in a multimodal way. On the one hand, the user can see the sheet music along with the currently played measure highlighted while listening to the musical work. On the other hand, the navigation within the sheet music representation yields an intuitive way to search for specific parts and to change the playback position in the audio representation.

3.3.2 Tempo Curves

In the previous application scenario, we showed how music synchronization techniques are of fundamental importance for realizing interfaces that allow users to browse, compare, or simply enjoy music in its various manifestations. We now indicate how synchronization results can also be used for automatically extracting temporal performance attributes from expressive music recordings. The motivation for such a task is that musicians give a piece of music their personal touch by continuously varying tempo, dynamics, and articulation. Instead of playing mechanically, they speed up at some places and slow down at others in order to shape a piece of music. Similarly, they continuously change the sound intensity and stress certain notes. Such performance issues are of fundamental importance for the understanding and perception of music.

The automated analysis of different interpretations, also referred to as **performance analysis**, has become an active field of research. Generally speaking, one may distinguish between two complementary goals. The first goal is to find commonalities between different interpretations, which allow for the derivation of general performance rules. For example, such a rule could be that most musicians tend to gradually slow down towards the end of a piece of music, thus closing off the composition. A second, even harder goal is to capture what exactly is characteristic for the style of a particular interpreter. For example, what makes Karajan so special? And how do his performances differ from recordings by other conductors? Before analyzing a specific performance, one requires the information about when and how the notes of the underlying piece of music are actually played. Therefore, as the first step of performance analysis, one has to annotate the performance by means of suitable attributes that make the exact timing and intensity of the various note events explicit. The extraction of such performance attributes from audio recordings constitutes a challenging problem.

Many researchers manually annotate the audio material by marking salient data points in the audio stream. However, being very labor-intensive, such a manual process is prohibitive in view of large audio collections. On the other hand, the automated extraction of performance aspects such as the precise timing of beat positions and tempo directly from a given music recording turns out to be an extremely hard problem, in particular for performed music with local tempo deviations. We will discuss the problem of tempo estimation, beat tracking, and related issues in Chapter 6.

Instead of trying to derive the tempo information only on the basis of a given music recording, we now present an alternative approach for deriving tempo-related information using synchronization techniques. Many pieces from the Western classical music literature are based on a musical score. The basic idea is to use this score

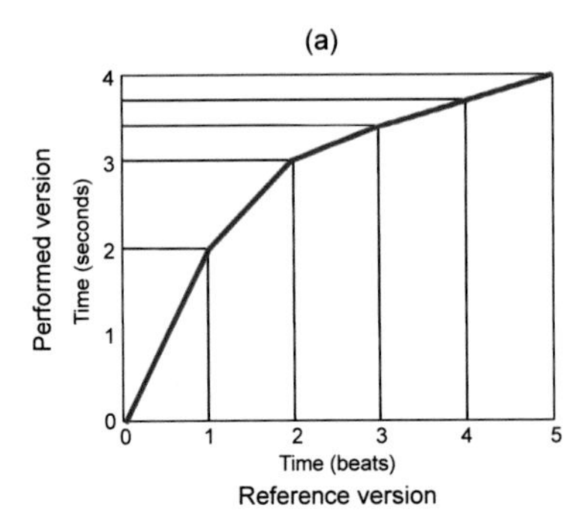

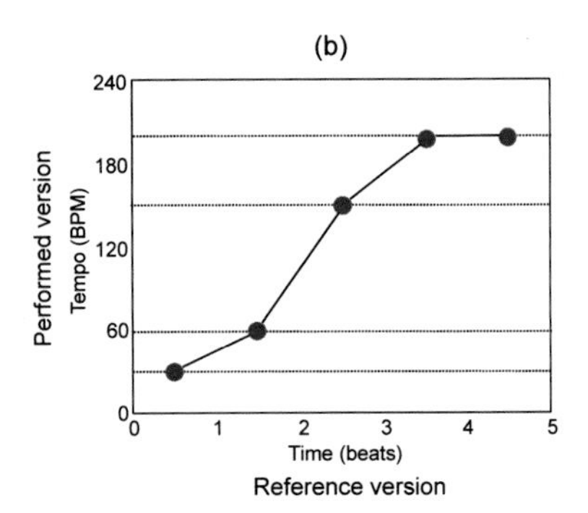

Fig. 3.23 Schematic illustration of **(a)** a warping path between a reference and a performed version and **(b)** the derived tempo curve.

as a reference version against which the performed version can be compared. In the reference version, the musical onset times as well as the pitch information of all occurring note events are known explicitly. In particular, the time axis is specified in beats per minute or BPM (see Section 1.1.2). In the first step, the score is converted into a kind of "neutral" piano-roll representation, where the notes are played with a known constant tempo in a purely mechanical way. Using music synchronization techniques, one can then temporally align the note events of the reference with their corresponding physical occurrences in the performed audio version. Now, the crucial observation is that the resulting warping path reveals the relative tempo differences between the actual performance and the neutral reference version (see Figure 3.23a). Knowing the absolute tempo of the reference version, one can then derive the tempo of the performed version at a certain position from the slope of the alignment path. More precisely, to derive a tempo value one needs to "observe" the performed version over a certain period of time. To this end, we fix a suitable time window. For example, let us fix a window on the reference time axis having a duration of one beat. From the alignment path one can read off the corresponding physical duration (given in seconds) of the performed version. In Figure 3.23a, the first beat of the reference version is aligned to a section of the performed version lasting two seconds. In other words, the performed version has in this section an average tempo of one beat every two seconds, resulting in 30 BPM. Similarly, the second beat lasts one second corresponding to 60 BPM, the third beat lasts 0.4 seconds corresponding to 150 BPM, and the fourth and fifth beats last 0.3 seconds corresponding to 200 BPM. This yields a tempo curve for the performed version specified for each beat interval of the reference version (see Figure 3.23b).

As an illustration, Figure 3.24 shows the tempo curves for three performances of the first eight measures of the "Träumerei" by Robert Schumann. Despite the significant differences in the overall tempo, there are also noticeable similarities in the relative shaping of the curves. For example, at the beginning of the second measure (region marked by the rectangular frame), all three pianists slow down. The musical reason is that there is a climax of the ascending melodic line in the first

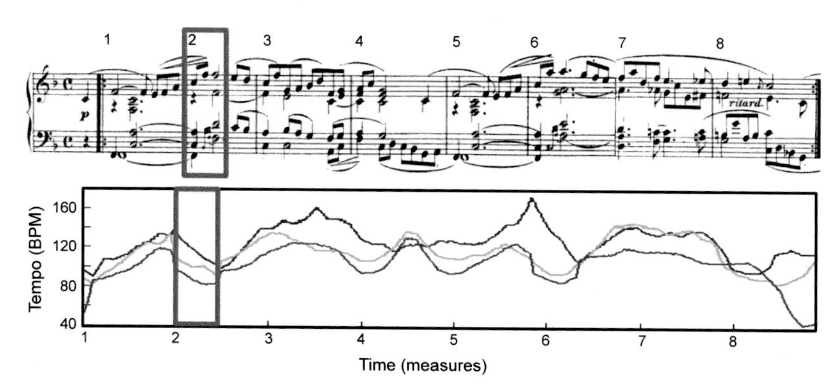

Fig. 3.24 Tempo curves for three different performances of Schumann's "Träumerei." The tempo is given in beats per minutes (BPM).

measure culminating in the marked subdominant chord (in B$^\flat$ major). This climax is further highlighted by a preceding ritardando. After the climax, the tension is released, and one can notice a considerable speed up in all three performances to return to the original tempo. As mentioned before, deriving such general rules is one of the goals of performance analysis.

We close this section with a more philosophical and critical discussion on the limitations of automated tempo extraction approaches. Generally speaking, the feeling of pulse and rhythm is one of the central components of music, closely related to what one commonly refers to as tempo. In order to define some notion of tempo, one requires a proper reference to measure against. Western music, for example, is often structured in terms of measures and beats, which allows for organizing and sectioning musical events over time. Based on a fixed time signature, one can then define the tempo as the number of beats per minute. Obviously, this definition requires a regular and steady musical beat or pulse over a certain period of time. Also, measurement itself is not as obvious as one may think. Which musical entities characterize a pulse? Are these only note onsets? How precisely can these entities be measured before getting drowned in noise? How many pulses or beats are needed to obtain a meaningful tempo estimation? With these questions, we want to indicate that the notion of tempo is far from being evident. Furthermore, due to discretization and synchronization errors, one needs numerically robust procedures to extract the tempo information by using average values over suitable time windows. Here, the window size constitutes a delicate trade-off between susceptibility to synchronization errors and sensibility towards timing nuances of the performance. In practice, it becomes a difficult problem to determine whether a given change in the tempo curve is caused by synchronization errors or whether it is the result of an actual tempo change in the performance.

3.4 Summary and Further Readings

In this chapter, we considered the task of music synchronization with the goal of identifying and linking semantically corresponding events present in different versions of the same underlying musical work. Using this task as a motivating scenario, we studied two problems that are of fundamental importance to music processing: feature extraction and sequence alignment. As for the first problem, we discussed how different music representations can be converted into common mid-level feature representations that capture relevant characteristics of the data while being invariant to aspects irrelevant for the given task. To this end, we introduced in Section 3.1 pitch- and chroma-based audio features, which capture melodic and harmonic properties of music. Then, closely following [25, Chapter 4], we studied in Section 3.2 the technique of dynamic time warping (DTW), which finds a cost-minimizing global alignment between two given sequences. Applied to sequences of chroma features, this alignment was used to account for the relative tempo differences that occur in the underlying music representations. In subsequent chapters of this book, we will encounter many more feature extraction and matching problems that are related to the techniques we have learned in this chapter.

Audio Features

As an important signal processing technique, we discussed log-frequency spectrograms based on the equal-tempered scale. In some way, such feature representations simulate the human pitch perception, which approximates to the pitch categorization underlying the equal-tempered scale. We refer to [21] for a detailed account on cognitive aspects of musical pitch. Identifying pitches that differ by one or several octaves, we then derived chroma-based audio features—sometimes also referred to as **pitch class profiles** [15]. As illustrated by Figure 3.25, there are many ways for computing and enhancing chroma features. In this chapter, we presented a simple strategy based on short-time Fourier transforms in combination with pooling strategies. To obtain better frequency resolutions, one popular alternative to using a single spectrogram is to construct a bank of bandpass filters, each corresponding to a pitch with an appropriately tuned bandwidth [25, Section 3.1]. Admittedly, this leads to the loss of the famed computational efficiency of the fast Fourier transform. However, some of this efficiency may be regained by processing the highest octave with a given frequency analysis method, then downsampling by a factor of two, and repeating for as many octaves as are desired [3]. This results in different sampling rates for each octave of the analysis, which decreases the overall computational complexity. A toolkit for such an analysis, also referred to as the constant-Q transform, has been created by Schörkhuber and Klapuri [33]. Further pre- and postprocessing steps are frequently applied to adjust the features' properties that concern spectral, temporal, and dynamical aspects. For an overview, we refer to [15, 27]. For example, preprocessing steps in the chroma computation based on logarithmic compression or related spectral whitening techniques are important to give small, yet relevant coef-

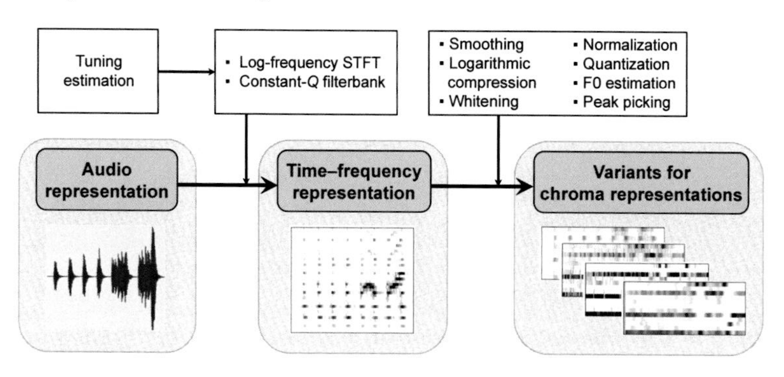

Fig. 3.25 Overview of feature extraction pipeline and chroma variants.

ficients a larger weight. Other techniques such as the estimation of the instantaneous frequency [11] (see also Section 8.2.1) or peak picking of the spectrum's local maxima often reduce the noise level significantly [15]. Generalized chroma representations with 24, 36, or even more dimensions (instead of the usual 12 dimensions) allow for dealing with tuning issues and operating with music beyond the twelve-tone equal-tempered scale [15]. Adding a further degree of abstraction by considering short-time statistics over energy distributions within the chroma bands, one obtains an entire family of scalable and robust audio features [29] (see also Section 7.2.1). To boost the degree of timbre invariance, a family of chroma-based audio features has been introduced where timbre-related information similar to that expressed by certain mel-frequency cepstral coefficients (MFCCs) has been discarded [26]. The idea of cyclically shifting chroma features to simulate transpositions has been formulated in [16] (see also Section 4.2.2.3). In recent years, deep learning techniques have been increasingly used to learn and adapt feature representations from training examples [20]. In conclusion, as we have mentioned already in Section 3.1.2, one should keep in mind that there is no "best" chroma variant and that the results of a specific music analysis task may crucially depend on the chroma type used. Besides music synchronization and alignment, we will encounter in subsequent chapters of this book various chroma variants for applications including chord recognition (Chapter 5), music structure analysis (Chapter 4), and content-based audio retrieval such as cover song and version identification (Chapter 8).

Dynamic Time Warping

In Section 3.2 on DTW, we closely followed the explanations from [25, Chapter 4]. Originating from speech processing, DTW has become a well-established method to account for temporal variations in the comparison of related time series. A classical and comprehensive account on DTW and related pattern recognition techniques is given by Rabiner and Juang [32] in the context of speech recognition. Furthermore, DTW has found numerous applications in a wide range of fields including data mining, information retrieval, bioinformatics, chemical engineering, signal processing, robotics, or computer graphics. Basically any data that can be transformed into a (linear) sequence of features can be analyzed with DTW, which includes data types such as text, video, audio, or general time series. Extensive research has been performed on how to accelerate DTW computations. For example, Keogh has suggested in the last two decades various indexing methods that make retrieval of time-warped time series feasible even for large datasets (see [19] for an early account and subsequent work). Closely related to DTW is the **edit distance** (see Exercise 3.19), which is sometimes also referred to as the **Levenshtein distance** [24]. The edit distance is used to compute a distance between strings, i.e., one-dimensional sequences consisting of discrete symbols (rather than sequences consisting of continuous features). The edit distance is used in fields such as text retrieval (spell checkers, plagiarism detection) or molecular biology to compute a distance between DNA sequences (see Exercise 3.20). For a detailed account on the edit distance with its applications to bioinformatics, we refer to standard textbooks such as [2]. A variant of the edit distance that is more robust to noise and outliers is known as the **longest common subsequence** (LCS) [6]. The computation of DTW, edit, as well as LCS distances can be done efficiently by means of dynamic programming—a widely used programming paradigm in computer science [6].

Music Synchronization

Music synchronization and related alignment tasks have been studied extensively within the field of music information retrieval. For an overview and pointers to the literature, we refer to [1, 8, 25]. Depending upon the respective types of music representations, one can distinguish between various synchronization scenarios. For example, **audio–audio** synchronization refers to the task of temporally aligning two different audio recordings of a piece of music. Similarly, the goal of **score–audio** and **MIDI–audio** synchronization is to align note and MIDI events with audio data. The task of automatically aligning scanned images of sheet music with audio or MIDI data was first tackled in [23]. Significant progress for cross-modal music matching and alignment could be achieved using deep learning techniques to learn joint feature representations [10]. Another cross-modal synchronization task is to align given lyrics to an audio recording of the underlying song [14], which turns out to be a hard problem. Again major progress could be achieved using deep learning with a loss function that involves some alignment strategy [34].

In the synchronization scenarios discussed in this chapter, the two data streams to be aligned are assumed to be known prior to the actual alignment. This assumption is exploited by DTW, which yields an optimal warping path by considering the two entire data streams. As opposed to such an **offline** scenario, one often has to deal with scenarios where the data streams are to be processed **online**. One prominent online scenario is known as **score following**, where a musician is performing a piece according to a given musical score. The goal is then to identify the currently played musical events depicted in the score with high accuracy and low latency [1, 5]. As

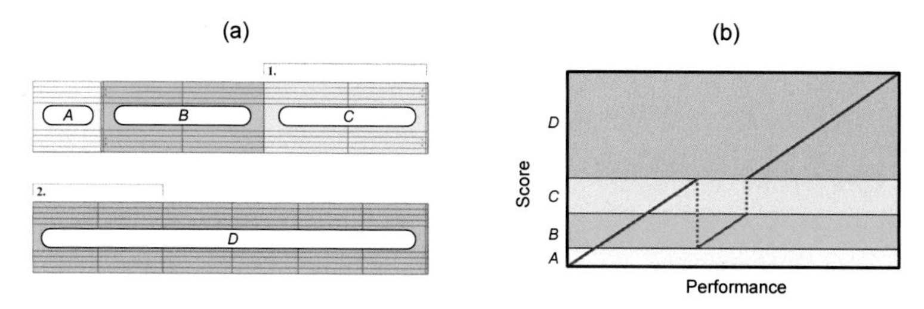

Fig. 3.26 **(a)** Schematic representation of a score with structural blocks *ABCD*. **(b)** Performance of the score as *ABCBD* (where the musician follows the repetition instructions). The resulting alignment path between the score and the performance is not monotonic and reveals sudden jumps.

opposed to classical DTW, such an online synchronization procedure inherently has a running time that is linear in the duration of the performed version. However, as a main disadvantage, an online strategy is very sensitive to local tempo variations and deviations from the score—once the procedure is out of sync, it is very hard to recover and return to the right track. In [1] a comprehensive music tracking system is described, which combines an online score-following strategy with music matching techniques for recovery. A further online synchronization problem is known as **automatic accompaniment**, where the task of the computer is to accompany a solo musician [8].

There are many more issues that need to be considered in the music synchronization context, including robustness, accuracy, and efficiency. Regarding efficiency, Dixon et al. [9] describe a linear-time DTW approach to audio synchronization based on forward path estimation. Even though the proposed algorithm is very efficient, the risk of missing the optimal alignment path is still relatively high. As an offline alternative, Prätzlich et al. [31] propose an efficient but robust music synchronization approach employing a multiscale strategy. Regarding the robustness and accuracy of the alignment, the choice of features is of crucial importance, and several feature modifications have been proposed to improve the overall synchronization quality [12, 18]. To obtain high temporal accuracy while keeping the robustness of the overall procedure, Ewert et al. [12] describe a hierarchical synchronization approach combining chroma-based features with features that capture note onset information. A further challenge is that the assumption of having a global correspondence between the sequences to be aligned is often violated. For example, recordings of the same piece of music may have a different number of stanzas, repetitions notated in the score may be omitted, or additional parts such as a solo may be added by the performer (see Figure 3.26 for an example). Various strategies have been proposed to handle such structural variations (see, e.g., [1, 13]).

Applications

As a first application of music synchronization, we described possible user interfaces that facilitate novel ways of accessing, listening to, viewing, comparing, and simply enjoying music in its various forms. An entire system for managing heterogeneous music collections, which also includes a variety of interfaces for multimodal music access and browsing, is described in [7]. For a more comprehensive overview of user interfaces, we refer to [17]. As a second application, we discussed how to derive tempo curves from audio recordings by aligning and comparing the performance with a score-like reference representation. Such a procedure is described in [28], where one also finds a discussion on how the tempo curves depend on the size of the analysis window and on the presence of synchronization inaccuracies. When it comes to expressive performances, there are many more parameters (including tempo, timing, dynamics, intonation, and articulation) that are shaped by a performer to bring out emotional and dramatic qualities of a piece of music [4]. Estimating, analyzing, and understanding such parameters in a systematic and quantitative way as well as automatically generating expressive performances have been long-standing research questions [35] which have experienced a renaissance in recent years thanks to data-driven deep learning techniques [4].

3.5 FMP Notebooks

In **Part 3** of the FMP notebooks [30], we provide and discuss Python code examples of all the components that are required to realize a basic music synchronization pipeline. In the first notebooks, we consider fundamental feature design techniques such as frequency binning, logarithmic compression, feature normalization, feature smoothing, tuning, and transposition. Then, we provide an implementation of the basic DTW procedure and introduce several experiments for exploring this technique in further depth. Finally, we close **Part 3** with a more comprehensive experiment on extracting tempo curves from music recordings, which nicely illustrates the many design choices, their impact, and the pitfalls one faces when dealing with a complex audio processing task.

We start with the **FMP Notebook *Log-Frequency Spectrogram and Chromagram***, which provides a step-by-step implementation for computing the log-frequency spectrogram as described in Section 3.1.1. Even though logarithmic frequency pooling as used in this approach has some drawbacks, it is instructive from an educational point of view. By studying the pitch-dependent sets $P(p)$ from (3.3), students gain a better understanding on how to interpret the frequency grid introduced by a discrete STFT. In particular, the pooling strategy reveals the problems associated with the insufficient frequency resolution for low pitches. In the notebook we make this problem explicit by considering a log-frequency spectrogram with empty pitch bins, leading to horizontal artifacts in our chromatic scale example. In the next step, we convert a log-frequency spectrogram into a chromagram by identifying pitches that share the same chroma. In a music processing course, it is an excellent exercise to let students compute and analyze the properties of chromagrams for music recordings of their own choice. This exploration can be done either visually, as demonstrated by our Burgmüller example, or acoustically using suitable sonification procedures as provided by the **FMP Notebook *Sonification*** of **Part B**. We close the notebook by discussing alternative variants of log-frequency spectrograms and chromagrams. In the subsequent notebooks, we often employ more elaborate chromagram implementations as provided by the Python package `librosa`.

Using spectrograms and chromagrams as instructive examples, we explore in the subsequent notebooks the effect of standard feature processing techniques. In the **FMP Notebook *Logarithmic Compression***, the discrepancy between large and small magnitude values is reduced by applying a suitable logarithmic function. To understand the effects of logarithmic compression, it is instructive to experiment with sound mixtures that contain several sources at different sound levels (e.g., a strong drum sound superimposed with a soft violin sound). In the **FMP Notebook *Feature Normalization***, we introduce different strategies for normalizing a feature representation including the Euclidean norm (ℓ^2), the Manhattan norm (ℓ^1), the maximum norm ($\ell^{\max}$), and the standard score (using mean and variance). Furthermore, similar to (3.10), we discuss different strategies for how one may handle small values (close to zero) in the normalization. This notebook is also well suited for practicing the transition from mathematical formulas to implementations. While logarithmic compression and normalization increase the robustness to variations in timbre or sound intensity, we study in the **FMP Notebook *Temporal Smoothing and Downsampling*** postprocessing techniques that can be used for making a feature sequence more robust to variations in aspects such as local tempo, articulation, and note execution. We consider two feature smoothing techniques, one based on local averaging and the other on median filtering. Using chroma representations of different recordings of Beethoven's Fifth Symphony (one of our favorite examples), we study smoothing effects and the role of the filter length (see also Figure 3.9). Finally, we introduce downsampling as a simple means to decimate the feature rate of a smoothed representation.

The **FMP Notebook *Transposition and Tuning*** covers central aspects of great musical and practical importance. As we discussed in Section 3.1.2.2, a musical transposition of one or several semitones can be simulated on the chroma level by a simple cyclic shift. We demonstrate this in the notebook using a C-major scale played on a piano (see also Figure 3.8). While transpositions are pitch shifts on the semitone level, we next discuss global frequency deviations on the sub-semitone level. Such deviations may be the result of instruments that are tuned lower or higher than the expected reference pitch A4 with center frequency 440 Hz. In the case that the tuning deviation is known, one can use this information to adjust the center and cutoff frequencies of the MIDI pitches for computing the log-frequency spectrogram and the chromagram (see Exercise 3.6). Estimating the tuning deviation, however, can be quite tricky. One way to introduce this topic in a music processing class is to let students perform, record, and analyze their own music. What is the effect when detuning a guitar or violin? How does strong vibrato affect the perception of

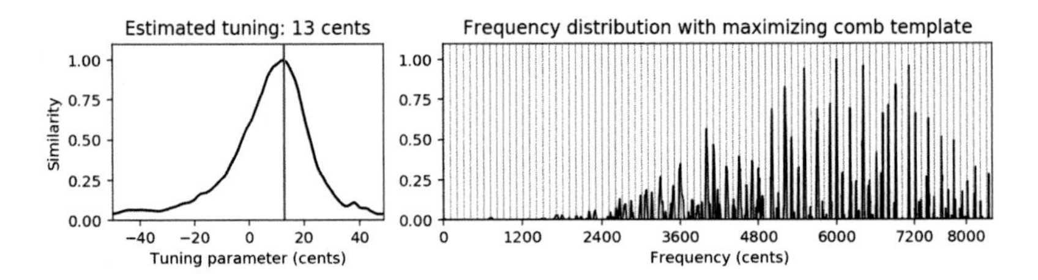

Fig. 3.27 Illustration of tuning procedure using a comb-filter approach. **Left:** Similarity function with maximizing tuning parameter at 13 cents. **Right:** Frequency distribution (with logarithmic frequency axis) and maximizing comb template (shown as red vertical lines).

pitch and tuning? What happens if the tuning changes throughout the performance? Having such issues in mind, developing and implementing a tuning estimation system can be part of an exciting and instructive student project. In this notebook, we present such a system that outputs a single number θ between -50 and $+50$ yielding the global frequency deviation (given in cents) on the sub-semitone level. In our approach, as illustrated by Figure 3.27, we first compute a frequency distribution from the given music recording, where we use different techniques such as the STFT, logarithmic compression, interpolation, local average subtraction, and rectification. The resulting distribution is then compared with comb-like template vectors, each representing a specific tuning. The template vector that maximizes the similarity to the distribution yields the tuning estimate. Furthermore, we conduct in the notebook various experiments that illustrate the benefits and limitations of our approach, while confronting the student with the various challenges one encounters when dealing with real music data.

In the next notebooks, we cover the second main topic of the chapter, dealing with alignment techniques (see Section 3.2). In the **FMP Notebook *Dynamic Time Warping (DTW)***, we provide an implementation of the basic DTW algorithm, closely following the description as presented in Table 3.2. This is a good opportunity for pointing out an issue one often faces in programming. In mathematics and some programming languages (e.g., MATLAB), one uses the convention that indexing starts with the index 1. In other programming languages such as Python, however, indexing starts with the index 0. Neither convention is good or bad. In practice, one needs to make adjustments in order to comply with the respective convention. Implementing the DTW algorithm is a good exercise to make students aware of this issue, which is often a source of programming errors. The **FMP Notebook *DTW Variants*** investigates the role of the step size condition, local weights, and global constraints (see Section 3.2.2). Rather than implementing all these variants from scratch, we employ a function from the Python package `librosa` and discuss various parameter settings. Finally, in the **FMP Notebook *Music Synchronization***, we apply the DTW algorithm in the context of our music synchronization scenario. Considering two performances of the beginning of Beethoven's Fifth Symphony (first 20 measures), we first convert the music recordings into chromagrams,

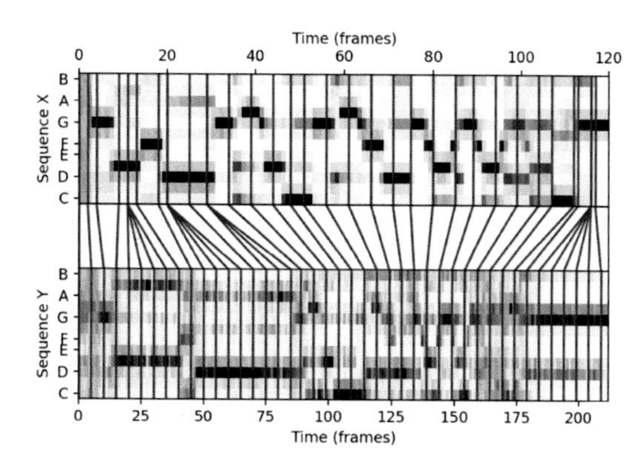

Fig. 3.28 Music synchronization result obtained for two input chromagrams.

which are then used as input to the DTW algorithm. The resulting warping path constitutes our synchronization result, as shown in Figure 3.28

Concluding **Part 3** of the FMP notebooks, we provide additional material for the applications described in Section 3.3. In the **FMP Notebook *Application: Music Navigation***, one finds two videos that illustrate the main functionalities of the Interpretation Switcher and Score Viewer Interface. Then, in the **FMP Notebook *Application: Tempo Curves***, we present an extensive experiment for extracting tempo information from a given music recording, following the overall ideas described in Section 3.3.2. Besides the recorded performance, one requires a score-based reference version, which we think of as a piano-roll representation with a musical time axis (given in measures and beats). On the basis of chroma representations, we apply DTW to compute a warping path between the performance and the score. Then, the idea is to compute the slope of the warping path and to take its reciprocal to derive the local tempo. In practice, however, this becomes problematic when the warping path runs horizontally (slope is zero) or vertically (slope is infinite). In the notebook, we solve this issue by thinning out the warping path to enforce strict monotonicity in both dimensions and then continue as indicated before. To make the overall procedure more robust, we also apply a local smoothing strategy in the processing pipeline. Our overall processing pipeline not only involves many steps with a multitude of parameters, but is also questionable from a musical point of view. Using the famous romantic piano piece "Träumerei" by Robert Schumann as a concrete real-world example, we discuss two conflicting goals. On the one hand, the tempo estimation procedure should be robust to local outliers that are the result of computational artifacts (e.g., inaccuracies of the DTW alginment). On the other hand, the procedure should be able to adapt to continuous tempo fluctuations and sudden tempo changes, being characteristic features of expressive performances. Through studying tempo curves and the way they are computed, one can learn a lot about the music as well as computational approaches. Also, this topic leads students to challenging and interdisciplinary research problems.

References

1. A. Arzt, *Flexible and Robust Music Tracking*, PhD thesis, Universität Linz, 2016.
2. H.-J. Böckenhauer and D. Bongartz, *Algorithmische Grundlagen der Bioinformatik: Modelle, Methoden und Komplexität*, Teubner, 2003.
3. J. C. Brown and M. S. Puckette, *An efficient algorithm for the calculation of a constant Q transform*, Journal of the Acoustic Society of America (JASA), 92 (1992), pp. 2698–2701.
4. C. E. C. Chacón, M. Grachten, W. Goebl, and G. Widmer, *Computational models of expressive music performance: A comprehensive and critical review*, Frontiers Digit. Humanit., 5 (2018), p. 25.
5. A. Cont, *A coupled duration-focused architecture for real-time music-to-score alignment*, IEEE Transactions on Pattern Analysis and Machine Intelligence, 32 (2010), pp. 974–987.
6. T. H. Cormen, C. E. Leiserson, R. L. Rivest, and C. Stein, *Introduction to Algorithms*, McGraw-Hill Higher Education, 2001.
7. D. Damm, C. Fremerey, V. Thomas, M. Clausen, F. Kurth, and M. Müller, *A digital library framework for heterogeneous music collections: from document acquisition to cross-modal interaction*, International Journal on Digital Libraries: Special Issue on Music Digital Libraries, 12 (2012), pp. 53–71.
8. R. B. Dannenberg and C. Raphael, *Music score alignment and computer accompaniment*, Communications of the ACM, Special Issue: Music Information Retrieval, 49 (2006), pp. 38–43.
9. S. Dixon and G. Widmer, *MATCH: A music alignment tool chest*, in Proceedings of the International Society for Music Information Retrieval Conference (ISMIR), London, UK, 2005, pp. 492–497.
10. M. Dorfer, J. Hajič Jr., A. Arzt, H. Frostel, and G. Widmer, *Learning audio-sheet music correspondences for cross-modal retrieval and piece identification*, Transactions of the International Society for Music Information (TISMIR), 1 (2018), pp. 22–31.
11. D. P. Ellis and G. E. Poliner, *Identifying 'cover songs' with chroma features and dynamic programming beat tracking*, in Proceedings of the IEEE International Conference on Acoustics, Speech, and Signal Processing (ICASSP), vol. 4, Honolulu, Hawaii, USA, 2007, pp. 1429–1432.
12. S. Ewert, M. Müller, and P. Grosche, *High resolution audio synchronization using chroma onset features*, in Proceedings of IEEE International Conference on Acoustics, Speech, and Signal Processing (ICASSP), Taipei, Taiwan, Apr. 2009, pp. 1869–1872.
13. C. Fremerey, M. Müller, and M. Clausen, *Handling repeats and jumps in score-performance synchronization*, in Proceedings of the International Society for Music Information Retrieval Conference (ISMIR), Utrecht, The Netherlands, 2010, pp. 243–248.
14. H. Fujihara and M. Goto, *Lyrics-to-audio alignment and its application*, in Multimodal Music Processing, M. Müller, M. Goto, and M. Schedl, eds., vol. 3 of Dagstuhl Follow-Ups, Schloss Dagstuhl – Leibniz-Zentrum für Informatik, Dagstuhl, Germany, 2012, pp. 23–36.
15. E. Gómez, *Tonal Description of Music Audio Signals*, PhD thesis, Universitat Pompeu Fabra, Barcelona, Spain, 2006.
16. M. Goto, *A chorus-section detecting method for musical audio signals*, in Proceedings of the IEEE International Conference on Acoustics, Speech, and Signal Processing (ICASSP), Hong Kong, China, 2003, pp. 437–440.
17. M. Goto and R. B. Dannenberg, *Music interfaces based on automatic music signal analysis: New ways to create and listen to music*, IEEE Signal Processing Magazine, 36 (2019), pp. 74–81.
18. C. Joder, S. Essid, and G. Richard, *Optimizing the mapping from a symbolic to an audio representation for music-to-score alignment*, in Proceedings of the IEEE Workshop on Applications of Signal Processing to Audio and Acoustics (WASPAA), New Paltz, NY, USA, 2011, pp. 121–124.
19. E. Keogh, *Exact indexing of dynamic time warping*, in Proceedings of the VLDB Conference, Hong Kong, 2002, pp. 406–417.

20. F. KORZENIOWSKI AND G. WIDMER, *Feature learning for chord recognition: The deep chroma extractor*, in Proceedings of the International Society for Music Information Retrieval Conference (ISMIR), New York City, USA, 2016, pp. 37–43.
21. C. L. KRUMHANSL, *Cognitive Foundations of Musical Pitch*, Oxford University Press, Oxford, UK, 1990.
22. F. KURTH, D. DAMM, C. FREMEREY, M. MÜLLER, AND M. CLAUSEN, *A framework for managing multimodal digitized music collections*, in ECDL, 2008, pp. 334–345.
23. F. KURTH, M. MÜLLER, C. FREMEREY, Y. CHANG, AND M. CLAUSEN, *Automated synchronization of scanned sheet music with audio recordings*, in Proceedings of the International Society for Music Information Retrieval Conference (ISMIR), Vienna, Austria, 2007, pp. 261–266.
24. V. I. LEVENSHTEIN, *Binary codes capable of correcting deletions, insertions, and reversals*, Soviet Physics Doklady, 10 (1966), pp. 707–710.
25. M. MÜLLER, *Information Retrieval for Music and Motion*, Springer Verlag, 2007.
26. M. MÜLLER AND S. EWERT, *Towards timbre-invariant audio features for harmony-based music*, IEEE Transactions on Audio, Speech, and Language Processing, 18 (2010), pp. 649–662.
27. ———, *Chroma Toolbox: MATLAB implementations for extracting variants of chroma-based audio features*, in Proceedings of the International Society for Music Information Retrieval Conference (ISMIR), Miami, Florida, USA, 2011, pp. 215–220.
28. M. MÜLLER, V. KONZ, A. SCHARFSTEIN, S. EWERT, AND M. CLAUSEN, *Towards automated extraction of tempo parameters from expressive music recordings*, in Proceedings of the International Society for Music Information Retrieval Conference (ISMIR), Kobe, Japan, Oct. 2009, pp. 69–74.
29. M. MÜLLER, F. KURTH, AND M. CLAUSEN, *Chroma-based statistical audio features for audio matching*, in Proceedings of the IEEE Workshop on Applications of Signal Processing (WASPAA), New Paltz, NY, USA, Oct. 2005, pp. 275–278.
30. M. MÜLLER AND F. ZALKOW, *FMP Notebooks: Educational material for teaching and learning fundamentals of music processing*, in Proceedings of the International Society for Music Information Retrieval Conference (ISMIR), Delft, The Netherlands, 2019, pp. 573–580.
31. T. PRÄTZLICH, J. DRIEDGER, AND M. MÜLLER, *Memory-restricted multiscale dynamic time warping*, in Proceedings of the IEEE International Conference on Acoustics, Speech, and Signal Processing (ICASSP), Shanghai, China, March 2016, pp. 569–573.
32. L. RABINER AND B.-H. JUANG, *Fundamentals of Speech Recognition*, Prentice Hall Signal Processing Series, 1993.
33. C. SCHÖRKHUBER AND A. P. KLAPURI, *Constant-Q transform toolbox for music processing*, in Proceedings of the Sound and Music Computing Conference (SMC), Barcelona, Spain, 2010.
34. D. STOLLER, S. DURAND, AND S. EWERT, *End-to-end lyrics alignment for polyphonic music using an audio-to-character recognition model*, in Proceedings of the IEEE International Conference on Acoustics, Speech, and Signal Processing (ICASSP), Brighton, UK, 2019, pp. 181–185.
35. G. WIDMER, S. DIXON, W. GOEBL, E. PAMPALK, AND A. TOBUDIC, *In search of the Horowitz factor*, AI Magazine, 24 (2003), pp. 111–130.

Exercises

Exercise 3.1. In Section 3.1.1, we computed a log-frequency spectrogram based on a semitone resolution using (3.3) and (3.4). In this exercise, we want to specify a log-frequency spectrogram with a resolution of half a semitone (resulting in 24 bands per octave). Write a small computer program that calculates the corresponding center frequencies, the cutoff frequencies, and the bandwidths for the various log-frequency bands, each corresponding to a half semitone (as in Table 3.1). Output all numbers for the resulting 25 bands between C4 and C5. Then, do the same for a log-frequency spectrogram with a resolution of a third semitone (resulting in 36 bands per octave). Again, output all numbers for the resulting 37 bands between C4 and C5.

Exercise 3.2. Assuming a sampling rate of $F_\mathrm{s} = 44100$ Hz and a window length of $N = 4096$, determine the largest pitch p for which the set $P(p)$ defined in (3.3) is empty. What are the center frequency, the cutoff frequencies, and the bandwidth of the corresponding log-frequency band?

Exercise 3.3. Let $\mathrm{BW}(p) = F_\mathrm{pitch}(p+0.5) - F_\mathrm{pitch}(p-0.5)$ be the bandwith for a pitch p as defined in (3.5). What is the relation between the bandwidths $\mathrm{BW}(p+12)$ and $\mathrm{BW}(p)$ of two pitches that are one octave apart? Give a mathematical proof for your claim. Similarly, determine the relation between the bandwidths $\mathrm{BW}(p+1)$ and $\mathrm{BW}(p)$ of two neighboring pitches.

Exercise 3.4. Given an audio signal at a sampling rate of $F_\mathrm{s} = 22050$ Hz, we want to compute a log-frequency spectrogram as in (3.4). As a requirement, all sets $P(p)$ (as defined in (3.3)) for all pitches corresponding to the notes C2 ($p = 36$) to C3 ($p = 48$) should contain at least four Fourier coefficients. To meet this requirement, what is the minimal window length N (assuming that N is a power of two) to be used in the STFT? For this N, determine the elements of the set $P(36)$ explicitly.

Exercise 3.5. The tuning of musical instruments is usually based on a fixed reference pitch. In Western music, one typically uses the **concert pitch** A4 having a frequency of 440 Hz (see Section 1.3.2). To estimate the deviation from this ideal reference, a musician is asked to play the note A4 on his or her instrument over the duration of four seconds. Describe a simple FFT-based procedure for estimating the tuning deviation of the instrument used. How would you choose the parameters (sampling rate, window size) to obtain an accuracy of at least 1 Hz in this estimation?

Exercise 3.6. Assume that an orchestra is tuned 20 cents upwards compared with the standard tuning. What is the center frequency of the tone A4 in this tuning? How can a chroma representation be adjusted to compensate for this tuning difference?

Exercise 3.7. Show that the DTW distance as defined in (3.21) is symmetric (i.e., $\mathrm{DTW}(X,Y) = \mathrm{DTW}(Y,X)$ for any two given sequences $X = (x_1, x_2, \ldots, x_N)$ and $Y = (y_1, y_2, \ldots, y_M)$) in the case that the local cost measure c is symmetric.

Exercise 3.8. Let $P = (p_1, p_2, \ldots, p_L)$ be an arbitrary (N,M)-warping path. Specify the smallest possible lower bound as well as the largest possible upper bound for the length L of P in terms of N and M.

Exercise 3.9. In this exercise, we show that there is a large number of theoretically possible warping paths. Let $\mu(N,M)$ be the number of possible (N,M)-warping paths for some given N and M. Obviously, in the case $N = 1$ or $M = 1$, there is only one possible warping path, i.e., $\mu(1,M) = \mu(N,1) = 1$. Show that $\mu(2,2) = 3$, $\mu(2,3) = 5$, and $\mu(3,3) = 13$. Derive a general recursive formula for $\mu(N,M)$ for $N > 1$ and $M > 1$. Compute $\mu(N,M)$ for $(N,M) \in [1:6] \times [1:6]$.

Exercise 3.10. Let $F = \mathbb{R}$ be a feature space and $c : \mathcal{F} \times \mathcal{F} \to \mathbb{R}_{\geq 0}$ a local cost measure defined by $c(x,y) = |x - y|$ for $x, y \in \mathbb{R}$. Compute $\mathrm{DTW}(X,Y)$ for the following sequences X and Y as well as all optimal warping paths. Also specify the cost matrix $\mathbf{C}$ and the accumulated cost matrix $\mathbf{D}$.

(a) $X = (1,7,4,4,6)$ and $Y = (1,2,2,7)$.
(b) $X = (1,2,2,1)$ and $Y = (1,0,0,1)$.

Exercise 3.11. In this excercise, we show that the DTW distance generally does not satisfy the triangle inequality. Let $\mathcal{F} := \{\alpha,\beta,\gamma\}$ be an abstract feature space consisting of three different elements. Define a cost measure $c : \mathcal{F} \times \mathcal{F} \to \{0,1\}$ by setting $c(x,y) := 1 - \delta_{xy}$ for $x,y \in \mathcal{F}$. In other words, $c(x,y) := 0$ if $x = y$ and $c(x,y) := 1$ if $x \neq y$ for $x,y \in \mathcal{F}$. Note that c defines a metric on $\mathcal{F}$ and, in particular, satisfies the triangle inequality. Now, consider the three sequences $X := (\alpha,\gamma,\gamma)$, $Y := (\alpha,\beta,\gamma)$, and $Z := (\alpha,\beta,\beta,\gamma)$ over $\mathcal{F}$. Compute $\mathrm{DTW}(X,Y)$, $\mathrm{DTW}(Y,Z)$, and $\mathrm{DTW}(X,Z)$. Furthermore, show that the triangle inequality does not hold in this example.

Exercise 3.12. Let $\mathcal{F} = \{\alpha,\beta,\gamma\}$ and $c : \mathcal{F} \times \mathcal{F} \to \{0,1\}$ be as in Exercise 3.11. Specify the DTW distances $\mathrm{DTW}(X,Y)$, $\mathrm{DTW}(X,Z)$, and $\mathrm{DTW}(Y,Z)$ for the sequences $X = (\gamma,\alpha,\beta)$, $Y = (\alpha,\alpha,\gamma,\alpha)$, and $Z = (\alpha,\beta,\gamma,\alpha,\beta,\gamma)$. Instead of using the dynamic programming approach, try to "guess" the DTW distances by specifying suitable warping paths. Then, argue that the specified warping paths are indeed optimal.

Exercise 3.13. Extend the accumulated cost matrix $\mathbf{D}$ from Section 3.2.1.3 by an additional row and column indexed by 0. Define $\mathbf{D}(n,0) := \infty$ for $n \in [1:N]$, $\mathbf{D}(0,m) := \infty$ for $m \in [1:M]$, and $\mathbf{D}(0,0) := 0$. Show that one obtains the original accumulated cost matrix when applying the recursion of (3.25) for $n \in [1:N]$ and $m \in [1:M]$.
[**Hint:** When computing with the value ∞, we assume that the sum of the value ∞ with a finite value is defined to be ∞. Furthermore, the minimum over a set containing finite values as well as the value ∞ is defined to be the minimum over the finite values.]

Exercise 3.14. In this exercise, we consider DTW with the step size condition $\Sigma = \{(2,1),(1,2),(1,1)\}$ (see (3.30)). As in Exercise 3.13, we extend the accumulated cost matrix $\mathbf{D}$, this time by two additional rows and columns indexed by -1 and 0. Then we set $\mathbf{D}(1,1) := \mathbf{C}(1,1)$, $\mathbf{D}(n,-1) := \mathbf{D}(n,0) := \infty$ for $n \in [-1:N]$, and $\mathbf{D}(-1,m) := \mathbf{D}(0,m) := \infty$ for $m \in [-1:M]$. $\mathbf{D}$ is then computed using the recursion of (3.31) for $n \in [1:N]$ and $m \in [1:M]$. Specify the cells $(n,m) \in [-1:N] \times [-1:M]$ for which one obtains $\mathbf{D}(n,m) = \infty$. Furthermore, describe some meaningful constraints for the lengths N and M in this alignment scenario.

Exercise 3.15. Let $F = \mathbb{R}$ be a feature space and $c : \mathcal{F} \times \mathcal{F} \to \mathbb{R}_{\geq 0}$ a local cost measure defined by $c(x,y) = |x-y|$ for $x,y \in \mathbb{R}$ (see Exercise 3.10). Compute $\mathrm{DTW}(X,Y)$ for the sequences $X = (1,7,4,4,6)$ and $Y = (1,2,2,7)$ as well as all optimal warping paths using the step size condition $\Sigma = \{(2,1),(1,2),(1,1)\}$ from (3.30). Also specify the cost matrix $\mathbf{C}$ and the accumulated cost matrix $\mathbf{D}$ using two additional rows and columns initialized with ∞ (see Exercise 3.14).

Exercise 3.16. In software such as MATLAB, an operation expressed as a matrix product can often be computed more efficiently than, e.g., using nested loops over the matrix indices. This motivates the following exercise. Let $c : \mathcal{F} \times \mathcal{F} \to \mathbb{R}$ be the cosine distance for $\mathcal{F} = \mathbb{R}^{12} \setminus \{0\}$ (see (3.14)). Given two feature sequences $X = (x_1,x_2,\ldots,x_N)$ and $Y = (y_1,y_2,\ldots,y_M)$ over $\mathcal{F}$, let $\mathbf{C}(n,m) := c(x_n,y_m)$ be the resulting cost matrix for $n \in [1:N]$ and $m \in [1:M]$ (see (3.13)). Show how $\mathbf{C}$ can be computed using matrix products (instead of a nested loop over the indices n and m to compute the individual entries $\mathbf{C}(n,m)$).

Exercise 3.17. Assume that, for two given sequences $X = (x_1,\ldots,x_N)$ and $Y = (y_1,\ldots,y_M)$, there is exactly one optimal (N,M)-warping path denoted by P^*. Furthermore, let $R \subseteq [1:N] \times [1:M]$ be a global constraint region (see Section 3.2.2.3). Show that the constrained optimal warping path P^*_R coincides with P^* if and only if P^* is contained in R.

Exercise 3.18. In this exercise, we analyze the multiscale approach to DTW (MsDTW) as outlined in Section 3.2.2.4. Let $X = (x_1,x_2,\ldots,x_N)$ and $Y = (y_1,y_2,\ldots,y_M)$ be sequences of length N and M, respectively. For simplicity, we assume that $N = M = 2^K$ for a natural number $K \in \mathbb{N}$.

Let $A^{\mathrm{DTW}}(N) = N^2$ denote the number of evaluations of the local cost measure that are required in the classical DTW algorithm. Furthermore, we assume that we have a coarsening and downsampling procedure for computing the coarsened sequences $X_1, X_2, \ldots, X_K$ and $Y_1, Y_2, \ldots, Y_K$, where the sampling rates are successively reduced by factors $f_1 = f_2 = \ldots = f_K = 2$. In the subsequent analysis, we neglect the operations required for the coarsening and downsampling procedure. Let $A^{\mathrm{MsDTW}}(N)$ denote the number of evaluations of the local cost measure that are required in the MsDTW algorithm. Specify a recursive equation for $A^{\mathrm{MsDTW}}(N)$. Derive from this equation an upper bound for $A^{\mathrm{MsDTW}}(N)$.
[**Hint:** Look at an upper bound for the length of a warping path at level k, $1 \leq k \leq K$.]

Exercise 3.19. In computer science, the **edit distance** (sometimes also referred to as the **Levenshtein distance**) is a string metric for measuring the difference between two sequences $X = (x_1, x_2, \ldots, x_N)$ and $Y = (y_1, y_2, \ldots, y_M)$ over an alphabet $\mathcal{F}$. The sequences are also often called **words**, and the elements of the alphabet are called **characters**. The edit distance $\mathrm{Edit}(X,Y)$ between X and Y is defined to be the minimum number of single-character edits required to change one sequence into the other. One allows three kinds of single-character edits referred to as **insertion** (including an additional character), **deletion** (omitting a character of a word), and **substitution** (replacing a character of a word by another character). Develop an algorithm based on dynamic programming (as in Table 3.2) that computes the edit distance between two given sequences X and Y.
[**Hint:** Define an accumulated cost matrix using (3.22). Let ε denote the empty word of length zero. Use this empty word as a recursion start to compute the accumulated cost matrix. For an example application, see Exercise 3.20.]

Exercise 3.20. The edit distance as introduced in Exercise 3.19 finds applications in biochemistry to compare the primary structures of biological molecules. In this exercise, we consider the case of **deoxyribonucleic acid** or DNA, which is a molecule that encodes the genetic instructions used in the development and functioning of living organisms. The primary structure of DNA can be specified by a sequence of simpler units called **nucleotides**, which are associated to base components referred to as adenine (A), cytosine (C), guanine (G), and thymine (T). Therefore, the primary structure of a DNA molecule can be specified by a sequence over the alphabet $\mathcal{F} := \Sigma := \{\mathtt{A}, \mathtt{C}, \mathtt{G}, \mathtt{T}\}$. In evolutionary biology, **homology** is the similarity between attributes of organisms (e.g., genes) that results from their shared ancestry. In genetics, homology is measured by comparing DNA sequences. A high sequence similarity between two DNA sequences is an indicator for a high probability of being homologous (e.g., sharing a common ancestor). Typical differences of homologous sequences caused by **mutation** are **substitutions** (e.g., $\mathtt{TGAT} \rightsquigarrow \underline{\mathtt{G}}\mathtt{GAT}$), **insertions** (e.g., $\mathtt{TGAT} \rightsquigarrow \mathtt{T}\dot{\mathtt{C}}\mathtt{GAT}$), and **deletions** (e.g., $\mathtt{TGAT} \rightsquigarrow \mathtt{T}\cancel{\mathtt{G}}\mathtt{AT}$). This illustrates why the edit distance is suitable for comparing the distance (or similarity) of DNA sequences.

By applying Exercise 3.19, compute the edit distance, the accumulated cost matrix, as well as the sequence of edits for the two sequences $X = \mathtt{TGAT}$ and $Y = \mathtt{CGAGT}$.

Exercise 3.21. Another problem related to DTW and the edit distance is known as the **longest common subsequence** (LCS) problem. Given two sequences $X = (x_1, x_2, \ldots, x_N)$ and $Y = (y_1, y_2, \ldots, y_M)$ over an alphabet $\mathcal{F}$, the goal is to find a longest subsequence common to both sequences. For example, the sequences $X = (b,a,b,c,b)$ and $Y = (a,b,b,c,c,b)$ over the alphabet $\mathcal{F} = \{a,b,c\}$ have the longest common subsequence (a,b,c,b). Develop an algorithm based on dynamic programming (as in Table 3.2) for determining the length $\mathrm{LCS}(X,Y)$ of a longest common subsequence of X and Y. Then, determine a longest common subsequence via backtracking. Finally, apply the algorithm to the two sequences $X = (b,a,b,c,b)$ and $Y = (a,b,b,c,c,b)$.
[**Hint:** Define an accumulated similarity matrix as in (3.22). Let ε denote the empty sequence of length zero. Use this empty sequence as a recursion start for computing the accumulated similarity matrix.]

Chapter 4
Music Structure Analysis

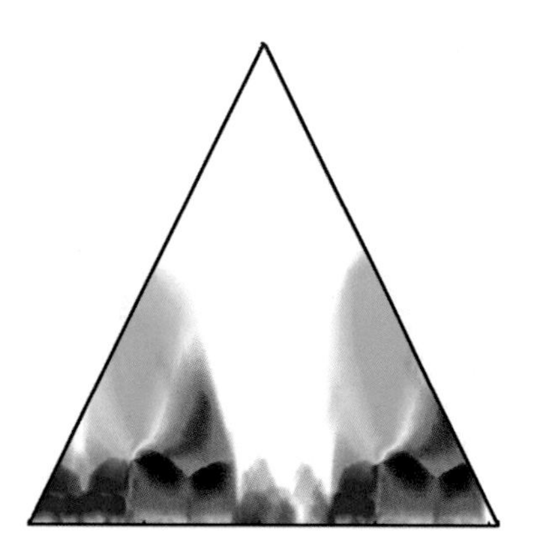

One of the attributes distinguishing music from random sound sources is the hierarchical structure in which music is organized. At the lowest level, one has events such as individual notes, which are characterized by the way they sound, their timbre, pitch, and duration. Combining various sound events, one obtains larger structures such as motifs, phrases, and sections, and these structures again form larger constructs that determine the overall layout of the composition. This higher structural level is also referred to as the musical structure of the piece, which is specified in terms of musical parts and their mutual relations. For example, in popular music such parts can be the intro, the chorus, and the verse sections of the song. Or in classical music, they can be the exposition, the development, and the recapitulation of a movement. The general goal of **music structure analysis** is to divide a given music representation into temporal segments that correspond to musical parts and to group these segments into musically meaningful categories.

Let us consider a concrete example. Figure 4.1a shows a sheet music representation of the Mazurka Op. 6, No. 4 by the Polish composer Frédéric Chopin. This piano piece can be subdivided into five sections, where the third and fifth sections are repetitions of the first section. Therefore, these sections belong to the same category denoted by the symbol A. Similarly, the fourth section is a repetition of the second one. These two sections belong to another group labeled by the symbol B. Hence, at an abstract level, the overall musical structure can be described by the sequence $A_1B_1A_2B_2A_3$ (see Figure 4.1d). Instead of using the musical score, one typical scenario is to derive structural information from a given audio recording

M. Müller, *Fundamentals of Music Processing*, https://doi.org/10.1007/978-3-030-69808-9_4

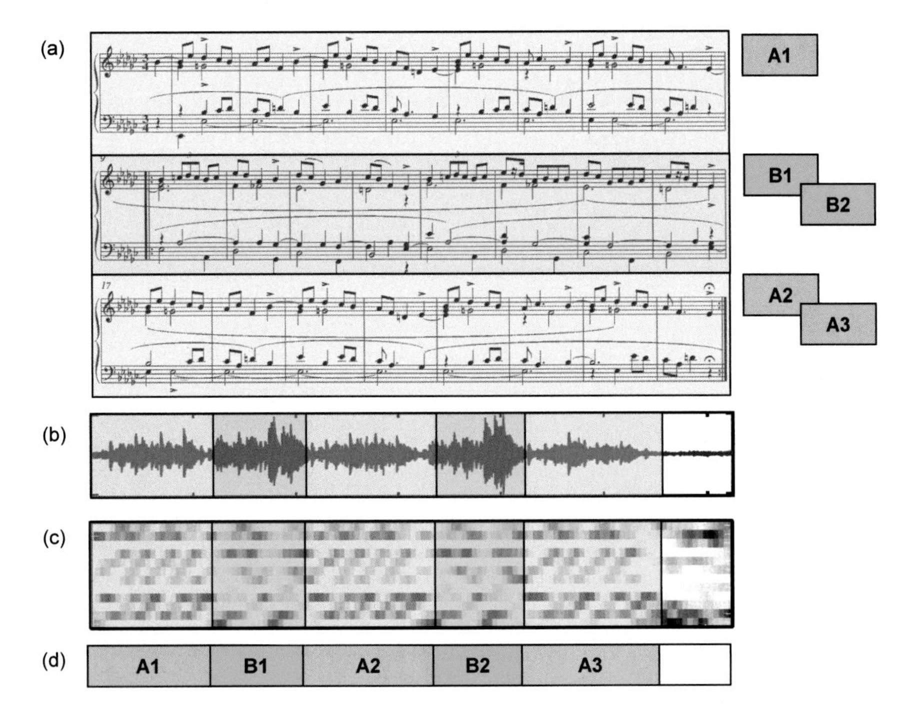

Fig. 4.1 Musical structure of the Mazurka Op. 6, No. 4 by Chopin. **(a)** Sheet music representation. **(b)** Waveform of an audio recording. **(c)** Chroma representation derived from (b). **(d)** Manually annotated segmentation of the audio recording.

(see Figure 4.1b). To this end, one needs to convert the waveform into a suitable feature representation that captures musical properties relevant for the structure of interest. In our example, as shown by Figure 4.1c, the repetition-based structure can be seen in a chroma representation that captures harmonic information.

As demonstrated by the previous example, the musical structure is often related to recurring patterns such as repeating sections. In general, however, there are many more criteria for segmenting and structuring music. For example, certain musical sections may be characterized by some homogeneity property such as a consistent timbre, the presence of a specific instrument, or the usage of certain harmonies. Furthermore, segment boundaries may go along with sudden changes in musical properties such as tempo, dynamics, or the musical key. These various segmentation principles require different methods, which may be loosely categorized into repetition-based, homogeneity-based, and novelty-based approaches.

In this chapter, we study general techniques for deriving structural information from a given music recording. In Section 4.1, we start by giving an overview of different segmentation principles, while introducing a working definition of the structure analysis problem as used in the subsequent sections. Furthermore, we discuss some feature representations that account for different musical dimensions. The concept of self-similarity matrices, which we study in Section 4.2, is of fundamental importance in computational music structure. In particular, we show how the various segmentation principles are reflected in such matrices and how this can be exploited for deriving structural information. As a first application of self-similarity matrices, we discuss in Section 4.3 a subproblem of music structure analysis known as audio thumbnailing. The goal of this problem is to determine the audio segment that best represents a given music recording. Providing a compact preview, such audio segments are useful for music navigation applications similar to visual thumbnails that help in organizing and accessing large photo collections. While we apply repetition-based principles for audio thumbnailing, we discuss in Section 4.4 some segmentation procedures that rely on novelty-based principles. The objective of such procedures is to specify points within a given audio recording where a human listener would recognize a change, a sudden event, or the transition between two contrasting parts. Finally, in Section 4.5, we address the issue of evaluating analysis results, which itself constitutes a nontrivial problem.

4.1 General Principles

Music structure analysis is a multifaceted and often ill-defined problem that depends on many different aspects. First of all, the complexity of the problem depends on the kind of music representation to be analyzed. For example, while it is comparatively easy to detect certain structures such as repeating melodies in sheet music, it is often much harder to automatically identify such structures in audio representations. Second, there are various principles including homogeneity, repetition, and novelty that a segmentation may be based on. While the musical structure of the piano piece shown in Figure 4.1 is based on repetition, musical parts in other music may be characterized by a certain instrumentation or tempo. Third, one also has to account for different musical dimensions, such as melody, harmony, rhythm, or timbre. For example, in Beethoven's Fifth Symphony the "fate motif" is repeated in various ways—sometimes the motif is shifted in pitch; sometimes only the rhythmic pattern is preserved. Finally, the segmentation and structure largely depend on the musical context and the temporal hierarchy to be considered. For example, the recapitulation of a sonata may be considered a kind of repetition of the exposition on a coarse temporal level even though there may be significant modifications in melody and harmony on a finer temporal level. Figure 4.2 gives an overview of various aspects that need to be considered when dealing with musical structures. In the following, we discuss these aspects in more detail. In particular, our goal is to raise the awareness that computational procedures as described in the subsequent sections are often based on simplifying model assumptions that only reflect certain aspects of the complex structural properties of music.

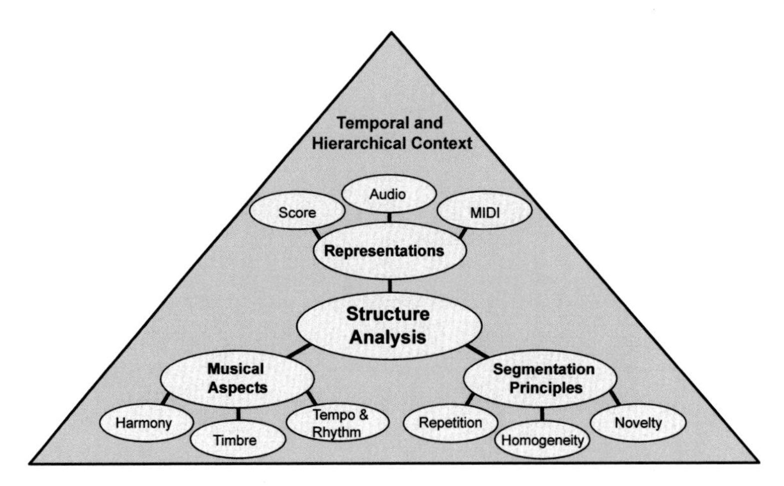

Fig. 4.2 Overview of various segmentation and structure principles.

4.1.1 Segmentation and Structure Analysis

The tasks of segmenting and structuring multimedia documents are of fundamental importance not only for the processing of music signals but also for general audio-visual content. **Segmentation** typically refers to the process of partitioning a given document into multiple segments with the goal of simplifying the representation into something that is more meaningful and easier to analyze than the original document. For example, in image processing the goal is to partition a given image into a set of regions such that each region is similar with respect to some characteristic such as color, intensity, or texture (see Figure 4.3 for an illustration). Region boundaries can often be described by contour lines or edges at which the image brightness or other properties change sharply and reveal discontinuities. In music, the segmentation task is to decompose a given audio stream into acoustically meaningful sections each corresponding to a continuous time interval that is specified by a start and end boundary. At a fine level, the segmentation may aim to find the boundaries between individual notes or to find the beat intervals specified by beat positions. At a coarser level, the goal may be to detect changes in instrumentation or harmony or to find the boundaries between verse and chorus sections. Also, discriminating between silence, speech, and music, finding the actual beginning of a music recording, or separating the applause at the end of a performance are typical segmentation tasks.

Going beyond mere segmentation, the goal of **structure analysis** is to also find and understand the relationships between the segments. For example, certain segments may be characterized by the instrumentation. There may be sections played only by strings. Sections played by the full orchestra may be followed by solo sections. The verse sections with a singing voice may be alternated with purely instrumental sections. Or a soft and slow introductory section may precede the main theme played in a much faster tempo. Furthermore, sections are often repeated. Most

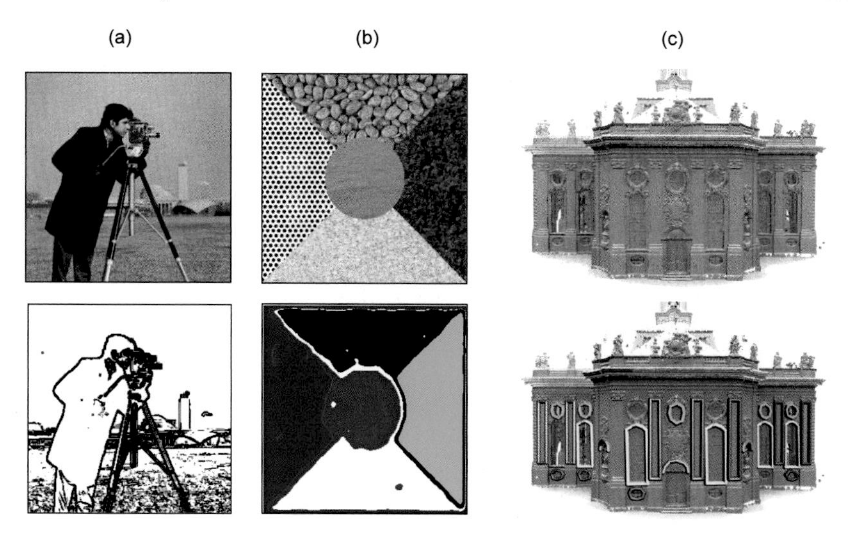

Fig. 4.3 Examples for segmentation results for image and 3D data. **(a)** Novelty-based image segmentation using edge detection. **(b)** Homogeneity-based texture segmentation. **(c)** Repetition-based segmentation of 3D geometry (from [35]).

events of musical relevance are repeated in a musical work in one way or another. However, repetitions are rarely identical copies of the original section, but undergo modifications in aspects such as the lyrics, the instrumentation, or the melody. One main task of structure analysis is to not only segment the given music recording, but to also group the segments into musically meaningful categories (e.g., intro, chorus, verse, outro).

The challenge in computational music structure analysis is that structure in music arises from many different kinds of relationships including repetition, contrast, variation, and homogeneity [27]. As we have already noted, **repetitions** play a particularly important role in music, where sounds or sequences of notes are often repeated [19]. Recurrent patterns can be of rhythmic, harmonic, or melodic nature. On the other hand, **contrast** is the difference between successive musical sections of different character. For example, a quiet passage may be contrasted by a loud one, a slow section by a rapid one, or an orchestral part by a solo. A further principle is that of **variation**, where motifs and parts are picked up again in a modified or transformed form. Finally, a section is often characterized by some sort of inherent **homogeneity**; for example, the instrumentation, the tempo, or the harmonic material may be similar within the section. All these principles need to be considered in the temporal context. Music happens in time (as opposed to, say, a painting), and it is the **temporal order** of events that is essential for building up musically and perceptually meaningful entities such as melodies or harmonic progressions [3].

In view of the various principles that crucially influence the musical structure, a large number of different approaches to music structure analysis have been developed. In this chapter, we want to roughly distinguish three different classes of methods. First, **repetition-based** methods are used to identify recurring patterns. Second, **novelty-based** methods are employed to detect transitions between contrasting parts. Third, **homogeneity-based** methods are used to determine passages that are consistent with respect to some musical property. Note that novelty-based and homogeneity-based approaches are two sides of a coin: novelty detection is based on observing some surprising event or change after a more homogeneous segment. While the aim of novelty detection is to locate the changes' time positions, the focus of homogeneity analysis lies in the identification of longer passages that are coherent with respect to some musical property. In the following section, we will study various procedures for structure analysis following one or several of these paradigms.

4.1.2 Musical Structure

As already mentioned in the introduction of this chapter, our focus is to analyze a given music recording on a rather coarse structural level. This level corresponds to what is often referred to as the **musical structure**, which describes the overall structural layout of a piece of music. In particular for Western classical music, one also encounters the term **musical form**, which refers to specific structural categories exploiting the principles of contrast and variety in one way or another. In this chapter, we use the term "musical structure" loosely, including with it the concept of musical form.

To specify musical structures, we now introduce some terminology as used in the remainder of this book. First of all, we want to distinguish between a piece of music (in an abstract sense) and a particular audio recording (an actual performance) of the piece. The term **part** is used in the context of the abstract music domain, whereas the term **segment** is used for the audio domain. Furthermore, we use the term **section** in a rather vague way for both domains to denote either a segment or a part. Musical parts are typically denoted by the capital letters $A, B, C, \ldots$ in the order of their first occurrence, where numbers (often written as subscripts) indicate the order of repeated occurrences. For example, the sequence $A_1B_1A_2B_2A_3$ describes the musical structure of the piano piece shown in Figure 4.1, which consists of three repeating A-parts and two repeating B-parts. Hence, given a recording of this piece of music, the goal of the structure analysis problem (as considered in this chapter) is to find the segments within the recording that correspond to the A- and B-parts.

In Western music, the musical structure often follows certain structural patterns (see Figure 4.4). The simplest of these patterns is the **strophic form**, which basically consists of a sequence of a part being repeated over and over again. The form $A_1A_2A_3A_4\ldots$ is, for example, used in folk songs or nursery rhymes, where the A-parts correspond to the stanzas of the underlying poem. Another structural pattern is referred to as **chain form**, which is simply a sequence of self-contained and unrelated parts ($ABCD\ldots$), sometimes with repeats ($A_1A_2B_1B_2C_1C_2D_1D_2\ldots$). This form is often used in a composition that consists of a concatenation of favorite tunes from

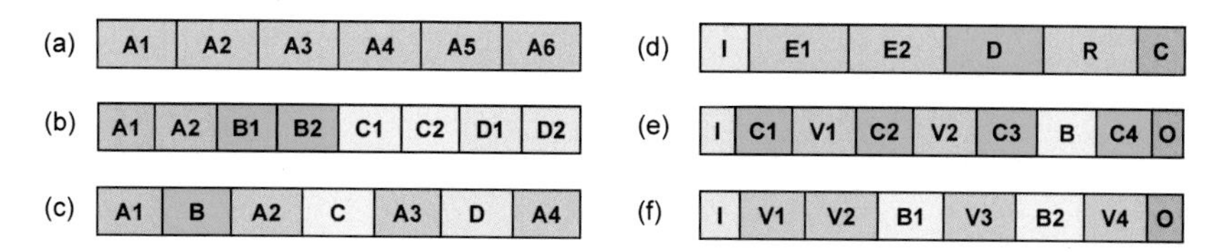

Fig. 4.4 Examples for musical structures as encountered in Western music. **(a)** Strophic form. **(b)** Chain form with repetitions. **(c)** Rondo form. **(d)** Sonata form. **(e)** Beatles song "Tell Me Why." **(f)** Beatles song "Yesterday."

popular songs, dances, or operettas. Examples are **medleys** or **potpourris**, which are pieces composed from parts of existing pieces that are simply juxtaposed with no strong connection or relationship. Another form is the **rondo form**, where a recurring theme alternates with contrasting sections, yielding the musical structure $A_1BA_2CA_3DA_4\ldots$.

In Western classical music, one of the most important musical structures is known as the **sonata form**, which is a large-scale musical structure typically used in the first movements of sonatas and symphonies. The basic sonata form consists of an **exposition** (E), a **development** (D), and a **recapitulation** (R), where the exposition is repeated once. Sometimes, one can find an additional **introduction** (I) and a closing **coda** (C), thus yielding the form IE_1E_2DRC. In particular, the exposition and the recapitulation stand in close relation to each other, both containing two subsequent contrasting subject groups (often simply referred to as the first and second theme) connected by some transition. As previously noted, at least at a coarse level, the recapitulation can be regarded as a kind of repetition of the exposition. However, at a finer level, there are significant differences. For example, the subject groups and transition in the recapitulation are musically altered and can be quite different from their corresponding occurrences in the exposition. Finally, we want to discuss some typical structural elements one finds in popular music. As with the sonata form, one sometimes uses generic names to denote the musical parts instead of using capital letters. The most important parts of a pop song are the **verse** (V) and the **chorus** (C) sections. Each verse usually employs the same melody (possibly with slight modifications), while the lyrics change for each verse. The chorus (sometimes also called the **refrain**) typically consists of a melodic and lyrical phrase which is repeated. Sometimes, pop songs may start with an **intro** (I) and close with an **outro** (O). Finally, verse and chorus sections may be connected by an additional part called a **bridge** (B). The verse and chorus are usually repeated throughout a song, while the intro and the outro appear only once. Some pop songs may have a **solo** section, where one or more instruments play a melodic line, typically following the melody previously introduced by the singer.

We have presented only a small selection of musical structures. In practice, there are many more structures as well as variations and deviations from standard forms as illustrated by the last two examples of Figure 4.4. A musical structure can be rather vague, and even music experts may argue about the construction of a given compo-

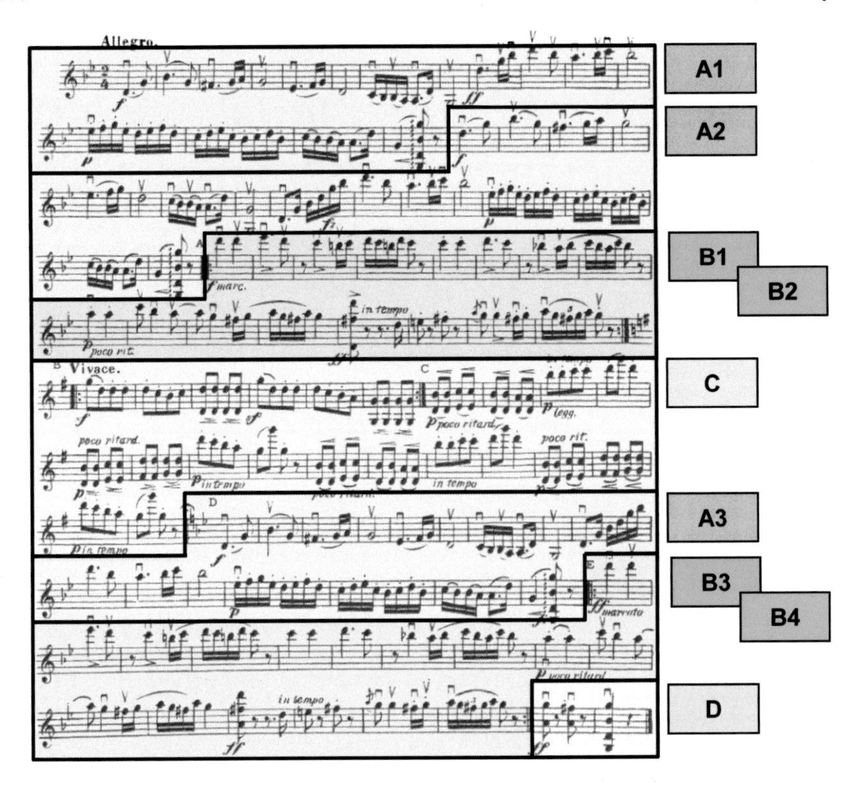

Fig. 4.5 Sheet music representation and musical structure of the Hungarian Dance No. 5 by Johannes Brahms. Only the voice for the violin of an arrangement for full orchestra is shown.

sition. In particular, what we call a repetition of a musical section is often far from being an exact copy. Segments that are considered to correspond to the same musical part may differ in instrumentation and tempo, or a segment may be transposed to another key, the melody may be changed while only the underlying harmonic progression is kept, and so on. Furthermore, musical structure is typically ordered in hierarchies, and it is often not clear which level should be considered when specifying the musical structure. For example, in the piece shown in Figure 4.1, the A-part can be further subdivided into substructures consisting of two or even four subparts. Similarly, the B-part can be regarded as a repetition of two subparts. These repeating substructures also become visible in the chroma representation derived from the music recording (see Figure 4.1c). In music notation, such subparts are often indicated using small letters $a, b, c, \ldots$.

As a final example, we want to consider the Hungarian Dance No. 5 by Johannes Brahms, which will also serve as our running example in the next sections. This piece is part of a set of 21 dance tunes composed by Brahms up to 1869 and based mostly on traditional Hungarian themes. Each dance has been arranged for a wide variety of instruments and ensembles, ranging from piano versions to versions for full orchestra. Figure 4.5 shows a sheet music representation for the violin voice of an arrangement for full orchestra. The musical structure as indicated in the figure is $A_1A_2B_1B_2CA_3B_3B_4D$, which consists of three repeating A-parts, four repeating B-parts, as well as a C-part and a short closing D-part. The A-part has a substructure consisting of two more or less repeating subparts. Furthermore, as becomes apparent when looking at the musical score, the middle C-part may be further subdivided into a substructure that may be described by $d_1d_2e_1e_2e_3e_4$ (see Figure 4.28).

The overall musical structure of this piece can be explained in terms of repeating elements. However, there are also many other musical cues that reinforce the musical structure. For example, the C-part stands in contrast to the remaining parts. First, there is a change of the musical key in the C-part (changing from G minor to G major). Then, there is a change in the notated tempo (changing from 'Allegro' to 'Vivace'). While the A- and B-parts have catchy tunes, there is no such melody in the C-part. Instead, the entire C-part is rather homogeneous with regard to harmony. However, this does not hold for other musical properties such as dynamics and tempo. For example, while the d-part segments are played in forte, the e-part segments are played in piano. Also there are many sudden tempo changes within the C-part. Therefore, in this case, a novelty-based segmentation procedure using tempo cues may be used to reveal the substructures of the C-part, whereas a homogeneity-based segmentation procedure using harmonic properties may be suited to distinguish the C-part from the other parts. We further develop this example in the next sections.

4.1.3 Musical Dimensions

We have already seen that the applicability of the different segmentation principles very much depends on the musical and acoustic properties of the audio signal to be analyzed. Since the sampled waveform of an audio signal is relatively uninformative by itself, the first step in automated structure analysis is to transform the given music recording into a suitable feature representation. As explained in the music synchronization scenario (Section 3.1), finding such a representation constitutes a delicate trade-off between robustness and expressiveness. Also, it is often unclear which musical properties are actually relevant for the given music signal and the considered segmentation scenario. For example, structural boundaries may be based on changes in harmony, timbre, or tempo. One major task in music processing is to transform a given audio signal into feature representations that correlate to the various musical aspects. In the following, we discuss this issue in more detail by considering three conceptually different feature representations (see Figure 4.6 for an overview).

As a first representation, we consider chroma features as introduced in Section 3.1.2. Recall that a normalized chroma vector describes the signal's local energy distribution over an analysis window (frame) across the twelve pitch classes of the equal-tempered scale (ignoring octave information). Capturing pitched content, a chroma-based feature sequence relates to harmonic and melodic properties

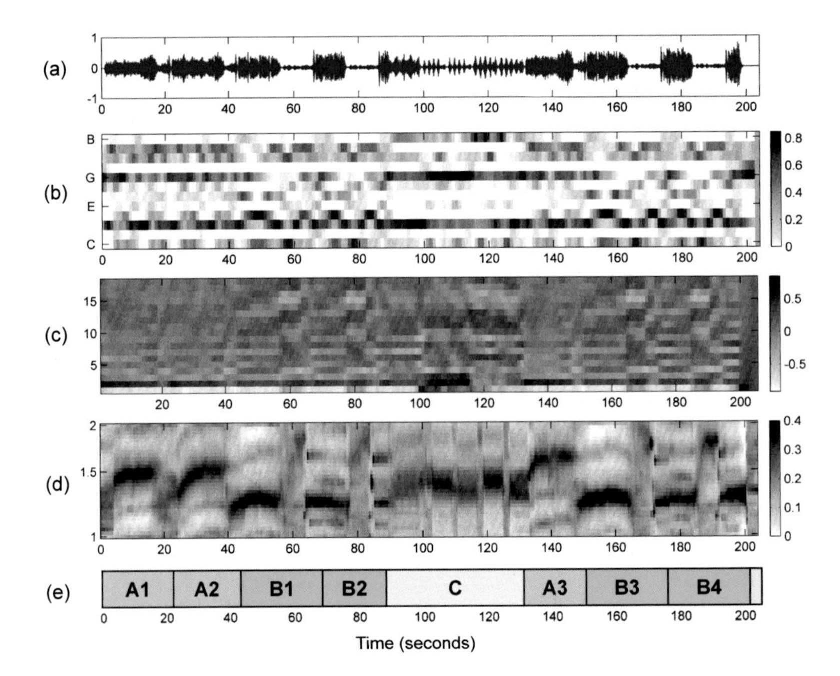

Fig. 4.6 Feature representations for a recording of the Hungarian Dance No. 5 by Johannes Brahms. **(a)** Waveform. **(b)** Chroma-based features. **(c)** MFCC-based features. **(d)** Tempo-based features. **(e)** Manually generated annotation.

of the music recording. Figure 4.6b shows a chroma representation derived from a recorded performance of our Brahms example, the Hungarian Dance No. 5. The patterns visible in the chromagram reveal important structural information. For example, the four repeating *B*-part segments are clearly visible as four similar characteristic subsequences in the chromagram. Furthermore, the *C*-part segment stands out in the chromagram by showing a high degree of homogeneity throughout the entire section. Indeed, for all chroma features of this segment, most of the signal's energy is contained in the G-, B-, and D-bands (which is not surprising since the *C*-part is in G major). In contrast, as for the *A*-part segments, many chroma vectors have dominant entries in the G-, $B^{\flat}$-, and D-bands (which nicely reflects that this part is in G minor).

Besides melody and harmony, the instrumentation and timbral characteristics are of great importance for the human perception of music structure. As we have discussed in Section 1.3.4, timbre is a rather vaguely defined perceptual property of sound, which is hard to describe and to extract from a music recording. For example, the automated recognition of musical instruments within polyphonic music signals is an extremely difficult problem. In applications such as structure analysis, it is often unnecessary to determine such information explicitly. Instead, mid-level representations that somehow correlate to aspects such as instrumentation and timbre may be sufficient. In the context of timbre-based structure analysis, one often uses **mel-frequency cepstral coefficients** (MFCCs), which were originally developed for automated speech recognition. Parametrizing the rough shape of the spectral envelope, MFCC-based features capture timbral properties of the signal. At this point, we do not want to give a technical description on how these features are computed. Instead, let us have a look at Figure 4.6c, which shows an MFCC-based feature representation for our Brahms example. One can recognize that MFCC features within the A-part segments are different from the ones in the B-part and C-part segments. For many music recordings such as pop songs, where sections with singing voice alternate with purely instrumental or percussive sections, MFCC-based feature representations are well suited for novelty-based and homogeneity-based segmentation.

As a third musical dimension, we consider properties that are related to beat, tempo, and rhythmic information. Estimation of the tempo and beat positions is one of the central topics in music processing, which we cover in Chapter 6. In the music segmentation context, such techniques are often applied to derive **beat-synchronous** feature representations, where the time axis is segmented according to musically meaningful beat positions. Such beat-synchronous representations are very useful to compensate for tempo changes in repeating parts. On the downside, beat tracking errors introduced by automated procedures may have negative consequences for the subsequent music processing tasks to be solved (see Section 6.3.3 for more details).

In music structure analysis, tempo and beat information may also be used in combination with homogeneity-based segmentation approaches. Instead of extracting such information explicitly, a mid-level feature representation that correlates to tempo and rhythm may suffice for deriving a meaningful segmentation at a higher structural level. As an example, Figure 4.6d shows such a mid-level representation, a **tempogram**, which encodes local tempo information. More precisely, a cyclic variant of a tempogram is shown, where tempi differing by a power of two are identified—similar to cyclic chroma features, where pitches differing by octaves are identified. Technical details on how to compute such tempograms can be found in Section 6.2.4. Having a look at Figure 4.6d, one can notice that the different musical parts are played in different tempi (even though the representation does not reveal the exact tempi). Furthermore, there are sections where the tempogram features do not have any dominating entries, which may indicate that there is no clear notion of a tempo in the recording. This kind of information is also important and can be used for segmentation purposes. As this example indicates, a tempogram may yield information that is complementary to the information obtained by chroma-based or MFCC-based feature representations.

Besides the various musical dimensions, there is another aspect one should keep in mind when looking for suitable feature representations: the temporal dimension. In all of the above-mentioned feature representations, an analysis window is shifted over the music signal. As we have already seen for the STFT in Section 2.5.2, the length of the analysis window as well as the hop size parameter have a crucial influence on the quality of the feature representation. For example, long window sizes and large hop sizes may be beneficial for smoothing out irrelevant local variations, which is often a desired property in homogeneity-based segmentation. On the downside, the temporal resolution decreases and important details may get lost, which can lead to problems when locating the exact segmentation boundaries.

In summary, a suitable choice of feature representations and parameter settings very much depends on the application context. Humans constantly and often unconsciously adapt themselves to the musical and acoustic characteristics of what they listen to. The richness and variety of musical structures make computational structure analysis a challenging problem.

4.2 Self-Similarity Matrices

We have seen that the principles of repetition, homogeneity, and novelty are fundamental for partitioning a given audio recording into musically meaningful structural elements. To study musical structures and their mutual relations, one general idea is to convert the music signal into a suitable feature sequence and then to compare each element of the feature sequence with all other elements of the sequence. This results in a **self-similarity matrix** (SSM), a tool which is of fundamental importance not only for music structure analysis but also for the analysis of many kinds of time series. In this section, we look at these matrices in detail. As we will see, one crucial property of self-similarity matrices is that repetitions typically yield path-like structures, whereas homogeneous regions yield block-like structures. These structural elements are exploited by most algorithms for visualizing, analyzing, and computing musical structures in one way or another. In Section 4.2.1, we introduce the concept of self-similarity matrices and discuss their basic structural properties. For applications, the improvement of these properties at an early state of the processing pipeline is of great importance, which is the topic of Section 4.2.2.

4.2.1 Basic Definitions and Properties

As said before, the concept of self-similarity matrices is fundamental for capturing structural properties of music recordings. Generally, one starts with a feature space $\mathcal{F}$ containing the elements of the feature sequence under consideration as well as with a similarity measure

$$s : \mathcal{F} \times \mathcal{F} \to \mathbb{R} \tag{4.1}$$

that makes it possible to compare these elements. Typically, the value $s(x,y)$ is high in case the elements $x, y \in \mathcal{F}$ are similar and small otherwise. Given a feature sequence $X = (x_1, x_2, \ldots, x_N)$, the idea is to compare all elements of the sequence with each other. This results in an N-square **self-similarity matrix** $\mathbf{S} \in \mathbb{R}^{N \times N}$ defined by

$$\mathbf{S}(n,m) := s(x_n, x_m), \tag{4.2}$$

where $x_n, x_m \in \mathcal{F}$, $n, m \in [1:N]$. In the following, a tuple $(n,m) \in [1:N] \times [1:N]$ is also called a **cell** of $\mathbf{S}$, and the value $\mathbf{S}(n,m)$ is referred to as the **score** of the cell (n,m).

Obviously, the concept of self-similarity matrices is closely related to the concept of cost matrices, which we have already encountered in Section 3.2.1. However, instead of a cost measure c as in (3.12), we now use a similarity measure s. And instead of comparing two sequences X and Y with each other, we now compare a single sequence X with itself. Depending on the application context and notion that is used to compare the data, there are many related concepts known under different names such as recurrence plot or self-distance matrix just to name a few. In this chapter, we only consider self-similarity matrices, but the techniques to be explained can easily be transferred to other types of matrices.

In the following discussion, we assume that the feature space is a Euclidean space $\mathcal{F} = \mathbb{R}^D$ of some dimension $D \in \mathbb{N}$. For simplicity and illustration purposes, we use as similarity measure s the inner product defined by

$$s(x,y) := \langle x|y \rangle \tag{4.3}$$

for two vectors $x, y \in \mathcal{F}$ (see (2.37)). With this similarity measure, the score between two orthogonal feature vectors is zero and otherwise it is nonzero. In the case that the feature vectors are normalized with respect to the Euclidean norm, the similarity values $s(x,y)$ lie in the interval $[-1,1]$. Obviously, there are many more possibilities to define a similarity measure (see Exercise 4.1). The suitability of a similarity measure depends on the properties of the considered features and vice versa.

Given a feature sequence $X = (x_1, x_2, \ldots, x_N)$, it seems reasonable to require that an element x_n should be maximally similar to itself. Using normalized features and the similarity measure from (4.3), the similarity measure assumes its maximal value $s(x_n, x_n) = 1$ for all $n \in [1:N]$. Therefore, the resulting SSM has a diagonal with large values. More generally, recurring patterns of the given feature sequence become visible in the SSM in the form of structures with large similarity values. The two most prominent structures induced by such patterns are often referred to as blocks and paths (see Figure 4.7a for an illustration). First, if the feature sequence captures musical properties that stay somewhat constant over the duration of an entire musical part, each of the feature vectors is similar to all other feature vectors within this segment. As a result, an entire **block** of large values appears in the SSM. In other words, homogeneity properties correspond to block-like structures. Second, if the feature sequence contains two repeating subsequences (e.g., two segments corresponding to the same musical part), the corresponding elements of the two subsequences are similar to each other. As a result, a **path** (or **stripe**) of high similarity running parallel to the main diagonal becomes visible in the SSM. In other words, repetitive properties correspond to path-like structures.

Before we further formalize these properties, let us have a look at Figure 4.7, which shows different self-similarity matrices for our Brahms example. Figure 4.7a shows an idealized SSM. For example, assuming that the three repeating A-part segments are homogeneous, the SSM has a quadratic block relating the segment

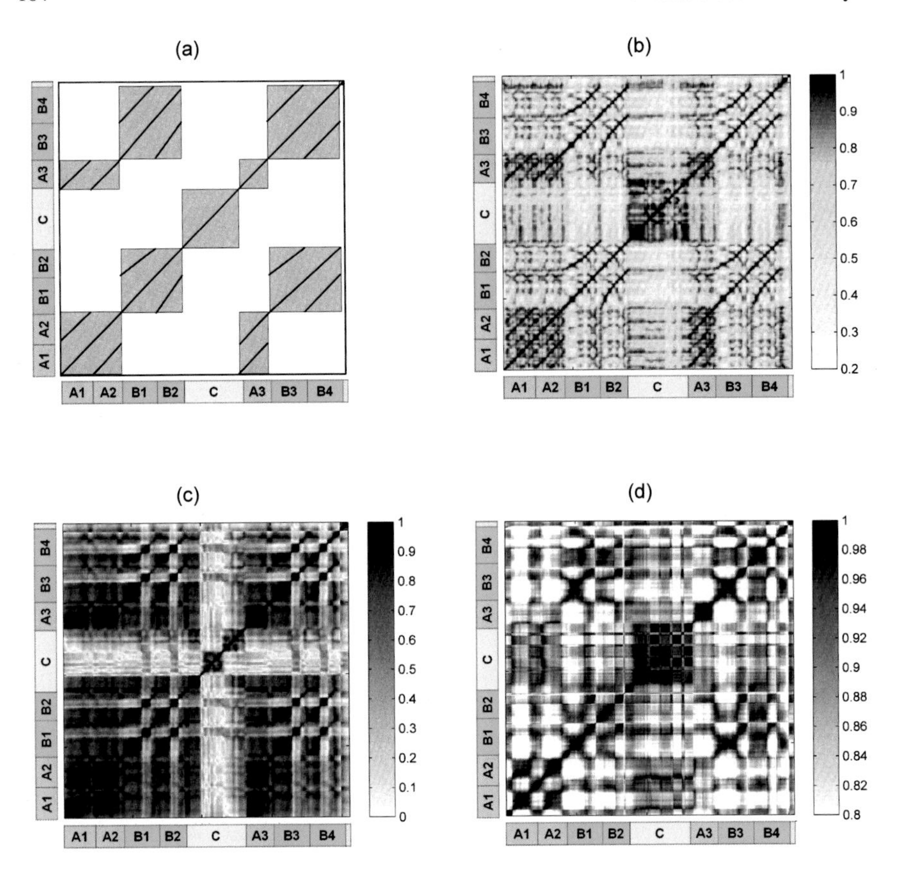

Fig. 4.7 Self-similarity matrices for the Hungarian Dance No. 5 by Johannes Brahms derived from various feature representations shown in Figure 4.6. **(a)** Idealized SSM. **(b)** SSM using chroma-based features. **(c)** SSM using MFCC-based features. **(d)** SSM using tempo-based features.

corresponding to A_1A_2 to itself and another quadratic block relating the A_3-part segment to itself. Furthermore, there are two rectangular blocks, one relating the A_1A_2-part segment to the A_3-part segment and the other relating the A_3-part segment to the A_1A_2-part segment. In case that the three repeating A-part segments are not homogeneous, the SSM reveals path structures that run (more or less) parallel to the main diagonal. For example, there is a path with large similarity values relating A_1 with A_2 and one relating A_1 with A_3.

How are such structures reflected in the case of "real" SSMs? Besides the idealized SSM, Figure 4.7 shows different self-similarity matrices for our Brahms example obtained from the three conceptually different feature sequences of Figure 4.6. In the visualization, large values of $\mathbf{S}$ are indicated by dark gray and small values by light gray. First, one can notice that properties of a self-similarity matrix crucially depend on the respective feature type. The SSM in Figure 4.7b, which is obtained from chroma-based features, resembles the idealized SSM to a large extent. The block-like structures corresponding to A-part segments indicate that these segments are quite homogeneous with respect to harmony. The same holds for the C-part segment. Furthermore, the small similarity values outside the C-part block (i.e., all cells relating the C-part frames to frames of other segments) show that the C-part segment is harmonically more or less unrelated to all other parts. For the B-part segments, there are path-like structures and no block-like structures. This shows that the B-part segments share the same harmonic progression (i.e., are repetitions with regard to harmony), but are not homogeneous with respect to harmony. An interesting observation is that, even though repeating, the B-part segments are played in different tempi and therefore have different lengths. For example, the shorter B_2-section is played faster than the B_1-section. As a result, the corresponding path does not run exactly parallel to the main diagonal. The gradient of the path indicates the relative tempo difference between the two related segments. Recall that we have discussed a similar issue already in the music synchronization context, where we derived a tempo curve from a warping path (see Section 3.3.2).

Looking at the other two self-similarity matrices the structures are not so clear. The SSM of Figure 4.7c, which results from MFCC-based features, mainly possesses block-like structures. In particular, the C-part segment has a low similarity to all other segments, which indicates a difference in timbre or instrumentation. Now, let us have a look at the tempogram-based SSM shown in Figure 4.7d. Again the C-part segment stands out, thus emphasizing its contrasting role. Furthermore, the SSM indicates the many tempo changes occurring in this music recording. In summary, the musical structure of the Brahms example can be best explained by the repetitive structure of the chroma-based SSM. Since this is the case with many musical works, in particular for melodic and harmonic Western music, we will mainly focus on this type of SSM in the subsequent sections.

We now formalize the concept of paths and blocks (see Figure 4.8). Let $X = (x_1, x_2, \ldots, x_N)$ be a feature sequence and $\mathbf{S}$ the resulting self-similarity matrix. We formally define a segment to be a set $\alpha = [s:t] \subseteq [1:N]$ specified by its starting point s and its end point t (given in terms of feature indices). Let

$$|\alpha| := t - s + 1 \tag{4.4}$$

denote the length of α. Next, a **path** over α of length L is a sequence

$$P = ((n_1, m_1), \ldots, (n_L, m_L)) \tag{4.5}$$

of cells $(n_\ell, m_\ell) \in [1:N]^2$, $\ell \in [1:L]$, satisfying $m_1 = s$ and $m_L = t$ (boundary condition) and $(n_{\ell+1}, m_{\ell+1}) - (n_\ell, m_\ell) \in \Sigma$ (step size condition), where Σ denotes a set of admissible step sizes. Note that this definition is very similar to the one of a warping path (see Section 3.2.1.1). In the case of $\Sigma = \{(1,1)\}$, one obtains paths that are strictly diagonal. In the following, we typically use the set

$$\Sigma = \{(2,1), (1,2), (1,1)\}, \tag{4.6}$$

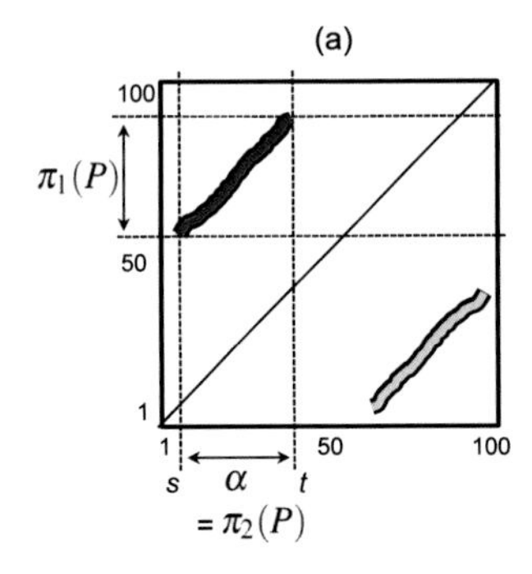

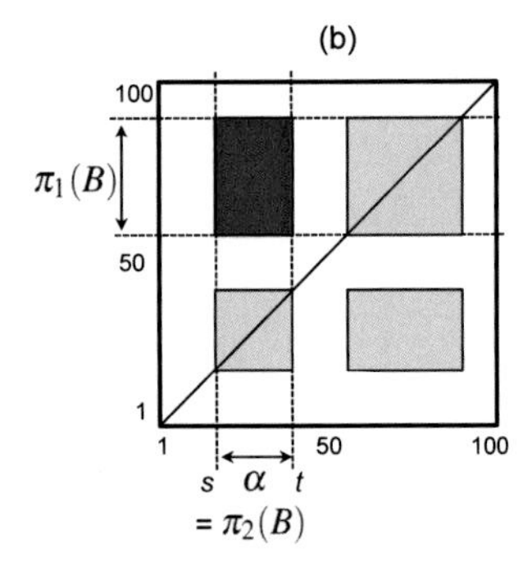

Fig. 4.8 Schematic view of self-similarity matrix with **(a)** a path and **(b)** a block.

which is the step size condition introduced in (3.30). For a path P, one can associate two segments defined by the projections

$$\pi_1(P) := [n_1 : n_L] \quad \text{and} \quad \pi_2(P) := [m_1 : m_L], \tag{4.7}$$

respectively (see Figure 4.8a). The boundary condition enforces $\pi_2(P) = \alpha$. The other segment $\pi_1(P)$ is referred to as the **induced segment**. The **score** $\sigma(P)$ of P is defined as

$$\sigma(P) := \sum_{\ell=1}^{L} \mathbf{S}(n_\ell, m_\ell). \tag{4.8}$$

Note that each path over the segment α encodes a relation between α and an induced segment, where the score $\sigma(P)$ yields a quality measure for this relation.

For blocks, we also introduce corresponding notions. A **block** over a segment $\alpha = [s : t]$ is a subset

$$B = \alpha' \times \alpha \subseteq [1 : N] \times [1 : N] \tag{4.9}$$

for some segment $\alpha' = [s' : t']$. Similar as for a path, we define the two projections $\pi_1(B) = \alpha'$ and $\pi_2(B) = \alpha$ for the block B and call α' the **induced segment** (see Figure 4.8b). Furthermore, we define the score of block B by

$$\sigma(B) = \sum_{(n,m)\in B} \mathbf{S}(n, m). \tag{4.10}$$

Based on paths and blocks, we can now consider different kinds of similarity relations between segments. We say that a segment α_1 is **path-similar** to a segment α_2, if there is a path P of high score with $\pi_1(P) = \alpha_1$ and $\pi_2(P) = \alpha_2$. Similarly, α_1 is **block-similar** to α_2, if there is a block B of high score with $\pi_1(B) = \alpha_1$ and $\pi_2(B) = \alpha_2$. Obviously, in case that the similarity measure s is symmetric, both the self-similarity matrix $\mathbf{S}$ and the above-defined similarity relations between segments are symmetric as well. Another important property of a similarity relation is **transitivity**, i.e., if a segment α_1 is similar to a segment α_2 and segment α_2 is similar to a segment α_3, then α_1 should also be similar to α_3 (at least to a certain degree). Also this property holds for path- and block-similarity in case that the similarity measure s has this property. As a consequence, path and block structures often appear in groups that fulfill certain symmetry and transitivity properties—at least in the ideal case. For example, if there is a block $B = \alpha' \times \alpha$ of high score, then the symmetry property implies that there is also a block $\alpha \times \alpha'$ of high score. Furthermore, if every frame belonging to α is similar to every other frame of α', then also the frames within the segments α and α' are similar to each other. This leads to additional blocks $\alpha \times \alpha$ and $\alpha' \times \alpha'$ (see Figure 4.8b). Figure 4.7 shows that such groups of similarity relations also appear in "real" SSMs.

Most computational approaches to music structure analysis exploit path- and block-like structures of SSMs in one way or another, and the overall algorithmic pipelines typically contain the following general steps:

1. The music signal is transformed into a suitable feature sequence.
2. A self-similarity matrix is computed from the feature sequence based on a similarity measure.
3. Blocks and paths of high overall score are derived from the SSM. Each block or path defines a pair of similar segments.
4. Entire groups of mutually similar segments are formed from the pairwise relations by applying a clustering step.

The last step can be considered as forming a kind of transitive closure of the pairwise segment relations induced by block and path structures. For example, in the case of Brahms' Hungarian Dance No. 5 (see Figure 4.7), the objective of the last step would be to find one group that contains all A-part segments and another group that contains all B-part segments.

In practice, this general processing pipeline leaves a lot of freedom and needs to be adjusted to account for particular properties of the underlying type of music and the requirements of the intended application. Furthermore, as mentioned before, major challenges arise from the fact that musical parts are rarely repeated in precisely the same way. Instead, audio segments that are considered as repetitions may differ significantly in aspects such as dynamics, orchestration, articulation, tempo, harmony, melody, or any combination of these. As a result, structure analysis becomes a hard and often ill-posed task. In particular, musical and acoustic variations may cause significant deteriorations in the path and block structures and their induced relations. This makes both steps, i.e., the block and path extraction step as well as the grouping step, error-prone and fragile. In the following, we discuss various strategies to cope with such challenges, e.g., by enhancing structural properties of SSMs (Section 4.2.2) or by jointly performing the two error-prone steps of path extraction and grouping within a joint optimization scheme (Section 4.3).

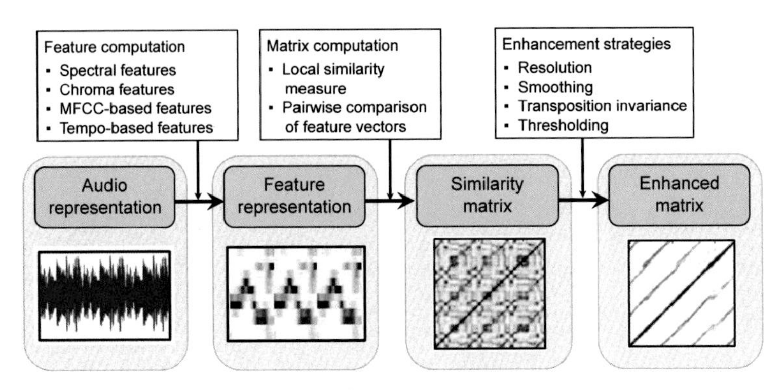

Fig. 4.9 Overview of the similarity matrix computation.

4.2.2 Enhancement Strategies

In this section, we describe various strategies for enhancing structural properties of self-similarity matrices (see Figure 4.9 for an overview). In particular, we focus on augmenting path-like structures, which play a central role in repetition-based structure analysis. Even though all the enhancement strategies are described for self-similarity matrices, similar strategies can be applied for more general similarity or cost matrices.

4.2.2.1 Feature Representation

In the first step, the given waveform-based audio recording is transformed into a suitable feature representation, which captures specific acoustic and musical properties. As we have already discussed in Section 4.2.1 and as illustrated by Figure 4.7, the structural properties of an SSM decisively depend on the feature type used. For example, MFCC-based and related spectral-based features may be suitable to capture aspects such as instrumentation and timbre. Other features based on onset information or tempograms are used to capture beat, tempo, and rhythmic information. In the following, we only consider the case of chroma-based audio features, which relate to harmonic and melodic properties as discussed in Section 3.1.2.

By considering a family of modified chroma representations similar to the ones used in Figure 3.9, we now demonstrate the influence of different parameter settings on the properties of the resulting SSM. Starting with a chroma representation of a given feature rate, this family comes along with two parameters: a length parameter $\ell \in \mathbb{N}$ (given in frames), which is used to smooth or average the feature values over ℓ consecutive frames, as well as a downsampling parameter d, which reduces the feature rate by a factor of d. For a more detailed description of such a procedure, we refer to Section 7.2.1 and Figure 7.10.

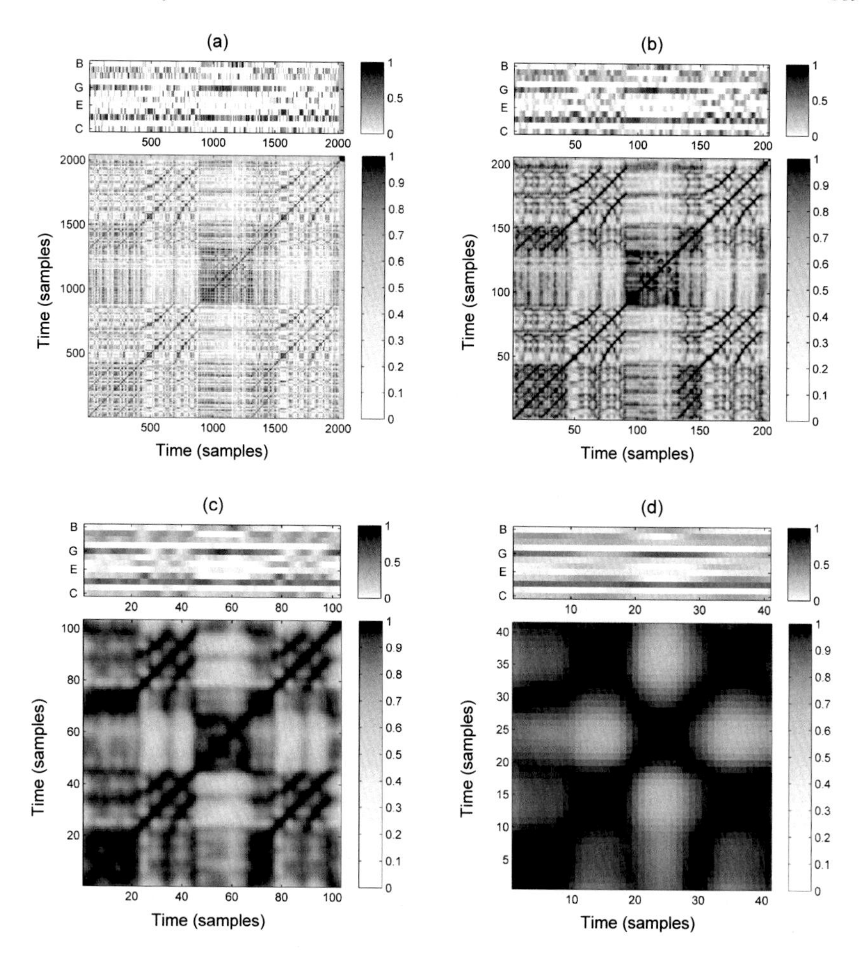

Fig. 4.10 Various chroma representations and resulting SSMs for the Hungarian Dance No. 5 by Johannes Brahms. **(a)** Usage of original normalized chroma features (10 Hz). **(b)** Applying $\ell = 40$ and $d = 10$ (1 Hz). **(c)** Applying $\ell = 160$ and $d = 20$ (0.5 Hz). **(d)** Applying $\ell = 480$ and $d = 50$ (0.2 Hz).

As an example, we start with normalized chroma features with a feature rate of 10 Hz. Figure 4.10a shows the resulting SSM, which yields a very detailed description of repetitive structures. Even though the path structures that correspond to the repeating *A*-part and *B*-part segments are visible, the SSM looks quite noisy and many of the shown details are irrelevant when only the overall musical structure is of interest.

Using a smoothing length of $\ell = 40$ (corresponding to four seconds of audio) and a downsampling by $d = 10$ (resulting in a feature rate of 1 Hz), one obtains the SSM shown in Figure 4.10b. Many of the details have been smoothed out, and some of the structurally relevant path and block structures have become more prominent.

In particular, this holds for the paths that relate to the B-part segments. Moreover, reducing the feature rate improves the computational efficiency for subsequent processing steps.

Further increasing the smoothing length and reducing the feature rate results in an emphasis of the rough harmonic content. In particular, neighboring elements in the feature sequence come closer together, which leads to an enhancement of block-like structures. For example, Figure 4.10c shows the SSM when using $\ell = 160$ (16 seconds) and $d = 20$ (feature rate of 0.5 Hz) and Figure 4.10d the SSM using $\ell = 480$ (48 seconds) and $d = 50$ (feature rate of 0.2 Hz). Using large smoothing windows, relevant path structures may be smeared out and lost for the subsequent steps. For other applications such as homogeneity-based structure analysis, however, averaging over large windows may be beneficial.

In summary, this example shows the importance not only of the feature type but also of the size of the analysis window and the feature rate. Knowing the temporal level of the music processing task is of great help for choosing suitable parameters. For example, for tasks such as extracting the musical structure from a given audio recording, smoothing and downsampling already on the feature level can lead to substantial improvements, not to speak of computational benefits in subsequent analysis steps. In particular, running time and memory requirements are important issues when employing concepts such as SSMs, which are quadratic in the length of the input feature sequence. As already mentioned in Section 4.1.3, another important strategy for adjusting and reducing the feature rate is based on **adaptive windowing**, where the analysis windows are determined by previously extracted onset and beat positions. This strategy will be discussed in more detail in Section 6.3.3.

4.2.2.2 Path Smoothing

We have seen that important structural elements of similarity matrices are paths of high similarity that run parallel to the main diagonal. Even though it is often easy for humans to recognize these structures, the automated extraction of paths constitutes a difficult problem due to significant distortions that are caused by variations in parameters such as dynamics, timbre, execution of note groups (e.g., grace notes, trills, arpeggios), modulation, articulation, or tempo progression. As an example, let us have a look at Figure 4.11a, which shows the SSM of a recording of the Waltz No. 2 from Dimitri Shostakovich's Suite for Variety Orchestra No. 1. This piece has the (rough) musical structure $A_1A_2BC_1C_2A_3A_4D$, where the theme, represented by the A-part, appears four times. However, there are significant variations in the four A-parts concerning instrumentation, articulation, as well as dynamics. For example, in A_1 the theme is played by a clarinet, in A_2 by strings, in A_3 by a trombone, and in A_4 by the full orchestra. As is illustrated by Figure 4.11a, these variations result in a rather poor and fragmented path structure. This makes it hard to identify the musically similar segments $\alpha_1 = [4:40]$, $\alpha_2 = [43:78]$, $\alpha_3 = [145:179]$, and $\alpha_4 = [182:217]$ corresponding to A_1, A_2, A_3, and A_4, respectively. In particular, as

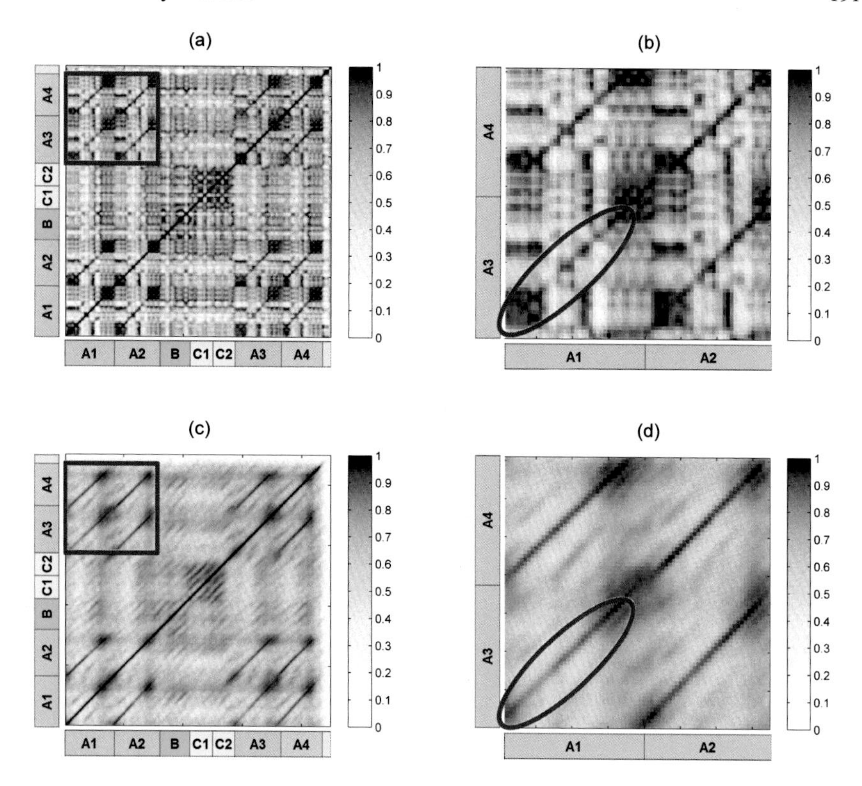

Fig. 4.11 Variants of SSMs for a recording of the Waltz No. 2 from Dimitri Shostakovich's Suite for Variety Orchestra No. 1. **(a)** Original SSM using chroma features (resolution of 1 Hz). **(b)** Enlargement of the submatrix indicated by the rectangular frame in (a). The path corresponding to segments α_1 (part A_1) and α_3 (part A_3) is highlighted by the oval. **(c)** SSM after applying diagonal smoothing. **(d)** Enlargement of the submatrix indicated by the rectangular frame in (c).

can be seen in the enlargement shown in Figure 4.11b, the path corresponding to the segments α_1 and α_3 is quite problematic.

To some extent, as we have seen above, structural properties of the SSM may be augmented by using longer analysis windows in the feature computation step. This, however, may also smooth out important details. As an alternative, we now show how to enhance the path structure of an SSM by applying image processing techniques. Recall that the relevant paths run along the direction of the main diagonal in the case that repeating parts are played in the same tempo. Therefore, in order to augment such paths, the general idea is to apply an averaging filter (or low-pass filter) in the direction of the main diagonal, which results in an emphasis of diagonal information and a softening of other, nondiagonal structures.

We now give a mathematical description of this procedure. Let $\mathbf{S}$ be an SSM of size $N \times N$ and let $L \in \mathbb{N}$ be a length parameter. Then we define the smoothed self-similarity matrix $\mathbf{S}_L$ by setting

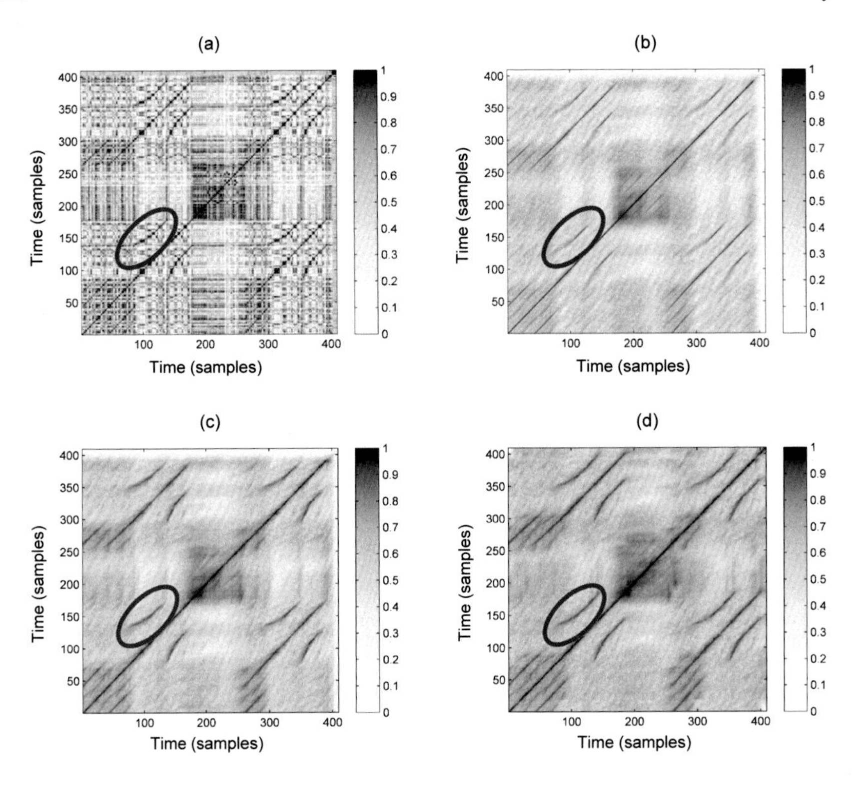

Fig. 4.12 Variants of SSMs for the Hungarian Dance No. 5 by Johannes Brahms. The path corresponding to the B_1-part and B_2-part segments is highlighted. **(a)** Original SSM using chroma features (resolution of 2 Hz). **(b)** SSM after applying diagonal smoothing. **(c)** SSM after applying tempo-invariant smoothing. **(d)** SSM after applying forward–backward smoothing.

$$\mathbf{S}_L(n,m) := \frac{1}{L}\sum_{\ell=0}^{L-1}\mathbf{S}(n+\ell,m+\ell) \tag{4.11}$$

for $n,m \in [1:N-L+1]$. In other words, the value $\mathbf{S}_L(n,m)$ is obtained by averaging the similarity values of two subsequences of length L, one starting at index n and the other at index m. By suitably extending $\mathbf{S}$ (e.g., by **zero-padding** where zero columns and rows are added), we may assume in the following that $\mathbf{S}_L(n,m)$ is defined for $n,m \in [1:N]$.

The averaging procedure results in a smoothing effect along the main diagonal, which is also illustrated by our Shostakovich example of Figure 4.11. Using the length parameter $L = 10$, the resulting self-similarity matrix $\mathbf{S}_{10}$ (Figure 4.11c) reveals the desired path structure much better than the original matrix $\mathbf{S}$ (Figure 4.11a). For example, the enhanced path highlighted in Figure 4.11d reveals the relation between the segments α_1 and α_3 much better than before (see Figure 4.11b).

A simple filtering along the main diagonal only works well if there are no relative tempo differences between the segments to be compared. However, this assumption is violated when a part is repeated with a faster or slower tempo. We have seen such a case in our Brahms example from Figure 4.7, where the shorter B_2-section is played much faster than the B_1-section. It is only the beginning of the B_2-section that is played much faster than the beginning of the B_1-section, whereas the two sections have roughly the same tempo towards the end of the part. This results in a path that does not run exactly parallel to the main diagonal (in particular at the beginning), so that applying an averaging filter in the direction of the main diagonal destroys some of the path structure (see Figure 4.12b). To deal with such relative tempo differences, one idea is to apply a multiple filtering approach, where the SSM is smoothed along various directions that lie in a neighborhood of the direction defined by the main diagonal. Each such direction corresponds to a tempo difference and results in a separate filtered matrix. The final self-similarity matrix is obtained by taking the cell-wise maximum over all these matrices. In this way, the path structure is also enhanced in the presence of local tempo variations as illustrated in Figure 4.12c.

To better understand the details of this procedure, first assume that we have two repeating segments α_1 and α_2 played at the same tempo. Then the direction of the resulting path is given by the gradient $(1,1)$. Next, assume that the second segment α_2 is played at half the tempo compared with α_1. Then the direction of the resulting path is given by the gradient $(1,2)$. In general, if the tempo difference between the two segments is given by a real number $\theta > 0$ (the second segment played θ times slower than the first one), the resulting gradient is $(1,\theta)$. We define the self-similarity matrix smoothed in the direction of $(1,\theta)$ by

$$\mathbf{S}_{L,\theta}(n,m) := \frac{1}{L}\sum_{\ell=0}^{L-1}\mathbf{S}(n+\ell, m+[\ell\cdot\theta]), \tag{4.12}$$

where $[\ell\cdot\theta]$ denotes the integer closest to the real number $\ell\cdot\theta$. Again, by suitably zero-padding the matrix $\mathbf{S}$, we may assume that $\mathbf{S}_{L,\theta}$ is defined for $n,m \in [1:N]$. Now, in practice, one does not know the local tempo difference that may occur in a given music recording. Also, the relative tempo difference between two repeating sections may change over time (as is the case with our Brahms example). Therefore, the idea is to consider a (finite) set Θ consisting of tempo parameters $\theta \in \Theta$ for different relative tempo differences. Then, we compute for each such θ a matrix $\mathbf{S}_{L,\theta}$ and obtain a final matrix $\mathbf{S}_{L,\Theta}$ by a cell-wise maximization over all $\theta \in \Theta$:

$$\mathbf{S}_{L,\Theta}(n,m) := \max_{\theta\in\Theta}\mathbf{S}_{L,\theta}(n,m). \tag{4.13}$$

In practice, one can use prior information on the expected relative tempo differences to determine the set Θ. For example, it rarely happens that the relative tempo difference between repeating segments is larger than 50 percent, so that Θ can be chosen to cover tempo variations of roughly -50 to $+50$ percent. Furthermore, in practice, the tempo range can be covered well by considering only a relatively small number of tempo parameters. For example, a typical choice could be

$\Theta = \{0.66, 0.81, 1.00, 1.22, 1.50\}$ (see Exercise 4.4). Note that choosing $\Theta = \{1\}$ reduces to the case $\mathbf{S}_{L,\Theta} = \mathbf{S}_L$.

This smoothing procedure works in the forward direction, which results in a fading out of the paths, particularly when using a large length parameter. To avoid this fading out, one idea is to additionally apply the averaging filter in a backward direction. The final self-similarity matrix is then obtained by taking the cell-wise maximum over the forward-smoothed and backward-smoothed matrices (see Exercise 4.2). The effect is illustrated in Figure 4.12d by means of the Brahms example.

4.2.2.3 Transposition Invariance

It is often the case that certain musical parts are repeated in a transposed form, where the melody is moved up or down in pitch by a constant interval. As an example, let us consider the song "In the Year 2525" by Zager and Evans, which has the musical structure $IV_1V_2V_3V_4V_5V_6V_7BV_8O$. The song starts with a slow intro, which is represented by the I-part. The verse of the song, which is represented by the V-part, is repeated eight times. While the first four verse sections are in the same musical key, V_5 and V_6 are transposed by one semitone upwards, and V_7 and V_8 are transposed by two semitones upwards. Figure 4.13b shows a path-enhanced version of the resulting self-similarity matrix based on some chroma feature representation. This matrix shows path structures that relate the first four V-sections with each other as well as V_5 with V_6 and V_7 with V_8. Because of the transpositions, however, the relation between the first four sections and the last four sections is not reflected in the SSM.

In the following, we show how repetitive structures can be made visible in the SSM even in the presence of key transpositions. We have already seen in Section 3.1.2 that such transpositions can be simulated by cyclically shifting chroma features. Mathematically, we modeled such shifts by the cyclic shift operator $\rho : \mathbb{R}^{12} \to \mathbb{R}^{12}$ defined in (3.11). Now, let $X = (x_1, x_2, \ldots, x_N)$ be the chroma feature sequence. We then define the **i-transposed self-similarity matrix** $\rho^i(\mathbf{S})$ by

$$\rho^i(\mathbf{S})(n,m) := s(\rho^i(x_n), x_m) \tag{4.14}$$

for $n, m \in [1:N]$ and $i \in \mathbb{Z}$. Obviously, one has $\rho^{12}(\mathbf{S}) = \mathbf{S}$. Intuitively, $\rho^i(\mathbf{S})$ describes the similarity relations between the original music recording (represented by $X = (x_1, x_2, \ldots, x_N)$) and the music recording transposed by i semitones upwards (represented by $\rho^i(X) = (\rho^i(x_1), \rho^i(x_2), \ldots, \rho^i(x_N))$). Since one does not know in general the kind of transpositions occurring in the music recording, we apply a similar strategy as before when dealing with relative tempo deviations. Taking a cell-wise maximum over the twelve different cyclic shifts, we obtain a single **transposition-invariant self-similarity matrix** $\mathbf{S}^{\mathrm{TI}}$ defined by

$$\mathbf{S}^{\mathrm{TI}}(n,m) := \max_{i \in [0:11]} \rho^i(\mathbf{S})(n,m). \tag{4.15}$$

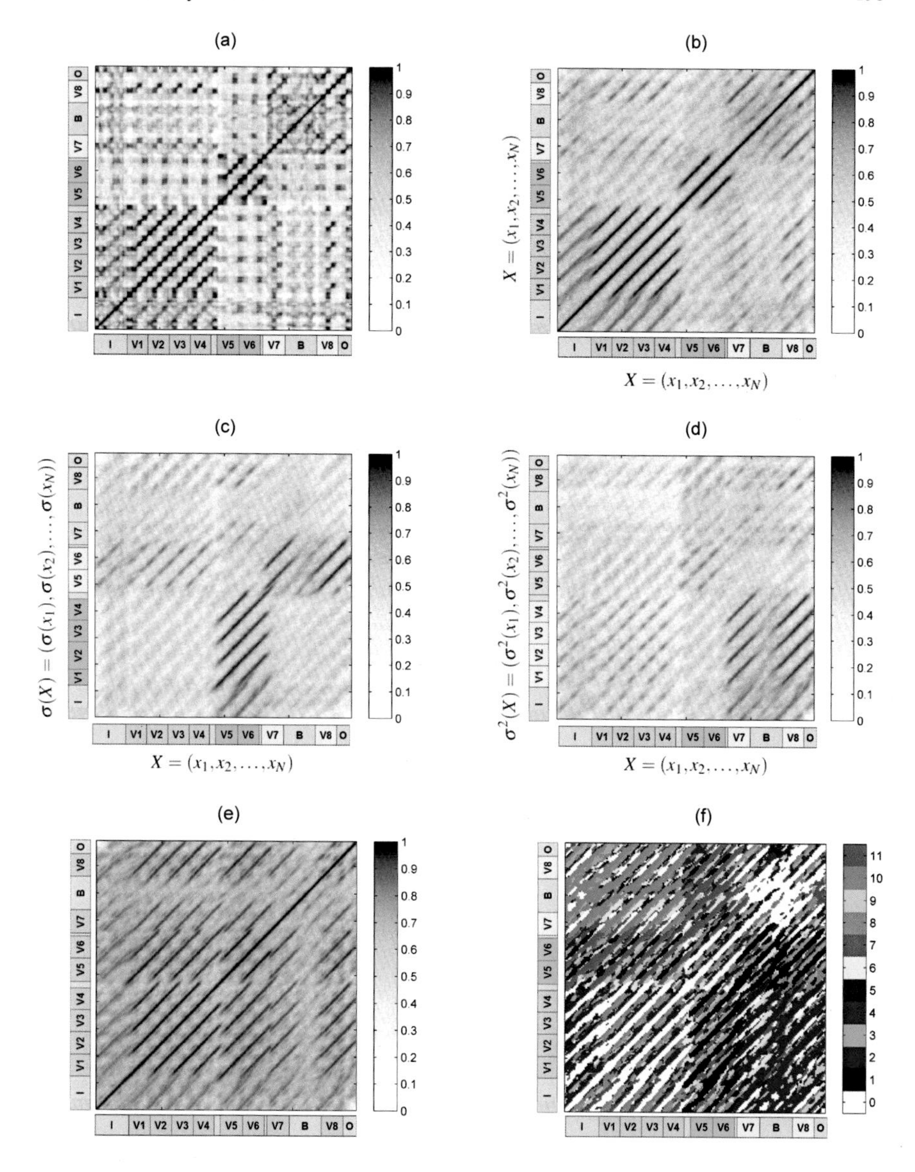

Fig. 4.13 Variants of SSMs for the song "In the Year 2525" by Zager and Evans. **(a)** Original SSM using chroma features (resolution of 1 Hz). **(b)** Path-enhanced SSM. **(c)** 1-transposed SSM. **(d)** 2-transposed SSM. **(e)** Transposition-invariant SSM. **(f)** Transposition index matrix.

Furthermore, we store the maximizing shift indices in an additional N-square matrix $\mathbf{I}$, which we refer to as the **transposition index matrix**:

$$\mathbf{I}(n,m) := \underset{i\in[0:11]}{\operatorname{argmax}}\, \rho^i(\mathbf{S})(n,m). \tag{4.16}$$

We illustrate the definitions by continuing the example shown in Figure 4.13 (see Exercise 4.3). Recall from above that shifting the sections V_1 to V_4 by one semitone upwards makes them similar to the original sections V_5 and V_6. This fact is revealed by the 1-transposed self-similarity matrix shown in Figure 4.13c. Similarly, shifting the sections V_1 to V_4 by two semitones upwards makes them similar to the original sections V_7 and V_8 (see Figure 4.13d). Putting together the information of all i-transposed self-similarity matrices by the maximization in (4.15), one obtains the transposition-invariant self-similarity matrix $\mathbf{S}^{\mathrm{TI}}$ shown in Figure 4.13e, where all pairwise similarity relations between the eight V-part segments become visible.

The resulting transposition index matrix is shown in Figure 4.13f in a color-coded form. We first discuss the case that the matrix $\mathbf{I}$ assumes the value $i = 0$ (white color in Figure 4.13f). The value $i = 0$ for a cell (n,m) indicates that $s(\rho^i(x_n),x_m)$ assumes a maximal value for $i = 0$. In other words, the chroma vector x_m is closer to x_n than to any other shifted version of x_n. Note, however, that this does not necessarily mean that x_m is close to x_n in absolute terms. As may be expected, the maximizing index is $i = 0$ at all positions where the conventional self-similarity matrix shown in Figure 4.13b reveals paths of low cost. Next, we consider the case that the matrix $\mathbf{I}$ assumes the value $i = 1$ (black color in Figure 4.13f). The value $i = 1$ for a cell (n,m) indicates that x_n becomes most similar to x_m when shifted one semitone upwards. Thus the strong path relations shown in Figure 4.13c correspond to cells assuming the value $i = 1$, and so on.

At this point, we want to note that introducing transposition invariance by cell-wise maximization over several matrices may increase the noise level in the resulting similarity matrix. Therefore, the transposition-invariant matrix should be computed on the basis of smoothed matrices, since the smoothing typically goes along with a suppression of unwanted noise. The definitions in (4.14) and (4.15) can be easily combined with the averaging approaches described by (4.11) and (4.12) to yield matrices $\rho^i_{L,\Theta}(\mathbf{S})$ and $\mathbf{S}^{\mathrm{TI}}_{L,\Theta}$. Such matrices are shown in Figure 4.13.

4.2.2.4 Thresholding

In many music analysis applications, self-similarity matrices are further processed by suppressing all values that fall below a given threshold. On the one hand, such a step often leads to a substantial reduction of unwanted noise-like components while leaving only the most significant structures. On the other hand, weaker but still relevant information may be lost. The thresholding strategy used may have a significant impact on the final result and has to be carefully chosen in the context of the considered application. Figure 4.14 shows some examples obtained by different thresholding settings as explained below.

The simplest strategy is to apply **global thresholding**, where all values $\mathbf{S}(n,m)$ of a similarity matrix $\mathbf{S}$ below a given threshold parameter $\tau > 0$ are set to zero:

$$\mathbf{S}_\tau(n,m) := \begin{cases} \mathbf{S}(n,m) & \text{if } \mathbf{S}(n,m) \geq \tau, \\ 0, & \text{otherwise.} \end{cases} \tag{4.17}$$

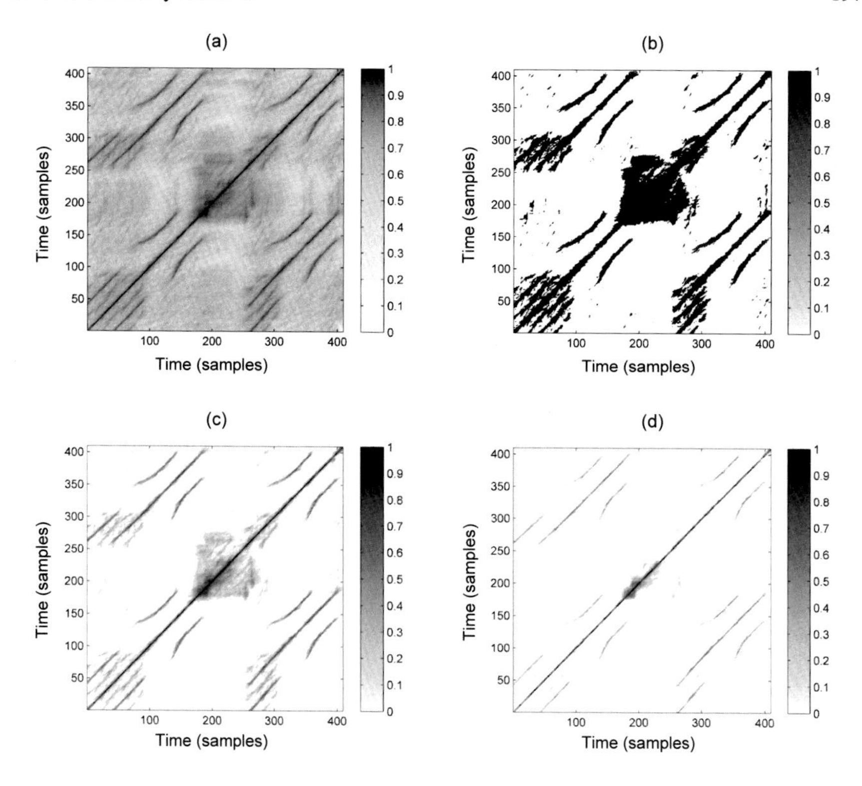

Fig. 4.14 Thresholding strategies applied to an SSM for the Hungarian Dance No. 5 by Johannes Brahms. **(a)** SSM from Figure 4.12d. **(b)** SSM after thresholding and binarization ($\tau = 0.75$). **(c)** SSM after thresholding and scaling ($\rho = 0.2$). **(d)** SSM after thresholding and scaling ($\rho = 0.05$).

Also, binarization of the similarity matrix can be applied by setting all values above or equal to the threshold to one and all others to zero. Instead of binarization, one may perform a scaling where the range $[\tau, \mu]$ is linearly scaled to $[0, 1]$ in the case that $\mu := \max_{n,m}\{\mathbf{S}(n,m)\} > \tau$, otherwise all entries are set to zero. Sometimes it may be beneficial to introduce an additional penalty parameter $\delta \leq 0$, setting all original values below the threshold to the value δ (see Section 4.3 for an application of this variant).

The global threshold τ can also be chosen in a **relative** fashion by keeping $\rho \cdot 100\%$ of the cells with the highest values using a relative threshold parameter $\rho \in [0, 1]$. Finally, thresholding can also be performed using a more **local** strategy by thresholding in a column- and rowwise fashion. To this end, for each cell (n, m), the value $\mathbf{S}(n, m)$ is kept if it is among the $\rho \cdot 100\%$ of the largest cells in row n and at the same time among the $\rho \cdot 100\%$ of the largest cells in column m, all other values being set to zero (see Exercise 4.5). As said before, the suitability of a thresholding setting depends on the respective music material and the application in mind. Often, suitable thresholds are learned and optimized using supervised learning procedures.

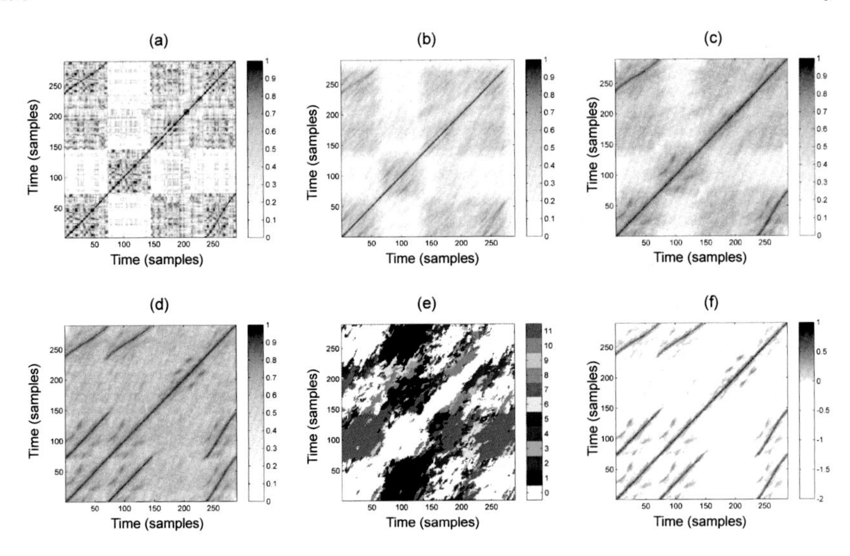

Fig. 4.15 Variants of similarity matrices for the same audio recording. **(a)** Original SSM using chroma features of 2 Hz resolution. **(b)** SSM after applying diagonal smoothing. **(c)** SSM after applying tempo-invariant and forward–backward smoothing. **(d)** Transposition-invariant SSM. **(e)** Transposition index matrix. **(f)** SSM after thresholding with penalty and scaling ($\rho = 0.2$, $\delta = -2$).

To conclude this section, Figure 4.15 summarizes the various enhancement and processing steps applied to a music recording having the musical structure $A_1A_2BA_3$. In this example, A_2 is a modulation of A_1 transposed by one semitone upwards, whereas A_3 is a repetition of A_1, however played much faster. Figure 4.15 shows a typical processing pipeline for computing an SSM as used in structure analysis applications. First, the music recording is converted into a sequence of normalized and smoothed chroma features as in Figure 3.9. Then, based on the similarity measure (4.3), an enhanced transposition-invariant self-similarity matrix $\mathbf{S}^{\mathrm{TI}}_{L,\Theta}$ is computed (see Figure 4.15d). In the next step, global thresholding is applied using a threshold parameter τ and a penalty parameter δ. Furthermore, the range $[\tau, 1]$ is linearly scaled to $[0, 1]$. As a result, the relevant path structure tends to lie in the positive part of the resulting SSM, whereas all other cells are given a negative score. Finally, setting $\mathbf{S}(n,n) = 1$ for $n \in [1:N]$, one can introduce a normalization property, which may have been lost in the smoothing process due to boundary effects. The SSM shown in Figure 4.15f is obtained in this way using a feature rate of 2 Hz. Settings for the enhancement are $L = 20$ for the length parameter and $\Theta = \{0.50, 0.63, 0.79, 1.26, 1.59, 2.00\}$ for the set of relative tempo differences (see Exercise 4.4). In this example, the threshold is chosen in a relative fashion by using the relative threshold $\rho = 0.2$ and the penalty parameter is set to $\delta = -2$.

4.3 Audio Thumbnailing

In this section, we deal with a prominent subproblem of music structure analysis commonly known as **audio thumbnailing**. Given a music recording, the objective is to automatically determine the most representative section, which may serve as a kind of "preview" giving a listener a first impression of the song or piece of music. Based on such previews, the user should be able to quickly decide if he or she would like to listen to the song or to move on to the next recording. Thus, audio thumbnails are an important browsing and navigation aid for finding interesting pieces in large music collections.

Often sections such as the chorus or the main theme of a song are good candidates for audio thumbnails. Such parts are typically repeated several times throughout the recording. Therefore, to determine a thumbnail automatically, most procedures try to identify a section that has on the one hand a certain minimal duration and on the other many (approximate) repetitions. As we have seen before, one challenge is that such repeating sections may show significant acoustic and musical differences in aspects that concern dynamics, instrumentation, articulation, and tempo.

We now describe a typical thumbnailing procedure for extracting the most repetitive segment from a given music recording. In particular, we show how enhanced self-similarity matrices as well as time warping techniques are applied for dealing with multiple variabilities. As the main technical tool, we introduce in Section 4.3.1 a fitness measure that assigns a fitness value to each audio segment. This measure simultaneously captures two aspects. First, it indicates **how well** a given segment explains other related segments, and second, it indicates **how much** of the overall music recording is covered by all these related segments. The audio thumbnail is then defined to be the segment of maximal fitness. In the computation of the fitness measure, one important concept is to avoid hard decisions and error-prone steps in an early stage of the algorithmic pipeline. To this end, an optimization scheme is applied for jointly performing path extraction and grouping—two error-prone steps that are often performed successively. Furthermore, we also have a look at an efficient algorithm based on dynamic programming for computing the fitness measure. In Section 4.3.2, we then introduce the concept of a scape plot representation that shows the fitness values over all possible audio segments. A visualization of this fitness scape plot yields a compact high-level view on the structural properties of the entire music recording. Finally, in Section 4.3.3, we discuss several explicit examples to indicate the potential as well as the limitations of the presented thumbnailing approach.

4.3.1 Fitness Measure

The idea of the fitness measure to be introduced is to simultaneously establish all relations between a given segment and its repetitions. To this end, a self-similarity matrix is required as described at the end of Section 4.2.2 and illustrated

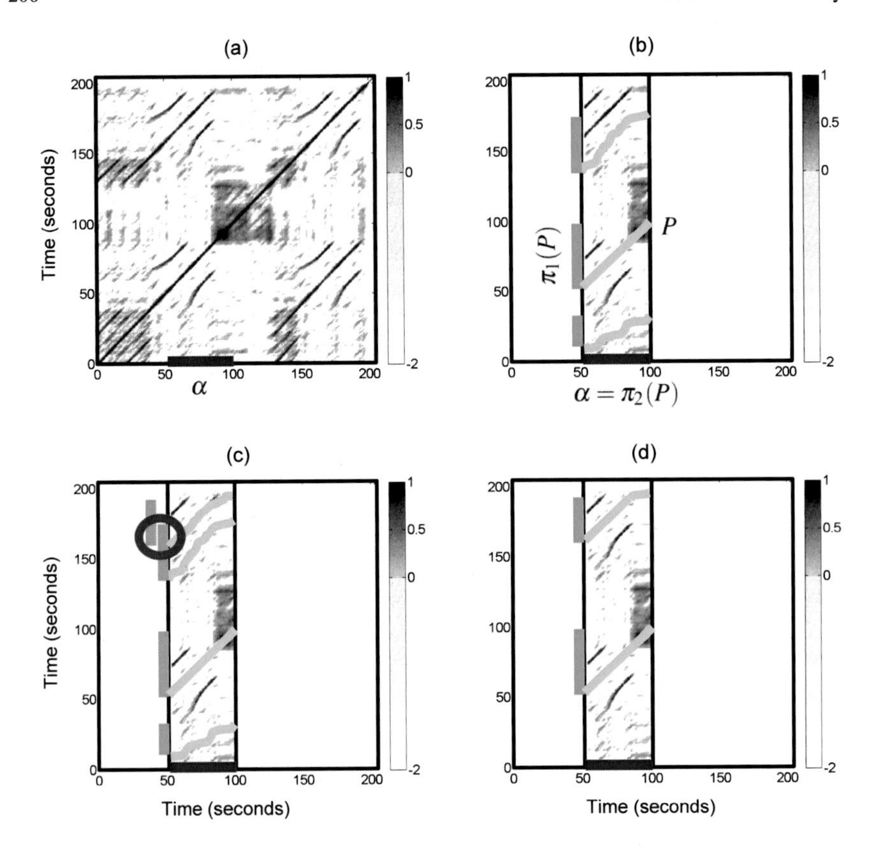

Fig. 4.16 SSM of our Brahms example with various paths over the segment $\alpha = [50:100]$. The induced segments are indicated on the vertical axis. **(a)** SSM. **(b)** Paths forming a path family. **(c)** Paths not forming a path family (induced segments overlap). **(d)** Paths forming an optimal path family.

by Figure 4.15f. Our Brahms example (see Figure 4.16) will serve as a running example for the subsequent steps. The following description of the fitness measure is generic in the sense that it works with general self-similarity matrices that only fulfill some basic normalization properties. From a technical point of view, only the properties

$$\mathbf{S}(n,m) \leq 1 \tag{4.18}$$

for all $n,m \in [1:N]$ and

$$\mathbf{S}(n,n) = 1 \tag{4.19}$$

for all $n \in [1:N]$ are required.

4.3.1.1 Path Family

Recall from Section 4.2.1 that a path P over a given segment $\alpha = [s:t] \subseteq [1:N]$ encodes a relation between $\alpha = \pi_2(P)$ and the induced segment $\pi_1(P)$. The score $\sigma(P)$ defined in (4.8) yields a quality measure for this relation. Extending the notion of a path, we now introduce the concept of a path family, which allows us to capture relations between α and several other segments in the music recording. To this end, we first define a **segment family** of size K to be a set

$$\mathcal{A} := \{\alpha_1, \alpha_2, \ldots, \alpha_K\} \tag{4.20}$$

of pairwise disjoint segments, i.e., $\alpha_k \cap \alpha_j = \emptyset$ for all $k, j \in [1:K]$ with $k \neq j$. Let

$$\gamma(\mathcal{A}) := \sum_{k=1}^{K} |\alpha_k| \tag{4.21}$$

be the **coverage** of $\mathcal{A}$ (see (4.4)). A **path family** over α is defined to be a set

$$\mathcal{P} := \{P_1, P_2, \ldots, P_K\} \tag{4.22}$$

of size K, consisting of paths P_k over α for $k \in [1:K]$. Furthermore, as an additional condition, we require that the induced segments are pairwise disjoint. In other words, the set $\{\pi_1(P_1), \ldots, \pi_1(P_K)\}$ is required to be a segment family. This definition is illustrated by Figure 4.16b, which shows a path family over the segment $\alpha = [50:100]$ consisting of $K = 3$ paths P_1, P_2, and P_3. The induced segments are $\pi_1(P_1) = [10:35]$, $\pi_1(P_2) = [50:100]$, and $\pi_1(P_3) = [136:174]$, which are pairwise disjoint. In contrast, the example shown in Figure 4.16c does not yield a path family, since the disjointness condition of the induced segments is violated. Extending the definition in (4.8), the **score** $\sigma(\mathcal{P})$ of the path family $\mathcal{P}$ is defined as

$$\sigma(\mathcal{P}) := \sum_{k=1}^{K} \sigma(P_k). \tag{4.23}$$

As indicated by Figure 4.16, there are in general a large number of possible path families over α. Among these path families, let

$$\mathcal{P}^* := \underset{\mathcal{P}}{\operatorname{argmax}}\, \sigma(\mathcal{P}) \tag{4.24}$$

denote an optimal path family of maximal score (see Figure 4.16d for an example). In the following, the family consisting of the segments induced by the paths of $\mathcal{P}^*$ will be referred to as the **induced segment family** (of $\mathcal{P}^*$ or of α). Intuitively, the induced segment family contains the (nonoverlapping) repetitions of the segment α. Next, we show how an optimal path family $\mathcal{P}^*$ can be computed efficiently using dynamic programming and then explain how the fitness measure is derived from the score $\sigma(\mathcal{P}^*)$ and the induced segment family of $\mathcal{P}^*$.

4.3.1.2 Optimization Scheme

We now describe an efficient algorithm for computing an optimal path family for a given segment in a running time that is linear in the product of the length of the segment and the length of the entire music recording. The algorithm is based on a modification of dynamic time warping (DTW) as discussed in Section 3.2. Recall that, given two sequences, say $X = (x_1, x_2, \ldots, x_N)$ and $Y = (y_1, y_2, \ldots, y_M)$, the objective of DTW is to compute an optimal path that **globally** aligns X and Y, where the first elements as well as the last elements of the two sequences are to be aligned. The step size condition as specified by the set Σ constrains the slope of the path. In particular, using $\Sigma = \{(2,1),(1,2),(1,1)\}$, as specified in (3.30) and (4.6), each element of X is aligned to at most one element of Y (and vice versa).

Now, when computing an optimal path family over a given segment $\alpha = [s:t] \subseteq [1:N]$, the role of Y is taken over by the segment α, and the conditions change compared with classical DTW. In particular, α can be simultaneously aligned to several (nonoverlapping) subsequences of X. However, for each such subsequence, the entire segment α is to be aligned. Furthermore, certain sections of X may be left completely unconsidered in the alignment. Finally, instead of finding a **cost-minimizing** warping path, we are now looking for a **score-maximizing** path family. To account for the new constraints, we need to introduce additional steps that allow us to skip certain sections of X and to jump from the end to the beginning of the given segment α. The following procedure is also illustrated by Figure 4.17.

First, considering paths over the segment $\alpha = [s:t]$ with $M := |\alpha|$, we only consider the $N \times M$ submatrix $\mathbf{S}^\alpha$, which consists of the columns s to t of the self-similarity matrix $\mathbf{S}$. Next, we specify an accumulated score matrix $\mathbf{D} \in \mathbb{R}^{N,M+1}$ by a recursive procedure (similar to (3.25) for the accumulated cost matrix). The rows of $\mathbf{D}$ are indexed by $[1:N]$, and the columns are indexed by $[0:M]$, where the role of the column indexed by $m = 0$ is explained later. For a given cell (n,m), we consider a **set of predecessors** denoted by $\Phi(n,m)$, which contains all cells that may precede (n,m) in a valid path family. For $n \in [2:N]$ and $m \in [2:M]$ this set is given by

$$\Phi(n,m) = \{(n-i, m-j) \mid (i,j) \in \Sigma\} \cap [1:N] \times [1:M], \tag{4.25}$$

and the accumulated score matrix is defined by

$$\mathbf{D}(n,m) = \mathbf{S}^\alpha(n,m) + \max\{\mathbf{D}(i,j) \mid (i,j) \in \Phi(n,m)\}. \tag{4.26}$$

So far, this is similar to the recursion of the DTW algorithm summarized in Table 3.2. The constraint conditions and additional steps are realized by the definition of the values of $\mathbf{D}$ for the remaining index pairs (n,m) with $n = 1$ or $m \in \{0,1\}$.

As said before, the first column of $\mathbf{D}$ indexed by $m = 0$ plays a special role. We define this first column recursively by $\mathbf{D}(1,0) = 0$ and

$$\mathbf{D}(n,0) = \max\{\mathbf{D}(n-1,0), \mathbf{D}(n-1,M)\} \tag{4.27}$$

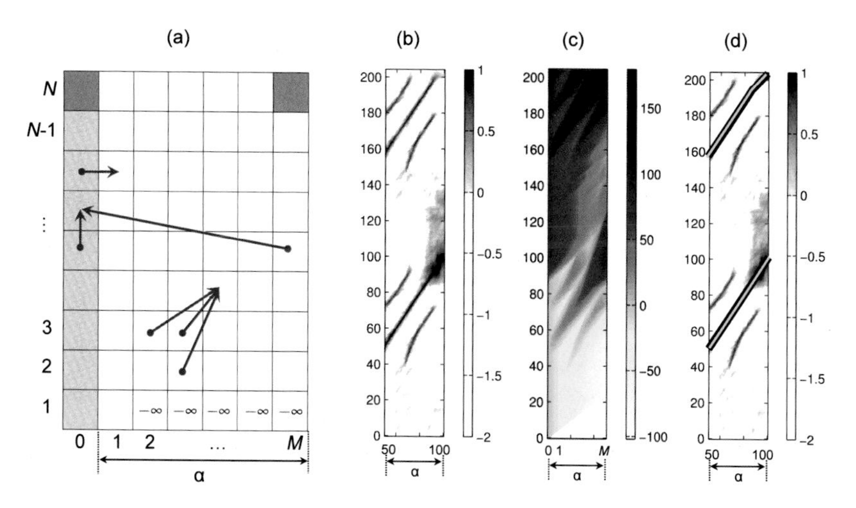

Fig. 4.17 **(a)** Illustration of the various predecessors in computing the accumulated score matrix. **(b)** Submatrix $\mathbf{S}^\alpha$ with $\alpha = [50:100]$ of the SSM shown in Figure 4.16a. **(c)** Accumulated score matrix $\mathbf{D}$. **(d)** Optimal path family.

for $n \in [2:N]$. The first term $\mathbf{D}(n-1,0)$ enables the algorithm to move upwards without accumulating any (possibly negative) score, thus realizing the condition that sections of X may be skipped without penalty (negative score). The second term $\mathbf{D}(n-1,M)$ closes up a path (ensuring that the entire segment α is aligned to a subsequence of X), while ensuring that the next possible segment does not overlap with the previous segment. Intuitively, the column indexed by $m = 0$ may be thought of as a kind of "elevator" column that makes it possible to skip arbitrary sections of X and to initialize new path components.

Next, we define the second column of $\mathbf{D}$ indexed by $m = 1$ by

$$\mathbf{D}(n,1) = \mathbf{D}(n,0) + \mathbf{S}^\alpha(n,1) \tag{4.28}$$

for $n \in [1:N]$. This definition makes it possible to start a new path component at cell $(n,1)$ coming from any position of the "elevator" column. Finally, to complete the initialization, we set $\mathbf{D}(1,m) = -\infty$ for $m \in [2:M]$. This forces the first path to come from the elevator column, thus starting with the first element of α. The score of an optimal path family is then given by

$$\sigma(\mathcal{P}^*) = \max\{\mathbf{D}(N,0), \mathbf{D}(N,M)\}. \tag{4.29}$$

The first term $\mathbf{D}(N,0)$ reflects the case that the final section of X may be skipped, and the second term $\mathbf{D}(N,M)$ ensures that in the other case the entire segment α is aligned to a suffix of X. The associated optimal path family $\mathcal{P}^*$ can be constructed from $\mathbf{D}$ using a backtracking algorithm as in the DTW algorithm (see Table 3.2). As the only modification, the cells of $\mathbf{S}^\alpha$ that belong to the first auxiliary column (indexed by $m = 0$) are to be omitted to obtain the final path family. In Exercise 4.6, we show that the recursive procedure for computing $\mathbf{D}$ has a computational complexity (in terms of memory requirements and running time) of $O(MN)$.

4.3.1.3 Definition of Fitness Measure

We have seen how to efficiently compute for a given segment α an optimal path family $\mathcal{P}^* = \{P_1, \ldots, P_K\}$, which reveals the repetition relations of α. In view of our intended fitness measure, one first idea is to simply use the total score $\sigma(\mathcal{P}^*)$ as defined in (4.23) as the fitness value for α. However, this measure does not yet have the desired properties, since it not only depends on the lengths of α and the paths, but also captures trivial self-explanations. For example, the segment $\alpha = [1:N]$ explains the entire sequence X perfectly, which is a trivial fact. More generally, each segment α explains itself perfectly, information that is encoded by the main diagonal of a self-similarity matrix. Therefore, one idea in defining the fitness measure is to disregard such trivial self-explanations. Assuming the normalization properties (4.18) and (4.19) of the underlying self-similarity matrix $\mathbf{S}$, this step can be done by simply subtracting the length $|\alpha|$ from the score $\sigma(\mathcal{P}^*)$. For example, in the case $\alpha = [1:N]$ this leads to the value zero. Furthermore, we normalize the score with regard to the lengths $L_k := |P_k|$ of the paths P_k contained in the optimal path family $\mathcal{P}^*$. This yields the **normalized score** $\bar{\sigma}(\alpha)$ defined by

$$\bar{\sigma}(\alpha) := \frac{\sigma(\mathcal{P}^*) - |\alpha|}{\sum_{k=1}^{K} L_k}. \tag{4.30}$$

From the assumption $\mathbf{S}(n,n) = 1$, we obtain $\bar{\sigma}(\alpha) \geq 0$ (see Exercise 4.7). Furthermore, note that, when using $\Sigma = \{(1,2),(2,1),(1,1)\}$, we get $\sum_k L_k \leq N$. This together with $\mathbf{S}(n,m) \leq 1$ implies the property $\bar{\sigma}(\alpha) \leq 1 - |\alpha|/N$. Intuitively, the value $\bar{\sigma}(\alpha)$ expresses the **average score** of the optimal path family $\mathcal{P}^*$ (minus a proportion for the self-explanation).

The normalized score indicates **how well** a given segment explains other segments, where the normalization eliminates the influence of segment lengths. This makes the normalized score a fair measure when comparing segments of different lengths. Besides repetitiveness, another issue is **how much** of the underlying music recording is covered by the thumbnail and its related segments. To capture this property, we define a **coverage** measure for a given α. To this end, let $\mathcal{A}^* := \{\pi_1(P_1), \ldots, \pi_1(P_K)\}$ be the segment family induced by the optimal path family $\mathcal{P}^*$, and let $\gamma(\mathcal{A}^*)$ be its coverage as defined in (4.21). Similar to the normalized score, we define the **normalized coverage** $\bar{\gamma}(\alpha)$ by

$$\bar{\gamma}(\alpha) := \frac{\gamma(\mathcal{A}^*) - |\alpha|}{N}. \tag{4.31}$$

As above, the length $|\alpha|$ is subtracted to compensate for trivial coverage. Obviously, one has $\bar{\gamma}(\alpha) \leq 1 - |\alpha|/N$. In other words, the value $\bar{\gamma}(\alpha)$ expresses the ratio be-

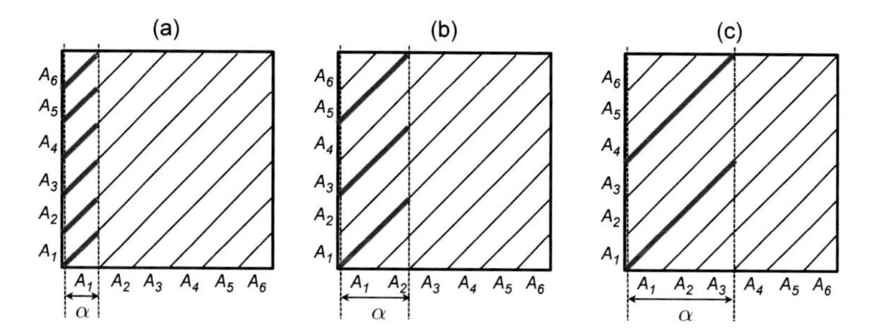

Fig. 4.18 Idealized SSM corresponding to the musical structure $A_1A_2 \ldots A_6$ with optimal path families for various segments α corresponding to **(a)** A_1, **(b)** A_1A_2, and **(c)** $A_1A_2A_3$.

tween the union of the induced segments of α and the total length of the original recording (minus a proportion for the self-explanation).

Having a high average score and a high coverage are both desirable properties for defining a thumbnail segment. However, these two properties are sometimes hard to satisfy at the same time. Shorter segments often have a higher average score, but a lower coverage, whereas longer segments tend to have a lower average score, but a higher coverage. To balance out these two trends, we combine the score and coverage measure by taking a suitable average. There are many ways for combining two values including the arithmetic, the geometric, and the harmonic mean. In the following, we use the harmonic mean, which (compared with the arithmetic mean) tends towards the smaller element and mitigates the impact of large differences between the two numbers to be averaged (see Exercise 4.8). We define the **fitness** $\varphi(\alpha)$ of the segment α to be the **harmonic mean**

$$\varphi(\alpha) := 2 \cdot \frac{\bar{\sigma}(\alpha) \cdot \bar{\gamma}(\alpha)}{\bar{\sigma}(\alpha) + \bar{\gamma}(\alpha)} \tag{4.32}$$

between the normalized score and normalized coverage. The fitness measure inherits the property $\varphi(\alpha) \leq 1 - |\alpha|/N$ from $\bar{\sigma}(\alpha)$ and $\bar{\gamma}(\alpha)$. The effect of combining score and coverage is illustrated by Figure 4.21 and will be further discussed in Section 4.3.2.

As an example, Figure 4.18 shows an idealized SSM of a piece having the musical structure $A_1A_2 \ldots A_6$, where we assume that each part is played in exactly the same way. Furthermore, we assume that the SSM has the value one on the indicated paths and otherwise the value zero. Let us first consider the segment α corresponding to A_1. The optimal path family consists of six paths over α (see Figure 4.18a). Since trivial self-explanations are left unconsidered, one obtains a normalized score of $\bar{\sigma}(\alpha) = 5/6$ and a normalized coverage of $\bar{\gamma}(\alpha) = 5/6$, which results in a fitness of $\varphi(\alpha) = 5/6$. Similarly, one obtains $\varphi(\alpha) = 2/3$ for the segment α corresponding to A_1A_2 (Figure 4.18b), and $\varphi(\alpha) = 1/2$ for the segment α corresponding to $A_1A_2A_3$ (Figure 4.18c). Obviously, the fitness is $\varphi(\alpha) = 0$ in case α corresponds to the entire music recording. In conclusion, the fitness measure allows for comparing segments of different length while slightly favoring shorter segments (since self-explanations are neglected). Further examples are discussed in Section 4.3.3 (see Exercise 4.9).

4.3.1.4 Thumbnail Selection

Based on the fitness measure, we define the audio thumbnail to be the segment of maximal fitness:

$$\alpha^* := \underset{\alpha}{\operatorname{argmax}}\, \varphi(\alpha). \tag{4.33}$$

By construction of the fitness measure, this segment has nonoverlapping repetitions that cover a possibly large portion of the audio recording. Furthermore, these repetitions are given by the induced segments obtained by the optimal path family of α^* yielding a segmentation of the audio recording into pairwise disjoint segments.

To account for prior knowledge and to remove spurious estimates, one can impose additional requirements on the thumbnail solution. In particular, introducing a lower bound θ for the minimal possible thumbnail length allows us to reduce the effect of noise scattered in the underlying self-similarity matrix. Extending the above definition, we define

$$\alpha_\theta^* := \underset{\alpha, |\alpha| \geq \theta}{\operatorname{argmax}}\, \varphi(\alpha). \tag{4.34}$$

In the next sections, we discuss and illustrate the properties of the fitness measure and the thumbnailing procedure in more detail.

4.3.2 Scape Plot Representation

The fitness measure assigns to each possible segment a fitness value that expresses a certain property. We now introduce a representation by which this segment-dependent property can be visualized in a compact and hierarchical way. Recall that a segment $\alpha = [s:t] \subseteq [1:N]$ is uniquely determined by its starting point s and its end point t. Since any two numbers $s, t \in [1:N]$ with $s \leq t$ define a segment, there are $(N+1)N/2$ different segments (see Exercise 4.10). Now, instead of considering start and end points, each segment can also be uniquely described by its center

$$c(\alpha) := (s+t)/2 \tag{4.35}$$

and its length $|\alpha|$. Using the center to parameterize a horizontal axis and the length to parameterize the height, each segment can be represented by a point in a triangular representation (see Figure 4.19). This way, the set of segments are ordered from bottom to top in a hierarchical way according to their length. In particular, the top of this triangle corresponds to the unique segment of maximal length N and the bottom points of the triangle correspond to the N segments of length one (where the start point coincides with the end point). Furthermore, all segments $\alpha' \subseteq \alpha$ contained in

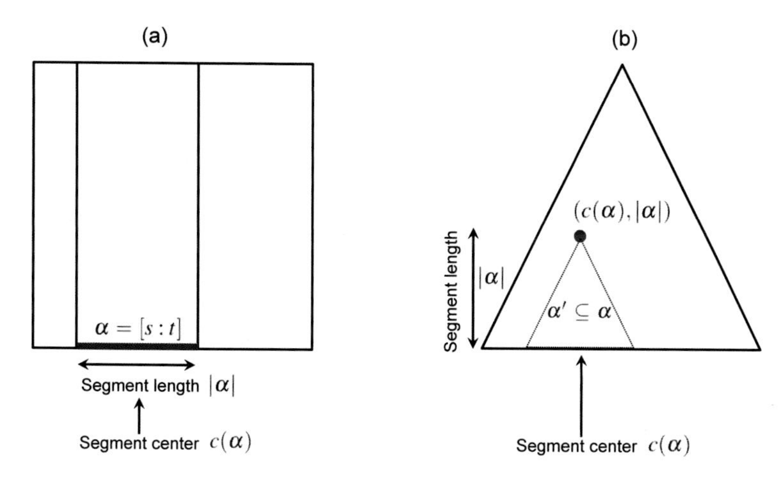

Fig. 4.19 Definition of scape plot representation. **(a)** Schematic SSM with segment. **(b)** Schematic scape plot with segment.

a given segment α correspond to points in the triangular representation that lie in a subtriangle below the point given by α (see Figure 4.19b and Exercise 4.12).

The triangular representation can be used as a grid for indicating the fitness values of all segments, which we also refer to as a **scape plot** representation of the fitness measure. More precisely, we define a scape plot Δ by setting

$$\Delta(c(\alpha), |\alpha|) := \varphi(\alpha) \tag{4.36}$$

for segment α. For our Brahms example, Figure 4.20 shows a scape plot representation in a color-coded form. Note that the maximal entry of Δ corresponds to the maximal fitness value, thus defining the thumbnail α^*.

4.3.3 Discussion of Properties

We now discuss some examples to illustrate the properties of the introduced fitness measure, the scape plot representation, and the induced segmentation. In our first example, we continue with our Brahms example. Recall that this piece has the musical structure $A_1A_2B_1B_2CA_3B_3B_4D$ (see Figure 4.5). Figure 4.16a shows a self-similarity matrix obtained from a given audio recording of this piece. Based on this SSM, the fitness measure is evaluated for all segments. The resulting fitness scape plot, which is shown in Figure 4.20a, reflects the musical structure in a hierarchical way. First note that the fitness-maximizing segment is $\alpha^* = [68:89]$. The coordinates in the scape plot are specified by the center $c(\alpha) = 78.5$ and the length $|\alpha| = 22$. Musically, this segment corresponds to the B_2-part, which is indeed the most repetitive part. The induced segment family consists of the four B-

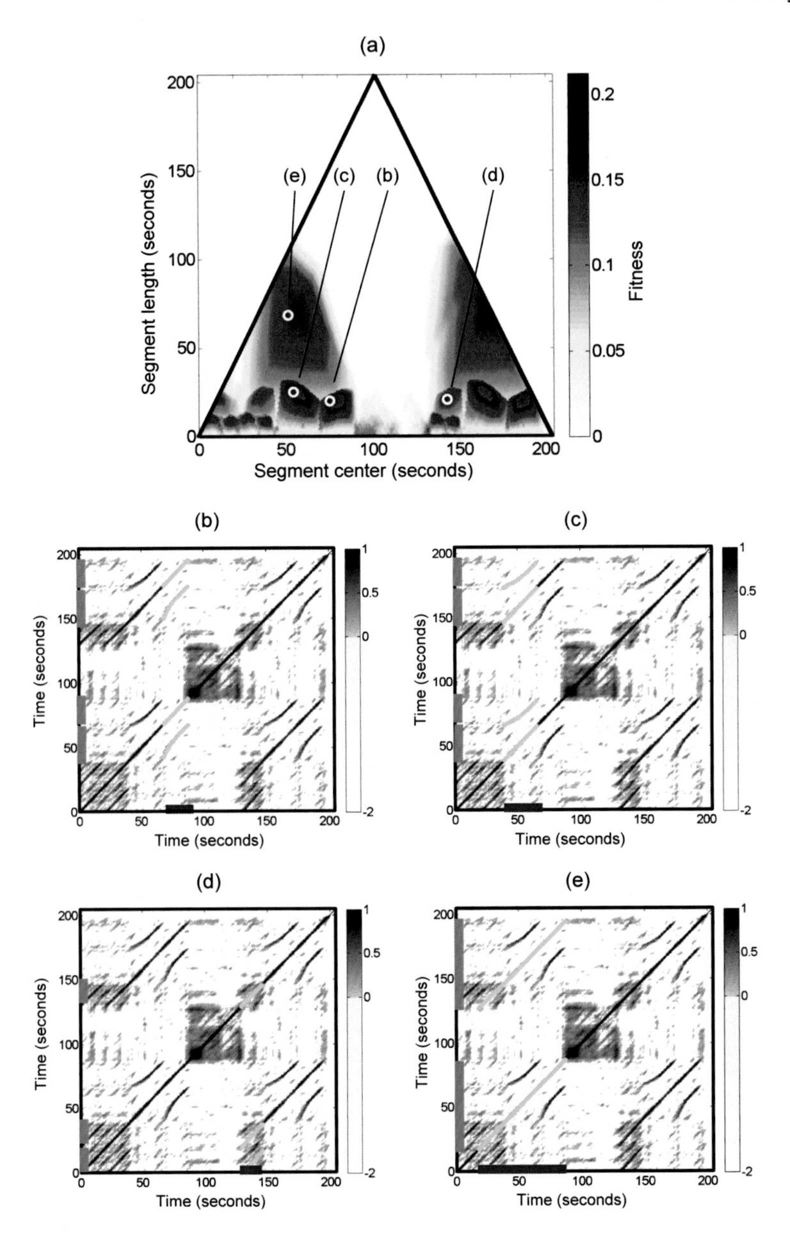

Fig. 4.20 Scape plot representation of fitness measure as well as different optimal path families and induced segment families over different segments α for our Brahms example. **(a)** Fitness scape plot. **(b)** $\alpha = \alpha^* = [68:89]$ (the thumbnail segment of maximal fitness corresponding to B_2). **(c)** $\alpha = [41:67]$ (corresponding to B_1). **(d)** $\alpha = [131:150]$ (corresponding to A_3). **(e)** $\alpha = [21:89]$ (corresponding to $A_1B_1B_2$).

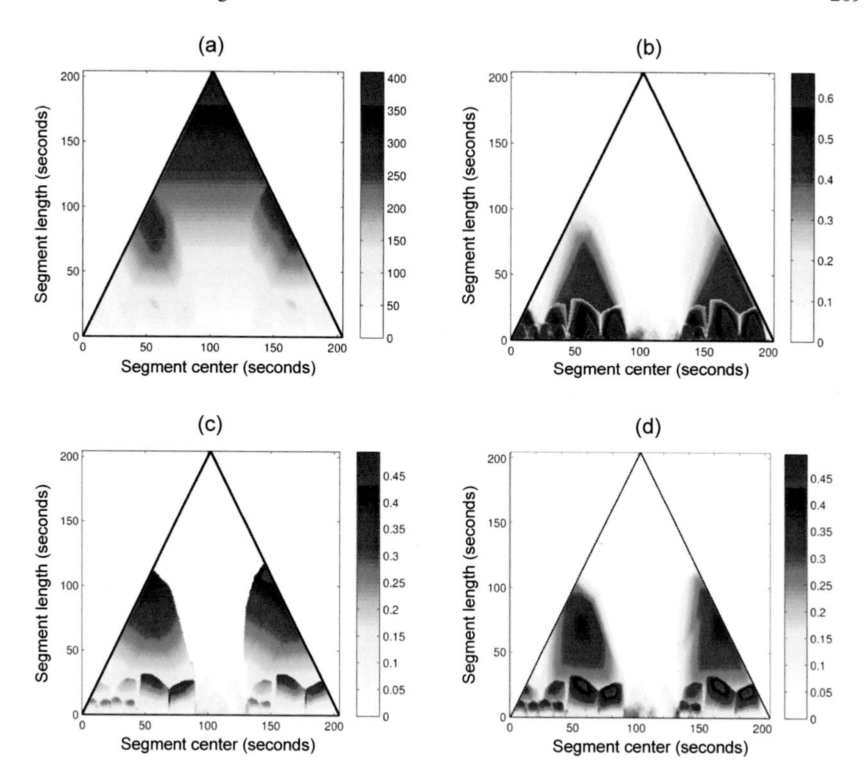

Fig. 4.21 Various scape plot representations. **(a)** Score. **(b)** Normalized score. **(c)** Normalized coverage. **(d)** Fitness measure (harmonic mean of (b) and (c)).

part segments (see Figure 4.20b). Note that all four B-part segments have almost the same fitness and lead to more or less the same segment family. For example, Figure 4.20c shows the induced segment family obtained from the B_1-part segment. This reflects the fact that each of the B-part segments may serve equally well as the thumbnail.

Recall that the introduced fitness measure slightly favors shorter segments (see Exercise 4.9). Therefore, since in this recording the B_2-part is played faster than the B_1-part, the fitness measure favors the B_2-part segment over the B_1-part segment. The scape plot also reveals other local maxima of musical relevance. For example, the local maximum corresponding to segment $\alpha = [131 : 150]$ ($c(\alpha) = 140.5$, $|\alpha| = 20$) corresponds to the A_3-part, and the induced segment family reveals the three A-parts (see Figure 4.20d). Furthermore, the local maximum assumed for segment $\alpha = [21 : 89]$ ($c(\alpha) = 55$, $|\alpha| = 69$) corresponds to $A_2B_1B_2$, which is repeated as $A_3B_3B_4$ (see Figure 4.20e). Again, note that, because of the normalization where self-explanations are disregarded, the fitness of the rather long segment $\alpha = [21 : 89]$ is well below that of the thumbnail $\alpha^* = [68 : 89]$.

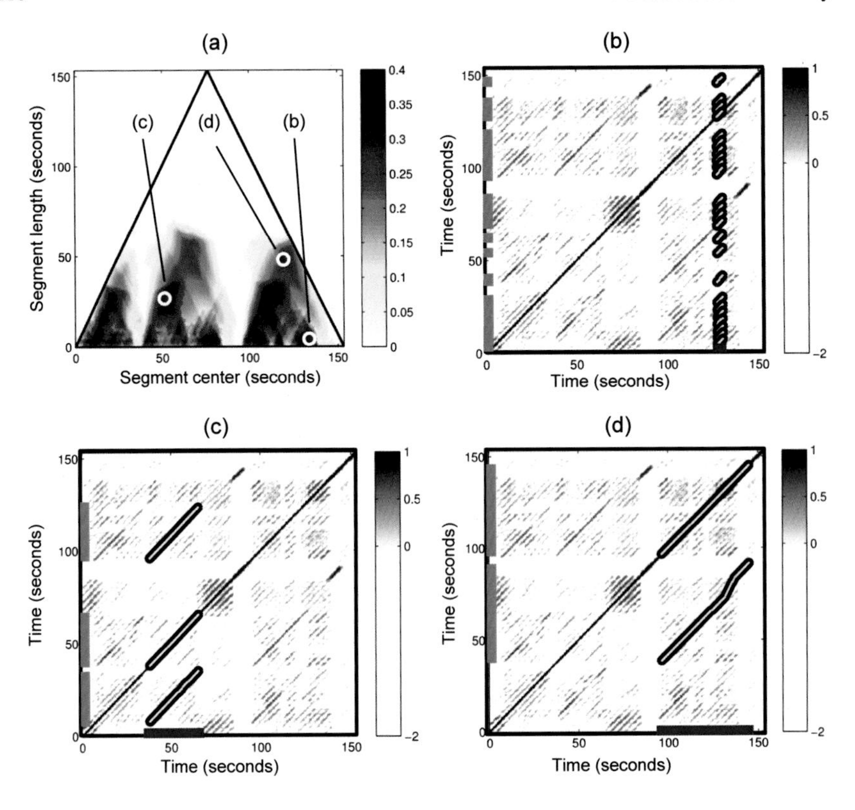

Fig. 4.22 Various optimal path families and induced segment families over different segments α for the Beatles song "Twist and Shout" having the musical structure $IV_1V_2B_1V_3B_2O$. **(a)** Fitness scape plot. **(b)** $\alpha = \alpha^* = [127:130]$. **(c)** $\alpha = \alpha^*_\theta = [38:65]$ using $\theta = 10$ (corresponding to V_2). **(d)** $\alpha = \alpha^*_\theta = [97:145]$ using $\theta = 40$ (corresponding to V_3B_2).

Next, we illustrate that in the definition (4.32) of the fitness measure the combination of the normalized score (4.30) and coverage (4.31) is of crucial importance. Figure 4.21b shows the scape plot when only using the normalized score. Since this measure expresses the average score of a path family without expressing how much of the audio material is actually covered, many of the small segments have a relatively high score. Using such a measure would typically result in false-positive segments of small length. In contrast, using only the normalized coverage would typically favor longer segments (see Figure 4.21c). The corresponding path families often contain components of rather low overall score (just above zero), which may result in rather weak repetitions. By combining score and coverage, the fitness measure balances out the two conflicting principles of having strong repetitions (high score) and of explaining possibly large portions of the recording (high coverage). Finally, we illustrate the importance of the normalization step by looking at the score $\sigma(\mathcal{P}^*)$ of the optimizing path family $\mathcal{P}^*$ over a segment α (see (4.23)). Figure 4.21d shows the resulting scape plot representation. Without normalization, longer segments typically dominate the shorter segments, with the entire recording having maximal score.

As a second example, Figure 4.22 shows the scape plot and various induced segment families for the Beatles song "Twist and Shout." This song has the musical structure $IV_1V_2B_1V_3B_2O$ consisting of a short intro (I-part), three verses (V-part), two bridges (B-part), and an outro (O-part). Interestingly, the fitness-maximizing segment $\alpha^* = [127:130]$ is very short and leads to a large number of spurious induced segments (see Figure 4.22b). The reason is that the song contains a short harmonic phrase, a so-called **riff**, which is repeated over and over again. As a consequence, the self-similarity matrix contains many repeated spurious path fragments which, as a whole family, lead to a high score as well as to a high coverage. To circumvent such problems, one can consider the segment α_θ^* as defined in (4.34) to enforce a minimal length for the thumbnail. In our example, setting $\theta = 10$ (given in seconds) one obtains the segment $\alpha_\theta^* = [38:65]$, which corresponds to the verse V_2 (see Figure 4.22c). This indeed yields a musically meaningful thumbnail. By further increasing the lower bound, one obtains superordinate repeating parts such as $\alpha_\theta^* = [97:145]$ corresponding to V_3B_2 (when using $\theta = 40$) (see Figure 4.22d).

4.4 Novelty-Based Segmentation

While the audio thumbnailing approach described in the previous section was based on the principle of repetition, we now discuss some segmentation procedures that are based on the principle of novelty. Recall from Section 4.1.1 that segment boundaries are often accompanied by a change in instrumentation, dynamics, harmony, tempo, or some other characteristics. It is the objective of novelty-based structure analysis to locate points in time where such musical changes occur, thus marking the transition between two subsequent structural parts. There are numerous approaches for novelty detection described in the literature. In the following, we present the main ideas of two of these approaches while introducing some general concepts that are also useful for the analysis of general time series. We start with a classical procedure where local changes are detected by correlating a checkerboard-like kernel along the main diagonal of a self-similarity matrix (Section 4.4.1). This procedure works particularly well when the underlying SSM has block-like structures. Then we introduce an approach for novelty detection that is based on structure features that encapsulate both local and global properties of the audio recording (Section 4.4.2). This procedure also highlights how various segmentation principles can be applied jointly within a single segmentation framework.

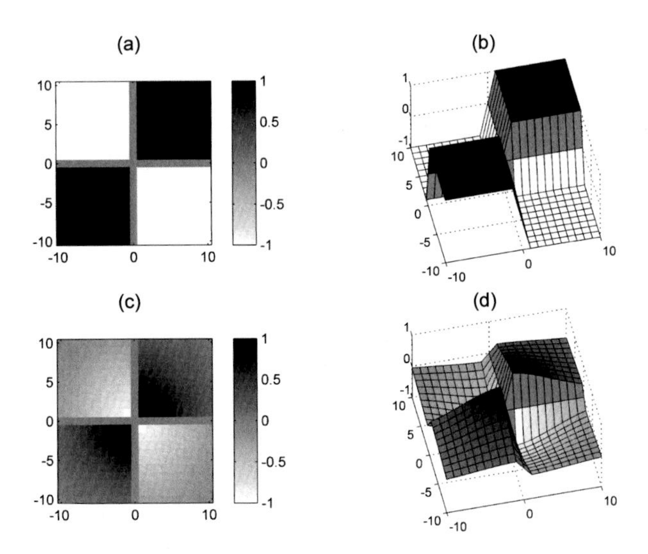

Fig. 4.23 Checkerboard kernel functions of size $M = 21$ ($L = 10$). **(a,b)** Box-like checkerboard kernel and 3D plot. **(c,d)** Gaussian checkerboard kernel and 3D plot.

4.4.1 Novelty Detection

As we have seen in Section 4.2.1, a self-similarity matrix reveals block-like structures in the case that the underlying feature sequence stays somewhat constant over the duration of an entire section. Often such a homogeneous segment is followed by another homogeneous segment that stands in contrast to the previous one. For example, a section played by strings may be followed by a section played by brass. Or there may be two contrasting sections each being homogeneous with respect to harmony, where the boundary between these sections is characterized by a change in the musical key. We have encountered such a case in our Brahms example, where one has homogeneous A-part segments in G minor and homogeneous C-part segments in G major (Figure 4.5).

One idea in novelty detection is to identify the boundary between two homogeneous but contrasting segments by correlating a checkerboard-like kernel function along the main diagonal of the SSM. This yields a **novelty function**. The peaks in this function indicate instances where significant changes occur in the audio signal. For example, using MFCCs, these peaks are good indicators for changes in timbre or instrumentation. Similarly, using chroma-based features, one obtains indicators for changes in harmony.

We now explain this procedure in more detail. As before, let $X = (x_1, x_2, \ldots, x_N)$ be a feature sequence and $\mathbf{S}$ a self-similarity matrix of size $N \times N$ derived from X. Let us first consider an audio recording that consists of two homogeneous but contrasting sections. When visualized, the resulting SSM looks like a 2×2 checkerboard as shown in Figure 4.23a. The two dark blocks on the main diagonal correspond to the regions of high similarity within the two sections. In contrast, the light regions outside these blocks express that there is a low cross-similarity between the sections. Thus, to find the boundary between the two sections one needs to identify the crux of the checkerboard. This can be done by correlating $\mathbf{S}$ with a kernel that itself looks like a checkerboard. The simplest such kernel is the (2×2)-unit kernel defined by

$$\mathbf{K} = \begin{bmatrix} -1 & 1 \\ 1 & -1 \end{bmatrix} = \begin{bmatrix} 0 & 1 \\ 1 & 0 \end{bmatrix} - \begin{bmatrix} 1 & 0 \\ 0 & 1 \end{bmatrix}. \tag{4.37}$$

This kernel can be written as the difference between a "coherence" and an "anti-coherence" kernel. The first kernel measures the self-similarity on either side of the center point and will be high when each of the two regions is homogeneous. The second kernel measures the cross-similarity between the two regions and will be high when there is little difference across the center point. The difference between the two values estimates the **novelty** of the feature sequence at the center point. The novelty is high when the two regions are self-similar but different from each other.

In audio structure analysis, where one is typically interested in changes on a larger time scale, kernels of larger size are used. Furthermore, since in this book we adopt a centered view (where a physical time position is associated to the center of a window or kernel), we assume that the size of the kernel is odd given by $M = 2L+1$ for some $L \in \mathbb{N}$. A box-like checkerboard kernel $\mathbf{K}_{\mathrm{Box}}$ of size M is an $(M \times M)$ matrix, which is indexed by $[-L:L] \times [-L:L]$. The matrix is defined by

$$\mathbf{K}_{\mathrm{Box}} = \mathrm{sgn}(k) \cdot \mathrm{sgn}(\ell), \tag{4.38}$$

where $k, \ell \in [-L:L]$ and "sgn" is the sign function (being -1 for negative numbers, 0 for zero, and 1 for positive numbers). For example, in the case $L = 2$, one obtains

$$\mathbf{K}_{\mathrm{Box}} = \begin{bmatrix} -1 & -1 & 0 & 1 & 1 \\ -1 & -1 & 0 & 1 & 1 \\ 0 & 0 & 0 & 0 & 0 \\ 1 & 1 & 0 & -1 & -1 \\ 1 & 1 & 0 & -1 & -1 \end{bmatrix} \tag{4.39}$$

(see Figure 4.23a). Note that the zero row and the zero column in the middle have been introduced more for theoretical reasons to ensure the symmetry of the kernel matrix. The checkerboard kernel can be smoothed to avoid edge effects using windows that taper towards zero at the edges. For this purpose, one may use a radially symmetric Gaussian function $\phi : \mathbb{R}^2 \to \mathbb{R}$ defined by

$$\phi(s,t) = \exp(-\varepsilon^2(s^2+t^2)), \tag{4.40}$$

where the parameter $\varepsilon > 0$ allows for adjusting the degree of tapering. Then the kernel $\mathbf{K}_{\mathrm{Gauss}}$ tapered by the Gaussian function is given by pointwise multiplication:

$$\mathbf{K}_{\mathrm{Gauss}}(k,\ell) = \phi(k,\ell) \cdot \mathbf{K}_{\mathrm{Box}}(k,\ell), \tag{4.41}$$

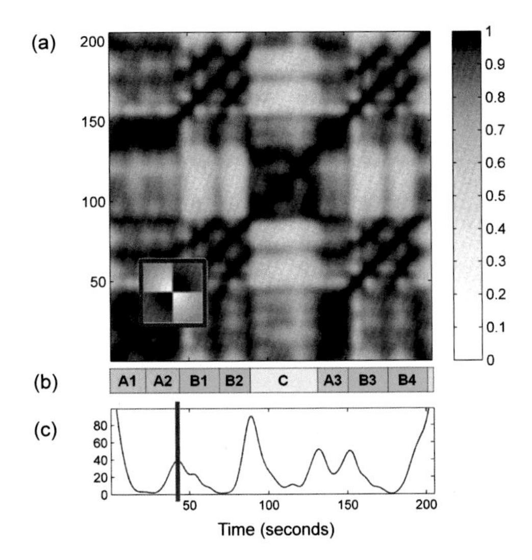

Fig. 4.24 Novelty function obtained by correlating an SSM with a Gaussian checkerboard kernel for a recording of the Hungarian Dance No. 5 by Johannes Brahms. **(a)** SSM similar to the one of Figure 4.10c. **(b)** Manually generated annotation of the musical structure. **(c)** Novelty function.

$k,\ell \in [-L:L]$ (see Figure 4.23c). Finally, to compensate for the influence of the actual kernel size and of the tapering, one may normalize the kernel. This can be done by dividing the kernel by the sum over the absolute values of the kernel matrix:

$$\mathbf{K}_{\mathrm{Norm}}(k,\ell) = \frac{\mathbf{K}_{\mathrm{Gauss}}(k,\ell)}{\sum_{k,\ell\in[-L:L]} |\mathbf{K}_{\mathrm{Gauss}}(k,\ell)|}. \tag{4.42}$$

The normalization becomes important when combining and fusing novelty information that is obtained from kernels of different size.

Now, to detect 2D corner points between adjoining blocks, the idea is to locally compare the SSM with a suitable checkerboard kernel. To this end, we slide a suitable checkerboard kernel $\mathbf{K}$ along the main diagonal of the SSM and sum up the element-wise product of $\mathbf{K}$ and $\mathbf{S}$:

$$\Delta_{\mathrm{Kernel}}(n) := \sum_{k,\ell\in[-L:L]} \mathbf{K}(k,\ell)\mathbf{S}(n+k,n+\ell) \tag{4.43}$$

for $n \in [L+1:N-L]$. Extending the matrix $\mathbf{S}$ on the boundaries by zero-padding (i.e., by setting $\mathbf{S}(k,\ell)=0$ for $(k,\ell)\in\mathbb{Z}\times\mathbb{Z}\setminus[1:N]\times[1:N]$), one may assume $n\in[1:N]$. This defines a function $\Delta_{\mathrm{Kernel}}:[1:N]\to\mathbb{R}$, also referred to as the **novelty function**, which specifies for each index $n\in[1:N]$ of the feature sequence a measure of novelty $\Delta_{\mathrm{Kernel}}(n)$. When the kernel $\mathbf{K}$ is positioned within a relatively uniform region of $\mathbf{S}$, the positive and negative values of the product tend to sum to zero and $\Delta_{\mathrm{Kernel}}(n)$ becomes small. Conversely, when the kernel $\mathbf{K}$ is positioned exactly at the crux of a checkerboard-like structure of $\mathbf{S}$, the values of the product are all positive and sum up to a large value $\Delta_{\mathrm{Kernel}}(n)$. Figure 4.24c shows a novelty function for our Brahms example using a chroma-based self-similarity matrix. The local maxima of the novelty function nicely indicate changes of harmony, which

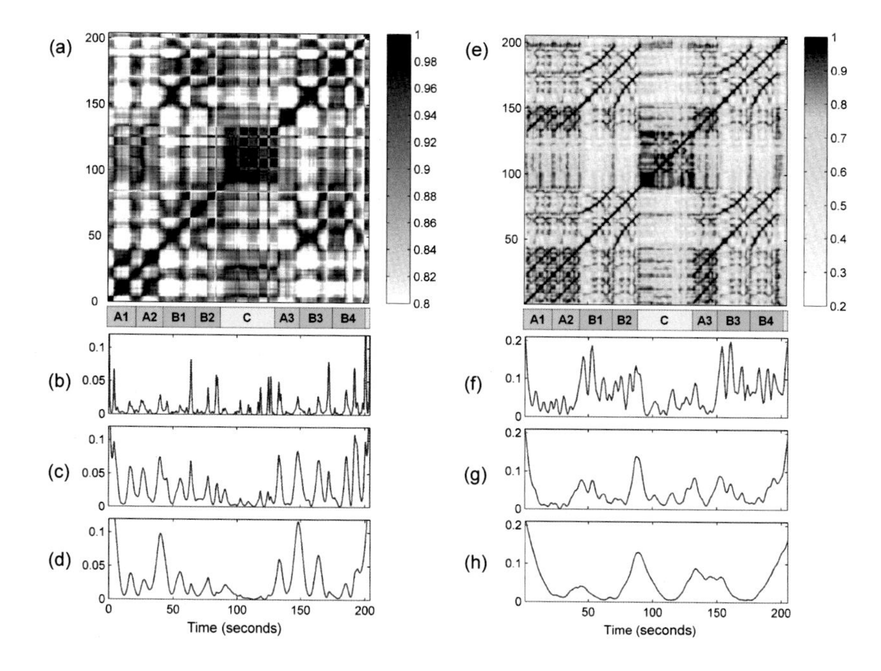

Fig. 4.25 Dependency of novelty functions on the feature representation and the kernel size. **(a)** SSM from Figure 4.7d using tempo-based features. **(b–d)** Novelty functions derived from (a) using a kernel of small/medium/large size. **(e)** SSM from Figure 4.7b using chroma-based features. **(f–h)** Novelty functions derived from (e) using a kernel of small/medium/large size.

particularly occur at boundaries between segments corresponding to different musical parts.

The size of the kernel has a significant impact on the properties of the novelty function. A small kernel may be suitable for detecting novelty on a short time scale, whereas a large kernel is suited for detecting boundaries and transitions between coarse structural sections. The suitability of a given kernel very much depends on the respective application and also on the properties of the underlying self-similarity matrix. This fact is illustrated by Figure 4.25, which shows novelty functions using different sizes and SSMs based on different features. Using a small kernel size may lead to a rather noisy novelty function with many spurious peaks. This particularly holds when the underlying SSM contains not only blocks but also path-like structures as is the case with the SSM shown in Figure 4.25e. Using a larger kernel averages out local fluctuations and results in a smoother novelty function. Note that a similar effect may be achieved by smoothing the SSM, which often leads to an enhancement of the block and an attenuation of the path structure. This effect becomes evident when comparing Figure 4.24 and Figure 4.25e.

In conclusion, the local maxima or peaks of a novelty function correspond to changes in the audio recording. These points often serve as good candidates for boundaries of neighboring segments that correspond to contrasting musical parts. In practice, there are many ways for computing novelty functions and for finding the relevant peaks. Besides the size of the kernel, a novelty function crucially depends on the characteristics of the underlying self-similarity matrix. In particular, the proposed novelty detection approach is only meaningful when the SSM has block-like structures, which are important in homogeneity-based structure analysis. Moreover, the peak selection strategy is also a delicate step that may have a substantial influence on the quality of the final result. Often, adaptive thresholding strategies where a peak is only selected when its value exceeds a local average of the novelty function are applied. To further reduce the number of spurious peaks, another strategy is to impose a constraint on the minimal distance between two subsequent peak positions.

In the following section, we describe a different approach for novelty detection, which makes it possible to identify structural changes as occurring in repetition-based structure analysis.

4.4.2 Structure Features

Most approaches for novelty detection are performed on the basis of features that capture local characteristics of the given music signal. For example, MFCC-based or chroma-based features capture local characteristics related to timbre or harmony, respectively. Then, a measure of novelty is computed by applying a local kernel or a type of derivative operator based on such feature representations. Computing local differences based on localized features makes such approaches vulnerable to more or less random noise-like fluctuations. We now describe a novelty detection procedure that incorporates global structural properties that go beyond local musical aspects such as harmony or timbre. To this end, we introduce **structure features** on the basis of which various structure analysis principles can be integrated within a unifying framework. The idea behind structure features is to jointly consider local and global aspects by measuring for each frame of a given feature sequence the relations to all other frames of the same feature sequence. This yields a frame-wise, i.e., **local**, feature representation that captures **global** structural characteristics of a feature sequence. The resulting structure features can then be used in combination with standard novelty detection procedures.

We start by introducing the concept of **time-lag matrices**, which is the main technical ingredient for defining the structure features. Let $\mathbf{S}$ be a self-similarity matrix derived from a feature sequence $X = (x_1, x_2, \ldots, x_N)$. Recall that two repeating segments, say $\alpha_1 = [s_1 : t_1]$ and $\alpha_2 = [s_2 : t_2]$, are revealed by a path of high similarity in $\mathbf{S}$ starting at (s_1, s_2) and ending at (t_1, t_2). Furthermore, if there is no relative tempo difference between the two segments, then the path runs exactly parallel to the main diagonal. One may also express this property by saying that segment α_1 is repeated after some time lag corresponding to $\ell = s_2 - s_1$ frames. This observation leads us to the notion of a time-lag representation of an SSM, where one time axis is replaced by a lag axis. To simplify notation, we assume in the following that

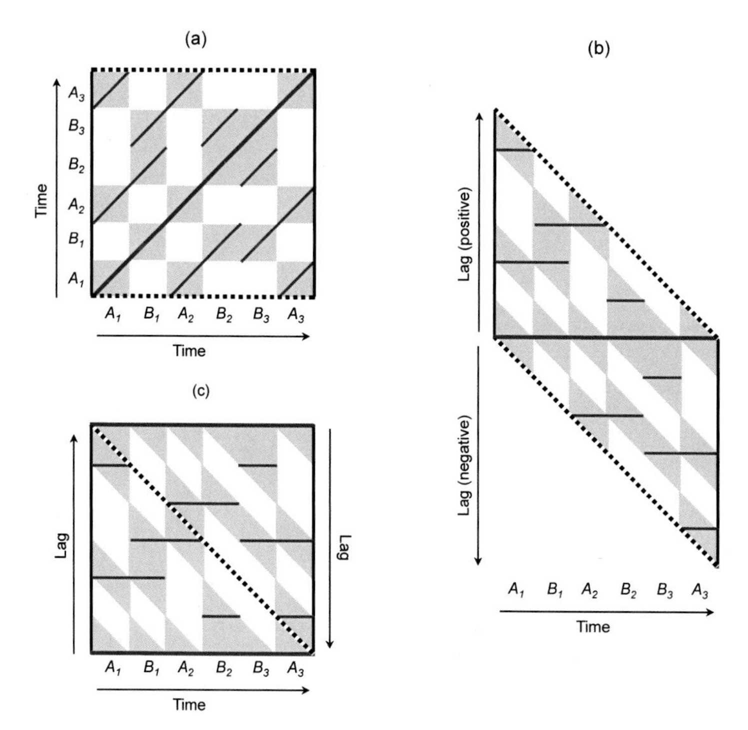

Fig. 4.26 **(a)** Self-similarity matrix **S**. **(b)** Time-lag representation **L**. **(c)** Cyclic time-lag representation of $\mathbf{L}^\circ$.

the frames are indexed starting with the index $n = 0$. Thus, $X = (x_0, x_1, \ldots, x_{N-1})$ and the self-similarity matrix **S** is indexed by $[0 : N-1] \times [0 : N-1]$. The **time-lag representation** of **S** is defined by

$$\mathbf{L}(\ell, n) = \mathbf{S}(n + \ell, n) \tag{4.44}$$

for $n \in [0 : N-1]$ and $\ell \in [-n : N-1-n]$. Note that the range for the lag parameter ℓ depends on the time parameter n. The lag index must be chosen in such a way that the sum $n + \ell$ lies in the range $[0 : N-1]$. For example, for time index $n = 0$ one can only look into the future with $\ell \in [0 : N-1]$, whereas for time index $n = N-1$ one can only look into the past with $\ell \in [-N+1 : 0]$. As an example, Figure 4.26a shows a self-similarity matrix **S** and Figure 4.26b its time-lag representation **L**, which is obtained by shearing the original matrix parallel to the horizontal axis. As a result, lines that are parallel to the main diagonal in **S** become horizontal lines in **L**. In other words, diagonal structures are transformed into horizontal structures. To simplify notation, we also introduce the **circular time-lag representation** $\mathbf{L}^\circ$ by defining

$$\mathbf{L}^\circ(\ell, n) = \mathbf{S}((n + \ell) \bmod N, n) \tag{4.45}$$

for $n \in [0:N-1]$ and $\ell \in [0:N-1]$. As also illustrated by Figure 4.26c, a negative time-lag parameter $\ell \in [-n:-1]$ as used in $\mathbf{L}$ is identified with $\ell + N$ in $\mathbf{L}^\circ$. Doing so, the time-lag representation $\mathbf{L}^\circ$ again becomes a matrix indexed by $[0:N-1] \times [0:N-1]$ as for the matrix $\mathbf{S}$.

What have we gained by considering a time-lag representation of a self-similarity matrix? In the following, let $\mathbf{S}^{[n]}$ denote the nth column of $\mathbf{S}$ for a given time frame $n \in [0:N-1]$. Recall that the vector $\mathbf{S}^{[n]} \in \mathbb{R}^N$ reveals the kind of relations that exist for time frame n. In the case that $\mathbf{S}^{[n]}(m)$ is large for some $m \in [0:N-1]$, then time frame n is related to time frame m. In the case that the value is small, the two frames are unrelated. In other words, $\mathbf{S}^{[n]}$ reveals the global structural relations of frame n. The same interpretation holds for the nth column of the time-lag matrix $\mathbf{L}^{\circ[n]}$. However, there is a crucial difference between $\mathbf{S}$ and $\mathbf{L}^\circ$. In the case that two subsequent frames n and $n+1$ have the same structural properties, the two vectors $\mathbf{S}^{[n]}$ and $\mathbf{S}^{[n+1]}$ are **cyclically shifted** versions of each other, whereas the two vectors $\mathbf{L}^{\circ[n]}$ and $\mathbf{L}^{\circ[n+1]}$ are **identical**.

Based on this observation, we define the **structure features** to be the columns $y_n := \mathbf{L}^{\circ[n]} \in \mathbb{R}^N$ for $\mathbf{L}^\circ$, $n \in [0:N-1]$. By this process, we have converted the original sequence $X = (x_0, x_1, \ldots, x_{N-1})$ of features x_n that capture local (acoustic, musical) characteristics into a sequence $Y = (y_0, y_1, \ldots, y_{N-1})$ of features y_n that capture global (structural) characteristics. As a result, boundaries of the global structural parts can be identified by looking for local changes in the feature sequence Y. There are many ways to capture such local changes. A simple strategy is to compute the difference between successive structure features based on a suitable distance function. For example, using the Euclidean norm of $\mathbb{R}^N$ (see (2.38)), one obtains a novelty function

$$\Delta_{\text{Structure}}(n) := \|y_{n+1} - y_n\| = \|\mathbf{L}^{\circ[n+1]} - \mathbf{L}^{\circ[n]}\| \tag{4.46}$$

for $n \in [0:N-2]$. Again, by zero-padding one may assume $n \in [0:N-1]$. The positions of local maxima or peaks of this function yield candidates for structural boundaries. The overall procedure depends on many design choices and parameter settings including the feature type used for the original sequence X or the way $\mathbf{S}$ is computed. Also, in practice, one often uses more involved derivative operators and applies suitable preprocessing steps (e.g., further enhancing the matrix $\mathbf{L}^\circ$) and postprocessing steps (e.g., normalizing the novelty function $\Delta_{\text{Structure}}$). Finally, as already mentioned in Section 4.4.1, the peak selection strategy may have a crucial influence on the final result.

We close this section by considering the example shown in Figure 4.27, which illustrates the overall procedure for structure-based novelty detection. The underlying piece of music is the Mazurka Op. 24, No. 1 by Frédéric Chopin, which has the musical structure $A_1A_2B_1B_2A_3A_4B_3B_4CA_5A_6$. Figure 4.27a shows a path-enhanced and binarized SSM computed from a chroma-based feature representation of an audio recording. The resulting circular time-lag representation $\mathbf{L}^\circ$ and novelty function $\Delta_{\text{Structure}}$ are shown in Figure 4.27b and Figure 4.27c, respectively. Note that the peak positions of $\Delta_{\text{Structure}}$ coincide well with the (joint) start and end positions of path components, which in turn concur with boundaries of the musical sections.

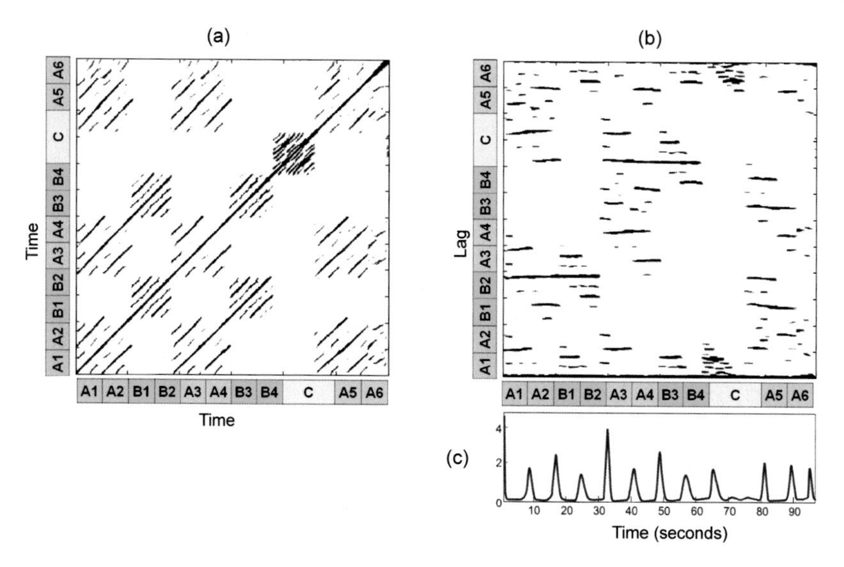

Fig. 4.27 Novelty-based segmentation using structure features for a recording of the Mazurka Op. 24, No. 1 by Frédéric Chopin. **(a)** Path-enhanced and binarized self-similarity matrix $\mathbf{S}$. **(b)** Circular time-lag representation $\mathbf{L}^{\circ}$. **(c)** Novelty function $\Delta_{\mathrm{Structure}}$.

The structure features work particularly well in this example because of two reasons. First, there are many repeating parts, resulting in a rich path structure. Second, the various repeating musical parts occur in different chronological orders, resulting in characteristic path discontinuations that are captured well by the structure features; That is, structure-based novelty detection does not work for a piece with musical structure $A_1A_2A_3A_4$ or $A_1B_1A_2B_2$, but works well for a piece with musical structure $A_1B_1A_2A_3$ or $A_1A_2B_1B_2$ (see Exercise 4.14).

4.5 Evaluation

We have described various procedures for extracting structural information from a given music recording. However, we have not yet discussed the issue of measuring **how well** a given procedure performs the task at hand. In this section, we address the problem of automatically evaluating structure analysis algorithms and explain why the evaluation itself constitutes a nontrivial task.

A general evaluation approach in structure analysis is to compare an **estimated result** obtained by some automated procedure against some **reference result**. To realize such a general approach, one needs to find answers to the following questions: How is a structure analysis result actually modeled? How should the estimated result be compared against the reference result? Where does the reference result come from and is it reliable? In particular the last question easily leads to philosophical

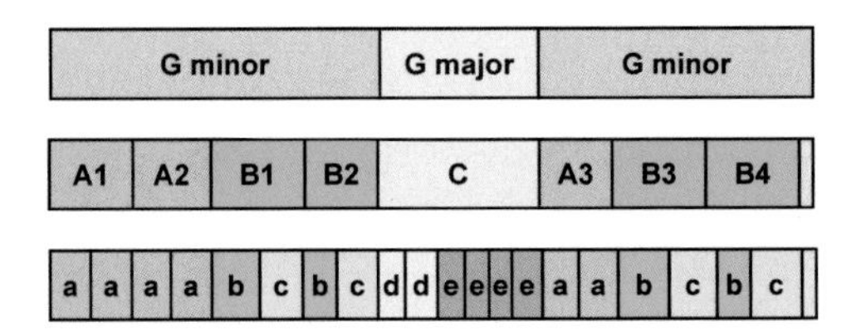

Fig. 4.28 Structure annotation on various scales of the Hungarian Dance No. 5 by Johannes Brahms.

considerations on the nature and meaning of musical structures. As we have already discussed in Section 4.1 and illustrated by Figure 4.2, music structure analysis is an ill-posed problem that depends on many different factors, not to mention the musical and acoustic variations that occur in real-world music recordings. Since a structure analysis result largely depends on the musical context and the considered temporal level, even two human experts may disagree in their analysis of a given piece of music. In the case of our Brahms example, as we have already discussed in Section 4.1.2 and as illustrated by Figure 4.28, one expert may annotate the structure on a larger scale resulting in the musical structure $A_1A_2B_1B_2CA_3B_3B_4D$, while another expert may consider a smaller scale, where the parts are further subdivided.

For the moment, we do not dwell on the latter issue any further. Instead, we assume that a valid reference structure annotation has been provided by a human expert, even though this is a simplistic and sometimes problematic assumption. Such an annotation is also often referred to as **ground truth**. The objective of the automated procedure is to estimate a structure annotation that is as close to the reference as possible. After introducing some general notions (Section 4.5.1), we discuss some evaluation metrics often used for comparing structure analysis results (Section 4.5.2).

4.5.1 Precision, Recall, F-Measure

Many evaluation measures are based on some notion of precision, recall, and F-measure—a concept that has been borrowed from the fields of information retrieval and pattern recognition. We now introduce this general concept in the context of binary classification (see Figure 4.29 for an overview). First, let $\mathcal{I}$ be a finite set of so-called **items**. For this set, one has a reference annotation that is the result of a binary classification. Each item $i \in \mathcal{I}$ is assigned either a label '$+$' (item is **positive** or **relevant**) or a label '$-$' (item is **negative** or **not relevant**). Let $\mathcal{I}_+^{\mathrm{Ref}}$ be the set of positive items, and $\mathcal{I}_-^{\mathrm{Ref}}$ be the set of negative items. Furthermore, one has an automated procedure that estimates the annotation for each item. Let $\mathcal{I}_+^{\mathrm{Est}}$ be the set of items being estimated as positive, and $\mathcal{I}_-^{\mathrm{Est}}$ be the set of items being estimated as negative. An item $i \in \mathcal{I}_+^{\mathrm{Est}}$ estimated as positive is called a **true positive** (TP) if it belongs to $\mathcal{I}_+^{\mathrm{Ref}}$, i.e., if $i \in \mathcal{I}_+^{\mathrm{Est}} \cap \mathcal{I}_+^{\mathrm{Ref}}$. Otherwise, if $i \in \mathcal{I}_+^{\mathrm{Est}} \cap \mathcal{I}_-^{\mathrm{Ref}}$, it is called a **false positive** (FP). Similarly, an item $i \in \mathcal{I}_-^{\mathrm{Est}}$ estimated as negative is called a **false negative** (FN) if it belongs to $\mathcal{I}_+^{\mathrm{Ref}}$, and **true negative** (TN) otherwise.

		Reference annotation ("Ground truth")		
		Positive	Negative	
Estimated annotation ("Algorithm")	Positive	True positive (TP)	False positive (FP)	$\mathrm{P} = \frac{\#\mathrm{TP}}{\#\mathrm{TP} + \#\mathrm{FP}}$
	Negative	False negative (FN)	True negative (TN)	
		$\mathrm{R} = \frac{\#\mathrm{TP}}{\#\mathrm{TP} + \#\mathrm{FN}}$		$\mathrm{F} = \frac{2\mathrm{PR}}{\mathrm{P} + \mathrm{R}}$

Fig. 4.29 Definition of precision, recall, and F-measure.

The **precision** P of the estimation is defined as the number of true positives divided by the total number of items estimated as positive:

$$\mathrm{P} = \frac{|\mathcal{I}_+^{\mathrm{Est}} \cap \mathcal{I}_+^{\mathrm{Ref}}|}{|\mathcal{I}_+^{\mathrm{Est}}|} = \frac{\#\mathrm{TP}}{\#\mathrm{TP} + \#\mathrm{FP}}. \tag{4.47}$$

In contrast, the **recall** R is defined as the number of true positives divided by the total number of positive items:

$$\mathrm{R} = \frac{|\mathcal{I}_+^{\mathrm{Est}} \cap \mathcal{I}_+^{\mathrm{Ref}}|}{|\mathcal{I}_+^{\mathrm{Ref}}|} = \frac{\#\mathrm{TP}}{\#\mathrm{TP} + \#\mathrm{FN}}. \tag{4.48}$$

Note that both precision and recall have values in the interval $[0,1]$. A perfect precision $\mathrm{P} = 1$ means that every item estimated as positive is indeed positive. In this case, there is no false positive, but there may exist some false negatives. In contrast, a perfect recall $\mathrm{R} = 1$ means that every positive item was also estimated as positive. In this case, there is no false negative, but there may exist some false positives. Only in the case $\mathrm{P} = 1$ and $\mathrm{R} = 1$ does the estimated annotation coincide with the reference annotation. Precision and recall are often combined by taking their harmonic mean to form a single measure, often referred to as the **F-measure**:

$$\mathrm{F} = \frac{2 \cdot \mathrm{P} \cdot \mathrm{R}}{\mathrm{P} + \mathrm{R}}. \tag{4.49}$$

The harmonic mean is further discussed in Exercise 4.8. One main property is that $\mathrm{F} \in [0,1]$ with $\mathrm{F} = 1$ if and only if $\mathrm{P} = 1$ and $\mathrm{R} = 1$.

4.5.2 Structure Annotations

With these formal definitions at hand, let us come back to the structure analysis scenario. Since there are many different analysis tasks and aspects to be considered, it is not at all clear how a mathematical model for the analysis result has to be specified. Let us start with the general task of deriving the musical structure from a given audio recording. In the following, we consider the discrete-time case, where the sampled time axis is indexed by $[1:N]$. We call the result of a structure analysis a **structure annotation**, which consists of a segmentation of the time axis together with a labeling of the segments. The segmentation is modeled by a segment family $\mathcal{A} = \{\alpha_1, \alpha_2, \ldots, \alpha_K\}$ of some size K as introduced in (4.20). Note that at this stage we make the assumption that segments are disjoint. On the one hand, this is a convenient restriction, which simplifies the comparison of different structure annotations. On the other hand, this assumption may not always be appropriate. For example, it does not allow to capture hierarchical and nested structures. For the labeling, let Λ be a set of possible labels. For example, Λ may be the set $\{A,B,C,\ldots,a,b,c,\ldots\}$, it may consist of a set of suitable strings such as $\{\text{Chorus}, \text{Verse}, \text{Bridge}, \ldots\}$, or it may simply be a subset of $\mathbb{N}$. Then the **labeling** can be modeled by assigning to each segment α_k a label $\lambda_k \in \Lambda$, $k \in [1:K]$. To simplify notation, we additionally assume that the segment family covers the entire time axis (i.e., $\bigcup_{k=1}^{K} \alpha_k = [1:N]$) so that each frame index $[1:N]$ is assigned to exactly one label. In Exercise 4.15, we show that this assumption does not lead to any loss of generality.

There are many ways to compare an estimated structure annotation with a reference annotation and for deriving some kind of "success" measure. In its strictest form, one could simply say that either the two annotations are identical ("success") or not ("fail"). In practice, however, such a binary measure is not very meaningful. Instead, one requires measures that indicate the degree of similarity of two given annotations and that is insensitive towards small differences in the annotations. For example, such differences may be due to small shifts in segment boundaries or local deviations in the labeling. Furthermore, even though two annotations may be based on the same segmentation and the same grouping of segments, they may differ in the naming of the labels. For example, in the reference annotation, segments may be labeled by strings such as "verse" or "chorus" whereas in the estimated annotation corresponding parts may be labeled by letters such as A or B. In many applications, such a mismatch in the label naming is not considered to be a failure of the algorithm. Having these issues in mind, we now discuss some evaluation measures in more detail.

4.5.3 Labeling Evaluation

We start with some purely frame-based evaluation measures, which are referred to as **pairwise** precision, recall, and F-measure. In these measures, the segment boundaries are left unconsidered and only the labeling information is used. For a given structure annotation, we define a **label function** $\varphi : [1:N] \to \Lambda$ by setting

$$\varphi(n) := \lambda_k \tag{4.50}$$

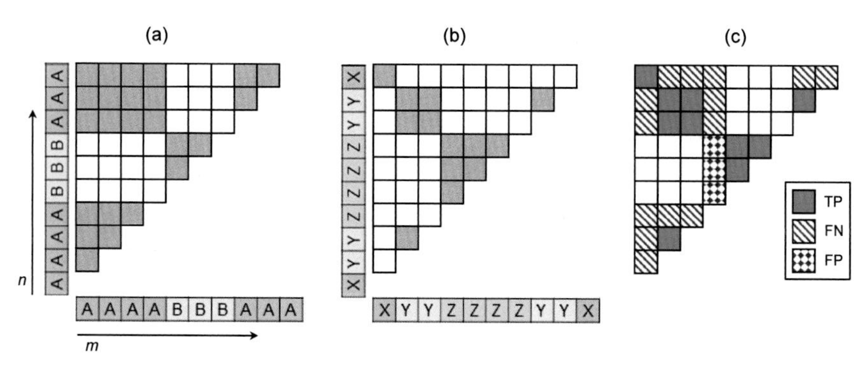

Fig. 4.30 Illustration of pairwise precision, recall, and F-measure. **(a)** Positive items (indicated by gray boxes) with regard to the reference annotation. **(b)** Positive items (indicated by gray boxes) with regard to the estimated annotation. **(c)** True positive (TP), false positive (FP), and false negative (FN) items.

for $n \in \alpha_k$ (assuming that the segment family covers the entire time axis). Let φ^{Ref} and φ^{Est} be the label functions for the reference and estimated structure annotation, respectively. In order to become independent of the actual label naming, the main idea is to not directly look at the labels, but to look for label co-occurrences. To this end, we consider pairs of frames that are assigned to the same label. More precisely, we define the set

$$\mathcal{I} = \{(n,m) \in [1:N] \times [1:N] \mid m < n\}, \tag{4.51}$$

which serves as a set of items as described in Section 4.5.1. For the reference and estimated annotations, we define the positive items by

$$\mathcal{I}_+^{\mathrm{Ref}} = \{(n,m) \in \mathcal{I} \mid \varphi^{\mathrm{Ref}}(n) = \varphi^{\mathrm{Ref}}(m)\}, \tag{4.52}$$

$$\mathcal{I}_+^{\mathrm{Est}} = \{(n,m) \in \mathcal{I} \mid \varphi^{\mathrm{Est}}(n) = \varphi^{\mathrm{Est}}(m)\}, \tag{4.53}$$

whereas $\mathcal{I}_-^{\mathrm{Ref}} = \mathcal{I} \setminus \mathcal{I}_+^{\mathrm{Ref}}$ and $\mathcal{I}_-^{\mathrm{Est}} = \mathcal{I} \setminus \mathcal{I}_+^{\mathrm{Est}}$. In other words, an item (n,m) is considered to be positive with regard to an annotation if the frames n and m have the same label. Now, the **pairwise precision** is defined to be the precision of this binary classification scheme. Similarly, the **pairwise recall** is the recall and the **pairwise F-measure** is the F-measure of this scheme.

The definitions are illustrated by Figure 4.30, where the sampled time interval $[1:N]$ consists of $N = 10$ samples. The reference structure annotation consists of three segments labeled with A, B, and A, respectively. As shown by Figure 4.30a, 24 out of the 45 items are positive with regard to the reference annotation. Similarly, Figure 4.30b shows an estimated structure annotation and the resulting 13 positive items. In Figure 4.30c, the true positives (#TP = 10), false positives (#FP = 3), and false negatives (#FN = 14) are indicated. From this, one obtains

$$\mathrm{P} = \#\mathrm{TP}/(\#\mathrm{TP}+\#\mathrm{FP}) = 10/13 \approx 0.769, \tag{4.54}$$

$$\mathrm{R} = \#\mathrm{TP}/(\#\mathrm{TP}+\#\mathrm{FN}) = 10/24 \approx 0.417, \tag{4.55}$$

$$\mathrm{F} = 2\mathrm{PR}/(\mathrm{P}+\mathrm{R}) \approx 0.541. \tag{4.56}$$

In this example, the precision of nearly 77% is relatively high, whereas the recall of 42% is relatively low. The F-measure is between these two values with a bias towards the smaller one. Further examples are discussed in the exercises.

4.5.4 Boundary Evaluation

The pairwise precision, recall, and F-measure are solely based on label information, whereas segment boundaries are treated implicitly by the presence of label changes. For other structure analysis tasks such as novelty-based segmentation, the precise detection of boundaries is the focus. To evaluate such procedures, one measures the deviation of the estimated segment boundaries from the boundaries of a reference annotation. To mathematically model this scenario, we introduce the notion of a **boundary annotation**, which is given by a sequence $B = (b_1, b_2, \ldots, b_K)$ of increasing indices $b_k \in [1:N]$, $k \in [1:K]$. For example, such a boundary annotation may be derived from a structure annotation by taking the start and possibly the end indices of the annotated segments. In the following, let B^{Ref} be the reference boundary annotation and B^{Est} the estimated boundary annotation. There are many ways to compare B^{Est} against B^{Ref}. For example, using $\mathcal{I} = [1:N]$ as a set of items, one can define $\mathcal{I}_+^{\mathrm{Ref}} := B^{\mathrm{Ref}}$ and $\mathcal{I}_+^{\mathrm{Est}} := B^{\mathrm{Est}}$. From this, the precision, recall, and F-measure can be computed in the usual way. In this case, an estimated boundary is considered correct only if it agrees with a reference boundary.

For certain applications small deviations in the boundary positions are acceptable. Therefore, one generalizes the previous measures by introducing a tolerance parameter $\tau \geq 0$ for the maximal acceptable deviation. An estimated boundary $b^{\mathrm{Est}} \in B^{\mathrm{Est}}$ is then considered **correct** if it lies within the τ-neighborhood of a reference boundary $b^{\mathrm{Ref}} \in B^{\mathrm{Ref}}$:

$$|b^{\mathrm{Est}} - b^{\mathrm{Ref}}| \leq \tau. \tag{4.57}$$

In this case, the sets $\mathcal{I}_+^{\mathrm{Ref}}$ and $\mathcal{I}_+^{\mathrm{Est}}$ can no longer be used for defining precision and recall. Instead, we generalize the notions of true positives, false positives, and false negatives. The **true positives** (TP) are defined to be the items $b^{\mathrm{Est}} \in B^{\mathrm{Est}}$ that are correct, and the **false positives** (FP) are the items $b^{\mathrm{Est}} \in B^{\mathrm{Est}}$ that are not correct. Furthermore, the **false negatives** (FN) are defined to be the items $b^{\mathrm{Ref}} \in B^{\mathrm{Ref}}$ with no estimated item in a τ-neighborhood. Based on these definitions, one can compute precision, recall, and F-measure from #TP, #FP, and #FN using the formulas of Figure 4.29.

However, this generalization needs to be taken with care. Because of the tolerance parameter τ, several estimated boundaries may be contained in the τ-neighborhood of a single reference boundary. Conversely, a single estimated bound-

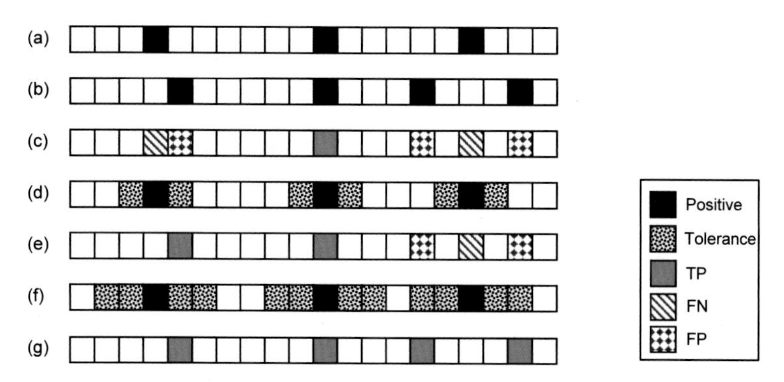

Fig. 4.31 Illustration of boundary evaluation. **(a)** Reference boundary annotation. **(b)** Estimated boundary annotation. **(c)** Evaluation of (b) with regard to (a). **(d)** τ-Neighborhood of (a) using the tolerance parameter $\tau = 1$. **(e)** Evaluation of (b) with regard to (d). **(f)** τ-Neighborhood of (a) using the tolerance parameter $\tau = 2$. **(g)** Evaluation of (b) with regard to (f).

ary may be contained in the τ-neighborhood of several reference boundaries. As a result, one may obtain a perfect F-measure even in the case that the sets B^{Est} and B^{Ref} contain a different number of boundaries. From a semantic point of view, this is not meaningful. To avoid such anomalies, one may introduce an additional assumption in the definition of a boundary annotation by requiring

$$|b_{k+1} - b_k| > 2\tau \tag{4.58}$$

for $k \in [1 : N-1]$. This is also a meaningful requirement from a musical point of view: a musical section (determined by two subsequent boundaries) should be much longer than the size of the tolerance parameter.

Figure 4.31 illustrates the boundary evaluation measures by means of a simple example. Using the tolerance parameter $\tau = 0$, one obtains #TP $= 1$, #FP $= 3$, and #FN $= 2$ (see Figure 4.31c). This yields P $= 1/4$, R $= 1/3$, and F $= 2/7$. In the case $\tau = 1$, one obtains #TP $= 2$, #FP $= 2$, and #FN $= 1$, which results in P $= 1/2$, R $= 2/3$, and F $= 4/7$ (see Figure 4.31e). Finally, when using $\tau = 2$, one obtains a perfect F-measure. However, in this case the condition (4.58) is violated and the meaning of the evaluation measure is questionable.

4.5.5 Thumbnail Evaluation

As a third scenario, we now discuss some evaluation measures for audio thumbnailing, which is a prominent subtask of general music structure analysis. In Section 4.3, we introduced an automated procedure for identifying the most representative section from a given audio recording. Mathematically, this section and its repetitions

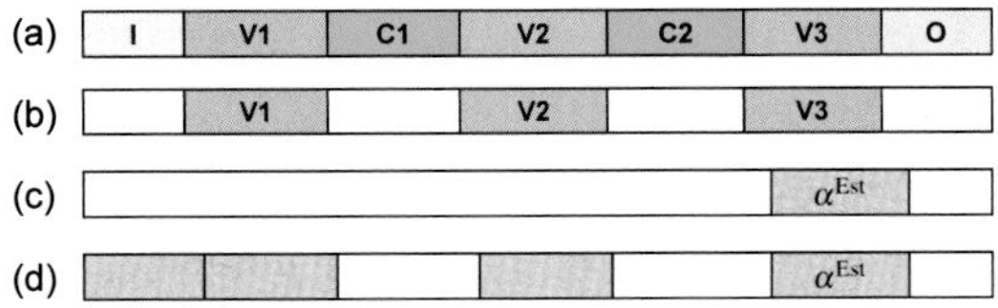

Fig. 4.32 Thumbnail evaluation. **(a)** Reference structure annotation. **(b)** Reference thumbnail family. **(c)** Estimated thumbnail. **(d)** Resulting segment family.

are modeled by a segment family $\mathcal{A} = \{\alpha_1, \alpha_2, \ldots, \alpha_K\}$ of some size K (see (4.20)). Each of the segments of this family may serve equally well as the thumbnail.

In the thumbnailing scenario, one does not need an entire structure annotation, but only the label and the associated segment family that represents the thumbnail and its repetitions. For example, for a popular song, the thumbnail may be the verse part of the song so that the associated segment family would consist of all verse sections of the recording (see Figure 4.32b). As in the previous evaluation scenarios, we assume that a suitable reference annotation is available. This annotation is given in the form of a segment family of the audio thumbnail, which is denoted by $\mathcal{A}^{\mathrm{Ref}}$ and also referred to as the **reference thumbnail family**. Furthermore, let α^{Est} be the estimated thumbnail segment. Since every segment $\mathcal{A}^{\mathrm{Ref}}$ can serve equally well as the reference thumbnail, we consider the estimated thumbnail α^{Est} to be correct if it agrees (at least to a large degree) with one of these segments. Therefore, to measure how well the estimated thumbnail α^{Est} corresponds to the reference thumbnail family, we compute

$$\mathrm{P}^{\alpha} = \frac{|\alpha^{\mathrm{Est}} \cap \alpha|}{|\alpha^{\mathrm{Est}}|}, \tag{4.59}$$

$$\mathrm{R}^{\alpha} = \frac{|\alpha^{\mathrm{Est}} \cap \alpha|}{|\alpha|}, \tag{4.60}$$

$$\mathrm{F}^{\alpha} = \frac{2\mathrm{P}^{\alpha}\mathrm{R}^{\alpha}}{\mathrm{P}^{\alpha} + \mathrm{R}^{\alpha}} \tag{4.61}$$

for each $\alpha \in \mathcal{A}^{\mathrm{Ref}}$ and define the **thumbnail F-measure** by

$$\mathrm{F}^{\mathrm{Thumb}} = \max\left\{\mathrm{F}^{\alpha} \mid \alpha \in \mathcal{A}^{\mathrm{Ref}}\right\}. \tag{4.62}$$

In other words, the thumbnail F-measure expresses to what extent α^{Est} maximally agrees with one of the reference thumbnails contained in $\mathcal{A}^{\mathrm{Ref}}$.

As an example, Figure 4.32 shows a song having the musical structure $IV_1C_1V_2C_2V_3O$, where the verse part is considered to be the thumbnail. Hence, the reference thumbnail family consists of the three segments corresponding to V_1, V_2, and V_3 (see Figure 4.32b). Figure 4.32c shows the estimated thumbnail segment α^{Est}. Since α^{Est} has no overlap with the V_1-part and V_2-part segments, the corresponding F-measures are zero. However, the F-measure between α^{Est} and the V_3-part segment α is $\mathrm{F}^{\alpha} = 0.8$, thus $\mathrm{F}^{\mathrm{Thumb}} = 0.8$ follows. Even though not needed for the evaluation, Figure 4.32d shows the segment family of the estimated thumb-

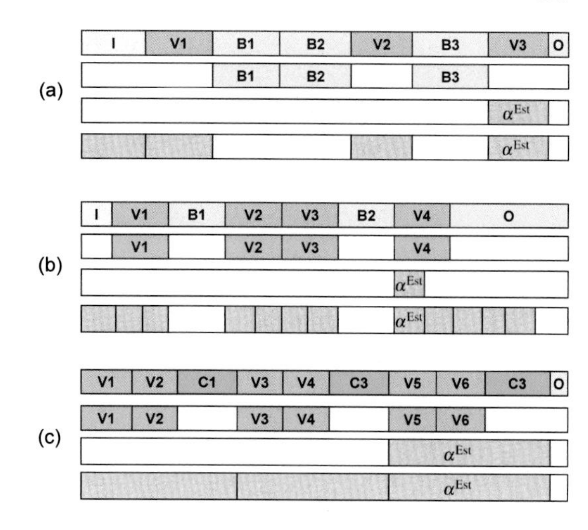

Fig. 4.33 Illustration of typical error sources in thumbnailing and music structure analysis (see Figure 4.32 for an explanation of the annotations). **(a)** Confusion problem for Beatles song "Martha My Dear." **(b)** Substructure (oversegmentation) problem for Beatles song "While My Guitar Gently Weeps." **(c)** Superordinate structure (undersegmentation) problem for Beatles song "For No One."

nail segment α^{Est}, which is obtained when using the fitness-based thumbnailing approach from Section 4.3. This family, which reveals all estimated repetitions of α^{Est}, contains four segments. In this example, it turns out that the intro (the section labeled as I) is harmonically very similar to the three verse sections (labeled as V_1, V_2, and V_3).

We close this section by discussing some typical error sources in audio thumbnailing. As a first example, let us consider the Beatles song "Martha My Dear" shown in Figure 4.33a. The annotated bridge segments were chosen as reference thumbnail family, whereas the estimated thumbnail corresponds to a verse segment (which is actually quite similar to the intro). For this song, the V-part and B-part segments both appear three times and have roughly the same duration. As a result, it is hard to decide whether to use the verse or the bridge for defining the reference segment family.

A second problem occurs when the thumbnail has a substructure. For example, in the Beatles song "While My Guitar Gently Weeps," the verse has a substructure basically consisting of two repeating subparts (see Figure 4.33b). Therefore, a segment that corresponds to the first or second half of the V-part may also serve as a meaningful thumbnail. Such a segment was chosen by the automated procedure as the estimated thumbnail. This is a typical example of a problem generally referred to as **oversegmentation**, where meaningful annotations exist on various scales. We have encountered this phenomenon already in our Brahms example shown in Figure 4.28.

Finally, as illustrated by the Beatles song "For No One" in Figure 4.33c, superordinate repeating parts may also have a high fitness, thus being selected as estimated thumbnails. In this example, the automated procedure identified the superordinate structure VVC (consisting of two verses and a chorus section) as thumbnail. This problem is generally referred to as **undersegmentation**, which is the counterpart to oversegmentation.

These three examples are typical for the kind of problems one has to face when dealing with ill-posed tasks such as music structure analysis. An automated procedure may yield a result that does not coincide with a reference annotation, but is still meaningful from a musical point of view. In such cases, it is less that the procedure has failed and more that the problem is ambiguous.

4.6 Summary and Further Readings

In this chapter, we studied various related research problems commonly subsumed under the name of music structure analysis. The general objective is to segment an audio recording with regard to various musical aspects, for example, identifying recurrent themes or detecting temporal boundaries between contrasting musical parts. Being organized in a hierarchical way, structure in music arises from various relationships between its basic constituent elements. The principles used to create such relationships include repetition, contrast, variation, and homogeneity. As a consequence, many different approaches to derive musical structures have been developed (see [5, 25, 27] for an overview and further references). Following [27], we distinguished between three different classes of methods. First, we looked at repetition-based methods, which are used to identify recurring patterns. As an important application, we applied such methods in Section 4.3 for audio thumbnailing. Second, we discussed novelty-based methods, which aim at detecting transitions and points of novelty. In Section 4.4, we studied two such approaches for finding structural boundaries between musical parts. Third, we considered homogeneity-based methods, which are used to determine passages that are consistent with respect to some musical property. In all three cases, one has to account for different musical dimensions such as melody, harmony, rhythm, or timbre [11].

In particular, the importance of repetitions in music has been emphasized in the literature [15]. Repetition is closely related to notions of coherence, intelligibility, and enjoyment in its perception, and studies show that, for a large variety of music, more than 90% of all musical passages longer than a few seconds in duration are repeated in some way or another at some point in the work [10]. One main challenge in structure analysis is that the notion of repetition can be quite ambiguous. What we refer to as repeating musical sections may include significant variations in the musical content. The principle of variation, where motifs and parts are picked up again in a modified or transformed form [13], is a central aspect of music. In this chapter, we only scratched the surface of the kind of structures that exist in music—not to speak of our bias towards Western music. From a computational perspective, using music structure analysis as a motivating scenario, we introduced basic notions and general techniques that may also be applicable for studying structural properties of other types of sequential data and time series.

Self-Similarity Matrices

As one essential tool, we discussed the concept of self-similarity matrices [5, 27]. These matrices are of great importance for the analysis of music signals and also general time series. For example, such matrices have been employed under the name "recurrence plot" for the analysis of dynamical systems [16, 32]. We have already encountered the related concept of cost matrices in the context of music synchronization in Chapter 3. Such matrices will also play an important role in Chapter 7 in the context of content-based music retrieval and version identification.

The first step for computing an SSM is to convert the given audio recording into a suitable feature representation. We saw that the properties of the resulting SSM crucially depend on the respective feature type [11, 27]. As one example, we considered MFCC-based features that correlate to the aspect of timbre [1]. Other features referred to as tempogram, rhythmogram, or beat spectrogram are used to capture beat, tempo, and rhythmic information—a topic that will be addressed in Chapter 6. In particular, we considered chroma-based audio features (as introduced in Section 3.1.2), which are particularly suited for analyzing the structure of repeating melodies and harmonies.

One important property of similarity matrices is the appearance of block- and path-like structures of high similarity. In Section 4.2.2, we studied several strategies to enhance such structural properties. To augment path-like structures, most enhancement procedures apply some kind of smoothing filter along the direction of the main diagonal [2, 23, 28, 32]. Such a filtering process is closely related to the concept of **time-delay embedding**, which has been previously used for the analysis of dynamical systems [16, 32]. The multiple filtering approach to deal with relative tempo differences between repeating parts was originally suggested in [23]. Also, morphological operations used in image processing to enhance contours and edges [6] can be applied for augmenting path-like structures. Relative and local thresholding is another essential concept for reducing the noise level in SSMs [32], which makes subsequent processing steps (e.g., path extraction) easier.

Besides the feature type, the window size and the temporal resolution used for feature extraction also crucially determine whether blocks or stripes are formed in an SSM. Block-enhanced SSMs have been used for structure analysis based on matrix factorization [12], a technique that we will encounter in Section 8.3 in a different context. In [9, 17], procedures for converting path structures into block structures are proposed. Such conversions make it possible for algorithms previously designed for homogeneity-based structure analysis to be applied to repetition-based structure analysis. Finally, we discussed that a musical part might be repeated in another key. Using chroma features, Goto [8] has suggested simulating transpositions by cyclic chroma shifts. Based on this idea, transposition-invariant SSMs were originally introduced in [20].

Audio Thumbnailing

Finding the repetitive structure of a music recording is a widely studied subtask within music structure analysis (see the overview articles [5, 27] and the references therein). One application of repetition-based structure analysis is audio thumbnailing, where the objective is to find the most representative and repetitive segment of a given music recording (see, e.g., [2, 4, 22]). To identify repetitions, most approaches extract the path structure from an SSM and apply a clustering step to the pairwise relations obtained from the paths to derive entire groups of mutually similar segments. Because of noisy and fragmented path structures due to variations, both steps—path extraction as well as grouping—are error-prone and fragile. In [8], a grouping process is described that balances out inconsistencies in the path relations by exploiting a constant tempo assumption. However, when dealing with varying tempo, the grouping process constitutes a challenging research problem. In the approach [22], which we discussed in Section 4.3, one main idea is to jointly perform the path extraction and grouping steps. This idea is realized by assigning a fitness value to a given segment in such a way that all existing relations within the entire recording are simultaneously accounted for. Instead of extracting individual paths, entire groups of paths (encoded by the concept of path families) are extracted, whereby the construction automatically enforces consistency properties within a group.

Cooper and Foote [4] have already formulated the general idea of assigning a fitness value to each segment of the audio recording. In this early work, the authors calculate the fitness of a given segment as the normalized sum of the self-similarity between the segment and the entire recording (however, the fitness measure does not take any specific path relations into account). In [28], Peeters introduced a fitness measure based on a binary-valued diagonal path structure extracted from an SSM. For visualizing the fitness values in a compact and hierarchical way, we have presented in Section 4.3.2 the concept of scape plots. In the music context, these plots were initially used by Sapp [30] to represent harmony in musical scores hierarchically. In [21], a refinement of the fitness scape plot is described, where some suitable color encoding indicates the relations between different segments. The frontispiece of this chapter shows such a refined scape plot representation for our Brahms example.

Segmentation Approaches

In Section 4.4 we addressed the topic of novelty-based segmentation, where the goal was to find boundaries between subsequent musical parts. The kernel-based approach (see Section 4.4.1) was originally described in the classical paper by Foote [7]. There are many other approaches for boundary detection. For example, Tzanetakis and Cook [36] calculate a Mahalanobis distance between successive frames to yield a novelty function. Using an optimization approach, Jensen [11] performs the boundary detection by minimizing the average distance within blocks (defined by neighboring segment boundaries), while keeping the number of segments small. More recent approaches based on deep learning aim at learning a novelty function (or activation function) from example snippets of input spectrograms with annotated segment boundaries (see, e.g., [37]). For further references on boundary detection and other structure analysis methods, we refer to [25].

In Section 4.4.2, we introduced the concept of structure features as a basis for novelty detection [31]. For computing these features, one idea was to transform an SSM into a time-lag representation, a concept that has been used in various structure analysis approaches [8, 28]. A critical aspect of the approach in [31] is that it integrates different structure analysis principles within a unifying framework: the structure features capture (global) repetition-based information, which is then analyzed using a (local) novelty-based procedure. However, to obtain a full structure annotation, the grouping needs to be done in a separate postprocessing step. The approach by Paulus and Klapuri [26] also combines different segmentation principles by introducing a cost function for structure annotations that considers block-like as well as path-like structures. The final structure annotation is obtained by minimizing the cost function over all possible annotations. However, this approach requires solving a combinatorial optimization task that is computationally prohibitive. To make the computations feasible, the number of candidate annotations is reduced drastically by applying a novelty-based boundary detection procedure in a separate preprocessing step.

Evaluation

In Section 4.5, we addressed the topic of evaluating automated structure analysis procedures. Many evaluation measures involve some kind of precision and recall rate. As typical examples, we adapted such metrics for several structure analysis scenarios. An overview of further evaluation measures can be found in [14]. For an implementation of the most common metrics used in general MIR research, we refer to the open-source Python library `mir_eval` [29]. To account for the fact that music exhibits structure at multiple scales, McFee et al. [18] introduced an evaluation metric which can compare hierarchical descriptions of musical structure.

Even though music is highly structured and obeys some general rules, what is interesting about any individual piece of music tends to be how it expands or breaks these rules. As a consequence, evaluating the performance of automated procedures is not as easy as one may think. Even so-called reference annotations made by different human experts may differ significantly [26, 33]. Therefore, to better reflect the ambiguity and richness of musical structures, the evaluation should be based on several annotations which have been generated by several human experts and are provided on different temporal scales [33]. An automated procedure could then be treated just as another "expert," and the estimated results could be compared against the entire pool of different reference annotations (instead of using only a single reference annotation). Rather than automatically extracting a structure annotation from scratch, another exciting research direction is to develop automated procedures that somehow explain an existing annotation. A procedure in this direction is described in [34], where the relevance of various features is determined in relation to a given annotation.

4.7 FMP Notebooks

We have seen that, due to the many different principles for musical relationships, music structure analysis constitutes a research area that offers a bouquet of fascinating and challenging subtasks. In **Part 4** of the FMP notebooks [24], we deepen some of the chapter's core concepts and algorithms, which are applicable beyond the music domain. In particular, we have a detailed look at the properties and variants of self-similarity matrices (SSMs). Then, considering some more specific music structure analysis tasks, we provide and discuss implementations of—as we think—some beautiful and instructive approaches for repetition and novelty detection. Using real-world music examples, we draw attention to the algorithms' strengths and weaknesses, while indicating the problems that typically arise from violations of the underlying model assumptions. We close **Part 4** by implementing and discussing evaluation metrics, which we take up again in other chapters of the book.

We start with the **FMP Notebook *Music Structure Analysis: General Principles***, where we create the general context of the subsequent notebooks of this part. In particular, we introduce our primary example used throughout these notebooks: Brahms' famous Hungarian Dance No. 5. Based on this example, we introduce implementations for parsing, adapting, and visualizing reference annotations for musical structures. Furthermore, we provide some Python code examples for converting music recordings into MFCC-, tempo-, and chroma-based feature representations (see also Figure 4.6). In a music processing course, we consider it essential to make students aware that such representations crucially depend on parameter settings and design choices. This fact can be made evident by suitably visualizing the representations. In the FMP notebooks in general, we attach great importance to a visual representation of results, which sharpens one's intuition and provides a powerful tool for questioning the results' plausibility.

One general idea to study musical structures and their mutual relations is to convert the music signal into a suitable feature sequence and compare each element of the feature sequence with all other sequence elements. This results in an SSM, a tool that is of fundamental importance not only for music structure analysis but also for analyzing many kinds of time series. Closely following Section 4.2, we cover this fundamental topic in the subsequent notebooks. The **FMP Notebook *Self-Similarity Matrix (SSM)*** explains the general ideas of SSMs and discusses basic notions such as paths and blocks. Furthermore, continuing our Brahms example, we provide Python code examples for computing and visualizing SSMs using different feature representations. It is an excellent exercise to turn the tables and to start with a structural description of a piece of music and then to transform this description into an SSM representation. This is what we do in the **FMP Notebook**

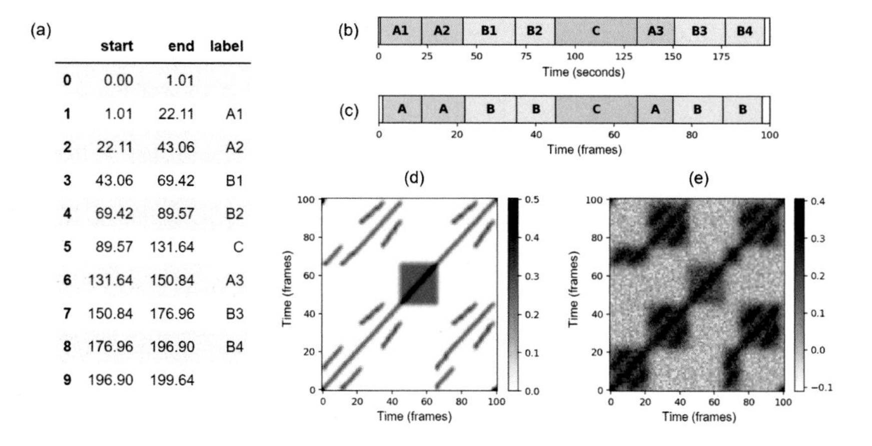

	start	end	label
0	0.00	1.01	
1	1.01	22.11	A1
2	22.11	43.06	A2
3	43.06	69.42	B1
4	69.42	89.57	B2
5	89.57	131.64	C
6	131.64	150.84	A3
7	150.84	176.96	B3
8	176.96	196.90	B4
9	196.90	199.64	

Fig. 4.34 **(a)** Structure annotation for our Brahms example given as a CSV file. **(b)** Visualization of structure annotation (time axis given in seconds). **(c)** Visualization of converted structure annotation (time axis given in frames). **(d),(e)** Different SSMs generated synthetically from the structure annotation.

SSM: Synthetic Generation, where we provide a function for converting a reference annotation of a music recording into an SSM. In this function, one can specify if the structural parts fulfill path-like (being repetitive) or block-like (being homogeneous) relations. Further parameters allow for modifying the SSM by applying a Gaussian smoothing filter or adding Gaussian noise (see Figure 4.34 for examples). Synthetically generating and visualizing SSMs is a very instructive way to gain a deeper understanding of these matrices' structural properties and their relation to musical annotations. Furthermore, synthetic SSMs are useful for debugging and testing automated procedures for music structure analysis. However, synthetic SSMs should not replace evaluation based on real music examples. In practice, SSMs computed from music and audio representations are typically far from being ideal—a painful experience that every student should have.

In Section 4.2.2, we described various strategies for enhancing the structural properties of SSMs. The notebooks implement these techniques step by step. In the **FMP Notebook *SSM: Feature Smoothing***, we study how feature smoothing affects the structural properties of an SSM, using our Brahms example as an illustration (see Figure 4.10). For example, starting with a chroma representation and increasing the smoothing length, one may observe an increase in homogeneity reflecting the rough harmonic content. As an alternative to average filtering, we also discussed median filtering. In the **FMP Notebook *SSM: Path Enhancement***, we discuss a strategy for enhancing path structures in SSMs. We show that simple filtering along the main diagonal works well if there are no relative tempo differences between the segments to be compared. Rather than directly implementing the equation (4.11) using nested loops, we provide a much faster matrix-based implementation, which exploits efficient array computing concepts provided by the Python package `numpy`. In a music processing course, this is an excellent opportunity for discussing efficiency and implementation issues. In this context, one may also discuss Python packages such as `numba` that translate specific Python code into fast machine code. After this little excursion on efficiency, we come back to our Brahms example, where the shorter B_2-section is played much faster than the B_1-section, leading to nondiagonal path structures. Here, diagonal smoothing fails, and we introduce a multiple filtering approach that preserves specific nondiagonal structures. Again, rather than implementing the equation (4.12) directly, we employ a matrix-based implementation using a tricky resampling strategy. Finally, we introduce a forward–backward smoothing approach that attenuates fading artifacts in particular at the end of path structures.

In Section 4.2.2.3, looking at the song "In the Year 2525" by Zager and Evans, we saw that certain musical parts may be repeated in a transposed form. In the **FMP Notebook *SSM: Transposition Invariance***, we provide an implementation for computing a transposition-invariant SSM. In particular, we show how the resulting transposition index matrix can be visualized (see Figure 4.12f). Such visualizations are—as we think—esthetically beautiful and say a lot about the harmonic relationships within a song. We close our studies on SSMs with the **FMP Notebook *SSM: Thresholding***, where we discuss global and local thresholding strategies, which are applicable to a wide range of matrix representations. The effect of different thresholding techniques can be nicely illustrated by small toy examples, which can also be integrated well into a music processing course in the form of small handwritten and programming exercises.

We now turn our attention to more concrete subtasks of music structure analysis. In the comprehensive **FMP Notebook *Audio Thumbnailing***, we provide a step-by-step implementation of the procedure described in Section 4.3. This is more of a notebook for advanced students who want to see how the mathematically rigorous description of an algorithm is put into practice. By interleaving theory, implementation details, and immediate application to a specific example, we hope that this notebook gives a positive example of making a complex algorithm more accessible. As the result of our audio thumbailing approach, we obtain a fitness measure that assigns to each possible segment a fitness value. The **FMP Notebook *Scape Plot Representation*** introduces a concept for visualizing the fitness values of all segments using a triangular image. This concept is an esthetically pleasing and powerful way to visualize segment properties in a compact and hierarchical form. Applied to our fitness measure, we deepen the understanding of our thumbnailing procedure by providing scape plot representations for the various measures involved (e.g., score, noramlized score, coverage, normlized coverate, and fitness; see also Figure 4.21). From a programming perspective, this notebook also demonstrates how to create elaborate illustrations using the Python library `matplotlib`—a task on which one can spend a lot of time.

Next, following Section 4.4, we deal with the music structure analysis subtask often referred to as novelty detection. In the **FMP Notebook *Novelty-Based Segmentation***, we cover the classical and widely used approach originally suggested by Foote [7]. We provide Python code examples for generating box-like and Gaussian checkerboard kernels, which are then shifted along the main diagonal of an SSM to detect 2D corner points. We strongly recommend that this simple, beautiful, explicit, and instructive approach be used as a baseline for any research in novelty-based segmentation before applying more intricate approaches. Of course, as we also demonstrate in the notebook, the procedure crucially depends on design choices and parameters such as the underlying SSM and the kernel size.

While most approaches for novelty detection use features that capture local characteristics, we consider in the **FMP Notebook *Structure Feature*** the concept of structure features that capture global structural properties. These features are basically the columns of an SSM's cyclic time–lag representation. In the notebook, we provide an implementation for converting an SSM into a time–lag representation. We also offer Python code examples that students can use to explore this conversion by experimenting with explicit toy examples. Again it is crucial to also apply the techniques to real-world music recordings, which behave completely differently than synthetic examples. In practice, one often obtains significant improvements by applying median filtering to remove undesired outliers or by applying smoothing filters to make differentiation less vulnerable to small deviations.[1]

We close **Part 4** with the **FMP Notebook *Evaluation***, where we discuss standard metrics based on precision, recall, and F-measure. Even though there are Python libraries such as `mir_eval` [29] that provide a multitude of metrics commonly used in MIR research, it is essential to exactly understand how these metrics are defined. Furthermore, requiring knowledge in basic data structures and data handling, students may improve their programming skills when implementing, adapting, and applying some of these metrics. In our notebook, one finds Python code examples for the standard precision, recall, and F-measure as well as adaptions of these measures for labeling and boundary evaluation. Again, we recommend using suitable toy examples and visualizations to get a feel for what the metrics actually express.

[1] Even though recent deep learning approaches try to circumvent such explicit processing steps, the "good old engineering" is at least instructive, yielding valuable insights into the task, the algorithm's potential, and the data at hand.

References

1. J.-J. AUCOUTURIER AND F. PACHET, *Improving timbre similarity: How high's the sky*, Journal of Negative Results in Speech and Audio Sciences, 1 (2004).
2. M. A. BARTSCH AND G. H. WAKEFIELD, *Audio thumbnailing of popular music using chroma-based representations*, IEEE Transactions on Multimedia, 7 (2005), pp. 96–104.
3. M. A. CASEY AND M. SLANEY, *The importance of sequences in musical similarity*, in Proceedings of the IEEE International Conference on Acoustics, Speech, and Signal Processing (ICASSP), Toulouse, France, 2006.
4. M. COOPER AND J. FOOTE, *Automatic music summarization via similarity analysis*, in Proceedings of the International Society for Music Information Retrieval Conference (ISMIR), Paris, France, 2002, pp. 81–85.
5. R. B. DANNENBERG AND M. GOTO, *Music structure analysis from acoustic signals*, in Handbook of Signal Processing in Acoustics, D. Havelock, S. Kuwano, and M. Vorländer, eds., vol. 1, Springer, New York, NY, USA, 2008, pp. 305–331.
6. E. R. DOUGHERTY, *An Introduction to Morphological Image Processing*, SPIE Optical Engineering Press, Bellingham, WA, USA, 1992.
7. J. FOOTE, *Automatic audio segmentation using a measure of audio novelty*, in Proceedings of the IEEE International Conference on Multimedia and Expo (ICME), New York, NY, USA, 2000, pp. 452–455.
8. M. GOTO, *A chorus section detection method for musical audio signals and its application to a music listening station*, IEEE Transactions on Audio, Speech, and Language Processing, 14 (2006), pp. 1783–1794.
9. H. GROHGANZ, M. CLAUSEN, N. JIANG, AND M. MÜLLER, *Converting path structures into block structures using eigenvalue decompositions of self-similarity matrices*, in Proceedings of the International Society for Music Information Retrieval Conference (ISMIR), Curitiba, Brazil, 2013, pp. 209–214.
10. D. B. HURON, *Sweet Anticipation: Music and the Psychology of Expectation*, The MIT Press, 2006.
11. K. JENSEN, *Multiple scale music segmentation using rhythm, timbre, and harmony*, EURASIP Journal on Advances in Signal Processing, (2007).
12. F. KAISER AND T. SIKORA, *Music structure discovery in popular music using non-negative matrix factorization*, in Proceedings of the International Society for Music Information Retrieval Conference (ISMIR), Utrecht, The Netherlands, 2010, pp. 429–434.
13. F. LERDAHL AND R. JACKENDOFF, *A Generative Theory of Tonal Music*, MIT Press, 1983.
14. H. LUKASHEVICH, *Towards quantitative measures of evaluating song segmentation*, in Proceedings of the International Society for Music Information Retrieval Conference (ISMIR), Philadelphia, USA, 2008, pp. 375–380.
15. E. H. MARGULIS, *On Repeat: How Music Plays the Mind*, Oxford University Press, 2014.
16. N. MARWAN, M. C. ROMANO, M. THIEL, AND J. KURTHS, *Recurrence plots for the analysis of complex systems*, Physics Reports, 438 (2007), pp. 237–329.
17. B. MCFEE AND D. ELLIS, *Analyzing song structure with spectral clustering*, in Proceedings of the International Society for Music Information Retrieval Conference (ISMIR), Taipei, Taiwan, 2014, pp. 405–410.
18. B. MCFEE, O. NIETO, M. M. FARBOOD, AND J. P. BELLO, *Evaluating hierarchical structure in music annotations*, Frontiers in Psychology, 8 (2017).
19. R. MIDDLETON, *Form*, in Key Terms in Popular Music and Culture, B. Horner and T. Swiss, eds., Wiley-Blackwell, 1999, pp. 141–155.
20. M. MÜLLER AND M. CLAUSEN, *Transposition-invariant self-similarity matrices*, in Proceedings of the International Society for Music Information Retrieval Conference (ISMIR), Vienna, Austria, Sept. 2007, pp. 47–50.
21. M. MÜLLER AND N. JIANG, *A scape plot representation for visualizing repetitive structures of music recordings*, in Proceedings of the International Society for Music Information Retrieval Conference (ISMIR), Porto, Portugal, 2012, pp. 97–102.

22. M. MÜLLER, N. JIANG, AND P. GROSCHE, *A robust fitness measure for capturing repetitions in music recordings with applications to audio thumbnailing*, IEEE Transactions on Audio, Speech, and Language Processing, 21 (2013), pp. 531–543.
23. M. MÜLLER AND F. KURTH, *Enhancing similarity matrices for music audio analysis*, in Proceedings of the International Conference on Acoustics, Speech, and Signal Processing (ICASSP), Toulouse, France, May 2006, pp. 437–440.
24. M. MÜLLER AND F. ZALKOW, *FMP Notebooks: Educational material for teaching and learning fundamentals of music processing*, in Proceedings of the International Society for Music Information Retrieval Conference (ISMIR), Delft, The Netherlands, 2019, pp. 573–580.
25. O. NIETO AND J. P. BELLO, *Systematic exploration of computational music structure research*, in Proceedings of the International Society for Music Information Retrieval Conference (ISMIR), New York City, USA, 2016, pp. 547–553.
26. J. PAULUS AND A. P. KLAPURI, *Music structure analysis using a probabilistic fitness measure and a greedy search algorithm*, IEEE Transactions on Audio, Speech, and Language Processing, 17 (2009), pp. 1159–1170.
27. J. PAULUS, M. MÜLLER, AND A. KLAPURI, *Audio-based music structure analysis*, in Proceedings of the International Society for Music Information Retrieval Conference (ISMIR), Utrecht, The Netherlands, 2010, pp. 625–636.
28. G. PEETERS, *Sequence representation of music structure using higher-order similarity matrix and maximum-likelihood approach*, in Proceedings of the International Society for Music Information Retrieval Conference (ISMIR), Vienna, Austria, 2007, pp. 35–40.
29. C. RAFFEL, B. MCFEE, E. J. HUMPHREY, J. SALAMON, O. NIETO, D. LIANG, AND D. P. W. ELLIS, *MIR_EVAL: A transparent implementation of common MIR metrics*, in Proceedings of the International Society for Music Information Retrieval Conference (ISMIR), Taipei, Taiwan, 2014, pp. 367–372.
30. C. S. SAPP, *Harmonic visualizations of tonal music*, in Proceedings of the International Computer Music Conference (ICMC), La Habana, Cuba, 2001, pp. 423–430.
31. J. SERRÀ, M. MÜLLER, P. GROSCHE, AND J. L. ARCOS, *Unsupervised music structure annotation by time series structure features and segment similarity*, IEEE Transactions on Multimedia, 16 (2014), pp. 1229–1240.
32. J. SERRÀ, X. SERRA, AND R. G. ANDRZEJAK, *Cross recurrence quantification for cover song identification*, New Journal of Physics, 11 (2009).
33. J. B. L. SMITH, J. A. BURGOYNE, I. FUJINAGA, D. D. ROURE, AND J. S. DOWNIE, *Design and creation of a large-scale database of structural annotations*, in Proceedings of the International Society for Music Information Retrieval Conference (ISMIR), Miami, Florida, USA, 2011, pp. 555–560.
34. J. B. L. SMITH AND E. CHEW, *Using quadratic programming to estimate feature relevance in structural analyses of music*, in Proceedings of the ACM International Conference on Multimedia, 2013, pp. 113–122.
35. M. SUNKEL, S. JANSEN, M. WAND, E. EISEMANN, AND H.-P. SEIDEL, *Learning line features in 3D geometry*, Computer Graphics Forum, 30 (2011), pp. 267–276.
36. G. TZANETAKIS AND P. COOK, *Multifeature audio segmentation for browsing and annotation*, in Proceedings of the IEEE Workshop on Applications of Signal Processing to Audio and Acoustics (WASPAA), New Paltz, NY, USA, 1999, pp. 103–106.
37. K. ULLRICH, J. SCHLÜTER, AND T. GRILL, *Boundary detection in music structure analysis using convolutional neural networks*, in Proceedings of the International Society for Music Information Retrieval Conference (ISMIR), Taipei, Taiwan, 2014, pp. 417–422.

Exercises

Exercise 4.1. Let $\mathcal{F} = \mathbb{R}^D$ be the real vector space of dimension $D \in \mathbb{N}$. Typical similarity measures are based on the Euclidean norm (also referred to as the ℓ^2-norm) defined by

$$\|x\|_2 := \left(\sum_{i=1}^{D} |x(i)|^2\right)^{1/2}$$

for a vector $x = (x(1), x(2), \ldots, x(D))^\top$. From this norm, one can derive the similarity measures $s^{a,b} : \mathcal{F} \times \mathcal{F} \to \mathbb{R}$ for constants $a \in \mathbb{R}$ and $b \in \mathbb{N}$ by setting

$$s^{a,b}(x,y) = a - \|x-y\|_2^b$$

for $x, y \in \mathcal{F}$. In the following, we consider the case $a = 2$ and $b = 2$. Furthermore, assume that x and y are normalized with respect to the ℓ^2-norm. Show that, in this case, the measure $s^{a,b}$ is simply twice the inner product $\langle x|y\rangle$, which measures the cosine of the angle between x and y.

Exercise 4.2. In (4.11), we have introduced a forward smoothing procedure. This procedure results in a fading out of the paths, in particular when using a large length parameter. To avoid this fading out, one idea is to additionally apply the averaging filter in backward direction. The final self-similarity matrix is then obtained by taking the cell-wise maximum over the forward-smoothed and backward-smoothed matrices. Formalize this procedure by giving a mathematical description. Furthermore, show how the backward smoothing can be realized by forward smoothing considering the time-reversed feature sequence.
[**Hint:** To avoid boundary considerations, assume that **S** is suitably zero-padded. The effect of the forward–backward smoothing procedure is illustrated by Figure 4.12d. Another example is shown in Figure 4.15c.]

Exercise 4.3. Let $\mathcal{F} = \mathbb{R}^D$ as in Exercise 4.1 and $s : \mathcal{F} \times \mathcal{F} \to \mathbb{R}$ be the similarity measure defined by $s(x,y) := |\langle x|y\rangle|$ for $x, y \in \mathcal{F}$ (see (4.3)). Show that the transposition-invariant self-similarity matrix $\mathbf{S}^{\mathrm{TI}}$ (see (4.15)) is symmetric. Is the transposition index matrix $\mathbf{I}$ (see (4.16)) symmetric? Describe the relation between the matrix $\mathbf{I}$ and its transposed matrix $\mathbf{I}^\top$.

Exercise 4.4. For computing the matrix $\mathbf{S}_{L,\Theta}$ in (4.13), a set Θ of relative tempo differences needs to be specified. Assume that θ_{min} is a lower bound and θ_{max} is an upper bound for the expected relative tempo differences. For a given number $K \in \mathbb{N}$, determine a set

$$\Theta = \{\theta_1 = \theta_{\mathrm{min}}, \theta_2, \ldots, \theta_{K-1}, \theta_K = \theta_{\mathrm{max}}\}$$

consisting of increasing tempo values that are logarithmically spaced. Write a small computer program for computing this set for the parameters $\theta_{\mathrm{min}} = 0.66$, $\theta_{\mathrm{max}} = 1.5$, and $K = 5$, as well as for $\theta_{\mathrm{min}} = 0.5$, $\theta_{\mathrm{max}} = 2$, and $K = 7$.
[**Hint:** Convert the tempo bounds θ_{min} and θ_{max} into the log domain by applying a logarithm. Then, linearly sample the resulting interval using K samples and apply an exponential function to the samples.]

Exercise 4.5. In this exercise, we look at the various thresholding strategies introduced in Section 4.2.2.4. Given the matrix

$$\mathbf{S} = \begin{bmatrix} 1 & 1 & 2 & 2 \\ 4 & 3 & 4 & 3 \\ 1 & 1 & 2 & 2 \\ 5 & 6 & 6 & 5 \end{bmatrix},$$

compute the matrices that are obtained by applying the following thresholding operations:

(a) Global thresholding using $\tau = 4$
(b) Global thresholding using $\tau = 4$ as in (a) with subsequent linear scaling of the range $[\tau, \mu]$ to $[0,1]$ using $\mu := \max\{\mathbf{S}(n,m) \mid n,m \in [1:4]\}$
(c) Global thresholding with subsequent linear scaling as in (b) and applying the penalty parameter $\delta = -1$
(d) Relative thresholding using the relative threshold parameter $\rho = 0.5$
(e) Local thresholding in a column- and rowwise fashion using $\rho = 0.5$

Exercise 4.6. Let $X = (x_1, x_2, \ldots, x_N)$ be a sequence and $\alpha = [s:t] \subseteq [1:N]$ a segment of length $M := |\alpha|$. Show that the optimization procedure for computing an optimal path family over α (as described in Section 4.3.1.2) has a complexity of $O(MN)$ regarding the memory requirements as well as the running time.

Exercise 4.7. Let $X = (x_1, \ldots, x_N)$ be a feature sequence and $\mathbf{S}$ the resulting SSM satisfying the normalization properties (4.18) and (4.19). Let $\mathcal{P}^*$ be an optimal path family over a given segment α. Show that $|\alpha| \leq \sigma(\mathcal{P}^*) \leq N$. In particular, this shows that $\sigma(\mathcal{P}^*) = N$ for $\alpha = [1:N]$.

Exercise 4.8. For two given real numbers $a, b \in \mathbb{R}$, the arithmetic mean is defined by $A(a,b) = (a+b)/2$, the geometric mean by $G(a,b) = \sqrt{ab}$, and the harmonic mean by $H(a,b) = 2ab/(a+b)$. Show that $H(a,b) \leq G(a,b) \leq A(a,b)$, i.e., the geometric mean always lies between the harmonic mean and the arithmetic mean. Furthermore, compute $A(a,b)$, $G(a,b)$, and $H(a,b)$ for the numbers $a = 1$ and $b \in \{1,2,3,4\}$.

Exercise 4.9. (a) Let us consider a piece of music having the musical structure $A_1B_1B_2A_2A_3$, where we assume that corresponding parts are repeated in exactly the same way. Furthermore, assume that the A-part and B-part segments are completely unrelated to each other and that a B-part segment has exactly twice the length of an A-part segment. Sketch an idealized SSM for this piece (as in Figure 4.18). Furthermore, determine the fitness values of the segments corresponding to A_1 and B_1, respectively.
(b) Next, consider a piece having the musical structure $A_1A_2A_3A_4$, where the four parts are repeated with increasing tempo. Assume that A_1 lasts 20 seconds, A_2 lasts 15 seconds, A_3 lasts 10 seconds, and A_4 lasts 5 seconds. Again sketch an idealized SSM and determine the fitness values of the four segments corresponding to the four parts.

Exercise 4.10. Let $[1:N]$ be a sampled time axis. Show that the number of different segments $\alpha = [s:t]$ with $s,t \in [1:N]$ and $s \leq t$ is $(N+1)N/2$.

Exercise 4.11. Determine the overall computational complexity of calculating the fitness scape plot as introduced in Section 4.3.2 for a feature sequence $X = (x_1, x_2, \ldots, x_N)$ of length N.
[**Hint:** Use Exercise 4.6 and Exercise 4.10.]

Exercise 4.12. Given a triangular representation of all segments within $[1:N]$ as in Figure 4.19b, visually indicate the following sets of segments:

(a) All segments having a minimal length above a given threshold $\theta \geq 0$
(b) All segments that contain a given segment α
(c) All segments that are disjoint to a given segment α
(d) All segments that contain the center $c(\alpha)$ of a given segment α

Exercise 4.13. Sketch the similarity matrix $\mathbf{S}$ and the circular time-lag matrix $\mathbf{L}^{\circ}$ as in Figure 4.26c for pieces with the following musical structure:

(a) $AB_1B_2B_3$, where all segments have the same length
(b) AB_1B_2, where the A-part and B_1-part segments have the same length and the B_2-part segment has twice the length (played with half the tempo of B_1)

Exercise 4.14. Sketch the similarity matrix $\mathbf{S}$, the circular time-lag matrix $\mathbf{L}^{\circ}$, and the resulting novelty function $\Delta_{\mathrm{Structure}}$ for pieces with the following musical structure (assuming that all segments corresponding to a musical part have the same length and that the kernel size used for computing the novelty function is much smaller than this length):

(a) $A_1A_2A_3A_4$
(b) $A_1B_1A_2B_2$
(c) $A_1B_1A_2A_3$
(d) $A_1A_2B_1B_2$

Exercise 4.15. Let $\mathcal{A} = \{\alpha_1, \alpha_2, \ldots, \alpha_K\}$ be a segment family together with a labeling $\lambda_k \in \Lambda$, $k \in [1:K]$. Let $\mu(\mathcal{A}) := \bigcup_{k=1}^{K} \alpha_k$ be the union of all segments. Show that one may assume $\mu(\mathcal{A}) = [1:N]$ by suitably extending the segment family, the label set Λ, and the labeling.

Exercise 4.16. In (4.51), we defined the set $\mathcal{I} = \{(n,m) \in [1:N] \times [1:N] \mid n < m\}$ to serve as a set of items for defining the pairwise evaluation measure. Determine the size of $\mathcal{I}$. Furthermore, let $\varphi : [1:N] \to \Lambda$ be a label function, and let $\mathcal{I}_{+}^{\mathrm{Ref}} = \{(n,m) \in \mathcal{I} \mid \varphi(n) = \varphi(m)\}$ be the set of positive items with regard to φ. Derive a general formula for the size of $\mathcal{I}_{+}^{\mathrm{Ref}}$.
[**Hint:** Note that the size of $\mathcal{I}_{+}^{\mathrm{Ref}}$ does not depend on the original order of the frames. Given a specific label, consider the number of frames assigned to that label. To derive a formula for the size of $\mathcal{I}_{+}^{\mathrm{Ref}}$, one needs to consider all possible labels assumed by φ.]

Exercise 4.17. In this exercise, we investigate how the pairwise labeling evaluation behaves with respect to under- and oversegmentation. To this end, let us consider the following structure annotations of a piece of music (similar to our Brahms example shown in Figure 4.28):

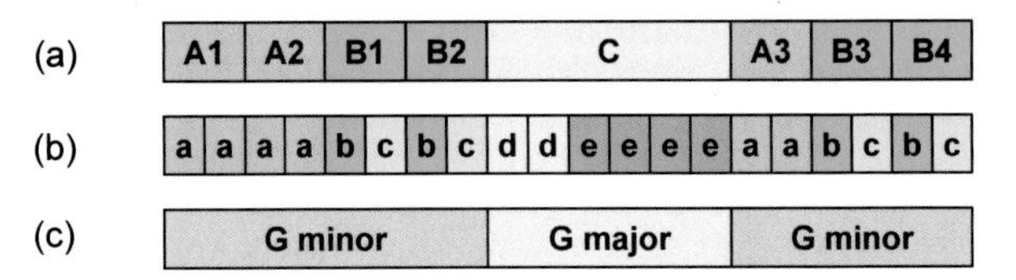

Compute the size $|\mathcal{I}_{+}|$ for each of the three annotations. Then, assume that (a) is the reference annotation. Compute the pairwise precision, recall, and F-measure for the case that (b) is the estimated annotation ("oversegmentation") and for the case that (c) is the estimated annotation ("undersegmentation").
[**Hint:** Use the results of Exercise 4.16.]

Exercise 4.18. Let $[1:N]$ be a sampled time axis with $N = 50$. Furthermore, let $B^{\mathrm{Ref}} = \{7, 13, 19, 28, 40, 44\}$ be a reference boundary annotation and $B^{\mathrm{Est}} = \{6, 12, 21, 29, 42\}$ be an estimated boundary annotation. Compute the boundary evaluation measures (precision, recall, F-measure) as in Section 4.5.4 for the tolerance parameter $\tau = 0$, $\tau = 1$, and $\tau = 2$, respectively. Why is the case $\tau = 2$ problematic for this example?

Exercise 4.19. Let $[1:N]$ be a sampled time axis with $N = 100$. Furthermore, let $\mathcal{A}^{\mathrm{Ref}} = \{[16:26], [40:49], [50:60], [75:84]\}$ be a reference thumbnail family. Compute the thumbnail F-measure as introduced in Section 4.5.5 for the following estimated thumbnail segments:

(a) $\alpha^{\mathrm{Est}} = [18:27]$
(b) $\alpha^{\mathrm{Est}} = [45:54]$
(c) $\alpha^{\mathrm{Est}} = [60:75]$

Chapter 5
Chord Recognition

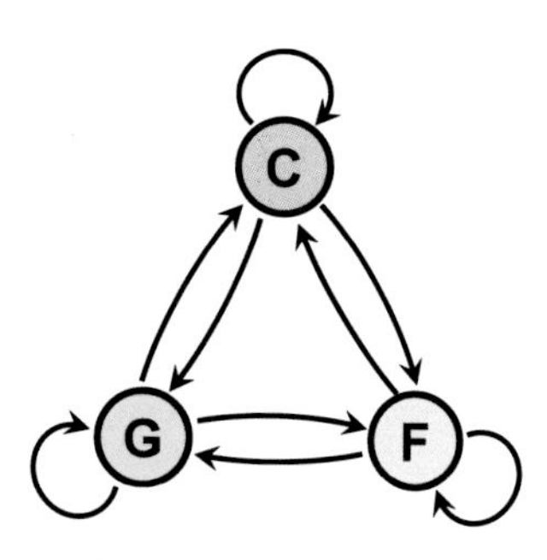

In music, **harmony** refers to the simultaneous sound of different notes that form a cohesive entity in the mind of the listener. The main constituent components of harmony, at least in the Western music tradition, are **chords**, which are musical constructs that typically consist of three or more notes. Harmony analysis may be thought of as the study of the construction, interaction, and progression of chords. The progression of chords over time closely relates to what is often referred to as the harmonic content of a piece of music. These progressions are of musical importance for composing, describing, and understanding Western tonal music including popular, jazz, and classical music. Therefore, features that capture the harmonic content are widely applied for music processing tasks including music structure analysis (Chapter 4) and music retrieval (Chapter 7).

In the analysis of harmony, there are many aspects that need to be examined, including the type of music representation, the temporal resolution, the level of abstraction, and the chords to be considered in the analysis. Harmony is a rather vague concept. Harmonic entities are formed by the human brain, which integrates sounds that may even come from notes played in succession rather than simultaneously. The musical context, as well as the listener's knowledge and expectation, have a crucial influence on how a specific harmony or chord is actually perceived and interpreted. The concept of harmony and chords is enriched by the existence of nonharmonic tones, which are notes that are not part of the implied chord within the harmonic framework. Such tones are used as a combining element to create smooth melody lines, as a passing element to prepare the transition from one chord to the next one, or as a dissonant element to create musical tension, thus adding some "spice" to the music. In other words, not all sounding notes need to be part of the underlying chord, which considerably complicates the harmonic analysis of music.

M. Müller, *Fundamentals of Music Processing*, https://doi.org/10.1007/978-3-030-69808-9_5

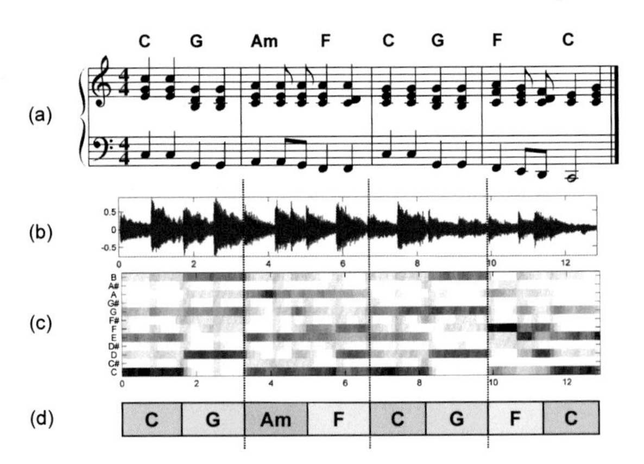

Fig. 5.1 Chord recognition task illustrated by the first measures of the Beatles song "Let It Be." **(a)** Score of the first four measures. **(b)** Waveform of an audio recording of these measures. **(c)** Chroma representation. **(d)** Chord recognition result.

In this chapter, we discuss computational approaches for the harmonic analysis of music. Rather than giving an overview of this wide area of research, we study a subproblem referred to as **chord recognition**, where we consider only a small number of the most important chords as occurring in Western music. Furthermore, we assume that the piece of music is given in the form of an audio recording. The resulting chord recognition task consists in splitting up the recording into segments and assigning a chord label to each segment. The segmentation specifies the start and end time of a chord, and the chord label specifies which chord is played during this time period.

As an example, let us have a look at the Beatles song "Let It Be" of Figure 5.1. The musical score as well as the manually annotated chord labels are shown in Figure 5.1a. In our scenario, we start with a waveform of the recorded song as indicated by Figure 5.1b. Most of the chord recognition procedures encountered in the literature proceed in two main steps. In the first step, the recording is converted into a sequence of audio features that capture harmony-related information. In particular, chroma-based features as introduced in Section 3.1.2 have proven to be a suitable representation (see Figure 5.1c). In the second step, pattern matching techniques are applied to map the audio features to chord labels, which yields the final recognition result (see Figure 5.1d).

In Section 5.1, we start with a short tutorial on intervals, chords, and scales as used in Western tonal music. Since the main focus of this book is on computational rather than musicological aspects, we adopt a rather simplistic approach by focusing on the major and minor triads as chords. The technically oriented reader, who is not interested in the musical background, should at least have a look at Figure 5.6 and may then skip most parts of this section. In Section 5.2, we discuss a basic procedure for chord recognition, where chroma features are compared with fixed templates that correspond to the chords to be considered. This simple procedure is very instructive, since it not only yields a baseline for more complex chord recognizers, but also highlights the role of the feature computation step. In particular, we give examples of how the recognition results can be influenced by modifying the audio features, by applying temporal smoothing techniques, and by adapting the underlying chord templates. One important idea on which many modern chord recognition systems are based is to jointly perform the pattern matching and temporal filtering steps within one optimization procedure. In this context, hidden Markov models (HMMs), which were originally developed for automatic speech recognition and are now widely used for a variety of music processing tasks, have proven to be a powerful tool. We introduce the main ideas of HMMs in Section 5.3 and then apply this concept for chord recognition.

5.1 Basic Theory of Harmony

In this section, we introduce some important concepts from music theory. Most of these notions are not needed for understanding the technical details presented in later sections, where (in technical terms) a chord may simply be thought of as a specific chroma distribution that has some musical meaning. However, a good understanding of the musical concepts becomes important when actually designing and evaluating chord recognition systems. In the following, we first discuss the notion of intervals (Section 5.1.1) and then introduce the more complex musical constructs of chords and scales (Section 5.1.2). Some of the material presented in this section has been inspired by Wikipedia articles and textbooks on harmony theory [8, 15, 28], where one can find further explanations and examples.

5.1.1 Intervals

In music, an **interval** may be loosely defined as the difference between two pitches. This definition is problematic in the sense that the underlying notion of pitch is already a rather vague one. Recall from Section 1.1.1 that pitch is a perceptual property that allows the ordering of sounds on a frequency-related logarithmic scale. When playing a note of a certain pitch on an instrument, the resulting musical tone is dominated by certain frequencies referred to as partials (Section 1.3.2). The frequency of the lowest partial is called the fundamental frequency, which is the frequency typically associated to the pitch. For harmonic sounds, the partials are (close to) integer multiples, the harmonics, of the fundamental frequency.

The most basic interval in music is the **octave**, which is defined as the distance between a pitch and another one with half or double its fundamental frequency. Starting with this basic interval, one can define notions of other intervals by considering frequency relations of harmonics (physical approach), geometric relations (mathematical approach), or note relations (musical approach). These approaches lead to slightly different notions of intervals, which are however referred to by the same interval names. In the following, we will discuss some of these notions and the resulting inconsistencies.

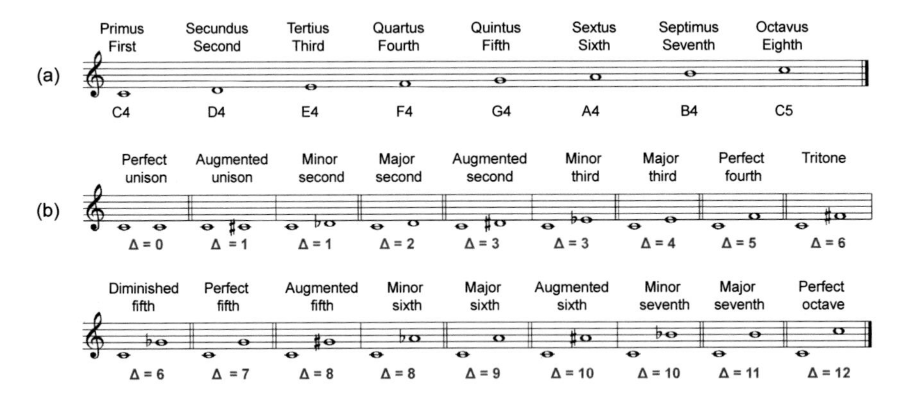

Fig. 5.2 **(a)** C major scale and enumeration of its constituent notes using ordinal numbers (in Latin and English). **(b)** Score representation of various intervals. Because of enharmonic equivalence, there may be musically different intervals having the same distance Δ (given in semitones).

5.1.1.1 Semitone Differences

Let us start with the musical approach as used for Western music, where one assumes a **twelve-tone equal-tempered scale**. Recall from Section 1.1.1 that, in this model, an octave is subdivided into twelve scale steps that are equally spaced on a logarithmic frequency axis. The smallest possible interval in this scale is called a **semitone**, which is the difference between two subsequent scale steps. In terms of frequencies, a semitone describes a ratio (rather than a difference) (see (1.2)).

Based on the notion of a semitone, one can now specify other intervals that are used in Western music theory. The naming conventions for these intervals are based on historical practice. In particular, an interval name may not only describe the difference in semitones between the lower and upper notes, but also how the interval is specified in score notation. As we have already seen in Section 1.1.1, different note symbols may refer to the same note, which is known as **enharmonic equivalence**.

Similarly, there exists enharmonic equivalence between musically different intervals (referred to by different interval names) that correspond to the same distance when measured in semitones (assuming the equal-tempered scale). For example, the two intervals named "augmented unison" and "minor second" both express the difference of one semitone, while specified differently when using score notation (see Figure 5.2b). In the following, even though being an oversimplification, we think of an interval in terms of a distance between pitches (given in semitones) while disregarding their music notation.

Figure 5.3 shows the most common interval names and their meaning in terms of semitone differences. For example, an interval consisting of two identical pitches is called a **unison**. In this case, the difference in semitones is zero. We have already seen the **octave**, which refers to a difference of twelve semitones and a frequency ratio of 1 : 2. The **fifth** denotes an interval that encodes the difference of seven semitones. To understand this naming convention, one needs to know that intervals are

Fig. 5.3 Names of intervals. The columns from left to right indicate the difference given in semitones (Δ), the name of the interval, the interval assuming C4 as the root note, the ratios with respect to just intonation (JI), and the Pythagorean ratios.

Δ	Interval name	Interval	JI ratio	Pyt. ratio
0	(Perfect) unison	C4 – C4	1:1	1:1
1	Minor second	C4 – D♭4	15:16	$3^5{:}2^8$
2	Major second	C4 – D4	8:9	$2^3{:}3^2$
3	Minor third	C4 – E♭4	5:6	$3^3{:}2^5$
4	Major third	C4 – E4	4:5	$2^6{:}3^4$
5	(Perfect) fourth	C4 – F4	3:4	$3{:}2^2$
6	Tritone	C4 – F♯4	32:45	$2^9{:}3^6$ or $3^6{:}2^{10}$
7	(Perfect) fifth	C4 – G4	2:3	2:3
8	Minor sixth	C4 – A♭4	5:8	$3^4{:}2^7$
9	Major sixth	C4 – A4	3:5	$2^4{:}3^3$
10	Minor seventh	C4 – B♭4	5:9	$3^2{:}2^4$
11	Major seventh	C4 – B4	8:15	$2^7{:}3^5$
12	(Perfect) octave	C4 – C5	1:2	1:2

traditionally named on the basis of a major scale, which consists of seven notes and an eighth note one octave apart from the first one (see Section 5.1.2.3). As an example, we consider the C major scale, which is shown in Figure 5.2a. Enumerating the notes of the major scale using ordinal numbers, the first and fifth note are a fifth interval (seven semitones) apart. Other intervals have names such as **major third** or **minor third**. Among these two third intervals, the term "major" refers to the larger interval spanning four semitones and the term "minor" to the smaller interval spanning three semitones. The **tritone** interval plays a special role in the sense that it divides the octave into two equal parts, each consisting of six semitones.

5.1.1.2 Frequency Ratios

As mentioned above, the concept of intervals can also be approached from a physical point of view by considering frequency relations that naturally occur in the harmonic partials of a pitched sound. Recall from Section 1.3.2 that the harmonics are the integer multiples of a fundamental frequency forming the **harmonic series** of a tone. As illustrated by Figure 1.20, the notes of the equal-tempered scale can be associated to partials contained in this series. However, there is a slight deviation between the fundamental frequencies of the notes and the frequency of the partials. In the physical approach, one derives the intervals from frequency relations between partials that occur within the same harmonic series. This is illustrated by Figure 5.4, which shows the harmonic series of the note C2. For example, the octave occurs as an interval between the first two harmonics, the fifth as an interval between the second and third partial, the fourth as an interval between the third and fourth partial, and so on. This observation leads to a definition of intervals that is based on ratios

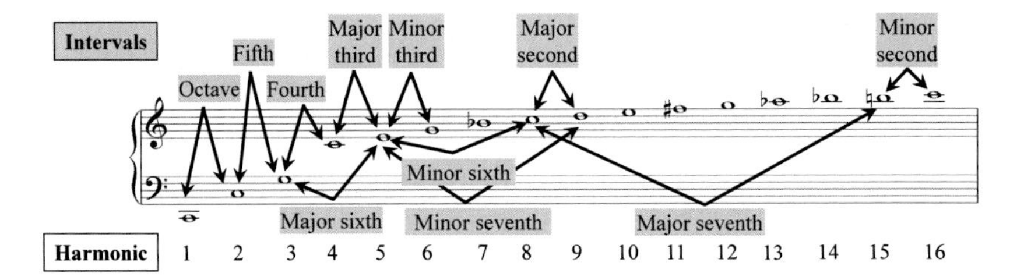

Fig. 5.4 Harmonic series in music notation starting with the root note C2 (see also Figure 1.20). The various intervals can be determined by the frequency ratios of suitable partials.

of small whole numbers. Any interval defined in this way is also called a **pure** or **just interval**. Similarly, the musical tuning based on harmonics is known as **pure** or **just intonation**. The frequency ratios of the various intervals with regard to just intonation are indicated in Figure 5.3. The definition of intervals based on these ratios leads to slightly different interval sizes compared with the equal-tempered case (see Exercise 5.4).

Besides just intonation, there are many more tuning systems that may be used for defining intervals in terms of frequency ratios. The oldest known tuning system was introduced by the Greek philosopher and mathematician Pythagoras (sixth century BC). The geometrically motivated **Pythagorean tuning** is based only on the frequency ratio 1 : 2 of the octave and the ratio 2 : 3 of the fifth. All other intervals are derived from these ratios by suitably adding and subtracting fifths and octaves. This results in intervals that can be expressed by frequency ratios that involve only powers of two or powers of three (see Figure 5.3). Further details on this tuning system are discussed in Exercise 1.10.

5.1.1.3 Consonance and Dissonance

Using the just intonation based on harmonic partials, we have seen that certain intervals can be described by ratios of small integers such as 1 : 1 (unison), 1 : 2 (octave), 2 : 3 (fifth), or 3 : 4 (fourth). Such intervals, which appear in the lower part of the harmonic series in a natural way, are usually perceived as coherent and pleasant. Used more generally, the term **consonance** refers to a combination of notes that sound pleasant to most people when being played simultaneously. In contrast, the term **dissonance** is used to refer to a combination of notes that sound harsh or unpleasant. Consonant sounds are considered to be stable (being at rest), as opposed to dissonant sounds that are considered to be unstable (having a transitional character).

There are various definitions of consonance and dissonance based on musical, physical, and perceptual criteria. As for intervals, the unisons, octaves, fifths and fourths are usually considered to be in **perfect** consonance. Therefore, these intervals are sometimes also called the **perfect intervals**. The major and minor thirds, as well as major and minor sixths, are still perceived as somehow consonant (**imperfect consonance**)—however, to a lower degree. The other intervals are typically considered as dissonant. This particularly holds for the tritone interval, the most dissonant interval. As shown by Figure 5.3, the frequency ratio of the tritone in just intonation involves the largest integers. Also, the tritone is the only interval that does not appear in the lower part of the harmonic series of Figure 5.4. When playing notes simultaneously, the degree of consonance relates to how many of the harmonics of the played notes coincide. From this perspective, consonance does not only depend on the size of the interval between two notes, but also on the combined spectral distribution of the resulting sound (see Exercise 5.5).

5.1.2 Chords and Scales

Based on musical notes and intervals, we now come to more complex musical constructs known as chords and scales. As already mentioned in the introduction, a **chord** can be loosely defined as a group of several notes that sound simultaneously. While most researchers agree that a chord should contain at least three notes, others also regard a combination of two notes as a chord. Depending on the number of distinct notes contained in a chord, one also speaks of a **dyad** (two notes, corresponding to intervals), a **triad** (three notes), a **tetrad** (four notes), and so on. In harmony analysis, notes that are one or several octaves apart are often considered to belong to the same "sound quality." Therefore, when defining the concept of a chord, it may be more precise to speak of distinct **pitch classes** (see Section 1.1.1), rather than distinct notes. For example, even though the chord shown in Figure 5.7d consists of five different notes, it is usually regarded as a triad consisting of the pitch classes C, E, and G.

5.1.2.1 Triads

Despite being an oversimplification, we restrict our considerations in the following to a small selection of chords. In Western music, the most important **triads** consist of three notes that can be stacked in thirds. When stacked in thirds, the lowest note is referred to as the **root note**. Since there are minor thirds (three semitones) and major thirds (four semitones), one can distinguish between four types of such triads (see Figure 5.5).

The first type is referred to as a **major triad**, where the interval between the root note and the second note is a major third, and the interval between the second note and the third note is a minor third (Figure 5.5a). As a result, the interval between the root note and the third note is a perfect fifth. The second type is the **minor triad**, where a minor third is followed by a major third (Figure 5.5b). Again the interval between the root note and the third note is a perfect fifth. The third type is called a **diminished triad**, which consists of two minor thirds (Figure 5.5c). Now, the interval between the root note and the third note consists of six semitones, a

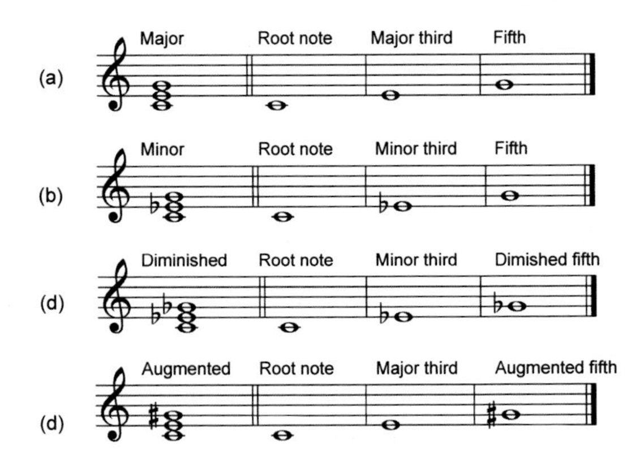

Fig. 5.5 Various types of triads over the root note C4. **(a)** Major triad. **(b)** Minor triad. **(c)** Diminished triad. **(d)** Augmented triad.

tritone (up to enharmonic equivalence) according to Figure 5.3. In the context of a diminished chord, the same interval is correctly referred to as a **diminished fifth**, since it is obtained by reducing the size of a perfect fifth. The fourth type is the **augmented triad**, which consists of two major thirds (Figure 5.5d). In this case, the interval between the root note and the third note consists of eight semitones, a minor sixth according (up to enharmonic equivalence) to Figure 5.3. The same interval is correctly referred to as an **augmented fifth**, since it can be obtained by increasing the size of a perfect fifth. These definitions again illustrate the complexity and ambiguity when speaking of intervals in a musical context.

Each of the chord types can be played based on different root notes. Regardless of the given root note, the four triad types have a different quality in how they are perceived by a listener. Consisting of a major third and a perfect fifth, the major triads comprise consonant intervals. Furthermore, there is a high agreement among the partials of the three constituent notes. As a result, the sound of a major triad is often described as being coherent, pleasant, and happy. By reducing the pitch of the second note by one semitone, a major triad transforms into a minor triad. Still regarded as consonant and coherent, minor chords are often perceived as sad, gloomy, or somber. For the diminished and augmented triads, there is little overlap in the partials of the constituent notes, and these triads are typically perceived as dissonant and unstable. Often, diminished and augmented triads are used in transitional passages to move between more stable harmonies based on major and minor chords. Finally, we want to note that all of the discussed triads also appear as subsets of more complex chords consisting of four, five or even more notes. For example, a major triad together with a minor seventh forms a tetrad known as a **dominant seventh chord**. Since they are beyond the scope of this book, we do not further discuss such more complex chords and their musical relations, but refer to the literature on music theory [8, 15, 28].

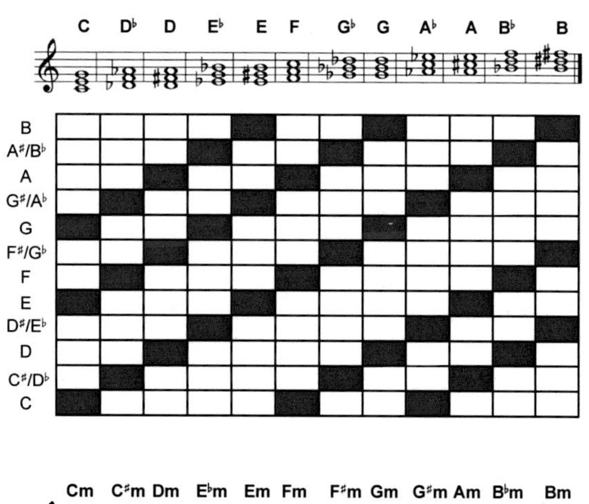

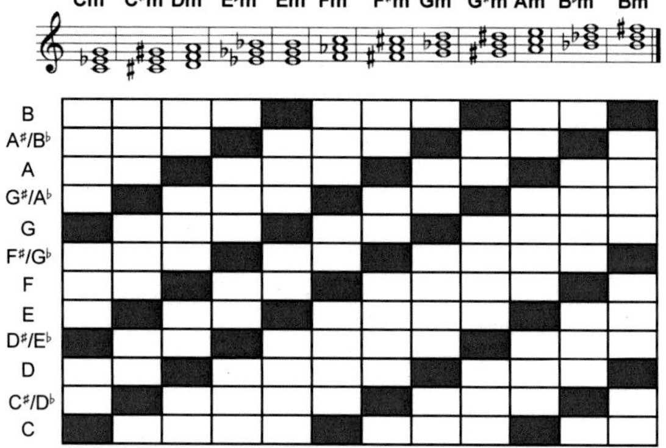

Fig. 5.6 Overview of all major chords (top) and all minor chords (bottom) up to enharmonic equivalence. One possible score notation is given for each chord, as well as a chroma pattern encoding the respective three-element subset (indicated by the colored cells).

5.1.2.2 Major and Minor Chords

Let us have a closer look at the major and minor triads. Since there are twelve different root notes (up to enharmonic equivalence and octave shifts), one can basically form twelve major and twelve minor triads. Figure 5.6 shows a score representation of these 24 triads, where the lowest note of each chord is the root note. A major chord is usually denoted with the same symbol used for the pitch class of its root note. In the following, we use bold characters to distinguish the concept of a chord from the concept of a pitch class. For example, the major chord with the root C is denoted by **C**. This chord, which is also called the C major chord, consists of three notes belonging to the pitch classes C, E, and G. For the minor chords, we use the same notation as for major chords except for adding a letter "**m**" that refers to "minor." For example, the C minor chord denoted by **Cm** consists of the three notes with pitch classes C, E$^\flat$, and G. Assuming the equal-tempered scale as for the case of pitch classes, we do not distinguish between chords such as **C**$^\sharp$ and **D**$^\flat$ or **G**$^\sharp$**m** and **A**$^\flat$**m**, even though these chords play a different role from a music theory perspective.

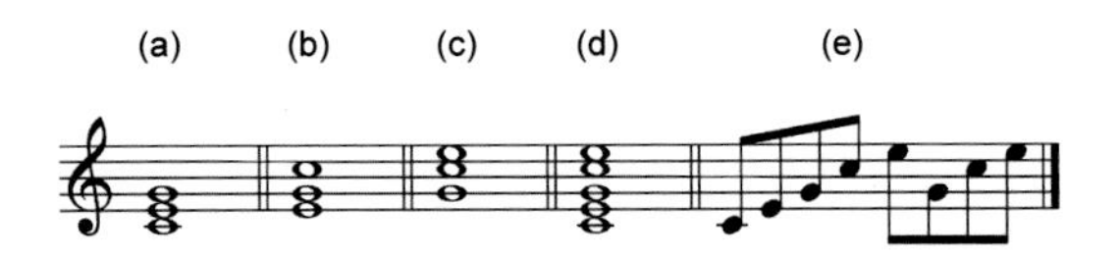

Fig. 5.7 Variants of the C major chord. **(a)** Root position. **(b)** First inversion. **(c)** Second inversion. **(d)** Octave doubling. **(e)** Broken chord.

As one may have noticed from the above considerations, chords can be seen from different perspectives. In this book, we adopt a rather simplistic view, where a major or minor chord is determined by the pitch classes or chroma values of its constituent notes. From a mathematical point of view, each of the triads can then be regarded as a three-element subset of the set $\{\mathrm{C},\mathrm{C}^\sharp,\mathrm{D},\ldots\mathrm{B}\}$ that consists of the twelve chroma attributes. Equivalently, a three-element subset can be regarded as a binary chroma vector with three entries of value 1 at the chroma positions encoded by the subset. The resulting chroma patterns of the 24 major and minor triads are shown in Figure 5.6. Based on this mathematical model, the twelve major chords can be obtained by cyclically shifting the major triad **C** in twelve different ways (see Section 3.1.2.2). Similarly, one obtains the twelve minor chords from **Cm**. Note that each of the 24 major and minor triads leads to a different three-element subset. In other words, on the pitch-class level, the major and minor triads are uniquely defined.

On the note level, there are generally many alternatives for realizing a given chord. As an example, Figure 5.7 shows some variants of the C major chord. When a chord's lowest note (the bass note) is its root, the chord is said to be in **root position** or in **normal form** (see Figure 5.7a). When the root is not the lowest note played in a chord, it is said to be **inverted**. The **first inversion** of the C major chord is shown in Figure 5.7b, and the **second inversion** in Figure 5.7c. Besides various types of inversions, the chord's notes may be further rearranged into different octaves. Additionally, notes of the same pitch class may be added (octave doubling) (see Figure 5.7d). Finally, the notes of a chord may be played one after another, resulting in what is referred to as a **broken chord** (see Figure 5.7e). As an entity that spans a certain period of time, such a group of notes may still be perceived as a single chord.

5.1.2.3 Musical Scales

Besides intervals and chords, we now consider another important musical construct that is referred to as a **musical scale**. Again, adopting a rather simplistic view, a scale can be regarded as a set of notes, where the elements are typically ordered by ascending pitch. While a chord may be thought of as a vertical structure, scales are usually associated to horizontal structures. Assuming the principle of octave equivalence, scales typically span a single octave, with higher or lower octaves simply repeating the pattern. In this way, a musical scale can be regarded as a division of the octave space into a certain number of **scale steps**, where each scale step is an interval between two successive notes.

Fig. 5.8 Two different spellings and score notations for the chromatic scale.

So far, we have already encountered a number of different scales. For example, in Section 1.1.1, we considered the twelve-tone equal-tempered scale, also referred to as the **chromatic scale**, where an octave is subdivided into twelve scale steps. In this case, all scale steps correspond to the same interval having a size of one semitone (or 100 cents). Due to enharmonic equivalence, there are various spellings and score notations to represent a chromatic scale; two of them are shown in Figure 5.8.

In the following, we only consider scales that are subsets of the chromatic scale, where the scale steps can be specified in semitones. In the context of scales, the minor second (one semitone) is also referred to as a **half step** and the major second (two semitones) as a **whole step**.

As with chords, there are two scale types that are of particular importance in Western music theory. The first scale type is known as a **major scale**, which is made up of seven notes and a repeated octave. The first note of a major scale is called the **key note** of the scale. Starting with the key note, the sequence of intervals between the successive notes of a major scale is

$$\text{whole, whole, half, whole, whole, whole, half.} \tag{5.1}$$

The chroma name of the key note also determines the name of the scale. For example, the major scale starting with a C is called C major (see Figure 5.9a). Sometimes, the symbol **C**, as used for chords, is also used as an abbreviation to refer to the scale. The other major scales are then obtained by cyclically shifting the C major scale. The notes of a major scale are given names, also known as **scale degrees**, to specify their positions relative to the key note. The key note is also called the **tonic**, which is the main note of the scale. The fourth note of the scale is called the **subdominant** and the fifth note the **dominant**. The remaining names are indicated in Figure 5.9b.

The second scale type we consider is known as the (natural) **minor scale**. Similar to a major scale, a minor scale consists of seven notes and a repeated octave. This time, however, the sequence of intervals between the notes is

$$\text{whole, half, whole, whole, half, whole, whole.} \tag{5.2}$$

Again, there are twelve minor scales with naming conventions similar to those of the major chords. As an example, Figure 5.9c shows the notes of a C minor scale.

Both major and minor scales can be subsumed under the general term **diatonic scale**, which is (by definition) a seven-pitch scale with five whole steps and two half steps for each octave. There are many more scales used in Western music and beyond. For example, besides the minor scale introduced above (also referred to as

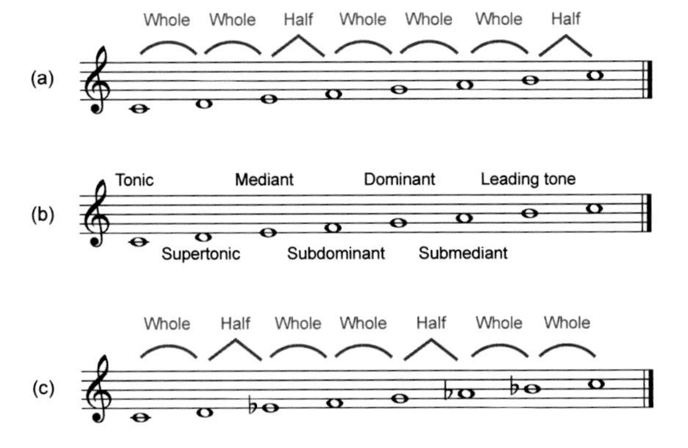

Fig. 5.9 Diatonic scales. **(a)** C major scale. **(b)** Names of scale degrees as used for major and minor chords. **(c)** C minor scale.

the **natural minor**), there are other types of minor scales called the **harmonic minor** and **melodic minor**. The notes of the harmonic minor scale are the same as the natural minor except that the seventh degree is raised by one semitone, making an augmented second between the sixth and seventh degrees. In this case, the sequence of intervals between the notes becomes

$$\text{whole, half, whole, whole, half, whole-and-a-half, half.} \tag{5.3}$$

We have already mentioned the chromatic scale, which involves twelve pitches. There are other scales such as the **pentatonic scale**, consisting of five pitches, or the **whole tone scale**, consisting of six pitches. A discussion of general musical scales and their relations is beyond the scope of this book, and we refer to the literature on music theory [8, 15, 28].

5.1.2.4 Circle of Fifths

One characteristic property of diatonic scales is that they can be obtained from a chain of six successive perfect fifths. For example, the C major scale is obtained from an ascending chain of six perfect fifths starting with F:

$$\mathrm{F}-\mathrm{C}-\mathrm{G}-\mathrm{D}-\mathrm{A}-\mathrm{E}-\mathrm{B}. \tag{5.4}$$

Being the most consonant nonoctave interval, the fifth interval plays a particularly important role when relating notes, chords, and scales. The property (5.4) means that all notes of a diatonic scale are related by fifths, which gives the scale a degree of coherence and balance.

Furthermore, considering fifth relationships also makes it possible to relate entire musical scales. This leads us to the famous **circle of fifths** shown in Figure 5.10. The circle of fifths is a visual representation of the relationships among the twelve tones of the chromatic scale and the associated major and minor scales. More precisely, the circle of fifths represents the relations between musical **keys**—a concept

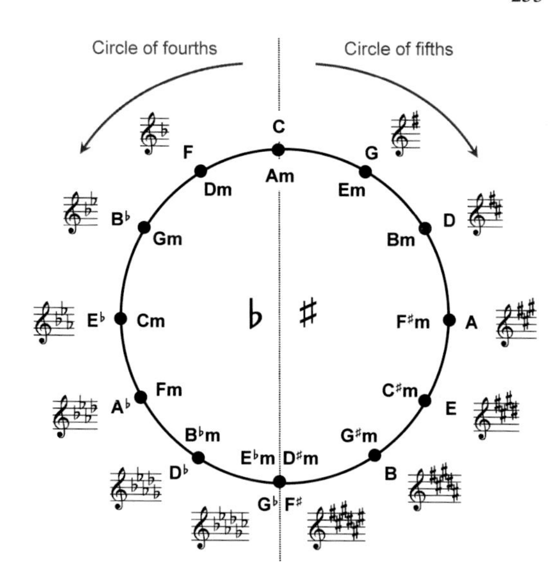

Fig. 5.10 Circle of fifths and fourths.

that is closely connected to major and minor scales. A piece's musical key usually identifies the tonic note as well as the major or minor triad that represents the harmonic center (giving a subjective sense of arrival and rest) of the piece. In general, the notes of a major (minor) scale constitute the basic material for the major (minor) key sharing the same name. Although many musicians confuse key with scale, as mentioned in [1], a scale is an ordered set of notes typically used in a key, while the key is the center of gravity, established by particular chord progressions. In this book, the distinction of these two concepts will not play an important role, and we may be a bit sloppy when using these terms.

After these remarks, let us come back to the circle of fifths. At the top of the circle, there is the C major key. The notes of the corresponding scale are C, D, E, F, G, A, and B, the so-called **natural notes**. These notes, which correspond to the white keys of a piano keyboard, do not require any **accidental** (♯, ♭) when encoded in Western music notation (see Section 1.1.1). As a result, the key signature of C major has no flats or sharps. Similarly, the A minor key, whose corresponding scale also consists of the seven natural notes, shares the same key signature with C major. In general, for each major key there is a minor key whose corresponding scales share the same notes. This relationship is referred to as a **relative relationship**. For example, the A minor key is denoted as the relative minor of the C major key and vice versa. In the circle of fifths shown in Figure 5.10, the major keys are noted outside the circle, whereas the corresponding relative minor keys are noted inside the circle.

Starting with C major at the top of the circle, one obtains the other keys proceeding clockwise by ascending fifths. The next key is the G major key. We have already seen that the key notes C and G are closely related by sharing many partials. What is more, the entire scales are closely related. First, the last four notes of the C major scale coincide with the first four notes of the G major scale. Then, the two scales share six out of seven notes; only the F in C major becomes an F$^\sharp$ in G major. This introduces a sharp in the key signature for G major. The same kind of relations hold between any two subsequent keys or scales along the circle of fifths. Proceeding one fifth upwards changes one note of the scale and introduces one additional sharp in the key signature. Repeating this process twelve times in the equal-tempered case, one returns to the original C major, thus closing the circle.

Similarly, one can travel along the circle in a counterclockwise fashion by descending perfect fifths. Because of octave equivalence, this amounts to ascending perfect fourths, resulting in a **circle of fourths** (see Exercise 5.2). Proceeding one step in this reverse direction again changes one note of the scale. This time, however, an additional flat is introduced in the resulting key signature. For example, C major becomes F major, where the note B becomes a B$^\flat$.

The circle of fifths reflects the degree of "musical" similarity between different scales; the closer two scales are located on the circle, the more they share in terms of tonal material. In music, **modulations** are used to move from one scale or musical key to another. Since six out of seven notes are shared by adjacent scales, a modulation by a perfect fifth can be accomplished in a very smooth fashion by only changing one note by a semitone. Intuitively, the chromatic scale may be regarded as a "global world" that contains all available tonal material. The major and minor scales can then be regarded as "local regions" of this world, each having its own "harmonic" flavor. The circle of fifths provides an orientation guide for the music to smoothly travel (if desired) from one region to another region.

5.1.2.5 Functional Relation of Chords

Many of the chords used in Western tonal music find a natural home within the system of diatonic scales. Recall that the positions of the notes of a given scale relative to the key note can be expressed in terms of scale degrees (see Figure 5.9). Similarly, one can consider the relative positions of chords in the context of a given diatonic scale. This leads us to the study of what is referred to as **functional** harmony theory, which tries to explain the principles of harmonic relationships evolving around the main scale degrees of the tonic, dominant, and subdominant. We now look again at the major and minor chords (Section 5.1.2.2) from a functional point of view.

Let us start with the C major scale, which consists of the seven notes C, D, E, F, G, A, and B. Using each of these notes as a root note, we can construct seven different triads that only consist of notes from the given scale. The resulting chords are shown in Figure 5.11a. For example, the triad over the key note C of the scale consists of the notes C, E, and G, which is the C major chord **C**. Or, the triad over the second note D consists of the notes D, F, and A, which is the D minor chord **Dm**. It turns out that the triads over the first, fourth, and fifth note of the C major scale are major chords, and the triads over the second, third, and sixth note are minor chords. Only the triad over the seventh note B is a diminished chord, denoted by **B$^\mathbf{o}$**. All of the major and minor chords that appear in this way in the C major scale are closely

Fig. 5.11 Roman numerals for the chords within a major scale. **(a)** C major scale. **(b)** G major scale. **(c)** D major scale.

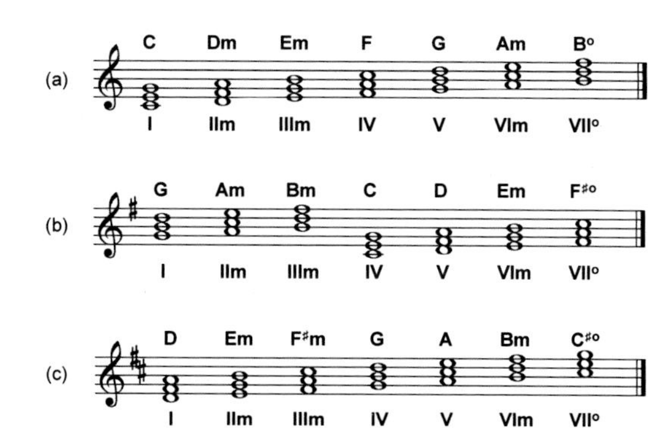

related, which is also reflected by their immediate proximity in the circle of fifths (see Figure 5.10).

Since chord types are defined in terms of intervals (semitone differences), they do not change when applying the same semitone shift to all notes. For example, a major chord remains a major chord when shifting all notes by seven semitones. The triad **C** becomes **G**, the triad **F** becomes **C**, the triad **G** becomes **D**, and so on. Therefore, the sequence of chord types is the same for all of the twelve major scales. This is also demonstrated by Figure 5.11b, which shows the triads over the G major scale, and by Figure 5.11c, which shows the triads over the D major scale. In other words, the relationships between the chords of a given major scale do not change when moving to another major scale.

The objective of functional harmony theory is to represent chords and to study their relationships independently of the selected scale or key. To this end, the chords appearing in a scale are represented in terms of the **scale degrees** as introduced in Section 5.1.2.3. Recall that we used the scale degrees to refer to note positions within a scale *relative* to the key note of the scale. Besides the names as specified in Figure 5.9b, Roman numerals (**I**, **II**, **III**, **IV**, ...) are often used to denote chords on scale degrees.[1] Furthermore, in the case of minor chords, a suffix **m** is added to the numeral. For example, **I** denotes the major chord over the key note, **IIm** the minor chord over the second note of the scale, and so on. Furthermore, the names of the scale degrees are also used to refer to the corresponding chord. For example, one speaks of the tonic meaning **I**, the subdominant meaning **IV**, and the dominant meaning **V**.

These conventions are illustrated by Figure 5.11, which shows the chords for the C major, G major, and D major scales. Note that, in functional harmony theory, the meaning of a chord depends on the context specified by the given scale. For example, the chord **C** is the tonic (**I**) within C major, but the subdominant (**IV**) within G major. Or the chord **G** is the dominant (**V**) within C major, but the tonic (**I**) within G major, and the subdominant (**IV**) within D major.

---

[1] This kind of scale-based chord analysis is also referred to as **Roman numeral analysis**. Another theory emphasizing functional chord names traces back to Hugo Riemann and is mainly used in the German music tradition. In this book, the differences between these theories are negligible.

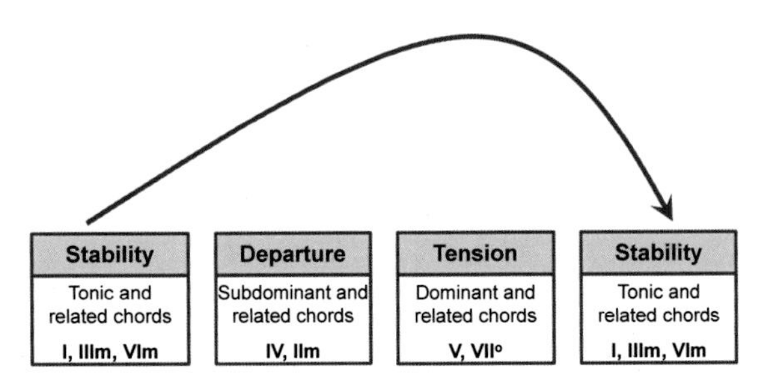

Fig. 5.12 Curve of suspense (exemplarily for major keys) and its elements as often encountered in music.

5.1.2.6 Chord Progressions

Music is rarely static, but rather moves along. It transports a message and aims for a goal. Musicians use their intuition and experience to generate a harmonic flow by suitably arranging and combining chords. Such an arrangement of chords over time is also referred to as a **chord progression**. Functional harmony theory yields a language for analyzing and describing such progressions.

Even though there seems to be an endless number of possible chords and chord progressions, there are often simple rules that govern typical combinations for certain types of music. One general principle, which is applied not only in music but also in literature or films, is based on the interplay between tension and release. For example, starting with a harmonically stable passage, the music departs from this feeling, creates tension, and then finally returns to the original feeling of stability (see Figure 5.12). Within a given scale, certain chords provide stability, while other chords generate a tension requiring some kind of resolution. In particular, the tonic chord (**I**) is considered as stable, whereas the dominant chord (**V**) and its variants cause the feeling of tension. Of course, these are only very coarse tendencies; the perception of such qualities also very much depends on the listener's experience and expectation, as well as on the musical context.

The progression of chords provides the harmonic foundation for many types of music. The Roman numerals as introduced in Figure 5.11 can be used to specify chord progressions in a generic form. Being independent of the underlying musical key, such descriptions emphasize the chords' function rather than their specific realization. For example, jazz musicians are used to Roman numeral notation and can realize the notated chords in any given musical scale. The essence of a song or piece of music can often be described by a characteristic chord progression. For example, many songs use only three chords based on the progression **I** – **IV** – **V**. Another well-known progression is the 12-bar Blues scheme, which (in its basic) form can be represented as: **I** – **I** – **I** – **I** – **IV** – **IV** – **I** – **I** – **V** – **IV** – **I** – **I**. This scheme is used in Blues music and related styles, where this basic progression is typically enriched by the use of seventh chords.

To conclude this short tutorial on music theory, we want to emphasize that harmony is one of the fundamental characteristics of music, along with other aspects including rhythm, melody, bass lines, instruments, dynamics, articulation, and lyrics. In the following sections of this chapter, we do not assume a detailed knowledge of

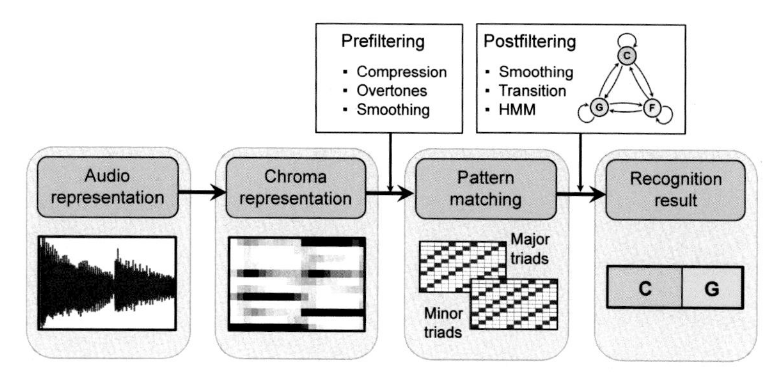

Fig. 5.13 Overview of the components of a typical processing pipeline for automated chord recognition.

harmony theory. However, the reader should keep in mind that harmony is closely related to chords and their relationships. Chords can be described in various ways including chroma-like patterns or in terms of functional properties. In particular, we have seen that the meaning and perception of a chord critically depend on the musical context. Also, for a given musical genre, there are certain chord progressions that are more likely to occur than others. For example, a C major chord **C** will most likely be followed by **G** or **Am**, rather than by $\mathbf{C}^\sharp$ or **Gm**. As we will see, this fact can be exploited for automated chord recognition by modeling a chord not as an isolated event, but as an element of an entire chord progression.

5.2 Template-Based Chord Recognition

As mentioned in the introduction, a typical chord recognition system consists of two main steps. This is again illustrated by Figure 5.13. In the first step, the given audio recording is cut into frames, and each frame is transformed into an appropriate feature vector. Most recognition systems rely on chroma-based audio features, which correlate to the underlying tonal information contained in the audio signal. In the second step, pattern matching techniques are used to map each feature vector to a set of predefined chord models. The best fit determines the chord label assigned to the given frame. To improve the chord recognition results, additional enhancement techniques are applied either before the pattern matching step (referred to as **prefiltering**) or after/within the pattern matching step (referred to as **postfiltering**) (see [4, 11]).

In Section 5.2.1, we introduce a first chord recognition procedure that employs a simple template-based matching strategy. Using suitable evaluation measures (Section 5.2.2), we then discuss some typical problems one has to cope with in audio-based chord recognition (Section 5.2.3). Finally, in Section 5.2.4, we describe how some of these problems can be alleviated by applying enhancement and prefiltering techniques. Later, in Section 5.3, we introduce a more involved postfiltering approach based on hidden Markov models, which reduces the number of performed chord changes by incorporating contextual information.

5.2.1 Basic Approach

Given an audio recording of a piece of music, the goal of our chord recognition task is to find out which chords are played at which time. The first step is to transform the recording into a sequence $X = (x_1, x_2, \ldots, x_N)$ of feature vectors $x_n \in \mathcal{F}$, $n \in [1:N]$, where $\mathcal{F}$ denotes a suitable feature space. Then, each feature vector x_n is to be mapped to a chord label $\lambda_n \in \Lambda$, where Λ denotes a set of possible chord labels. For example, one may consider the set

$$\Lambda = \{\mathbf{C}, \mathbf{C}^\sharp, \ldots, \mathbf{B}, \mathbf{Cm}, \mathbf{C}^\sharp\mathbf{m}, \ldots, \mathbf{Bm}\} \tag{5.5}$$

consisting of the twelve major and minor triads. In this case, each frame $n \in [1:N]$ is assigned to a major chord or a minor chord specified by λ_n. There are many ways for transforming the music recording into a feature sequence and for performing the pattern matching step. Let us start with the most basic procedure.

For the feature extraction step, basically every chord recognition procedure relies on some type of chroma features. This is because chroma-based features capture a signal's short-time tonal content, which is closely correlated to the harmonic progression of the underlying piece. Recall from Section 3.1.2 that, assuming the equal-tempered scale, the chroma values correspond to the set $\{\mathrm{C}, \mathrm{C}^\sharp, \mathrm{D}, \ldots, \mathrm{B}\}$. This set consists of twelve chroma attributes, which are determined up to enharmonic equivalence. As in Section 3.1.2, we identify this set with the set $[0:11]$ by enumerating the chroma attributes such that 0 corresponds to C, 1 to $\mathrm{C}^\sharp$, and so on. A chroma feature can then be expressed as a 12-dimensional vector $x = (x(0), x(1), \ldots, x(11))^\top \in \mathcal{F}$ with $\mathcal{F} = \mathbb{R}^{12}$. In the following, we start with a simple chroma variant as defined in (3.6) of Section 3.1.2. Furthermore, in our examples, we use a window size that corresponds to 200 ms and a hop size of half the window length. As a result, our feature sequence $X = (x_1, x_2, \ldots, x_N)$ has a feature rate of 10 Hz. Later in this section, we introduce various feature enhancement strategies and smoothing steps and discuss their influence on the final chord recognition results.

For the pattern matching step, let us start with a simple template-based chord recognition strategy (see Figure 5.14). The idea is to precompute a set $\mathcal{T} \subset \mathcal{F} = \mathbb{R}^{12}$ of templates denoted by $\mathbf{t}_\lambda \in \mathcal{T}$, $\lambda \in \Lambda$. Intuitively, each template can be thought of as a prototypical chroma vector that represents a specific musical chord. Furthermore, we fix a similarity measure $s : \mathcal{F} \times \mathcal{F} \to \mathbb{R}$ that allows for comparing different chroma vectors. Then, the template-based procedure consists in assigning the chord label that maximizes the similarity between the corresponding template

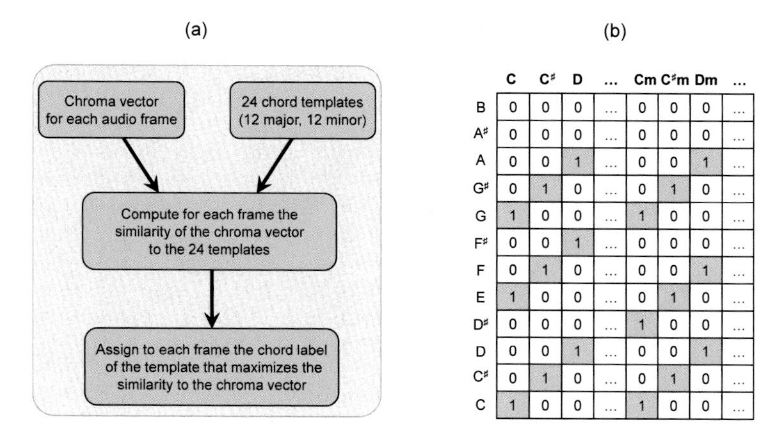

	C	C♯	D	...	Cm	C♯m	Dm	...
B	0	0	0	...	0	0	0	...
A♯	0	0	0	...	0	0	0	...
A	0	0	1	...	0	0	1	...
G♯	0	1	0	...	0	1	0	...
G	1	0	0	...	1	0	0	...
F♯	0	0	1	...	0	0	0	...
F	0	1	0	...	0	0	1	...
E	1	0	0	...	0	1	0	...
D♯	0	0	0	...	1	0	0	...
D	0	0	1	...	0	0	1	...
C♯	0	1	0	...	0	1	0	...
C	1	0	0	...	1	0	0	...

Fig. 5.14 **(a)** Overview of a template-based chord recognition procedure. **(b)** Binary chord templates for major and minor triads.

and the given feature vector x_n:

$$\lambda_n := \underset{\lambda \in \Lambda}{\operatorname{argmax}}\, s(\mathbf{t}_\lambda, x_n). \tag{5.6}$$

In this procedure, there are many design choices that crucially influence the performance of a chord recognizer. Which chords should be considered in $\mathcal{T}$? How are the chord templates defined? What is a suitable similarity measure to compare the feature vectors with the chord templates?

For the chord label set Λ, we choose the twelve major and minor triads as in (5.5). We will see that the restriction to these 24 chord classes is often problematic from a musical point of view. However, since this choice is simple, convenient, and instructive, it is often made in the chord recognition literature. We have already seen in Figure 5.6 how one may define chord templates. Considering chords up to enharmonic equivalence and up to octave shifts, each chord is specified by a subset of the set $\{\mathrm{C}, \mathrm{C}^\sharp, \mathrm{D}, \ldots, \mathrm{B}\}$ or, equivalently, of the set $[0:11]$. For example, a C major chord **C** corresponds to the three-element subset $\{\mathrm{C}, \mathrm{E}, \mathrm{G}\}$ or to the subset $\{0,4,7\}$. Each subset, in turn, can be identified with a binary twelve-dimensional chroma vector $x = (x(0), x(1), \ldots, x(11))^\top$, where $x(i) = 1$ if and only if the chroma value $i \in [0:11]$ is contained in the chord. For example, in the case of the C major chord **C**, the resulting chroma vector is

$$\mathbf{t}_{\mathbf{C}} := x = (1,0,0,0,1,0,0,1,0,0,0,0)^\top \tag{5.7}$$

(see Figure 5.14b). Recall from Section 5.1.2.2 that, when using a chroma-based encoding, the twelve major chords and twelve minor chords can be obtained by cyclically shifting the binary vectors for the C major and the C minor triads, respectively.

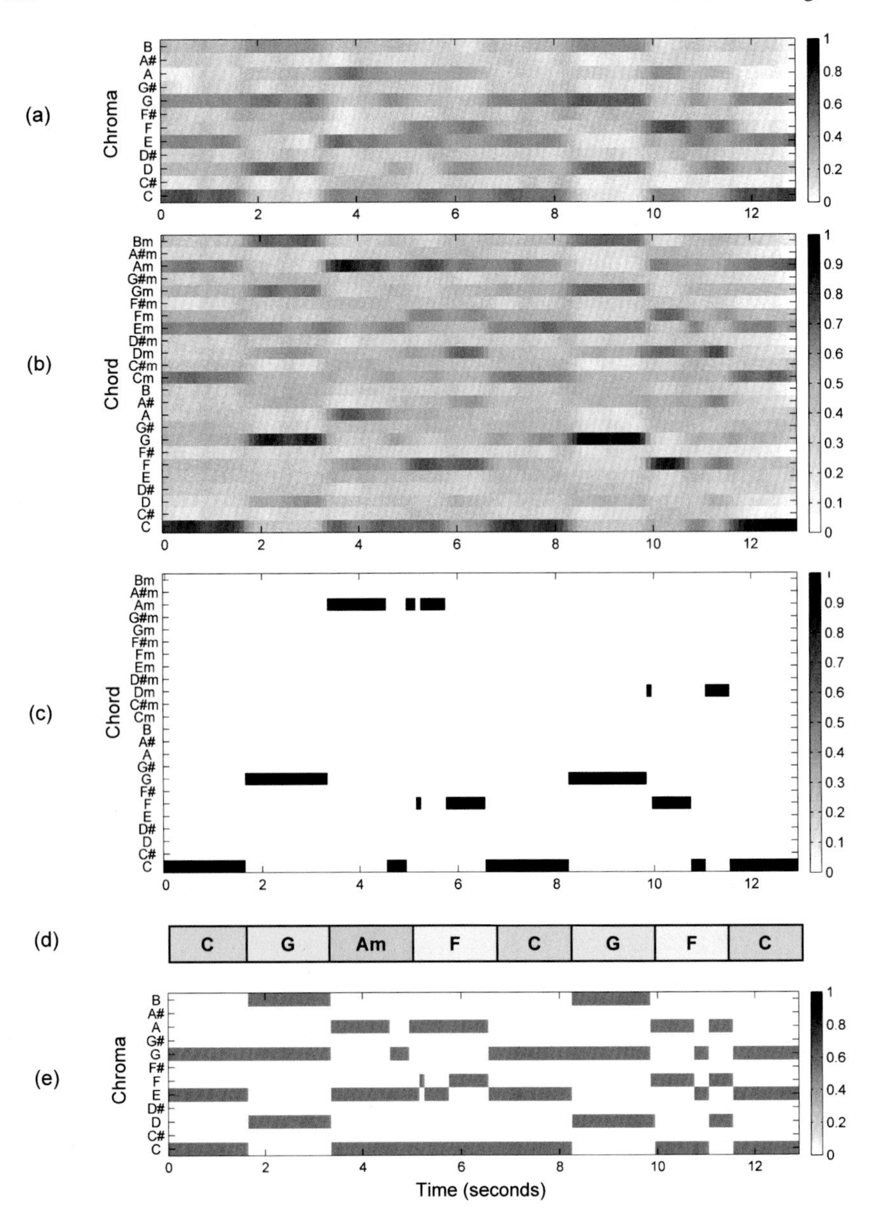

Fig. 5.15 Template-based chord recognition using binary templates for the 24 major and minor triads. The audio recording consists of the first four measures of the Beatles song "Let It Be" (see Figure 5.1). **(a)** Chroma representation. **(b)** Similarity values between the chroma vectors and the 24 chord templates. **(c)** Chord recognition result. **(d)** Manually specified chord annotations. **(e)** Normalized binary templates of the chord recognition result.

Furthermore, there are many ways for comparing chroma features and chord templates. Because of its simplicity, we use the inner product of normalized vectors as a similarity measure s (see (2.37)). This measure has already been used in Chapter 4 for the task of music structure analysis. For convenience, we repeat its definition:

$$s(x,y) = \frac{\langle x|y\rangle}{\|x\| \cdot \|y\|} \tag{5.8}$$

for $x, y \in \mathcal{F} \setminus \{0\}$. In the case $\|x\| = 0$ or $\|y\| = 0$, we set $s(x,y) = 0$. Note that this measure always yields a value $s(x,y) \in [-1,1]$. In the case that the vectors x and y only have positive entries, one has $s(x,y) \in [0,1]$.

To obtain a better understanding of this procedure, we continue our Beatles example (the beginning of the song "Let It Be") from Figure 5.1. In the feature extraction step, the audio recording is converted into the chroma representation shown in Figure 5.15a. In the next step, each chroma vector is compared with each of the 24 binary chord templates, which yields 24 similarity values per frame. These similarity values are visualized in Figure 5.15b in the form of a **time–chord representation**. For example, this visualization shows that the chroma vectors at the beginning of the song are most similar to the template for the C major chord **C**. Furthermore, there is also a higher degree of similarity to the templates for **Cm**, **Em**, and **Am**. We analyze this phenomenon in more detail later in Section 5.2.2 (see Figure 5.17).

According to (5.6), we select for each frame the chord label λ_n of the template that maximizes the similarity value over all 24 chord templates. This yields our final chord recognition result (see Figure 5.15c). Comparing this result with a reference annotation generated by a music expert (Figure 5.15d), one can observe that the result obtained from the automated procedure agrees with the reference labels for most of the frames. Finally, the similarity-maximizing chord templates, which have been normalized with respect to the Euclidean norm as used in (5.8), are shown in Figure 5.15e. In a way, this sequence of chord templates may be thought of as a musically informed quantization of the original chroma representation of Figure 5.15a.

5.2.2 Evaluation

In order to evaluate the quality of a chord recognition procedure, a general approach is to compare the computed result against a reference annotation. We have already seen in the context of music structure analysis (Section 4.5) that such an evaluation often gives rise to various questions. How should the agreement between the computed result and the reference annotation be quantified? Is the reference annotation well defined? Is it reliable? Are the model assumptions in the formalization of the chord recognition task appropriate? To what extent do violations of these assumptions influence the final result? For all these questions there are no definite answers, and evaluation results need to be taken with care. Still, quantitative evaluations are useful indicators. They generally reflect the overall performance of automated pro-

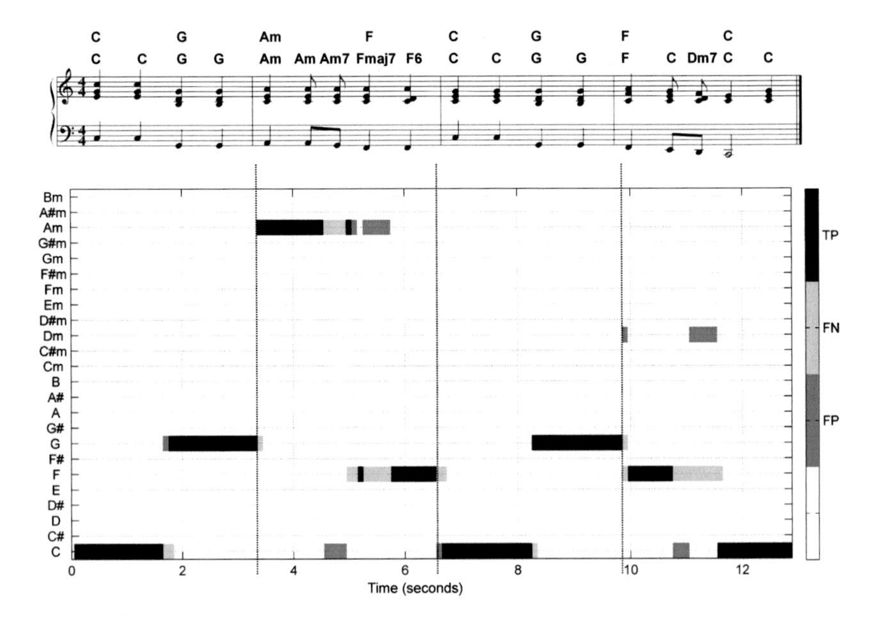

Fig. 5.16 Evaluation of a chord recognition result for the first four measures of the Beatles song "Let It Be". The top shows a score representation along with two different chord annotations provided by music experts. One annotation is specified on the half-measure level (every two quarter notes) and the other annotation on a finer temporal level. The bottom shows the chord recognition result from Figure 5.15c evaluated on the basis of the half-measure annotation.

cedures and give valuable insights into the characteristics of the underlying music data.

In the following, we introduce a typical evaluation approach while discussing various challenges one has to cope with in the modeling, annotation, evaluation, and computation stage. As in Section 4.5, we assume that there exists a reference annotation, which is also called **ground truth**. The objective of the automated procedure is to estimate a chord recognition result that is as close to the reference annotation as possible.

5.2.2.1 Manual Annotation

The reference annotation is usually generated by music experts, who often perform harmony analysis on a piece of music, based on a score representation. The expert partitions the score into sections and assigns to each section a chord label that describes the predominant harmony. The sections may have different lengths ranging from a quarter note or even a shorter duration, up to several measures. Depending on the temporal granularity, the suggested annotations may differ significantly.

As an example, Figure 5.16 shows two annotations of the first four measures of the Beatles song "Let It Be." The first annotation has been specified on the half-measure level (every two quarter notes) and reflects the rough harmonic progression of the piece. On this level, an annotated section may contain notes that, strictly speaking, do not belong to the selected chord. These notes may serve as **passing notes** to prepare the transition to the next chord or as **suspended notes** to intensify the tension that is to be resolved in the following chords. The second annotation shown in Figure 5.16 provides chord labels on a finer temporal level, which corresponds to a more detailed analysis. For example, on the coarse level, the fourth half-measure may be regarded as an F major section, even though it contains notes beyond the triad such as the E or the D. On the finer level, more advanced chord models are employed to precisely describe the harmonies in this section.

While manually generated annotations are often based on a musical score, the automated chord recognition procedure to be evaluated may work on the basis of an audio recording of the piece. In this case, one requires methods for converting chord annotations specified on the score's musical time axis into annotations specified on the physical time axis of the given recording. This process is often done manually—a tedious and time-consuming task. Instead, one may also employ music synchronization techniques as described in Chapter 3 (see also Figure 3.1). Such automated transfer, however, may bear the risk of introducing additional synchronization errors.

Most audio-based chord recognition procedures work in a frame-wise fashion. To match the reference annotation to the frame windows used in the feature extraction step, one needs an additional quantization step. Furthermore, the chord models used for the manual annotation may disagree with the chord models used in the automated procedure. In this case, one needs a way to make the chord models comparable, e.g., by transforming all chords to the 24 major and minor triads. As also illustrated by Figure 5.16, such a reduction is problematic due to oversimplifications, and often introduces additional ambiguities.

In summary, we have seen that the generation and usage of so-called ground-truth annotations involves several issues. First of all, there is—in general—no well-defined ground truth, and even music experts may disagree on how to annotate a given piece of music. Second, the annotations may depend on the employed temporal granularity. Third, one needs to adapt the manual annotations to make them comparable with the computed results.

5.2.2.2 Precision, Recall, F-measure

Following Section 4.5.1, we now introduce evaluation measures for comparing estimated results computed by a chord recognition procedure with manually generated reference annotations (even though these annotations and measures are to be treated with caution as discussed before). While considering a frame-wise chord recognition scenario, we do not want to assume that every frame needs to be annotated. For example, the recording may start with silence or end with applause. In these cases, there is no meaningful chord annotation, and the corresponding frames should be left unconsidered in the evaluation. To model this, we extend our label set Λ by an additional symbol **N**, which we refer to as **nonchord label**. In the following, we assume that we have for each frame $n \in [1:N]$ an estimated chord label $\lambda_n \in \Lambda \cup \{\mathbf{N}\}$ and a reference chord label $\lambda_n^{\mathrm{Ref}} \in \Lambda \cup \{\mathbf{N}\}$. Adapting the notions from Section 4.5.1, we define the set of items to be $\mathcal{I} = [1:N] \times \Lambda$. Note that, when using this definition, the nonchord label **N** is left unconsidered. Then,

$$\mathcal{I}_+^{\mathrm{Ref}} := \{(n, \lambda_n) \in \mathcal{I} \mid n \in [1:N]\} \tag{5.9}$$

are the **positive items** (or **relevant items**) and

$$\mathcal{I}_+^{\mathrm{Est}} := \{(n, \lambda_n^{\mathrm{Ref}}) \in \mathcal{I} \mid n \in [1:N]\}, \tag{5.10}$$

are the **items estimated as positive**. From this, following Section 4.5.1, one can derive the notions of true positive (TP), false positive (FP), and false negative (FN) items as well as the precision (P), recall (R), and F-measure (F).

Let us have a closer look at the chord recognition result from Figure 5.15, whose evaluation is illustrated by Figure 5.16. Using a feature rate of 10 Hz we obtain $N = 130$ frames. As reference, the annotations on the half-measure level are used. These annotations are transferred to the physical time axis of the recording and further quantized to match the audio frames. Note that some frames may be left unannotated. For the evaluation measures, we obtain $\mathrm{P} = 0.84$, $\mathrm{R} = 0.79$, and $\mathrm{F} = 0.82$. In other words, most of the computed chord labels agree with the reference labels. However, there are also some deviations. In particular, these deviations occur for frames with chord ambiguities due to additional passing or suspended notes. This becomes clear when looking at the annotation of the finer level. For example, the chord **Am7** in the second measure consists of the four chroma A, C, E, and G. Although musically close to **Am**, three of the four chroma also occur in the chord **C**. As a result, the automated procedure erroneously labeled most of these frames as **C**. A second source for deviations are transition regions between chords, where the ending sound of one chord may be present in the same analysis window as the beginning sound of the next chord. For example, one encounters some misclassified frames in the transition from the chord **C** to the next chord **G** in the first measure.

5.2.3 Ambiguities in Chord Recognition

Let us delve into more detail and try to get a better understanding of the various kinds of ambiguities one encounters in chord recognition.

5.2.3.1 Chord Ambiguities

First of all, as the discussion of the previous example showed, different chords may be closely related by sharing some of their notes. This fact is again illustrated by Figure 5.17. For example, as shown by Figure 5.17a, the C major chord shares two

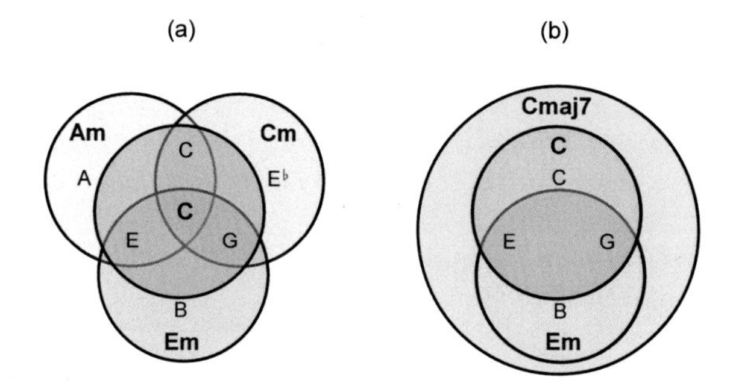

Fig. 5.17 Ambiguity of chords. **(a)** Among the 24 major and minor triads, the chord **C** shares two notes with the chords **Am**, **Cm**, and **Em**, respectively. **(b)** The chord **Cmaj7**, which consists of the four notes C, E, G, and B, includes the chords **C** and **Em**.

of its three notes with the triads **Am**, **Cm**, and **Em**. Therefore, the degree of similarity between **C** and these three chords is relatively high and may cause confusion in the classification stage, particularly in the presence of additional notes. This is also demonstrated by Figure 5.17b. The shown chord **Cmaj7** consists of four notes C, E, G, and B, thus containing the two triads **C** and **Em**. When using the similarity measure s as defined in (5.8), one obtains the following equality of similarity values:

$$s(\mathbf{t}_{\mathbf{Cmaj7}}, \mathbf{t}_{\mathbf{C}}) = s(\mathbf{t}_{\mathbf{Cmaj7}}, \mathbf{t}_{\mathbf{Em}}). \tag{5.11}$$

In other words, when only considering the 24 major and minor triads, a chroma vector corresponding to **Cmaj7** may be mapped either to the major chord **C** or to the minor chord **Em**. Most of the misclassifications in Figure 5.16 stem from such chord ambiguities due to an oversimplification of the chord models. This problem may be mitigated by extending the chord label set. For example, besides the major and minor triads, one may also introduce chord templates that correspond to major seventh chords. However, on the downside, increasing the number of possible chords also increases the confusion probability in the classification stage.

5.2.3.2 Acoustic Ambiguities

In our chord recognition scenario, we start with an audio representation of a given musical work rather than with a score representation. Therefore, besides chord ambiguities due to musical reasons, one also has to deal with ambiguities that are introduced by acoustic properties of the recorded music. In particular, the presence of partials may have a significant influence on the results of a chord recognizer. Recall from Section 1.3.2 that playing even a single note on an instrument produces a complex sound mixture. The main frequency components are the fundamental frequency and its harmonics, the integer multiples of the fundamental frequency. As illustrated by Figure 1.20 and Figure 5.4, the first harmonic of a harmonic series can be approximated by the notes of the equal-tempered scale. For example, in the case of the note C2, the first eight harmonics correspond to the notes

$$\mathrm{C2}, \mathrm{C3}, \mathrm{G3}, \mathrm{C4}, \mathrm{E4}, \mathrm{G4}, \mathrm{B^{\flat}4}, \mathrm{C5}. \tag{5.12}$$

On the chroma level, these harmonics cover the chroma C, E, G, and B$^\flat$. In other words, because of harmonics, the musical tone of the note C2 may possess substantial energy not only in the chroma band C, but also in the chroma bands G, E, and possibly B$^\flat$ (among others). Since the energy usually decreases with higher harmonics, most of it is typically contained in the C and G bands.

What does this observation imply for our template-based chord recognition procedure based on binary chord templates for the 24 major and minor triads? When comparing the chroma pattern of a musical tone for a single note C2 with the binary chord templates, the template for the chord **C** will yield the highest similarity. From a musical point of view, this is not a meaningful or desirable property, since the note C2 may serve different purposes. For example, it could also be the root note of the chord **Cm** or the dominant of the chord **F**.

Now, when playing an entire chord, one obtains a superposition of the harmonics of all involved notes. The above effects become quite complex and confusing. For example, let us consider the minor chord **Cm**, which consists of the chroma C, E$^\flat$, and G. Besides having energy in these three chroma bands, the acoustic sound of this chord may also have substantial frequency components in the chroma band E (coming from the fifth harmonic of C). When comparing the resulting chroma vector with the binary major and minor chord templates, the energy in the E-band as well as in the E$^\flat$-band may cause some confusion between the chords **Cm** and **C**. This kind of problem, which is also known as **major–minor confusion**, often occurs in automated chord recognition. The effect is further aggravated by the fact that the notes of a chord may be played with different intensities. In particular, when the third of a minor chord is played weaker, the resulting sound may be classified as a major chord.

The effect of harmonics is also illustrated by Figure 5.18, which shows the similarity values between various chroma patterns and the binary chord templates for the 24 triads (see (5.5)). In Figure 5.18a, we start with chroma patterns that are obtained by simply considering normalized binary chord templates. In this case, the similarity values shown in Figure 5.18c reflect the number of notes that are shared by the respective chroma patterns and chord templates. In Figure 5.18b, the chroma patterns were modified by considering the first eight harmonics for each involved note. Contrary to the previous scenario, the energy is now spread over more chroma bands with a bias to certain bands. For example, in the pattern for **C** (first column), the chroma band G contains most of the energy. This is because G appears not only as one of the three notes in the chord, but also as the third harmonic of the root note C. As shown by Figure 5.18d, the modified chroma pattern for **C** is now closer to the binary template for **G** than to the template for **C**.

5.2.3.3 Tuning

Often, a music recording is not tuned as specified by the center frequencies of the equal-tempered scale (see (1.1) of Section 1.3.2). For example, as we have already mentioned in Section 3.1.2.1, orchestras are sometimes deliberately tuned with a

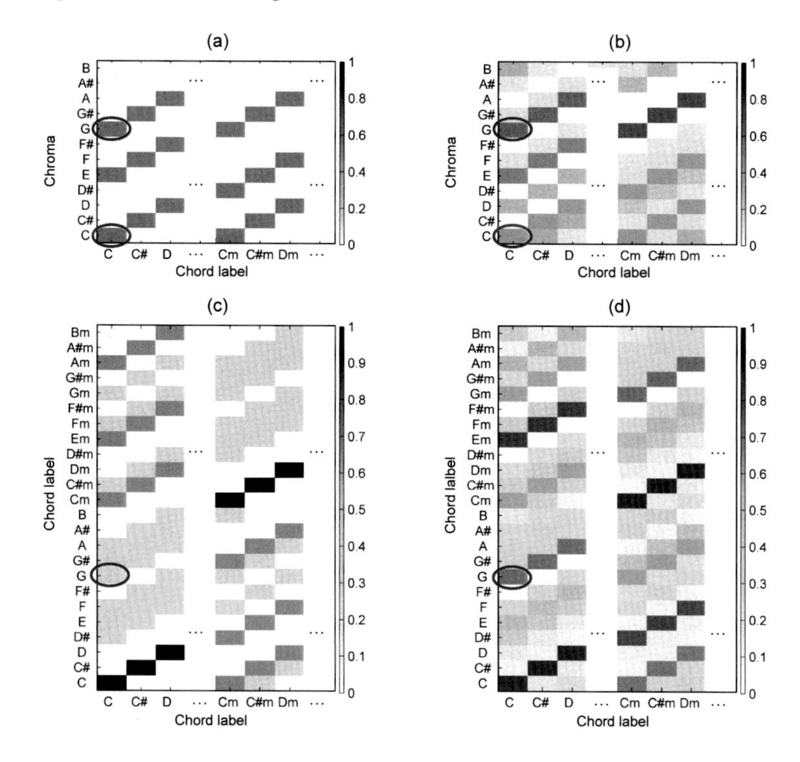

Fig. 5.18 Similarity values between chroma patterns and binary chord templates considering the 24 triads (indexed by chord labels) of (5.5). The marked entries in the figure are discussed in the text. **(a)** Normalized binary chroma patterns. **(b)** Normalized chroma patterns considering the first eight harmonics for each note involved. **(c)** Similarity values between the chroma patterns of (a) (horizontal axis) and binary chord templates (vertical axis). **(d)** Similarity values between the chroma patterns of (b) (horizontal axis) and binary chord templates (vertical axis).

tuning frequency that lies above or below the usual 440 Hz. Furthermore, the tuning may differ from the expected values due to a modification of the playback speed or the application of other postprocessing operations. In Figure 3.8 we have seen that a deviation from the assumed center frequencies may introduce severe degradations in the quality of musically informed audio features such as chroma-based features.

In chord recognition, tuning issues can lead to fatal errors. Such an example is shown in Figure 5.19, where the audio recording has been modified by shifting the frequencies half a semitone (50 cents) upwards. In the resulting chromagram (Figure 5.19a), one can observe that the notes' energy is spread across neighboring chroma bands. This leads to chroma patterns that do not fit well within the binary chord templates. While some of the computed chord labels correspond to the expected chord labels up to a semitone shift, most of the classification results are somewhat chaotic.

The compensation of tuning effects is of great importance in audio analysis tasks such as chord recognition. In Section 3.1.2.1 and Exercise 3.5, we discussed how to

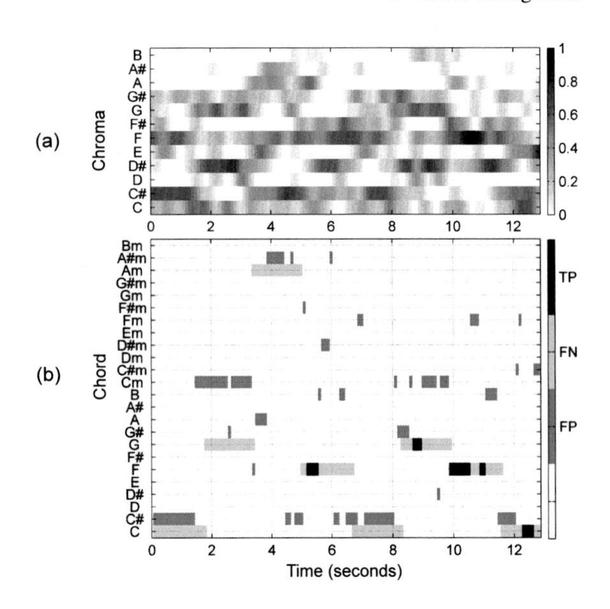

Fig. 5.19 Chord recognition result for the Beatles example from Figure 5.16, where the audio recording has been tuned half a semitone (50 cents) upwards. **(a)** Chroma representation. **(b)** Chord recognition result.

approach such tuning issues by adjusting the frequency binning used in the computation of the pitch-based log-frequency spectrogram.

5.2.3.4 Segmentation Ambiguities

Besides ambiguities in the chroma patterns and oversimplification issues introduced by the chord models, one also has to deal with ambiguities in the temporal dimension. To illustrate this problem, let us start with an example: the famous C major prelude by Johann Sebastian Bach. The first four measures, along with their chord annotations, are shown in Figure 5.20a. As demonstrated by Figure 5.20b, our chord recognizer exhibits many classification errors, in particular in the second and third measure. Let us analyze the reasons for this behavior.

In this example, each half-measure starts with a bass note. Then the other notes join in and gradually build up the sound of an entire chord. We have already encountered such **broken chords** in Figure 5.7e. Even though the notes are not played simultaneously, a broken chord as a whole may be perceived as a single harmonic unit.

In our basic chord recognition procedure, we have chopped up the signal into short frames and classified each frame separately. For example, in Figure 5.20b, we have used an analysis window with a duration of 200 ms (and a hop size corresponding to 100 ms, which yields a feature resolution of 10 Hz). In other words, the recognized chord label of a frame only represents the harmonic content of a 200 ms section of the music recording.

Let us come back to our Bach example. In the recording, a whole measure has a duration of about 2400 ms. Thus we have 600 ms per quarter note and 150 ms per

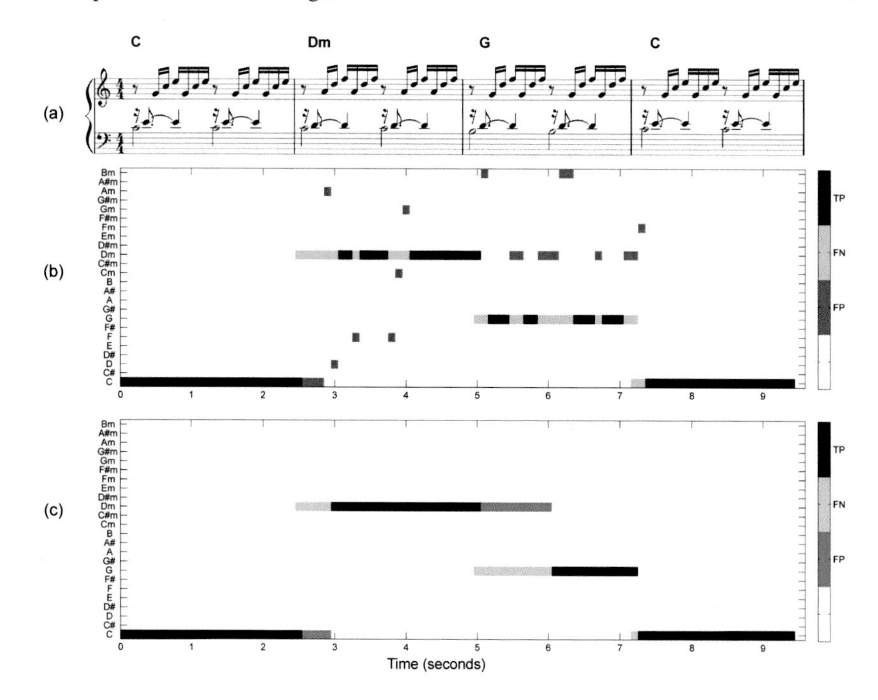

Fig. 5.20 Evaluation of a chord recognition result for the first four measures of the Prelude BWV 846 in C major by Johann Sebastian Bach. **(a)** Musical score and reference annotations. **(b)** Chord recognition result using a frame length of 200 ms and a hop size of half the window length (yielding a feature rate of 10 Hz). **(c)** Chord recognition result after applying prefiltering using 20 frames.

sixteenth note. Therefore, in each analysis frame of 200 ms, one finds the onsets of at most one note. Even though the sound of each note may last much longer than the notated duration, the harmonic content of each frame is dominated by only one or two notes. This explains the many misclassifications and chord label changes in the recognition result of Figure 5.20b. Interestingly, these errors only occur in the second and third measures, while the first and fourth measures are classified correctly (see Exercise 5.6).

An obvious strategy for improving the chord recognition result in our Bach example is to use larger frame sizes that better correspond to the half-measure or measure level of the annotations. Alternatively, one may merge or smooth over the analysis frames by applying a filter prior to the pattern matching step. For example, applying a smoothing procedure (similar to Figure 3.9) to merge the information over 20 consecutive frames (covering in total roughly 2 sec of the audio), one can significantly improve the classification result (see Figure 5.20c). We discuss this strategy in more detail in Section 5.2.4.4.

5.2.4 Enhancement Strategies

We have seen that the challenges in automated chord recognition arise from a multitude of factors that concern musical, acoustic, and temporal aspects. Due to the vagueness of the chord recognition task, even humans may disagree and argue about how to harmonically interpret certain musical phenomena. Therefore, automated methods alone will never be enough to "solve" chord recognition. However, there are certain strategies for "improving" automated methods in the sense that they come closer to what humans may have agreed on. In the following, we discuss some of these strategies.

5.2.4.1 Templates with Harmonics

As a first strategy, the overall procedure may be improved by refining the chord templates that are used in the pattern matching step. So far, as illustrated by Figure 5.14, we have used idealized binary chord templates that indicate the presence or absence of notes in the given chord. For real music recordings, however, the presence of harmonics and other sound components leads to chroma features where the energy is spread over the chroma bands in a more unstructured, nonbinary fashion. This motivates the modification of the chord templates by introducing a model that accounts for harmonics.

Let us have a look at a typical construction. As in Section 5.2.3.2, we consider the first eight harmonics and their note correspondences. In the case of a note with chroma C, we have seen in (5.12) that the first eight harmonics correspond to the chroma values

$$\mathrm{C},\mathrm{C},\mathrm{G},\mathrm{C},\mathrm{E},\mathrm{G},\mathrm{B}^{\flat},\mathrm{C}. \tag{5.13}$$

Often, the energy in the harmonics decays in an exponential fashion. This can be modeled by assuming that the energy of the kth partial is α^{k-1} for some $\alpha \in [0,1]$, $k \in \mathbb{N}_0$. Form this, one can define a template with harmonics for the chroma C by setting

$$\mathbf{t}^{\mathrm{h}}_{\mathrm{C}} = (1+\alpha+\alpha^3+\alpha^7,0,0,0,\alpha^4,0,0,\alpha^2+\alpha^5,0,0,\alpha^6,0)^{\top}. \tag{5.14}$$

A chord template with harmonics for the major chord **C** is obtained by summing up the templates over the chord's chroma classes:

$$\mathbf{t}^{\mathrm{h}}_{\mathbf{C}} = \mathbf{t}^{\mathrm{h}}_{\mathrm{C}} + \mathbf{t}^{\mathrm{h}}_{\mathrm{E}} + \mathbf{t}^{\mathrm{h}}_{\mathrm{G}}. \tag{5.15}$$

Similarly, one obtains chord templates for the other major and minor triads. The chroma patterns shown in Figure 5.18b have been computed this way using the parameter $\alpha = 0.9$. The chord template in (5.15) can be further refined by introducing additional parameters that weight the notes of a given chord in different ways (see Exercise 5.8).

5.2.4.2 Templates from Examples

Instead of explicitly modeling the harmonics, a conceptually different approach is to "learn" chroma patterns for the chord templates from labeled training data. The input of such a learning procedure, which is also known as **supervised learning**, consists of a set of training examples. Each example, in turn, is a pair consisting of an **input object** and an **output value**. In our case, the input objects are chroma vectors and the output values are chord labels. The objective of supervised learning is to infer a classification scheme that correctly determines the chord labels for a number of unknown chroma vectors, thus generalizing from the training data to unseen situations in a reasonable way.

In the chord recognition context, the simplest way to incorporate knowledge from labeled chroma vectors into the recognition pipeline is to replace the chord templates by averages. To this end, we assume that we have for each $\lambda \in \Lambda$ a set $\mathbf{T}_\lambda$ of chroma vectors labeled with λ. Then, we define the average template $\mathbf{t}_\lambda^{\mathrm{a}}$ by taking the entrywise average:

$$\mathbf{t}_\lambda^{\mathrm{a}}(i) := \frac{1}{|\mathbf{T}_\lambda|} \sum_{x \in \mathbf{T}_\lambda} x(i) \tag{5.16}$$

for $i \in [0:11]$, where $|\mathbf{T}_\lambda|$ denotes the number of elements contained in the set $\mathbf{T}_\lambda$.

The main advantage of such a data-driven approach is that the learned chord templates naturally inherit the musical and acoustic properties from the training vectors (see Figure 5.21c). In other words, the models automatically adapt to the underlying data. On the downside, supervised learning requires suitably labeled training data, which should accurately represent the properties of the unseen data to be classified. In particular, one needs for each possible output value a sufficient amount of training material. Such labeled data are often hard to come by.

In the chord recognition context, one can alleviate this problem by exploiting the close relationships within the major and minor triads. Recall that the chroma patterns of the twelve major triads are related by cyclic shifts (see Section 5.1.2.2). This also holds for the twelve minor triads. Therefore, instead of learning 24 chord templates, it suffices to learn the chord templates only for the C major and the C minor chords. The other templates can then be derived from these two prototype patterns via cyclic shifts. A similar trick can be used to increase the training material for learning the C major and the C minor patterns. To this end, all training chroma vectors labeled with a major chord are cyclically shifted to correspond to C major, while the chroma vectors labeled with a minor chord are transformed into a C minor pattern.

Taking the average templates is a simple way to adapt a template-based chord recognizer to given training data. More involved approaches are often based on statistical models that capture not only the averages but also the variances in the training data. In such approaches, the templates are replaced by chord models that are specified by, e.g., Gaussian distributions given in terms of a mean vector and a covariance matrix. The similarity of a given chroma vector to a chord model is then expressed by a Gaussian probability value and the assigned label is determined

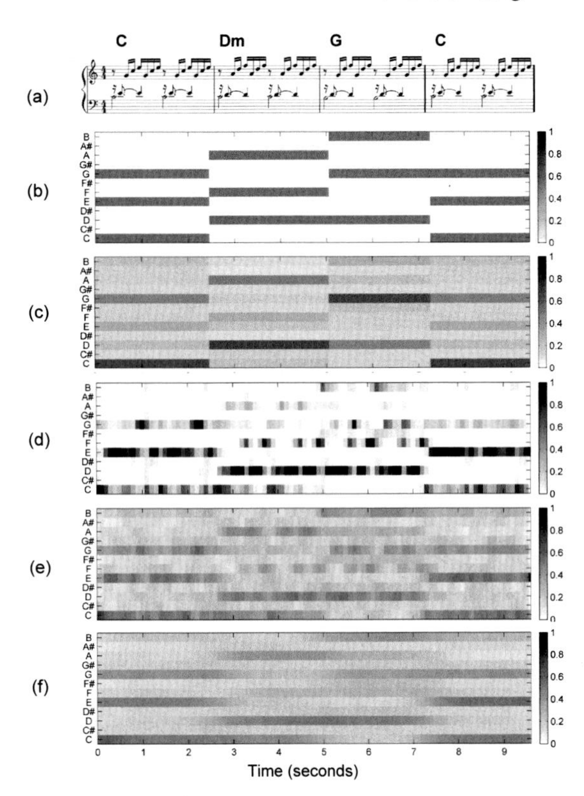

Fig. 5.21 Various types of enhanced chroma features. All chroma vectors are normalized with respect to the Euclidean norm and have a 10 Hz resolution. **(a)** Musical score and reference annotations. **(b)** Binary chord templates (corresponding to reference annotations). **(c)** Average chord templates obtained from training data (corresponding to reference annotations). **(d)** Original chroma features obtained from an audio recording. **(e)** Chroma features from (d) after applying logarithmic compression (using $\gamma = 10$). **(f)** Chroma features from (e) after applying smoothing (using $L = 20$).

by the probability-maximizing chord model. The discussion of such statistical approaches is beyond the scope of this book, and we refer the reader to the literature such as [4].

In practice, training and evaluation are often performed on the basis of the same manually annotated dataset. One typically splits up the available dataset into disjoint subsets called the **training set** and **validation set** (sometimes also called the **testing set**). In a first stage, the training set is used to learn the chord templates. Then, in the second stage, the validation set is used to evaluate the resulting chord recognition procedure (see Section 5.2.2). The goal of such a **cross-validation** approach is to obtain insights into how well the procedure generalizes when applied to unseen data.

To reduce the dependency of the overall procedure on the partitioning of the dataset, one often performs multiple rounds of cross-validation. One way is to randomly partition the dataset into K subsets of equal size. One of the subsets is used as the validation set, whereas the union of the other $K-1$ subsets is used as the training set. The cross-validation is repeated K times (the so-called **folds**) with each of the K subsets serving exactly once as the validation set. The K results are then combined, e.g., by taking the average of the considered evaluation measure, to form a single estimate. This approach, which is commonly known as **K-fold cross-validation**, is frequently used to evaluate chord recognition procedures that involve some kind of training.

As an illustrative example, let us consider a concrete scenario. A dataset widely used in automated chord recognition consists of a collection of 180 Beatles songs along with publicly available reference chord annotations [7, 16]. The original annotations, which go beyond the 24 major and minor triads, need to be reduced to the 24 chord labels. This may be done by considering only the first two intervals of each chord, where augmented chords are mapped to major chords and diminished chords to minor chords. In the case that no meaningful reduction is possible or that no meaningful chord information exists, the label **N** is used (see Section 5.2.2.2).

Although this dataset is limited to only one rock band, the results still show certain tendencies of the chord recognition accuracies. Let us denote this collection by $\mathcal{D}$. In a 3-fold cross-validation, one randomly partitions $\mathcal{D}$ into three subcollections $\mathcal{D}_k$, $k \in \{1,2,3\}$, each consisting of 60 recordings. Then, two of the three subcollections are used to train the recognizer, which is then tested on the remaining one. In each of the three folds, we obtain an evaluation measure. For example, using the F-measure as introduced in Section 5.2.2.2, one may obtain $\mathrm{F} = 0.61$ in the first fold, $\mathrm{F} = 0.66$ in the second fold, and $\mathrm{F} = 0.65$ in the third fold. On average, this yields $\mathrm{F} = 0.64$, which is the result of the cross-validation. We will come back to this experimental setup in the following sections.

5.2.4.3 Spectral Enhancement

Besides refining and adapting the chord templates, another general strategy is to modify and enhance the chroma features extracted from the audio recordings to be analyzed. We have already seen in Section 3.1.2 that there are many chroma variants with quite different properties. The chroma type used has a strong influence on the chord recognition results, as has been demonstrated by [4, 11, 17]. In the following, we consider two typical enhancement strategies.

As a first strategy, we apply **logarithmic compression** as introduced in Section 3.1.2.1. Recall that this strategy makes the signal's chroma distribution more uniform, thus giving smaller components a larger weight relative to the stronger components. Starting with the original chromagram $\mathcal{C}$, the logarithmically compressed chromagram is defined by

$$\mathcal{C}_\gamma := \Gamma_\gamma \circ \mathcal{C} \tag{5.17}$$

(see (3.8)). The constant $\gamma \in \mathbb{R}_{>0}$ determines the degree of compression. The effect of logarithmic compression has already been demonstrated by Figure 3.7. Similar effects become visible in Figure 5.21e, which shows the compressed chromagram using $\gamma = 10$ for our Bach example.

To illustrate the kind of dependencies of the chord recognition result on the underlying feature type, we now conduct a small experiment based on the Beatles dataset introduced in Section 5.2.4.2. Besides the template-based approach using

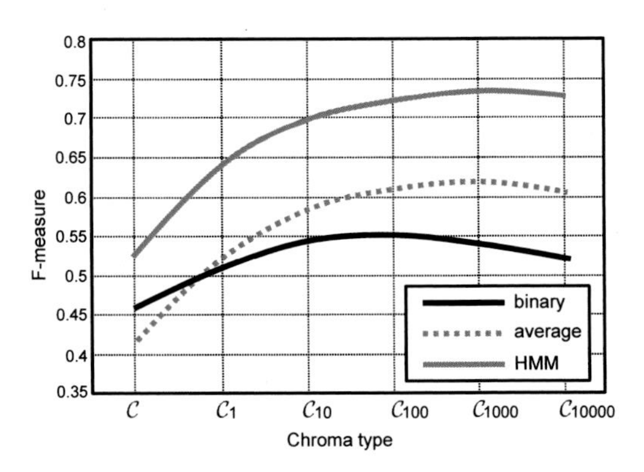

Fig. 5.22 Dependency of the chord recognition accuracy on the compression parameter γ used in the chroma feature $\mathcal{C}_\gamma$. The figure shows the average F-measures obtained from a 3-fold cross-validation conducted on the Beatles dataset. The results are shown for three different chord recognizers based on binary templates, average templates, and HMMs.

binary as well as average templates, we also test a third recognizer that employs hidden Markov models (HMMs). We will discuss this more advanced approach, which incorporates some temporal context in the classification stage, in Section 5.3. For each classification strategy, we build chord recognizers that are based on different chroma types by changing the constant γ in the logarithmic compression. Then, each of the recognizers is evaluated by performing a 3-fold cross-validation using the F-measure as explained in Section 5.2.4.2. The results are shown in Figure 5.22.

Let us discuss the results in more detail, first considering the case of a template-based chord recognizer using binary templates. Figure 5.22 shows the chord recognition accuracy for $\mathcal{C}$ (no compression) and $\mathcal{C}_\gamma$ for $\gamma \in \{1, 10, 100, 1000, 10000\}$. Starting with an F-measure of $\mathrm{F} = 0.46$ for $\mathcal{C}$, the recognition rate first increases, reaching a maximum of $\mathrm{F} = 0.55$ for $\mathcal{C}_{100}$, before it drops again when further increasing γ. This tendency becomes even more distinct when using the chord recognizer based on average templates, where the F-measure rises from $\mathrm{F} = 0.42$ for $\mathcal{C}$ to $\mathrm{F} = 0.61$ for $\mathcal{C}_{1000}$, before it drops again. In the HMM-based approach, which also involves a training step, one can recognize the same trend.

One reason for the increased chord recognition accuracy is that weak spectral components, which are often perceptually relevant, are enhanced by the compression. For very large compression factors, however, the chroma patterns become less characteristic. The amplification of irrelevant noise-like components may outweigh the effect of balancing out the harmonically relevant components, thus leading to a decrease in the overall chord recognition accuracy. In the case of binary chord templates, the templates are independent of the chroma feature type used. In particular, higher harmonics are not taken into account by the idealized binary templates. This may be one of the reasons why the strategy based on binary templates does not benefit from logarithmic compression to the same extent as the other methods do, where the templates and chord models are adapted to the underlying chroma type in the training stage.

Rather than optimizing a given chord recognition procedure, we wanted to demonstrate how small modifications at the feature extraction stage may have a significant influence on the chord recognition quality—independently of the classification strategy used. In particular, our experiments showed that logarithmic compression is an important step in most chord recognition procedures.

5.2.4.4 Prefiltering

Logarithmic compression can also be thought of as a type of *spectral* smoothing. As a second feature enhancement strategy, we now introduce *temporal* smoothing. As discussed in Section 3.1.2.3, such smoothing operations can be beneficial for reducing the effect of irrelevant local variations.

We start with a chromagram given by some sequence $X = (x_1, x_2, \ldots, x_N)$, which consists of chroma features $x_n = (x_n(0), x_n(1), \ldots, x_n(11))^\top \in \mathbb{R}^{12}$, $n \in [1:N]$. The easiest smoothing strategy is to apply an averaging filter on each of the twelve components of the sequence. To this end, we fix a number $L \in \mathbb{N}$ that determines the length of the averaging filter. We then define

$$x_n^L(i) := \frac{1}{L} \sum_{\ell=0}^{L-1} x_{n+\ell-\lfloor (L-1)/2 \rfloor}(i) \tag{5.18}$$

for each component $i \in [0:11]$ and each frame $n \in [1:N]$. To avoid boundary problems in (5.18), we assume that x_n is defined for $n \in \mathbb{Z}$ by setting $x_n := 0$ for $n \in \mathbb{Z} \setminus [1:N]$. This procedure is often referred to as **zero-padding**. Furthermore, note that we apply in (5.18) the averaging in a "centered" way. This preserves the correspondence between the frame index and the physical time position (given in seconds) in the audio recording.

As a result of this procedure, we obtain a sequence $X^L = (x_1^L, x_2^L, \ldots, x_N^L)$, which is a smoothed version of the original sequence X. The parameter L determines the degree of smoothing applied in this procedure. The case $L = 1$ yields the original sequence X. The effect of smoothing is illustrated by Figure 5.21f, which shows the chromagram after applying an averaging filter of length $L = 20$ (measured in frames). Having used a feature rate of 10 Hz, the length of the smoothing window corresponds to two seconds of the original audio recording. Note that in our Bach example, this window roughly corresponds to the duration of a measure, thus nicely covering all notes of a broken chord. However, in the transitions between two subsequent measures, the note information from two different chords is merged into a single smoothing window.

We now investigate the effect of temporal smoothing on the chord recognition accuracy by conducting the same kind of experiment as in Section 5.2.4.3. This time, however, we use the fixed chroma type C_{10} and then apply smoothing for different choices of the parameter L. Since smoothing is applied before the pattern matching step, this step is also referred to as **prefiltering**. Figure 5.23 shows the resulting F-measures for a 3-fold cross-validation using three different classification strategies (binary, average, HMM) and $L \in \{1, 2, \ldots, 26\}$. For the chord recognizer based on binary templates, the F-measure is $\mathrm{F} = 0.54$ for $L = 1$. The recognition rates first improve with increasing L and reach a maximum of $\mathrm{F} = 0.64$ for L between 17 and

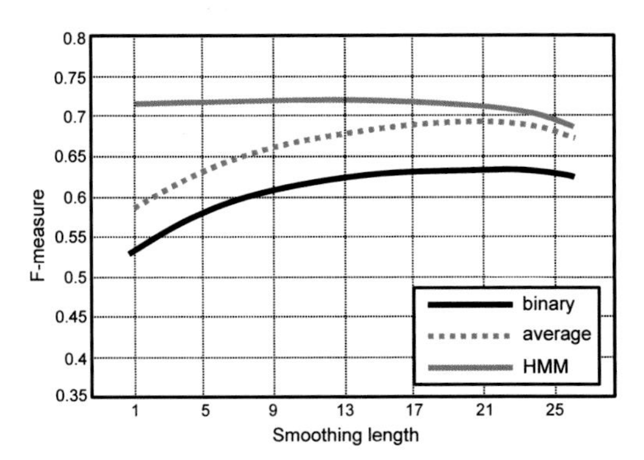

Fig. 5.23 Dependency of the chord recognition accuracy on the smoothing length used in the prefiltering (based on a fixed chroma type). The figure shows the average F-measures obtained from a 3-fold cross-validation conducted on the Beatles dataset. The results are shown for three different chord recognizers based on binary templates, average templates, and HMMs.

23. Then, the rates decrease again when further increasing L. A similar trend can be observed for the chord recognizer based on average templates. For both strategies, smoothing removes temporal fluctuations and local outliers in the features, thus improving the frame-wise classification result. Also, the smoothing integrates nonsynchronous notes that musically belong to the same chord—an effect we have seen in Figure 5.21f for the broken chords. On the other hand, smoothing reduces the temporal resolution and may prevent a recognizer from detecting chords of short duration. For the particular Beatles dataset, a smoothing window corresponding to roughly two seconds of the audio turns out to be the best trade-off between increased robustness to outliers and decreased temporal resolution. This trade-off, however, is highly dependent on the chord change rate of the underlying audio material used for the evaluation.

In the case of the HMM-based recognizer, temporal smoothing of the features has a less significant effect on the chord recognition accuracy. This recognizer, as we discuss in Section 5.3, already incorporates context-aware smoothing within the classification stage. Therefore, additional smoothing prior to the classification does not further improve the overall recognition result.

In summary, one can say that temporal feature smoothing is an easy way to improve the overall evaluation measure of chord recognizers that work in a purely frame-wise fashion. However, the "optimal" smoothing length very much depends on the data distribution. Rather than using an averaging filter of fixed length, an alternative approach is to employ musically informed adaptive segmentation techniques. For example, chord changes often go along with beat positions. Therefore, filtering could be performed in a **beat-synchronous** fashion, where each analysis window is determined by two consecutive beat positions (see Section 6.3.3). However, finding beat positions by automated methods, as will be seen in Chapter 6, is a challenging task by itself. Instead of modifying the features prior to the classification stage, another general approach is to perform some kind of filtering within the pattern matching step (see Figure 5.13). This leads us to context-aware classification schemes as discussed in the next section.

5.3 HMM-Based Chord Recognition

So far, our template-based chord recognizer worked in a purely frame-wise fashion: each frame was classified separately without regarding the previous or future frames. In music, however, chord progressions are not arbitrary, but follow certain rules. Some harmonic progressions are more probable than others. For example, as illustrated in Figure 5.12, one often encounters the harmonic progression **I** – **IV** – **V** – **I**, where tension is built up by traveling from the tonic via the subdominant to the dominant, which resolves again to the tonic. Starting with the C major chord **C** (tonic), it is more likely to find **F** (subdominant) or **G** (dominant) rather than come across harmonically unrelated chords such as **Fm** or $\mathbf{D}^\flat$. Furthermore, one often finds the tendency to stay within the same harmonic context for a certain duration. In other words, chord changes are relatively rare events given the high temporal resolution typically used in the feature extraction step.

We have seen that some of these issues can be alleviated by applying a prefiltering step, which basically removes local outliers and effectively increases the analysis window. On the downside, prefiltering introduces a fixed time grid and washes out characteristic information, thus resulting in a loss of flexibility. As an alternative, we now study a more refined strategy that performs the filtering in a context-sensitive fashion. The main idea is to introduce a transition model that expresses the likelihood of passing over from one chord to another. This leads us to a concept known as the **hidden Markov model** (HMM). The concept of HMMs has been widely used in applications such as speech recognition [25] and also constitutes the de facto standard method in most automated procedures for chord recognition [29]. In the following, we discuss the basic ideas behind HMMs and their application to chord recognition.

5.3.1 Markov Chains and Transition Probabilities

As said before, certain transitions from one chord to another are more likely than others. To capture such likelihoods, we now introduce the concept of Markov chains, closely following the tutorial by Rabiner [25]. Abstracting from our chord recognition scenario, we assume that the chords to be considered are represented by a set

$$\mathcal{A} := \{\alpha_1, \alpha_2, \ldots, \alpha_I\} \tag{5.19}$$

of size $I \in \mathbb{N}$ consisting of distinct elements α_i for $i \in [1:I]$. The elements α_i are also referred to as **states**. A progression of chords is realized by a system that can be described at any time instance $n = 1, 2, 3, \ldots$ as being in one of the states $s_n \in \mathcal{A}$. The change from one state to another is specified according to a set of probabilities associated with each state. In general, a probabilistic description of such a system can be quite complex. To simplify the model, one often makes the assumption that the probability of a change from the current state s_n to the next state s_{n+1} only depends on the current state, and not on the events that preceded it. In terms of conditional probabilities, this property is expressed by

$$P[s_{n+1}=\alpha_j \mid s_n=\alpha_i, s_{n-1}=\alpha_k, \ldots] = P[s_{n+1}=\alpha_j \mid s_n=\alpha_i]. \tag{5.20}$$

At this point, we have used the notion of conditional probability without giving a proper definition. An introduction to the required concepts from probability theory is beyond the scope of this book, and we need to refer to a standard textbook such as [6]. Intuitively, a **conditional probability** expresses the probability of an event given that another event has occurred. If the events are denoted by E_1 and E_2, respectively, the conditional probability is commonly denoted by $P[E_1 \mid E_2]$ and called "the probability of E_1 given E_2." For example, in the case $P[s_{n+1}=\alpha_j \mid s_n=\alpha_i]$ (see (5.20)), the event E_2 expresses that the system is in state α_i at the current time point n. Furthermore, E_1 means that the system is in state α_j at the next time point $n+1$. In the following, we will use similar terms without further explanations.

Let us come back to the property expressed by the conditional probabilities of (5.20), which make the system "memoryless" with regard to events that took place before the current time point n. Besides this property, we also assume that the system is invariant under time shifts. By definition, the following coefficients become independent of the index n:

$$a_{ij} := P[s_{n+1}=\alpha_j \mid s_n=\alpha_i] \in [0,1] \tag{5.21}$$

for $i,j \in [1:I]$. These coefficients, which are also called **state transition probabilities**, obey the standard stochastic constraint

$$\sum_{j=1}^{I} a_{ij} = 1 \tag{5.22}$$

for $i \in [1:I]$. These coefficients can be expressed by an $(I \times I)$ matrix, which we denote by A. A system that satisfies these properties is also called a (discrete-time) **Markov chain**. The specific kind of "amnesia" introduced in (5.20) is called the **Markov property**.

To illustrate these definitions, let us consider Figure 5.24. In this example, the Markov chain consists of $I=3$ states α_1, α_2, and α_3, which correspond to the major chords **C**, **G**, and **F**, respectively. In the graph representation of Figure 5.24a, the states correspond to the nodes, the transitions to the edges, and the transition probabilities to the labels attached to the edges. For example, the transition probability to remain in the state $\alpha_1 = \mathbf{C}$ is $a_{11}=0.8$, whereas the transition probability of changing from $\alpha_1=\mathbf{C}$ to $\alpha_2=\mathbf{G}$ is $a_{12}=0.1$. The transition probability matrix A is shown in Figure 5.24b. Note that each row of this matrix sums up to one, which is the stochastic constraint (5.22). However, this property need not hold for the columns of A.

The model expresses the probability of all possible chord changes. To compute the probability of a given chord progression, we also need the information on how the model gets started. This information is specified by additional model param-

(a)

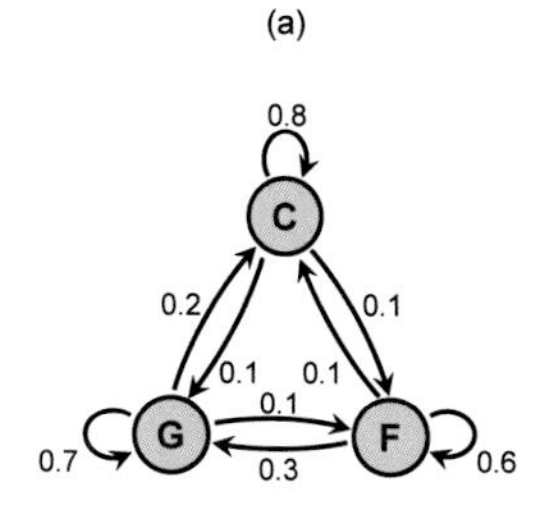

(b)

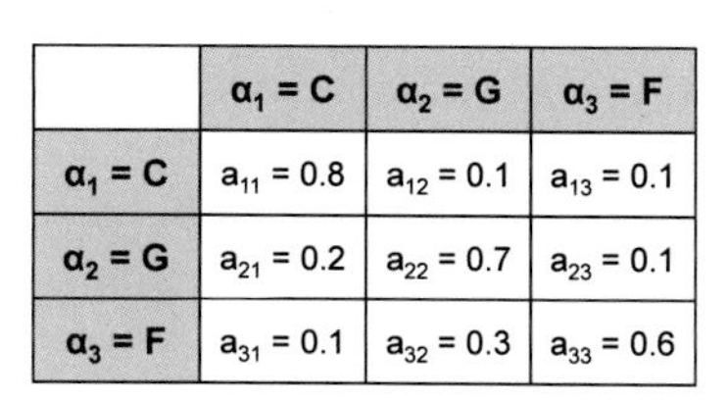

	α_1 = C	α_2 = G	α_3 = F
α_1 = C	$a_{11} = 0.8$	$a_{12} = 0.1$	$a_{13} = 0.1$
α_2 = G	$a_{21} = 0.2$	$a_{22} = 0.7$	$a_{23} = 0.1$
α_3 = F	$a_{31} = 0.1$	$a_{32} = 0.3$	$a_{33} = 0.6$

Fig. 5.24 Illustration of a Markov chain with three states. **(a)** Graph representation of the Markov chain. **(b)** Transition probability matrix.

eters referred to as **initial state probabilities**. For a general Markov chain, these probabilities are specified by the numbers

$$c_i := P[s_1 = \alpha_i] \in [0,1] \tag{5.23}$$

for $i \in [1:I]$. These coefficients, which sum up to one, can be expressed by a vector of length I denoted by C.

Continuing our example from Figure 5.24, let us assume that $C = (c_1, c_2, c_3)^\top = (0.6, 0.2, 0.2)^\top$. Furthermore, let us assume that we are given a state sequence $S = (s_1, \ldots, s_N)$, which is in our case a chord progression. For example, let us consider the sequence

$$S = (\mathbf{C}, \mathbf{C}, \mathbf{C}, \mathbf{G}, \mathbf{G}, \mathbf{F}, \mathbf{F}, \mathbf{C}, \mathbf{C}) \tag{5.24}$$

having length $N = 9$. Then, using the transition probabilities from Figure 5.24, the probability of the chord progression with regard to the given model can be computed via

$$\begin{aligned} P[S \mid \text{Model}] &= c_1 \cdot a_{11} \cdot a_{11} \cdot a_{12} \cdot a_{22} \cdot a_{23} \cdot a_{33} \cdot a_{31} \cdot a_{11} \\ &= 0.6 \cdot 0.8 \cdot 0.8 \cdot 0.1 \cdot 0.7 \cdot 0.1 \cdot 0.6 \cdot 0.1 \cdot 0.8 \\ &\approx 0.000129 = 1.29 \cdot 10^{-4}. \end{aligned} \tag{5.25}$$

This probability is quite a small number. However, rather than its *absolute* size, we will see that the size *relative* to the probabilities of other sequences becomes important. For example, the chord progression $S' = (\mathbf{F}, \mathbf{C}, \mathbf{F}, \mathbf{C}, \mathbf{F}, \mathbf{C}, \mathbf{F}, \mathbf{C}, \mathbf{F})$ has a probability of $P[S' \mid \text{Model}] = 2 \cdot 10^{-9}$, which is much smaller compared with the one in (5.25). Thus, with regard to our model, the observation S is much more likely to happen than S'.

5.3.2 *Hidden Markov Models*

Based on a Markov chain, we can compute a probability for a given observation consisting of a sequence of states or chords. In our chord recognition scenario, however, this is not what we need. Rather than observing a sequence of chords, we observe a sequence of chroma vectors that are somehow related to the chords. In other words, the state sequence is not directly visible, but only a fuzzier observation sequence that is generated based on the state sequence. Furthermore, rather than computing a probability of the observation sequence, the goal is to uncover the relation between the observed feature vectors and the underlying chords.

In the following, we extend the concept of Markov chains to a statistical model referred to as a **hidden Markov model** (HMM). The idea is to represent the relation between the observed feature vectors and the chords (the states) using a probabilistic framework. Each state is equipped with a probability function that expresses the likelihood for a given chord to output or emit a certain feature vector. As a result, we obtain a two-layered process consisting of a hidden layer and an observable layer. The hidden layer produces a state sequence that is not observable ("hidden"), but generates the observation sequence on the basis of the state-dependent probability functions.

Before we formally define this concept, we illustrate the main ideas by continuing our example from Figure 5.24. Recall that the Markov chain with its parameters (the state transition and the initial state probabilities) can be used to generate a state sequence. In the case of a hidden Markov model, this state sequence is no longer directly visible, which also explains the term "hidden." In Figure 5.25, this property is illustrated by the "cloud" covering the Markov model. Instead, each state emits certain entities, which are visible or observable to the outside world. Therefore, these entities are also referred to as **observations** of the process. The observations are emitted according to emission probabilities associated to each of the states. In our example, the observations are chroma vectors. In Figure 5.25, the space of possible observations consists of three different chroma vectors. The first observation vector is emitted with probability 0.7 by state $\alpha_1 = \mathbf{C}$, with probability 0.1 by state $\alpha_2 = \mathbf{G}$, and with probability 0 by state $\alpha_3 = \mathbf{F}$. This probability distribution reflects the fact that the first observation vector, which has most of its energy in the C, E, and G bands, is most likely produced by the state corresponding to the C-major chord model.

We now formally specify the components of a hidden Markov model. The first layer of an HMM is a Markov chain as introduced in Section 5.3.1. It is specified by a set $\mathcal{A} = \{\alpha_1, \alpha_2, \ldots, \alpha_I)$ of states (5.19), a matrix $A = (a_{ij})_{i,j\in[1:I]}$ of state transition probabilities (5.21), and a vector $C = (c_1, c_2, \ldots, c_I)^\top$ of initial state probabilities (5.23).

To define the second layer of an HMM, we need to specify a space of possible output values and a probability function for each state. In general, the output space can be any set including the real numbers, a vector space, or any kind of feature space. For example, in the case of chord recognition, this space may be modeled as

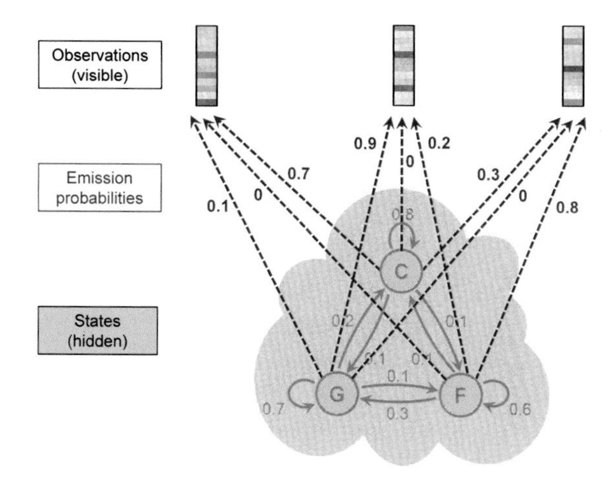

Fig. 5.25 Illustration of a hidden Markov model extending Figure 5.24. The state-dependent emission probabilities are indicated by the labels of the dashed arrows.

the feature space $\mathcal{F} = \mathbb{R}^{12}$ consisting of all possible 12-dimensional chroma vectors. For the sake of simplicity, we only want to consider the case of a **discrete HMM**, where the output space is assumed to be discrete and even finite. In this case, the space can be modeled as a finite set

$$\mathcal{B} = \{\beta_1, \beta_2, \ldots, \beta_K\} \tag{5.26}$$

of size $K \in \mathbb{N}$ consisting of distinct output elements β_k, $k \in [1:K]$. The elements β_k are also referred to as **observation symbols**.

As indicated before, an HMM associates with each state a probability function, which is also referred to as the **emission probability** or **output probability**. In the discrete case, the emission probabilities are specified by coefficients

$$b_{ik} \in [0,1] \tag{5.27}$$

for $i \in [1:I]$ and $k \in [1:K]$. Each coefficient b_{ik} expresses the probability of the system to output the observation symbol β_k when in state α_i. Similarly to the state transition probabilities, the emission probabilities are required to satisfy the stochastic constraint

$$\sum_{k=1}^{K} b_{ik} = 1 \tag{5.28}$$

for $i \in [1:I]$ (thus forming a probability distribution for each state). The coefficients can be expressed by an $(I \times K)$ matrix, which we denote by B.

In summary, an HMM is specified by a tuple

$$\Theta := (\mathcal{A}, A, C, \mathcal{B}, B) \tag{5.29}$$

(see Figure 5.26). The sets $\mathcal{A}$ and $\mathcal{B}$ are usually considered to be fixed components of the model, while the three probability measures A, C, and B are the free parame-

Component	Meaning	Reference
$\mathcal{A}$	Set of states α_i for $i \in [1:I]$	(5.19)
A	State transition probabilities a_{ij} for $i, j \in [1:I]$	(5.21)
C	Initial state probabilities c_i for $i \in [1:I]$	(5.23)
$\mathcal{B}$	Set of observation symbols β_k for $k \in [1:K]$	(5.26)
B	Emission probabilities b_{ik} for $i \in [1:I]$ and $k \in [1:K]$	(5.27)

Fig. 5.26 Overview of the components $(\mathcal{A}, A, C, \mathcal{B}, B)$ used to specify an HMM.

Algorithm: HMM-based generation of observations

Input: HMM specified by $(\mathcal{A}, A, C, \mathcal{B}, B)$
Output: Observation sequence $O = (o_1, o_2, \ldots, o_N)$

Procedure:
Compute in a loop for $n = 1, 2, \ldots, N$:

If $n = 1$: Choose an initial state s_1 according to the initial state distribution C.
If $n > 1$: Transit to the new state s_n according to the state transition probability (specified by the ith row of A when $s_{n-1} = \alpha_i$ for some $i \in [1:I]$).
Choose $o_n = \beta_k$ for some $k \in [1:K]$ according to the emission probabilities (specified by the jth row of B when $s_n = \alpha_j$ for some $j \in [1:I]$).

Table 5.1 Algorithm for generating a sequence of observations of length N based on an HMM.

ters to be determined. This can be done explicitly by an expert based on his or her musical knowledge or by employing a learning procedure based on suitably labeled training data.

Once an HMM has been specified by $(\mathcal{A}, A, C, \mathcal{B}, B)$, it can be used for various analysis and synthesis applications. In Section 5.3.3.1, we will see how an HMM can be used for explaining a given sequence of observations. Since it is very instructive, we now discuss how to (artificially) generate an observation sequence

$$O := (o_1, o_2, \ldots, o_N) \tag{5.30}$$

of length $N \in \mathbb{N}$ on the basis of a given HMM. As said before, the elements $o_n \in \mathcal{B}$ are called **observations**. The basic idea is to choose a first state according to the initial state distribution, to generate an observation according to the emission probabilities, to move on to a next state according to the transition probabilities, and to iterate this process (see Table 5.1 for a more detailed description).

In our chord recognition scenario, we encounter the inverse problem. Instead of generating an observation sequence, we are given an observation sequence in the form of a sequence of chroma vectors extracted from an audio recording. Then our goal is to associate with each frame a chord label that best "explains" the corresponding chroma vector. Based on a suitable HMM, where the states correspond to the considered chords, this problem can be viewed as finding or "uncovering" the hidden state sequence that generates the given observation sequence with the highest probability. In the following section, we introduce several problems for HMMs that become important for real-world applications and present an algorithm for solving the specific problem at hand.

5.3.3 *Evaluation and Model Specification*

Following the tutorial by Rabiner [25], we now discuss some general problems for HMMs that concern, on the one hand, the specification of the free model parameters, and on the other hand, the evaluation of observation sequences.

5.3.3.1 Evaluation Problem

In Section 5.3.1, we discussed how to compute the probability of a state sequence with regard to a given Markov chain. We now consider the same kind of **evaluation problem** for the case of HMMs. In this case, we are given an HMM specified by $\Theta = (\mathcal{A}, A, C, \mathcal{B}, B)$ as in (5.29) and an observation sequence $O = (o_1, o_2, \ldots, o_N)$. The task is to compute the probability

$$P[O \mid \Theta] \tag{5.31}$$

of the observation sequence given the model. From a slightly different viewpoint, this probability can be regarded as a score value that expresses how well a given model matches a given observation sequence. This interpretation becomes useful in the case where one is trying to choose among several competing models. The solution would then be to choose the model which best matches the observation sequence.

A typical application scenario is the task of **isolated word recognition**. For example, let us assume that we want to build a system for recognizing the words "zero," "one," "two," "three," and so on for the ten digits. A typical procedure is to build a separate HMM for each of the words. Then, given a speech signal of a word, the signal is transformed into a suitable feature representation used as the observation sequence. Each of the ten HMMs is evaluated on this observation sequence, yielding a probability value. Finally, the speech signal is classified according to the word represented by the probability-maximizing HMM.

Computing the overall probability for an HMM to output a given observation sequence is not as straightforward as in the case of a Markov chain. The reason is that the hidden state sequence is not known and that every possible state sequence may contribute to the overall probability. To be more precise, let $O = (o_1, o_2, \ldots, o_N)$ be an observation sequence with $o_n = \beta_{k_n} \in \mathcal{B}$ for some suitable $k_n \in [1:K]$, $n \in [1:N]$. Furthermore, let $S = (s_1, s_2, \ldots, s_N)$ be a fixed state sequence with $s_n = \alpha_{i_n} \in \mathcal{A}$ for some suitable $i_n \in [1:I]$, $n \in [1:N]$. Then the probability $P[O, S \mid \Theta]$ for generating the state sequence S as well as the observation sequence O can be obtained as follows: First, the HMM needs to jump into the state α_{i_1} (with probability c_{i_1}), then it emits o_1 (with probability $b_{i_1k_1}$), moves on to state α_{i_2} (with probability $a_{i_1i_2}$), and so on. This yields

$$P[O,S \mid \Theta] = c_{i_1} \cdot b_{i_1k_1} \cdot a_{i_1i_2} \cdot b_{i_2k_2} \cdot \ldots \cdot a_{i_{N-1}i_N} \cdot b_{i_Nk_N}. \tag{5.32}$$

Next, to obtain the overall probability $P[O \mid \Theta]$, one needs to consider all possible state sequences of length N:

$$P[O \mid \Theta] = \sum_{S=(s_1,s_2,\ldots,s_N)} P[O,S \mid \Theta] \tag{5.33}$$

$$= \sum_{i_1=1}^{I} \sum_{i_2=1}^{I} \ldots \sum_{i_N=1}^{I} c_{i_1} \cdot b_{i_1k_1} \cdot a_{i_1i_2} \cdot b_{i_2k_2} \cdot \ldots \cdot a_{i_{N-1}i_N} \cdot b_{i_Nk_N}. \tag{5.34}$$

This leads to I^N summands, a number that is exponential in the length N of the observation sequence. Therefore, in practice, this calculation is computationally infeasible even for a small N. The good news is that there is a more efficient way to compute $P[O \mid \Theta]$ using an algorithm that is based on the dynamic programming paradigm. This procedure requires a number of operations on the order of I^2N (instead of I^N). Since we do not require this procedure in our chord recognition scenario, we do not discuss it in this book and refer to [25] for a detailed description.

5.3.3.2 Uncovering Problem

Next, we discuss the so-called **uncovering problem**, which is the problem relevant for our chord recognition scenario. Again, we are given an HMM specified by $\Theta = (\mathcal{A}, A, C, \mathcal{B}, B)$ and an observation sequence $O = (o_1, o_2, \ldots, o_N)$. Instead of finding the overall probability for O, where one needs to consider *all* possible state sequences, the goal of the uncovering problem is to find the *single* state sequence $S = (s_1, \ldots, s_N)$ that "best explains" the observation sequence. The uncovering problem stated so far is not well defined since—in general—there is not a single "correct" state sequence generating the observation sequence. Indeed, one needs a kind of optimization criterion that specifies what is meant when talking about a best possible explanation. There are several reasonable choices for such a criterion, and the actual choice will depend on the intended application.

One possible optimization criterion is to choose the state sequence S^* that yields the highest probability Prob^* when evaluated against the observation sequence O:

$$\mathrm{Prob}^* = \max_{S=(s_1,s_2,\ldots,s_N)} P[O,S \mid \Theta], \tag{5.35}$$

$$S^* = \operatorname*{argmax}_{S=(s_1,s_2,\ldots,s_N)} P[O,S \mid \Theta]. \tag{5.36}$$

Note that the probability $P[O,S \mid \Theta]$ is one of the summands contributing to the overall probability $P[O \mid \Theta]$ in (5.33). To find the sequence S^* using the naive approach, one would have to compute the probability value $P[O,S \mid \Theta]$ for each of the I^N possible state sequences of length N and then look for the maximizing argument. Fortunately, there is a technique known as the **Viterbi algorithm** for finding the optimizing state sequence in a much more efficient way.

The Viterbi algorithm, which is similar to the DTW algorithm (Section 3.2), is based on **dynamic programming** (Section 3.2.1.3). The idea is to recursively compute an optimal (i.e., probability-maximizing) state sequence from optimal solutions for subproblems, where one considers truncated versions of the observation sequence. Let $O = (o_1, o_2, \ldots, o_N)$ be the observation sequence. As in the case of DTW (see (3.22)) we define the prefix $O(1:n) := (o_1, \ldots, o_n)$ of length $n \in [1:N]$ and set

$$\mathbf{D}(i,n) := \max_{(s_1,\ldots,s_n)} P[O(1:n),(s_1,\ldots,s_{n-1},s_n = \alpha_i) \mid \Theta] \tag{5.37}$$

for $i \in [1:I]$. In other words, $\mathbf{D}(i,n)$ is the highest probability along a single state sequence $(s_1,\ldots,s_n)$ that accounts for the first n observations and ends in state $s_n = \alpha_i$. The state sequence yielding the maximal value

$$\mathrm{Prob}^* = \max_{i \in [1:I]} \mathbf{D}(i,N) \tag{5.38}$$

is the solution to our uncovering problem.

The $(I \times N)$ matrix $\mathbf{D}$ can be computed recursively along the column index $n \in [1:N]$ (see Figure 5.27). For the case $n = 1$, the prefix observation sequence consists of the single element $o_1 = \beta_{k_1}$. Therefore, the value $\mathbf{D}(i,1)$ for some $i \in [1:I]$ is the probability to start (and end) with state α_i and to emit the element $o_1 = \beta_{k_1}$:

$$\mathbf{D}(i,1) = c_i b_{ik_1}. \tag{5.39}$$

This constitutes the initialization of our recursion. Now, let us assume that we want to compute $\mathbf{D}(i,n)$ for some $n \in [2:N]$ and $i \in [1:I]$. Let $(s_1,\ldots,s_{n-1},s_n)$ be the optimal state sequence yielding $\mathbf{D}(i,n)$. Then, by the definition in (5.37), we have $s_n = \alpha_i$ and the observation $o_n = \beta_{k_n}$ is emitted with probability b_{ik_n}. Next, let us look at the truncated sequence $(s_1,\ldots,s_{n-1})$ and suppose that it ends with the state $s_{n-1} = \alpha_{j^*}$ for some $j^* \in [1:I]$. Note that the probability of changing from the state α_{j^*} at time $n-1$ to state α_i at time n is specified by a_{j^*i}. Therefore, we obtain

$$\mathbf{D}(i,n) = b_{ik_n} \cdot a_{j^*i} \cdot P[O(1:n-1),(s_1,\ldots,s_{n-1} = \alpha_{j^*}) \mid \Theta]. \tag{5.40}$$

Furthermore, note that the truncated sequence must be optimal for $\mathbf{D}(j^*,n-1)$ (otherwise, the sequence $(s_1,\ldots,s_{n-1},s_n)$ would not be optimal for $\mathbf{D}(i,n)$). Therefore, we obtain

$$\mathbf{D}(j^*,n-1) = P[O(1:n-1),(s_1,\ldots,s_{n-1} = \alpha_{j^*}) \mid \Theta]. \tag{5.41}$$

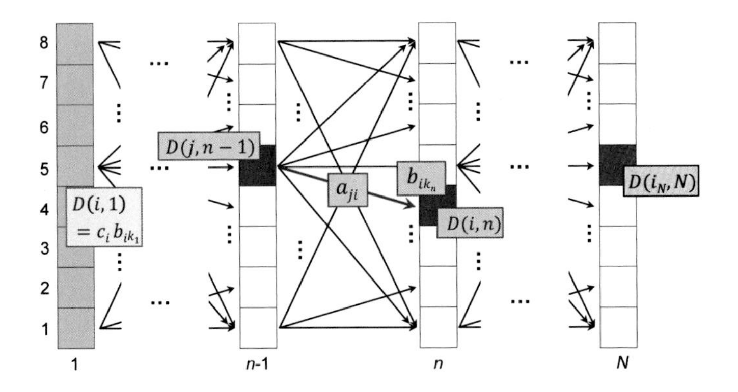

Fig. 5.27 Illustration of the Viterbi algorithm. The blue cells indicate the entries $\mathbf{D}(i,1)$ initialized by (5.39). The red cells illustrate the computation of (5.42). The black cell indicates the index computed in (5.43).

In the computation, we do not know the truncated sequence $(s_1,\ldots,s_{n-1})$, neither do we know the index $j^* \in [1:I]$. However, it is not hard to see that the index j^* must be the index that maximizes the product $\mathbf{D}(j,n-1)\cdot a_{ji}$ (otherwise $(s_1,\ldots,s_{n-1},s_n)$ would not be optimal). Using (5.40) and (5.41), this implies the following recursion:

$$\mathbf{D}(i,n) = b_{ik_n} \cdot \max_{j\in[1:I]} \big(a_{ji}\cdot\mathbf{D}(j,n-1)\big). \tag{5.42}$$

From the $(I\times N)$ matrix $\mathbf{D}$, we obtain the maximal probability via (5.38). However, we do not yet know the optimal state sequence. Again, as with finding the optimal warping path in the case of DTW, we need to apply a **backtracking** procedure that constructs the optimal state sequence in reverse order. Let $S^* = (\alpha_{i_1},\ldots,\alpha_{i_N})$ denote the optimal state sequence to be constructed. Then, by (5.38), the last element α_{i_N} is determined by

$$i_N := \operatorname*{argmax}_{j\in[1:I]} \mathbf{D}(j,N). \tag{5.43}$$

Furthermore, the element α_{i_n} for $n = N-1, N-2,\ldots,1$ is determined by the maximizing argument in the recursion (5.42):

$$i_n = \operatorname*{argmax}_{j\in[1:I]} \big(a_{ji_{n+1}}\cdot\mathbf{D}(j,n)\big). \tag{5.44}$$

Table 5.2 summarizes the entire procedure for computing the maximizing probability and the optimal state sequence. As indicated in Table 5.2, the backtracking can be simplified by introducing an $(I\times(N-1))$ matrix $\mathbf{E}$ that keeps track of the maximizing argument in (5.42). Note that the 'argmax' in the backtracking may not be unique, thus opening up the possibility of having more than one optimal state se-

Algorithm: VITERBI

Input: HMM specified by $\Theta = (\mathcal{A}, A, C, \mathcal{B}, B)$
Observation sequence $O = (o_1 = \beta_{k_1}, o_2 = \beta_{k_2}, \ldots, o_N = \beta_{k_N})$

Output: Optimal state sequence $S^* = (s_1^*, s_2^*, \ldots, s_N^*)$

Procedure: Initialize the $(I \times N)$ matrix $\mathbf{D}$ by $\mathbf{D}(i,1) = c_i b_{ik_1}$ for $i \in [1:I]$. Then compute in a nested loop for $n = 2, \ldots, N$ and $i = 1, \ldots, I$:

$$\begin{aligned}\mathbf{D}(i,n) &= \max_{j\in[1:I]} \left(a_{ji} \cdot \mathbf{D}(j,n-1)\right) \cdot b_{ik_n}\\ \mathbf{E}(i,n-1) &= \operatorname{argmax}_{j\in[1:I]} \left(a_{ji} \cdot \mathbf{D}(j,n-1)\right)\end{aligned}$$

Set $i_N = \operatorname{argmax}_{j\in[1:I]} \mathbf{D}(j,N)$ and compute for decreasing $n = N-1, \ldots, 1$ the maximizing indices

$$i_n = \operatorname{argmax}_{j\in[1:I]} \left(a_{ji_{n+1}} \cdot \mathbf{D}(j,n)\right) = \mathbf{E}(i_{n+1}, n).$$

The optimal state sequence $S^* = (s_1^*, \ldots, s_N^*)$ is defined by $s_n^* = \alpha_{i_n}$ for $n \in [1:N]$.

Table 5.2 Viterbi algorithm based on dynamic programming.

quence. To obtain a uniquely determined sequence, one may take, for example, the lexicographically smallest index in case 'argmax' is not unique.

Continuing our example from Figure 5.25, we illustrate the principle of the Viterbi algorithm. The HMM used in this example is specified by Figure 5.28a. In particular, this model consists of $I = 3$ states $\mathcal{A} := \{\alpha_1, \alpha_2, \alpha_3\}$ and $K = 3$ observation symbols $\mathcal{B} = \{\beta_1, \beta_2, \beta_3\}$. Let us consider the observation sequence

$$O = (o_1, o_2, \ldots, o_6) = (\beta_1, \beta_3, \beta_1, \beta_3, \beta_3, \beta_2) \tag{5.45}$$

of length $N = 6$. The matrices $\mathbf{D}$ and $\mathbf{E}$, which are computed as in Table 5.2, are shown in Figure 5.28b. For example, one obtains $\mathbf{D}(1,1) = c_1 b_{11} = 0.6 \cdot 0.7 = 0.42$. As another example, the entry $\mathbf{D}(1,2)$ is obtained by $\mathbf{D}(1,2) = a_{11} \cdot \mathbf{D}(1,1) \cdot b_{12} = 0.8 \cdot 0.42 \cdot 0.3 = 0.1008$, where $\mathbf{E}(1,1) = i_1 = 1$ is the maximizing argument. The probability of the optimal state sequence is $\text{Prob}^* = 0.0006$ (rounded up to four decimal points) with $i_6 = 2$ being the maximizing argument. Starting with this index, backtracking yields the index sequence $(1,1,1,3,3,2)$, which is highlighted by the rectangles in the matrix $\mathbf{E}$ shown in Figure 5.28b. This corresponds to the optimal state sequence

$$S^* = (\alpha_1, \alpha_1, \alpha_1, \alpha_3, \alpha_3, \alpha_2). \tag{5.46}$$

This example also illustrates the kind of context-sensitive smoothing introduced by the HMM. Without regarding the previous and subsequent observations, the second observation $o_2 = \beta_3$ is best explained by the state α_3 (with an emission probability of $b_{33} = 0.8$). However, being preceded by $o_1 = \beta_1$ as well as succeeded by $o_3 = \beta_1$, the HMM-based procedure favors the state α_1 to explain $o_2 = \beta_3$ (even though the emission probability is only $b_{13} = 0.3$). In other words, it is beneficial to stay in state α_1 for the first three time instances rather than to switch back and forth between state

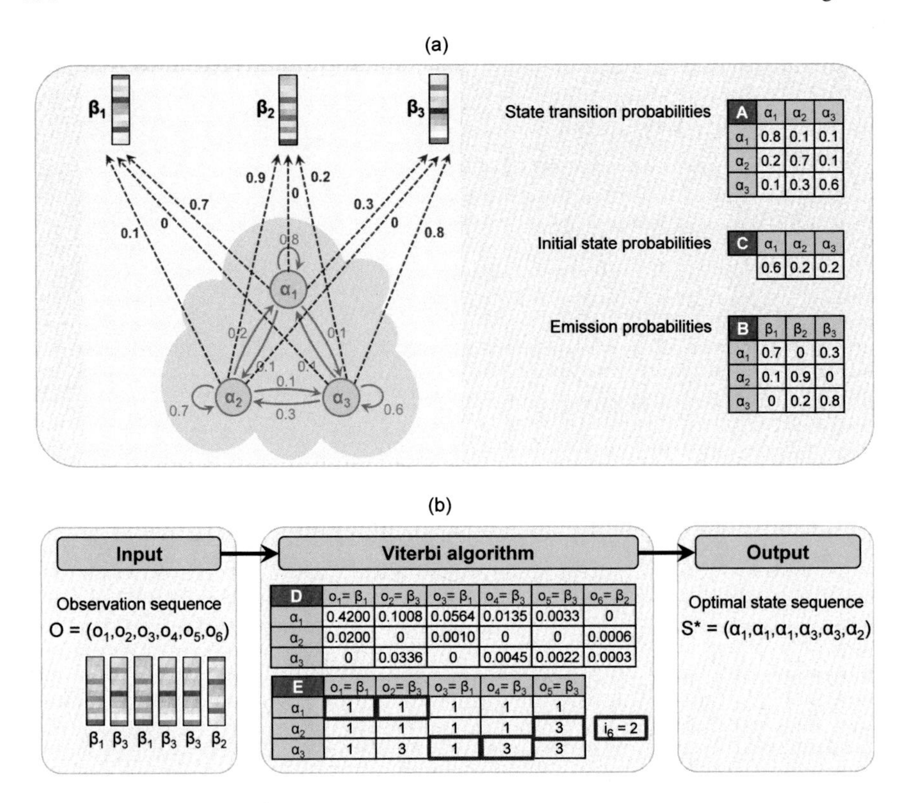

Fig. 5.28 Illustration of the uncovering problem. **(a)** Specification of an HMM. **(b)** Application of the Viterbi algorithm.

α_1 and α_3. The HMM trades off the loss in emission probability against the gain in transition probability, thus reducing the number of transitions between different states (see Exercise 5.10).

Finally, let us have a look at the overall computational complexity of the Viterbi algorithm. For the memory requirements, one basically needs to store the $(I \times N)$ matrix $\mathbf{D}$ along with other data including the HMM parameters (requiring $O(I^2)$ numbers) and the observation sequence (requiring $O(N)$ numbers). The number of operations in the Viterbi algorithm is dominated by the recursion (5.42). In this recursion, one needs to maximize a product over $j \in [1:I]$ for each $n \in [2:N]$ and each $i \in [1:I]$. This requires $O(I^2N)$ operations. The number of remaining operations has an order smaller than $O(I^2N)$. This shows that the overall complexity of the Viterbi algorithm is $O(I^2N)$, which is much better than $O(I^N)$ required for the naive approach.

5.3.3.3 Estimation Problem

Besides the evaluation and uncovering problems, the third basic problem for HMMs is referred to as the **estimation problem**. Given an observation sequence O, the objective is to determine the three probability measures specified by A, C, and B of Θ that maximize the probability $P[O \mid \Theta]$. In other words, the free model parameters are to be estimated so as to best describe the observation sequence. This is a typical instance of an optimization problem where the observation sequence serves as "training material" for adjusting the HMM. As stated by Rabiner [25], the estimation problem is by far the most difficult problem of HMMs. In fact, there is no known way to explicitly solve the given optimization problem. However, iterative procedures that find locally optimal solutions have been suggested. One of these procedures is known as the **Baum–Welch algorithm**. In the following, we give an overview of this procedure. For the technical details and proper definitions of the involved variables we refer to [9, 25].

The Baum–Welch algorithm is a reestimation procedure, which iteratively updates and improves the free model parameters A, C, and B. In order to describe this procedure, we need to define some variables that depend on a given observation sequence O and a given HMM specified by Θ. First, let

$$\gamma_n(i) = P[s_n = \alpha_i \mid O, \Theta] \tag{5.47}$$

be the probability of being in state α_i, $i \in [1:I]$, at a given time instance $n \in [1:N]$. Similarly, we define

$$\gamma_n(i,j) = P[s_n = \alpha_i, s_{n+1} = \alpha_j \mid O, \Theta] \tag{5.48}$$

to be the probability of being in state α_i at time n and in state α_j at time $n+1$ for $i,j \in [1:I]$ and $n \in [1:N-1]$. The two variables are related by

$$\gamma_n(i) = \sum_{j=1}^{I} \gamma_n(i,j). \tag{5.49}$$

Summing $\gamma_n(i)$ over the time index n, we obtain a quantity that can be interpreted as the expected number of times a state α_i has been visited. Excluding the last time index $n = N$ from the summation, the resulting quantity can also be interpreted as the number of transitions that start from state α_i and end in an arbitrary state α_j for $i \in [1:I]$:

$$\sum_{n=1}^{N-1} \gamma_n(i) = \text{expected number of transitions from } \alpha_i. \tag{5.50}$$

Similarly, summation of $\gamma_n(i,j)$ over the time index n yields a quantity that expresses the expected number of transitions from α_i to α_j:

$$\sum_{n=1}^{N-1} \gamma_n(i,j) = \text{expected number of transitions from } \alpha_i \text{ to } \alpha_j. \tag{5.51}$$

The computation of the described variables is not trivial. The good news is that there exist efficient algorithms based on dynamic programming, which we have already mentioned in Section 5.3.3.1. For a description of these procedures, we refer to the literature [9, 25].

In the Baum–Welch algorithm, one starts with an HMM specified by parameters $\Theta = (\mathcal{A}, A, C, \mathcal{B}, B)$ and an observation sequence $O = (o_1, o_2, \ldots, o_N)$. From this, one computes

$$\begin{aligned}\hat{c}_i &= \text{expected number of transitions from } \alpha_i \text{ at time } (n=1)\\ &= \gamma_1(i)\end{aligned} \tag{5.52}$$

$$\begin{aligned}\hat{a}_{ij} &= \frac{\text{expected number of transitions from } \alpha_i \text{ to } \alpha_j}{\text{expected number of transitions from } \alpha_i}\\ &= \frac{\sum_{n=1}^{N-1} \gamma_n(i,j)}{\sum_{n=1}^{N-1} \gamma_n(i)}\end{aligned} \tag{5.53}$$

$$\begin{aligned}\hat{b}_{ik} &= \frac{\text{expected number of transitions from } \alpha_i \text{ and observing } \beta_k}{\text{expected number of times in state } \alpha_i}\\ &= \frac{\sum_{n=1}^{N} \gamma_n(i) \cdot \delta_{o_n,\beta_k}}{\sum_{n=1}^{N} \gamma_n(i)}\end{aligned} \tag{5.54}$$

for $i, j \in [1:I]$ and $k \in [1:K]$. The function δ_{o_n,β_k}, which is also referred to as the Kronecker delta, assumes a value 1 if $o_n = \beta_k$ and a value 0 otherwise. These numbers define a transition probability matrix $\hat{A} = (\hat{a}_{ij})_{i,j\in[1:I]}$, an initial state probability vector $\hat{C} = (\hat{c}_1, \ldots, \hat{c}_I)^\top$, and an emission probability matrix $\hat{B} = (\hat{b}_{ik})_{i\in[1:I],k\in[1:K]}$, thus yielding a new HMM specified by $\hat{\Theta} = (\mathcal{A}, \hat{A}, \hat{C}, \mathcal{B}, \hat{B})$. One central result in the theory of HMMs, which we state here without proof, is that

$$P[O \mid \hat{\Theta}] \geq P[O \mid \Theta]. \tag{5.55}$$

In other words, replacing the free parameters of the original HMM by the estimated parameters $(\hat{A}, \hat{C}, \hat{B})$ is either insignificant, or increases the probability of the model outputting the given observation sequence. In the case $P[O \mid \hat{\Theta}] = P[O \mid \Theta]$, the model Θ is called a **critical point**. In the case $P[O \mid \hat{\Theta}] > P[O \mid \Theta]$, one says that the model $\hat{\Theta}$ is more likely than the model Θ with regard to the observation O.

Based on this mathematical fact, one can iteratively use $\hat{\Theta}$ in place of Θ and repeat the reestimation calculation defined by (5.52), (5.53), and (5.54). This procedure is guaranteed to converge to a critical point. The final result of this reestimation procedure is called a **maximum-likelihood** estimate of the parameters of the HMM. It is important to note that, in general, this critical point only yields a local maximum of the likelihood function, which assigns to each parameter setting (A, C, B) the value $P[O \mid \Theta]$. In practice, this local maximum may be far from the global maximum. Furthermore, the final estimate delivered by the Baum–Welch algorithm crucially depends on the initialization of the HMM to start with. For further details, proofs, and links to the literature, we refer to [9, 25].

5.3.4 Application to Chord Recognition

We now show how the concept of HMMs can be applied to our chord recognition scenario. In particular, we have a look at the transition model, which introduces a kind of context-aware postfiltering. This leads, as we will see in Section 5.3.4.3, to substantial improvements in the recognition results beyond the simple prefiltering approach discussed in Section 5.2.4.4.

First of all, we need to create an HMM that suitably models our chord recognition problem. Recall from (5.29) that an HMM is specified by a tuple $\Theta := (\mathcal{A}, A, C, \mathcal{B}, B)$. The states of the HMM are used to model the various chords that are allowed in the recognition problem. In the following, we consider the twelve major and minor triads as in (5.5), thus setting

$$\mathcal{A} = (\alpha_1, \ldots, \alpha_I) := \{\mathbf{C}, \mathbf{C}^\sharp, \ldots, \mathbf{B}, \mathbf{Cm}, \mathbf{C}^\sharp\mathbf{m}, \ldots, \mathbf{Bm}\}. \tag{5.56}$$

In this case, the HMM consists of $I = 24$ states, which we enumerate as indicated by (5.56). For example, α_1 corresponds to **C** and α_{13} to **Cm**.

5.3.4.1 Specification of Emission Probabilities

In our chord recognition scenario, the observations are chroma vectors that have previously been extracted from the given audio recording. In other words, the observations are 12-dimensional real-valued vectors which are elements of the continuous feature space $\mathcal{F} = \mathbb{R}^{12}$. So far, we have only considered the case of **discrete HMMs**, where the observations are discrete symbols coming from a finite output space $\mathcal{B}$ (see (5.26)). To make discrete HMMs applicable to our scenario, one possible procedure is to introduce a finite set of prototype vectors, a so-called **codebook**. Such a codebook can be regarded as a discretization of the continuous feature space $\mathcal{F} = \mathbb{R}^{12}$, where each prototype vector represents an entire range of feature vectors. The process of mapping an arbitrary feature vector to one of the prototype vectors is also referred to as **quantization**.

There are many ways for determining a codebook and for performing the quantization. Many approaches are based on **clustering** techniques, where a set of given vectors is grouped in such a way that vectors in the same group (called a cluster) are more similar to each other than to those in other clusters. The prototype vectors of the codebook are then defined as, e.g., the centroids of the resulting clusters. Furthermore, in the quantization, an arbitrary vector is mapped to the most similar (or closest) prototype vector.

Finding a suitable codebook as well as a suitable quantization function is in general not an easy problem, and there are many design choices to be made. Furthermore, the quantization may introduce serious degradations, in particular for vectors that are not represented well by the codebook. An alternative is to use **continuous HMMs**, where the finite output space $\mathcal{B}$ is replaced by a continuous output space. In this case, the emission probabilities specified by the matrix B are to be replaced by continuous **probability density functions** (PDFs). Also the reestimation procedure needs to be modified by considering values of the density functions instead of discrete probabilities. The PDFs used in practice are typically based on Gaussian functions or mixtures thereof. Such PDFs can be compactly described by a few parameters (means, variances, mixture coefficients). Similar to the procedure described in Section 5.3.3.3, one can derive an iterative estimation procedure that adjusts the PDF parameters (instead of the emission probabilities as in the case of discrete HMMs) on the basis of given training sequences. For details, which go beyond the scope of this book, we refer to [9, 25].

5.3.4.2 Specification of Transition Probabilities

In music, certain chord transitions are more likely than others. This observation was our main motivation for introducing HMMs, where the first-order temporal relationships between the various chords are captured by the transition probability matrix A. We introduce the notation $\alpha_i \rightarrow \alpha_j$ referring to the transition from state α_i to state α_j, $i,j \in [1:I]$. For example, the coefficient $a_{1,2}$ expresses the probability for the transition $\alpha_1 \rightarrow \alpha_2$ (corresponding to $\mathbf{C} \rightarrow \mathbf{C}^\sharp$), whereas $a_{1,8}$ expresses the probability for $\alpha_1 \rightarrow \alpha_8$ (corresponding to $\mathbf{C} \rightarrow \mathbf{G}$). In real music, the change from a tonic to the dominant is much more likely than transposing by one semitone, so that the probability $a_{1,8}$ should be much larger than $a_{1,2}$. The coefficients a_{ii} express the probability of staying in state α_i, i.e., $\alpha_i \rightarrow \alpha_i$, $i \in [1:I]$. These coefficients are also referred to as **self-transition** probabilities.

A transition probability matrix can be specified in many ways. For example, the matrix may be defined manually by a music expert based on rules from harmony theory. The most common approach is to generate such a matrix automatically by estimating the transition probabilities from labeled data. In the following, we describe such an automated approach.

To this end, one needs suitably labeled training data. For example, one may use the annotated Beatles dataset described in Section 5.2.4.2. In the following, we assume that each audio recording of the training dataset is represented by a sequence of frames and that each frame is labeled with one of the states (the 24 major and minor triads (see (5.56))). Next, we count how often each of the 24×24 possible chord transitions occurs in the training data. To this end, we consider adjacent elements, so called **bigrams**, in the labeled frame sequences. Let $\mu(i,j)$ be the number of transitions $\alpha_i \rightarrow \alpha_j$, $i,j \in [1:I]$. In other words, $\mu(i,j)$ is the number of bigrams in the training data, where the first frame is labeled by α_i and the next frame by α_j. Finally, we define the transition probability matrix $A = (a_{ij})_{i,j\in[1:I]}$ by setting

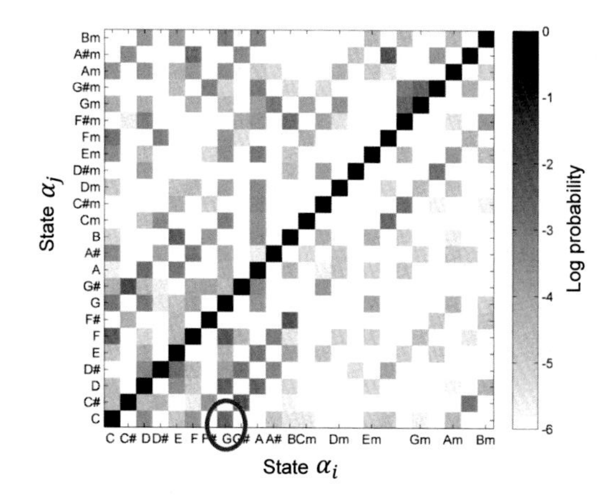

Fig. 5.29 Estimated transition probability matrix (using a log probability scale for visualization purposes). The matrix was computed based on (5.57) using the Beatles dataset from Section 5.2.4.2. As an example, the coefficient $a_{1,8}$ (corresponding to the transition $\mathbf{C} \rightarrow \mathbf{G}$) has been highlighted.

$$a_{ij} = \frac{\mu(i,j)}{\sum_{k \in [1:I]} \mu(i,k)}. \tag{5.57}$$

The numerator of this fraction is the total number of transitions $\alpha_i \rightarrow \alpha_j$, whereas the denominator counts the total number of transitions starting at α_i and going to an arbitrary state. Note that this definition corresponds to the update rule (5.53).

As an illustration, let us have a look at the transition probability matrix of Figure 5.29, which has been computed using the Beatles dataset described in Section 5.2.4.2 (disregarding the frames annotated by **N**). The feature rate is 10 Hz as in the previous experiments of this chapter. For the sake of visibility, Figure 5.29 shows the log probability values $\log(a_{ij})$. First of all, note that the matrix is dominated by the coefficients on the main diagonal, which correspond to self-transitions. Why is this the case? Most of the chord durations occurring in the training data are much longer (often around one second) than the frame length in use (100 ms corresponding to a feature rate of 10 Hz). As a result, the chord remains stable for several subsequent frames, thus making the self-transition probabilities much higher than the probabilities of moving to a different chord.

Next, one can observe in Figure 5.29 that certain transition probabilities such as $a_{1,8}$ are much higher than others. This reflects the fact that certain transitions such as $\mathbf{C} \rightarrow \mathbf{G}$ occur more often than others. This observation is not surprising, since this transition expresses the musically important relation between **C** and its dominant **G**. More generally, transitions such as $\mathbf{I} \rightarrow \mathbf{V}$ lead to high transition probabilities between major chords that differ by five or seven semitones, respectively. This explains some of the secondary diagonals visible in Figure 5.29.

In general, structures of secondary diagonals express the fact that chord transition probabilities depend on functional relations, which are independent of the underlying musical key (Section 5.1.2.5), between chords. For example, the probability of moving from the tonic to the dominant is independent of playing a piece of music in the C major scale or in the C$^\sharp$ major scale. The transition probability matrix shown in Figure 5.29, however, is more irregular than expected. The reason is that

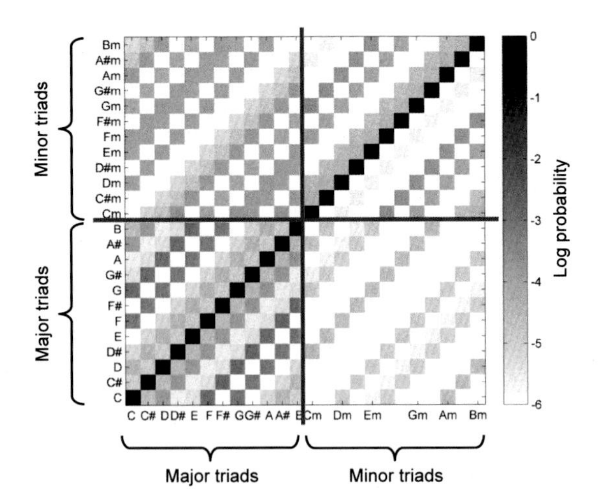

Fig. 5.30 Transposition-invariant transition probability matrix (using a log probability scale for visualization purposes) starting with the matrix from Figure 5.29.

the training set used is not balanced with regard to the various existing musical keys. For example, the Beatles dataset may contain many songs in the C major or A minor scales, but rarely a song in the D$^\flat$ major or B$^\flat$ minor scales. As a result, there is not sufficient training data for obtaining robust estimates of certain transition probabilities.

To make the training set more balanced and invariant to the underlying musical key, one can apply a similar trick as in Section 5.2.4.2, where we cyclically shifted the chroma features. This time, we apply a cyclic chroma shift to the considered bigrams so that the first chord of each bigram corresponds either to **C** or to **Cm**. In this case, we say that the bigrams have been normalized with respect to the chroma C. For example, after normalization, the transitions $\mathbf{C}^\sharp \rightarrow \mathbf{G}^\sharp$ or $\mathbf{D} \rightarrow \mathbf{A}$ become $\mathbf{C} \rightarrow \mathbf{G}$. Afterwards, one takes twelve versions of each of the resulting C-normalized bigrams. These twelve versions are then normalized with respect to the twelve different chroma values. As a result, the size of the training data used for deriving the matrix A has been increased by a factor of twelve. Furthermore, the statistics on the number of transitions have become invariant under transpositions (cyclic chroma shifts). Therefore, we call the resulting matrix a **transposition-invariant** transition probability matrix. Figure 5.30 shows such a matrix when applying this construction to the matrix shown in Figure 5.29.

5.3.4.3 Effect of HMM-Based Postfiltering

We now demonstrate the effect of applying HMMs to our chord recognition scenario and continue the discussion of the Beatles experiments in Section 5.2.4. As described in Section 5.2.4.2, we use a 3-fold cross-validation on the basis of 180 Beatles songs. The free parameters of the HMM-based and template-based chord recognizers are either learned automatically from the training set or set manually using musical knowledge. In either case, the validation set is kept independent of the free parameters of the respective chord recognizer model. In all experiments, we use a feature rate of 10 Hz.

In Figure 5.22, we have studied the dependency of the chord recognition accuracy on the underlying chroma type. In particular, we have seen that the accuracy crucially depends on the compression parameter. The same figure shows that the HMM-based chord recognizer clearly outperforms both considered template-based chord recognizers—independently of the feature type used. For example, using the feature type $\mathcal{C}$, the template-based chord recognizer using binary templates achieves an F-measure of $\mathrm{F} = 0.46$, while the HMM-based approach yields $\mathrm{F} = 0.53$. The improvements become even more apparent when considering logarithmically compressed features. For example, for the best performing feature type $\mathcal{C}_{1000}$, the template-based chord recognizer using average templates achieves an F-measure of $\mathrm{F} = 0.61$, while the HMM-based approach yields $\mathrm{F} = 0.74$.

The improvements in the HMM-based approach come specifically from the transition model that introduces context-sensitive smoothing (see also the example in (5.46) of Section 5.3.3.2). In the case of high self-transition probabilities, a chord recognizer tends to stay in the current chord rather than change to another one. As a result, many of the noise-like outliers and ambiguous chroma vectors are skipped, which can be regarded as a kind of smoothing.

This effect is also demonstrated in Figure 5.31. The broken chords cause many chord ambiguities of short duration. This leads to many random-like chord changes when using a simple template-based chord recognizer (see also Figure 5.31b). Using an HMM-based approach, chord changes are only performed when the relatively low transition probabilities are compensated by a substantial increase of emission probabilities. Consequently, only the dominant chord changes remain (see Figure 5.31d).

We have seen in Section 5.2.4.4 that a similar smoothing effect can be achieved by applying a prefiltering step. This effect is again demonstrated by Figure 5.31c. However, there are some crucial differences. In prefiltering, which is performed prior to the pattern matching step, a smoothing filter is applied to the feature representation. This not only smooths out noise-like frames, but also washes out characteristic chroma information and blurs transitions. As opposed to prefiltering, the HMM-based approach leaves the feature representation untouched. Furthermore, the smoothing is performed in combination with the pattern matching step. For this reason, we also call this approach **postfiltering**. As a result, the original chroma information is preserved and transitions are kept sharp. The resulting effect on the recognition accuracy is also illustrated by comparing Figure 5.31d with Figure 5.31c, where the prefiltering introduces recognition errors, in particular at the chord boundaries.

The effect of prefiltering and postfiltering is also demonstrated by Figure 5.23, where we have studied the dependency of the chord recognition accuracy on the smoothing length used in the prefiltering. In the HMM-based approaches, the accuracy is always higher than the one in the template-based approaches—even when using the best performing prefiltering strategy. The figure also demonstrates that the

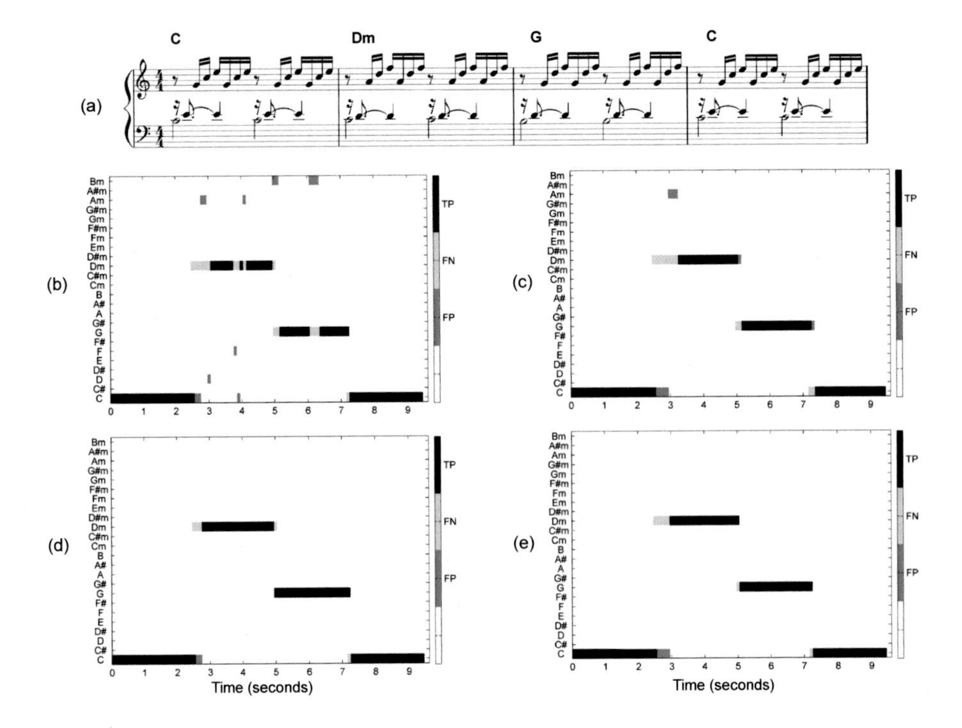

Fig. 5.31 Different chord recognition procedures applied to the first four measures of the Prelude BWV 846 in C major by Johann Sebastian Bach (see Figure 5.20). In all procedures, the same chroma type C_{10} at a feature rate of 10 Hz is used. **(a)** Musical score and reference annotations. **(b)** Template-based chord recognition. **(c)** Template-based chord recognition using prefiltering (20 frames). **(d)** HMM-based chord recognition. **(e)** HMM-based chord recognition using prefiltering (20 frames).

combination of postfiltering with prefiltering does not further improve the recognition results; the two filtering approaches lead to similar improvements that do not add up, compare also Figure 5.31d with Figure 5.31e.

In conclusion, we want to mention that there is a complex interaction between the various enhancement and smoothing strategies that have been suggested and applied for automated chord recognition. For a detailed analysis of these components, we refer to the excellent study by Cho and Bello [4]. One main result of this study is the importance of self-transitions in HMM-based chord recognizers. Even though the strengthening of transitions such as $\mathbf{I} \rightarrow \mathbf{IV}$ or $\mathbf{I} \rightarrow \mathbf{V}$ leads to a boost of certain musically meaningful chord changes, the main improvements come from high self-transition probabilities that essentially reduce the number of chord changes. This result is supported by an experiment using a transition matrix in which all transitions are assigned the same (relatively small) probability value, except for the self-transitions which are assigned a much larger value (see Figure 5.32). Even when using this uniform transition probability matrix, which only reduces the number of

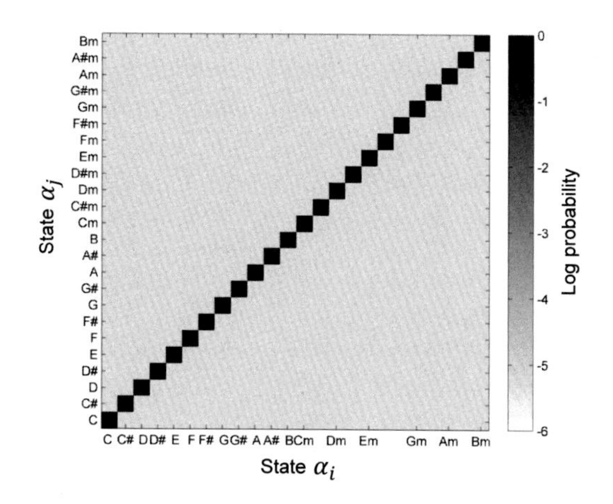

Fig. 5.32 Uniform transition probability matrix with a large value on the main diagonal (self-transitions) and a much smaller value at all remaining positions.

chord transitions without considering musical context, one obtains similar improvements to the ones when using more complex transition probability matrices.

5.4 Summary and Further Readings

In this chapter, we studied the problem of chord recognition with the objective of automatically extracting chord labels from a given music recording. This is only a restricted scenario within the much wider field of harmony analysis, where one studies the construction of and relationships between chords and their progressions—not to mention further aspects such as the underlying musical key and the role of non-chord tones. In Section 5.1, we gave some background on notions such as intervals, chords, and scales with a focus on Western tonal music. In particular, we discussed the relation between the perceived consonance of sound mixtures and frequency ratios of the constituent tonal sound components. For the sake of simplicity, we focused on the twelve major and minor triads and their (functional) relations within a given musical scale. For real music, harmony analysis can be very complex and ambiguous. There are many textbooks on harmony theory, ranging from basic to advanced material. For introductory texts, we refer to, e.g., [8, 15, 28].

Chord Recognition

Automatic chord recognition, as considered in this chapter, is one of the central tasks in the field of music information retrieval (MIR). Starting with the paper by Fujishima [5], numerous approaches have been suggested over the last two decades, moving from knowledge-driven to data-driven systems. For an overview of this development and links to further references, we refer to [24]. Most of the traditional model-based chord recognition strategies proceed in a similar fashion, as illustrated by Figure 5.13. In the first step, the given music recording is converted into a sequence of chroma-based audio features. These features are often further processed, for example, by applying suitable smoothing filters to even out temporal outliers or by applying logarithmic compression procedures to enhance small yet perceptually relevant spectral components. In the next step, pattern matching techniques are applied to map the chroma features to chord labels that correspond to the various musical chords to be considered. Further postfiltering techniques may be applied to smooth out local misclassifications. Often, hidden Markov models, which jointly perform the pattern matching and temporal filtering steps within one optimization procedure, are used. In Section 5.2, we discussed the main ideas of template-based classifiers and then introduced in Section 5.3 a typical HMM-based approach.

One of this chapter's main objectives was to illustrate the delicate interplay of the various feature extraction, filtering, and pattern matching components composing a chord recognition system—a phenomenon that is not limited to chord recognition but occurs in most music analysis and retrieval tasks. The situation is complicated because the components' behavior may critically depend on various parameters used for adjusting temporal, spectral, or dynamical aspects. For example, the chroma representation used as input to the pattern matching step may already have a significant impact on the chord recognition accuracy [11, 17]. A detailed analysis of the interrelation of different chord recognition components can be found in the article by Cho and Bello [4]. We highly recommend this excellent article, which has also been a source of inspiration for this book chapter. As reviewed by Pauwels et al. [24], there has been a paradigm shift from model-based approaches (as described in this chapter) to data-driven approaches. In recent years, automatic chord recognition has been dominated by deep learning approaches thanks to their capability of performing feature extraction and classification within a single optimization procedure [10, 13, 19, 24]. By exploiting relationships with other musical properties, various approaches have been proposed for jointly analyzing chords and musical attributes such as key, bass notes, and metric positions (see, e.g., [18, 21, 22, 26]). Further references to such multi-task learning approaches can be found in [24].

To overcome the substantial simplifications that go along when considering only the 24 major and minor triads, large-vocabulary automatic chord recognizers with up to 170 chord labels have been proposed [19]. Although making much more sense from a musical perspective, large-vocabulary systems are often hard to train, since, for real music, class distributions often become highly skewed. Furthermore, such systems often ignore the structural similarity between related chords [19]. Despite substantial progress in automatic chord recognition and the integration of such techniques in commercial products, there is still room for improvement [24]. In the article [10], the authors give some insights into obstacles and limitations in chord estimation. One question is whether standard dictionaries and comparison methods, where all chords are treated independently, reflect the natural relationships between chord models (see also [19]). Another issue is that chords' perception in real music can be highly subjective, thus making the very notion of "ground truth" annotations tenuous [10]. One research direction to remedy this chord ambiguity problem is pre-

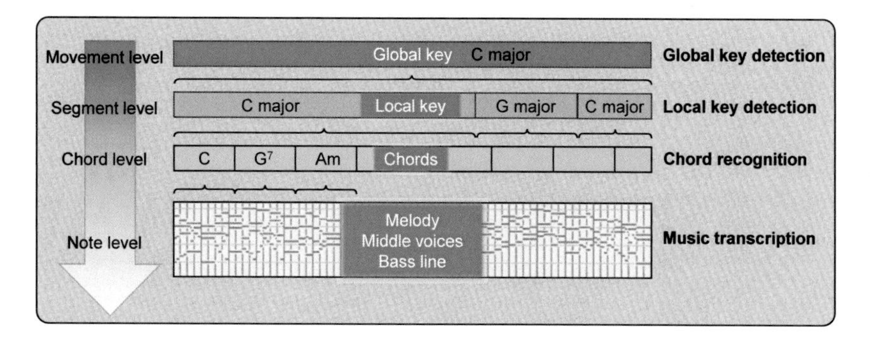

Fig. 5.33 Overview of different concepts for tonal analysis and related MIR tasks. (Figure adapted from [31].)

dicting context-dependent interpretations and the personalization of chord recognition [12].

There are several music processing tasks that are related to chord recognition and may be subsumed under the term "tonal analysis" [31]. These tasks, as illustrated by Figure 5.33, refer to different types of tonal structures that hierarchically depend on each other. For example, the object of **global key detection** is to assign a single key label (e.g., tonic note and mode) for the whole song or movement. In contrast, **local key analysis** aims at resolving key changes (modulations) that may occur throughout a song or movement. While some approaches to local key estimation try to partition a piece into key segments [3, 34], others propose hierarchical visualizations that better account for possible segmentation and key ambiguities [27]. Similar to chord recognition, there has been a shift in the last years from model-based to data-driven approaches based on deep learning [14, 33]. Typically, a key segment can be further subdivided into several chord segments. Thus, chord recognition refers to a finer temporal level. Further descending the temporal hierarchy, one reaches music processing tasks related to **music transcription** with the goal of extracting note-level information [2]. Several methods address the problem of detecting chords and local keys at the same time [18, 23, 26]. Apart from concrete tonal items, several researchers introduce methods for measuring more abstract concepts such as musical and tonal complexity [30, 32]. For an overview on tonal analysis tasks and further references to the literature, we refer to [31, 33].

Hidden Markov Models

The chord recognition scenario served as motivation for considering hidden Markov models (HMMs), one of the essential concepts for modeling time-dependent data streams. Initially introduced in the late 1960s, HMMs have become mainly known for their application to speech recognition. Following the classical tutorial by Rabiner [25], which we strongly recommend for further reading, we first introduced the

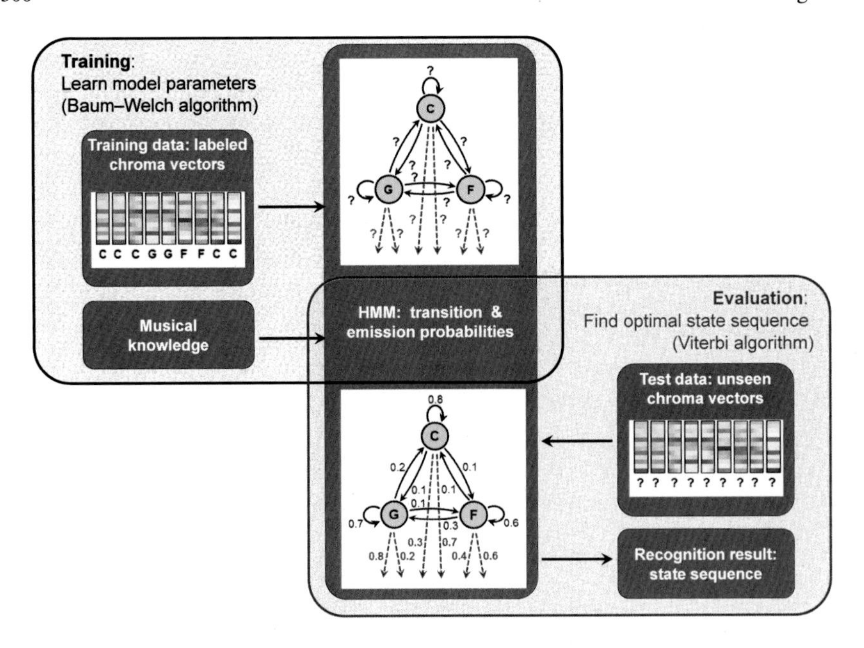

Fig. 5.34 Overview of a typical HMM-based chord recognition approach consisting of a training and an evaluation stage.

concept of Markov chains, which was then extended to the more powerful concept of hidden Markov models. In Markov chains, the states are directly visible to the observer, and therefore the state transition probabilities are the only parameters. In a *hidden* Markov model, the states are not directly visible. Instead, an HMM emits output entities according to a given state-dependent probability distribution.

Following Rabiner's tutorial [25], we introduced three fundamental problems for HMMs: the evaluation problem (computing the probability of an observation sequence given a specific HMM), the uncovering problem (finding the best sequence of model states), and the estimation problem (adjusting the model parameters to best account for an observed sequence). In particular, we discussed the uncovering problem in more detail and introduced the efficient Viterbi algorithm for solving this problem. The Viterbi algorithm is based on dynamic programming—a paradigm we have already encountered in Section 3.2 for computing DTW distances and optimal warping paths. In the application to chord labeling, which is summarized by Figure 5.34, an HMM was used to uncover the most likely chord labeling sequence that generates a given sequence of chroma features—an idea originally introduced by [29].

In this book, we have only considered the most basic variant of discrete HMMs. There are many more variants and extensions of HMMs, including continuous HMMs, autoregressive HMMs, and HMMs with specific state transition topologies (see [25]). The estimation of the model parameters can become very intricate, leading to challenging and deep mathematical problems. For an excellent textbook on the classical theory of HMMs, including the discrete as well as the continuous case, we refer to [9].

5.5 FMP Notebooks

In **Part 5** of the FMP notebooks [20], we provide and discuss Python code examples of all the components that are required to realize a template-based and an HMM-based chord recognizer. Based on evaluation metrics and suitable time–chord representations, we quantitatively and qualitatively discuss how the various components and their parameters affect the chord recognition results. To this end, we consider real-world music recordings, which expose the weaknesses of the automatic procedures, the problem modeling, and the evaluation metrics.

The first notebooks of **Part 5** mainly provide sound examples of the basic musical concepts introduced in Section 5.1. In the **FMP Notebook *Intervals***, we provide Python code examples for generating sinusoidal sonifications of intervals. We then generate sound examples of the various music intervals with respect to equal temperament, just intonation, and Pythagorean tuning. Besides a mathematical specification of deviations (given in cents), the sound examples allow for an acoustic comparison of intervals generated based on the different intonation schemes. Similarly, in the **FMP Notebook *Chords***, we give sound examples for different chords. In particular, we provide a piano recording as well as a synthesized version of each of the twelve major and minor triads. Finally, in the **FMP Notebook *Musical Scales and Circle of Fifths***, we cover the notions of musical scales and keys. In particular, we look at diatonic scales, which are obtained from a chain of six successive perfect fifth intervals and can be arranged along the circle of fifths. In summary, these three notebooks show how simple sonifications may help to better understand musical concepts. In a music processing course, one may develop small tools for ear training in basic harmony analysis as part of small student projects.

After the musical warm-up in the previous notebooks, we introduce in the **FMP Notebook *Template-Based Chord Recognition*** a simple yet instructive chord recognizer. For illustration, we use the first measures of the Beatles song "Let It Be" (see Figure 5.1), which is converted into a chroma representation. As we already discussed in the context of music synchronization (see Section 3.4), there are many different ways of computing chroma features. As examples, we compute and visualize three different chroma variants as provided by the Python package `librosa`. Furthermore, we provide Python code examples to generate chord templates, compare the templates against the recording's chroma vectors in a frame-wise fashion, and visualize the resulting similarity values in the form of a time–chord representation. By looking at the template that maximizes the similarity value, one obtains the frame-wise chord estimate, which we visualize in the form of a binary time–chord representation. Finally, we discuss these results by visually comparing them with manually generated chord annotations. We recommend that students also use the functionalities provided by the **FMP Notebook *Sonification*** of **Part B** to complement the visual inspections by acoustic ones. We think that qualitative inspections—based on explicit music examples and using visualizations and sonifications of all intermediate results—are essential for students to understand the technical components, to become aware of the model assumptions and their implications, and to sharpen their intuition of what to expect from computational approaches.

Besides a qualitative investigation using explicit examples and visualizations, one also requires quantitative methods to evaluate an automatic chord recognizer's performance. To this end, one typically compares the computed result against a reference annotation. Such an evaluation, as we discuss in the **FMP Notebook *Chord Recognition Evaluation***, gives rise to several questions (see also Section 5.2.2). How should the agreement between the computed result and the reference annotation be quantified? Is the reference annotation reliable? Are the model assumptions appropriate? To what extent do violations of these assumptions influence the final result? Such issues should be kept in mind before turning to specific metrics. Our evaluation focuses on some simple metrics based on precision, recall, and F-measure, as we already encountered in the **FMP Notebook *Evaluation*** of **Part 4**. Before the evaluation, one needs to convert the reference annotation into a suitable format that conforms with the respective metric and the automatic approach's format. Our notebook demonstrates how one may convert a segment-wise reference annotation (where segment boundaries are specified in seconds) into a frame-wise format. Furthermore, one may need to adjust the chords and naming conventions. All these conversion steps are, by far, not trivial and often require simplifying design choices. Continuing our Beatles example, we discuss such issues and make them explicit using suitable visualizations. Furthermore, we address some of the typical evaluation problems that stem from chord ambiguities (e.g., due to an oversimplification of the chord models) or segmentation ambiguities (e.g., due to broken chords). We hope that this notebook is a source of inspiration for students to conduct experiments with their own music examples.

Motivated by the chord recognition problem, the **FMP Notebook *Hidden Markov Model (HMM)*** deepens the topic (closely following Section 5.3). We start by providing a Python function that generates a state and observation sequence from a given discrete HMM (see Table 5.1). Conversely, knowing an observation sequence as well as the underlying state sequence it was generated from (which is normally hidden), we show how one can estimate the state transition and output probability matrices. The general problem of estimating HMM parameters only on the basis of observation sequences is much harder—a topic beyond the scope of this textbook. The uncovering problem of HMMs (see Section 5.3.3.2) is discussed in the **FMP Notebook *Viterbi Algorithm***. We first provide an implementation of the Viterbi algorithm as presented in Table 5.2. In practice, however, this multiplicative version of the algorithm is problematic since the product of probability values decreases exponentially with the number of factors, which may finally lead to a numerical underflow. To remedy this problem, one applies a logarithm to all probability values and replaces multiplication by summation. Our notebook also provides this log-variant implementation of the Viterbi algorithm and compares it against the original version using a toy example.

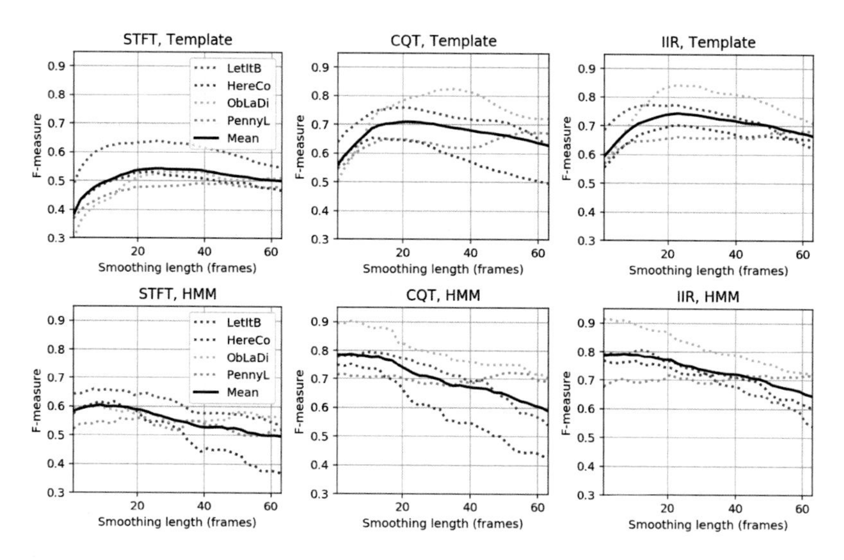

Fig. 5.35 Prefiltering experiments for template-based and HMM-based chord recognizers and three different chroma representations (`STFT`, `CQT`, `IIR`). The evaluation is performed on the basis of four Beatles songs (`LetItB`, `HereCo`, `ObLaDi`, `PennyL`).

In the **FMP Notebook *HMM-Based Chord Recognition***, we apply the HMM concept to chord recognition. Rather than learning the HMM parameters from training examples, we fix all the HMM parameters using musical knowledge. In this way, besides keeping the technical requirements low (not to speak of the massive training data required for the learning procedure), the HMM-based chord recognizer can be regarded as a direct extension of the template-based procedure. Note that we considered in this textbook only the case of discrete HMMs, where the observations are discrete symbols coming from a finite output space (see Section 5.3.2). In our application, however, the observations are real-valued chroma vectors. Therefore, we use an HMM variant in our notebook, where we replace the discrete output space by a continuous feature space $\mathbb{R}^{12}$. Furthermore, we replace a given state's emission probability by a normalized similarity value defined as the inner product of a state-dependent normalized template and a normalized observation (chroma) vector. As for the transition probabilities, we use different models as illustrated by Figure 5.29, Figure 5.30, and Figure 5.32, respectively. Based on this HMM variant, we implement an adapted Viterbi algorithm using a numerically stable log version. Considering real-world music examples, we finally compare the resulting HMM-based chord recognizer with the template-based approach, showing the evaluation results in the form of time–chord visualizations, respectively.

As said before, chord recognition has always been and still is one of the central tasks in MIR. Besides chords being a central concept in particular for Western music, another reason for the topic's popularity is the availability of a dataset known as the Beatles Collection, which we already quickly mentioned in Section 5.2.4.2.

This dataset is based on twelve Beatles albums comprising 180 audio tracks. While being a well-defined, medium-sized collection of musical relevance, the primary value of the dataset lies in the availability of high-quality reference annotations for chords, beats, key changes, and music structures [7, 16]. In the **FMP Notebook *Experiments: Beatles Collection***, we take the opportunity to present a few systematic studies in the context of chord recognition. To keep the notebook slim and efficient, we only use from the collection the following four representative Beatles songs: “Let It Be” (`LetItB`), “Here Comes the Sun” (`HereCo`), “Ob-La-Di, Ob-La-Da” (`ObLaDi`), and “Penny Lane” (`PennyL`). The provided experimental setup and implementation can be easily extended to an arbitrary number of examples. We provide the full processing pipeline in the notebook, starting with raw audio and annotation files and ending with parameter sweeps and quantitative evaluations. First, the reference annotations are converted into a suitable format. Then, the audio files are transformed into chroma representations, where we consider three different chroma types (`STFT`, `CQT`, `IIR`). All this data is computed in a preprocessing step and stored in a suitable data structure for later usage. In our experiments, we consider two different pattern matching techniques (a template-based and an HMM-based approach) to map the chroma features to chord labels that correspond to the 24 major and minor triads. As for the quantitative evaluation, we use the standard precision, recall, and F-measure. After looking at some individual results using time–chord representations, we conduct a first comprehensive experiment to study the role of prefiltering (see Section 5.2.4.4). To this end, we consider a parameter $L \in \{1,3,\ldots,63\}$ that determines the smoothing length (applied to the input chromagram) and report on the resulting F-measure for each of the four songs and its mean over the four songs. The overall parameter sweep result is shown in Figure 5.35 for different chroma types and pattern matching techniques. Similarly, we conduct a parameter sweep experiment to study the role of self-transition probabilities used in HMM-based chord recognizers. Finally, we present two small experiments where we question the musical relevance of the results achieved. First, we discuss a problem related to an imbalance in the class distribution. As a concrete example, we consider a rather dull chord recognizer that, based on some global statistics of the song, decides on a single major or minor triad and outputs the corresponding chord label for all time frames. In the case of the song “Ob-La-Di, Ob-La-Da,” this dull procedure achieves an F-measure of $F = 0.551$—which does not seem bad for a classification problem with 24 classes. Second, we discuss a problem that comes from the reduction to only 24 chords and illustrates the role of the nonchord model. While these experiments nicely demonstrate some of the obstacles and limitations in chord estimation mentioned by [10], we see another main value of this notebook from an educational perspective. Giving concrete examples for larger-scale experiments, we hope that students get some inspiration from this notebook for conducting similar experiments in the context of other music processing tasks.

References

1. W. Apel, *Harvard Dictionary of Music*, Harvard University Press, 1969.
2. E. Benetos, S. Dixon, Z. Duan, and S. Ewert, *Automatic music transcription: An overview*, IEEE Signal Processing Magazine, 36 (2019), pp. 20–30.
3. W. Chai and B. Vercoe, *Detection of key change in classical piano music*, in Proceedings of the International Society for Music Information Retrieval Conference (ISMIR), London, UK, 2005, pp. 468–474.
4. T. Cho and J. P. Bello, *On the relative importance of individual components of chord recognition systems*, IEEE/ACM Transactions on Audio, Speech, and Language Processing, 22 (2014), pp. 477–492.
5. T. Fujishima, *Realtime chord recognition of musical sound: A system using common lisp music*, in Proceedings of the International Computer Music Conference (ICMC), Beijing, China, 1999, pp. 464–467.
6. A. Gut, *Probability: A Graduate Course*, Springer, New York, 2nd ed., 2013.
7. C. Harte, M. B. Sandler, S. Abdallah, and E. Gómez, *Symbolic representation of musical chords: A proposed syntax for text annotations*, in Proceedings of the International Society for Music Information Retrieval Conference (ISMIR), London, UK, 2005, pp. 66–71.
8. F. Haunschild, *The New Harmony Book*, AMA Verlag, 2000.
9. X. D. Huang, Y. Ariki, and M. A. Jack, *Hidden Markov Models for Speech Recognition*, Edinburgh University Press, 1990.
10. E. J. Humphrey and J. P. Bello, *Four timely insights on automatic chord estimation*, in Proceedings of the International Society for Music Information Retrieval Conference (ISMIR), Málaga, Spain, 2015, pp. 673–679.
11. N. Jiang, P. Grosche, V. Konz, and M. Müller, *Analyzing chroma feature types for automated chord recognition*, in Proceedings of the AES Conference on Semantic Audio, Ilmenau, Germany, 2011.
12. H. V. Koops, W. B. de Haas, J. Bransen, and A. Volk, *Automatic chord label personalization through deep learning of shared harmonic interval profiles*, Neural Computing and Applications, 32 (2020), pp. 929–939.
13. F. Korzeniowski and G. Widmer, *Feature learning for chord recognition: The deep chroma extractor*, in Proceedings of the International Society for Music Information Retrieval Conference (ISMIR), New York City, USA, 2016, pp. 37–43.
14. ———, *Genre-agnostic key classification with convolutional neural networks*, in Proceedings of the International Society for Music Information Retrieval Conference (ISMIR), Paris, France, 2018, pp. 264–270.
15. S. Kostka, D. Payne, and B. Almen, *Tonal Harmony*, McGraw-Hill, 7th ed., 2012.
16. M. Mauch, C. Cannam, M. E. P. Davies, S. Dixon, C. Harte, S. Kolozali, D. Tidhar, and M. B. Sandler, *OMRAS2 metadata project 2009*, in Late Breaking Demo of the International Society for Music Information Retrieval Conference (ISMIR), Kobe, Japan, 2009.
17. M. Mauch and S. Dixon, *Approximate note transcription for the improved identification of difficult chords*, in Proceedings of the International Society for Music Information Retrieval Conference (ISMIR), Utrecht, The Netherlands, 2010, pp. 135–140.
18. ———, *Simultaneous estimation of chords and musical context from audio*, IEEE Transactions on Audio, Speech, and Language Processing, 18 (2010), pp. 1280–1289.
19. B. McFee and J. P. Bello, *Structured training for large-vocabulary chord recognition*, in Proceedings of the International Society for Music Information Retrieval Conference (ISMIR), Suzhou, China, 2017, pp. 188–194.
20. M. Müller and F. Zalkow, *FMP Notebooks: Educational material for teaching and learning fundamentals of music processing*, in Proceedings of the International Society for Music Information Retrieval Conference (ISMIR), Delft, The Netherlands, 2019, pp. 573–580.

21. Y. Ni, M. McVicar, R. Santos-Rodriguez, and T. D. Bie, *An end-to-end machine learning system for harmonic analysis of music*, IEEE Transactions on Audio, Speech, and Language Processing, 20 (2012), pp. 1771–1783.
22. H. Papadopoulos and G. Peeters, *Joint estimation of chords and downbeats from an audio signal*, IEEE Transactions on Audio, Speech, and Language Processing, 19 (2011), pp. 138–152.
23. ———, *Local key estimation from an audio signal relying on harmonic and metrical structures*, IEEE Transactions on Audio, Speech, and Language Processing, 20 (2012), pp. 1297–1312.
24. J. Pauwels, K. O'Hanlon, E. Gómez, and M. B. Sandler, *20 years of automatic chord recognition from audio*, in Proceedings of the International Society for Music Information Retrieval Conference (ISMIR), Delft, The Netherlands, 2019, pp. 54–63.
25. L. R. Rabiner, *A tutorial on hidden Markov models and selected applications in speech recognition*, Proceedings of the IEEE, 77 (1989), pp. 257–286.
26. T. Rocher, M. Robine, P. Hanna, and L. Oudre, *Concurrent estimation of chords and keys from audio*, in Proceedings of the International Society for Music Information Retrieval Conference (ISMIR), Utrecht, The Netherlands, 2010, pp. 141–146.
27. C. S. Sapp, *Visual hierarchical key analysis*, ACM Computers in Entertainment, 3 (2005), pp. 1–19.
28. C. Schroeder and K. Wyatt, *Harmony and Theory: A Comprehensive Source for All Musicians*, Musicians Institute Press, 1998.
29. A. Sheh and D. P. W. Ellis, *Chord segmentation and recognition using EM-trained hidden Markov models*, in Proceedings of the International Society for Music Information Retrieval Conference (ISMIR), Baltimore, MD, USA, 2003, pp. 185–191.
30. S. Streich, *Music Complexity: A Multi-Faceted Description of Audio Content*, PhD thesis, University Pompeu Fabra, Barcelona, Spain, 2007.
31. C. Weiss, *Computational Methods for Tonality-Based Style Analysis of Classical Music Audio Recordings*, PhD thesis, Ilmenau University of Technology, Ilmenau, Germany, 2017.
32. C. Weiss and M. Müller, *Tonal complexity features for style classification of classical music*, in Proceedings of the IEEE International Conference on Acoustics, Speech, and Signal Processing (ICASSP), Brisbane, Australia, 2015, pp. 688–692.
33. C. Weiss, H. Schreiber, and M. Müller, *Local key estimation in music recordings: A case study across songs, versions, and annotators*, IEEE/ACM Transactions on Audio, Speech & Language Processing, 28 (2020), pp. 2919–2932.
34. Y. Zhu and M. S. Kankanhalli, *Key-based melody segmentation for popular songs*, in Proceedings of the International Conference on Pattern Recognition (ICPR), vol. 3, Cambridge, UK, 2004, pp. 862–865.

Exercises

Exercise 5.1. Determine, for each of the following intervals, the number of semitones and the interval name (as specified in Figure 5.3):

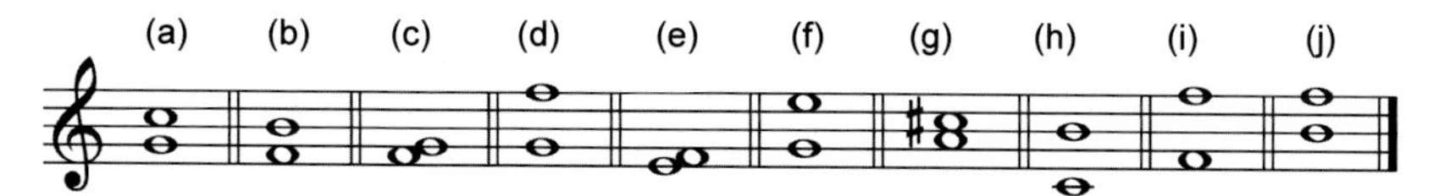

Exercise 5.2. The **complement** of an interval is the interval which, when added to the original interval, spans an octave in total. Specify the complement for each interval in Figure 5.3. In which way is the tritone interval special?

Exercise 5.3. Determine the chord symbol for each of the following chords (similar to Figure 5.6):

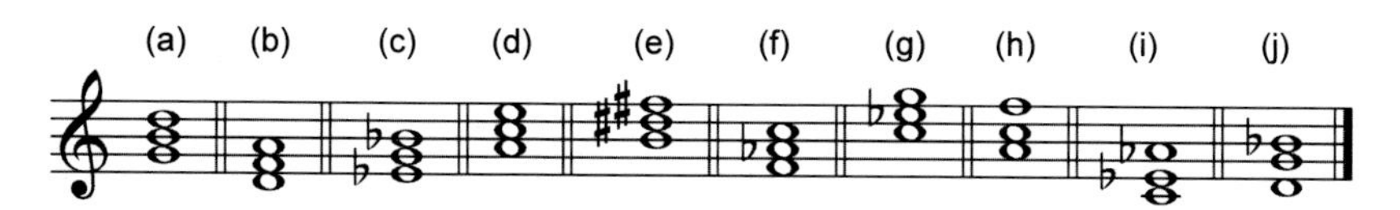

Exercise 5.4. In this exercise, we compare the size of the intervals obtained from different definitions. First, assuming the twelve-tone equal-tempered scale, determine the size (given in cents) and frequency ratios for each of the 13 intervals shown in Figure 5.3. Next, assuming just intonation, determine the size (given in cents) and the frequency ratios of the intervals (see Figure 5.3). Finally, compute the difference of the interval sizes (given in cents) between the just intonation and the equal-tempered case.
[**Hint:** Write a small computer program that helps you with the calculations.]

Exercise 5.5. In this exercise, we investigate the dependency between the degree of consonance of an interval and the coincidence of partials of the two notes underlying the interval. Assuming the twelve-tone equal-tempered scale, we look at the intervals that are formed by the root note C4 and each of the following seven notes: C4, E$^\flat$4, E4, F4, F$^\sharp$4, G4, and C5. Consider for each of these notes the first eight harmonics. Determine for each of the resulting harmonics the closest musical note along with the difference (given in cents) between the harmonic's actual frequency and the center frequency of the musical note (see also Figure 1.20). For example, the following table shows these results for the two notes C4 and E$^\flat$4 (with the differences being specified in brackets):

1	2	3	4	5	6	7	8
C4 [0]	C5 [0]	G5 [+2]	C6 [0]	E6 [-14]	G6 [+2]	B$^\flat$6 [-31]	C7 [0]
E$^\flat$4 [0]	E$^\flat$5 [0]	B$^\flat$5 [+2]	E$^\flat$6 [0]	G6 [-14]	B$^\flat$6 [+2]	D$^\flat$6 [-31]	E$^\flat$7 [0]
⋮	⋮	⋮	⋮	⋮	⋮	⋮	⋮

Then investigate, for each of the seven intervals, which of the harmonics of the two involved notes coincide (or, to be more precise, are close together with respect to frequency). For example, the coincidences of harmonics between the notes of the interval C4–C4 and the interval C4–E$^\flat$4 are indicated by frameboxes in the above table (where the first row represents the interval C4–C4 and the second one the interval C4–E$^\flat$4). Note that G6 appears as the sixth harmonic of C4 and as the fifth harmonic of E$^\flat$4. However, this coincidence is "tarnished" by the fact that the sixth harmonic of C4 deviates by $+2$ cents from the center frequency of G6, whereas the fifth harmonic of E$^\flat$4 deviates by -14 cents from G6. Similarly, discuss the results for the other intervals.

Exercise 5.6. In Figure 5.20b, one can observe many misclassifications and chord label changes in the recognition result. Explain why these errors only occur in the second and third measure, while the first and fourth measure have been classified correctly.

Exercise 5.7. Let Λ be the set of the major and minor triads (see (5.5)). Furthermore, for a given chord $\lambda \in \Lambda$, let $\mathbf{t}_\lambda^{\mathrm{h}}$ be the chord template with harmonics based on the first eight harmonics (see (5.14) and (5.15)). Compute $\mathbf{t}_\lambda^{\mathrm{h}}$ for $\lambda = \mathbf{C}$ and $\lambda = \mathbf{Cm}$, respectively, using the parameter $\alpha = 1$.

Exercise 5.8. In this exercise, we extend the chord template model as defined by (5.14) and (5.15) by introducing some additional weight parameters. For the C-major chord $\lambda = \mathbf{C}$, we define the template

$$\mathbf{t}_{\mathbf{C}}^{\mathrm{h,w}} = w_1 \cdot \mathbf{t}_{\mathrm{C}}^{\mathrm{h}} + w_2 \cdot \mathbf{t}_{\mathrm{E}}^{\mathrm{h}} + w_3 \cdot \mathbf{t}_{\mathrm{G}}^{\mathrm{h}}$$

for $w = (w_1, w_2, w_3)^\top \in \mathbb{R}^3$. Similarly, using the same weights, we define the chord templates $\mathbf{t}_\lambda^{\mathrm{h,w}}$ for the other major and minor triads $\lambda \in \Lambda$ (see (5.5)). We now compare these new chord templates with the original binary templates $\mathbf{t}_\lambda$ (see (5.7)) using the similarity measure s as defined in (5.8). Write a small computer program to compute the similarity values $s(\mathbf{t}_{\mathbf{C}}^{\mathrm{h,w}}, \mathbf{t}_\lambda)$ and $s(\mathbf{t}_{\mathbf{Cm}}^{\mathrm{h,w}}, \mathbf{t}_\lambda)$ for all 24 major and minor triads $\lambda \in \Lambda$ using the following parameters:

(a) $\alpha = 0$ and $w = (1,1,1)$
(b) $\alpha = 1$ and $w = (1,1,1)$
(c) $\alpha = 0$ and $w = (1,0.2,1)$
(d) $\alpha = 1$ and $w = (1,0.2,1)$

In which case is there a confusion between the C-major and C-minor chord? Explain the reason for this confusion in words.

Exercise 5.9. Let us consider a Markov chain with I states $\{\alpha_1, \alpha_2, \ldots, \alpha_I\}$ and transition probability coefficients a_{ij}, $i, j \in [1:I]$ (see (5.21)). The goal of this exercise is to determine how long the resulting system stays (on average) in a given state. To this end, consider an observation sequence $S = (\alpha_i, \ldots, \alpha_i, \alpha_j)$ of length $d+1$ consisting of d states α_i for some $i \in [1:I]$ and a final state α_j for some $j \neq i$. Compute the probability $P_i(d) := P[S \mid \text{Model}, s_1 = \alpha_i]$, where the condition $s_1 = \alpha_i$ means that the system is assumed to start with state α_i. From this, compute the expected duration $\overline{d}_i$ for state i, which is defined by $\overline{d}_i := \sum_{d=1}^{\infty} d \cdot P_i(d)$. Finally, determine the expected durations for the states α_1, α_2, and α_3 of the system specified in Figure 5.24.
[**Hint:** Use the fact that $\sum_{d=1}^{\infty} d \cdot a^{d-1} = 1/(1-a)^2$ for a number $a \in [0,1)$.]

Exercise 5.10. Let us consider the HMM as specified in Figure 5.28a. Compute the optimal state sequence and its probability for the observation sequence $O = (\beta_1, \beta_3, \beta_1, \beta_3, \beta_3)$, which is a prefix of the observation sequence used in Figure 5.28b. Compare the result with the one obtained in Figure 5.28b.

Exercise 5.11. Let us consider the HMM as specified in Figure 5.28a. Determine the optimal state sequence for the observation sequence $O = (\beta_1, \beta_3^{N-1})$ for each $N \in \mathbb{N}$. Argue why the respective state sequence is optimal.

Chapter 6
Tempo and Beat Tracking

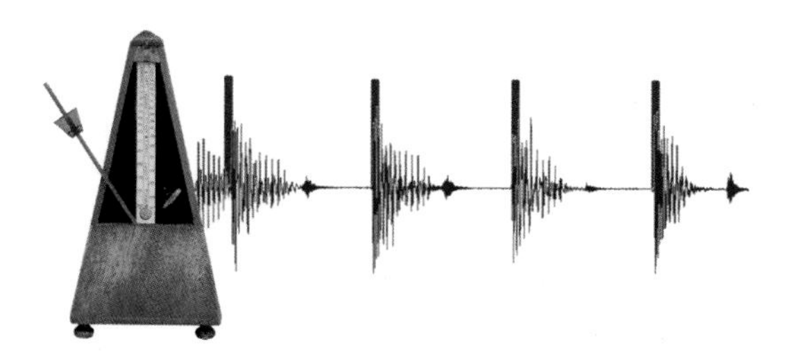

Temporal and structural regularities are perhaps the most important incentives for people to get involved and to interact with music. It is the **beat** that drives music forward and provides the temporal framework of a piece of music. Intuitively, the beat corresponds to the pulse a human taps along when listening to music. The beat is often described as a sequence of perceived pulse positions, which are typically equally spaced in time and specified by two parameters: the **phase** and the **period** (see Figure 6.1b). The term **tempo** refers to the rate of the pulse and is given by the reciprocal of the beat period. Tempo and beat are fundamental aspects of music, and the automated extraction of such information from audio recordings constitutes one of the central and well-studied research areas in music processing. In this chapter, we introduce some key techniques used in tempo estimation and beat tracking. Furthermore, we discuss some of the challenges one has to face when dealing with music where certain model assumptions are not fulfilled.

When listening to a piece of music, we as humans are often able to tap along with the musical beat without difficulty—sometimes, we even do this unconsciously. In the case that we lose track at some point in time, maybe because of a tempo change or rhythmic displacement, we are able to recover quickly and resume tapping. However, simulating this cognitive process with an automated beat tracking system is much harder than one may think. Recent beat tracking systems can cope well with modern pop and rock music that has a strong and steady beat. In deriving this information, most systems are based on the assumptions that beats correspond to note onsets (typically percussive in nature) and that beats are periodically spaced in time. However, there are many types of music where these assumptions are violated. For example, in string music a note may be played softly with a barely noticeable onset, or a musician may slightly lengthen certain notes to shape musical phrases. In general, musicians do not play mechanically at a fixed tempo, but slow down or accelerate at certain positions to create tension and release. As a consequence, the

M. Müller, *Fundamentals of Music Processing*, https://doi.org/10.1007/978-3-030-69808-9_6

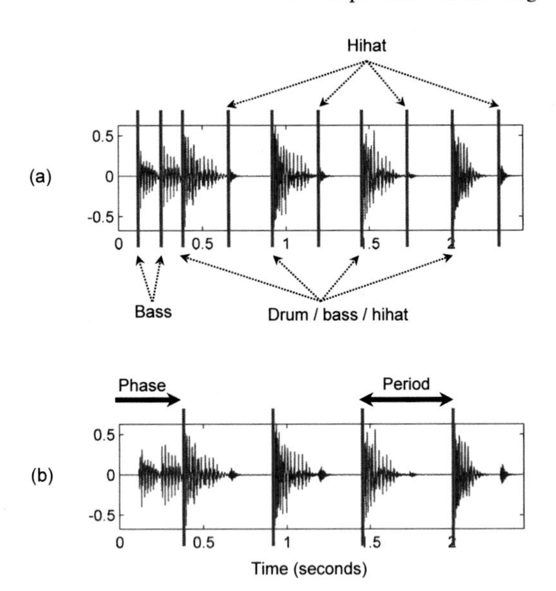

Fig. 6.1 Waveform representation of an excerpt of "Another One Bites the Dust" by Queen. **(a)** Note onsets. **(b)** Beat positions.

presence of such local tempo changes makes the extraction of beat positions a very challenging task. Still, at least when familiar with the type of music, humans are capable of anticipating local tempo changes and tracking the beats even for highly complex music.

In most approaches to automated tempo and beat tracking, the first step is to estimate the positions of note onsets within the music signal (see Figure 6.1a). This task, which is also referred to as **onset detection**, is discussed in Section 6.1. In particular, we show how to transform a given music signal into a novelty representation that captures certain changes in the signal's energy or spectrum. The peaks of such a representation yield good indicators for note onset candidates. We have seen a similar concept when applying novelty detection to music structure analysis (see Section 4.4). In Section 6.2, we introduce the notion of a tempogram, which represents local tempo information on different pulse levels. Such a time–tempo representation is obtained by analyzing a novelty representation with regard to reoccurring patterns and quasiperiodic pulse trains. In this context, we study two important methods for periodicity analysis, one using Fourier and the other using autocorrelation analysis techniques. We then continue in Section 6.3 with the topic of beat tracking. First, we introduce a mid-level representation that captures meaningful local pulse information even in the presence of significant tempo changes. Then, based on a dynamic programming approach, we discuss a robust beat tracking procedure, which assumes a roughly constant tempo throughout the recording.

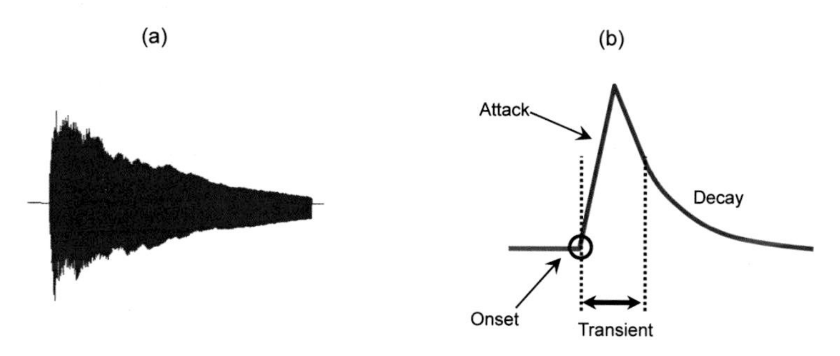

Fig. 6.2 Illustration of attack, transient, onset, and decay of a single note (based on [1]). **(a)** Note played on a piano. **(b)** Idealized amplitude envelope.

6.1 Onset Detection

Generally speaking, **onset detection** is the task of determining the starting times of notes or other musical events as they occur in a music recording. In practice, however, the notion of an onset can be rather vague and is related to other concepts such as attacks or transients. As discussed in Section 1.3.4, there is often a sudden increase of energy at the beginning of a musical tone (see Figure 6.2a). The **attack** of a note refers to the phase where the sound builds up, which typically goes along with a sharply increasing amplitude envelope. This is also reflected by the initial phase of the ADSR model shown in Figure 1.22. The concept of a **transient** is more difficult to grasp. As noted in Section 1.3.4, a transient may be described as a noise-like sound component of short duration and high amplitude typically occurring at the beginning of a musical tone or a more general sound event. However, the release or offset of a sustained note may also contain a transient-like component. In transient regions, the signal evolves quickly in an unpredictable and rather chaotic way. For example, in the case of a piano, the transient corresponds to the initial phase where a key is hit, the damper is raised, the hammer strikes the strings, the strings start to vibrate, and the vibrations are transmitted to the large soundboard that starts resonating to finally yield a steady and sustained sound. As opposed to the attack and transient, the **onset** of a note refers to the single instant (rather than a period) that marks the beginning of the transient, or the earliest time point at which the transient can be reliably detected (see Figure 6.2b).

To detect note onsets in the signal, the general idea is to capture sudden changes that often mark the beginning of transient regions. For notes that have a pronounced attack phase, onset candidates may be determined by locating time positions where the signal's amplitude envelope starts increasing. When this is not the case, such as for nonpercussive music with soft onsets and blurred note transitions, the detection of onsets is much more challenging. For example, the waveform of a violin sound, as shown in Figure 1.23b, may exhibit a slow energy increase rather than an abrupt change as in a piano sound. For soft sounds, it is hard to determine the exact onset position. The detection of individual note onsets becomes even harder when dealing with complex polyphonic music. Simultaneously occurring sound events may result in masking effects, where no significant changes in the signal's energy are measurable. In these cases, more refined onset detection methods are needed, e.g., by looking at changes in the signal's short-time spectrum or other statistical properties.

In this section, we study four different approaches for onset detection: an energy-based approach (Section 6.1.1), a spectral-based approach (Section 6.1.2), a phase-based approach (Section 6.1.3), and a complex-domain approach (Section 6.1.4). All approaches follow the same algorithmic pipeline, but differ in the signal properties that are exploited to derive onset candidates. In this pipeline, the signal is first converted into a suitable feature representation that better reflects the properties of interest. Then, a type of derivative operator is applied to the feature sequence and a novelty function is derived. Finally, a peak-picking algorithm is employed to locate the onset candidates. Note that this general procedure is exactly the same as for novelty detection in the context of music structure analysis (see Section 4.4.1). However, the features and, in particular, the temporal levels that are relevant in structure analysis and onset detection are quite different. While a tolerance window of 500 ms up to a couple of seconds may be used in the case of structural boundaries, the accuracy needed in onset detection is usually far below 100 ms, sometimes even on the order of 10 ms.[1]

6.1.1 Energy-Based Novelty

We have seen that playing a note on an instrument often coincides with a sudden increase of the signal's energy. For example, this holds when striking a key on a piano, plucking a string on a guitar, or hitting a drum with a stick. Based on this observation, a straightforward way to detect note onsets is to transform the signal into a local energy function that indicates the local energy of the signal for each time instance and then to look for sudden changes in this function. Mathematically, this procedure can be realized as follows: Let x be a DT-signal. As in the case of a discrete STFT (see Section 2.5.3), we fix a discrete window function $w:\mathbb{Z}\to\mathbb{R}$, which is shifted over the signal x to determine local sections. In particular, we assume that w is a bell-shaped function centered at time zero[2] and that $w(m)$ for $m\in[-M:M]$ comprises the nonzero samples of w for some $M\in\mathbb{N}$. The **local energy** of x with regard to w is defined to be the function $\mathrm{E}_w^x:\mathbb{Z}\to\mathbb{R}$ given by

$$\mathrm{E}_w^x(n) := \sum_{m=-M}^{M} |x(n+m)w(m)|^2 = \sum_{m\in\mathbb{Z}} |x(m)w(m-n)|^2 \tag{6.1}$$

[1] This is the range where the human ear is no longer capable of distinguishing between two subsequent transients [26].

[2] In Section 2.5.3, to simplify notation, we considered the noncentered case assuming that the nonzero window coefficients are $w(n)$ for $n\in[0:N-1]$.

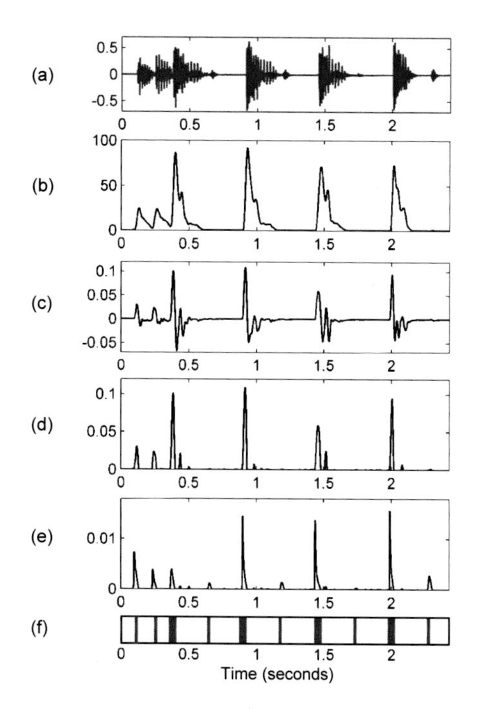

Fig. 6.3 Computation of an energy-based novelty function of the signal from Figure 6.1. **(a)** Waveform. **(b)** Local energy function. **(c)** Discrete derivative. **(d)** Novelty function Δ_{Energy} obtained after half-wave rectification. **(e)** Novelty function $\Delta_{\text{Energy}}^{\text{Log}}$ based on a logarithmic energy function. **(f)** Annotated note onsets (the four beat positions are marked by thick lines).

for $n \in \mathbb{Z}$. In other words, $\mathrm{E}_w^x(n)$ contains the energy (as defined in (2.41)) of the signal x multiplied with a window shifted by n samples. Let us have a look at the example shown in Figure 6.3b, which shows a local energy function for the beginning of "Another One Bites the Dust" by Queen. Starting with an offbeat consisting of two sixteenth notes played only by bass, four percussive beats (played by kick drum, snare drum, hihat, and bass) follow (see Figure 6.1). Furthermore, between each two subsequent beats, there is an additional hihat stroke. As the energy function shows, the percussive beats contain a lot of energy, whereas the low-energy hihat strokes are not as strongly captured.

Intuitively, to measure energy changes, we take a derivative of the local energy function. In the discrete case, the easiest way to realize such a derivative is to take the difference between two subsequent energy values (see Figure 6.3c). Furthermore, since we are interested in energy increases (and not decreases), we keep only the positive differences while setting the negative differences to zero. The latter step is known as **half-wave rectification** and is notated as:

$$|r|_{\geq 0} := \frac{r+|r|}{2} = \begin{cases} r, & \text{if } r \geq 0, \\ 0, & \text{if } r < 0 \end{cases} \tag{6.2}$$

for $r \in \mathbb{R}$. Altogether, we obtain an **energy-based novelty function** $\Delta_{\text{Energy}} : \mathbb{Z} \to \mathbb{R}$ given by

$$\Delta_{\text{Energy}}(n) := |\mathrm{E}_w^x(n+1) - \mathrm{E}_w^x(n)|_{\geq 0} \tag{6.3}$$

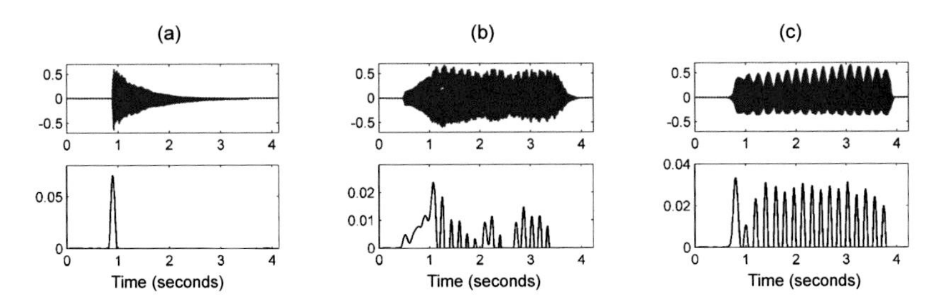

Fig. 6.4 Waveform and energy-based novelty function of the note C4 (261.6 Hz) played by different instruments (see Figure 1.23). **(a)** Piano. **(b)** Violin. **(c)** Flute.

for $n \in \mathbb{Z}$. The resulting function is shown in Figure 6.3d for our example "Another One Bites the Dust." The four quarter-note drum beats correspond to the four highest peaks. Therefore, these beats can be correctly detected by a simple peak-picking procedure. Also, the two beginning offbeats played by the bass are correctly identified by the first two peaks. However, the four hihat strokes between the beat positions do not show up in Δ_{Energy} (see Figure 6.3d). As mentioned before, these four hihat events contain relatively little energy and, when compared with the high-energy drum events, become invisible in the energy-based novelty function.

As we discussed in Section 1.3.3, the human perception of sound intensity is logarithmic in nature. Therefore, even musical events of rather low energy may still be perceptually relevant. For example, the hihat is clearly audible even at the beat positions where it is overlaid with the strong drum hits. To account for such phenomena, one often applies a logarithm to the energy values, for example, by switching to the logarithmic decibel scale (1.6) or by applying logarithmic compression (3.7). Note that, in the logarithmic case, the resulting novelty function corresponds to (the logarithm of) energy ratios rather than differences as shown by the following equation:

$$\Delta_{\mathrm{Energy}}^{\mathrm{Log}}(n) := |\log(\mathrm{E}_w^x(n+1)) - \log(\mathrm{E}_w^x(n))|_{\geq 0} = \left|\log\left(\frac{\mathrm{E}_w^x(n+1)}{\mathrm{E}_w^x(n)}\right)\right|_{\geq 0}. \tag{6.4}$$

As can be seen in Figure 6.3e, even the weak hihat onsets become visible in the logarithmic novelty function. On the downside, however, the logarithm may also amplify noise-like sound components, possibly leading to spurious peaks.

Another general problem in onset detection is energy fluctuation in nonsteady sounds as a result of vibrato or tremolo (see Section 1.3.4). Especially for purely energy-based procedures, amplitude modulations often lead to spurious peaks in the resulting novelty function. This is demonstrated by Figure 6.4, which shows the energy-based novelty function for the note C4 played by different instruments. While the novelty function shows a single clear peak in the case of a piano sound, there are many additional peaks in the case of a violin or flute sound. Furthermore, the relatively slow energy increase at the beginning of the violin sound leads to a smeared and temporally inaccurate onset peak.

To increase the robustness of onset detection, a typical approach is to first decompose the signal into several subbands that contain complementary frequency information. Then one computes a novelty function for each subband separately and suitably combines the individual functions to derive the onset information. For example, the subbands may correspond to musical pitches as discussed in Section 3.1.1, which results in pitch-based novelty functions. To exploit prior knowledge, one may use broader frequency bands that correspond to typical ranges of musical instruments (see Exercise 1.11). In the next section, we study an approach that decomposes a signal into subbands that correspond to the spectral coefficients. In this case, the resulting novelty function measures spectral changes, which yields more refined information than purely energy-based approaches.

6.1.2 Spectral-Based Novelty

Onset detection becomes a much harder problem for polyphonic music with simultaneously occurring sound events. A musical event of low intensity may be masked by an event of high intensity. Energy fluctuations (e.g., coming from vibrato) in the sustain phase of one instrument may be stronger than energy increases in the attack phase of other instruments. Therefore, in the case of multiple instruments playing at the same time, it is generally hard to detect all onsets when using purely energy-based methods. However, the characteristics of note onsets may strongly depend on the respective type of instrument. For example, for percussive instruments with an impulse-like onset, one can observe a sudden increase in energy that is spread across the entire spectrum of frequencies (see Figure 2.21a). Such noise-like broadband transients may be observable in certain frequency bands even in polyphonic mixtures. In particular, since the energy of harmonic sources is concentrated more in the lower part of the spectrum, transients are often well detectable in the higher-frequency region.

Motivated by such observations, the idea of spectral-based novelty detection is to first convert the signal into a time–frequency representation and then to capture changes in the frequency content. In the following, let $\mathcal{X}$ be the discrete STFT of the DT-signal x as defined in (2.26) or (2.148). For a discussion of the various parameters, including the sampling rate $F_s = 1/T$, the window length N of the discrete window w, and the hop size H, we refer to Section 2.5.3 or Section 3.1.1. For the moment, we only need to keep in mind that $\mathcal{X}(n,k) \in \mathbb{C}$ denotes the kth Fourier coefficient for frequency index $k \in [0:K]$ and time frame $n \in \mathbb{Z}$, where $K = N/2$ is the frequency index corresponding to the Nyquist frequency.

To detect spectral changes in the signal, one basically computes the difference between subsequent spectral vectors using a suitable distance measure. This results in a **spectral-based novelty function**, which is also known as the **spectral flux**. There are many different ways of computing such a novelty function, which depend not only on the parameters of the STFT and the distance measure, but also on pre- and postprocessing steps that are often applied.

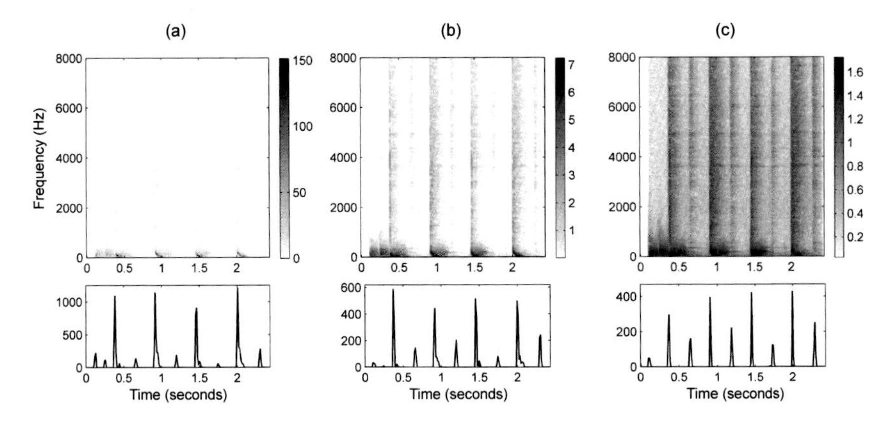

Fig. 6.5 Logarithmic compression (using the same audio excerpt as in Figure 6.1). The figure shows the respective magnitude spectrogram (top) and the resulting novelty function $\bar{\Delta}_{\mathrm{Spectral}}$ (bottom). **(a)** Magnitude spectrogram. **(b)** Compressed spectrogram using $\gamma = 1$. **(c)** Compressed spectrogram using $\gamma = 1000$.

In the following, we describe a typical procedure. First, to enhance weak spectral components, we apply a **logarithmic compression** to the spectral coefficients. Such a step, as we have already encountered in the context of chroma features (Figure 3.7), is often applied to account for the logarithmic sensation of sound intensity and to balance out the dynamic range of the signal. To obtain the compressed spectrogram, we apply the function Γ_γ of (3.7) to the magnitude spectrogram $|\mathcal{X}|$. This yields

$$\mathcal{Y} := \Gamma_\gamma(|\mathcal{X}|) = \log(1 + \gamma \cdot |\mathcal{X}|) \tag{6.5}$$

for a suitable constant $\gamma \geq 1$. In onset detection, logarithmic compression is particularly helpful for enhancing the comparatively weak high-frequency information. This is also illustrated by Figure 6.5, which continues our example "Another One Bites the Dust" from Figure 6.1. In the visualization of the original spectrogram $|\mathcal{X}|$ (Figure 6.5a), the harmonic components of the bass are visible in the low-frequency part. However, the transients at the beat positions can hardly be recognized. Using a compressed spectrogram with $\gamma = 1$ (Figure 6.5b), the vertical structures of the transients become more prominent—even the weak transients of the hihat between subsequent beats become visible. By increasing γ, the low-intensity values are further enhanced. On the downside, a large compression factor γ may also amplify nonrelevant noise-like components.

In the next step, we compute the discrete temporal derivative of the compressed spectrum $\mathcal{Y}$. Similarly to the energy-based novelty function, we only consider the positive differences (increase in intensity) and discard negative ones. This yields the spectral-based novelty function $\Delta_{\mathrm{Spectral}} : \mathbb{Z} \to \mathbb{R}$ defined by

$$\Delta_{\mathrm{Spectral}}(n) := \sum_{k=0}^{K} |\mathcal{Y}(n+1,k) - \mathcal{Y}(n,k)|_{\geq 0} \tag{6.6}$$

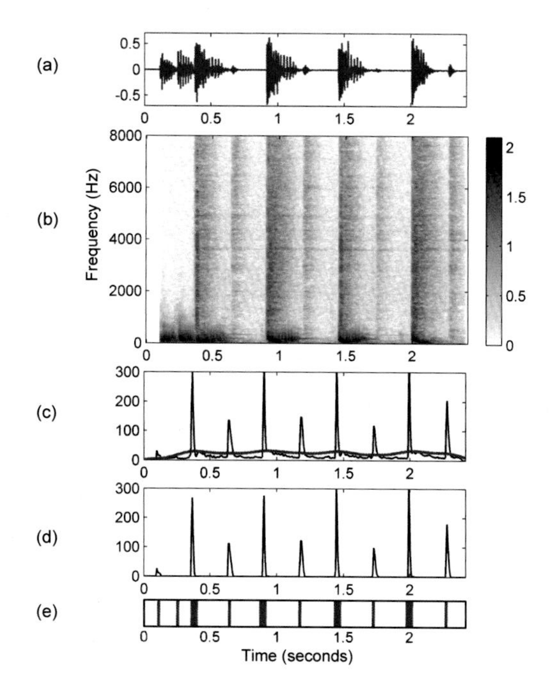

Fig. 6.6 Computation of a spectral-based novelty function for the signal from Figure 6.1. **(a)** Waveform. **(b)** Compressed spectrogram using $\gamma = 100$. **(c)** Novelty function $\Delta_{\mathrm{Spectral}}$ and local average function μ (in thick/red). **(d)** Novelty function $\bar{\Delta}_{\mathrm{Spectral}}$. **(e)** Annotated note onsets (the four beat positions are marked by thick lines).

for $n \in \mathbb{Z}$, where we use the half-wave rectification as introduced in (6.2). One can further enhance the properties of the novelty function by applying suitable postprocessing steps. For example, in view of a subsequent peak-picking step, one objective may be to enhance the peak structure of the novelty function, while suppressing small fluctuations. To this end, we introduce a local average function $\mu : \mathbb{Z} \to \mathbb{R}$ by setting

$$\mu(n) := \frac{1}{2M+1} \sum_{m=-M}^{M} \Delta_{\mathrm{Spectral}}(n+m), \tag{6.7}$$

$n \in \mathbb{Z}$, where the parameter $M \in \mathbb{N}$ determines the size of an averaging window. The enhanced novelty function $\bar{\Delta}_{\mathrm{Spectral}}$ is obtained by subtracting the local average from $\Delta_{\mathrm{Spectral}}$ and by only keeping the positive part (half-wave rectification):

$$\bar{\Delta}_{\mathrm{Spectral}}(n) := \left|\Delta_{\mathrm{Spectral}}(n) - \mu(n)\right|_{\geq 0} \tag{6.8}$$

for $n \in \mathbb{Z}$. Figure 6.6 illustrates the computational pipeline by means of our running example. As opposed to the energy-based novelty functions (Figure 6.3), the enhanced spectral-based novelty function $\bar{\Delta}_{\mathrm{Spectral}}$ (Figure 6.6d) not only indicates the onsets at the four beat positions, but also has significant peaks at the four weak hihat onsets between the beats. Even though the hihat sounds have a comparatively low intensity, they produce sharp transients, which are captured well by the compressed magnitude spectrogram (see also Figure 6.5c).

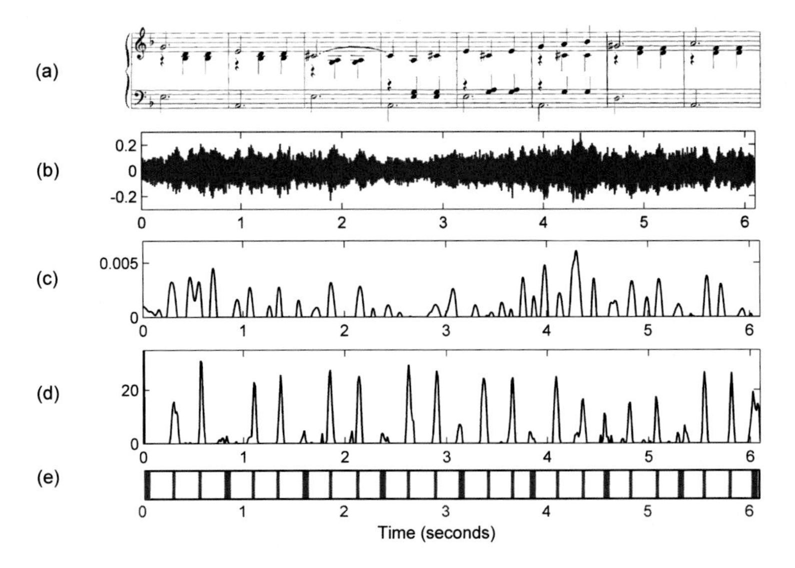

Fig. 6.7 Different novelty functions for an audio excerpt of Shostakovich's Waltz No. 2 from the "Suite for Variety Orchestra No. 1." **(a)** Score representation (in a piano reduced version). **(b)** Waveform. **(c)** Energy-based novelty function. **(d)** Spectral-based novelty function. **(e)** Annotated note onsets (downbeat positions are marked by thicker lines).

As a second example, let us have a look at an excerpt of an orchestra recording of the Waltz No. 2 from Dimitri Shostakovich's Suite for Variety Orchestra No. 1, an example we have already used in Figure 4.11. The first beats (downbeats) of the $3/4$ meter are played softly by nonpercussive instruments, leading to relatively weak and blurred onsets. In contrast, the second and third beats are played sharply ("staccato"), supported by percussive instruments. These properties are also reflected by the spectral-based novelty function shown in Figure 6.7d. The peaks that correspond to downbeats are hardly visible or even missing, whereas the peaks that correspond to the percussive beats are much more pronounced. The figure also shows the improvements one obtains for this example when using spectral-based methods (Figure 6.7d) compared with purely energy-based methods (Figure 6.7c).

As said before, there are many more approaches for computing spectral-based novelty functions. For example, as with the energy-based case, it may be beneficial to first split up the spectrum into several frequency bands (often five to eight logarithmically spaced bands are used). The resulting bandwise novelty functions are then weighted and summed up to yield the single overall novelty function (see Exercise 6.4).

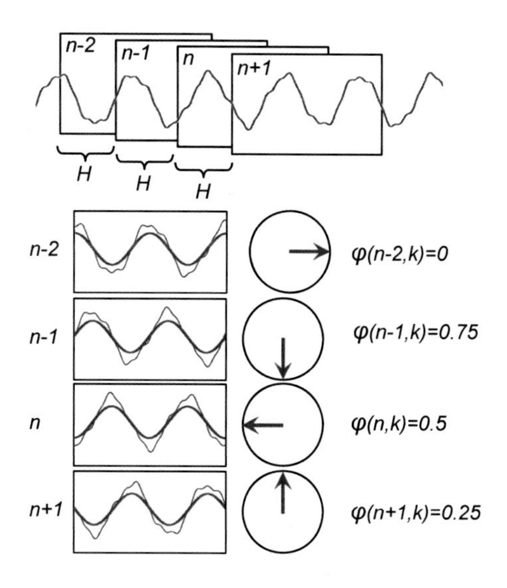

Fig. 6.8 Locally stationary signal and its correlation to a sinusoid corresponding to frequency index k for the frames $n-2$, $n-1$, n, and $n+1$. The angular representation of the phases is indicated by the circles.

6.1.3 Phase-Based Novelty

In the definition of the spectral-based novelty function, we have only used the magnitude of the spectral coefficients. However, the phases of the complex coefficients are also an important source of information for various audio analysis and synthesis tasks. In the following, we show how the phase information can be used for onset detection. In particular, we exploit the fact that stationary tones have a stable phase, while transients have an unstable phase. For another application of the phase information, we refer to Section 8.2.1.

As before, let $\mathcal{X}(n,k) \in \mathbb{C}$ be the complex-valued Fourier coefficient for frequency index $k \in [0:K]$ and time frame $n \in \mathbb{Z}$. Using the polar coordinate representation (2.9), this complex coefficient can be written as

$$\mathcal{X}(n,k) = |\mathcal{X}(n,k)| \exp(2\pi i \varphi(n,k)) \tag{6.9}$$

with the phase $\varphi(n,k) \in [0,1)$ (see also Section 2.3.2.2). Intuitively, as we explained in Section 2.1.1.1, the phase $\varphi(n,k)$ determines how the sinusoid of frequency $F_{\mathrm{coef}}(k) = F_{\mathrm{s}} \cdot k/N$ (see (2.28)) has to be shifted to best correlate with the windowed signal corresponding to the nth frame. Let us assume that the signal x has a high correlation with this sinusoid (i.e., $|\mathcal{X}(n,k)|$ is large) and shows a steady behavior in a region of a number of subsequent frames $\ldots, n-2, n-1, n, n+1, \ldots$ (i.e., x is locally stationary). Then the phases $\ldots, \varphi(n-2,k)$, $\varphi(n-1,k)$, $\varphi(n,k)$, $\varphi(n+1,k), \ldots$ increase from frame to frame in a fashion that is linear in the hop size H of the STFT (see Figure 6.8). Therefore, the frame-wise phase difference in this region remains approximately constant (possibly up to some integer, as we discuss shortly in this section):

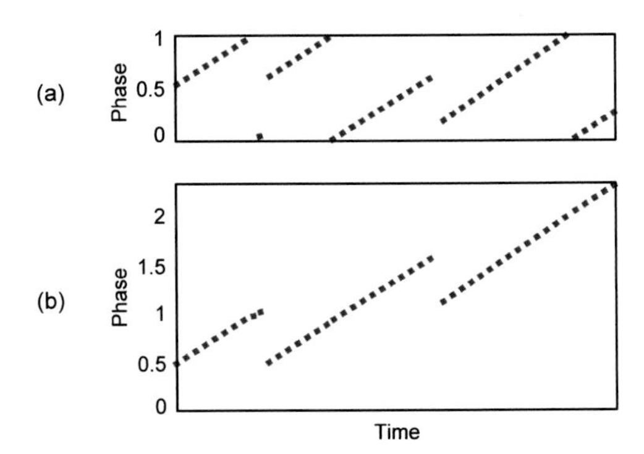

Fig. 6.9 Illustration of phase unwrapping. **(a)** Wrapped phase. **(b)** Unwrapped phase.

$$\varphi(n,k) - \varphi(n-1,k) \approx \varphi(n-1,k) - \varphi(n-2,k). \tag{6.10}$$

Let us define the first-order difference by

$$\varphi'(n,k) := \varphi(n,k) - \varphi(n-1,k) \tag{6.11}$$

and the second-order difference by

$$\varphi''(n,k) := \varphi'(n,k) - \varphi'(n-1,k). \tag{6.12}$$

Note that one obtains $\varphi''(n,k) \approx 0$ in steady regions of x. However, in transient regions, the phase behaves quite unpredictably across the entire frequency range. As a result, a simultaneous disturbance of the values $\varphi''(n,k)$ for $k \in [0:K]$ is a good indicator for note onsets. Motivated by this observation, we define the **phase-based novelty function** Δ_{Phase} by

$$\Delta_{\mathrm{Phase}}(n) = \sum_{k=0}^{K} |\varphi''(n,k)| \tag{6.13}$$

for $n \in \mathbb{Z}$.

At this point, we need to discuss a technical issue. Recall that the phase γ (in radians) of a complex number $c \in \mathbb{C}$ is defined only up to integer multiples of 2π (see (2.9)). Therefore, the phase is often constrained to the interval $[0,2\pi)$ and the number $\gamma \in [0,2\pi)$ is called the **principal value** of the phase. In the scenario of Fourier analysis, we are using the normalized phases $\varphi = \gamma/(2\pi)$. In this case, the interval $[0,1)$ represents the principal values. When considering a function or a time series of phase values (e.g., the phase values over the frames of an STFT as above), the choice of principal values may introduce unwanted discontinuities. These artificial phase jumps are the results of **phase wrapping**, where a phase value just below one is followed by a value just above zero (or vice versa). To avoid such discontinuities, one often applies a procedure called **phase unwrapping**, where the objective is to recover a possibly continuous sequence of (unwrapped) phase values (see Figure 6.9). Such a procedure, however, is in general not well defined since the

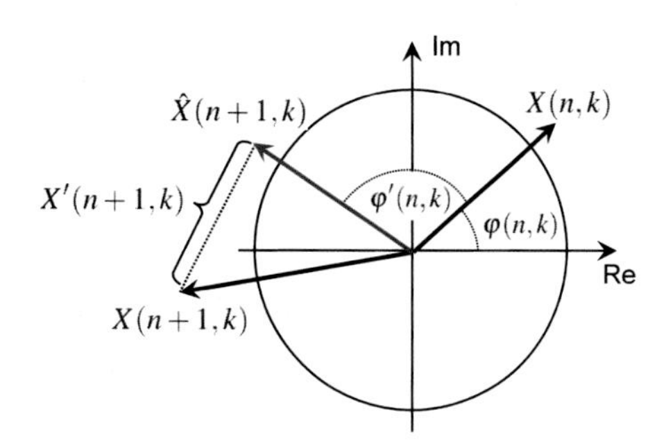

Fig. 6.10 Illustration of the complex-domain difference $\mathcal{X}'(n,k)$ between an estimated spectral coefficient $\hat{\mathcal{X}}(n+1,k)$ and the actual coefficient $\mathcal{X}(n+1,k)$.

original time series may possess "real" discontinuities that are hard to distinguish from "artificial" phase jumps. In the onset detection context, phase jumps due to wrapping may occur when computing the differences in (6.11) and (6.12). In these cases, one needs to use an unwrapped version of the phase. As an alternative, we introduce a **principal argument function**

$$\Psi : \mathbb{R} \to [-0.5, 0.5] \tag{6.14}$$

which maps phase differences into the range $[-0.5, 0.5]$. To this end, a suitable integer value is added to or subtracted from the original phase difference to yield a value in $[-0.5, 0.5]$. The differences as defined in (6.11) and (6.12) are then replaced by

$$\varphi'(n,k) := \Psi\big(\varphi(n,k) - \varphi(n-1,k)\big), \tag{6.15}$$
$$\varphi''(n,k) := \Psi\big(\varphi'(n,k) - \varphi'(n-1,k)\big). \tag{6.16}$$

Even though the principal argument function may cancel out large discontinuities in the phase differences, this effect is attenuated since we consider in (6.13) the sum of differences over all frequency indices.

6.1.4 Complex-Domain Novelty

We have seen that steady regions within a signal may be characterized by a phase-based criterion in the case that the sinusoid correlates well with the signal. However, if the magnitude of the Fourier coefficient $\mathcal{X}(n,k)$ is very small, the phase $\varphi(n,k)$ may exhibit a rather chaotic behavior due to small noise-like fluctuations that may occur even within a steady region of the signal. To obtain a more robust detector, one idea is to weight the phase information with the magnitude of the spectral coefficient. This leads to a complex-domain variant of the novelty function, which jointly considers phase and magnitude. The assumption of this variant is that the phase differences as well as the magnitude stay more or less constant in steady regions. Therefore, given the Fourier coefficient $\mathcal{X}(n,k)$, one obtains a steady-state estimate

$\hat{\mathcal{X}}(n+1,k)$ for the next frame by setting

$$\hat{\mathcal{X}}(n+1,k) = |\mathcal{X}(n,k)|\exp(2\pi i(\varphi(n,k)+\varphi'(n,k))) \tag{6.17}$$

(see Figure 6.10). Then, we can use the magnitude between the estimate $\hat{\mathcal{X}}(n+1,k)$ and the actual coefficient $\mathcal{X}(n+1,k)$ to obtain a measure of novelty:

$$\mathcal{X}'(n+1,k) = |\hat{\mathcal{X}}(n+1,k) - \mathcal{X}(n+1,k)|. \tag{6.18}$$

The complex-domain difference $\mathcal{X}'(n,k)$ quantifies the degree of nonstationarity for frame n and coefficient k. Note that this number does not discriminate between note onsets (energy increase) and note offsets (energy decrease). Therefore, we decompose $\mathcal{X}'(n,k)$ into a component $\mathcal{X}^+(n,k)$ of increasing magnitude and a component $\mathcal{X}^-(n,k)$ of decreasing magnitude:

$$\mathcal{X}^+(n,k) = \begin{cases} \mathcal{X}'(n,k) & \text{for } |\mathcal{X}(n,k)| > |\mathcal{X}(n-1,k)| \\ 0 & \text{otherwise,} \end{cases} \tag{6.19}$$

$$\mathcal{X}^-(n,k) = \begin{cases} \mathcal{X}'(n,k) & \text{for } |\mathcal{X}(n,k)| \leq |\mathcal{X}(n-1,k)| \\ 0 & \text{otherwise.} \end{cases} \tag{6.20}$$

A **complex-domain novelty function** Δ_{Complex} for detecting note onsets can then be defined by summing the values $\mathcal{X}^+(n,k)$ over all frequency coefficients:

$$\Delta_{\text{Complex}}(n,k) = \sum_{k=0}^{K} \mathcal{X}^+(n,k). \tag{6.21}$$

Similarly, for detecting general transients or note offsets, one may compute a novelty function using $\mathcal{X}'(n,k)$ or $\mathcal{X}^-(n,k)$, respectively.

6.2 Tempo Analysis

The extraction of tempo and beat information from audio recordings is a challenging problem in particular for music with weak note onsets and local tempo changes. For example, in the case of romantic piano music, the pianist often takes the freedom of speeding up and slowing down the tempo—an artistic means also referred to as **tempo rubato**. There is a wide range of music where the notions of tempo and beat remain rather vague or are even nonexistent. Sometimes, the rhythmic flow of music is deliberately interrupted or disturbed by **syncopation**, where certain notes outside the regular grid of beat positions are stressed. To make the problem of tempo and beat tracking feasible, most automated approaches rely on two basic assumptions. The first assumption is that beat positions occur at note onset positions, and the second assumption is that beat positions are more or less equally spaced—at least for a certain period of time. Even though both assumptions may be violated and

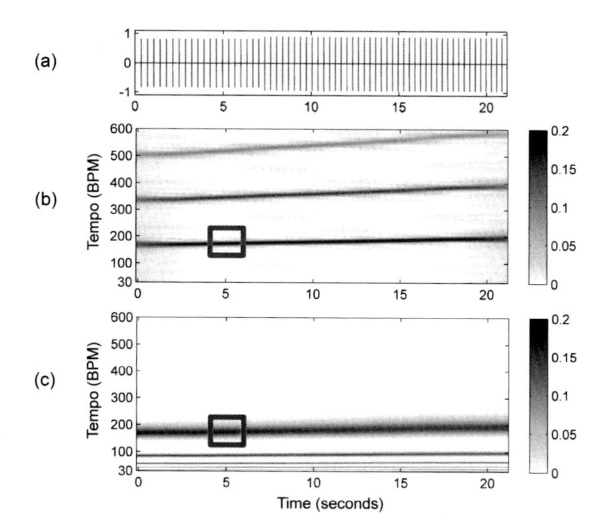

Fig. 6.11 Illustration of two different tempogram representations $\mathcal{T}$ of a click track with increasing tempo (170 to 200 BPM). The large values $\mathcal{T}(t,\tau)$ around $t = 5$ sec and $\tau = 180$ BPM are highlighted by the rectangular frames. **(a)** Novelty function of click track. **(b)** Tempogram with harmonics. **(c)** Tempogram with subharmonics.

inappropriate for certain types of music, they are convenient and reasonable for a wide range of music including most rock and popular songs.

Based on these two assumptions, we discuss in this section various time–tempo or tempogram representations, which capture local tempo characteristics of music signals (Section 6.2.1). To derive such representations, we study two methods for analyzing novelty functions with regard to reoccurring or quasiperiodic patterns. Using Fourier analysis, we show how to derive a tempogram by comparing the novelty function with templates that consist of windowed sinusoids, each representing a specific frequency or tempo (Section 6.2.2). For the second method, we discuss an autocorrelation approach where a tempogram is obtained by comparing a novelty function with localized time-shifted copies of itself (Section 6.2.3). Finally, we introduce robust mid-level representations referred to as cyclic tempograms (Section 6.2.4), which are the tempo-related counterpart of the harmony-related chroma representations. The properties of the tempogram representations are illustrated in the context of music segmentation.

6.2.1 Tempogram Representations

In Section 2.5.2, we studied the concept of a (magnitude) spectrogram, which represents the time–frequency content of a given signal. A large value $\mathrm{Spec}(t,\omega)$ of a spectrogram indicates that the signal contains at time instance t a periodic component that corresponds to the frequency ω (see (2.141)). We now introduce a similar concept referred to as a **tempogram**, which indicates for each time instance the local relevance of a specific tempo for a given music recording. Mathematically, we model a tempogram as a function

$$\mathcal{T}:\mathbb{R}\times\mathbb{R}_{>0}\to\mathbb{R}_{\geq 0} \tag{6.22}$$

depending on a time parameter $t\in\mathbb{R}$ measured in seconds and a tempo parameter $\tau\in\mathbb{R}_{>0}$ measured in beats per minute (BPM). Intuitively, the value $\mathcal{T}(t,\tau)$ indicates the extent to which the signal contains a locally periodic pulse of a given tempo τ in a neighborhood of time instance t. For example, the tempogram of Figure 6.11b has a large value $\mathcal{T}(5,180)$, thus indicating that the music signal has a dominant tempo of $\tau=180$ BPM around time position $t=5$ sec. Just as with spectrograms (Section 2.5.3), one computes a tempogram in practice only on a discrete time–tempo grid. As before, we assume that the sampled time axis is given by $[1:N]$. To avoid boundary cases and to simplify the notation in the subsequent considerations, we extend this axis to $\mathbb{Z}$. (The respective representations are then extended by, e.g., zero-padding.) Furthermore, let $\Theta\subset\mathbb{R}_{>0}$ be a finite set of tempi specified in BPM. Then, a **discrete tempogram** is a function

$$\mathcal{T}:\mathbb{Z}\times\Theta\to\mathbb{R}_{\geq 0}. \tag{6.23}$$

Most approaches for deriving a tempogram representation from a given audio recording proceed in two steps. Based on the assumption that pulse positions usually go along with note onsets, the music signal is first converted into a novelty function (see Section 6.1). This function typically consists of impulse-like spikes, each indicating a note onset position. In the second step, the locally periodic behavior of the novelty function is analyzed. To obtain a tempogram, one quantifies the periodic behavior for various periods $T>0$ (given in seconds) in a neighborhood of a given time instance. The rate $\omega=1/T$ (measured in Hz) and the tempo τ (measured in BPM) are related by

$$\tau=60\cdot\omega. \tag{6.24}$$

For example, a sequence of impulse-like spikes that are regularly spaced with period $T=0.5$ sec corresponds to a rate of $\omega=1/T=2$ Hz or a tempo of $\tau=120$ BPM.

One major problem in determining the tempo of a music recording arises from the fact that pulses in music are often organized in complex hierarchies that represent the rhythm. In particular, there are various levels that are presumed to contribute to the human perception of tempo and beat. For example, as illustrated by Figure 6.12, one may consider the tempo on the **tactus** level, which typically corresponds to the quarter note level and often matches the foot tapping rate. Thinking at a larger musical scale, one may also perceive the tempo at the **measure** level, in particular when listening to fast music or to highly expressive music with strong rubato. Finally, one may also consider the **tatum** (temporal atom) level, which refers to the fastest repetition rate of musically meaningful accents occurring in the signal.

Often the tempo ambiguity that arises from the existence of different pulse levels is also reflected in a tempogram $\mathcal{T}$. Higher pulse levels often correspond to integer multiples $\tau,2\tau,3\tau,\ldots$ of a given tempo τ. As with pitch (Section 1.3.2), we call such integer multiples **(tempo) harmonics** of τ. Furthermore, integer fractions $\tau,\tau/2,\tau/3,\ldots$ are referred to as **(tempo) subharmonics** of τ. Analogous to the

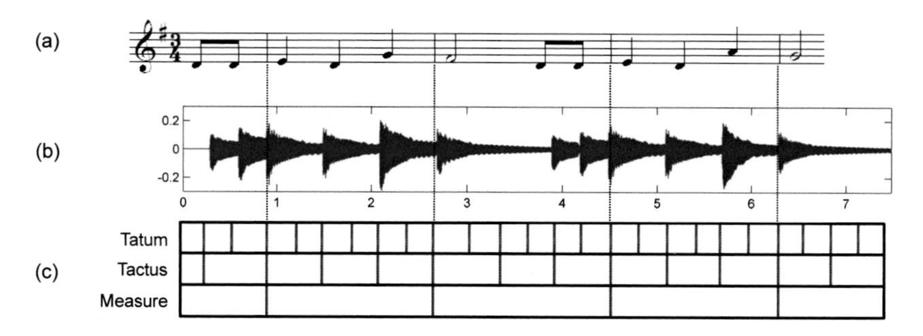

Fig. 6.12 Illustration of various pulse levels. In this example, the tactus level corresponds to the quarter note and the tatum level to the eighth note level. **(a)** Score representation. **(b)** Waveform of an audio excerpt of "Happy Birthday to you." **(c)** Annotation of pulse levels.

notion of an octave for musical pitches (see Section 1.1.1), the difference between two tempi with half or double the value is called a **tempo octave**. For an illustration, we refer to Figure 6.11, which shows two different types of tempograms for a click track of increasing tempo (raising from 170 to 200 BPM over the course of 20 sec). The tempogram of Figure 6.11b emphasizes tempo harmonics, whereas the tempogram of Figure 6.11c emphasizes tempo subharmonics. In the following, we will study two conceptually different methods that are used to derive these two tempograms.

6.2.2 Fourier Tempogram

As a first periodicity estimation method, we show how a short-time Fourier transform can be used to derive a tempogram from a given novelty function $\Delta : \mathbb{Z} \to \mathbb{R}$. Dealing with a discrete-time signal Δ, we consider the discrete version of the STFT as discussed in Section 2.5.3. To this end, we fix a window function $w : \mathbb{Z} \to \mathbb{R}$ of finite length centered at $n = 0$ (e.g., a sampled Hann window as defined in (2.140)). Then, for a frequency parameter $\omega \in \mathbb{R}_{\geq 0}$ and time parameter $n \in \mathbb{Z}$, the complex Fourier coefficient $\mathcal{F}(n, \omega)$ is defined by

$$\mathcal{F}(n,\omega) := \widetilde{\Delta^w}(n,\omega) = \sum_{m\in\mathbb{Z}} \Delta(m)\overline{w}(m-n)\exp(-2\pi i\omega m). \tag{6.25}$$

This definition corresponds to (2.143) when using a hop size $H = 1$. Converting frequency to tempo values based on (6.24), we define the (discrete) **Fourier tempogram** $\mathcal{T}^{\mathrm{F}} : \mathbb{Z} \times \Theta \to \mathbb{R}_{\geq 0}$ by

$$\mathcal{T}^{\mathrm{F}}(n,\tau) := |\mathcal{F}(n,\tau/60)|. \tag{6.26}$$

The Fourier-based analysis of the novelty function is also illustrated by Figure 6.13, which continues our Shostakovich example from Figure 6.7. As the Fourier tempogram $\mathcal{T}^{\mathrm{F}}$ (Figure 6.13b) reveals, the dominant tempo of this excerpt is between 200 and 300 BPM. Starting with roughly $\tau = 225$ BPM, the tempo slightly increases over time. An entry $\mathcal{T}^{\mathrm{F}}(n,\tau)$ of the tempogram is obtained by locally comparing the novelty function Δ in a neighborhood of n with a windowed sinusoid that represents the tempo τ. This kind of analysis is shown in Figure 6.13c for a time index n that corresponds to the physical time $t = 2$ sec and a frequency parameter ω that corresponds to the tempo $\tau = 230$ BPM. In this case, the positive parts of the windowed sinusoid nicely align with the impulse-like peaks of the novelty function Δ, whereas the negative parts of the sinusoid fall into the zero-regions of Δ. As a result, there is a high correlation between the windowed sinusoid and Δ, which leads to a large coefficient $\mathcal{T}^{\mathrm{F}}(n,\tau)$. In contrast, using a sinusoid that represents only half this tempo leads to a small coefficient, as illustrated by Figure 6.13d. In this case, every second peak of Δ falls into the positive parts of the sinusoid, whereas the remaining peaks of Δ fall into the negative parts of the sinusoid. Because of the resulting cancellations, the correlation between Δ and the sinusoid becomes small. Finally, Figure 6.13e illustrates that one obtains a high correlation when using a sinusoid that represents twice the main tempo. In this case, the peaks of Δ are aligned with every second positive part of the sinusoid, whereas all other parts of the sinusoid fall into the zero-regions of Δ. Our discussion shows that a Fourier tempogram generally indicates tempo harmonics, but suppresses tempo subharmonics. This fact is illustrated by Figure 6.11b, which shows the Fourier tempogram of a synthetic click track. Also, in our Shostakovich example, the second tempo harmonic starting at $\tau = 450$ BPM is clearly visible in $\mathcal{T}^{\mathrm{F}}$ (Figure 6.13b). Interestingly, because of the weak downbeats every third beat within the 3/4 meter (see our discussion of Figure 6.7), the tempogram $\mathcal{T}^{\mathrm{F}}$ also shows some larger coefficients that correspond to $1/3$ and $2/3$ of the main tempo (see Exercise 6.5)

For practical applications, $\mathcal{T}^{\mathrm{F}}$ is computed only for a small number of tempo parameters. For example, one may choose the set $\Theta = [30:600]$ covering the (integer) musical tempi between 30 and 600 BPM. The bounds are motivated by the assumption that only musical events showing a temporal separation between roughly 100 ms (600 BPM) and 2 sec (30 BPM) contribute to the perception of tempo. This tempo range requires a spectral analysis of high resolution in the lower frequency range. Therefore, a straightforward FFT as discussed in Section 2.4.3 is not suitable. However, since only relatively few frequency bands (tempo values) are needed for the tempogram, computing the required Fourier coefficients individually according to (6.25) still has a reasonable computational complexity. As for the temporal resolution, one can set w to be a sampled Hann window as defined in (2.140) of size $2N+1$ for some $N \in \mathbb{N}$. Depending on the respective application and the nature of the music recording, a window size corresponding to 4–12 sec of audio is a reasonable range. Finally, note that the feature rate of the resulting tempogram can be adjusted by introducing a hop size parameter H in (6.25) as used in (2.143).

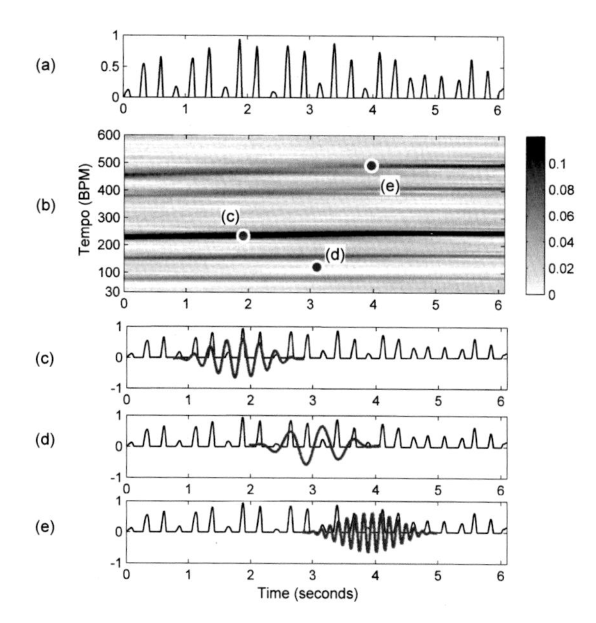

Fig. 6.13 Fourier-based tempo analysis for the Shostakovich example from Figure 6.7. **(a)** Novelty function Δ. **(b)** Fourier tempogram $\mathcal{T}^{\mathrm{F}}$. **(c–e)** Correlation of Δ and various analyzing windowed sinusoids.

6.2.3 Autocorrelation Tempogram

As a second periodicity estimation method, we now discuss an autocorrelation-based approach. Generally speaking, the **autocorrelation** is a mathematical tool for measuring the similarity of a signal with a time-shifted version of itself. Since the inner product as defined in (2.43) is used for this measurement, this technique is also known as the **sliding inner product**. In the following, we only consider the case of discrete-time and real-valued signals. Let $x \in \ell^2(\mathbb{Z})$ be such a signal having finite energy (see (2.41)). The autocorrelation $R_{xx} : \mathbb{Z} \to \mathbb{R}$ of the real-valued signal x is defined by

$$R_{xx}(\ell) = \sum_{m\in\mathbb{Z}} x(m)x(m-\ell), \tag{6.27}$$

which yields a function that depends on the time-shift or **lag** parameter $\ell \in \mathbb{Z}$. As shown in Exercise 6.6, the autocorrelation is well defined for signals in the space $\ell^2(\mathbb{Z})$. Furthermore, $R_{xx}(\ell)$ is maximal for $\ell = 0$ and symmetric in ℓ. Intuitively, if the autocorrelation is large for a given lag, then the signal contains repeating patterns that are separated by a time period as specified by the lag parameter.

We now apply the autocorrelation in a local fashion for analyzing a given novelty function $\Delta : \mathbb{Z} \to \mathbb{R}$ in the neighborhood of a given time parameter n. As in the case of the Fourier tempogram discussed in the last section, we fix a window function $w : \mathbb{Z} \to \mathbb{R}$ of finite length centered at $n = 0$. The windowed version $\Delta_{w,n} : \mathbb{Z} \to \mathbb{R}$ localized at point $n \in \mathbb{Z}$ is defined by

$$\Delta_{w,n}(m) := \Delta(m)w(m-n), \tag{6.28}$$

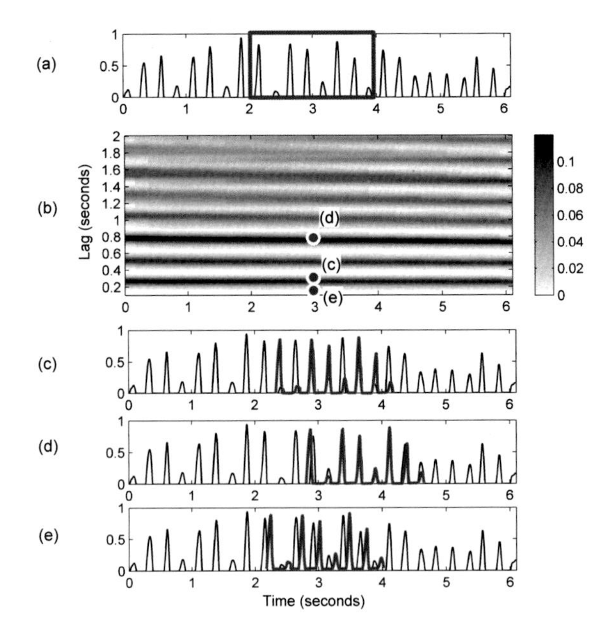

Fig. 6.14 Autocorrelation-based tempo analysis for the Shostakovich example from Figure 6.7. **(a)** Novelty function Δ. **(b)** Time-lag representation $\mathcal{A}$. **(c–e)** Correlation of Δ and various time-shifted windowed sections.

$m \in \mathbb{Z}$. Recall that we have used a similar definition when introducing the STFT (see (2.133)). To obtain the **short-time autocorrelation** $\mathcal{A} : \mathbb{Z} \times \mathbb{Z} \to \mathbb{R}$, we apply (6.27) to $\Delta_{w,n}$ and define

$$\mathcal{A}(n,\ell) := \sum_{m\in\mathbb{Z}} \Delta(m)w(m-n)\Delta(m-\ell)w(m-n-\ell). \tag{6.29}$$

When assuming that the window function w is of finite length, the autocorrelation of the localized novelty function is zero for all but a finite number of time lag parameters. In the following, let us assume that the support of the window function w lies in the interval $[-L:L]$ for some $L \in \mathbb{N}$. Then one has $\mathcal{A}(n,\ell) = 0$ for $|\ell| \geq 2L+1$ (see Exercise 6.7). Because of this property and the symmetry of the autocorrelation, one only needs to consider the time lag parameters $\ell \in [0:2L]$. Furthermore, because of the windowing, at most $2L+1-\ell$ of the summands in (6.29) are nonzero. To balance out the effect of the windowing, the value $\mathcal{A}(n,\ell)$ may be divided by a factor that depends on the window properties and the overlap $2L+1-\ell$ of the window and its time-shifted version.

Visualizing the short-time autocorrelation $\mathcal{A}$ leads to a **time-lag representation**. Before we discuss how this representation can be converted into a time–tempo representation, let us first have a look at Figure 6.14, which continues our Shostakovich example. The window w used in this example is a rectangular window that has a length corresponding to 2 sec of the original audio recording. Let us consider the time index n corresponding to the time instance $t = 3$ sec. To compute $\mathcal{A}(n,\ell)$, one only considers the section of the novelty function Δ between 2 sec and 4 sec (Figure 6.14a). We have seen that the tempo of our Shostakovich recording is roughly 230 BPM in this section. In other words, the duration of the interval between two subsequent beats is roughly $s = 0.26$ sec. Let us consider the lag parameter ℓ that corresponds to a time shift of $s = 0.26$ sec. Then, as illustrated by Figure 6.14c, the novelty function in this section nicely correlates with its time-shifted version: the peaks of the section fall onto peaks of the section shifted by one beat period. The same holds when shifting the section by two, three or more beat periods. For example, Figure 6.14d shows the case $s = 0.78$ sec (three beat periods). This period corresponds to a tempo of 77 BPM, which is the tempo on the measure level. In contrast, when using a lag ℓ that corresponds to half a beat period $s = 0.13$ sec (double tempo 461 BPM), the peaks of the section and the peaks of the shifted section miss each other, thus resulting in a coefficient $\mathcal{A}(n,\ell)$ close to zero. This case is illustrated by Figure 6.14e.

To obtain a time–tempo representation from the time-lag representation, one needs to convert the lag parameter into a tempo parameter. To this end, one requires the frame rate or time resolution of the novelty function. Suppose that each time frame corresponds to r seconds, then a time lag of ℓ (given in frames) corresponds to $\ell \cdot r$ seconds. Since a shift of $\ell \cdot r$ seconds corresponds to a rate of $1/(\ell \cdot r)$ Hz, one obtains from (6.24) the tempo

$$\tau = \frac{60}{r \cdot \ell} \text{ BPM}. \tag{6.30}$$

Based on this conversion, the lag axis can be interpreted as a tempo axis as illustrated by Figure 6.15b. This allows us to define the **autocorrelation tempogram** $\mathcal{T}^{\mathrm{A}}$ by setting

$$\mathcal{T}^{\mathrm{A}}(n,\tau) := \mathcal{A}(n,\ell) \tag{6.31}$$

for each tempo $\tau = 60/(r \cdot \ell)$, $\ell \in [1:L]$. Note that in this case, since the tempo values are reciprocal to the linearly sampled lag values, the tempo axis is sampled in a nonlinear fashion. To obtain a tempogram $\mathcal{T}^{\mathrm{A}} : \mathbb{Z} \times \Theta \to \mathbb{R}_{\geq 0}$ that is defined on the same tempo set Θ as the Fourier tempogram $\mathcal{T}^{\mathrm{F}}$, one can use standard resampling and interpolation techniques applied to the tempo domain. The result of such an interpolation step is shown in Figure 6.15c.

As another example, Figure 6.11c shows the autocorrelation tempogram of a click track. This figure illustrates that, as opposed to the Fourier tempogram, an autocorrelation tempogram exhibits tempo subharmonics, but suppresses tempo harmonics. We have already given the argument for this behavior when discussing Figure 6.14: a high correlation of a local section of the novelty function with the section shifted by ℓ samples also implies a high correlation with a section shifted by $k \cdot \ell$ lags for integers $k \in \mathbb{N}$. Assuming that ℓ corresponds to tempo τ, the lag $k \cdot \ell$ corresponds to the subharmonic τ/k.

This property is also evident in our Shostakovich example. Similar to the Fourier tempogram $\mathcal{T}^{\mathrm{F}}$ (Figure 6.13b), the autocorrelation tempogram $\mathcal{T}^{\mathrm{A}}$ (Figure 6.15c) reveals the dominant tempo at $\tau = 225$ BPM, which corresponds to the quarter note level. However, as opposed to $\mathcal{T}^{\mathrm{F}}$, the dominant tempo revealed by $\mathcal{T}^{\mathrm{A}}$ is at $\tau = 75$ BPM, which corresponds to the tempo on the measure level and is the third

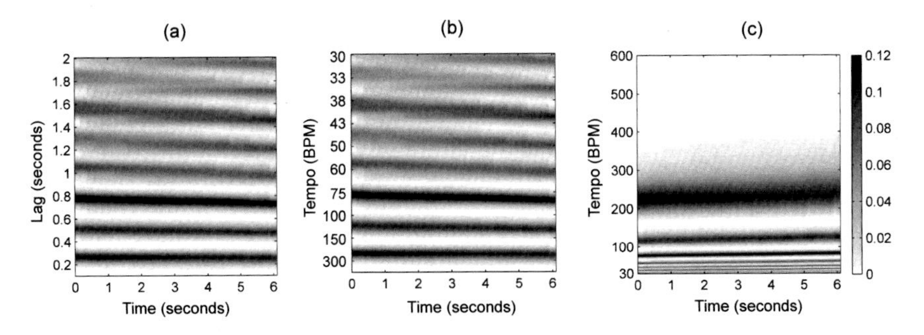

Fig. 6.15 Conversion from lag to tempo. **(a)** Time-lag representation with linear lag axis. **(b)** Representation from (a) with tempo axis. **(c)** Time–tempo representation with linear tempo axis.

subharmonic of $\tau = 225$ BPM. Reflecting the 3/4 meter of the waltz, the dominance of the tempi $\tau = 225$ BPM and $\tau = 75$ BPM is also of musical relevance. In conclusion, one may say that the Fourier tempogram and autocorrelation tempogram yield different types of tempo information and ideally complement each other.

Assuming a more or less steady tempo, it suffices to determine one **global** tempo value for the entire recording. Such a value may be obtained by averaging the tempo values obtained from a frame-wise periodicity analysis. For example, based on a tempogram representation, one can average the tempo values over all time frames to obtain a function $\mathcal{T}_{\text{Average}} : \Theta \to \mathbb{R}_{\geq 0}$ that only depends on $\tau \in \Theta$. Assuming that the relevant time positions lie in the interval $[1:N]$, one may define $\mathcal{T}_{\text{Average}}$ by

$$\mathcal{T}_{\text{Average}}(\tau) := \frac{1}{N} \sum_{n\in[1:N]} \mathcal{T}(n,\tau). \tag{6.32}$$

The maximum

$$\hat{\tau} := \max\{\mathcal{T}_{\text{Average}}(\tau) \mid \tau \in \Theta\} \tag{6.33}$$

of this function then yields an estimate for the global tempo of the recording. Of course, more refined methods for estimating a single tempo value may be applied. For example, instead of using a simple average in (6.32), we may apply median filtering, which is more robust to outliers and noise. Also, to alleviate the problem of tempo octave confusion, one may improve the result by a combined usage of the Fourier and autocorrelation tempograms.

When dealing with music that exhibits significant tempo changes, one needs to estimate the **local** tempo in the neighborhood of each time instance, which is a much harder problem than global tempo estimation. Having computed a tempogram, the frame-wise maximum yields a good indicator of the locally dominating tempo. In the case that the tempo is relatively steady over longer periods of time, one may increase the window size to obtain more robust and smoother tempo estimates. However, it then becomes harder to detect sudden tempo changes and local tempo fluctuations—the same trade-off we have already encountered in the case of the

STFT (see Section 2.5.2). Furthermore, instead of simply taking the frame-wise maximum—a strategy that is prone to local inconsistencies and outliers—global optimization techniques based on dynamic programming may be used to obtain smooth tempo trajectories. Such strategies will be discussed in Section 6.3 in the context of beat tracking. In both global and local tempo estimation, one often has to struggle with confusions of tempo harmonics and subharmonics, which are the result of the existence of various pulse levels such as measure, tactus, and tatum. In the following section, we introduce a robust mid-level representation that is impervious to tempo octave confusions while still capturing local tempo information.

6.2.4 Cyclic Tempogram

The various pulse levels mentioned above can be seen in analogy to the existence of harmonics in the pitch context (see Section 1.3.2). To reduce the effects of harmonics, we introduced in Section 3.1.2 the concept of chroma-based audio features. By identifying pitches that differ by one or several octaves, we obtained a cyclic mid-level representation that captures harmonic information while being robust to changes in timbre. Inspired by the concept of chroma features, we now introduce the concept of cyclic tempograms. The idea is to form tempo equivalence classes by identifying tempi that differ by a power of two. More precisely, we say that two tempi τ_1 and τ_2 are **octave equivalent**, if they are related by $\tau_1 = 2^k \tau_2$ for some $k \in \mathbb{Z}$. For a tempo parameter τ, we denote the resulting tempo equivalence class by $[\tau]$. For example, for $\tau = 120$ one obtains $[\tau] = \{\ldots, 30, 60, 120, 240, 480 \ldots\}$. Given a tempogram representation $\mathcal{T} : \mathbb{Z} \times \mathbb{R}_{>0} \to \mathbb{R}_{\geq 0}$, we define the **cyclic tempogram** by

$$\mathcal{C}(n, [\tau]) := \sum_{\lambda \in [\tau]} \mathcal{T}(n, \lambda). \tag{6.34}$$

Note that the tempo equivalence classes topologically correspond to a circle. Fixing a reference tempo τ_0, the cyclic tempogram can be represented by a mapping $\mathcal{C}_{\tau_0} : \mathbb{Z} \times \mathbb{R}_{>0} \to \mathbb{R}_{\geq 0}$ defined by

$$\mathcal{C}_{\tau_0}(n, s) := \mathcal{C}(n, [s \cdot \tau_0]) \tag{6.35}$$

for $n \in \mathbb{Z}$ and a **scaling parameter** $s \in \mathbb{R}_{>0}$. Note that $\mathcal{C}_{\tau_0}(n, s) = \mathcal{C}_{\tau_0}(n, 2^k s)$ for $k \in \mathbb{Z}$. In particular, $\mathcal{C}_{\tau_0}$ is completely determined by its values $s \in [1, 2)$.

These definitions are illustrated by Figure 6.16, which shows various tempograms for a click track of increasing tempo (110 to 130 BPM), similar to the one used in Figure 6.11. As demonstrated by Figure 6.16a, the Fourier tempogram $\mathcal{T}^{\mathrm{F}}$ indicates the tempo as well as its tempo harmonics. Using a reference tempo $\tau_0 = 60$ BPM, the resulting **cyclic Fourier tempogram**, which we denote by $\mathcal{C}^{\mathrm{F}}_{\tau_0}$, is shown in Figure 6.16c. In the pitch context, given a reference frequency ω, the frequency 3ω is an octave plus a fifth higher, and 3ω can be regarded as the **dominant** to the **tonic** ω. In analogy to the pitch context, we call the tempo class $[3\tau]$, which

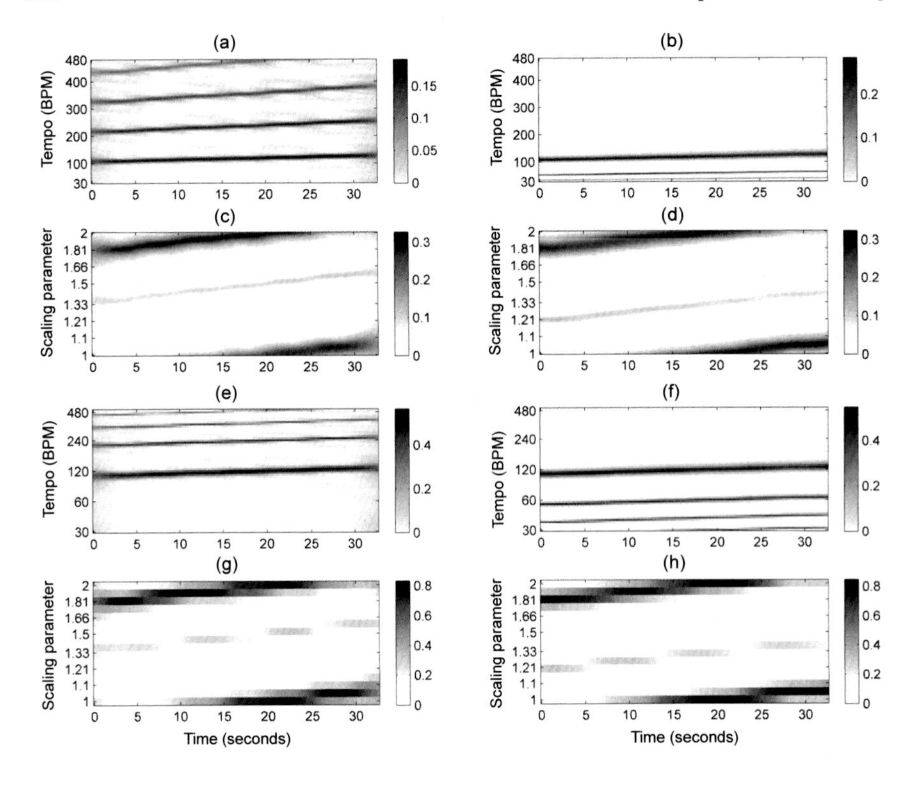

Fig. 6.16 Various tempogram representations for a click track of increasing tempo (110 to 130 BPM). **(a,b)** Fourier and autocorrelation tempogram representations. **(c,d)** Cyclic tempograms obtained from (a,b) using $\tau_0 = 60$. **(e,f)** Tempograms with logarithmic tempo axis using $M = 15$. **(g,h)** Cyclic tempograms from (e,f) using $\tau_0 = 60$.

corresponds to the third harmonic 3τ, the **tempo dominant** of $[\tau]$. In Figure 6.16c, the tempo dominant is visible as the weak increasing line starting with $s = 1.33$ at time $t = 0$. Similarly, the autocorrelation tempogram $\mathcal{T}^{\mathrm{A}}$ with its tempo subharmonics is displayed in Figure 6.16b. We denote the resulting **cyclic autocorrelation tempogram** by $\mathcal{C}^{\mathrm{A}}_{\tau_0}$ (see Figure 6.16d). Again inspired by the case of pitches, we call the tempo class $[\tau/3]$ of the third subharmonic $\tau/3$ the **tempo subdominant** of $[\tau]$. In Figure 6.16d, the tempo subdominant appears as a weak increasing line starting with $s = 1.2$ at time $t = 0$.

So far we have assumed that the space of tempo parameters is continuous. In practice, one can compute a cyclic tempogram $\mathcal{C}_{\tau_0}$ only for a finite number of parameters $s \in [1,2)$. Recall that for computing a value $\mathcal{C}_{\tau_0}(n,s)$ one needs to sum the values $\mathcal{T}(n,\tau)$ for tempo parameters $\tau \in \{s \cdot \tau_0 \cdot 2^k \mid k \in \mathbb{Z}\}$. In other words, the required tempo values are spaced exponentially on the tempo axis. Therefore, as with chroma features, where one uses a log-frequency axis, one requires a log-tempo axis for computing a cyclic tempogram. To this end, the tempo range is sampled in a logarithmic fashion such that each tempo octave contains M tempo samples

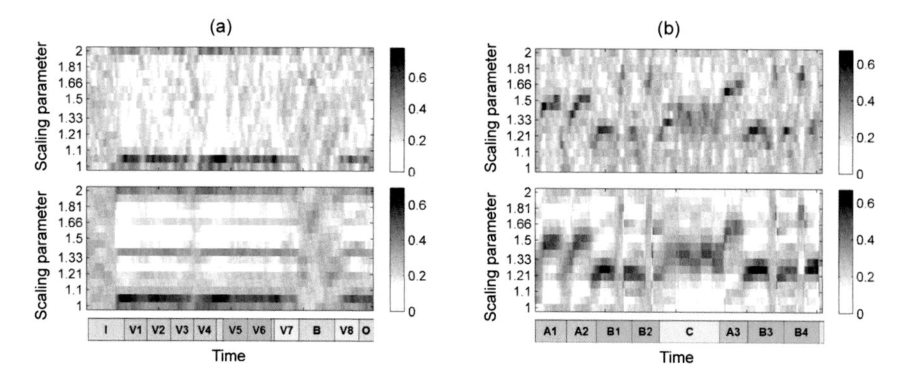

Fig. 6.17 Cyclic Fourier (**top**) and autocorrelation (**bottom**) tempogram representations used for homogeneity-based music segmentation. **(a)** "In the Year 2525" by Zager and Evans (see also Figure 4.13). **(b)** Hungarian Dance No. 5 by Johannes Brahms (see also Figure 4.6d).

for a given number $M \in \mathbb{N}$. Then one obtains a discrete cyclic tempogram $\mathcal{C}_{\tau_0}$ simply by adding up the corresponding values of the different octaves as described in (6.34). This yields an M-dimensional feature vector for every time frame $n \in \mathbb{Z}$, where the cyclic tempo axis is sampled at M positions (see Exercise 6.9) As an example, Figure 6.16e shows a Fourier tempogram with a logarithmic tempo axis. In this case, four tempo octaves ranging from $\tau = 30$ to $\tau = 480$ BPM are used, where each octave is logarithmically sampled using $M = 15$ tempo parameters. The resulting cyclic Fourier tempogram is shown in Figure 6.16g. Similarly, Figure 6.16f shows an autocorrelation tempogram with logarithmic tempo axis and Figure 6.16h the resulting cyclic tempogram.

As mentioned previously, cyclic tempogram representations are the tempo-based counterparts of harmony-based chromagram representations. Compared with standard tempograms, the cyclic versions are more robust to ambiguities that are caused by the various pulse levels. Furthermore, one can simulate changes in tempo by cyclically shifting a cyclic tempogram. Note that this is similar to the property of chromagrams, which can be cyclically shifted to simulate modulations in pitch. As one further advantage, even low-dimensional versions of discrete cyclic tempograms still bear valuable local tempo information of the underlying musical signal. We now indicate how cyclic tempograms can be used as a tool for audio segmentation.

Recall from Section 4.1 that there are many different strategies for segmenting music signals including novelty-based, repetition-based, and homogeneity-based approaches. In the latter, the idea is to partition the music signal into segments that are homogeneous with regard to a specific musical property. In this context, we have considered various feature representations that capture different musical properties such as timbre, harmony, and tempo (see Figure 4.6). We now indicate how cyclic tempograms may be useful for tempo-based segmentation. In the following examples, we use low-dimensional versions of $\mathcal{C}_{60}^{\mathrm{A}}$ and $\mathcal{C}_{60}^{\mathrm{F}}$ based on $M = 15$ different tempo classes. In our first example, we consider the song "In the Year 2525" by

Zager and Evans, a song we have already encountered in Figure 4.13. Recall that the song has a repetitive structure represented by $IV_1V_2V_3V_4V_5V_6V_7BV_8O$. The song starts with a slow intro (I-part), which has a contemplative character with a rather vague notion of tempo and rhythm. The music is dominated by a singing voice, which is accompanied mainly by constant strumming of a guitar. The bridge (B-part) towards the end of the song is played in the same style. As opposed to the intro and bridge, the eight repeating verse sections (V-parts) are played much faster with a clear notion of tempo and rhythm, which are supported by percussive instruments. As seen in Figure 6.17a, the slow parts can be easily discerned from the fast parts in both cyclic tempograms, $\mathcal{C}_{60}^{\mathrm{F}}$ and $\mathcal{C}_{60}^{\mathrm{A}}$. In the slow parts, the tempograms exhibit a noise-like character, where no clear tempo is visible. In contrast, in the fast parts, the tempograms have a dominating tempo corresponding to the scaling parameter value $s = 1.05$, which reflects the actual constant tempo $\tau = s \cdot 60 \cdot 2 = 126$ BPM of the verse sections.

As a second example, we consider a recording of Brahms' Hungarian Dance No. 5, which has already served as our running example in Chapter 4. The musical structure of this recording can be described by $A_1A_2B_1B_2CA_3B_3B_4D$. In this recording, the different musical parts are played in different tempi. Furthermore, there are numerous abrupt changes in tempo, even within some of the parts. Although the cyclic tempogram representations shown in Figure 6.17b do not reveal the exact tempi, they capture tempo-related information that may be useful for homogeneity-based structure analysis. In the two considered examples, the cyclic tempograms yield musically meaningful segmentations purely based on a low-dimensional representation of tempo. These segments cannot be recovered using MFCCs or chroma features, since the homogeneity assumption does not hold with regard to timbre or harmony (see Section 4.1.3).

Finally, we want to note that the tempogram representation shown in Figure 4.6d was generated using a Fourier tempogram with $M = 120$ different tempo classes. For tasks such as homogeneity-based audio segmentation, it may be beneficial to use a much coarser resolution (e.g., $M = 15$ in Figure 6.17b). Decreasing the dimension of the features makes them more robust to small tempo fluctuations. Furthermore, a low feature dimension has advantages for many analysis and retrieval tasks with regard to aspects that concern efficiency, indexing, and learning.

6.3 Beat and Pulse Tracking

The task of beat and pulse tracking can be seen as an extension of tempo estimation in the sense that, additionally to the rate, it also considers the phase of the pulses (see Figure 6.1b). Starting with a Fourier tempogram, we introduce in Section 6.3.1 a robust pulse representation that reveals the predominant local pulse occurring in a neighborhood of a certain time instance in the music signal. This yields a pulse tracker that can adjust to continuous and sudden changes in tempo as long as the underlying novelty function possesses locally periodic patterns. We will see that the pulse representation can be thought of as a kind of periodicity enhancement of the novelty function. The pulse representation does not aim at extracting pulses at a specific level, but locally switches to the dominating pulse level. If one is interested in the pulse positions that correspond to the beat level, one needs to exploit additional knowledge such as a rough estimate of the expected tempo. In Section 6.3.2, we discuss such a beat tracking procedure based on dynamic programming, which assumes a more or less constant tempo throughout the music recording.

6.3.1 Predominant Local Pulse

We now introduce a robust procedure for the extraction of musically meaningful local pulse information even in the case of complex music. Intuitively speaking, the idea is to construct a mid-level representation that explains the local periodic nature of a given (possibly noisy) onset representation. More precisely, starting with a novelty function, we determine for each time position a windowed sinusoid that best captures the local peak structure of the novelty function. Instead of looking at the windowed sinusoids individually, the crucial idea is to employ an overlap–add technique by accumulating all sinusoids over time. As a result, one obtains a single function that can be regarded as a local periodicity enhancement of the original novelty function. Revealing **predominant local pulse** (PLP) information, this representation is referred to as a **PLP function**. In this context, we use the term **predominant pulse** in a rather loose way to refer to the strongest pulse level that is measurable in the underlying novelty function. Since the PLP representation yields the predominant pulse in a (windowed) neighborhood of each time position, continuous tempo variations and local changes in the pulse level can be captured to a certain degree.

6.3.1.1 Definition of PLP Function

In the following construction, we start with a Fourier tempogram $\mathcal{T}^{\mathrm{F}}:\mathbb{Z}\times\Theta\to\mathbb{R}_{\geq 0}$ as defined in (6.26). For each time position $n\in\mathbb{Z}$, we compute the tempo parameter $\tau_n\in\Theta$ that maximizes $\mathcal{T}^{\mathrm{F}}(n,\tau)$:

$$\tau_n := \underset{\tau\in\Theta}{\operatorname{argmax}}\,\mathcal{T}^{\mathrm{F}}(n,\tau). \tag{6.36}$$

As an example, Figure 6.18b shows the maximizing tempo parameters τ_n at seven different time positions. We now make use of the phase information that is given by the complex Fourier coefficients $\mathcal{F}(n,\omega)$ defined in (6.25). Recall from (6.26) that we have $\mathcal{T}^{\mathrm{F}}(n,\tau) := |\mathcal{F}(n,\tau/60)|$. Therefore, the phase φ_n that belongs to the windowed sinusoid of tempo τ_n is given by

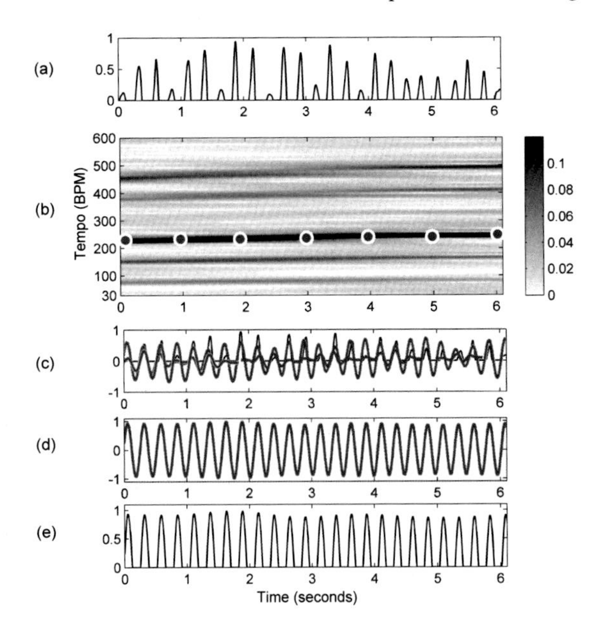

Fig. 6.18 Illustration of the PLP computation. **(a)** Novelty function Δ. **(b)** Tempogram $\mathcal{T}$ with frame-wise tempo maxima (indicated by circles) shown at seven time positions n. **(c)** Optimal windowed sinusoids κ_n (using a window size of 4 seconds) corresponding to the maxima (see also Figure 6.13 for an illustration of individual sinusoids). **(d)** Accumulation of all sinusoids (overlap–add). **(e)** PLP function Γ obtained after half-wave rectification.

$$\varphi_n = -\frac{1}{2\pi}\gamma_n, \tag{6.37}$$

where γ_n is the angle of the polar coordinate representation $\mathcal{F}(n,\tau/60) = |\mathcal{F}(n,\tau/60)|\exp(i\gamma_n)$ (see also (2.89)). Based on τ_n and φ_n, we define the optimal windowed sinusoid $\kappa_n : \mathbb{Z} \to \mathbb{R}$ by setting

$$\kappa_n(m) := w(m-n)\cos\Big(2\pi\big((\tau_n/60)\cdot m - \varphi_n\big)\Big) \tag{6.38}$$

for each time point $n \in \mathbb{Z}$, where we use the same window function w as for the Fourier tempogram. For example, Figure 6.13c shows such an optimal windowed sinusoid for the time index n that corresponds to $t = 2$ sec and the maximizing tempo parameter $\tau_n = 230$ BPM. Intuitively, the sinusoid κ_n best explains the local periodic nature of the novelty function at time position n with respect to the tempo set Θ. The period $60/\tau_n$ corresponds to the predominant periodicity of the novelty function, and the phase information φ_n takes care of accurately aligning the maxima of κ_n and the peaks of the novelty function. The relevance of the sinusoids κ_n depends not only on the quality of the novelty function, but also on the window size of w and the tempo set Θ. For example, as we discussed before, increasing the window size typically yields more robust estimates at the cost of temporal flexibility.

The estimation of optimal windowed sinusoids in regions with a strongly corrupt peak structure is problematic. This particularly holds in the case of small window sizes. To make the periodicity estimation more robust while keeping the temporal flexibility, the idea is to form a single function instead of looking at the sinusoids in a one-by-one fashion. To this end, we apply an overlap–add technique, where the optimal windowed sinusoids κ_n are accumulated over all time positions $n \in \mathbb{Z}$ (see Figure 6.18). Furthermore, we only consider the positive part of the resulting function. More precisely, we define a function $\Gamma : \mathbb{Z} \to \mathbb{R}_{\geq 0}$ as follows:

$$\Gamma(m) = |\sum_{n\in\mathbb{Z}} \kappa_n(m)|_{\geq 0} \tag{6.39}$$

for $n \in \mathbb{Z}$, where we use the half-wave rectification as introduced in (6.2). The resulting function is our mid-level representation referred to as a **PLP function**.

As an example, Figure 6.18d shows the accumulation for the seven optimal windowed sinusoids indicated in Figure 6.18c. Note how the maxima of the different windowed sinusoids align well not only with the peaks of the novelty function, but also with the maxima of neighboring sinusoids in the overlapping regions. This leads to constructive interferences—a phenomenon that we have already seen in Figure 2.19a. By suitably normalizing the window function w (in particular to compensate for overlaps of subsequent windows in the tempogram computation), one can achieve values in the accumulated sinusoids within the interval $[-1,1]$ as well as local maxima close to 1 if and only if the overlapping sinusoids align well. The final PLP function Γ is obtained through half-wave rectification (see Figure 6.18e). In the subsequent discussion, we show how to obtain suitable candidates for pulse and beat positions from the peaks of Γ.

6.3.1.2 Discussion of Properties

We now discuss various properties of the PLP concept by looking at some representative examples. As a first example, we continue with the Shostakovich excerpt introduced in Figure 6.7. Recall that in this recording the first beats (downbeats) of the $3/4$ meter are weak, whereas the second and third beats are strong. This property is also reflected by the peak structure of the novelty function shown in Figure 6.18a, where the peaks corresponding to downbeats are very low. As indicated by the Fourier tempogram (Figure 6.18b), the dominant tempo lies between 200 and 250 BPM throughout this excerpt with a slight tempo increase starting with roughly $\tau = 225$ BPM. The maximizing tempo values as well as the corresponding optimal windowed sinusoids are indicated for seven different time positions. Note that each of these windowed sinusoids tries to explain the locally periodic nature of the peak structure of the novelty function, where small deviations from the "ideal" periodicity and weak peaks are balanced out. The resulting PLP function Γ is shown in Figure 6.18e. Note that the predominant pulse positions are clearly indicated by the peaks of Γ even though some of these pulse positions are rather weak in the original novelty function. In this sense, the PLP function can be regarded as a local periodicity enhancement of the original novelty function, where the predominant pulse level is taken into account.

As a second example, we consider an orchestra recording of the Hungarian Dance No. 5 by Johannes Brahms, which was already used in Figure 6.17b. In the following, we only consider a small audio excerpt (the section between $t_1 = 35$ sec and

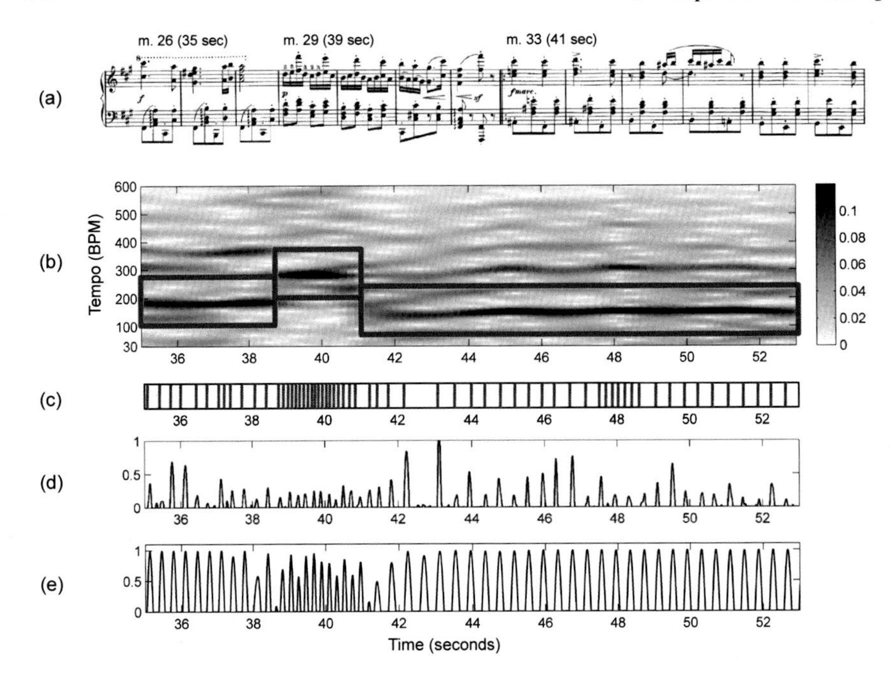

Fig. 6.19 Excerpt (corresponding to measures 26 to 38) of an orchestra recording conducted by Ormandy of Brahms' Hungarian Dance No. 5. **(a)** Score in a piano reduced version. **(b)** Fourier tempogram. The rough tempo range of the predominant pulse is highlighted by the rectangular frames. **(c)** Manual annotation of note onset positions. **(d)** Novelty function $\Delta_{\mathrm{Spectral}}$. **(e)** PLP function Γ.

$t_2 = 53$ sec of the recording), where measures 26 to 38 are played. Figure 6.19a shows the musical score of a piano reduced version of these measures. With respect to the overall musical structure $A_1A_2B_1B_2CA_3B_3B_4D$, the excerpt comprises the transition from the A_2-section to the B_1-section. Because of significant local tempo changes, this recording constitutes a great challenge for tempo estimation and pulse tracking. Considering a quarter-note pulse level, a manual inspection shows that the excerpt starts with a tempo of 90 BPM (measures 26–28, seconds 35–39), then abruptly changes to 140 BPM (measures 29–32, seconds 39–41), and continues with 75 BPM (measures 33–38, seconds 41–53). Many of the note onsets, which are indicated in Figure 6.19c, are poorly captured by the novelty function shown in Figure 6.19d. Furthermore, because of large differences in dynamics, there are some strong onsets that dominate the novelty function as well as some weak onsets that can hardly be distinguished from spurious peaks not related to any note onsets. As a result, the height of a peak is not necessarily the only indicator of its relevance. Despite these challenges, the tempo is reflected well by the Fourier tempogram on the eighth-note pulse level (the second harmonic of the quarter-note tempo); see Figure 6.19b. Although corrupt, the peak structure of the novelty function still possesses some local periodic regularities, which are captured by the windowed sinusoids corresponding to the predominant local tempo. The resulting PLP function Γ, as shown by Figure 6.19e, is capable of revealing the pulse positions on the eighth-note level.

The Brahms example illustrates another important property of the PLP function: it not only reveals positions of predominant pulses but also indicates a kind of confidence in the estimation. Note that the amplitudes of the optimal windowed sinusoids do not depend on the amplitude of the novelty function. This makes a PLP function invariant to changes in dynamics of the underlying music signal. Recall that we estimate the windowed sinusoids using a sliding window technique and add them up over all considered time positions. Since neighboring sinusoids overlap, constructive and destructive interference phenomena in the overlapping regions influence the amplitude of the resulting PLP function Γ. Consistent local tempo estimates result in sinusoids that produce constructive interferences in the overlap–add process. In such regions, the peaks of the PLP function assume a value close to one. In contrast, sudden random-like changes in the local tempo estimates result in inconsistencies in the overlap regions of subsequent sinusoids, which in turn cause destructive interferences and lower values of Γ. In Figure 6.19e, this effect is visible in measures 28–33 (seconds 38–42), where several sudden tempo changes occur within the window size of the analysis sinusoids. The confidence property of PLP functions can be used to detect problematic passages in music recordings.

Although the PLP concept is designed to automatically adapt to the predominant tempo, for some applications the local nature of these estimates might not be desirable. In particular, taking the frame-wise maximum to determine the predominant tempo as in (6.36) has both advantages and drawbacks. On the one hand, it allows the PLP function to quickly adjust—even to sudden changes in tempo. On the other hand, it may lead to unwanted jumps such as random switches between tempo octaves. This situation is illustrated by our third example, shown in Figure 6.20, where a piano recording of a piece by Burgmüller is analyzed. Using the tempo set $\Theta = [30:600]$ yields the PLP function of Figure 6.20b, where several changes between the quarter-note and eighth-note level occur. Such switches in the pulse level can be avoided when constraining the tempo set Θ in the maximization. For example, using a constrained set $\Theta = [60:200]$, one obtains the tempogram and PLP function shown in Figure 6.20c. In this case, the PLP function correctly reveals the quarter-note (tactus) pulse positions with a tempo of roughly 130 BPM. Similarly, using the set $\Theta = [200:340]$ reveals the positions on the eighth-note (Figure 6.20d) and using the set $\Theta = [450:600]$ the positions on the sixteenth-note pulse level. In other words, one can easily incorporate into the PLP framework prior knowledge on the expected tempo range to reveal the pulses on a specific level.

6.3.2 Beat Tracking by Dynamic Programming

There are many types of music with a strong and steady beat, where the tempo is more or less constant throughout the entire recording. In such cases, the flexibility

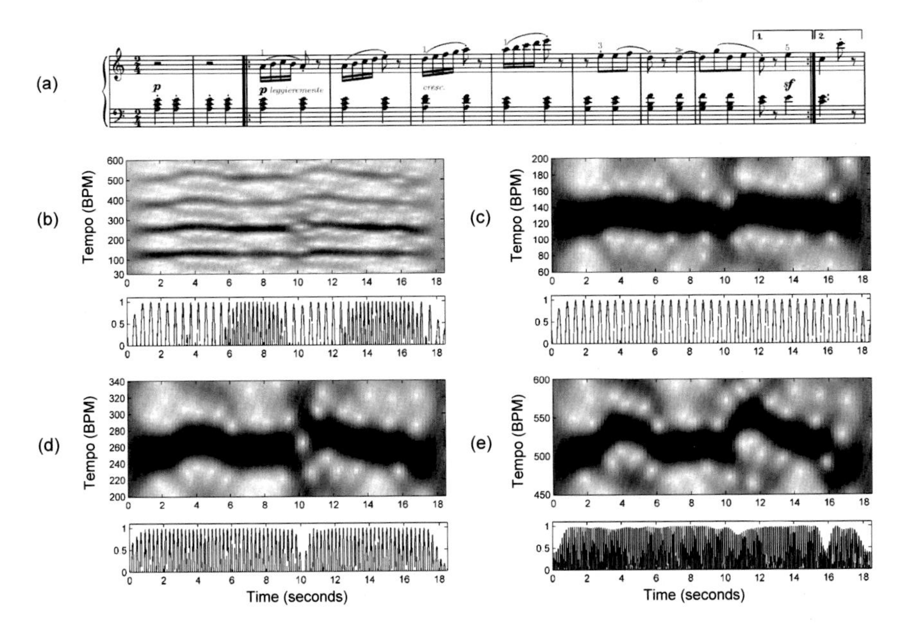

Fig. 6.20 Beat level adjustment by tempo range restriction based on a piano recording of the beginning of the Piano Etude Op. 100 No. 2 by Burgmüller. Tempograms and PLP functions (KS = 4 sec) are shown for various sets Θ specifying the tempo range used (given in BPM). **(a)** $\Theta = [30:600]$ (full tempo range). **(b)** $\Theta = [60:200]$ (tempo range around quarter-note level). **(c)** $\Theta = [200:340]$ (tempo range around eighth-note level). **(d)** $\Theta = [450:600]$ (tempo range around sixteenth-note level).

offered by the PLP concept is not needed. Following the approach originally suggested by Ellis [9], we now describe a beat tracking procedure which is based on the assumptions that beat positions go along with the strongest note onsets and that the tempo is roughly constant. The main idea is to construct a score function that measures how well an arbitrary beat sequence reflects these two assumptions. The score-maximizing beat sequence constitutes the final beat tracking result. We will see that such an optimal beat sequence can be computed efficiently using dynamic programming—an algorithmic paradigm we have introduced in Section 3.2.1.3.

The input of the beat tracking procedure consists of a novelty function $\Delta : [1:N] \to \mathbb{R}$ as well as a rough estimate $\hat{\tau} \in \mathbb{R}_{>0}$ of the global tempo. As usual, we assume that the interval $[1:N]$ represents the sampled time axis used for the novelty feature computation. The tempo estimate $\hat{\tau}$ may be specified manually or obtained by an automated procedure such as (6.33). From $\hat{\tau}$ and the feature rate, one can derive an estimate for the beat period. For simplicity, we assume that the beat period is specified in terms of samples or feature indices (rather than in seconds). Let $\hat{\delta} \in \mathbb{N}$ be this number. Assuming a roughly constant tempo, the difference δ of two consecutive beats should be close to $\hat{\delta}$. To account for the deviation of δ from the ideal beat period $\hat{\delta}$, we introduce a penalty function $P_{\hat{\delta}} : \mathbb{N} \to \mathbb{R}$ by setting

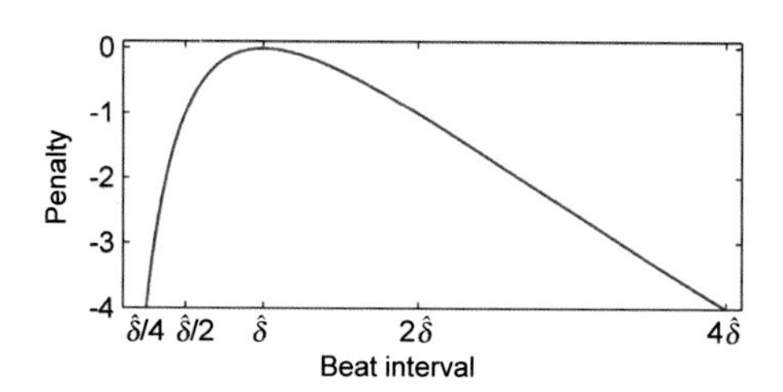

Fig. 6.21 Penalty function $P_{\hat{\delta}}$ measuring the deviation of a given beat period δ from the ideal beat period $\hat{\delta}$.

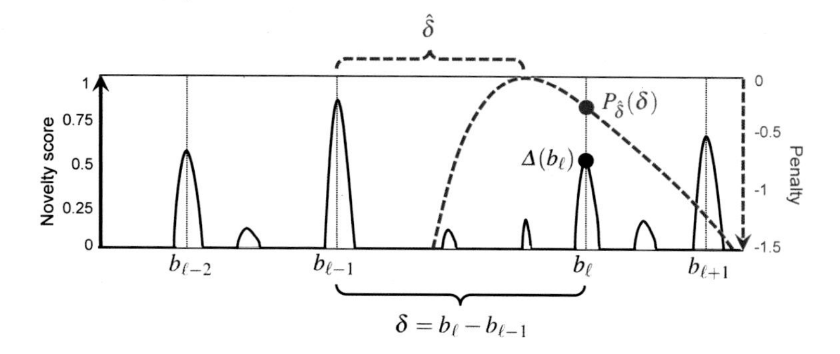

Fig. 6.22 Illustration of the score value $\mathbf{S}(B)$ from (6.41), which involves a (positive) novelty score $\Delta(b_\ell)$ and a (negative) deviation penalty $P_{\hat{\delta}}(b_\ell - b_{\ell-1})$.

$$P_{\hat{\delta}}(\delta) := -\big(\log_2(\delta/\hat{\delta})\big)^2 \tag{6.40}$$

for $\delta \in \mathbb{N}$. This function has a maximal value of zero at $\delta = \hat{\delta}$ and exhibits increasingly negative values for larger deviations (see Figure 6.21). Furthermore, since tempo deviations are relative in nature (doubling the tempo should be penalized to the same degree as halving the tempo), the penalty function is defined to be symmetric on a logarithmic axis.

For a given $N \in \mathbb{N}$, let $B = (b_1, b_2, \ldots, b_L)$ be a sequence of length $L \in \mathbb{N}_0$ consisting of strictly monotonically increasing beat positions $b_\ell \in [1:N]$ for $\ell \in [1:L]$. In the following, we refer to such a sequence as a **beat sequence**. By definition, the beat sequence of length $L = 0$ is the empty sequence. Furthermore, let $\mathcal{B}^N$ denote the set consisting of all possible beat sequences for a given parameter $N \in \mathbb{N}$. To measure the quality of a beat sequence $B \in \mathcal{B}^N$, we introduce a score value $\mathbf{S}(B)$ that includes positive values of the novelty function Δ and negative values of the penalty function $P_{\hat{\delta}}$ (see Figure 6.22). To this end, we set

$$\mathbf{S}(B) := \sum_{\ell=1}^{L} \Delta(b_\ell) + \lambda \sum_{\ell=2}^{L} P_{\hat{\delta}}(b_\ell - b_{\ell-1}). \tag{6.41}$$

In particular, we obtain $\mathbf{S}(B) = 0$ for the empty beat sequence with $L = 0$. Furthermore, in the case $L = 1$, we obtain $\mathbf{S}(B) = \Delta(b_1)$. In order to achieve a large score, the novelty values $\Delta(b_\ell)$ should be large, while the nonpositive penalty values $P_{\hat{\delta}}(b_\ell - b_{\ell-1})$ should be close to zero. The weight parameter $\lambda \in \mathbb{R}_{>0}$ is introduced to balance out these two conflicting conditions. The beat sequence

$$B^* := \mathrm{argmax}\{\mathbf{S}(B) \mid B \in \mathcal{B}^N\} \tag{6.42}$$

that has the maximal score among all possible beat sequences yields the solution of the beat tracking problem.

Note that the number of possible beat sequences is exponential in N (see Exercise 6.10). We now show how an optimal beat sequence can be computed efficiently using **dynamic programming**. We have already encountered this algorithmic paradigm in previous chapters for solving similar optimization problems. For example, in Section 3.2.1.3, we used dynamic programming to compute a cost-minimizing warping path or, in Section 5.3.3.2, to determine a probability-maximizing state sequence (Viterbi algorithm).

Recall from Section 3.2.1.3 that, in the context of DTW, we broke down the optimization problem into simpler subproblems by considering prefixes of the two sequences to be aligned. In the beat tracking context, we employ a similar idea by considering prefixes of the underlying novelty function. More precisely, let $\mathcal{B}_n^N \subset \mathcal{B}^N$ denote the set of all beat sequences that end in $n \in [0:N]$. In other words, for a beat sequence $B = (b_1, b_2, \ldots, b_L) \in \mathcal{B}_n^N$, we have $b_L = n$. Note that, in the case $n = 0$, the only possible beat sequence is the empty one. Let

$$\mathbf{D}(n) := \max\left\{\mathbf{S}(B) \mid B \in \mathcal{B}_n^N\right\} \tag{6.43}$$

denote the maximal score over all beat sequences ending with $n \in [0:N]$. The value $\mathbf{D}(n)$ is also referred to as the **accumulated score**. It is not hard to see that

$$\mathcal{B}^N = \bigcup_{n \in [0:N]} \mathcal{B}_n^N \tag{6.44}$$

(see Exercise 6.10). Therefore, the maximal score $\mathbf{S}(B^*)$ for an optimal beat sequence B^* as defined in (6.42) is obtained by looking for the largest value of $\mathbf{D}$:

$$\mathbf{S}(B^*) = \max_{n \in [0:N]} \mathbf{D}(n). \tag{6.45}$$

We now show how the values $\mathbf{D}(n)$ can be computed in an iterative fashion for $n = 0, 1, \ldots, N$. For a summary of the procedure, we also refer to Table 6.1.

First, we consider the case $n = 0$, which is used for initializing the procedure. In this case, we have $\mathbf{D}(n) = 0$, which is the accumulated score of the empty beat sequence of length $L = 0$. Next, let $n > 0$. Assuming that we already know the values $\mathbf{D}(m)$ for $m \in [0:n-1]$, we need to compute the value $\mathbf{D}(n)$. Let $B_n^* = (b_1, b_2, \ldots, b_L)$ with $b_L = n$ denote a score-maximizing beat sequence that yields the value $\mathbf{D}(n) = \mathbf{S}(B_n^*)$. Even though we may not know such a sequence explicitly, we know, at least, that the last beat $b_L = n$ contributes with the novelty value $\Delta(n)$. We distinguish between two cases. The first case is $L = 1$, where one

Algorithm: OPTIMAL BEAT SEQUENCE

Input: Novelty function $\Delta : [1:N] \to \mathbb{R}$
Estimate $\hat{\delta}$ for the beat period (given in samples)
Weight parameter $\lambda \in \mathbb{R}$

Output: Optimal beat sequence $B^* = (b_1, b_2, \ldots, b_L)$

Procedure: Initialize $\mathbf{D}(0) = 0$ and $\mathbf{P}(0) = 0$. Then compute in a loop for $n = 1, \ldots, N$:

$\mathbf{D}(n) = \Delta(n) + \max\{0, \max_{m\in[1:n-1]}\{\mathbf{D}(m) + \lambda P_{\hat{\delta}}(n-m)\}\}$
If $\mathbf{D}(n) = \Delta(n)$ then set $\mathbf{P}(n) = 0$,
otherwise set $\mathbf{P}(n) = \mathrm{argmax}_{m\in[1:n-1]}\{\mathbf{D}(m) + \lambda P_{\hat{\delta}}(n-m)\}$

Set $\ell = 1$ and $a_\ell = \mathrm{argmax}_{n\in[0:N]}\mathbf{D}(n)$. Then repeat the following steps until $\mathbf{P}(a_\ell) = 0$:

Increase ℓ by one.
Set $a_\ell = \mathbf{P}(a_{\ell-1})$.

If $a_\ell = 0$, then set $L = 0$ and return $B^* = \emptyset$.
Otherwise let $L = \ell$ and return $B^* = (a_L, a_{L-1}, \ldots, a_1)$.

Table 6.1 Computation of an optimal beat sequence using dynamic programming.

has a single beat and $\mathbf{D}(n) = \Delta(n)$. The second case is $L > 1$, where one has a beat $b_{L-1} \in [1:n-1]$ that precedes $b_L = n$. Using (6.41), the accumulated score $\mathbf{D}(n)$ is obtained by

$$\mathbf{D}(n) = \Delta(n) + \lambda P_{\hat{\delta}}(n - b_{L-1}) + \mathbf{D}(b_{L-1}). \tag{6.46}$$

In other words, the optimal score $\mathbf{D}(n)$ is obtained as the sum of the novelty value $\Delta(n)$, the (weighted) penalty of the beat period $\delta = b_L - b_{L-1}$, and the optimal score $\mathbf{D}(b_{L-1})$ of a beat sequence ending at b_{L-1}. Even though we do not know b_{L-1} explicitly so far, we have already computed all values $\mathbf{D}(m)$ for $m \in [0:n-1]$. From this and by considering the two cases ($L = 1$ and $L > 1$), we obtain the following recursion:

$$\mathbf{D}(n) = \Delta(n) + \max\begin{cases} 0, \\ \max_{m\in[1:n-1]}\{\mathbf{D}(m) + \lambda P_{\hat{\delta}}(n-m)\}. \end{cases} \tag{6.47}$$

This concludes the computation of the accumulated score $\mathbf{D}$. By (6.45), this also yields the maximal score of an optimal beat sequence. However, we do not yet know what an optimal beat sequence looks like. To determine such an optimal beat sequence, we need to apply a **backtracking** procedure—similar to the previous scenarios in Section 3.2.1.3 and Section 5.3.3.2.

While calculating $\mathbf{D}(n)$, we additionally store the information on the maximization process in (6.47) by means of a number $\mathbf{P}(n) \in [0:n-1]$. In the case that the maximum in (6.47) is 0, we have $L = 1$ and there is no preceding beat. Therefore, we set $\mathbf{P}(n) := 0$. Otherwise, one has $L > 1$, and there is a preceding beat. In this case we set

$$\mathbf{P}(n) := \underset{m \in [1:n-1]}{\operatorname{argmax}} \left\{ \mathbf{D}(m) + \lambda P_{\hat{\delta}}(n-m) \right\}. \tag{6.48}$$

The maximizing index $n^* \in [0:N]$ in (6.45) determines the last beat $b_L = n^*$ of an optimal beat sequence B^*. (Only in the case $n^* = 0$, there is no last beat and $B^* = \emptyset$.) The remaining beats of B^* can then be obtained by backtracking using the predecessor information supplied by $\mathbf{P}$. In the case $\mathbf{P}(n^*) = 0$, the backtracking is terminated and $L = 1$. Otherwise, $b_{L-1} = \mathbf{P}(b_L)$ determines the beat preceding the last beat $b_L = n^*$. This procedure is then iterated to determine $b_{L-2} = \mathbf{P}(b_{L-1})$, $b_{L-3} = \mathbf{P}(b_{L-2})$, and so on, until the condition $\mathbf{P}(b_1) = 0$ terminates the backtracking. Note that the length L is not known a priori and results from the backtracking. Table 6.1 summarizes the entire procedure (compare this with the DTW procedure of Table 3.2). Thanks to dynamic programming, the exponential number of operations needed to compute the score of all possible beat sequences is reduced to a number of operations that is quadratic in N (see Exercise 6.11). In practice, further savings can be achieved by restricting the search space in the maximization (6.47) (see Exercise 6.13).

As said before, the main limitation of the beat tracking procedure is its dependency on a single, predefined tempo $\hat{\tau}$. Using a small weighting parameter λ, the procedure may yield good beat tracking results even in the presence of local deviations from the ideal beat period $\hat{\delta}$. However, the presented procedure is not designed for handling music with slowly varying tempo (such as ritardando or accelerando) or abrupt changes in tempo. Despite these limitations, the simplicity and efficiency of the dynamic programming approach to beat tracking makes it an attractive choice for many types of music.

6.3.3 Adaptive Windowing

Onset and beat positions are expressive features that often segment a music recording into semantically meaningful units. We now indicate how such a segmentation can help to improve general feature extraction. One crucial step in practically all music analysis tasks consists of transforming the given audio signal into a suitable feature representation that captures certain musical properties while being invariant to other aspects. For example, as we have seen in Section 3.1.2, chroma features are a powerful representation for revealing harmonic properties of a music recording. Since most musical properties vary over time, the given audio signal is typically split up into segments or frames, which are then further processed individually. The underlying assumption is that the signal stays (approximately) stationary within each segment with regard to the property to be captured.

In practice, as is the case with the short-time Fourier transform (Section 2.5), a predefined window of fixed size is used for the time localization, where the size is determined empirically and optimized for the specific application in mind. Using **fixed-size windowing**, however, may lead to a violation of the homogeneity assumption: the boundaries of the resulting windowed sections often do not coincide

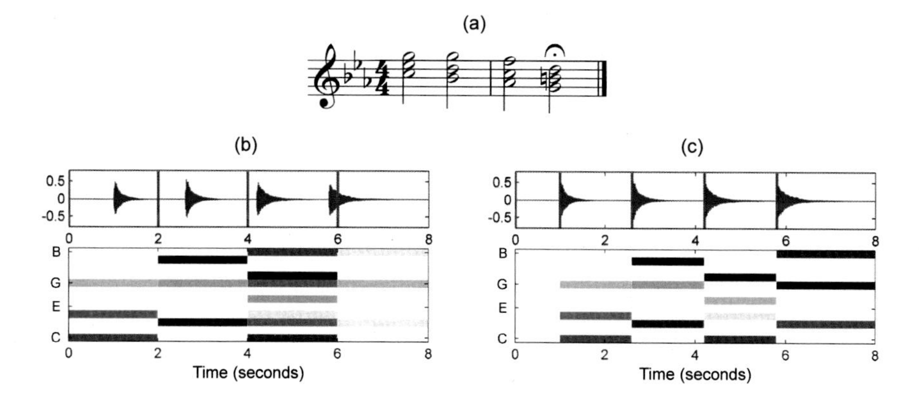

Fig. 6.23 Score, audio recording, and chroma representation of a sequence of four chords. **(a)** Musical score of the four chords. **(b)** Segmentation using a window of fixed size. **(c)** Adaptive segmentation resulting in beat-synchronized features.

with the positions where the changes of the signal occur. To illustrate this problem, Figure 6.23b shows a chroma representation for an audio excerpt with four subsequent chords, where fixed-size windowing has been used. Note that the third frame comprises a chord change leading to a rather "noisy" chroma feature where the chroma bands contain energy from two different chords. To attenuate the problem, one may decrease the window size at the cost of an increased feature rate and a poorer frequency resolution. As an alternative to fixed-size windowing, one can employ a more musically meaningful **adaptive windowing** strategy, where segment boundaries are induced by previously extracted onset and beat positions. Since musical changes typically occur at onset positions, this often leads to an increased homogeneity within the adaptively determined frames and a significant improvement in the resulting feature quality. In our chord example, as shown by Figure 6.23c, such an onset-based adaptive windowing leads to clean chroma features that nicely capture the characteristics of the four chords.

Adaptive windowing techniques based on beat information are of particular importance for many music analysis and retrieval applications. In this case, a windowed section is determined by two consecutive beat positions, which results in one feature vector per beat. Such **beat-synchronous** feature representations have the advantage of possessing a musical time axis (given in beats) rather than a physical time axis (given in seconds). This makes the feature representation robust to differences in tempo (given in BPM). In the context of music synchronization (Chapter 3), the usage of beat-synchronous features could make DTW-like alignment techniques obsolete—at least in the ideal case of having perfect beat positions. For example, knowing the beat positions already yields a beatwise synchronization of two different performances that follow the same musical score. However, in practice, such strategies have to be treated with caution, in particular when the beat positions are determined automatically. As we have discussed in this chapter, automated

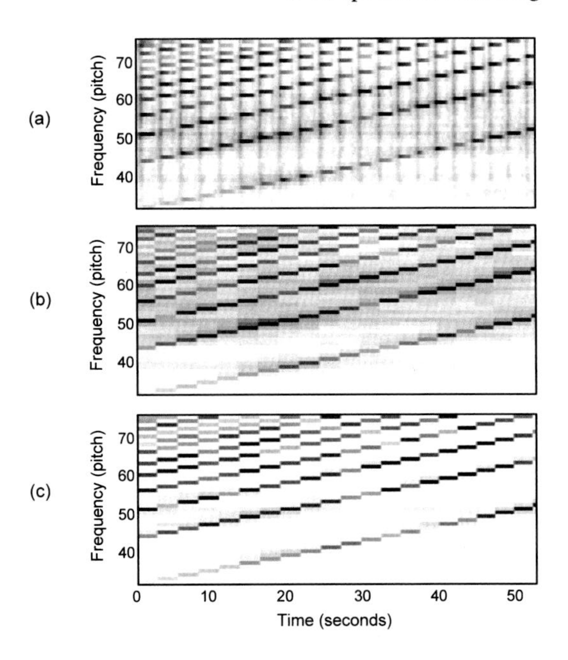

Fig. 6.24 Time–pitch representations for a piano recording of a chromatic scale (similar to Figure 3.3). **(a)** Original time–pitch representation using fixed-size windowing with a small window size. **(b)** Adaptive windowing using $\lambda = 1$. **(c)** Adaptive windowing using $\lambda = 0.5$.

beat tracking procedures work well for music with percussive onsets and a steady tempo. However, when dealing with weak note onsets and expressive music with local tempo changes, the automated generation of beat positions becomes an error-prone task, not to mention the problems related to tempo octave confusion. Using corrupt beat information at the feature extraction stage may have immense consequences for the subsequent music processing tasks to be solved. For example, to compensate for beat tracking errors in the music synchronization context, one may have to reintroduce error-tolerant techniques that are similar to DTW.

Next, we show how the knowledge of onset and beat positions can be used in another way for improving the quality of audio features. Recall that note onsets often go along with noise-like energy bursts spread over the entire spectrum, especially for instruments such as the piano, guitar, or percussion. This phenomenon is illustrated in Figure 6.24a, which shows a time–pitch representation of a chromatic scale played on a piano (similar to Figure 3.3). In this representation, the transients that go along with note onsets are clearly visible as vertical structures at the beginning of each note. While these transients are useful for the detection of note onset positions, they cause undesired artifacts in features that capture harmonic or melodic information. As an example, Figure 6.24b shows the time–pitch representation using adaptive windowing based on consecutive note onset positions. Each of the resulting feature vectors reveals the pitch content of a single note. However, the nonharmonic transient components also enter the analysis window and introduce noise-like artifacts in pitch bands that are not related to the underlying notes. To remove these noise-like artifacts while keeping the harmonic information, one idea is to exclude a neighborhood around each note onset position. To this end, we introduce a parameter $\lambda \in \mathbb{R}$, $0 < \lambda < 1$, which determines the size of the neighborhoods. Let $s,t \in [1:N]$ denote the start and end positions of a given adaptive window. Then we define

$$s_\lambda := s + \left\lfloor \frac{1-\lambda}{2}(t-s) \right\rfloor \quad \text{and} \quad t_\lambda := t - \left\lfloor \frac{1-\lambda}{2}(t-s) \right\rfloor, \tag{6.49}$$

which determine the start and end positions of the shortened window used for the feature computation. With this definition, the center of the adaptive window is preserved, while its size is reduced by a factor λ relative to its original size $(t-s)$. The effect of this procedure is illustrated by Figure 6.24c, where the factor $\lambda = 0.5$ has been used to remove the transients at the note onset positions. As a result, the harmonic information of the time–pitch representation has clearly been enhanced.

6.4 Summary and Further Readings

The automated extraction of onset, beat, and tempo information is one of the central tasks in music signal processing and constitutes a key element for a number of music analysis and retrieval applications. Tempo and beat are not only expressive descriptors per se, but also induce natural and musically meaningful segmentations of the underlying audio signals. In this chapter, we studied a number of key techniques and important principles that are used in this vibrant and well-studied area of research. Furthermore, we showed how these techniques and principles are applied in specific tempo and beat tracking procedures while discussing the benefits and limitations of automated methods.

Onset Detection

In general, the notions of onset, beat, and tempo are not as well defined as one may think at first sight. However, for most music, beat positions go along with note onsets or percussive events. Therefore, in typical tempo and beat tracking approaches, the first step consists in locating such events in the given signal—a task we referred to as onset detection or novelty detection. Following the excellent tutorial by Bello et al. [1], we studied in Section 6.1 different methods for computing novelty functions that capture changes in the signal's energy, spectrum, or phase. When playing a note, the onset often goes along with a sudden change of the signal's energy. In such a case, note onset candidates may be determined by locating time positions where the signal's amplitude envelope starts to increase. Much more challenging is the detection of onsets in the case of nonpercussive music, where one has to deal with soft onsets or blurred note transitions. As a result, more refined methods have to be used for computing a novelty function, e.g., by analyzing the signal's spectral content, pitch, harmony, or phase. For an overview and links to further literature, we refer to [1, 20]. In complex polyphonic mixtures of music, simultaneously occurring events of high intensities lead to masking effects that prevent any observation of an energy increase of a low-intensity onset. To circumvent these masking effects, detection functions were proposed that analyze the signal in a bandwise fashion to extract transients occurring in certain frequency regions of the signal [22, 31]. As a side-effect of a sudden energy increase, there appears an accompanying broadband noise burst in the signal's spectrum. This effect is mostly masked by the signal's energy in lower frequency regions, but is easily detectable in the higher-frequency regions of the spectrum. In this context, logarithmic compression and spectral whitening are techniques for enhancing the high-frequency information. Rather than explicitly modeling and capturing signal changes, more recent deep learning approaches try to learn novelty (or activation) functions from labeled training data [11, 32].

Tempo Analysis

The estimation of a music recording's tempo, which can be loosely defined as the frequency with which humans tap along, is a central and long-studied task in music information retrieval [7, 31]. To derive the beat period and tempo from a novelty function, one strategy is to explicitly determine note onset positions and then analyze the resulting interonset intervals (IOIs). Considering suitable histograms or probabilities of the occurring IOIs, one may derive hypotheses on the beat period and tempo [7]. The drawback of such approaches is that they rely on an explicit localization of a discrete set of note onsets—a fragile and error-prone step, particularly for weak and blurry onsets. Avoiding the explicit extraction of note onsets, novelty functions can be directly analyzed concerning reoccurring or quasiperiodic patterns. Generally speaking, one may distinguish between three different methods for measuring periodicities. The autocorrelation method aims at detecting periodic self-similarities by comparing a novelty function with time-shifted copies [5, 9, 15, 29]. Another widely used method is based on a bank of comb filter resonators, where a novelty function is compared with templates consisting of equally spaced spikes, with each template representing a specific tempo [23, 31]. Similarly, one can use a short-time Fourier transform [16, 29] to derive a frequency representation of the novelty functions. In this case, the novelty function is compared with sinusoidal templates, each corresponding to a specific frequency. Avoiding even the intermediate step of computing a novelty function, recent deep learning techniques try to estimate the tempo directly from a short patch of a signal's time–frequency representation. For example, framing tempo estimation as a multi-class classification problem, Schreiber et al. [33] present such a single-step tempo estimation system based on a convolutional neural network.

The estimation of a music recording's tempo is more challenging, as one may guess. First, the notions of tempo and beat are often vague and subjective, due to the complex hierarchical structure of rhythm [28]. In particular, as illustrated by Figure 6.12, there are various levels that contribute to the human perception of tempo and beat. Furthermore, the detection of locally periodic patterns becomes challenging when the music recording reveals significant tempo changes. This often occurs in performances of classical music as a result of ritardandi, accelerandi,

Table 6.2 Comparison of tempogram representations obtained by Fourier and autocorrelation analysis.

Fourier tempogram	Autocorrelation tempogram
Comparison of novelty curve with windowed sinusoids with each sinusoid representing a tempo.	Comparison of novelty curve with time-shifted windowed sections of itself with each lag representing a tempo.
Conversion of frequency (Hertz) into tempo (BPM).	Conversion of lag (seconds) into tempo (BPM).
Measurement of novelty periodicities.	Measurement of novelty self-similarities.
Emphasis of tempo harmonics.	Emphasis of tempo subharmonics.
Suitable to analyze tempo on tactus and tatum level.	Suitable to analyze tempo on tactus and measure level.

fermate, and so on [7]. Probably the biggest problem with tempo estimation, apart from fluctuating tempi, is the so-called octave error, which occurs when estimates are integer multiples or fractions of the reference tempo. Due to these challenges, the evaluation of tempo estimation procedures constitutes a research problem in itself. To conduct a basic evaluation of a tempo estimation system, one needs test recordings with well-defined (locally stable) tempi, reliable annotations, and suitable evaluation metrics. Starting with the work by Goto et al. [13] and Scheirer [31], the MIR research community has been conducting such evaluations for 25 years [15, 34, 38]. Despite technological improvements, however, many of the basic questions have remained the same. What are the applications tempo estimation is used for? Do datasets and metrics match the use cases? Are there music examples for which no system estimates the correct tempo, or recordings most systems estimate different tempi for? Does that mean the annotation is wrong, the tempo is hard to estimate, or the recording is not suitable for the task? For a discussion of these questions and further references to the literature, we refer to [15, 34].

Rather than explicitly determining the tempo, another strand of research is to convert a music recording into a feature representation that implicitly encodes information related to tempo, rhythm, and other periodic characteristics. Such representations may then be used for other MIR tasks such as genre classification or music structure analysis. In the literature, one finds various suggestions for spectrogram-like representations, including **tempograms** [17], **rhythmograms** [21], or **beat spectrograms** [12]. In Section 6.2, we studied two different techniques based on Fourier and autocorrelation analysis to derive such time–tempo representations. A summary of these two methods and their implications can be found in Table 6.2. Furthermore, following [17, 24], we derived a cyclic version of tempogram representations by identifying tempo octaves—as we did with chroma representations, where we identified pitch octaves (see Table 6.3 for a comparison). We indicated how these features, which are invariant to pulse level confusions, may be used for music segmentation applications.

Table 6.3 Steps for computing chromagram and cyclic tempogram representations.

Steps	Chromagram	Cyclic tempogram
1.	Analysis of waveform.	Analysis of novelty curve.
2.	Computation of spectrogram.	Computation of tempogram.
3.	Usage of log-frequency axis.	Usage of log-tempo axis.
4.	Cyclic projection.	Cyclic projection.

Beat Tracking

As a further central topic of this chapter, we discussed in Section 6.3 the problem of automated pulse and beat tracking following [16]. Using Fourier analysis, we exploited magnitude as well as phase information to derive for each time position a windowed sinusoid that best explains the local periodic nature of a novelty representation. Then, employing an overlap–add technique, a single function that reveals the predominant local pulse (PLP) was derived. The PLP function reveals the local pulse information even in the presence of continuous tempo changes. Furthermore, its amplitudes indicate a kind of confidence in the periodicity estimation. For modern pop and rock music with a strong beat and relatively steady tempo, one does not need the flexibility offered by the PLP concept. To handle such music, we studied a robust beat tracking procedure, which was originally proposed by Ellis [9]. To compute an optimal beat sequence, we discussed an efficient and elegant algorithm based on dynamic programming—a paradigm we have already encountered in the DTW context (Section 3.2).

There are numerous contributions within the area of beat tracking, and, in this textbook, we have only offered a glimpse into this exciting research area. We have seen that various pulse levels contribute to the human perception of tempo and beat [28]. Most work in the literature has focused on determining musical pulses on the tactus or foot tapping level [5, 9, 29], but few approaches exist for analyzing the signal on the measure level [23, 30] or finer tatum level [6, 35]. Many of these approaches handle different pulse levels simultaneously. The PLP concept introduced in [16] does not explicitly address the problem of extracting pulses at a specific level. Instead, a PLP function can be regarded as a kind of mid-level representation that captures the locally predominant pulse.

A recent trend in research aims at improving beat and pulse tracking through the use of neural networks (see, e.g., [3, 8, 10]). Like onset detection, many of these approaches learn an activation function encoding the probability of pulse positions (classifying a frame as pulse or nonpulse). The relevant pulse positions are then determined using some kind of peak picking applied in a post processing step. Motivated by the fact that music genres may correlate to typical beat patterns and tempo ranges, Böck et al. [3] introduce a system based on multiple recurrent neural networks, each being specialized on a particular musical style. The deep learning framework has also opened up new possibilities for multi-task learning approaches, e.g., estimating tempo and beat jointly [2].

Due to the challenges mentioned already in the context of tempo estimation, the evaluation of beat tracking procedures constitutes a research problem in itself (see, e.g., [4, 19, 37]). The evaluation measures used may be divided into two groups: firstly, measures that analyze each beat position separately, and secondly, measures that take the tempo and metrical levels into account [5, 23, 25]. Another interesting approach for evaluating and improving beat tracking systems is to measure the mutual agreement between beat sequences obtained by different beat tracking systems [19, 37]. In particular, looking at inconsistencies across the different beat sequences makes it possible to identify challenging music excerpts without the need for ground-truth annotations. Investigations of musical properties that influence the beat tracking quality have also been conducted in [7, 18].

Finally, we want to mention that the extraction of onset, beat, and tempo information is of fundamental importance for the determination of higher-level musical structures such as rhythm and meter [14, 28, 36]. Generally, the term **rhythm** is used to refer to a temporal patterning of event durations, which are specified by a regular succession of strong and weak stimuli. Furthermore, the perception of rhythmic patterns also depends on other cues, such as the dynamics and timbre of the sound events involved. Such repeating patterns of accents form characteristic **pulse groups**, which determine the **meter** of a piece of music. Here, each group typically starts with an accented beat and consists of all pulses until the next accent. In this sense, the term **meter** is often used synonymously with the term **time signature**, which specifies the beat structure of a musical measure or bar. It expresses a regular pattern of beat stresses continuing through a piece, thus defining a hierarchical grid of beats at various time scales.

6.5 FMP Notebooks

It is the beat that drives music forward and makes people move or tap along with the music. Thus the extraction of beat and tempo information from audio recordings constitutes a natural entry point into music processing and yields a multifaceted application for teaching and learning signal processing. In **Part 6** of the FMP notebooks [27], we demonstrate this by providing Python implementations of fundamental signal and music processing algorithms along with instructive music and audio examples.

As discussed in this chapter, most approaches to beat tracking are based on two assumptions: first, the beat positions correspond to note onsets (often percussive in nature), and, second, beats are periodically spaced in time. In the first notebooks, starting with the **FMP Notebook *Onset Detection***, we consider the problem of determining the starting times of notes or other musical events as they occur in a music recording. To get a feeling for this seemingly simple task, we look at various sound examples of increasing complexity, including a click sound, an isolated piano sound, an isolated violin sound, and a section of a complex string quartet recording. It is very instructive to look at such examples to demonstrate that the detection of individual note onsets can become quite tricky for soft onsets in the presence of vibrato, not to speak of complex polyphonic music. Furthermore, we introduce an excerpt of the song "Another One Bites the Dust" by Queen, which will serve as our running example throughout the subsequent notebooks (see also Figure 6.1). For later usage, we introduce some Python code for parsing onset and beat annotations and show how such annotations can be sonified via click tracks using a function from the Python package `librosa`.

In the **FMP Notebook *Energy-Based Novelty***, we implement the onset detector from Section 6.1.1 step by step, computing a local energy function, taking a discrete derivative, and applying half-wave rectification. In particular, we explain the role of the window function used in the first step and apply logarithmic compression as a way to enhance small energy values. Involving basic signal processing elements, this simple procedure is instructive from an educational point of view. However, for nonpercussive sounds, the approach has significant weaknesses.

This naturally leads us to the **FMP Notebook *Spectral-Based Novelty***, where we discuss a novelty representation that is commonly known as spectral flux (see Section 6.1.2). The idea is to convert the signal into a spectrogram and then measure spectral changes by taking the distance between subsequent spectral vectors. This technique is suited to recall a phenomenon from Fourier analysis: the energy of transient events is spread across the entire spectrum of frequencies, thus yielding broadband spectral structures. These structures can be detected well by the spectral-based novelty detection approach. Again, we highlight the role of logarithmic compression and further enhance the novelty function by subtracting its local average.

As an alternative to the spectral flux, we introduce in the **FMP Notebook *Phase-Based Novelty*** an approach that is well suited to study the role of the STFT's phase (see Section 6.1.3). We use this opportunity to discuss phase unwrapping and introduce the principal argument function—topics that beginners in signal processing often find tricky. In the onset detection context, the importance of the phase is highlighted by the fact that slight signal changes (e.g., caused by a weak onset) can hardly be seen in the STFT's magnitude, but may already introduce significant phase distortions. In the **FMP Notebook *Complex-Domain Novelty***, we discuss how phase and magnitude information can be combined (Section 6.1.4).

Each novelty detection procedure has its benefits and limitations, as demonstrated in the **FMP Notebook *Novelty: Comparison of Approaches***. Different approaches may lead to novelty functions with different feature rates. Therefore, we show how one may adjust the feature rate using a resampling approach. Furthermore, we introduce a matrix-based visualization that allows for easy comparison and averaging of different novelty functions. In summary, the notebooks on onset detection constitute an instructive playground for students to learn and explore fundamental signal processing techniques while gaining a deeper understanding of essential onset-related properties of music signals.

The novelty functions introduced so far serve as the basis for onset detection. The underlying assumption is that the positions of peaks (thought of as well-defined local maxima) of the novelty function are good indicators for onset positions. Similarly, in the context of music structure analysis, the peak positions of a novelty function were used to derive segment boundaries between musical parts (see Section 4.4). If the novelty function has a clear peak structure with impulse-like and well-separated peaks, the peaks' selection is a simple problem. However, in practice, one often has to deal with rather noisy novelty functions with many spurious peaks. In such situations, the strategy used for peak picking may substantially influence the quality of the final detection or segmentation result. In the **FMP Notebook *Peak Picking***, we cover this important, yet often underestimated topic. In particular, we present and discuss Python code examples that demonstrate how to use and adapt existing implementations of various peak picking strategies. Instead of advocating a specific procedure, we discuss various heuristics that are often applied in practice. For example, simple smoothing operations may reduce the effect of noise-like fluctuations in the novelty function. Furthermore, adaptive thresholding strategies, where a peak is only selected when its value exceeds a local average of the novelty function, can be applied. Another strategy is to impose a constraint on the minimal distance between two subsequent peak positions to reduce the number of spurious peaks further. In a music processing class, it is essential to note that there is no best peak picking strategy per se—the suitability of a peak picking strategy depends on the requirements of the application. On the one hand, unsuitable heuristics and parameter choices may lead to surprising and unwanted results. On the other hand, exploiting specific data statistics (e.g., minimum distance of two subsequent peaks) at the peak picking stage can lead to substantial improvements. Therefore, knowing the details of peak picking strategies and the often delicate interplay of their parameters is essential when building MIR systems.

While novelty and onset detection are in themselves important tasks, they also constitute the basis for other music processing problems such as tempo estimation, beat tracking, and rhythmic analysis. When designing processing pipelines, a general principle is to avoid intermediate steps based on hard and error-prone decisions. In the following notebooks, we apply this principle for tempo estimation, where we avoid the explicit extraction of note onset positions by directly analyzing a novelty representation concerning periodic patterns. We start with the introductory **FMP Notebook *Tempo and Beat***, where we discuss basic notions and assumptions on which most tempo and beat tracking procedures are based. As already noted before, one first assumption is that beat positions occur at note onset positions, and a second assumption is that beat positions are more or less equally spaced—at least for a certain period. These assumptions may be questionable for certain types of music, and we provide some concrete music examples that illustrate this. For example, in passages with syncopation, beat positions may not go along with any onsets, or the periodicity assumption may be violated for romantic piano music with strong tempo fluctuations. We think that the explicit discussion of such simplifying assumptions is at the core of researching and teaching music processing. In the notebook, we also introduce the notion of pulse levels (e.g., measure, tactus, and tatum level) and give audio examples to illustrate these concepts. Furthermore, we use the concept of tempograms (time–tempo representations) to illustrate tempo phenomena over time. To further deepen the understanding of beat tracking and its challenges, we sonify the beat positions with click sounds and mix them into the original audio recording—a procedure also described in the **FMP Notebook *Sonification*** of **Part**

B. At this point, we again advocate the importance of visualization and sonification methods to make teaching and learning signal processing an interactive pursuit.

Closely following the theory of Section 6.2, we study in the next notebooks the concept of tempograms, which reveal tempo-related phenomena. In the **FMP Notebook *Fourier Tempogram***, the basic idea is to analyze a novelty function using an STFT (see (6.25)) and to re-interpret frequency (given in Hertz) as tempo (given in BPM). We adopt a centered view in our implementation, where the novelty function is zero-padded by half the window length. The Fourier coefficients are computed frequency by frequency, allowing us to explicitly specify the tempo values and tempo resolution (typically corresponding to a nonlinear frequency spacing). Even though losing the FFT algorithm's efficiency, the computational complexity may still be reasonable, when only considering a relatively small number of tempo values. In the **FMP Notebook *Autocorrelation Tempogram***, we cover a second approach for capturing local periodicities of the novelty function. After a general introduction of autocorrelation and its short-time variant, we provide an implementation for computing the time–lag representation and visualization of its interpretation (similar to Figure 6.14). Furthermore, we show how to apply interpolation for converting the lag axis into a tempo axis. Next, in the **FMP Notebook *Cyclic Tempogram***, we provide an implementation of the procedure described in Section 6.2.4. Again we apply interpolation to convert the linear tempo axis into a logarithmic axis before identifying tempo octaves—similar to the approach for computing chroma features. We then study the properties of Fourier- and autocorrelation-based cyclic tempograms, focusing on the tempo discretization parameter. Finally, using real music recordings with tempo changes, we demonstrate the potential of tempogram features for music segmentation applications.

As we discussed in this chapter, the task of beat and pulse tracking extends tempo estimation in the sense that, additionally to the rate, it also considers the phase of the pulses. We described in Section 6.3 how one can obtain a local pulse representation from a Fourier-based tempogram along with its phase. In the **FMP Notebook *Fourier Tempogram***, we provide Python code examples to compute and visualize the optimal windowed sinusoids underlying the idea of Fourier analysis (see Figure 6.13). Then, in the **FMP Notebook *Predominant Local Pulse (PLP)***, we apply an overlap–add technique, where such optimal windowed sinusoids are accumulated over time, yielding the PLP function. Considering challenging music examples with continuous and sudden tempo changes, we explore the role of various parameters, including the sinusoidal length and tempo range. Although the techniques and their implementation are sophisticated, the results (presented in the form of visualizations and sonifications) are highly instructive and, as we find, esthetically pleasing.

Rather than being a beat tracker per se, the PLP concept should be seen as a tool for bringing out a locally predominant pulse track within a specific tempo range. In Section 6.3.2, we introduced a genuine beat tracking algorithm that aims at extracting a stable pulse track from a novelty function, given an estimate of the expected tempo. In the **FMP Notebook *Beat Tracking by Dynamic Programming***, we provide an implementation of this instructive algorithm, which can be solved using

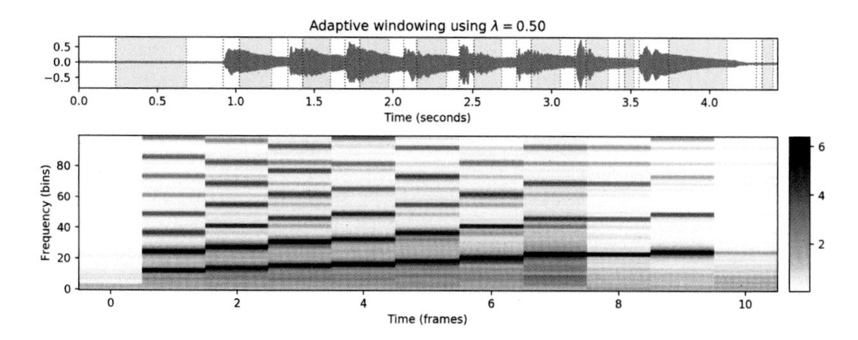

Fig. 6.25 Example of adaptive windowing using a parameter $\lambda \in \mathbb{R}$ to control the neighborhood's relative size to be excluded (see (6.49)). **Top:** Waveform and time grid (indicated by red vertical lines). **Bottom:** Feature representation with frames corresponding to the shaded segments of the signal.

dynamic programming. We apply this algorithm to a small toy example, which is something that is not only helpful for understanding the algorithm but should always be done to test one's implementation. We then move on to real music recordings to indicate the algorithm's potential and limitations.

Finally, in the **FMP Notebook *Adaptive Windowing***, we discuss another important application of beat and pulse tracking, following Section 6.3.3. Our algorithm's input is a feature representation based on fixed-size windowing and an arbitrary (typically nonuniform) time grid, e.g., consisting of previously extracted onset and beat positions. The output is a feature representation adapted according to the input time grid. In our implementation, an additional parameter allows for excluding a certain neighborhood around each time grid position (see Figure 6.25). This strategy may be beneficial when expecting signal artifacts (e.g., transients) around these positions, which degrade the feature representation.

References

1. J. P. Bello, L. Daudet, S. Abdallah, C. Duxbury, M. Davies, and M. B. Sandler, *A tutorial on onset detection in music signals*, IEEE Transactions on Speech and Audio Processing, 13 (2005), pp. 1035–1047.
2. S. Böck, M. E. P. Davies, and P. Knees, *Multi-task learning of tempo and beat: Learning one to improve the other*, in Proceedings of the International Society for Music Information Retrieval Conference (ISMIR), Delft, The Netherlands, 2019, pp. 486–493.
3. S. Böck, F. Krebs, and G. Widmer, *A multi-model approach to beat tracking considering heterogeneous music styles*, in Proceedings of the International Society for Music Information Retrieval Conference (ISMIR), Taipei, Taiwan, 2014, pp. 603–608.
4. M. E. P. Davies and S. Böck, *Evaluating the evaluation measures for beat tracking*, in Proceedings of the International Society for Music Information Retrieval Conference (ISMIR), Taipei, Taiwan, 2014, pp. 637–642.
5. M. E. P. Davies and M. D. Plumbley, *Context-dependent beat tracking of musical audio*, IEEE Transactions on Audio, Speech, and Language Processing, 15 (2007), pp. 1009–1020.
6. N. Degara, A. Pena, M. E. P. Davies, and M. D. Plumbley, *Note onset detection using rhythmic structure*, in Proceedings of the IEEE International Conference on Acoustics, Speech, and Signal Processing (ICASSP), Dallas, Texas, USA, 2010, pp. 5526–5529.
7. S. Dixon, *Automatic extraction of tempo and beat from expressive performances*, Journal of New Music Research, 30 (2001), pp. 39–58.
8. S. Durand, J. P. Bello, B. David, and G. Richard, *Robust downbeat tracking using an ensemble of convolutional networks*, IEEE/ACM Transactions on Audio, Speech, and Language Processing, 25 (2017), pp. 76–89.
9. D. P. Ellis, *Beat tracking by dynamic programming*, Journal of New Music Research, 36 (2007), pp. 51–60.
10. A. Elowsson, *Beat tracking with a cepstroid invariant neural network*, in Proceedings of the International Society for Music Information Retrieval Conference (ISMIR), New York City, USA, 2016, pp. 351–357.
11. F. Eyben, S. Böck, B. Schuller, and A. Graves, *Universal onset detection with bidirectional long short-term memory neural networks*, in Proceedings of the International Society for Music Information Retrieval Conference (ISMIR), Utrecht, The Netherlands, August 2010, pp. 589–594.
12. J. Foote and S. Uchihashi, *The beat spectrum: A new approach to rhythm analysis*, in Proceedings of the International Conference on Multimedia and Expo (ICME), Tokyo, Japan, August 2001, pp. 881–884.
13. M. Goto and Y. Muraoka, *A beat tracking system for acoustic signals of music*, in Proceedings of the ACM International Conference on Multimedia, San Francisco, CA, USA, 1994, pp. 365–372.
14. F. Gouyon and S. Dixon, *A review of automatic rhythm description systems*, Computer Music Journal, 29 (2005), pp. 34–54.
15. F. Gouyon, A. P. Klapuri, S. Dixon, M. Alonso, G. Tzanetakis, C. Uhle, and P. Cano, *An experimental comparison of audio tempo induction algorithms*, IEEE Transactions on Audio, Speech, and Language Processing, 14 (2006), pp. 1832–1844.
16. P. Grosche and M. Müller, *Extracting predominant local pulse information from music recordings*, IEEE Transactions on Audio, Speech, and Language Processing, 19 (2011), pp. 1688–1701.
17. P. Grosche, M. Müller, and F. Kurth, *Cyclic tempogram – a mid-level tempo representation for music signals*, in Proceedings of IEEE International Conference on Acoustics, Speech, and Signal Processing (ICASSP), Dallas, Texas, USA, Mar. 2010, pp. 5522–5525.
18. P. Grosche, M. Müller, and C. S. Sapp, *What makes beat tracking difficult? A case study on Chopin Mazurkas*, in Proceedings of the International Society for Music Information Retrieval Conference (ISMIR), Utrecht, The Netherlands, 2010, pp. 649–654.

19. A. HOLZAPFEL, M. E. P. DAVIES, J. R. ZAPATA, J. L. OLIVEIRA, AND F. GOUYON, *Selective sampling for beat tracking evaluation*, IEEE Transactions on Audio, Speech, and Language Processing, 20 (2012), pp. 2539–2548.
20. A. HOLZAPFEL, Y. STYLIANOU, A. C. GEDIK, AND B. BOZKURT, *Three dimensions of pitched instrument onset detection*, IEEE Transactions on Audio, Speech, and Language Processing, 18 (2010), pp. 1517–1527.
21. K. JENSEN, J. XU, AND M. ZACHARIASEN, *Rhythm-based segmentation of popular Chinese music*, in Proceedings of the International Society for Music Information Retrieval Conference (ISMIR), London, UK, 2005.
22. A. P. KLAPURI, *Sound onset detection by applying psychoacoustic knowledge*, in Proceedings of the IEEE International Conference on Acoustics, Speech, and Signal Processing (ICASSP), Washington, DC, USA, 1999, pp. 3089–3092.
23. A. P. KLAPURI, A. J. ERONEN, AND J. ASTOLA, *Analysis of the meter of acoustic musical signals*, IEEE Transactions on Audio, Speech, and Language Processing, 14 (2006), pp. 342–355.
24. F. KURTH, T. GEHRMANN, AND M. MÜLLER, *The cyclic beat spectrum: Tempo-related audio features for time-scale invariant audio identification*, in Proceedings of the International Society for Music Information Retrieval Conference (ISMIR), Victoria, Canada, Oct. 2006, pp. 35–40.
25. M. F. MCKINNEY, D. MOELANTS, M. E. P. DAVIES, AND A. P. KLAPURI, *Evaluation of audio beat tracking and music tempo extraction algorithms*, Journal of New Music Research, 36 (2007), pp. 1–16.
26. B. C. MOORE, *An Introduction to the Psychology of Hearing*, Brill Academic Publisher, 6th ed., 2013.
27. M. MÜLLER AND F. ZALKOW, *FMP Notebooks: Educational material for teaching and learning fundamentals of music processing*, in Proceedings of the International Society for Music Information Retrieval Conference (ISMIR), Delft, The Netherlands, 2019, pp. 573–580.
28. R. PARNCUTT, *A perceptual model of pulse salience and metrical accent in musical rhythms*, Music Perception, 11 (1994), pp. 409–464.
29. G. PEETERS, *Template-based estimation of time-varying tempo*, EURASIP Journal on Advances in Signal Processing, (2007).
30. G. PEETERS AND H. PAPADOPOULOS, *Simultaneous beat and downbeat-tracking using a probabilistic framework: Theory and large-scale evaluation*, IEEE Transactions on Audio, Speech, and Language Processing, 19 (2011), pp. 1754–1769.
31. E. D. SCHEIRER, *Tempo and beat analysis of acoustical musical signals*, Journal of the Acoustical Society of America, 103 (1998), pp. 588–601.
32. J. SCHLÜTER AND S. BÖCK, *Improved musical onset detection with convolutional neural networks*, in Proceedings of the IEEE International Conference on Acoustics, Speech, and Signal Processing (ICASSP), Florence, Italy, May 2014, pp. 6979–6983.
33. H. SCHREIBER AND M. MÜLLER, *A single-step approach to musical tempo estimation using a convolutional neural network*, in Proceedings of the International Society for Music Information Retrieval Conference (ISMIR), Paris, France, 2018, pp. 98–105.
34. H. SCHREIBER, J. URBANO, AND M. MÜLLER, *Global music tempo estimation: Are we done yet?*, Transactions of the International Society for Music Information Retrieval (TISMIR), 3 (2020), pp. 111–125.
35. J. SEPPÄNEN, *Tatum grid analysis of musical signals*, in Proceedings of the IEEE Workshop on Applications of Signal Processing to Audio and Acoustics (WASPAA), New Paltz, NY, USA, 2001, pp. 131–134.
36. W. A. SETHARES, *Rhythm and Transforms*, Springer, 2007.
37. J. R. ZAPATA, M. E. P. DAVIES, AND E. GÓMEZ, *Multi-feature beat tracking*, IEEE/ACM Transactions on Audio, Speech, and Language Processing, 22 (2014), pp. 816–825.
38. J. R. ZAPATA AND E. GÓMEZ, *Comparative evaluation and combination of audio tempo estimation approaches*, in Proceedings of the Audio Engineering Society (AES) Conference on Semantic Audio, Ilmenau, Germany, 2011.

Exercises

Exercise 6.1. Let $x : \mathbb{Z} \to \mathbb{R}$ be a signal with the nonzero samples $(x(0),\ldots,x(6)) = (0.1,-0.1,0.1,0.9,0.7,0.1,-0.3)$ (all other samples being zero). Furthermore, let $w : \mathbb{Z} \to \mathbb{R}$ be a rectangular window function with nonzero coefficients $w(-1) = w(0) = w(1) = 1$ (i.e., $M = 1$; see Section 6.1.1). Compute all nonzero coefficients of the energy-based novelty function $\Delta_{\mathrm{Energy}} : \mathbb{Z} \to \mathbb{R}$ as defined in (6.3).

Exercise 6.2. Let $x : \mathbb{Z} \to \mathbb{R}$ be a discrete signal. Furthermore, let $w : \mathbb{Z} \to \mathbb{R}$ be a rectangular window function of length $2M+1$ centered at time zero, i.e., $w(m) = 1$ for $m \in [-M : M]$ and $w(m) = 0$ otherwise. Then the local energy E_w^x (see (6.1)) is given by

$$\mathrm{E}_w^x(n) := \sum_{m=-M}^{M} x(n+m)^2$$

for $n \in \mathbb{Z}$. In the following, an operation refers to a multiplication, an addition, or a subtraction of two real-valued samples. Determine the overall number of operations that are required to compute $\mathrm{E}_w^x(n)$ for $n \in [0 : N-1]$ using a naive approach. Then, describe an improved procedure that reduces the overall number of required operations. How many operations are needed by your procedure?

Exercise 6.3. Let $\mathcal{Y}$ be an $(N \times (K+1))$ matrix with coefficients $\mathcal{Y}(n,k)$ indexed by $n \in [0 : N-1]$ and $k \in [0 : K]$. In the following, we consider the matrix $\mathcal{Y}$ defined by

$$\mathcal{Y}^\top = \begin{bmatrix} 0 & 0.1 & 0.1 & 0 & 0.2 & 0.1 \\ 0 & 0 & 0.1 & 0.1 & 0.2 & 0.1 \\ 0 & 0.8 & 0.7 & 0.5 & 0.6 & 0.4 \\ 0 & 0 & 0 & 0 & 0.8 & 0.7 \\ 0 & 0 & 0 & 0.1 & 0 & 0 \end{bmatrix},$$

where $N = 6$ and $K = 4$. (Note that the transposed matrix has been specified.) Interpreting this matrix as a magnitude spectrogram, compute the novelty function $\Delta_{\mathrm{Spectral}}$ as defined in (6.6). Furthermore, compute the local average function μ using $M = 1$ (see (6.7)) and the enhanced novelty function $\bar{\Delta}_{\mathrm{Spectral}}$ (see (6.8)).

Exercise 6.4. Realize a bandwise approach for spectral-based novelty detection as outlined at the end of Section 6.1.2. More concretely, let x denote the given music signal sampled at a rate of $F_s = 22050$ Hz and $\mathcal{Y}$ the resulting (possibly compressed) magnitude spectrogram (as in (6.5)) using an STFT window length of $N = 4096$ and a hop size of $H = N/2$. In a first step, divide the frequency range into bands with the first band covering 0–500 Hz, the second 500–1000 Hz, the third 1000–2000 Hz, the fourth 2000–4000 Hz, and the fifth band 4000–11025 Hz. Determine for each of the bands the set of spectral coefficients (similar to (3.3)). Then, compute a novelty function for each of the bands separately (similar to (6.6)). Finally, compute a single overall novelty function by considering a weighted sum over the bandwise novelty functions using the weighting factor $w_\ell \in \mathbb{R}_{>0}$ for the ℓth band, $\ell \in [1 : 5]$. Give a formal description of this procedure by specifying the mathematical details.

Exercise 6.5. In this exercise, we consider the novelty functions corresponding to the click tracks shown in the following figure:

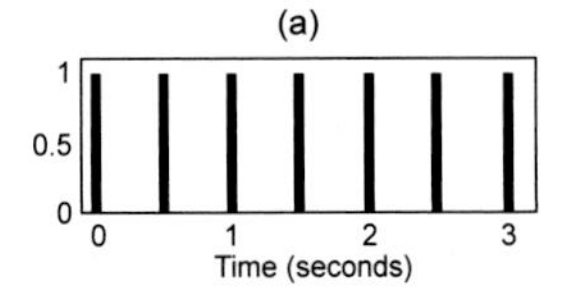

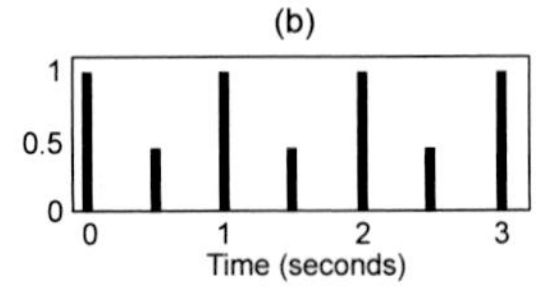

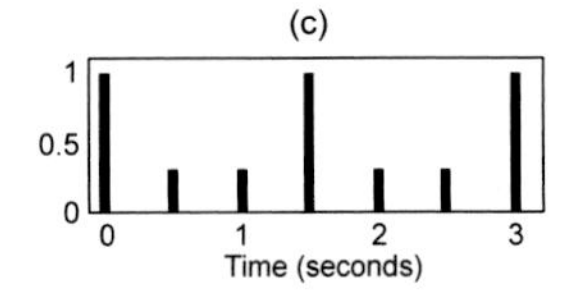

For each of these novelty functions, sketch the Fourier tempogram (see Section 6.2.2) in the tempo range between 20 and 250 BPM. In particular, specify the tempo parameters for which one expects large tempogram coefficients. What is the smallest such parameter (corresponding to the lowest relevant tempo) for each case? Finally, for each of the three novelty functions, indicate visually (as in Figure 6.13) the correlation with the analyzing sinusoid corresponding to this smallest, yet relevant tempo.

Exercise 6.6. Let $x \in \ell^2(\mathbb{Z})$ be a real-valued discrete-time signal. Furthermore, let R_{xx} be the autocorrelation of x, which is given by $R_{xx}(\ell) = \sum_{n\in\mathbb{Z}} x(n)x(n-\ell)$ for each lag parameter $\ell \in \mathbb{Z}$ (see (6.27)). Show that $R_{xx}(0) = \mathrm{E}(x)$ (see (2.41)) and $|R_{xx}(\ell)| \leq R_{xx}(0)$. Furthermore, show that R_{xx} is symmetric, i.e., $R_{xx}(\ell) = R_{xx}(-\ell)$.
[**Hint:** Use the Cauchy–Schwarz inequality $|\langle x|y\rangle| \leq \|x\|\|y\|$ from (2.40), which holds for any $x, y \in \ell^2(\mathbb{Z})$.]

Exercise 6.7. Let $x : \mathbb{Z} \to \mathbb{R}$ be a real-valued signal. Assume that the support of x lies in the interval $[-M : M]$ for some $M \in \mathbb{N}$. Let R_{xx} be the autocorrelation as defined in (6.27). Show that $R_{xx}(\ell) = 0$ for $|\ell| \geq 2M + 1$. Furthermore, show that at most $2M + 1 - |\ell|$ of the summands in (6.29) are nonzero.

Exercise 6.8. Let $\Delta : \mathbb{Z} \to \mathbb{R}$ be a novelty function with a feature rate of 10 Hz. Furthermore, let $\mathcal{T}^{\mathrm{A}}$ be the autocorrelation tempogram derived from Δ (see (6.31)). What is the maximal tempo that is captured by $\mathcal{T}^{\mathrm{A}}$?

Exercise 6.9. In this exercise, we consider a discrete cyclic tempogram representation $\mathcal{C}_{\tau_0}$ using a reference tempo $\tau_0 = 60$ BPM (see (6.35)). For computing $\mathcal{C}_{\tau_0}$, we use four tempo octaves ranging from $\tau = 30$ to $\tau = 480$ BPM, where each octave is logarithmically sampled using $M \in \mathbb{N}$ tempo parameters. Specify a formula for the tempo values that are needed to compute $\mathcal{C}_{\tau_0}$. Furthermore, using $M = 10$, determine the eleven tempo values between $\tau = 60$ and $\tau = 120$ BPM. Next, assume that $\mathcal{C}_{\tau_0}$ is derived from an autocorrelation tempogram based on a feature rate of 10 Hz. Determine the lag parameters corresponding to the eleven computed tempo values. Which problems arise? Make suggestions to alleviate these problems.

Exercise 6.10. For a given parameter $N \in \mathbb{N}$, let $\mathcal{B}^N$ be the space of all possible beat sequences within the interval $[1 : N]$ (see Section 6.3.2). Determine the number $|\mathcal{B}^N|$. Furthermore, given a length parameter $L \in [0 : N]$, determine the number of beat sequences of length L. Finally, let $\mathcal{B}_n^N \subset \mathcal{B}^N$ denote the subset of all beat sequences that end in $n \in [0 : N]$ (where the case $n = 0$ refers to the empty beat sequence). Determine the number $|\mathcal{B}_n^N|$. Finally, show that $\mathcal{B}^N = \cup_{n\in[0:N]} \mathcal{B}_n^N$ (see (6.44)).

Exercise 6.11. Given a novelty function $\Delta : [1 : N] \to \mathbb{R}$, analyze the computational complexity of the beat tracking procedure described in Section 6.3.2 (see also Table 6.1) in terms of memory requirements as well as in terms of the number of required operations. Assume that an operation is an addition, a multiplication, an evaluation of $P_{\hat{\delta}}$, or a maximization (where maximization over a set of $M \in \mathbb{N}$ elements counts as M operations).

Exercise 6.12. Apply the beat tracking procedure described in Section 6.3.2 (see also Table 6.1) to the novelty function $\Delta : [1 : N] \to \mathbb{R}$ with $N = 11$ given by the following values:

n	1	2	3	4	5	6	7	8	9	10	11
$\Delta(n)$	0.1	0.0	1.0	0.0	1.0	0.8	0.0	0.2	0.4	1.0	0.0

For the computations, use the weight parameter $\lambda = 1$ and the following values for the penalty function $P_{\hat{\delta}}$ which favors the beat period $\hat{\delta} = 3$ (note that, for the sake of simplicity, these values are not obtained from (6.40)):

n	1	2	3	4	5	6	7	8	9	10	11
$P_{\hat{\delta}}(n)$	-2	-0.2	1.0	0.5	-0.1	-1	-1.5	-3	-5	-8	-12

Compute the accumulated score values $\mathbf{D}(n)$ and the predecessors $\mathbf{P}(n)$ for $n \in [1:N]$. Furthermore, derive the optimal beat sequence B^*.

Exercise 6.13. The penalty function $P_{\hat{\delta}}$ defined in (6.40) (see also Figure 6.21) decreases rapidly with larger deviations from the ideal beat period $\hat{\delta}$. Therefore, it becomes unlikely that the predecessor m of some beat position n lies far from the position $n-\hat{\delta}$. This observation can be used to achieve significant savings by restricting the search space $m \in [1:n-1]$ in the maximization (6.47). For example, assuming that the next beat to be estimated has at least the distance $\hat{\delta}/2$ and at most the distance $2\hat{\delta}$ from its predecessor beat, one may replace the search space $m \in [1:n-1]$ by the constrained search space $m \in [1:n-1] \cap [n-2\hat{\delta}, n-\hat{\delta}/2]$. Analyze the computational complexity of the modified procedure (as in Exercise 6.11). Compare the result with the original procedure.

Exercise 6.14. Recall that a beat sequence $B = (b_1, b_2, \ldots, b_L)$ is a sequence of increasing indices $b_\ell \in [1:N]$, $\ell \in [1:L]$. Mathematically, this is identical to the notion of a boundary annotation, which we introduced for evaluating novelty-based segmentation procedures in the context of music structure analysis (see Section 4.5.4). Therefore, to evaluate a beat tracking procedure, one can use exactly the same evaluation measures as for novelty detection. Following Section 4.5.4, let B^{Ref} be a reference beat sequence and B^{Est} an estimated beat sequence. Furthermore, let $\tau \geq 0$ be a tolerance parameter for the maximal acceptable deviation. Similar to (4.57), an estimated beat $b^{\mathrm{Est}} \in B^{\mathrm{Est}}$ is considered **correct** if it lies within the τ-neighborhood of a reference beat $b^{\mathrm{Ref}} \in B^{\mathrm{Ref}}$:

$$|b^{\mathrm{Est}} - b^{\mathrm{Ref}}| \leq \tau.$$

Following Section 4.5.4, introduce the notions of true positives, false positives, and false negatives, and then derive the precision, recall, and F-measure. Furthermore, using $\tau = 1$, compute these measures for the following beat sequences:

$$\begin{aligned} B^{\mathrm{Ref}} &= (10,20,30,40,50,60,70,80,90) \\ B^{\mathrm{Est}} &= (10,19,26,34,42,50,61,70,78,89) \end{aligned}$$

Exercise 6.15. In the evaluation measure considered in Exercise 6.14, the beat positions were evaluated independently of each other. However, when tapping to the beat of music, a listener obviously requires the temporal context of several consecutive beats. Therefore, in evaluating beat tracking procedures, it seems natural to consider beats in the temporal context instead of looking at the beat positions individually. To account for these temporal dependencies, we now introduce a context-sensitive evaluation measure. Let $B^{\mathrm{Ref}} = (r_1, r_2, \ldots, r_M)$ be a reference beat sequence with $r_m \in [1:N]$, $m \in [1:M]$. Similarly, let $B^{\mathrm{Est}} = (b_1, b_2, \ldots, b_L)$ be an estimated beat sequence with $b_\ell \in [1:N]$, $\ell \in [1:L]$. Furthermore, let $K \in \mathbb{N}$ be a parameter that specifies the temporal context measured in beats, and let $\tau \geq 0$ be a tolerance parameter for the maximal acceptable deviation. Then, an estimated beat b_ℓ is considered a **K-correct detection** if there exists a subsequence $b_i, \ldots, b_{i+K-1}$ of B^{Est} containing b_ℓ (i.e., $\ell \in [i:i+K-1]$) as well as a subsequence $r_j, \ldots, r_{j+K-1}$ of B^{Ref} such that

$$|b_{i+k} - r_{j+k}| \leq \tau$$

for all $k \in [0:K-1]$. Intuitively, for a beat to be considered K-correct, one requires an entire track consisting of K consecutive estimated beats that match (up to the tolerance τ) a track of K consecutive reference beats. Note that a single outlier in the estimated beats voids this property. Let L_K be the number of K-correct estimated beats. Then, we define the context-sensitive precision $\mathrm{P}_K := L_K/L$, recall $\mathrm{R}_K := L_K/M$, and F-measure $\mathrm{F}_K := 2\mathrm{P}_K\mathrm{R}_K/(\mathrm{P}_K + \mathrm{R}_K)$. For B^{Ref} and B^{Est} as specified in Exercise 6.14, determine the set of K-correct beat sequences as well as the context-sensitive precision, recall, and F-measure for $\tau = 1$ and $K \in \{1,2,3,4\}$.

Chapter 7
Content-Based Audio Retrieval

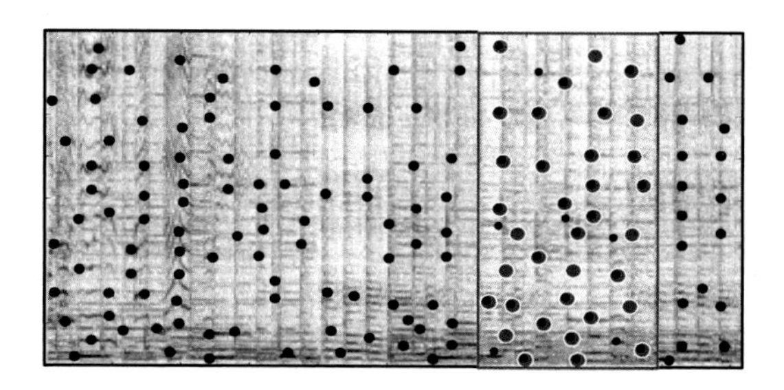

The revolution in music distribution and storage brought about by digital technology has fueled tremendous interest in and attention to the ways that information technology can be applied to this kind of content. The rapidly growing corpus of digitally available music data requires novel technologies that allow users to browse personal collections or discover new music on the world wide web, or to help music creators to manage and protect their rights. The general field of **information retrieval** (IR) is devoted to the task of organizing information and of making it accessible and useful. An information retrieval process begins when a user specifies his/her information needs by means of a **query**. The retrieval system should then deliver from a given data collection all **documents** or **items** that are somehow related to the query. For example, in the case of a typical web search, the query may consist of a text string of words and the search engine should deliver all text documents containing the specified words.

While ten years ago, most digital content was textual, it has now expanded to include audio, images, video, and other types of multimedia documents. This particularly holds for the music domain, where listeners enjoy ubiquitous access to huge music collections containing audio recordings, digitized images of sheet music, album covers, and an increasing number of video clips. Such huge amounts of readily available music require retrieval strategies that allow users to explore large music collections in a convenient and enjoyable way. Most of the available services for music recommendation and playlist generation rely on metadata and textual annotations of the actual audio content. For example, a music recording may be described by the name of the artist or composer, the title of the piece, or the song lyrics—editorial data that is typically created manually by domain experts. Typical query terms may be a title such as "Day Tripper" when searching for the song by The Beatles, or a composer's name such as "Beethoven" when looking for the Fifth

M. Müller, *Fundamentals of Music Processing*, https://doi.org/10.1007/978-3-030-69808-9_7

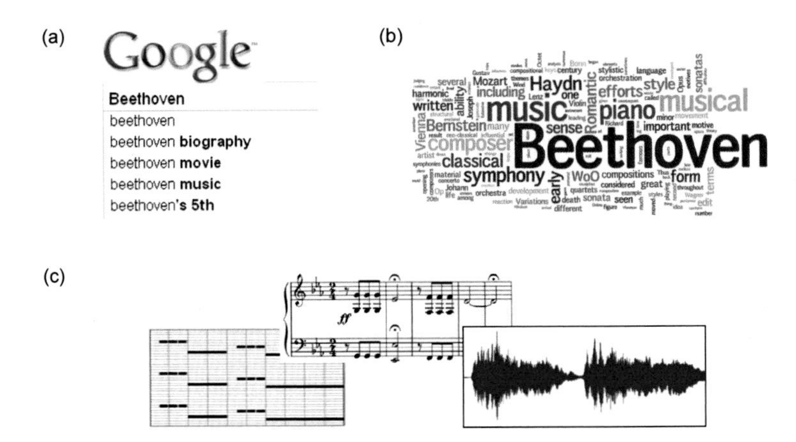

Fig. 7.1 Illustration of retrieval scenarios. **(a)** Traditional retrieval using textual metadata (e.g., artist, title) and a web search engine. **(b)** Retrieval based on rich and expressive metadata given by tags. **(c)** Content-based retrieval using audio, MIDI, or score information.

Symphony. In this scenario, a user needs to have a relatively clear idea of what he or she is looking for (see Figure 7.1a). To overcome these limitations, recent retrieval systems complement editorial metadata with general and expressive annotations, which are also referred to as **tags**. Such tags may describe the musical style or genre of a recording, and also include information about the mood, the musical key, or the tempo. Many music recommendation systems rely on a large number of tags that have been generated by different users, automatically extracted from music blogs, and enriched by statistical information on user behavior and music consumption. Even though such tags may be quite noisy, they still express certain general trends and describe the music content in a statistical and human-centered way.

While text-based retrieval systems can be very powerful, they require the audio material to be enriched with suitable metadata—an assumption that is often not valid, in particular for less popular music or music material that is scattered in unstructured data collections. Furthermore, not all retrieval scenarios can be handled by a purely text-based approach. How should a retrieval system be designed if the user's query consists of a short excerpt of a CD recording or a melody sung into a microphone? What can be done if only a few measures of a musical score are available? How can a user be satisfied if he or she looks for music with a specific rhythmic pattern or harmonic progression which have not been annotated? To handle such scenarios, one requires **content-based retrieval** systems that only make use of the raw music data, rather than relying on manually generated metadata. The term **content** loosely refers to any kind of information that can be directly derived from the music material to be queried, compared, and retrieved.

In this chapter, we present various content-based retrieval strategies that follow the **query-by-example** paradigm: given a music representation or a fragment of it (used as a query or example), the task is to automatically retrieve documents from a music collection containing parts or aspects that are similar to the query. Such strategies can be loosely classified according to their **specificity**, which refers to the degree of similarity between the query and the database documents. High specificity is related to a strict notion of similarity, whereas low specificity refers to a rather vague one. Starting with a retrieval task of high specificity, in Section 7.1 we deal with the problem of **audio identification**. Given a small audio fragment as query, the task of audio identification consists in identifying the particular audio recording that is the source of the query. At a lower specificity level, one finds the task of **audio matching**, which we discuss in Section 7.2. Given a query fragment, the goal of audio matching is to retrieve all audio excerpts that musically correspond to the query. In this scenario, one explicitly allows semantically motivated variations as they typically occur in different performances and arrangements of a piece of music. Further softening the notion of similarity, we finally introduce the task of **version identification** in Section 7.3. In this scenario, one deals not only with performance variations in instrumentation and tempo, but also with more extreme variations concerning the musical structure, key, or melody, as typically occur in remixes and cover songs. By means of these three scenarios, we introduce a number of key techniques and discuss the trade-off between requirements that become important when designing and implementing retrieval systems.

7.1 Audio Identification

Within the area of content-based music retrieval, the task of audio identification has received a lot of attention in both academic research and industry. Audio identification techniques have now been integrated into many commercial applications such as broadcast and copyright monitoring, or added-value services for delivering metadata and other content information. In Section 7.1.1, we explain the most important requirements for audio identification systems, including robustness, reliability, granularity, scalability, and efficiency. In the identification process, the audio material is compared by means of so-called **audio fingerprints**, which are compact and descriptive audio features. There are many different ways for designing and computing audio fingerprints, and the suitability of a specific type of fingerprint very much depends on the requirements imposed by the application in mind. Serving as an instructive and important example, we discuss the main ideas of the fingerprinting techniques that are used in the commercial Shazam[1] music identification service, which were developed by Wang [30]. First, in Section 7.1.2, we introduce fingerprints based on spectral peaks and discuss their properties. Then, in Section 7.1.3, we present retrieval and indexing techniques that are needed to scale up the fingerprinting system to huge audio collections.

---

[1] www.shazam.com

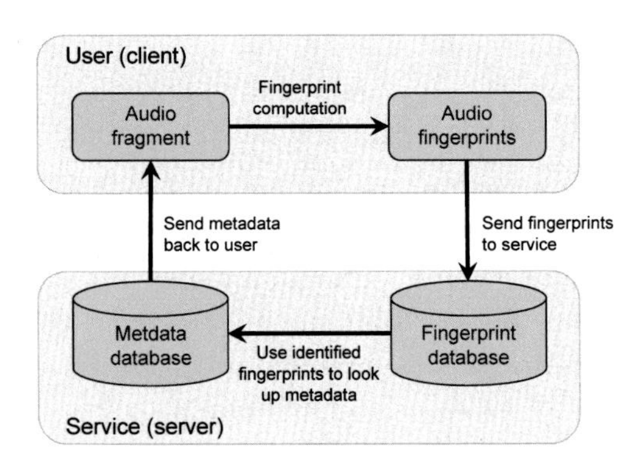

Fig. 7.2 Client–server model for an added-value service which delivers metadata linked to an identified query audio fragment.

7.1.1 General Requirements

Let us assume you hear a song in a restaurant, in a shopping mall, or in a car, and you want to learn more about it. For example, you want to know the song's title or the name of the performer or the artist. Recent music discovery services help users in such situations by identifying the audio recording and delivering suitable content information. A typical scenario is that a user, also called the **client**, records a short audio fragment of the unknown song using a smartphone. The audio fragment is then converted into a compact fingerprint representation, which is transmitted to the identification service, also called the **server**. The server hosts various data resources including a fingerprint database that covers all music recordings to be identified, as well as a metadata database that contains content information linked to these recordings. The server receives the query fingerprints sent by the client and compares them with the fingerprints contained in the database. This step is typically realized by an efficient database look-up supported by suitable index structures. In the case of a successful identification, the server retrieves the content information linked to the identified fingerprints and sends back the desired metadata to the client. Figure 7.2 presents a schematic overview of the underlying **client–server model** of the described metadata delivery service.

Real-world music recognition systems need to be robust and computationally efficient, which leads to a number of technical challenges to be solved. In particular, the audio fingerprints used in such systems need to fulfill certain requirements including high specificity, robustness, compactness, and scalability. We now discuss these requirements in more detail.

First of all, audio fingerprints should possess a high **specificity** so that even an audio fragment only a few seconds long suffices to reliably identify the corresponding recording and to distinguish it from millions of others. To this end, the fingerprints must retain a sufficient amount of acoustically relevant information to allow discrimination over a large number of fingerprints.

The discrimination requirement, however, is undermined by the fact that audio signals may be exposed to many different kinds of distortions. For example, the user may record the audio fragment in a noisy environment, where the music is superimposed with other sound sources such as people speaking, traffic, or engine noise. Additionally, there may be environmental factors such as reverberation and absorption. Furthermore, the audio signal to be identified may be modified and degraded by imperfect recording, transmission, or playback devices. For example, transmission through telephone equipment reduces the frequencies to a range from about 300 Hz to 4000 Hz. Other artifacts may be introduced by lossy compression, pitch shifting, time scaling, equalization, or dynamics compression. For a reliable identification, fingerprints have to possess a significant degree of **robustness** against such background noise and signal distortions.

As another important requirement, audio fingerprints should be as **compact** as possible. A small-sized fingerprint representation is required in view of the millions of recordings the server has to process and provide in its database. Furthermore, in smartphone-based applications, audio fingerprints need to be transmitted over channels with limited bandwidth. Thus the size of the query fingerprints should be kept to a minimum.

Beyond having compact fingerprints, a more general requirement for audio identification systems is referred to as **scalability**. In order to scale to millions of recordings, the computation of audio fingerprints should be simple and efficient—a requirement that is also needed when computing the fingerprints on mobile devices with limited processing power. Most importantly, for the design of large-scale audio identification systems, one requires efficient retrieval strategies to facilitate fast database look-ups. In this context, hash-based indexing techniques come into play. Such techniques also become crucial when the server needs to process hundreds or thousands of queries per second. Finally, in view of a constantly growing digital music catalog, one requires efficient update procedures for extending and maintaining the server's fingerprint database and the underlying index structures.

All these requirements are important for the design of large-scale audio identification systems. However, improving a certain requirement often implies losing performance in some other, and one has to face a delicate trade-off between contradicting principles. For example, boosting the robustness typically leads to an increase of wrong identifications (false positives), thus deteriorating the accuracy of the identification system. Similarly, even though beneficial for computational and compactness reasons, an excessive reduction of the fingerprint size negatively affects the discrimination capability. Conversely, fingerprints of high specificity and robustness may not be usable in practice if their computation requires extensive processing power. The balance between the different requirements very much depends on the respective application. For certain applications, one may only deal with rather mild signal distortions (e.g., only compression artifacts, but no background noise), which significantly lowers the robustness requirement. In other applications, even query fragments of long duration may be acceptable, which makes the discrimination between different items much easier.

7.1.2 Audio Fingerprints Based on Spectral Peaks

Closely following the original publication by Wang [30], we now present the main ideas of the audio fingerprint techniques that are used in the Shazam audio identification system. Recall that, in smartphone-based applications, the audio fragment used as a query may be severely distorted and superimposed with other sound sources. Furthermore, the duration of the fragment is typically short (a couple of seconds) and it may come from any portion of the original music recording. Therefore, besides robustness and specificity, temporal locality and translation invariance are further important properties that audio fingerprints should possess. **Temporal locality** means that each fingerprint should be calculated using audio samples only from a small neighborhood of a corresponding point in time, so that distant samples do not have any influence. The requirement of **translation invariance** means that the fingerprints should be reproducible regardless of the position of the audio fragment within the original music recording, as long as the neighborhood needed to compute the fingerprint is contained in the fragment. We now describe audio fingerprints that are based on the concept of spectral peaks. Being characteristic points in the time–frequency plane, such peaks turn out to fulfill most of the desired properties. In particular, the time–frequency coordinates of a spectral peak often remain unchanged even in the presence of noise and additional sound sources, which makes them highly suitable for the aforementioned smartphone-based application.

7.1.2.1 Design of Audio Fingerprints

Given an audio signal in the form of a sampled waveform, the first step in deriving the fingerprints consists in computing an STFT $\mathcal{X}$ as in (2.148) or (3.1). Recall that $\mathcal{X}(n,k)$ denotes the kth Fourier coefficient for the nth time frame, where $k \in [0:K]$ and $n \in \mathbb{Z}$. In the following, an element $k \in [0:K]$ is also referred to as a **frequency stamp** and an element $n \in \mathbb{Z}$ as a **time stamp**. Hence, the coordinates of a Fourier coefficient $\mathcal{X}(n,k)$ are specified by a time–frequency point $(n,k) \in \mathbb{Z} \times [0:K]$ consisting of a time stamp n and a frequency stamp k.

In the second step of the fingerprint computation, the STFT representation of the signal is reduced to a sparse set of time–frequency points. To this end, one uses a peak-picking strategy that identifies time–frequency points that have a higher magnitude than all their neighbors within a region around the respective points. More precisely, let $\tau > 0$ and $\kappa > 0$ be parameters that determine the size of the neighborhood in the time and frequency direction, respectively. Then a point (n_0,k_0) is selected as a peak if

$$|\mathcal{X}(n_0,k_0)| > |\mathcal{X}(n,k)| \tag{7.1}$$

for all $(n,k) \in \big([n_0 - \tau, n_0 + \tau] \times [k_0 - \kappa, k_0 + \kappa]\big) \cap \big(\mathbb{Z} \times [0:K]\big)$. These definitions are illustrated by Figure 7.3a, which shows a spectrogram and one extracted peak with its local neighborhood. Note that increasing the size of the neighborhood makes it harder for a point to be selected. Thus, the parameters τ and κ can be used to adjust

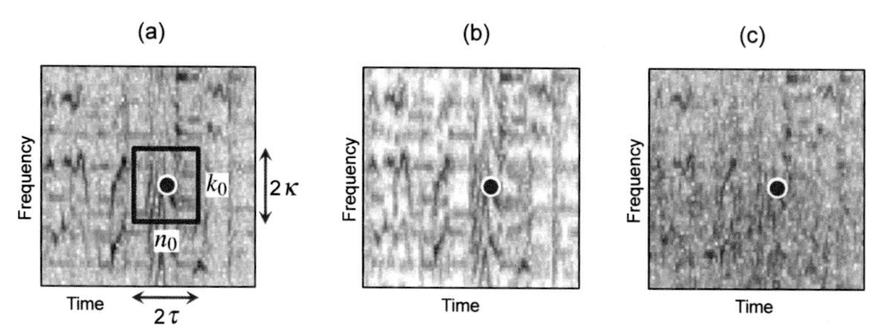

Fig. 7.3 Illustration of peak selection and signal distortions. **(a)** Original spectrogram of an undistorted signal with a peak at position (n_0, k_0) and a neighborhood specified by the parameters τ and κ. **(b)** Spectrogram of the signal after applying lossy audio data compression (e.g., MP3). **(c)** Spectrogram of the signal superimposed with strong background noise.

the density of the selected peaks and to yield a reasonably uniform coverage of the time–frequency plane.

By definition, the spectral peaks are local in nature with regard to time and frequency, thus fulfilling the locality requirement. Furthermore, when choosing a small hop size H (e.g., $H = 1$, which shifts the analysis window sample by sample) in the STFT computation, the peak features become invariant to signal translations. One major benefit of using spectral peaks is their robustness to even severe signal degradations. For example, Figure 7.3b shows a spectrogram obtained after applying lossy audio compression (e.g., MP3) using a very low bitrate. Even though the magnitudes of the spectral coefficients may have changed significantly, many of the time–frequency coordinates of the local maxima have not changed. Figure 7.3c shows an example where the music recording has been superimposed with strong noise and people speaking in the background. Still, some of the characteristic peaks can be found in the distorted spectrogram.

In general, one may distinguish between two different types of distortions. The first type of distortion affects all spectral coefficients in a similar fashion. For example, this is the case when changing the amplitude of the signal or when adding white noise. Even though the magnitudes of the spectral coefficients may change, the proportions of the magnitudes and therefore the peak coordinates remain more or less unaffected. For the second type, the distortions are concentrated in certain time–frequency regions, while other regions are left unchanged. For example, this happens when the signal is bandlimited to a certain frequency range or when more structured sound sources such as people speaking in the background are superimposed. Because of their local nature, peak positions are relatively independent from each other and many of them survive in the unchanged regions.

The peak selection step reduces a complicated spectrogram representation of the signal to a sparse set of coordinates (see Figure 7.4 for an illustration). Notice that the magnitudes of the peaks are no longer used—only the time and frequency stamps of the peaks are considered, which introduces a high degree of robustness. In the following, the representation consisting of all peak coordinates is also referred to as a

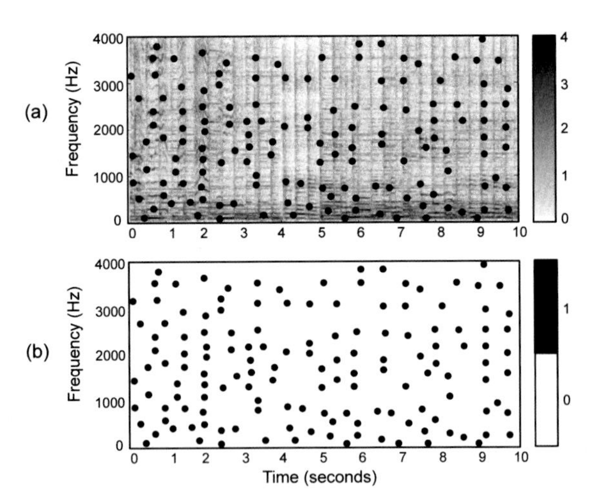

Fig. 7.4 Illustration of the peak-based audio fingerprints using a recording of "Day Tripper" by The Beatles as an example. **(a)** Spectrogram representation with extracted peak fingerprints. **(b)** Resulting constellation map.

constellation map. We will see how one can build robust yet efficient audio identification systems by using fingerprint representations based on such constellation maps.

7.1.2.2 Fingerprint Matching

Let us consider a short query audio fragment, called $\mathcal{Q}$, as well as a database recording, called $\mathcal{D}$. Furthermore, let $\mathcal{C}(\mathcal{Q})$ and $\mathcal{C}(\mathcal{D})$ denote the constellation maps of $\mathcal{Q}$ and $\mathcal{D}$, respectively. In the case that the query $\mathcal{Q}$ is contained in the recording $\mathcal{D}$, the constellation map $\mathcal{C}(\mathcal{Q})$ should more or less agree with the corresponding section within the constellation map $\mathcal{C}(\mathcal{D})$. Intuitively, the basic idea of audio identification is to put the constellation map of $\mathcal{D}$ on a strip chart and the constellation map of $\mathcal{Q}$ on a transparent piece of plastic. The latter is then shifted over the former, and when the number of matching points is significant, the query is considered to be contained in the document. The proper time offset of the matching position within $\mathcal{C}(\mathcal{D})$ is given by the shift. This matching process, which is similar to computing the cross-correlation between two time series, is illustrated by Figure 7.5. In our example, the time offset of the best matching segment in $\mathcal{D}$ corresponds to time position $t = 4.5$ sec (see Figure 7.5b). Even in the case of spurious and missing peaks due to signal degradations, the query may be correctly identified as long as the number of correctly matching peak coordinates is statistically significant. For example, in current music discovery services, a query audio fragment of 10 to 15 seconds of duration may still be identified correctly even in the case that less than 10 % of the peaks have survived [30].

We now formalize the outlined matching procedure. Recall that a peak position is specified by its coordinates (n,k) consisting of a time stamp $n \in \mathbb{Z}$ and a frequency stamp $k \in [0:K]$. A constellation map is then a finite set of such coordinates. As

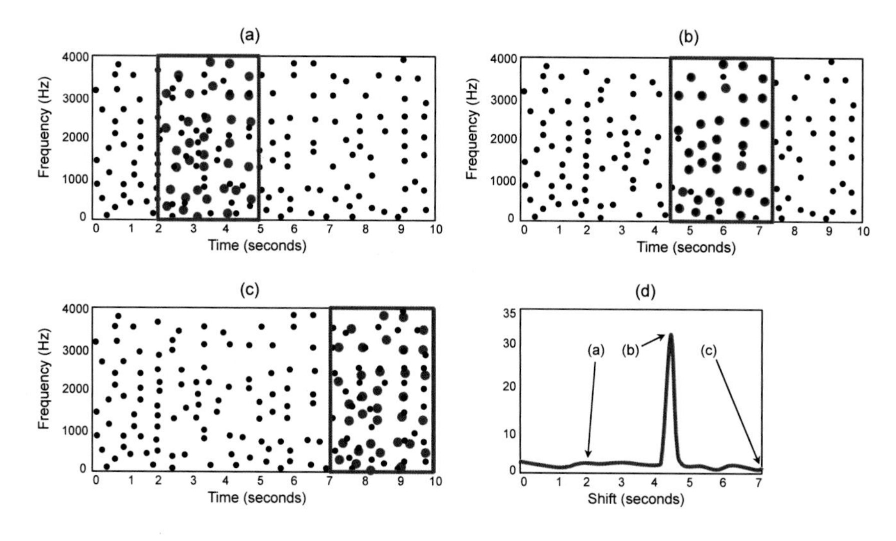

Fig. 7.5 Constellation maps of a database document $\mathcal{D}$ (black, small dots) and a query $\mathcal{Q}$ (red, large dots). The constellation map of $\mathcal{Q}$ (containing $|\mathcal{Q}| = 35$ points) is shifted and locally compared with the one of $\mathcal{D}$. **(a)** Shift corresponding to time offset $t = 2$ sec. **(b)** Shift corresponding to time offset $t = 4.5$ sec (matching position). **(c)** Shift corresponding to time offset $t = 7$ sec. **(d)** Matching function $\Delta_{\mathcal{C}}$.

before, let $\mathcal{C}(\mathcal{Q}) \subset \mathbb{Z} \times [0:K]$ and $\mathcal{C}(\mathcal{D}) \subset \mathbb{Z} \times [0:K]$ be the constellation maps of the query $\mathcal{Q}$ and the document $\mathcal{D}$, respectively. Shifting the query by $m \in \mathbb{Z}$ positions yields the constellation map $m + \mathcal{C}(\mathcal{Q})$ defined by

$$m + \mathcal{C}(\mathcal{Q}) := \{(m+n,k) \mid (n,k) \in \mathcal{C}(\mathcal{Q})\}. \tag{7.2}$$

To count the matching peak coordinates between a shifted query and a database document, we simply need to intersect the corresponding constellation maps and to determine the size of the resulting set. This yields a matching function $\Delta_{\mathcal{C}} : \mathbb{Z} \to \mathbb{N}_0$ defined by

$$\Delta_{\mathcal{C}}(m) := \big|(m + \mathcal{C}(\mathcal{Q})) \cap \mathcal{C}(\mathcal{D})\big| \tag{7.3}$$

for $m \in \mathbb{Z}$. In general, when the query and the database documents are unrelated, the number $\Delta_{\mathcal{C}}(m)$ of (coincidentally) matching peak positions is usually small compared with $|\mathcal{C}(\mathcal{Q})|$. Only if the query is contained in the database document will the matching function have a large value $\Delta_{\mathcal{C}}(m)$ for some shift index $m \in \mathbb{Z}$. This index indicates the time offset between the query $\mathcal{Q}$ and the matching section in $\mathcal{D}$. Figure 7.5d shows an example of a matching function $\Delta_{\mathcal{C}}$, which assumes its maximum at the matching position corresponding to the time offset $t = 4.5$ sec.

7.1.3 Indexing, Retrieval, Inverted Lists

In the matching processes described so far, the query needs to be compared against all sections (having the same duration as the query) of all documents contained in the database. Obviously, such an exhaustive search strategy, whose run-time linearly depends on the number and sizes of the documents, is not feasible for large databases containing millions of recordings. In view of scalability, one requires search strategies that facilitate fast information access without sacrificing the accuracy of the retrieval results. Such search strategies typically use **indexing** techniques, which optimize speed and performance by cutting down the search space through suitable look-up operations. An **index** is constructed similarly to a traditional book index, which consists of a collection of alphabetically ordered key words. For each key word, in turn, there is a list of increasing page numbers indicating the occurrences of the given key word in the book.

Before we come back to our fingerprinting scenario, we now introduce a general indexing and retrieval framework based on inverted lists. In the following, we assume that the data items to be indexed consist of a time stamp $n \in \mathbb{Z}$ as well as a hash $h \in \mathcal{H}$, where $\mathcal{H}$ denotes a finite set of possible hash values. In information retrieval, the notion of a **hash** is used to refer to a fixed-length identifier (e.g., a binary string consisting of a fixed number of bits) that acts as a shortened and compact reference to the original, more complex data entity. In analogy to a traditional book index, the hash values play the role of the key words and the time stamps the role of the page numbers.

In the following we assume that a document $\mathcal{D}$ can be represented by a finite set of data items. The resulting set $\mathcal{F}(\mathcal{D}) \subset \mathbb{Z} \times \mathcal{H}$ is also referred to as a **feature representation** of $\mathcal{D}$. In practice, one has to deal with an entire database containing a large number of documents. To simplify notation, we only regard the case that the database consists of a single document $\mathcal{D}$. This can be assumed without loss of generality since one may concatenate all database documents to form a single (possibly very large) document (keeping track of possible document boundaries in an additional data structure).

Extending the notion of a book index, we now introduce an index structure that provides a mapping from the hashes (corresponding to the key words) to their locations (corresponding to the page numbers). To this end, we construct for each hash $h \in \mathcal{H}$ an **inverted list** $L(h)$. For a fixed hash h, the list $L(h)$ consists of the time stamps $n \in \mathbb{Z}$ with $(n,h) \in \mathcal{F}(\mathcal{D})$, where the time stamps are sorted in increasing order. For an illustrative example, let us have a look at Figure 7.6a, which shows a feature representation $\mathcal{F}(\mathcal{D})$ of a document $\mathcal{D}$. Each of the plotted points indicates an element $(n,h) \in \mathcal{F}(\mathcal{D})$ consisting of a time stamp $n \in \mathbb{Z}$ and a hash $h \in \mathcal{H} = \{1,2,3,4\}$. Figure 7.6b shows the inverted lists for each of the four hashes. For example, the list $L(3) = (0,3,5)$ encodes that the set $\mathcal{F}(\mathcal{D})$ contains the three elements $(0,3)$, $(3,3)$, $(5,3)$ all having the hash $h = 3$.

Next, we show how the inverted lists can be used to accelerate the retrieval process. Let $\mathcal{Q}$ be a query with feature representation $\mathcal{F}(\mathcal{Q})$. As in (7.2), we define the shifted version $m + \mathcal{F}(\mathcal{Q})$ for $m \in \mathbb{Z}$ by

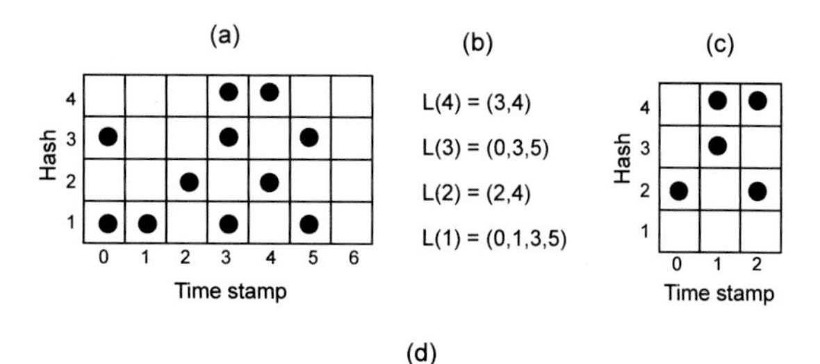

(d)

Query (n,h)	L(h) - n	Indicator functions									
		...	-1	0	1	2	3	4	5	6	...
(0,2)	(2,4)	0	0	0	0	1	0	1	0	0	0
(1,3)	(-1,2,4)	0	1	0	0	1	0	1	0	0	0
(1,4)	(2,3)	0	0	0	0	1	1	0	0	0	0
(2,2)	(0,2)	0	0	1	0	1	0	0	0	0	0
(2,4)	(1,2)	0	0	0	1	1	0	0	0	0	0
Matching function		**0**	**1**	**1**	**1**	**5**	**1**	**2**	**0**	**0**	**0**

Fig. 7.6 Illustrative example of index-based retrieval using inverted lists. **(a)** Feature representation $\mathcal{F}(\mathcal{D}) \subset \mathbb{Z} \times \mathcal{H}$ of a database document $\mathcal{D}$ with $\mathcal{H} = \{1,2,3,4\}$. **(b)** Inverted lists for $\mathcal{F}(\mathcal{D})$. **(c)** Feature representation $\mathcal{F}(\mathcal{Q})$ of a query $\mathcal{Q}$. **(d)** Computation of the matching function $\Delta_{\mathcal{F}}$ using the indicator functions of shifted inverted lists.

$$m + \mathcal{F}(\mathcal{Q}) := \{(m+n,h) \mid (n,h) \in \mathcal{F}(\mathcal{Q})\}. \tag{7.4}$$

Generalizing (7.3), we define the matching function $\Delta_{\mathcal{F}} : \mathbb{Z} \to \mathbb{N}_0$ by setting

$$\Delta_{\mathcal{F}}(m) := \big|(m + \mathcal{F}(\mathcal{Q})) \cap \mathcal{F}(\mathcal{D})\big| \tag{7.5}$$

for $m \in \mathbb{Z}$. Instead of scanning over the entire database document, we can use the inverted lists to reduce the search space by looking only at the hashes that actually occur in the query. For example, let us consider the query indicated by Figure 7.6c, where the hash $h = 1$ does not occur for any of the points contained in $\mathcal{F}(\mathcal{Q})$. Therefore, the database points encoded by the inverted list $L(h)$ with $h = 1$ are irrelevant for computing the matching function $\Delta_{\mathcal{F}}$ and can be left unconsidered.

How can the matching function $\Delta_{\mathcal{F}}$ be computed efficiently using the information supplied by the inverted lists? Let us first examine a single element $(n,h) \in \mathcal{F}(\mathcal{Q})$ of the query. How does this element need to be shifted to match a database point? By construction of the inverted list, each time stamp $\ell \in L(h)$ indicates that $(\ell,h) \in \mathcal{F}(\mathcal{D})$. Furthermore, the query point (n,h) needs to be shifted by $m = \ell - n$ positions to match the database point (ℓ,h). In other words, the shifted list $L(h) - n$ defined by

$$L(h) - n := \{\ell - n \mid \ell \in L(h)\} \tag{7.6}$$

contains exactly all the shifts that, when applied to the query point (n,h), result in a match with a database point having the same hash h. For example, when considering the query point $(n,h) = (1,3)$ (see Figure 7.6c), one obtains the shifted list $L(h) - n = L(3) - 1 = (-1,2,4)$ (see Figure 7.6b). Shifting the query point $(1,3)$ by $m = -1$, $m = 2$, or $m = 4$ yields the matching database points $(0,3)$, $(3,3)$, and $(5,3)$, respectively.

We now look at all elements $(n,h) \in \mathcal{F}(\mathcal{Q})$ of the query. For a given shift m, the value $\Delta_{\mathcal{F}}(m)$ counts the number of matching points between the m-shifted query and the database document. Therefore, a point $(n,h) \in \mathcal{F}(\mathcal{Q})$ contributes to this number if and only if $m \in (L(h)-n)$. In other words, looking at all $(n,h) \in \mathcal{F}(\mathcal{Q})$, one can derive the value $\Delta_{\mathcal{F}}(m)$ by counting how often the shift index m appears in the shifted lists $L(h)-n$.

To formalize this counting procedure, we introduce the notion of an **indicator function** (sometimes also called a **characteristic function**). Given an arbitrary set A and a subset $B \subseteq A$, the indicator function is a function $\mathbf{1}_B : A \to \{0,1\}$ defined by

$$\mathbf{1}_B(a) := \begin{cases} 1 & \text{if } a \in B, \\ 0 & \text{if } a \notin B. \end{cases} \tag{7.7}$$

We apply the concept of indicator functions for the case of inverted lists by regarding a list simply as a set of its list elements. Setting $A = \mathbb{Z}$ and $B = L(h)-n$, one obtains $\mathbf{1}_{L(h)-n}(m) = 1$ if and only if $m \in (L(h)-n)$. From the above argumentation, the matching function $\Delta_{\mathcal{F}}$ can be derived from the indicator functions in the following way:

$$\Delta_{\mathcal{F}}(m) = \sum_{(n,h)\in\mathcal{F}(\mathcal{Q})} \mathbf{1}_{L(h)-n}(m) \tag{7.8}$$

for $m \in \mathbb{Z}$. This equation yields a method for calculating $\Delta_{\mathcal{F}}$, which we illustrate by means of our example from Figure 7.6. The feature representation $\mathcal{F}(\mathcal{Q})$ of the query contains five points. For each of the points $(n,h) \in \mathcal{F}(\mathcal{Q})$, we consider the indicator function $\mathbf{1}_{L(h)-n}$ of the shifted list $(L(h)-n)$. Finally, all of the five resulting indicator functions are summed up to yield the matching function $\Delta_{\mathcal{F}}$ (see Figure 7.6d).

Let us analyze what we have gained by using the index-based computation of the matching function. First, note that the number $N := |\mathcal{F}(\mathcal{D})|$ of database items is in general very large, whereas the number $M := |\mathcal{F}(\mathcal{Q})|$ of query items is small. Let $L := |\mathcal{H}|$ be the number of different hashes. Assuming that the hash values of the database items are more or less evenly distributed over the set $\mathcal{H}$, each of the L inverted lists contains roughly N/L elements. Such a uniformity property is one of the main objectives when designing a hash function, which should map the entities of the database documents as evenly as possible over the range of available hash values.

Since each database item enters exactly one of the inverted lists, computing and storing the inverted lists has a computational complexity that is linear in N. Furthermore, this step is independent of the query so that the construction of the index can be done offline. In the query stage, only the information contained in the inverted lists corresponding to the M items of $\mathcal{F}(\mathcal{Q})$ is needed. These M lists need to be accessed, shifted, and further processed to derive the matching function—all operations that are linear in the sizes of the required lists. Therefore, the overall complexity in processing a query is linear in

$$\frac{M \cdot N}{L}. \tag{7.9}$$

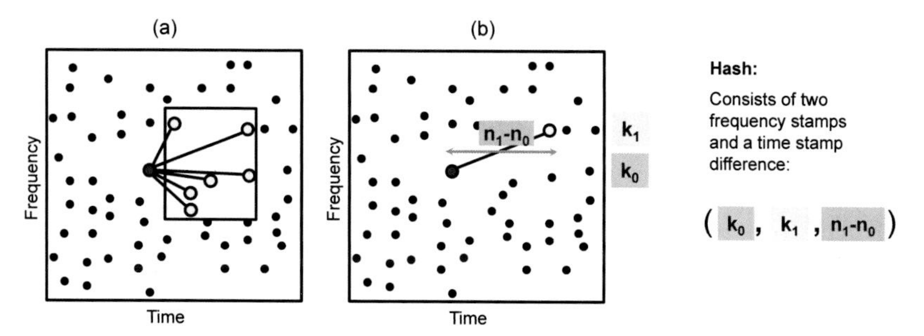

Fig. 7.7 Illustration of the peak pairing strategy to form fingerprint hashes. **(a)** Anchor peak and associated target zone. The fan-out in this example is $F = 6$. **(b)** Pairing of anchor peak and target peaks to form hashes.

Compared with the naive approach for computing the matching function, which requires $M \cdot N$ operations, we have gained an increase in efficiency by a factor of L. Even though, for a fixed number of hashes, the overall complexity for processing a query is still linear in the size N of the database items, the speed-up in practical applications can be drastic in the case that L is large. Therefore, in order to speed up the computations, one straightforward idea is to increase the number L of hashes. This, however, comes at a cost—as we will see in the next section when applying the index-based retrieval concept for audio identification.

7.1.4 Index-Based Audio Identification

The indexing and retrieval framework of the last section can be directly applied to our fingerprinting scenario based on spectral peaks. For the feature representations, we use the constellation maps $\mathcal{F}(\mathcal{Q}) = \mathcal{C}(\mathcal{Q})$ for the query and $\mathcal{F}(\mathcal{D}) = \mathcal{C}(\mathcal{D})$ for the database document. The set of possible hashes is then given by $\mathcal{H} = [0 : K]$, which consists of the different frequency stamps resulting from the STFT computation. With these settings, the matching function $\Delta_\mathcal{C}$ of (7.3) is given by $\Delta_\mathcal{C} = \Delta_\mathcal{F}$.

In practice, this simple indexing approach turns out to be problematic, since the number of frequency stamps may be too small to sufficiently speed up the retrieval process. For example, using a 1024-bin frequency axis yields only a speed-up factor of $L = 1024$ compared with the brute-force approach (see (7.9)). In view of the huge databases containing millions of recordings, much larger speed-up factors are needed. One idea could be to increase the number of hashes by using a finer frequency resolution in the peak computation, e.g., using $L = 16384$ frequency bins instead of $L = 1024$. This, however, would drastically affect the robustness of the overall system. Using a higher frequency resolution would significantly increase the probability that small spectral distortions in the query signal lead to modifications of the frequency stamps of spectral peaks. Since the frequency stamps serve as hash values, this would result in many mismatches when comparing the query peaks with the corresponding peaks in the database document.

How can one increase the specificity and the number of hashes without sacrificing the robustness of the fingerprints? We now present a possible approach, which was originally suggested in [30]. The main idea is to form fingerprint hashes by considering pairs of peaks instead of individual peaks (see Figure 7.7). To this end, one fixes a point $(n_0,k_0) \in \mathcal{C}(\mathcal{D})$ to serve as an **anchor point** as well as a **target zone** $T_{(n_0,k_0)} \subset \mathbb{Z} \times [0:K]$ associated to it (see Figure 7.7a). The target zone should be thought of as a small rectangular region in the time–frequency plane close to the anchor point. Then one considers pairs of points

$$\big((n_0,k_0),(n_1,k_1)\big) \tag{7.10}$$

consisting of the anchor point (n_0,k_0) and some target point $(n_1,k_1) \in T_{(n_0,k_0)} \cap \mathcal{C}(\mathcal{D})$. Each pair yields a triple

$$(k_0,k_1,n_1-n_0) \tag{7.11}$$

consisting of two frequency stamps and a difference of two time stamps. The idea is to use these triples as hashes instead of single frequency stamps.

Based on such triples, we redefine our feature representations for the database document and the query. Let

$$\mathcal{T}(n_0,k_0) := \{(k_0,k_1,n_1-n_0) \mid (n_1,k_1) \in T_{(n_0,k_0)} \cap \mathcal{C}(\mathcal{D})\} \tag{7.12}$$

be the set of all triples that can be formed by using a single anchor point $(n_0,k_0) \in \mathcal{C}(\mathcal{D})$. Then, we obtain the new feature representation for $\mathcal{D}$ by adding the time stamp n_0 of the anchor point to each of these triples and by considering **every** point of $\mathcal{C}(\mathcal{D})$ as an anchor:

$$\mathcal{F}(\mathcal{D}) := \bigcup_{(n_0,k_0)\in\mathcal{C}(\mathcal{D})} \{(n_0,h) \mid h \in \mathcal{T}(n_0,k_0)\}. \tag{7.13}$$

Similarly, we redefine the feature representation $\mathcal{F}(\mathcal{Q})$ for the query $\mathcal{Q}$. Note that, instead of $[0:K]$, the set of possible hashes is now a subset

$$\mathcal{H} \subset [0:K] \times [0:K] \times \mathbb{Z}. \tag{7.14}$$

With these newly defined feature representations, we can use exactly the same indexing and retrieval framework as introduced in Section 7.1.3 for computing a matching function $\Delta_{\mathcal{F}}$. This time, however, we count matching triples (consisting of two frequency stamps and a time stamp difference) between the shifted query and the database document instead of considering only matching frequency stamps. Still, a high value $\Delta_{\mathcal{F}}(m)$ is a good indicator that the query is likely to be part of the database document.

What have we gained by this construction? Recall that it was our goal to accelerate the retrieval process by reducing the length of the inverted lists without losing fingerprint robustness. To this end, we increased the specificity of the fingerprints by considering pairs of peaks along with their distance instead of individual peaks. Obviously, the number of items to be indexed and queried increases by this construction. Using all pairs of peaks would result in a number of items quadratic in the number of points contained in a constellation map. To avoid such a combinatorial explosion, the target zones come into play. Assuming that the peak coordinates are more or less evenly distributed in a constellation map and using suitably shifted target zones of the same fixed size for all anchor points, the number of constellation points per target zone is roughly the same. We call this number the **fan-out** of the target zone and denote it by $F \in \mathbb{N}$ (see also Figure 7.7a). For example, if $F = 10$, then the number of items (the pairs of peaks) contained in a feature representation $\mathcal{F}(\mathcal{D})$ (or $\mathcal{F}(\mathcal{Q})$) is approximately ten times the number of items (the peaks) in the original constellation map $\mathcal{C}(\mathcal{D})$ (or $\mathcal{C}(\mathcal{Q})$). In other words, the fan-out factor leads directly to a cost factor in terms of the number of items to be stored, indexed, and processed.

Despite this increase in data items, we gain a tremendous acceleration in the search process. In view of a more concrete explanation, let us assume that the two frequency stamps, as well as the difference of time stamps of a triple $(k_0, k_1, n_1 - n_0)$, can each be encoded using B bits. Then the triples yield a number of $L_{\mathrm{Triple}} = 2^{B+B+B}$ possible hash values, as opposed to the number of $L = 2^B$ hash values when using only a single frequency stamp. In other words, the specificity of the hashes has increased by a factor of 2^{B+B}. On the downside, the number $N_{\mathrm{Triple}} := |\mathcal{F}(\mathcal{D})|$ and $M_{\mathrm{Triple}} := |\mathcal{F}(\mathcal{Q})|$ contained in the new feature representations have both increased by a factor of F compared with $N = |\mathcal{C}(\mathcal{D})|$ and $M = |\mathcal{C}(\mathcal{Q})|$, respectively. By (7.9), the overall complexity of the new procedure is

$$\frac{M_{\mathrm{Triple}} \cdot N_{\mathrm{Triple}}}{L_{\mathrm{Triple}}} = \frac{F \cdot M \cdot F \cdot N}{L \cdot L \cdot L} = \frac{F^2}{L^2} \cdot \frac{M \cdot N}{L}. \tag{7.15}$$

Thus, the time to process a query has been reduced by a factor of L^2/F^2. For example, having $F = 10$ and using $B = 10$ bits, this results in a huge speed-up by a factor of $2^{20}/100 \approx 10000$.

What have we lost on the side of the overall procedure's robustness? First note that we have not changed the way the spectral peaks are computed—we only combined two peaks to form a new, more complex and therefore more specific item. However, being a combination of two individual peaks, it becomes more likely that this item will be modified due to signal distortions. To be more concrete, let us assume that the probability of an original spectral peak surviving the distortions to which the query audio fragment may be exposed is given by a number $p \in [0, 1]$. For example, $p = 0.2$ means that 20 % of the original peaks survive. Furthermore, let us simplistically assume that this probability is independent and identically distributed over all points involved. Then the probability that two specific points in the query survive is p^2. For example, in the case of $p = 0.2$, one obtains that this probability is only $p^2 = 0.04$.

Assuming that the distance between two surviving peaks is not affected by the distortion, the probability of a triple $(k_0, k_1, n_1 - n_0)$ surviving is also p^2. This reduction in item resilience is a trade-off against the tremendous amount of speed-up provided. Furthermore, the reduced probability of individual item survival is mitigated by the F times larger number of available items compared with the number of original constellation points. For a given anchor point, the probability of at least one item surviving is the joint probability of the anchor point and at least one target point in its target zone surviving. This probability is given by

$$p \cdot (1 - (1 - p)^F) \tag{7.16}$$

(see Exercise 7.3). Now, if F is not too small, e.g., $F = 10$, then even if the survival probability of a single peak is only $p = 0.2$, the probability of at least one target point surviving is $(1 - (1 - p)^F) = 1 - 0.8^{10} \approx 0.9$. Therefore, we are not much worse off than before. In summary, we have seen that, by using more complex items for indexing, we have traded off approximately $F = 10$ times the storage space for approximately 10000 times improvement in speed, and a small loss in probability of audio identification.

In recent years, many different fingerprinting and indexing techniques have been proposed and are now being used in commercial products. In this section, we have had a closer look at one of these techniques, which was originally developed for the Shazam audio identification system [30]. We have discussed the main ideas underlying this system, but there are many parameters that need to be adjusted in order to find a good trade-off between the various requirements including robustness, specificity, scalability, and compactness. Important aspects include the temporal and spectral resolutions used for the STFT computation, the peak-picking strategy, the size of the target zones, the fan-out parameter, and suitable data structures for processing the inverted lists.

Although robust to many kinds of signal distortions, the discussed fingerprinting approach is not designed for handling temporal deformations. The matching of the constellation maps as well as the time stamp differences in the peak pairs are both sensitive to relative tempo differences between the query and database document. Therefore, one needs other techniques to become invariant to time scale modifications.

The fingerprints using spectral peaks are designed to be highly sensitive to a particular version of a piece of music. For example, given a multitude of different performances of a song by the same artist, the fingerprinting system is likely to pick the correct one even if they are virtually indistinguishable by the human ear. In general, audio identification systems are designed to target identification of recordings that are already present in the database. Therefore, such techniques usually do not generalize to live recordings or performances that are not part of the database. In the following sections, we discuss retrieval tasks and techniques that aim at identifying different versions of a given recording including different performances of the same piece, arrangements, and cover songs.

7.2 Audio Matching

While significant progress has been made for highly specific retrieval scenarios such as audio identification, retrieval scenarios of lower specificity still pose many challenges. In this section, we address a retrieval task referred to as **audio matching**: given a short query audio clip, the goal is to automatically retrieve all excerpts from all recordings within a given audio database that **musically** correspond to the query [16, 15]. In this matching scenario, as opposed to classic audio identification, one allows semantically motivated variations as they typically appear in different performances and arrangements of a piece of music. For example, two different performances of the same piece may exhibit significant nonlinear global and local differences in tempo, articulation, and phrasing, which are due to the freedom an artist has in executing performance directives such as ritardandi, accelerandi, fermate, or ornamentations. Furthermore, one has to deal with considerable spectral variations, which are due to differences in instrumentation, dynamics, accentuation, and so on. In Section 7.2.1, we address in more detail the various challenges one has to face in audio matching and discuss the implications for the feature design step. In particular, we introduce a scalable class of chroma-based audio features. Based on these features, we discuss in Section 7.2.2 a basic matching procedure for locally comparing the query sequence with database subsequences. Finally, to better account for temporal deformations in this comparison, we discuss in Section 7.2.3 a more flexible DTW-based matching approach.

7.2.1 General Requirements and Feature Design

To further illustrate the audio matching scenario, let us again consider the Symphony No. 5 by Ludwig van Beethoven, which we have already encountered as a running example in the previous chapters. Being one of the most popular pieces in the Western classical music literature, there exist a large number of different performances and arrangements of Beethoven's Fifth Symphony. More than 100 recordings are commercially available, not to mention numerous nonprofessional live performances as may be found on video-sharing websites such as YouTube.

Now imagine you are sitting in a student orchestra concert at your university and listen to a performance of Beethoven's Fifth. You take your smartphone, record a few seconds, and send the audio fragments to an identification service. Since the live performance of the query is not part of the system's fingerprint database, the service will not be able to identify the recording when using traditional fingerprinting techniques.

This is exactly the scenario where audio matching techniques should step in. For example, let us assume that the query consists of a recording of the first theme of Beethoven's Fifth. Then the goal of audio matching is to find all audio fragments that musically correspond to the query in a given database. The retrieved matches should include the repetitions of the main theme in the exposition and recapitulation

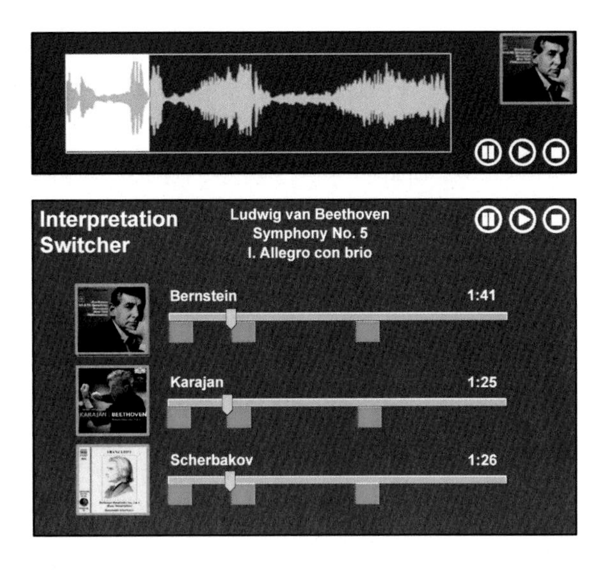

Fig. 7.8 User interface for content-based music retrieval and navigation. The user may specify a query by marking certain parts of a waveform. The system then retrieves all recordings that contain sections musically similar to the query. In the shown example, the query consists of the first 21 seconds of a Bernstein recording of Beethoven's Fifth. The system retrieved three recordings (Bernstein, Karajan, Scherbakov) along with three matching sections (represented by the rectangles) in each of the recordings.

within the same performance as well as the corresponding excerpts in other recordings, arrangements, and synthesized versions (e.g., obtained from MIDI files). All the retrieved matches can then be presented and made accessible to a user by means of suitable interfaces. Extending the functionality of the Interpretation Switcher introduced in Figure 3.21, all the matching sections within the retrieved recordings can be highlighted, as indicated by Figure 7.8. Based on this interface, the user can easily browse through and listen to the retrieved audio fragments, possibly starting a new retrieval process with a refined query or a retrieved item.

Note that variation across different performances of the same musical work may be quite significant. First of all, the various performances may be played in different tempi. For example, in a recording by Herbert von Karajan, the first theme has a duration of 18 seconds, as opposed to 21 seconds in the Bernstein version. There are also different instrumental versions of Beethoven's Fifth. The famous pianist and composer Franz Liszt scaled down the symphony to a piano solo, which significantly deviates from an orchestral version in timbre and dynamics. There are even more distant arrangements including arrangements for marching bands, disco adaptations, and swinging jazz versions.

The audio fingerprints based on spectral peaks as introduced in Section 7.1.2 are suitable features to characterize the local acoustic properties of a specific audio recording. Also, as we have seen, such features possess a high degree of robustness to certain signal distortions and superpositions. However, spectral peaks are not designed to handle musical variations. This fact is illustrated by the left part of Figure 7.9, which shows the spectrograms along with the resulting peak-based constellation maps for a Bernstein as well as for a Karajan version of the beginning of Beethoven's Fifth. The spectrogram peaks for the two different recordings are quite inconsistent with regard to their frequency stamps (due to differences in timbre) as well as with regard to their time stamps (due to differences in tempo). For the

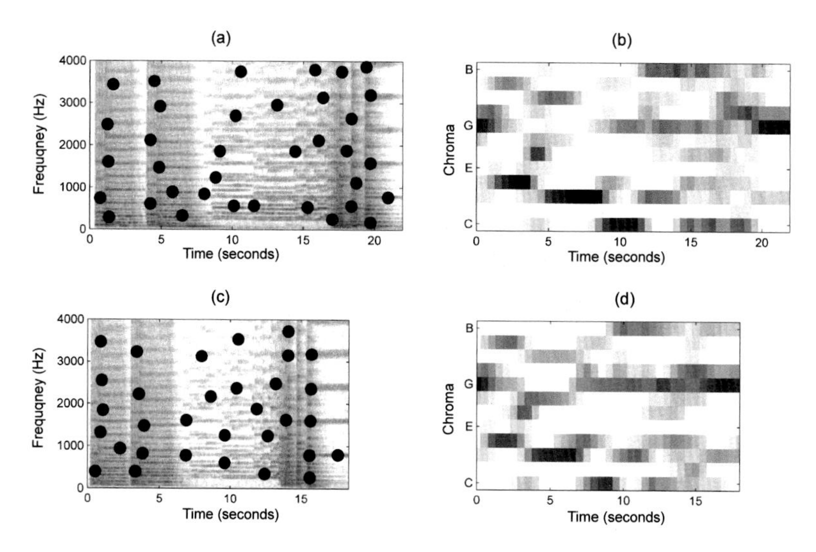

Fig. 7.9 Different representations derived from recordings of the beginning first theme of Beethoven's Fifth Symphony. **(a)** Spectrogram and spectral peaks for a Bernstein recording. **(b)** Chromagram for a Bernstein recording. **(c)** Spectrogram and spectral peaks for a Karajan recording. **(d)** Chromagram for a Karajan recording.

audio matching task, one requires descriptors that capture **musical** characteristics of the underlying piece of music rather than **acoustic** characteristics of a specific recording.

Thinking of different performances of the same musical work, all these versions are based on more or less the same note material. The same melodies are played within the same harmonic context. As we have seen in the music synchronization scenario (Chapter 3), chroma-based audio features are suitable mid-level representations for capturing this kind of information. This is again demonstrated by the right part of Figure 7.9, which shows chroma-based audio representations for the Bernstein and Karajan versions. The two chromagrams exhibit a much higher degree of similarity than the corresponding spectrograms, thus revealing a much higher invariance against performance variations.

Recall from Section 3.1.2 that chroma features are based on the twelve pitch spelling attributes C, $\mathrm{C}^\sharp$, D, ..., B as used in Western music notation, where each chroma vector indicates how the energy over a signal's frame is distributed across the twelve chroma bands. Measuring such distributions over time yields a time–chroma representation that closely correlates to the melodic and harmonic progression. Such progressions are often similar for different recordings of the same piece of music, thus making chroma features a suitable tool for our matching task. We have already seen in Section 3.1.2 that there are different ways of computing chroma features and that the properties of chroma features can be adjusted by applying suitable postprocessing steps such as logarithmic compression, normalization, or smoothing.

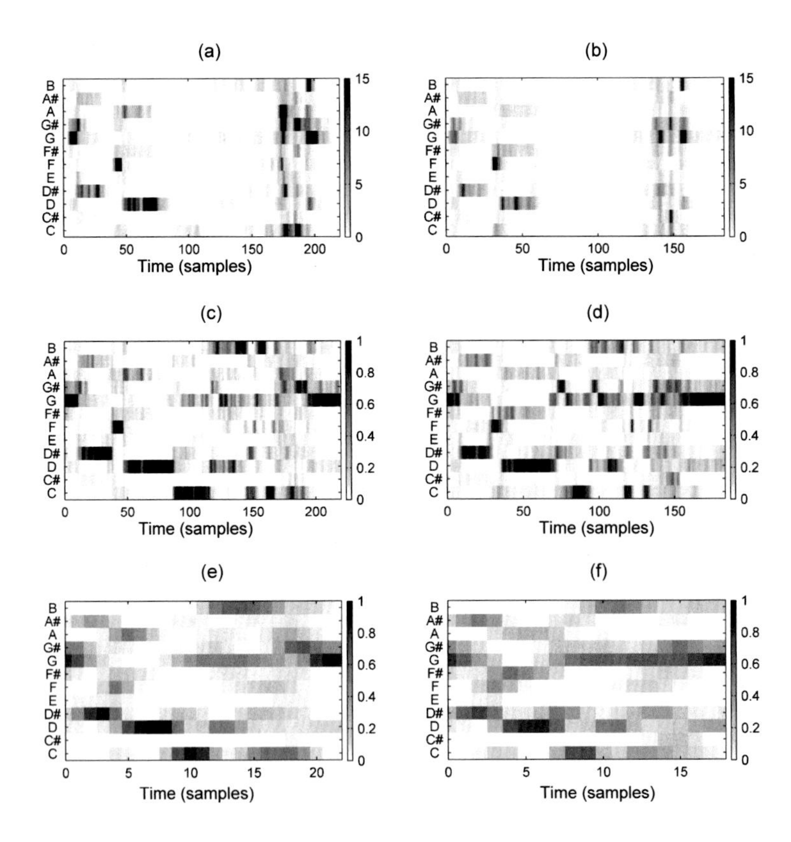

Fig. 7.10 Different chroma representations for two orchestra recordings (Bernstein, Karajan) of the beginning first theme of Beethoven's Fifth Symphony. **(a)/(b)** Basic chroma features with a 10 Hz feature rate. **(c)/(d)** Normalized chroma features. **(e)/(f)** Smoothed and downsampled CENS^{41}_{10}-features.

Which chroma variant is suitable for the task of audio matching? Let us have a look at Figure 7.10, which shows three chroma variants for a Bernstein as well as a Karajan recording of the beginning of Beethoven's Fifth. The first row of Figure 7.10 shows a basic chroma variant as computed in (3.6) using a feature rate of 10 Hz, where each chroma vector corresponds to a window of 200 ms with a window overlap of half the size. To balance out the huge differences in dynamics within and across the recordings, one can apply normalization techniques. For example, one may normalize each chroma vector with regard to the Euclidean norm or by using the more robust projection operator as defined in (3.10). The result of such a normalization is illustrated by the second row of Figure 7.10. Even though these normalized chromagram representations already reveal similar patterns across the two recordings, there are still many performance-specific differences. Therefore, one idea is to apply additional quantization and smoothing procedures to further reduce the effect of local fluctuations due to variations in local tempo, articulation, and note execution.

As an example, we now discuss a concrete postprocessing procedure as originally suggested in [17], which illustrates how these steps may be implemented in practice. We start with some basic chroma features, as illustrated by the first row of Figure 7.10, and normalize each chroma vector with respect to the Manhattan norm so that the twelve chroma values add up to one. Let $X = (x_1, x_2, \ldots, x_N)$ denote the resulting sequence of normalized chroma vectors $x_n \in [0,1]^{12}$, $n \in [1:N]$. Note that each of these vectors has only positive entries between zero and one.

Next, we define a quantization function $Q : [0,1] \to \{0,1,2,3,4\}$ by

$$Q(a) := \begin{cases} 0 \text{ for } & 0 \leq a < 0.05, \\ 1 \text{ for } & 0.05 \leq a < 0.1, \\ 2 \text{ for } & 0.1 \leq a < 0.2, \\ 3 \text{ for } & 0.2 \leq a < 0.4, \\ 4 \text{ for } & 0.4 \leq a \leq 1. \end{cases} \tag{7.17}$$

In the first step, we quantize each chroma vector $x_n = (x_n(0), \ldots, x_n(11))^\top \in [0,1]^{12}$ by applying Q to each component of x_n, yielding

$$Q(x_n) := (Q(x_n(0)), \ldots, Q(x_n(11)))^\top. \tag{7.18}$$

Intuitively, this quantization assigns a value of 4 to a chroma component if the corresponding chroma class contains more than 40% of the signal's total energy and so on. The chroma components below a 5% threshold are set to zero, which introduces robustness to noise. The thresholds are chosen in a logarithmic fashion to account for the logarithmic perception of sound intensity. For example, the vector $x_n = (0.02, 0.5, 0.3, 0.07, 0.11, 0, \ldots, 0)^\top$ is transformed into the vector $Q(x_n) := (0,4,3,1,2,0,\ldots,0)^\top$. In the second step, the quantized sequence $(Q(x_1), \ldots, Q(x_N))$ is further smoothed. To this end, we fix a number $\ell \in \mathbb{N}$ that determines the length of a smoothing window (e.g., a Hann window as defined in (2.140)) and then consider local averages (weighted by the window function) of each of the twelve components of the sequence $(Q(x_1), \ldots, Q(x_N))$. This again results in a sequence of 12-dimensional vectors with nonnegative entries. In the last step, this sequence is downsampled by a factor of d, and the resulting vectors are normalized with respect to the Euclidean norm.

The two steps, quantization and smoothing, amount to computing weighted statistics of the energy distribution over a window of ℓ consecutive vectors. Therefore, we call the resulting features CENS_d^ℓ (chroma energy normalized statistics). The main idea of CENS features is that taking statistics over relatively large windows smooths out local deviations in tempo, articulation, and execution of note groups such as trills or arpeggios. As an illustration of this effect, the third row of Figure 7.10 shows the sequences of CENS_{10}^{41}-features for the two Beethoven performances. Starting with a feature rate of 10 Hz for the original chroma sequence, the parameter $\ell = 41$ corresponds to a window size of 4100 ms. Furthermore, using the parameter $d = 10$ reduces the feature rate to 1 Hz (one feature per second). Com-

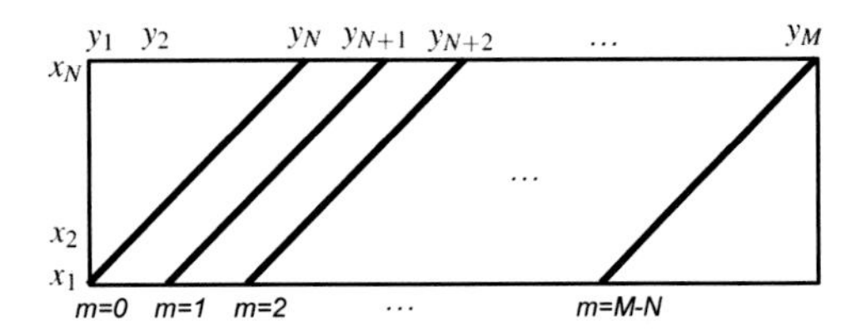

Fig. 7.11 Schematic illustration of the computation of the matching function Δ_{Diag} by summing up diagonals of the cost matrix $\mathbf{C}$.

pared with the original chroma sequences, the resulting CENS sequences of the two performances possess a much higher degree of similarity, while still capturing some characteristic musical information.

In summary, providing a family of chroma features depending on the two parameters ℓ and d, the described procedure is a flexible and computationally inexpensive way to adjust the feature specificity and resolution without repeating the cost-intensive spectral audio decomposition.

7.2.2 Diagonal Matching

As in the audio identification scenario, we are given a query audio fragment $\mathcal{Q}$ and a database recording $\mathcal{D}$. Instead of using constellation maps, however, the query and the database document are now compared on the basis of chroma features. Let $X = (x_1, x_2, \ldots, x_N)$ and $Y = (y_1, y_2, \ldots, y_M)$ be the chroma feature sequences for $\mathcal{Q}$ and $\mathcal{D}$, respectively. The length N of the query is typically short in comparison with the length M of the database recording. Intuitively, to test if and where the query $\mathcal{Q}$ is contained in $\mathcal{D}$, we shift the sequence X over the sequence Y and locally compare X with suitable subsequences of Y. Every subsequence of Y that is similar or, equivalently, has a small distance to X is considered a **match** for the query.

There are many ways for locally comparing X with subsequences of Y. Following [16], let us start with a basic procedure that we refer to as **diagonal matching**. First of all, we need to fix a local cost measure (or local distance measure) to compare the chroma vectors of the sequences X and Y. As in the context of music synchronization, we use the **cosine distance**, denoted by c (see (3.14)). Assuming that the chroma vectors are normalized with respect to the Euclidean norm, this distance is given by

$$c(x,y) = 1 - \langle x|y \rangle \tag{7.19}$$

for two vectors x and y with $\|x\| = \|y\| = 1$. One simple way for comparing two chroma sequences that share the same length is to compute the average distance between corresponding vectors of the two sequences. Doing so, we compare the query sequence $X = (x_1, \ldots, x_N)$ with all subsequences $(y_{1+m}, \ldots, y_{N+m})$ of Y having the same length N as the query, where $m \in [0 : M-N]$ denotes the shift index. This procedure, which is illustrated by Figure 7.11, yields a **matching function** $\Delta_{\text{Diag}} : [0 : M-N] \to \mathbb{R}$ defined by

$$\Delta_{\mathrm{Diag}}(m) := \frac{1}{N} \sum_{n=1}^{N} c(x_n, y_{n+m}). \tag{7.20}$$

We now slightly reformulate the way this matching function is computed. As in (3.13), let $\mathbf{C} \in \mathbb{R}^{N \times M}$ be the **cost matrix** given by

$$\mathbf{C}(n,m) := c(x_n, y_m) \tag{7.21}$$

for $n \in [1:N]$ and $m \in [1:M]$. Then the value $\Delta_{\mathrm{Diag}}(m)$ is obtained (up to the normalization by the query length) by summing up diagonals of the matrix $\mathbf{C}$ as illustrated by Figure 7.11. This explains why we denote this procedure as "diagonal" matching.

We now discuss how the matching function can be applied for retrieving all matches that are similar to the query fragment. As in Section 7.1.3, to simplify notation, we assume that the database is represented by a single document $\mathcal{D}$. To determine the best match between $\mathcal{Q}$ and $\mathcal{D}$, we simply look for the index $m^* \in [0 : M - N]$ that minimizes the matching function Δ_{Diag}:

$$m^* := \underset{m \in [0:M-N]}{\mathrm{argmin}} \; \Delta_{\mathrm{Diag}}(m). \tag{7.22}$$

The best match is then given by the audio clip corresponding to the subsequence

$$Y(1+m^* : N+m^*) := (y_{1+m^*}, \ldots, y_{N+m^*}). \tag{7.23}$$

To obtain further matches, we exclude a neighborhood of the best match from further considerations. For example, one may exclude a neighborhood of $\rho = N/2$ around m^* by (intuitively) setting $\Delta_{\mathrm{Diag}}(m) = \infty$ for $m \in [m^* - \rho, m^* + \rho] \cap [0 : M - N]$. This ensures that the subsequent matches do not overlap by more than half the query length. To find subsequent matches, the latter procedure is repeated until a certain number of matches is obtained or a specified distance threshold is exceeded.

As an illustration, let us consider a database that consists of four recordings: one recording of a waltz by Shostakovich, two recordings (Bernstein, Karajan) of the first movement of Beethoven's Fifth, and one recording of the Hungarian Dance No. 5 by Brahms. We obtain the document $\mathcal{D}$ by concatenating these four recordings. To avoid matches across different recordings, we keep track of the boundaries in an additional data structure. Technically, this can be done by inserting additional columns with ∞ value at the recordings' boundary positions in the feature representation Y of the database document $\mathcal{D}$ (see Exercise 7.5). As a query $\mathcal{Q}$, we again use the first 21 seconds of the Bernstein recording, which correspond to the first theme of Beethoven's Fifth. This theme appears again in the repetition of the exposition and once more, with slight modifications, in the recapitulation. Since $\mathcal{D}$ contains two different performances of Beethoven's Fifth, we expect six matches that musically correspond to the query. Now, let us have a look at Figure 7.12a, which shows the matching function Δ_{Diag} using CENS^{41}_{10}-features. The minimizing index m^* appears at the beginning of the Bernstein recording with a cost value of $\Delta_{\mathrm{Diag}}(m^*) \approx 0$. This

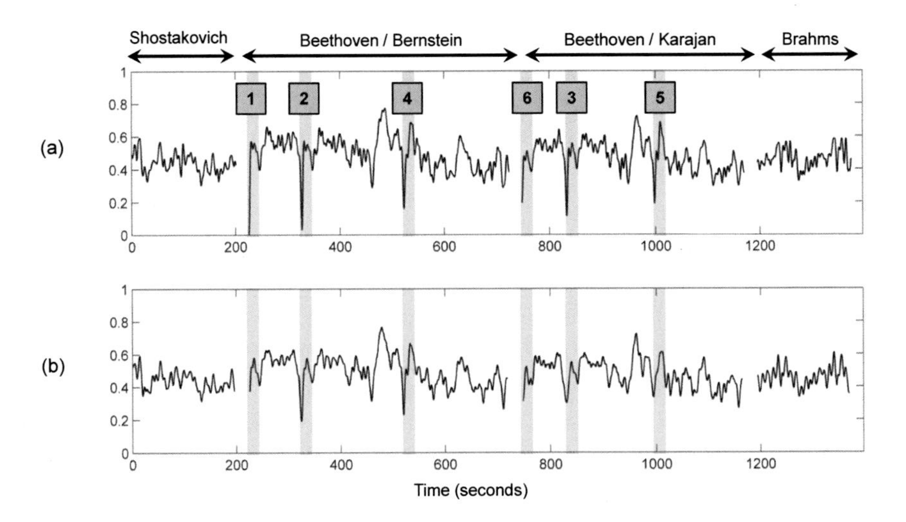

Fig. 7.12 Matching functions for a database consisting of four recordings including two recordings of the first movement of Beethoven's Fifth Symphony. **(a)** Matching function Δ_{Diag} using a query that consists of the first 21 seconds (first theme) of the Bernstein recording. The top six matches correspond to the six occurrences of the first theme. **(b)** Matching function Δ_{Diag} using a modified Bernstein query with a tempo reduced by 25 %.

is no surprise, since the query is part of the database. In other words, the query $\mathcal{Q}$ has been correctly identified within $\mathcal{D}$ by means of the first match. The matching function yields additional matches by looking at the positions of all local minima of Δ_{Diag} that are close to zero and fall below a certain cost threshold $\tau > 0$. For example, looking at all local minima below $\tau = 0.2$, we obtain six matching positions that indeed correspond to the six musically meaningful matches. For example, the second match corresponds to the theme in the repetition of the exposition in the Bernstein recording, the third match to the corresponding section in the Karajan recording, and so on. The cost value $\Delta_{\mathrm{Diag}}(m)$ at a position m indicates the distance between the respective matching section and the query.

This basic matching procedure works well in the case that the tempo of the query roughly coincides with the tempo within the sections to be matched. Using relatively coarse and smoothed features (such as the CENS^{41}_{10}-features in our example), the temporal blurring is capable of introducing robustness to local tempo variations. This explains why the themes—even in the Karajan recording—were found despite being played faster than in the Bernstein query. However, when the tempo difference between the query and a database section becomes too large, the diagonal matching procedure is doomed to fail. This fact is illustrated by Figure 7.12b, which shows the matching function Δ_{Diag} when using a modified Bernstein query with a tempo reduced by 25 %. Since the costs at the desired matching positions have increased significantly, most of these positions can no longer be distinguished from spurious matches.

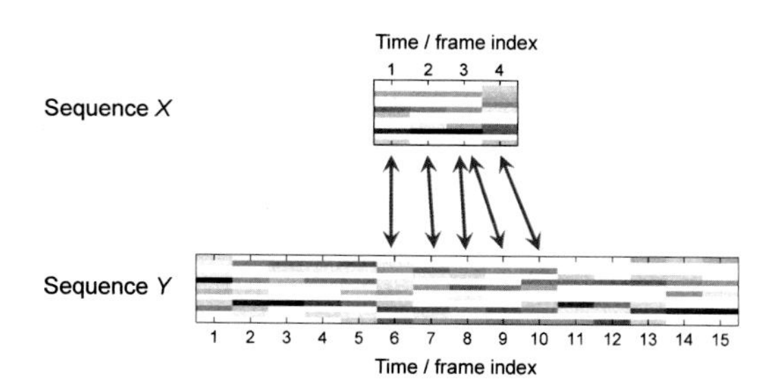

Fig. 7.13 Time alignment of a sequence X with a subsequence of Y (see Figure 3.10 for comparison). Aligned points or frames are indicated by the arrows.

7.2.3 DTW-Based Matching

In Section 3.2, we have studied how one can deal with tempo differences when comparing two feature sequences. Based on the notion of a warping path, we computed an optimal global alignment between the two sequences using dynamic time warping (DTW). In the audio matching scenario, the alignment task is slightly different. Instead of finding a global alignment between the two given sequences, the objective is to find a subsequence within the longer sequence that optimally fits the shorter sequence (see Figure 7.13). We now show how the problem of finding optimal subsequences can be solved by a variant of dynamic time warping.

In the following, we adopt the notation introduced in Section 3.2. Let $X = (x_1, x_2, \ldots, x_N)$ and $Y = (y_1, y_2, \ldots, y_M)$ be two feature sequences over the feature space $\mathcal{F}$, where we assume that the length M is much larger than the length N. Furthermore, let $c : \mathcal{F} \times \mathcal{F} \to \mathbb{R}$ be a local cost measure as in (3.12) and $\mathbf{C}$ the resulting cost matrix given by $\mathbf{C}(n,m) = c(x_n, y_m)$ for $n \in [1:N]$ and $m \in [1:M]$. For two indices $a, b \in [1:M]$ with $a \leq b$, we use the notation

$$Y(a:b) := (y_a, y_{a+1}, \ldots, y_b) \tag{7.24}$$

to denote a subsequence of Y. Based on the global DTW distance, our matching problem can be formulated as the following optimization task: find the subsequence of Y (over all possible subsequences of Y) that minimizes the DTW distance to X. In other words, the task is to determine the indices defined by

$$(a^*, b^*) := \underset{(a,b):1 \leq a \leq b \leq M}{\operatorname{argmin}} \ \mathrm{DTW}\big(X, Y(a:b)\big). \tag{7.25}$$

This task seems to involve two different kinds of optimization steps. First, one needs to consider all possible subsequences of Y to find the optimal one. Second, for each subsequence in turn, one needs to compute the DTW distance to X, which itself involves an optimization for determining the cost of an optimal warping path. The good news is that both the indices a^* and b^*, as well as an optimal alignment between X and the subsequence $Y(a^* : b^*)$, can be computed within a single optimization framework. Only a slight modification of the original DTW algorithm (described in Table 3.2) is necessary. The basic idea is to allow for omissions at the beginning and at the end of Y in the alignment with X. Instead of giving a formal proof of the entire procedure, we describe in the following only the necessary modifications.

As with the original DTW algorithm, one defines an $N \times M$ accumulated cost matrix denoted by $\mathbf{D}$. As in (3.23), the first column of this matrix is initialized by setting

$$\mathbf{D}(n,1) := \sum_{k=1}^{n} \mathbf{C}(k,1) \tag{7.26}$$

for $n \in [1:N]$. However, as opposed to (3.24), the first row of $\mathbf{D}$ is now initialized by

$$\mathbf{D}(1,m) := \mathbf{C}(1,m) \tag{7.27}$$

for $m \in [1:M]$. This initialization makes it possible to start at any position of the sequence Y without accumulating any cost, thus realizing the idea of skipping the beginning of Y when being matched to X. The remaining values of $\mathbf{D}$ are defined recursively as in (3.25) for $n \in [2:N]$ and $m \in [2:M]$. Finally, instead of looking only at the coefficient $\mathbf{D}(N,M)$ to obtain the global DTW cost, the second modification is to consider the entire last row $\mathbf{D}(N,m)$ for $m \in [1:M]$. From this row, the index b^* defined in (7.25) can be determined by

$$b^* = \underset{b \in [1:M]}{\mathrm{argmin}}\ \mathbf{D}(N,b). \tag{7.28}$$

Choosing the cost-minimizing index in this row (instead of taking the last index as is done in the original DTW approach) realizes the idea of skipping the end of Y when being matched to X.

The start index a^* defined in (7.25) cannot be directly read off from the matrix $\mathbf{D}(N,m)$. To determine a^*, one needs to apply a backtracking procedure as in classical DTW to construct an optimal warping path (see Table 3.2). This time, however, one starts with $q_1 = (N,b^*)$ (instead of $q_1 = (N,M)$) and stops as soon as the first row of $\mathbf{D}$ is reached by some element $q_L = (1,m)$, $m \in [1:M]$ (instead of $q_L = (1,1)$). The index $a^* \in [1:M]$ is then determined by this index m. Furthermore, the path $(q_L, q_{L-1}, \ldots, q_1)$ defines an optimal warping path between the sequence X and the subsequence $Y(a^*:b^*)$ (see also Exercise 7.6).

Let us have a look at an example while reflecting on the overall procedure from a more general perspective. Based on the cost matrix $\mathbf{C}$ of Figure 7.14a, one obtains the accumulated cost matrix $\mathbf{D}$ shown in Figure 7.14b. The optimal index $b^* = 24$ can be determined from the values of the top row of $\mathbf{D}$. The index $a^* = 13$ is derived via backtracking, which also yields the optimal alignment path between the sequence $Y(13:24)$ and X. Besides revealing the optimal index b^*, the top row of $\mathbf{D}$ provides more information. Each entry $\mathbf{D}(N,m)$ for an arbitrary $m \in [1:M]$ indicates the total cost of aligning X with an optimal subsequence of Y that ends at position m. This motivates us to define a **matching function** $\Delta_{\mathrm{DTW}} : [1:M] \to \mathbb{R}$ by setting

$$\Delta_{\mathrm{DTW}}(m) := \frac{1}{N}\,\mathbf{D}(N,m) \tag{7.29}$$

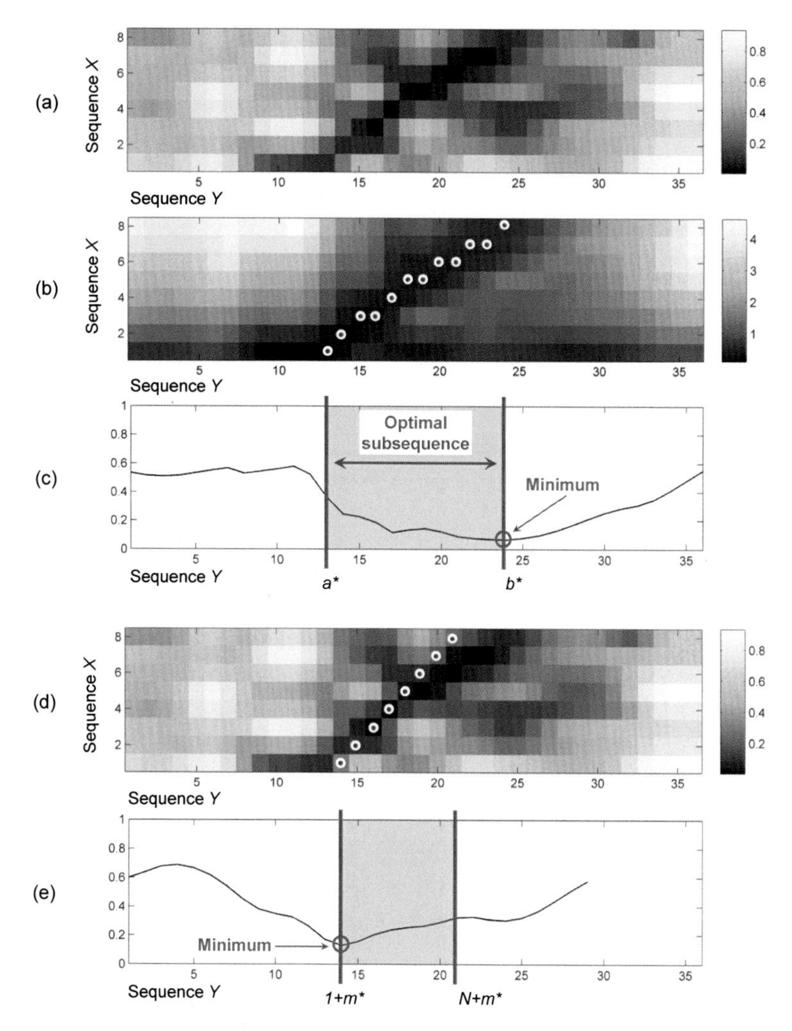

Fig. 7.14 Illustration of subsequence DTW for finding an optimal subsequence of Y that matches X. **(a)** Cost matrix **C**. **(b)** Accumulated cost matrix **D** with warping path between X and the optimal subsequence $Y(a^* : b^*)$. **(c)** Matching function Δ_{DTW}. **(d)** Cost matrix **C** with optimal diagonal match. **(e)** Matching function Δ_{Diag}.

for $m \in [1:M]$, where we have normalized the accumulated cost by the length N of the query. Each local minimum $b \in [1:M]$ of Δ_{DTW} that is close to zero indicates the end position of a subsequence $Y(a:b)$ that has a small DTW distance to X. The start index $a \in [1:M]$ as well as the optimal alignment between this subsequence and X are obtained by a backtracking procedure starting with the cell $q_1 = (N,b)$.

In summary, we have obtained a generalization of our diagonal matching procedure, which is illustrated by the lower part of Figure 7.14. However, instead of revealing **start** positions of matching sections as is the case with Δ_{Diag}, the matching function Δ_{DTW} indicates **end** positions of matching sections. In simple terms, in diagonal matching, the query X is processed in forward direction, whereas in the DTW-based approach it is processed in backward direction. Furthermore, instead of just diagonally aligning the matching section with the query, the DTW-based approach introduces warping operations that make it possible to handle temporal deviations between the match and the query. This effect becomes visible in our example of Figure 7.14, where the query sequence X has length $N = 8$, while the best matching subsequence $Y(13:24)$ has length 12. The DTW-based matching procedure, which is based on the step size set $\Sigma = \{(1,0),(0,1),(1,1)\}$, can account for such temporal differences as indicated by Figure 7.14b. By using diagonal matching, however, the length of the matching subsequence is forced to have the same length N as the query X. This leads to a poorer alignment and inaccuracies in the matching, as illustrated by Figure 7.14d.

As another example, let us again have a look at Figure 7.12b, where we have used a modified Bernstein query with a tempo reduced by 25 %. As a result of large tempo differences, the expected matches in the much faster Karajan recording could not be identified using diagonal matching. Using DTW-based matching, however, the matching positions clearly stand out in Δ_{DTW} as demonstrated by Figure 7.15c. Furthermore, Figure 7.15a illustrates how the matching sections lead to "corridors" of low cost in the matrix $\mathbf{C}$. Accumulating the costs in these corridors leads to comparatively small values in the matrix $\mathbf{D}$, as shown by Figure 7.15b.

The discussed subsequence variant of DTW can be modified in the same way as classical DTW (see Section 3.2.2). In particular, the step size condition may be changed by replacing the set $\Sigma = \{(1,0),(0,1),(1,1)\}$. For example, using the set $\Sigma = \{(1,1)\}$, DTW-based matching basically reduces to diagonal matching—except for the backward processing instead of the forward processing (see also Exercise 7.7). In general, using the set $\Sigma = \{(1,0),(0,1),(1,1)\}$ may lead to alignment paths that are highly deteriorated. In the extreme case, the sequence X may be assigned to a single element of Y. Therefore, in certain applications, it may be beneficial to use the set $\Sigma = \{(2,1),(1,2),(1,1)\}$, which yields a compromise between the strict diagonal matching and the DTW-based matching with the full flexibility (see also Figure 3.17). Further properties and variants of the DTW-based matching procedure are discussed in the exercises.

Finally, we want to discuss how one can handle possible transpositions between the query and matching database sections. As discussed in Section 3.1.2, one can simulate transpositions by cyclically shifting the chroma features along the 12-dimensional chroma axis. Let $\rho : \mathbb{R}^{12} \to \mathbb{R}^{12}$ be the **cyclic shift** operator as defined in (3.11) and let $\rho^i(X) = (\rho^i(x_1), \rho^i(x_2), \ldots, \rho^i(x_N))$, $i \in [0:11]$ be the shifted versions of the query X. Then, the idea is to use each of the twelve sequences $\rho^i(X)$ as a separate query to retrieve matches from the database. Similar to the concept of transposition-invariant self-similarity matrices as introduced in Section 4.2.2.3, one can also define a transposition-invariant matching function. To this end, one first computes a separate matching function, say Δ^i, for each $\rho^i(X)$ and Y. The **transposition-invariant matching function** Δ^{TI} is then obtained by setting

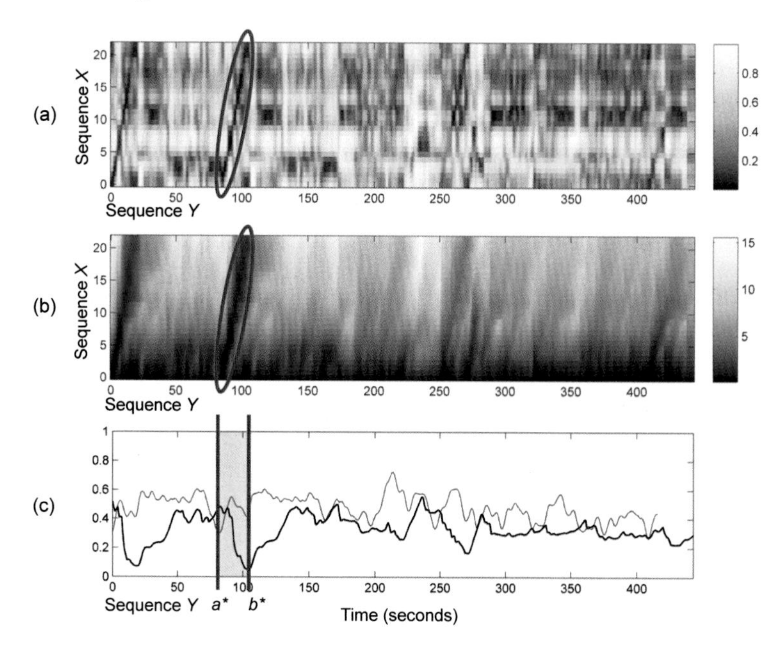

Fig. 7.15 Illustration of matching procedures for the modified Bernstein query from Figure 7.12b (sequence X) and the Karajan recording (sequence Y). **(a)** Cost matrix **C** with "corridor" of low cost (as highlighted by the oval in the image) corresponding to the top match. **(b)** Accumulated cost matrix **D**. **(c)** Diagonal matching function Δ_{Diag} (thin red curve) and DTW-based matching function Δ_{DTW} (thick black curve). The best matching subsequence $Y(a^* : b^*)$ is highlighted.

$$\Delta^{\mathrm{TI}}(m) := \min_{i\in[0:11]} \Delta^i(m) \tag{7.30}$$

for $m \in [1:M]$. This is exactly the same approach as used in the SSM context (see (4.15)).

This procedure is illustrated by Figure 7.16 by means of the song "In the Year 2525" by Zager and Evans, which we have already encountered in Section 4.2.2.3. Recall that the musical structure of this song is given by $IV_1V_2V_3V_4V_5V_6V_7BV_8O$, where the first four verse sections are in the same musical key, V_5 and V_6 are transposed by one semitone upwards, and V_7 and V_8 are transposed by two semitones upwards. Using a query X that corresponds to V_1 and a database sequence Y that consists of the entire song, the first four rows of Figure 7.16 show the matching functions Δ^i for $i = 0,1,2,3$ using the DTW-based matching approach (i.e., $\Delta^0 = \Delta_{\mathrm{DTW}}$). The resulting transposition-invariant matching function Δ^{TI}, as illustrated by Figure 7.16e, correctly indicates the end positions of all eight verse sections.

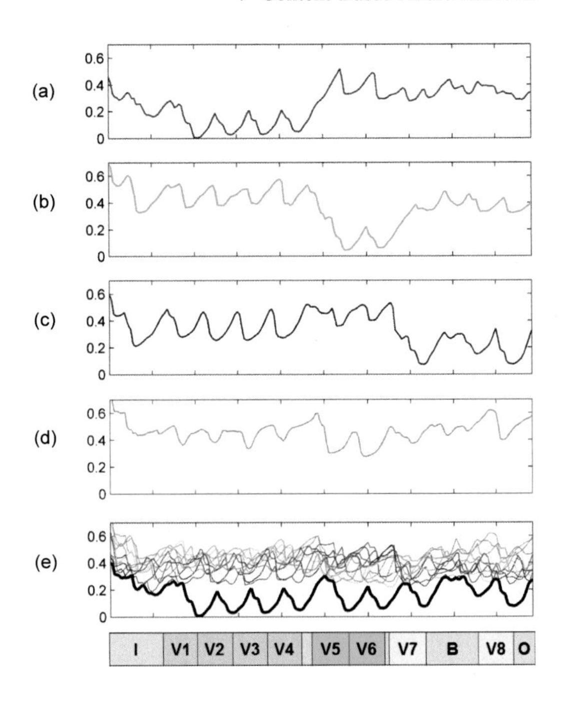

Fig. 7.16 Illustration of the concept of a transposition-invariant matching function by means of the song "In the Year 2525" by Zager and Evans. The query X corresponds to the verse section V_1 and the database sequence Y to the entire song. **(a)** Matching function $\Delta^0 = \Delta_{\mathrm{DTW}}$. **(b)** Δ^1. **(c)** Δ^2. **(d)** Δ^3. **(e)** All twelve matching functions Δ^i for $i \in [0:11]$ and the resulting transposition-invariant matching function Δ^{TI} (thick black curve).

7.3 Version Identification

In the previous tasks, we used an audio fragment as a query and looked for similar fragments contained in a database of music recordings. The degree of specificity considered was very high for audio identification and much lower for audio matching. In this section, we further relax the notion of similarity and deal with a task referred to as **version identification**. A version may differ from the original recording in many ways, possibly including significant changes in timbre, instrumentation, tempo, key, harmony, melody, lyrics, and musical structure. For example, when looking for versions of Beethoven's Fifth Symphony, one may be interested in retrieving a live performance played by a punk-metal band, where many notes have been modified and most of the original structure has been lost. Despite radical changes in tone and tempo, one may still recognize the original composition by means of characteristic melodic, harmonic, or rhythmic elements of the original composition that shine through in the modified version. In Section 7.3.1, we discuss in more detail the various ways an original musical work may be modified to create new versions such as cover songs, remixes, mashups, or medleys. We then describe the main ideas of a retrieval system to identify versions that share some characteristic melodic and harmonic progressions. As opposed to audio matching, where one compares a given short query fragment with local sections of other recordings, one typically compares entire recordings in version identification. In Section 7.3.2, we describe an algorithm for deriving a global similarity measure by finding potentially long matching sections within the query and database document. Finally, in Section 7.3.3, we address the topic of evaluating the quality of ranked retrieval results as typically obtained from version identification systems.

7.3.1 Versions in Music

In Western culture, when speaking of a **piece of music**, one typically thinks of a specific composition given in music notation or given in the form of a recorded track. Often, the origin of a piece of music can be traced back and associated to the name of a composer or a music group. For example, the musical score written down by Ludwig van Beethoven is considered the original version of the Fifth Symphony. Or the original version of the song "Yellow Submarine" is without question the recording by the English rock band "The Beatles." For music in general, however, this view is somewhat simplistic and using the term "piece of music"—not to mention the term "original version"—may be problematic. For example, what should be considered the original version of a folk song that has been orally transmitted over hundreds of years, undergoing numerous changes and existing in many different versions? In **jazz**, where improvisation and the resulting variations are key elements of the music, one often speaks of a **standard** rather than a piece of music. For a jazz piece there typically exists an entire family of performances, which only share certain variations on a melody or are in accordance with certain chord progressions. The terms are even more problematic for Indian classical music, where a **raga** only yields the tonal or melodic framework on which a composition or improvisation is based, rather than denoting a specific piece of music.

7.3.1.1 Types of Versions

It is beyond the scope of this book to discuss such musicological aspects in greater depth. Closely following [23], we adopt a simplistic point of view and use the terms "piece of music" and "version" in a rather loose way. Instead of trying to give a formal definition of these terms, let us consider some typical examples in the context of Western music.

In Western classical music, the original version of a piece of music is often given in the form of a musical score, which has been written down and authorized by the composer. All performances following the musical score are typically also considered original versions of the piece of music, even though there may exist significant differences between these versions, as we have seen in the context of audio matching. An **arrangement** refers to a reworking of a piece of music so that it can be played by instruments different from the ones notated in the original score. For example, a piece for violin might be arranged so that it can be played on a clarinet instead, or a piece written for piano solo might be arranged so that it can be played by a full orchestra. Conversely, symphonic and chamber music has often been arranged for one or two pianos, a reworking also referred to as **piano transcription**. Such transcriptions became popular in particular during the 19th century, since this was the only way symphonic music originally composed for orchestra could also be listened to by a wider audience.

There are many terms that refer to music that has been created by concatenating and mixing existing pieces and recordings. Already in the 15th century, one finds pieces of music called **quodlibet** (Latin for "what pleases"), where different melodies, usually popular folk tunes, were combined in an often humorous manner. Later, in the 19th century, such pieces also became known as **potpourri**, where individual sections of popular operas, operettas, or songs are simply juxtaposed with no strong connection or relationship. Similar techniques are also used nowadays in mainstream popular music. In this context, the term **medley** is used to refer to a piece composed from parts of existing pieces, typically the most memorable parts of songs, played one after another, sometimes overlapping. When selecting and playing back music, a disc jockey (DJ) tries to create smooth transitions between the individual tracks by suitably manipulating existing audio material, thus creating a new version often referred to as a **DJ mix**. Going beyond simple playback, DJs may act as performing musicians by manipulating, blending, and mixing existing audio material, e.g., by applying turntable scratching to create percussive sounds mixed into the original recordings. Mixing techniques are often used by songwriters and music producers to create backing instrumentals for new songs. In the music production context, **sampling** (not to be confused with sampling as used in the discretization context, see Section 2.2.2) refers to the technique of taking portions, or samples, of one recording and reusing them as a "new" instrument in a different piece.

For popular music, there exist many different terms that are used to refer to versions that have been derived from original recordings or songs. For example, a **remix** is a recording that has been edited or completely recreated to sound different from the original version. Such modifications may range from changes in dynamics, pitch, tempo, and playing time to complete rearrangements of voices and instrumental tracks. Thus, on the one hand, a remix may be close to a **remaster** with the goal of enhancing the sound quality of a previously existing recording. On the other hand, it may be a **mashup**, where a new composition has been created by overlaying the vocal track of one song seamlessly over the instrumental track of another. The term **sound collage** is used to refer to a composition obtained by combining portions of existing recordings or pieces of music—similar to a collage in the visual arts, where portions of other artwork or texts, photographs, and other found objects are glued to a piece of paper or canvas. The practice of directly using existing musical material such as a melody or theme in a new composition is called a **quotation**. This is different from a **variation**, where musical material is repeated in an altered form with changes that may involve harmony, melody, rhythm, timbre, orchestration or any combination of these. Also the notion of **parody**, where the objective is to imitate an original work in a ridiculing and trivializing way, is known in music.

Finally, we want to mention the notion of a **cover version** or **cover song**, which loosely refers to a new performance of a previously released song by someone other than the original artist. Typically within the pop and rock genres, this notion some-

Fig. 7.17 Examples of different versions of the painting "Mona Lisa" by Leonardo da Vinci (adopted from [23]).

times has a slightly negative connotation, in particular because the release of cover songs has been a strategy of the marketing industry to profit from the popularity of previously successful songs without remunerating the composer or original group. Cover versions of popular songs are now widespread, sometimes with a radical change in style.

This overview only gives a glimpse of the various notions that exist to refer to versions that can be associated to an original musical work. Nowadays, with the availability of personal digital technology for distributing, recording, and processing audio material, semiprofessional music bands and amateurs can easily produce their own music—often on the basis of existing songs and audio material. For many songs such as "Summertime" by George Gershwin or "Yesterday" by The Beatles one can find hundreds of versions on video-sharing websites such as YouTube—some of them having millions of clicks. Of course, the imitation, manipulation, and modification of existing artworks to create new versions is not limited to music. Principles based on quotation, parody, translation, and plagiarism can be found in literature, painting, sculpture or photography. Figure 7.17 illustrates this by means of the painting "Mona Lisa" by Leonardo da Vinci.

7.3.1.2 Types of Modifications

Different versions of a piece of music typically share some characteristic elements—the "essence" of the piece. However, versions may differ significantly with regard to a wider range of musical and acoustic properties. Following [23], we now discuss some of these properties in more detail.

First of all, versions of the same piece of music may differ substantially in their **tempo** (see Section 1.1.2). In particular for classical Western music, the tempo may only loosely be specified by terms such as **Largo** (very slow), **Andante** (at a walking pace), or **Presto** (extremely fast), and performers take the freedom to play the music at their own pace. As an extreme example, one may mention the two famous recordings by the Canadian classical pianist Glenn Gould of Johann Sebastian Bach's Goldberg Variations. In his 1955 debut album, Glenn Gould plays the beginning Aria with a tempo of 60 BPM. Shortly before his death, Gould re-recorded the Goldberg Variations in the year 1981 in a more introverted way, this time choosing a tempo of only 33 BPM for the Aria—nearly half the tempo compared with his first recording. Tempo changes can also be introduced by processing an existing music recording using techniques based on resampling or time scale modification. For example, DJs use beatmatching tools for speeding up or slowing down a recording in order to match the tempo of a previous track so both can be seamlessly mixed. The tempo-related aspect of **timing** deals with a much smaller temporal scale. Rather than strictly following a predefined tempo, musicians give a performance an individual touch by introducing subtle temporal fluctuations—a musical phenomenon also referred to as **expressive timing**.

Besides tempo and timing, one of the most typical differences between versions of a piece of music is the "sound color" or timbre (see Section 1.3.4). Such differences result from modifications in the instrumentation, the different playing styles of artists, room acoustics, the application of postproduction techniques such as equalization, dynamic compression, and so on.

In versions such as cover songs it is also quite common to have a change of the original **structure** of the piece (see Chapter 4). For example, the intro may be skipped, an instrumental solo section may be added, or a chorus section may be repeated more often than in the original song. Similarly, in a classical solo concerto, an improvised **cadenza** may be included by the soloist displaying his or her virtuosic skills.

Sometimes the **musical key** is changed when creating a new version of a piece of music. For example, in an arrangement, the pitch range may be adapted to a different singer or another instrument. Such key changes are typically applied to an entire piece, although they may also be restricted to a single musical part. Besides the main key, one may also find changes in the **harmonization**, the chordal accompaniment to a melody. Such reharmonization techniques are used to make the music sound more interesting or to create a different mood. Particularly in jazz, musicians often modify a well-known jazz standard by applying reharmonization techniques and then also adapt the melody to fit the new chord progression.

There are also other properties that may be changed when creating new versions. Often the **lyrics**, the set of words underlying a song, are replaced to adjust the song for a specific occasion such as a celebration. Also, popular songs and operas are translated into other languages to open up to new audiences including people from other countries. Last but not least, in versions such as live performances, the actual

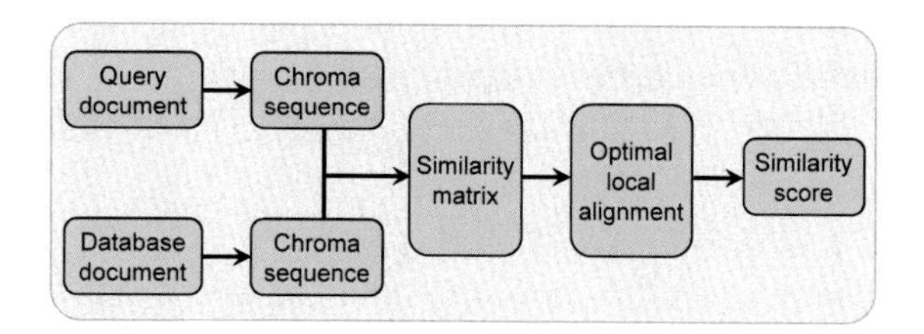

Fig. 7.18 Overview of a pipeline for a version identification system.

music may be superimposed by additional sound sources such as applause, rhythmic clapping, people shouting, and so on.

7.3.2 Identification Procedure

Given an audio recording (or another representation) of a piece of music, the objective of version identification is to retrieve from a given music database all recordings that can be considered a version of the given piece. In this retrieval scenario, the query typically consists of an entire recording—as opposed to audio identification and audio matching, where the query was only a small audio fragment. Therefore, version identification is usually considered a **document-level** retrieval task, where a single similarity measure is used to globally compare entire documents. As we have discussed before, the original and a derived version of a piece of music typically share some common characteristics. However, due to possible structural differences between these versions, it is not clear where these common elements occur. Therefore, when assessing the **global** similarity between a given query document and a database document, a general strategy in version identification is to look for **local** concurrences with regard to certain musical properties. In other words, the global comparison is performed on a local basis.

In view of the many possible kinds of modifications that may be applied when creating new versions, it is not realistic to assume that one can deal with all the resulting variations by using a single technique. In the following, we restrict ourselves to the scenario where the versions to be identified share a similar melodic or harmonic progression—at least in certain sections. On the other side, we allow differences in aspects such as tempo, instrumentation, timbre, and the overall musical structure. Following [23], we present the main ideas of a typical version identification procedure, which is tailored towards capturing tonal elements of music recordings while showing a high degree of invariance with regard to a wide range of modifications. Based on these assumptions, the overall strategy for comparing two given documents is as follows (see also Figure 7.18): First, to capture the tonal characteristics, we convert the recordings into sequences of chroma-based audio features. Second, we compute a similarity matrix by comparing the elements of these two sequences in a pairwise fashion. Third, we try to identify a potentially long path of high similarity. The presence of such a path indicates that the two chroma sequences share some related subsequences. Finally, from this information, we derive a similarity score between the two versions to be compared.

Many of the required techniques to realize such a version identification system are similar to the ones we have already used for audio matching (Section 7.2), for music structure analysis (Chapter 4), and for music synchronization (Chapter 3). Therefore, in the following, we only highlight how these techniques need to be adjusted for the given scenario. As in audio matching, we are interested in capturing the rough tonal progression while retaining invariance to dynamics and local tempo fluctuations. Therefore, the requirements for the feature representations are similar to the ones discussed in Section 7.2.1. Both the query and the database document are converted into chroma-based feature sequences, say $X = (x_1, x_2, \ldots, x_N)$ and $Y = (y_1, y_2, \ldots, y_M)$. Since we want to blend out nuances, the usage of a smoothed and normalized chroma variant is beneficial. In the following examples, we use the CENS_5^{21}-features as introduced in Section 7.2.1. These features, which have a rate of 2 Hz (two chroma vectors per second), constitute a good trade-off between robustness and specificity.

In the audio matching scenario, we looked for subsequences of Y that are similar to the full query sequence X. Now, in the version identification scenario, the situation is different. The assumption is that the query document and the database document may share a similar tonal progression in certain parts, but we know neither the strength and the duration nor the locations where these concurrences occur. In this context, the matching task can be formulated as follows: given the sequences X and Y, we are looking for a subsequence within X and a subsequence within Y such that these two subsequences are as similar as possible. Furthermore, in the comparison of these two subsequences, we want to be able to deal with temporal deformations.

As in the case of audio matching, this task can be expressed in the form of an optimization problem, which can be solved efficiently by using dynamic programming. However, we need to assume a different viewpoint on the given problem. In audio matching, we looked for a cost-minimizing subsequence of Y that matched the query X (see (7.25)). This optimization criterion worked since the **entire** query sequence X was forced to be matched against a subsequence of Y. Now, when we assume that only a **subsequence** of X needs to be matched against one of Y, there is a trivial solution: the empty subsequences. Indeed, matching two empty sequences results in an overall cost of zero, which is an optimal solution when we assume that the cost matrix $\mathbf{C}$ has no negative values. This is definitely not what we are looking for.

Intuitively speaking, we are not only looking for subsequences that can be matched with minimal cost, but also for preferably long subsequences that have a certain relevance. In other words, we need to balance two principles at the same time: minimizing the overall matching cost on the one hand, and maximizing the lengths of the subsequences on the other hand. To remedy this problem, we assume a "positive" viewpoint by using a similarity matrix (or score matrix) instead of a cost matrix. Then, instead of identifying cost-minimizing subsequences, we are looking for score-maximizing subsequences. Further constraints for the comparison of the

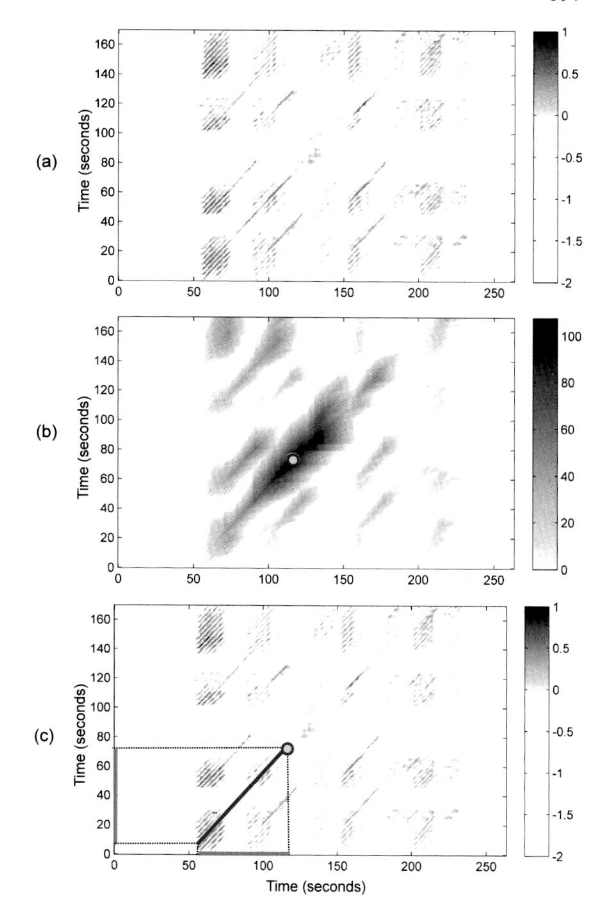

Fig. 7.19 Comparison of the original version of the song "Day Tripper" by The Beatles (vertical axis) with a cover version by the band Ocean Colour Scene (horizontal axis). **(a)** Similarity matrix **S** computed in the same fashion as the one shown in Figure 4.15f. **(b)** Accumulated score matrix **D**. **(c)** **S** with score-maximizing path and induced segments.

subsequences are imposed by introducing negative penalty values as well as suitable step size conditions. Note that we have already employed similar strategies in the context of audio thumbnailing (see Section 4.3). We now quickly review these concepts and apply them to our version identification task.

Fixing a similarity measure $s : \mathcal{F} \times \mathcal{F} \rightarrow \mathbb{R}$ as in (4.3), we compute an $N \times M$ similarity matrix by setting

$$\mathbf{S}(n,m) := s(x_n, y_m) \tag{7.31}$$

for $n \in [1:N]$ and $m \in [1:M]$. The properties of this similarity matrix can be further improved by using the same enhancement strategies as described for the case of self-similarity matrices (see Section 4.2.2). In particular, we apply some thresholding with respect to a threshold parameter $\tau > 0$ and a penalty parameter $\delta \leq 0$ as discussed in Section 4.2.2.4 and illustrated by Figure 4.15f. The resulting matrix is constructed in such a way that the cells that may express relevant similarity relations tend to have a positive score, whereas all other cells are given a negative score. This property is crucial in view of the following procedure, which tries to find a path component that accumulates a possibly large score. Such a path mainly lies in the positive part of $\mathbf{S}$ while avoiding running through cells of negative score (see also Figure 7.19c for an example).

To formalize the optimization task, we need again the notion of a path to account for temporal deformations when comparing two feature sequences. As in (4.5), a path is defined to be a sequence $P = ((n_1,m_1),\ldots,(n_L,m_L))$ of cells $(n_\ell,m_\ell) \in [1:N] \times [1:M]$, $\ell \in [1:L]$, satisfying $(n_{\ell+1},m_{\ell+1}) - (n_\ell,m_\ell) \in \Sigma$ for some set Σ of admissible step sizes. Thus, choosing $\Sigma = \{(0,1),(1,0),(1,1)\}$, this definition is the same as the one for a warping path in Section 3.2.1.1 except for having omitted the boundary condition. The reason for this is that we do not want to globally align the sequences X and Y, but only subsequences. As in (4.7), we associate for a path P the two **induced segments** $\pi_1(P) := [n_1 : n_L]$ and $\pi_2(P) := [m_1 : m_L]$. The **score** $\sigma(P)$ of P is defined as $\sigma(P) := \sum_{\ell=1}^{L} \mathbf{S}(n_\ell,m_\ell)$ (see (4.8)). With these definitions at hand, our optimization task is to find the score-maximizing path

$$P^* := \underset{P}{\operatorname{argmax}}\, \sigma(P) \tag{7.32}$$

over all possible paths (with arbitrary start and end positions). The two best matching subsequences of X and Y are then given by the induced segments $\pi_1(P^*)$ and $\pi_2(P^*)$, respectively.

A score-maximizing path can be computed via dynamic programming similar to the DTW algorithm. To this end, we define an $N \times M$ accumulated score matrix $\mathbf{D}$ by

$$\mathbf{D}(n,m) := \max\{\sigma(P) \mid P \text{ is path ending at } (n,m)\ \} \tag{7.33}$$

for $n \in [1:N]$ and $n \in [1:M]$. In other words, $\mathbf{D}(n,m)$ is the maximal possible score that can be accumulated by a path that starts at some arbitrary cell but ends with cell (n,m). In this definition, the empty path $P = \emptyset$ is considered to be a path of score $\sigma(P) = 0$, which belongs to the paths considered in (7.33). For cells (n,m) with $n = 1$ or $m = 1$, one obtains

$$\mathbf{D}(1,1) = \max\{0,\mathbf{S}(1,1)\}, \tag{7.34}$$
$$\mathbf{D}(n,1) = \max\{0,\mathbf{D}(n-1,1)+\mathbf{S}(n,1)\} \quad \text{for } n \in [2:N], \tag{7.35}$$
$$\mathbf{D}(1,m) = \max\{0,\mathbf{D}(1,m-1)+\mathbf{S}(1,m)\} \quad \text{for } m \in [2:M], \tag{7.36}$$

constituting the boundary cases for our recursion. For $n \in [2:N]$ and $m \in [2:M]$, one can compute $\mathbf{D}$ via

$$\mathbf{D}(n,m) = \max \begin{cases} 0, \\ \mathbf{D}(n-1,m-1)+\mathbf{S}(n,m), \\ \mathbf{D}(n-1,m)+\mathbf{S}(n,m), \\ \mathbf{D}(n,m-1)+\mathbf{S}(n,m). \end{cases} \tag{7.37}$$

Note that this recursion differs from the one in (3.25) in two ways. First, instead of minimizing over costs, we now maximize over (possibly negative) scores. Even

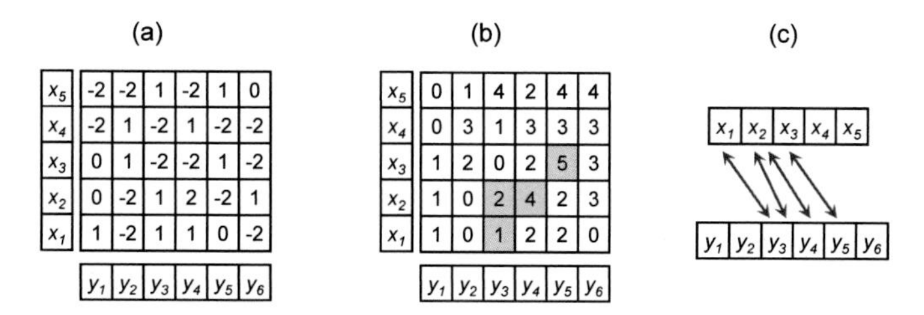

Fig. 7.20 **(a)** Similarity matrix **S** for two sequences $X = (x_1, \ldots, x_5)$ and $Y = (y_1, \ldots, y_6)$. **(b)** Accumulated score matrix **D** with a score-maximizing path. **(c)** Resulting local alignment.

more important, the second difference is that the maximization includes a zero value. This modification makes it possible to start a path at any position without having accumulated any potentially negative score. This realizes the idea of skipping the beginnings of the sequences X and Y when being compared.

How do we obtain the score-maximizing path P^* as defined in (7.32)? Since we do not impose any boundary constraints on the considered paths, P^* may end at any cell. Therefore, to obtain the maximal score over all possible paths, we need to look at the maximal entry of $\mathbf{D}$:

$$\mathbf{D}^{\max} := \sigma(P^*) = \max_{(n,m)\in[1:N]\times[1:M]} \mathbf{D}(n,m). \tag{7.38}$$

In general, there may be several entries of $\mathbf{D}$ having maximal value. We start with one of these entries, say q_1, which defines the end position of an optimal path P^*. The full path is then obtained via backtracking—as we did with DTW (see Table 3.2). This time, however, the stop condition of the backtracking is different. Let $q_1, q_2, \ldots, q_\ell$ be the iteratively determined cells, then the backtracking is stopped as soon as a cell $q_\ell = (a,b)$ with $\mathbf{D}(a,b) = 0$ or $q_\ell = (1,1)$ is reached. In the first case, the cell $q_\ell = (a,b)$ is omitted since it has a nonpositive score $\mathbf{S}(a,b) \leq 0$. This yields the optimal path $P^* = (q_{\ell-1}, \ldots, q_1)$. In the second case the path starts with the cell $q_\ell = (1,1)$ if $\mathbf{S}(1,1) > 0$ or with the cell $q_{\ell-1}$ if $\mathbf{S}(1,1) \leq 0$.

For an example, let us have a look at Figure 7.20, which shows a similarity matrix $\mathbf{S}$ and the resulting accumulated score matrix $\mathbf{D}$. The matrix $\mathbf{D}$ assumes its maximal value at the cell $(3,5)$. Backtracking yields the score-maximizing path $P^* = ((1,3),(2,3),(2,4),(3,5))$ with $\sigma(P^*) = 5$. The two induced segments are $\pi_1(P^*) = [1:3]$ and $\pi_2(P^*) = [3:5]$, which yield the subsequences (x_1, x_2, x_3) and (y_3, y_4, y_5) (see Figure 7.20c).

An example obtained from real audio recordings is shown in Figure 7.19, where the original Beatles version of the song "Day Tripper" is compared with a cover version by the band Ocean Colour Scene. Figure 7.19a shows an enhanced and thresholded similarity matrix $\mathbf{S}$. In this matrix, one can clearly notice a number of path components, which lie in the positive part of $\mathbf{S}$. On the other hand, there are no path-like structures at the beginning and the end of the cover version, which is longer than the original version. Listening to the cover song reveals that in the first fifty seconds the band members interact and talk to the audience so that there are no tonal relations to the original Beatles version. The accumulated score matrix $\mathbf{D}$, which is shown in Figure 7.19b, assumes a maximal value of $\sigma(P^*) = 107$ at the cell (n,m) that corresponds to the end position $(72,116)$ (given in seconds) of a score-maximizing optimal path. The entire path as well as the induced segments are shown in Figure 7.19c.

In summary, given two audio recordings, we computed chroma-based feature sequences X and Y and derived an enhanced similarity matrix $\mathbf{S}$. In the design of this matrix it was important that the cells expressing tonal concurrences lie in the positive part of $\mathbf{S}$, while the irrelevant cells lie in the negative part. The threshold parameter τ and penalty parameter δ can be used to balance out the positive and negative parts. We then computed the highest value $\mathbf{D}^{\max}$ of an accumulated score matrix $\mathbf{D}$, which amounts to the total score of the two best matching subsequences of X and Y (see (7.38)). By construction, the value $\mathbf{D}^{\max}$ is high if and only if the recordings share some common tonal progression of a certain strength and duration. The duration of possible **gaps** (cells of negative score within an alignment path) can be controlled by the penalty parameter δ. The concept of an alignment path allows for temporal deviations between these progressions, while the chroma features used yield some robustness to differences in timbre and dynamics. Furthermore, using a local alignment, the specific locations of the progression within the versions are not relevant, which accounts for possible structural differences of the two versions.

Because of these properties, we define the overall similarity score $\gamma(\mathcal{Q},\mathcal{D})$ for the document-level comparison between a query document $\mathcal{Q}$ and a database document $\mathcal{D}$ by setting

$$\gamma(\mathcal{Q},\mathcal{D}) := \mathbf{D}^{\max}. \tag{7.39}$$

This definition implements the idea of performing the global comparison on the basis of local concurrences. For the version identification task, a given query document is compared with all database documents. The database documents are then ranked according to the computed similarity score. In the next section, we discuss this retrieval process in more detail and address the topic of evaluating the retrieval quality.

7.3.3 Evaluation Measures

In audio identification and audio matching we dealt with **fragment-level** retrieval tasks, where the objective was to identify specific sections of a database document that are similar to the given query fragment. In these scenarios, to simplify notation, we assumed that the database consists of a single document $\mathcal{D}$ (by concatenating all database documents). Now, in version identification, we deal with a **document-level** retrieval task, where the objective is to retrieve entire documents rather than fragments (even though the document-level similarity score is based on a fragment-level comparison). We now give a mathematical description of this retrieval scenario and introduce some evaluation measures to assess the quality of a document-level retrieval result.

In the following, we assume that the database consists of a set of K documents, $\{\mathcal{D}_1, \mathcal{D}_2, \ldots, \mathcal{D}_K\}$, where each document $\mathcal{D}_k$ is associated to an identifier $k \in [1:K]$. Furthermore, given a query document $\mathcal{Q}$, we suppose that we have a similarity measure that yields a value $\gamma(\mathcal{Q}, \mathcal{D}_k) \in \mathbb{R}$ for each $k \in [1:K]$. Based on these values, we can **rank** (i.e., sort) the database documents in descending order of their similarity values. In this way (imagining a Google-like retrieval scenario), the retrieval results that are most similar to the given query appear early in the result list displayed to a user. The **top rank** is the retrieval result that is ranked at the first position (i.e., the document with the highest similarity value). Furthermore, we say that the document $\mathcal{D}_k$ has a **higher rank** than the document $\mathcal{D}_\ell$ if

$$\gamma(\mathcal{Q}, \mathcal{D}_k) > \gamma(\mathcal{Q}, \mathcal{D}_\ell) \tag{7.40}$$

for $k, \ell \in [1:K]$. To break ties in the case $\gamma(\mathcal{Q}, \mathcal{D}_k) = \gamma(\mathcal{Q}, \mathcal{D}_\ell)$, we simply say that $\mathcal{D}_k$ has a higher rank than $\mathcal{D}_\ell$ if $k < \ell$. Mathematically, a ranking can be specified by a permutation

$$\rho_\mathcal{Q} : [1:K] \to [1:K] \tag{7.41}$$

that sorts the database documents in descending order of rank:

$$\gamma(\mathcal{Q}, \mathcal{D}_{\rho_\mathcal{Q}(1)}) \geq \gamma(\mathcal{Q}, \mathcal{D}_{\rho_\mathcal{Q}(2)}) \geq \ldots \geq \gamma(\mathcal{Q}, \mathcal{D}_{\rho_\mathcal{Q}(K)}). \tag{7.42}$$

Note that, using this notation, the top rank (highest rank) is represented by the index value 1 (corresponding to the document with identifier $\rho_\mathcal{Q}(1)$) and the lowest rank by the index value K (corresponding to the document with identifier $\rho_\mathcal{Q}(K)$).

The version identification system described in Section 7.3.2 yields such a ranked list of documents based on the similarity score of (7.39). Depending on the application in mind, one can think of many different ways for presenting the final retrieval result and assessing its quality. In certain applications, only the top-ranked match may be of interest, and will be further processed by the retrieval system. For example, in audio identification, only the top match is needed to identify the audio recording and to retrieve suitable metadata, which is then presented to the user (see Figure 7.2). In other scenarios, the system may show the top ten matches, a typical standard setting as used in web search systems such as Google. When looking only at the similarity values, the top matches may include just documents that are nearly identical to the query document. Therefore, some retrieval systems try to find a good balance between document similarity, relevance, and diversity when presenting retrieval results to the user.

In the design and evaluation of retrieval systems, such issues are of great importance and constitute a separate area of research [13]. In the following, we introduce the main ideas of some general evaluation measures based on precision and recall. Using the notions of Section 4.5.1, the **items** of our current scenario are the database documents or its identifiers. Let $\mathcal{I} := [1:K]$ be the set of items. For the evaluation, we assume that a reference annotation of the relevant (or positive) items is available.

Let

$$\mathcal{I}_{\mathcal{Q}} := \mathcal{I}_{+}^{\mathrm{Ref}} \subseteq \mathcal{I} \tag{7.43}$$

be the set of the relevant items, which depends on the given query document $\mathcal{Q}$. In Section 4.5.1, we assumed that the retrieval system returns a set $\mathcal{I}_{+}^{\mathrm{Est}} \subseteq \mathcal{I}$, which consists of the items being estimated as positive. Based on this fixed set of items, we introduced precision and recall measures. Now, the situation is different: instead of having a fixed and unordered set of retrieved items, the system returns an ordered list of documents. One possibility for evaluating ranked retrieval results is to consider an entire family of precision and recall values depending on the ranking parameter $r \in [1:K]$. For later convenience, we define a **relevance function** $\chi_{\mathcal{Q}} : [1:K] \to \{0,1\}$ that assumes the value 1 if the document at rank r is relevant; otherwise it assumes the value 0. Mathematically, this definition is expressed by

$$\chi_{\mathcal{Q}}(r) := \begin{cases} 1 & \text{if } \rho_{\mathcal{Q}}(r) \in \mathcal{I}_{\mathcal{Q}}, \\ 0 & \text{if } \rho_{\mathcal{Q}}(r) \in \mathcal{I} \setminus \mathcal{I}_{\mathcal{Q}}. \end{cases} \tag{7.44}$$

We then define the precision $\mathrm{P}_{\mathcal{Q}}(r)$ and recall $\mathrm{R}_{\mathcal{Q}}(r)$ at rank $r \in [1:K]$ by setting

$$\mathrm{P}_{\mathcal{Q}}(r) := \frac{1}{r} \sum_{k=1}^{r} \chi_{\mathcal{Q}}(k), \tag{7.45}$$

$$\mathrm{R}_{\mathcal{Q}}(r) := \frac{1}{|\mathcal{I}_{\mathcal{Q}}|} \sum_{k=1}^{r} \chi_{\mathcal{Q}}(k). \tag{7.46}$$

Note that these definitions agree with (4.47) and (4.48) when $\mathcal{I}_{+}^{\mathrm{Est}}$ is defined to be the set of the top r items of the ranked list (see Exercise 7.11). The family $\{(\mathrm{P}_{\mathcal{Q}}(r), \mathrm{R}_{\mathcal{Q}}(r)) \mid r \in [1:K]\}$ can be visualized in a two-dimensional plane with one axis referring to precision and the other to recall. Combining subsequent points yields a so-called **precision–recall curve** or **PR curve**.

Let us have a look at the example of Figure 7.21, where we consider a database consisting of $K = 10$ documents. The similarity scores and the resulting rank positions of these documents are shown in Figure 7.21a. Furthermore, in Figure 7.21b, the ten documents have been sorted according to their rank. For the permutation $\rho_{\mathcal{Q}}$ of (7.41), which specifies this reordering, one obtains $\rho_{\mathcal{Q}}(1) = 9$, $\rho_{\mathcal{Q}}(2) = 2$, $\rho_{\mathcal{Q}}(3) = 6$, and so on. In our example, the set of relevant items is $\mathcal{I}_{\mathcal{Q}} = \{2,7,8,9\}$. These documents occur at the rank positions $r \in \{1,2,4,8\}$, which are also the positions of the 1-values of the relevance function $\chi_{\mathcal{Q}}$ (see (7.44)). The last two columns of Figure 7.21b indicate the precision $\mathrm{P}_{\mathcal{Q}}(r)$ and recall $\mathrm{R}_{\mathcal{Q}}(r)$ at rank $r \in [1:K]$. For example, the top three matches contain two of the four relevant documents, which yields $\mathrm{P}_{\mathcal{Q}}(3) = 2/3$ and $\mathrm{R}_{\mathcal{Q}}(3) = 2/4$.

The precision–recall curve of our example is shown in Figure 7.21c. In general, a PR curve has a characteristic sawtooth shape: if the rth document is relevant, both the precision and the recall increase. If it is nonrelevant, the precision drops while the recall remains the same. Furthermore, a value $\mathrm{P}_{\mathcal{Q}}(r) = 1$ means that all of the top r matches are relevant.

(a)

ID	Score	Rank
1	8	10
2	52	2
3	22	5
4	10	9
5	12	7
6	34	3
7	11	8
8	27	4
9	72	1
10	18	6

(b)

Rank	ID	Rel.	P(r)	R(r)
1	9	+	1.00	0.25
2	2	+	1.00	0.50
3	6	-	0.67	0.50
4	8	+	0.75	0.75
5	3	-	0.60	0.75
6	10	-	0.50	0.75
7	5	-	0.43	0.75
8	7	+	0.50	1.00
9	4	-	0.44	1.00
10	1	-	0.40	1.00

(c)

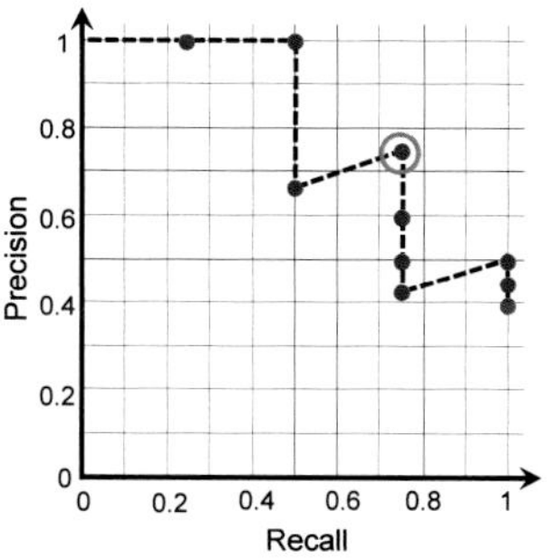

Fig. 7.21 Precision and recall values for a ranked retrieval result for $K = 10$ documents and a given query $\mathcal{Q}$. **(a)** Table indicating the document identifier, the similarity scores, and the rank $r \in [1:K]$. **(b)** Table indicating the rank, the document identifier, the relevance of the document, and the precision values $\mathrm{P}_{\mathcal{Q}}(r)$ and recall values $\mathrm{R}_{\mathcal{Q}}(r)$ at rank $r \in [1:K]$. **(c)** Plot of the resulting precision–recall curve with break-even point indicated by the circle.

For a certain query, the PR curve gives a good impression of the overall quality of the ranked retrieval result. However, dealing with an entire family of precision and recall values can be quite cumbersome. Therefore, one often reduces this family of PR values to a single evaluation measure that still bears some characteristic information on the overall retrieval performance. In the following, we discuss some prominent examples.

As noted above, for applications such as a web search, a user typically looks only at the retrieval results presented on the first or second page. Therefore, rather than looking at the entire ranked list, only the quality of the first ten or twenty top matches is what matters to the user. To measure the retrieval performance in this setting, one can simply look at the precision $\mathrm{P}_{\mathcal{Q}}(r)$ at a specific rank r, for example, $\mathrm{P}_{\mathcal{Q}}(r)$ at rank $r = 10$ or at $r = 20$. Using this measure has the advantage that one does not need to know the entire set of relevant documents. On the downside, however, such a single precision value does not reflect well the overall behavior of the retrieval system. Furthermore, precision at some fixed rank r does not average well when considering different queries. The reason for this is that the total number of relevant documents for a query has a strong influence on $\mathrm{P}_{\mathcal{Q}}(r)$. For example, if there are only five relevant documents in total, one obtains $\mathrm{P}_{\mathcal{Q}}(10) \leq 0.5$ even in the case of a perfect retrieval result. On the other side, if the number of relevant documents is large, $\mathrm{P}_{\mathcal{Q}}(r)$ also tends to be large.

A more robust alternative is known as the **break-even point** of the PR curve, which is defined to be the positive value where the precision equals the recall. In the example of Figure 7.21, the break-even point is $\mathrm{P}_{\mathcal{Q}}(4) = \mathrm{R}_{\mathcal{Q}}(4) = 0.75$. For further details, see also Exercise 7.13.

As another measure, one may consider the "best point" of the PR curve in terms of the F-measure. To this end, let $\mathrm{F}_{\mathcal{Q}}(r)$ denote the F-measure of $\mathrm{P}_{\mathcal{Q}}(r)$ and $\mathrm{R}_{\mathcal{Q}}(r)$ (see (4.49)). In the case that $\mathrm{P}_{\mathcal{Q}}(r) = 0$ and $\mathrm{R}_{\mathcal{Q}}(r) = 0$, we set $\mathrm{F}_{\mathcal{Q}}(r) = 0$. Then, the

maximal F-measure of the PR curve is defined as

$$\mathrm{F}_{\mathcal{Q}}^{\max} := \max_{r\in[1:K]} \mathrm{F}_{\mathcal{Q}}(r). \tag{7.47}$$

In the example of Figure 7.21, the maximal F-measure is $\mathrm{F}_{\mathcal{Q}}^{\max} = 0.75$ and equals the break-even point. In general, however, the two measures do not need to agree (see Exercise 7.14).

As a third example, we consider the **average precision** $\overline{\mathrm{P}}_{\mathcal{Q}}$ defined by

$$\overline{\mathrm{P}}_{\mathcal{Q}} := \frac{1}{|\mathcal{I}_{\mathcal{Q}}|}\sum_{r=1}^{K} \mathrm{P}_{\mathcal{Q}}(r)\chi_{\mathcal{Q}}(r). \tag{7.48}$$

In this definition, the precision $\mathrm{P}_{\mathcal{Q}}(r)$ is only considered when $\chi_{\mathcal{Q}}(r) = 1$. In other words, the average is computed only at the ranks where the recall level changes (see also Exercise 7.15). For example, for the retrieval result of Figure 7.21 one obtains an average precision of

$$\overline{\mathrm{P}}_{\mathcal{Q}} = \frac{1}{4}(1+1+0.75+0.5) = 0.8125. \tag{7.49}$$

As for the break-even point and the maximal F-measure, the average precision ranges between zero and one, where the value one is assumed if and only if all relevant documents are ranked at the top. One advantage of the average precision is that it takes into account the entire ranked list, contrary to the other two measures. For example, interchanging in Figure 7.21 the values $\chi_{\mathcal{Q}}(8)$ and $\chi_{\mathcal{Q}}(9)$ (thus ranking a relevant document one position lower) leaves the break-even point and the maximal F-measure unchanged, whereas the average precision decreases to $\overline{\mathrm{P}}_{\mathcal{Q}} = (1+1+0.75+0.44)/4 = 0.7975$.

So far, we have considered evaluation measures for a single query document $\mathcal{Q}$. In practice, when evaluating the performance of a retrieval system, one should use many different queries that reflect the typical information a user needs within a given application scenario. To obtain a single evaluation number, one often combines the query-dependent values by taking the mean over all these values. As an example, let us consider the case of average precision. Let $\{\mathcal{Q}_1,\ldots,\mathcal{Q}_J\}$ be the set of query documents to be considered in the evaluation. Then we obtain a value $\overline{\mathrm{P}}_{\mathcal{Q}_j}$ for each $j \in [1:J]$. The **mean average precision** or MAP is defined by

$$\overline{\mathrm{P}} := \frac{1}{J}\sum_{j=1}^{J} \overline{\mathrm{P}}_{\mathcal{Q}_j}. \tag{7.50}$$

Having good discrimination and stability properties, the mean average precision is widely employed in information retrieval for different multimedia domains including text, image, audio, and music. In particular, the MAP has been used as a standard measure for evaluating the performance of version identification systems. State-of-the-art procedures, which use techniques similar to those described in Section 7.3.2,

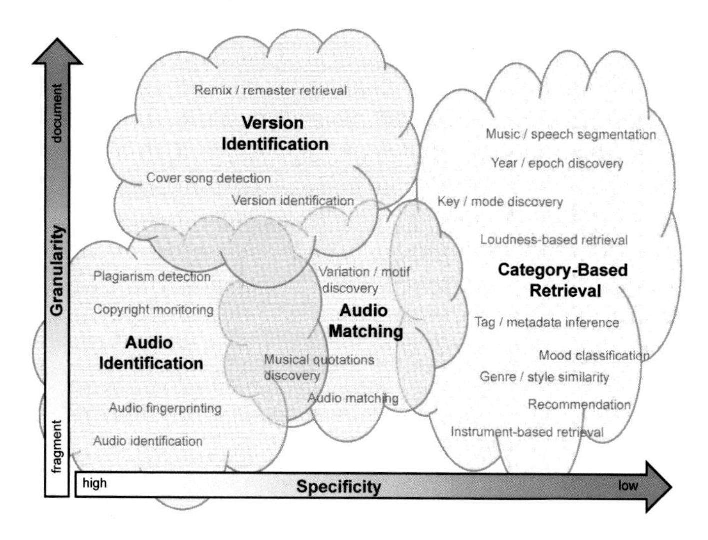

Fig. 7.22 Various content-based music retrieval tasks arranged within a specificity–granularity plane (adopted from [12]).

reach MAP values between 0.7 and 0.8 or even slightly above (see [23]). Of course, such numbers need to be treated with caution, since they also critically depend on the properties of the evaluation datasets used.

7.4 Summary and Further Readings

In this chapter, we discussed various content-based audio retrieval scenarios that follow the query-by-example paradigm. Given an audio recording or a fragment of it (used as a query), the task was to automatically retrieve documents from an audio database containing parts or aspects similar to the query. Retrieval systems based on this paradigm do not require any textual descriptions in the form of metadata or tags. However, the notion of similarity used to compare different audio recordings (or fragments) is of great importance and largely depends on the respective application as well as the user requirements.

There are many different scenarios for content-based audio retrieval which follow different strategies and aim at different applications. We had a look at three such scenarios: audio identification, audio matching, and version identification. Generally, such retrieval tasks can be characterized by various aspects, such as the notion of similarity, the underlying matching principles, or the query format. Following [5, 12], we consider two aspects: **specificity** and **granularity** (see also Figure 7.22). The **specificity** of a retrieval system refers to the degree of similarity between the query and the database documents to be retrieved. Highly specific retrieval systems return exact or near copies of the query, whereas low-specific re-

Table 7.1 Overview of the three different retrieval scenarios and feature types as considered in this chapter. (Note that there are many more approaches that may use other feature types as well.)

Retrieval task	Audio identification	Audio matching	Version identification
Identification	Specific audio recording	Different interpretations	Different versions
Query	Short fragment (5–10 seconds)	Audio clip (10–40 seconds)	Entire recording
Retrieval level	Fragment	Fragment	Document
Specificity	High	Medium	Medium / low
Features	Spectral peaks (abstract)	Chroma (harmony)	Chroma (harmony)

trieval systems return semantically related matches that, from a numerical point of view, may be quite different from the original query. The **granularity** refers to the temporal level considered in the retrieval scenario. In **fragment-level** retrieval scenarios, the query consists of a short fragment of an audio recording, and the goal is to retrieve all related fragments that are contained in the database documents. For example, such fragments may cover a few seconds of audio content or may correspond to a motif, a theme, or a musical part. In contrast, in **document-level** retrieval, the query may consist of an entire document which is compared with entire documents of the database. In this case, the notion of similarity tends to be coarser, even though the global similarity score between documents may still be based on a local, fragment-level comparison.

The various content-based retrieval scenarios may be loosely classified and arranged according to their specificity and granularity (see Figure 7.22). In the specificity–granularity plane shown, the three scenarios considered in this chapter (audio identification, audio matching, and version identification) are represented by clouds, which enclose several related retrieval scenarios. A fourth cloud that represents **category-based** retrieval scenarios has been added. In these scenarios, the similarity relationships are somewhat vague and express cultural or musicological categories, such as genre [29] or emotion [32]. A more in-depth discussion of such retrieval scenarios is outside the scope of this book, and we refer to [19] for further entry points into this area of research. Even though the taxonomy indicated by Figure 7.22 may be too simplistic, it gives an intuitive overview of various retrieval paradigms while illustrating their subtle but crucial differences. The main differences between the three scenarios considered in this chapter are again summarized by Table 7.1. In the following, we give pointers to relevant literature for further readings.

Audio Identification

Among the three considered scenarios, the task of audio identification (also referred to as audio fingerprinting) is of the highest specificity. Given a short, potentially distorted audio fragment, the objective is to identify the original recording in which the fragment is contained. Being a fragment-level task, the query may be as short as a couple of seconds. As discussed in Section 7.1.1, one main challenge in this scenario is that the audio signals may be affected by noise, artifacts from lossy audio compression, pitch shifting, time scaling, equalization, or dynamics compression [3]. As a result, many different strategies for the design and computation of fingerprints have been suggested (see, e.g., [2, 3, 14, 30] for some earlier approaches). As an example, we have described in Section 7.1.2 a peak-based fingerprinting strategy, where we have closely followed the original publication by Wang [30]. In Section 7.1.3, we turned to the topic of data indexing with the objective of improving the speed of the retrieval process without sacrificing accuracy. Motivated by our fingerprinting scenario, we introduced and applied the concept of inverted lists—a standard technique used in information retrieval [31]. One central issue for obtaining a good index structure is the design of suitable hashes that serve as key words for the inverted lists. Following [30], we discussed in Section 7.1.4 an approach (using pairs of peaks) that balances out the conflicting goals of high specificity, robustness, and efficiency. Sonnleitner et al. [28] extended this approach (using quadruples of peaks) to obtain a fingerprinting system that is robust to time scale modifications and pitch shifting. For further pointers to the literature, we refer to the references discussed in [1, 12, 22, 25, 28].

Audio identification is never performed as an end in itself—it always serves a specific purpose, an application (see [2, 3, 14]). Following [22], we want to mention four important application scenarios. A first application is **broadcast monitoring**, where audio identification techniques are used to monitor radio stations, e.g., to ensure that advertisements were actually broadcast, to automatically create playlists based on what has been broadcast, or to compile charts. Audio fingerprinting is also applied for **copyright monitoring**, with the advantage that one can identify copyrighted material without having to rely on filenames, embedded metadata, or watermarks. Such techniques are now an integral part of video-sharing platforms such as YouTube, where users can freely upload video and audio content. A third kind of application is subsumed under the heading **connected audio**. Typically, such an application is able to identify a song played in the environment, e.g., on the radio or in a club, and looks up metadata for the user. To monetize, consumers are then often given the opportunity to purchase the recognized song from one of the major online music stores. Finally, audio fingerprinting techniques are used for **file metadata correction**. Often, when ripped from CD, audio recordings are given semi-anonymous names such as "`01_Track.mp3`" without any additional information. To fix this issue, metadata services identify badly labeled tracks and import or update the missing or wrong metadata.

Audio Matching

While the problem of audio identification can be regarded as largely solved even for vast music collections, less specific retrieval tasks still pose many open problems regarding robustness and scalability. In Section 7.2, we highlighted the differences between high-specific audio identification and mid-specific audio matching. Furthermore, we presented various strategies to cope with musically motivated variations. Instead of using rather abstract features such as spectral peaks, we used chroma features that capture musical (tonal, harmonic, melodic) information. We have already encountered these features in the context of music synchronization (Chapter 3), music structure analysis (Chapter 4), and chord recognition (Chapter 5). To make the features more robust to local temporal fluctuations, we discussed a chroma variant obtained by applying suitable postprocessing techniques, an approach originally introduced in [17]. For a detailed discussion of relevant literature on chroma features, we refer to Section 3.4.

Instead of using sparse peak representations as with audio identification, we employed a subsequence search based on time–chroma representations. By locally comparing a query chromagram with all database chromagrams, we obtained a matching function that indicated the matching subsequences. Diagonal matching (see Section 7.2.2) yields a simple matching procedure that can be computed efficiently but becomes problematic in the presence of tempo differences. To deal with nonlinear tempo differences, we introduced in Section 7.2.3 a subsequence variant of DTW. However, DTW-based approaches become more cumbersome when it comes to indexing and scalability to huge datasets. An alternative approach is to apply a multiple query strategy similar to the tempo-invariant smoothing strategy introduced in Section 4.2.2.2. The idea is to generate multiple versions of a query by applying scaling operations that simulate different tempi, compute a separate matching function for each of the scaled versions using diagonal matching, and then minimize over all resulting matching functions [16].

The matching procedures described in Section 7.2 are exhaustive in the sense that the query is shifted over all database documents. This is not practical for large-scale retrieval scenarios with millions of recordings. Similar to audio fingerprinting, Kurth et al. [15] use inverted file indexing, where the hashes are defined by suitably quantized chroma vectors. However, to deal with the variations, one requires various fault-tolerance mechanisms, which partly undermine the speed-up obtained by this method. As an alternative to considering long feature sequences, one may split up the audio material into small overlapping **shingles** that consist of short feature subsequences. Having a fixed size, the shingles can then be retrieved efficiently using index-based nearest-neighbor search [4, 11]. To cope with temporal variations and to obtain low-dimensional shingles, each shingle may cover only a small portion of the audio material, with the result that queries typically consist of many shingles [4]. Such an approach, as a downside, may lead to a high number of table lookups and to many intermediate results that need to be merged in a post processing step. Increasing the shingle duration (e.g., covering 15 to 25 seconds of the audio) may alleviate this problem but requires a combination of local feature smoothing and global query scaling techniques to handle temporal variations [11]. In this context, feature embedding strategies have been successfully applied to make feature representations more robust to musical variations while significantly reducing the shingle dimension [34].

Version Identification

In Section 7.3, we discussed the task of version identification, which is an instance of document-level retrieval at a lower specificity level compared with audio matching. In version identification, one has to deal not only with changes in instrumentation, tempo, and tonality, but also with more extreme variations concerning the musical structure, key, or melody, as typically occur in remixes and cover songs. This requires document-level similarity measures that globally compare entire documents. Closely following [23, 24], we presented a typical approach for version identification that consists of two ingredients. First, since versions such as cover songs typically share similar harmonic and melodic progressions, most procedures are based on one or another variant of chroma-based feature representations. Second, as we discussed, the global similarity measure is often inferred from locally comparing only parts of the documents—a strategy that allows for dealing with nontrivial structural changes. Therefore, as for the audio matching task, dynamic programming algorithms are a standard choice for dealing with tempo variations, this time applied in a local fashion to identify matching subsequences or local alignments. In particular, inspired by [24, 23], we presented in Section 7.3.2 an approach based on finding possibly long matching subsequences. In [9], beat-synchronous feature representations are used as a strategy to cope with tempo variations. However, as discussed in Section 6.3.3, the required beat tracking step is often error-prone for certain types of music, which may have a negative impact on the final retrieval result. Circumventing the need for an explicit temporal alignment, recent deep neural networks have been applied to learn a nonlinear embedding function that maps a song's high-dimensional feature representation into a low-dimensional metric space. The retrieval of cover songs can then be performed in the embedding space using simple nearest-neighborhood search (see, e.g., [7, 33]). Furthermore, we want to draw attention to the work by Raffel and Ellis [20, 21], which combines traditional alignment and indexing techniques with recent deep learning techniques for content-based matching of MIDI and audio files.

Finally, in Section 7.3.3, we addressed the issue of measuring the performance of version identification and other retrieval systems [23]. In particular, we looked at the general scenario, where one submits a query and obtains a ranked list of matches retrieved from a given collection. We introduced various standard evaluation measures for ranked lists (using notions of precision and recall as introduced in Section 4.5). At this point, we also want to mention that systematic evaluations have been conducted for the task of audio cover song identification in the MIREX (Music Information Retrieval Evaluation eXchange) initiative. For further details, we refer to [8, 23].

Alignment Scenarios

We want to conclude this section by reflecting on the various alignment scenarios that have crossed our way so far. The objective of sequence alignment is to identify

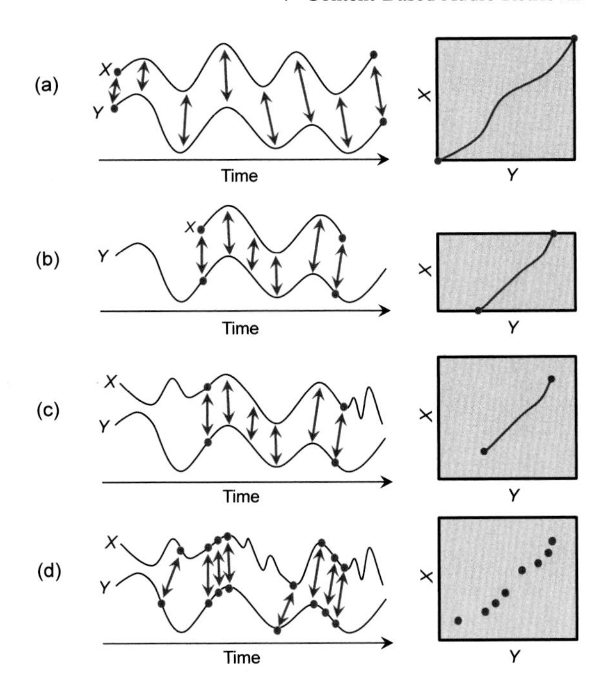

Fig. 7.23 Illustration of different alignment scenarios. **(a)** Global alignment. **(b)** Mixed global/local alignment. **(c)** Local alignment. **(d)** Partial matching.

regions of similarity that are shared by two given sequences $X = (x_1, x_2, \ldots, x_N)$ and $Y = (y_1, y_2, \ldots, y_M)$. At the same time, the elements of the matching regions are brought into correspondence. Generally speaking, one may distinguish between two categories of alignments: **global** alignments and **local** alignments. In the global case, one forces the alignment to span the entire length of the two sequences. By contrast, in the local case, one tries to identify regions of similarity within longer sequences that are often widely divergent overall.

In Section 3.2, we have encountered a first alignment technique known as dynamic time warping, where the alignment was mathematically modeled by the notion of a warping path (see Section 3.2.1.1). The boundary condition in this definition enforced that the two given sequences X and Y were aligned globally, as illustrated by Figure 7.23a. Furthermore, one could adjust the degree of continuity in the alignment by suitably modifying the step size condition.

In the context of audio matching (Section 7.2.3), the two sequences X and Y were treated in a different way. While the sequence X had to be aligned as a whole, only a subsequence of the sequence Y was sufficient as a matching counterpart (see Figure 7.23b). Therefore, this scenario required a mixed global/local alignment approach. Technically, the idea was to relax the boundary conditions for the sequence Y, where a suffix and a prefix could be left out in the alignment without any cost.

In Section 7.3.2, we then encountered a local alignment scenario, where for both sequences X and Y only matching subsequences were to be identified (see Figure 7.23c). To compute the matching, the boundary condition for the sequence X was also dropped. Furthermore, we needed a second technical modification. In the previous two scenarios, having a global constraint at least on the side of the sequence X, the alignment was computed by considering a cost-minimizing path. Now, in the third scenario with no global constraints, such an approach would lead to an empty alignment. Therefore, we introduced a different optimization criterion by looking at a score-maximizing alignment based on a similarity matrix containing cells with a positive score (encoding relevant information) and cells with a negative score (encoding irrelevant information). Of course, the procedure and the threshold used to determine the positive and negative score are crucial for the final optimization result. Instead of using negative score values, one finds in the literature many related approaches that use other kinds of penalty terms to account for the introduction of gaps and other discontinuities [23]. Such an instance is the famous **Smith–Waterman algorithm**, which is widely used for the identification of common molecular subsequences [27].

As discussed in Section 3.4, there are many more approaches for aligning and matching sequences—many of them inspired by applications in bioinformatics to align sequences of DNA, RNA, or proteins. Besides the boundary conditions, other factors, such as the cost (or similarity) measure and the step size condition, crucially influence the alignment result. For example, in the extreme case, one may completely drop the step size condition and replace it with a much weaker monotonicity condition. One such example is the **longest common subsequence** (LCS) problem whose objective is to find the longest subsequence shared by two (or more) sequences (see [6]). A similar problem, referred to as **partial matching**, is discussed in Exercise 7.10 and illustrated by Figure 7.23d.

In all the alignment scenarios we have considered, an optimal solution could be computed efficiently by employing dynamic programming [6]. One fundamental property that is needed for the recursion to work is the monotonicity condition in the alignments. We have seen an example in the context of audio thumbnailing where monotonicity for only one of the sequences still suffices for applying dynamic programming. In Section 4.3.1.2, we described such an approach when computing an optimal path family, where additional steps were introduced to allow for nonmonotonic jumps (from the end to the start) in one of the sequences. For a more unifying description of the various alignment scenarios, we also refer to [10].

7.5 FMP Notebooks

Motivated by content-based audio retrieval tasks, we studied in this chapter fundamental concepts for comparing music documents based on local similarity cues. In particular, we introduced efficient algorithms for globally and locally aligning feature sequences—concepts useful for handling temporal deformations in general time series. In **Part 7** of the FMP notebooks [18], we provide Python implementations of the core algorithms and explain how they work using instructive and explicit toy examples. Furthermore, using real music recordings, we show how the algorithms

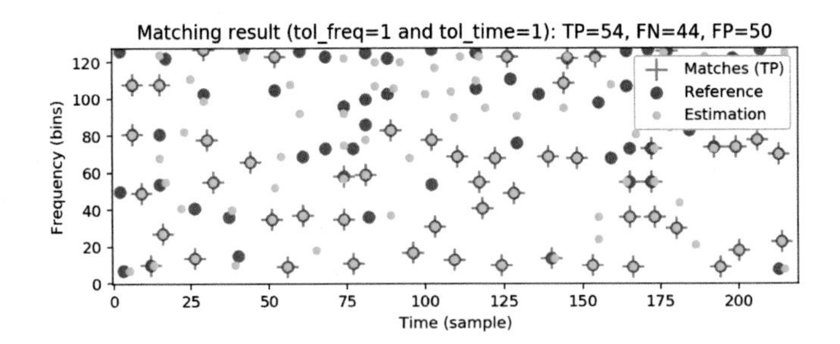

Fig. 7.24 Evaluation measures that indicate the agreement between two constellation maps computed for an original version (Reference) and a noisy version (Estimation).

are used in the respective retrieval application. We close **Part 7** with an implementation and discussion of metrics for evaluating retrieval results given in the form of ranked lists.

While giving a brief outline of the various music retrieval aspects considered in this chapter, the primary purpose of the **FMP Notebook *Content-Based Audio Retrieval*** is to provide concrete music examples that highlight typical variations encountered. In particular, we work out the differences in the objectives of audio identification, audio matching, and version identification by looking at different versions of Beethoven's Fifth Symphony. Furthermore, providing cover song excerpts of the song "Knockin' On Heaven's Door" by Bob Dylan, we indicate some of the most common modifications as they appear in different versions of the original song. In a lecture, we consider it essential to let students listen to, discuss, and find their own music examples, which they can then use as a basis for subsequent experiments.

In the **FMP Notebook *Audio Identification***, we discuss the requirements placed on a fingerprinting system by looking at specific audio examples. In particular, using a short excerpt from the Beatles song "Act Naturally," we provide audio examples with typical distortions a fingerprinting system needs to deal with. Then, we introduce the main ideas of an early fingerprinting approach originally developed by Wang [30] and successfully used in the commercial `Shazam` music identification service. In this system, the fingerprints are based on spectral peaks and the matching is performed using constellation maps that encode peak coordinates (see Figure 7.24). Closely following Section 7.1.2.1, we provide a naive Python implementation for computing a constellation map by iteratively extracting spectral peaks. While being instructive, looping over the frequency and time axis of a 2D spectrogram representation is inefficient in practice—even when using a high-performance Python compiler as provided by the `numba` package. As an alternative, we provide a much faster implementation using 2D filtering techniques from image processing. Comparing running times of different implementations should leave a deep impression on students—an essential experience everyone should have in a computer science lecture. We test the robustness of constellation maps towards signal degradations by considering our Beatles example. To this end, we introduce overlay visualizations of constellation maps for qualitative analysis and information retrieval metrics for quantitative analysis (see Figure 7.24 for an example). Extending the procedure of Section 7.1.2.2, we then provide an implementation of a matching function with additional tolerance parameters to account for small deviations of spectral peak positions. Again we use our modified Beatles excerpts to illustrate the behavior of the matching function under signal distortions. In particular, we demonstrate that the overall fingerprinting procedure is robust to adding noise or other sources while breaking down when changing the signal using time-scale modification or pitch shifting. The concept of indexing (as discussed in Section 7.1.3 and Section 7.1.4) is not covered in our FMP notebooks. For a Python implementation of a full-fledged fingerprinting system, we refer to [26].

Motivated by the audio matching application, the next notebooks provide Python implementations of all the required algorithmic components, closely following Section 7.2. Rather then using abstract features such as spectral peaks, audio matching requires features that capture musical (e.g., tonal, harmonic, melodic) properties. In the **FMP Notebook *Feature Design (Chroma, CENS)***, we consider a family of scalable and robust chroma-related audio features (called CENS), originally proposed in [17]. Using different performances of Beethoven's Fifth Symphony, we study the effects introduced by the quantization, smoothing, normalization, and downsampling operations used in the CENS computation. In the **FMP Notebook *Feature Design (Chroma, CENS)***, we provide a reference implementation of the CENS concept, which yields a family of scalable and robust chroma-related audio features (see Section 7.2.1 and [17]). Using different performances of Beethoven's Fifth Symphony, we study the effects introduced by the quantization, smoothing, normalization, and downsampling operations. The CENS concept can be applied to various chromagram implementations as introduced in the **FMP Notebook *Log-Frequency Spectrogram and Chromagram*** of **Part 3** and the **FMP Notebook *Template-Based Chord Recognition*** of **Part 5**. From an educational viewpoint, these notebooks should make students aware that one may change a feature's properties considerably by applying a little post processing. When trying out some complicated techniques, one should keep an eye on the simple and straightforward approaches (which often yield profound insights into the task and data at hand and may serve as baselines to compare against).

Next, the **FMP Notebook *Diagonal Matching*** provides a step-by-step implementation of the retrieval procedure described in Section 7.2.2. Using a toy example, we discuss the matching function's behavior under signal distortions (including stretching and compressing). We then introduce a function that iteratively extracts local minima under certain neighborhood constraints. In some sense, this procedure can be regarded as a simple peak picking strategy, which should be compared with the more involved alternatives, as discussed in the **FMP Notebook *Peak Picking***. Finally, we cover the multiple-query strategy, where we generate multiple versions of a query by applying scaling operations that simulate different tempi. We illustrate the effect of this procedure by continuing our toy example from above. Providing suitable functions for visualizing the results of all intermediate steps (including feature representations, cost matrices, matching functions, and retrieved matches) is a central feature of the notebook, which allows students to analyze the results and create their own illustrations.

As an alternative to diagonal matching, we studied in Section 7.2.3 a DTW-based matching approach, which is covered in the **FMP Notebook** ***Subsequence DTW***. The Python code of this notebook closely follows the implementation of the original DTW approach (see the **FMP Notebook** ***Dynamic Time Warping (DTW)*** of **Part 3**), which allows students to recognize the differences between the two approaches immediately. Again we draw attention to indexing conventions used in Python (where indexing starts with the index 0) and go through easy-to-understand toy examples. Furthermore, we highlight conceptual differences between the matching functions obtained by diagonal matching and subsequence DTW and discuss their relation to different step size conditions. Finally, we compare our implementation with the one provided by the Python package `librosa` and discuss various parameter settings.

In the **FMP Notebook** ***Audio Matching***, we put the individual components together to create a complete audio matching system. We apply our implementation to several real-world music examples starting with three performances (two orchestral and one piano version) of Beethoven's Fifth Symphony. Then, we consider two performances of the second waltz of Shostakovich's Jazz Suite No. 2, which contains repeating parts with different instrumentation. This example is very instructive when using one of these parts as a query since it illustrates to what extent the matching procedure is capable of identifying the other parts across different performances. We also present an experiment, which shows how the quality of the matching results crucially depends on the length of the query: queries of short duration (having low specificity) will generally lead to a large number of spurious matches, while enlarging the query length (thus increasing its specificity) will generally reduce the number of such matches. Finally, using the song "In the Year 2525" by Zager and Evans, we implement the transposition-invariant matching function and provide a visualization function that reproduces Figure 7.16.

Turning to the task of version identification, we introduce in the **FMP Notebook** ***Common Subsequence Matching*** another sequence alignment variant that drops the boundary condition for both sequences. In our implementation, we follow the same line as with the original DTW and subsequence DTW, thus facilitating an easy comparison of the different algorithms. Furthermore, to round off the alignment topic, we also provide an implementation of the partial matching algorithm, which replaces the step size with a weaker monotonicity condition (see Exercise 7.10). As we discussed in this chapter, the four alignment problems illustrated by Figure 7.23 can all be solved efficiently using dynamic programming. Studying the subtle differences in the algorithmic approaches is an ideal exercise in a computer science curriculum to deepen algorithmic understanding.

In the **FMP Notebook** ***Version Identification***, we present a baseline system that integrates common subsequence matching as a main algorithmic component. We illustrate how the system works by using the original recording and a cover version of the Beatles song "Day Tripper" as input documents. Using chromagrams of the two recordings, we first create a score matrix that encodes potential relations

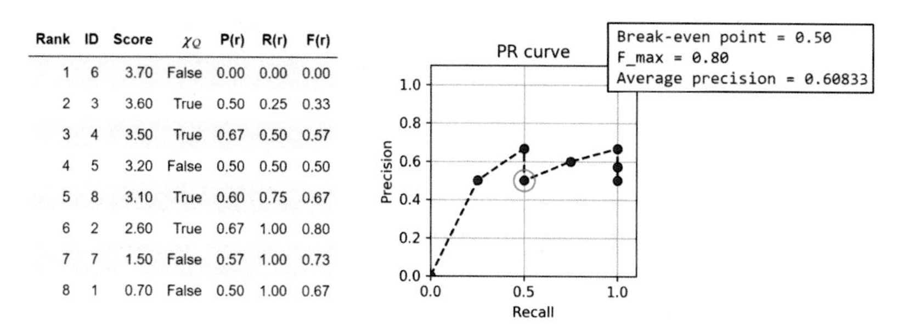

Rank	ID	Score	χ_Q	P(r)	R(r)	F(r)
1	6	3.70	False	0.00	0.00	0.00
2	3	3.60	True	0.50	0.25	0.33
3	4	3.50	True	0.67	0.50	0.57
4	5	3.20	False	0.50	0.50	0.50
5	8	3.10	True	0.60	0.75	0.67
6	2	2.60	True	0.67	1.00	0.80
7	7	1.50	False	0.57	1.00	0.73
8	1	0.70	False	0.50	1.00	0.67

Fig. 7.25 Evaluation metrics for the toy example of Exercise 7.12.

between the two input sequences. In the notebook, we provide an implementation for computing such a score matrix using path-enhancement and thresholding techniques, as introduced in the **FMP Notebook *Audio Thumbnailing*** of **Part 4**. We then apply common subsequence matching for computing a potentially long path of high similarity, show this path for our Beatles example, and provide the audio excerpts that correspond to the two induced segments. The Python functions provided in this notebook may serve as a suitable basis for mini-projects within a music processing curriculum. Understanding the influence of the feature representation, the score matrix, and the matching strategy is crucial before students move on to apply more involved techniques such as supervised deep learning techniques. In particular, listening to the audio excerpts encoded by the alignment path says a lot about the versions' musical relationships.

In the final **FMP Notebook *Evaluation Measures***, we provide an implementation of some evaluation metrics that are useful for document-level retrieval scenarios (see Section 7.3.3). This continues our discussion of general evaluation metrics from the **FMP Notebook *Evaluation*** of **Part 4**. We start with an implementation for computing and visualizing a PR curve and its characteristic points. Then, we turn to the average precision and mean average precision. We test our implementations using toy example, where the evaluation measures can be computed manually. We strongly advise students to perform such sanity checks to verify the correctness of the implementation, but also to deepen the understanding of the measures. The notebook also provides Python scripts (e.g., based on Python data manipulation tools such as `pandas`) that show how one may generate nice-looking tables and figures of evaluation results. One such example is shown in Figure 7.25, which presents the solution of Exercise 7.12.

References

1. B. Agüera y Arcas, B. Gfeller, R. Guo, K. Kilgour, S. Kumar, J. Lyon, J. Odell, M. Ritter, D. Roblek, M. Sharifi, and M. Velimirović, *Now playing: Continuous low-power music recognition*, CoRR, abs/1711.10958 (2017).
2. E. Allamanche, J. Herre, O. Hellmuth, B. Fröba, and M. Cremer, *AudioID: Towards content-based identification of audio material*, in Proceedings of the Audio Engineering Society (AES) Convention, Amsterdam, The Netherlands, 2001.
3. P. Cano, E. Batlle, T. Kalker, and J. Haitsma, *A review of audio fingerprinting*, The Journal of VLSI Signal Processing, 41 (2005), pp. 271–284.
4. M. A. Casey, C. Rhodes, and M. Slaney, *Analysis of minimum distances in high-dimensional musical spaces*, IEEE Transactions on Audio, Speech, and Language Processing, 16 (2008), pp. 1015–1028.
5. M. A. Casey, R. Veltkap, M. Goto, M. Leman, C. Rhodes, and M. Slaney, *Content-based music information retrieval: Current directions and future challenges*, Proceedings of the IEEE, 96 (2008), pp. 668–696.
6. T. H. Cormen, C. E. Leiserson, R. L. Rivest, and C. Stein, *Introduction to Algorithms*, The MIT Press, 3rd ed., 2009.
7. G. Doras and G. Peeters, *A prototypical triplet loss for cover detection*, in Proceedings of the IEEE International Conference on Acoustics, Speech, and Signal Processing (ICASSP), Barcelona, Spain, 2020, pp. 3797–3801.
8. J. S. Downie, M. Bay, A. F. Ehmann, and M. C. Jones, *Audio cover song identification: MIREX 2006–2007 results and analyses*, in Proceedings of the International Society for Music Information Retrieval Conference (ISMIR), Philadelphia, USA, 2008, pp. 468–474.
9. D. P. Ellis and G. E. Poliner, *Identifying 'cover songs' with chroma features and dynamic programming beat tracking*, in Proceedings of the IEEE International Conference on Acoustics, Speech, and Signal Processing (ICASSP), vol. 4, Honolulu, Hawaii, USA, 2007, pp. 1429–1432.
10. S. Ewert, M. Müller, V. Konz, D. Müllensiefen, and G. A. Wiggins, *Towards cross-version harmonic analysis of music*, IEEE Transactions on Multimedia, 14 (2012), pp. 770–782.
11. P. Grosche and M. Müller, *Toward characteristic audio shingles for efficient cross-version music retrieval*, in Proceedings of the IEEE International Conference on Acoustics, Speech, and Signal Processing (ICASSP), Kyoto, Japan, 2012, pp. 473–476.
12. P. Grosche, M. Müller, and J. Serrà, *Audio content-based music retrieval*, in Multimodal Music Processing, M. Müller, M. Goto, and M. Schedl, eds., vol. 3 of Dagstuhl Follow-Ups, Schloss Dagstuhl – Leibniz-Zentrum für Informatik, Dagstuhl, Germany, 2012, pp. 157–174.
13. F. Guillet and H. Hamilton, eds., *Quality Measures in Data Mining*, Springer, 2007.
14. J. Haitsma and T. Kalker, *A highly robust audio fingerprinting system*, in Proceedings of the International Society for Music Information Retrieval Conference (ISMIR), Paris, France, 2002, pp. 107–115.
15. F. Kurth and M. Müller, *Efficient index-based audio matching*, IEEE Transactions on Audio, Speech, and Language Processing, 16 (2008), pp. 382–395.
16. M. Müller, F. Kurth, and M. Clausen, *Audio matching via chroma-based statistical features*, in Proceedings of the International Society for Music Information Retrieval Conference (ISMIR), London, UK, 2005, pp. 288–295.
17. ———, *Chroma-based statistical audio features for audio matching*, in Proceedings of the IEEE Workshop on Applications of Signal Processing (WASPAA), New Paltz, NY, USA, Oct. 2005, pp. 275–278.
18. M. Müller and F. Zalkow, *FMP Notebooks: Educational material for teaching and learning fundamentals of music processing*, in Proceedings of the International Society for Music Information Retrieval Conference (ISMIR), Delft, The Netherlands, 2019, pp. 573–580.

19. J. NAM, K. CHOI, J. LEE, S. CHOU, AND Y. YANG, *Deep learning for audio-based music classification and tagging: Teaching computers to distinguish rock from bach*, IEEE Signal Processing Magazine, 36 (2019), pp. 41–51.
20. C. RAFFEL AND D. P. W. ELLIS, *Large-scale content-based matching of MIDI and audio files*, in Proceedings of the International Society for Music Information Retrieval Conference (ISMIR), Málaga, Spain, 2015, pp. 234–240.
21. ———, *Optimizing DTW-based audio-to-MIDI alignment and matching*, in Proceedings of the IEEE International Conference on Acoustics, Speech, and Signal Processing (ICASSP), Shanghai, China, 2016, pp. 81–85.
22. H. SCHREIBER AND M. MÜLLER, *Accelerating index-based audio identification*, IEEE Transactions on Multimedia, 16 (2014), pp. 1654–1664.
23. J. SERRÀ, *Identification of Versions of the Same Musical Composition by Processing Audio Descriptions*, PhD thesis, Universitat Pompeu Fabra, Barcelona, Spain, 2011.
24. J. SERRÀ, E. GÓMEZ, P. HERRERA, AND X. SERRA, *Chroma binary similarity and local alignment applied to cover song identification*, IEEE Transactions on Audio, Speech, and Language Processing, 16 (2008), pp. 1138–1151.
25. J. SIX, F. BRESSAN, AND M. LEMAN, *A case for reproducibility in MIR: Replication of 'A Highly Robust Audio Fingerprinting System'*, Transactions of the International Society for Music Information Retrieval (TISMIR), 1 (2018), pp. 56–67.
26. J. SIX AND M. LEMAN, *Panako – A scalable acoustic fingerprinting system handling time-scale and pitch modification*, in Proceedings of the International Society for Music Information Retrieval Conference (ISMIR), Taipei, Taiwan, 2014, pp. 259–264.
27. T. F. SMITH AND M. S. WATERMAN, *Identification of common molecular subsequences*, Journal of Molecular Biology, 147 (1981), pp. 195–197.
28. R. SONNLEITNER AND G. WIDMER, *Robust quad-based audio fingerprinting*, IEEE Transactions on Audio, Speech, and Language Processing, 24 (2016), pp. 409–421.
29. G. TZANETAKIS AND P. COOK, *Musical genre classification of audio signals*, IEEE Transactions on Speech and Audio Processing, 10 (2002), pp. 293–302.
30. A. WANG, *An industrial strength audio search algorithm*, in Proceedings of the International Society for Music Information Retrieval Conference (ISMIR), Baltimore, MD, USA, 2003, pp. 7–13.
31. I. H. WITTEN, A. MOFFAT, AND T. C. BELL, *Managing Gigabytes: Compressing and Indexing Documents and Images*, Morgan Kaufmann, 1999.
32. Y.-H. YANG AND H. H. CHEN, *Music Emotion Recognition*, CRC Press, 2011.
33. F. YESILER, J. SERRÀ, AND E. GÓMEZ, *Accurate and scalable version identification using musically-motivated embeddings*, in Proceedings of the IEEE International Conference on Acoustics, Speech, and Signal Processing (ICASSP), Barcelona, Spain, 2020, pp. 21–25.
34. F. ZALKOW AND M. MÜLLER, *Learning low-dimensional embeddings of audio shingles for cross-version retrieval of classical music*, Applied Sciences, 10 (2020).

Exercises

Exercise 7.1. Consider the constellation maps $\mathcal{C}(\mathcal{D})$ (left) and $\mathcal{C}(\mathcal{Q})$ (right) as specified by the figure below. Determine the resulting matching function $\Delta_{\mathcal{C}} : \mathbb{Z} \to \mathbb{N}_0$ as defined in (7.3) by shifting $\mathcal{C}(\mathcal{Q})$ over $\mathcal{C}(\mathcal{D})$ (see Figure 7.5).

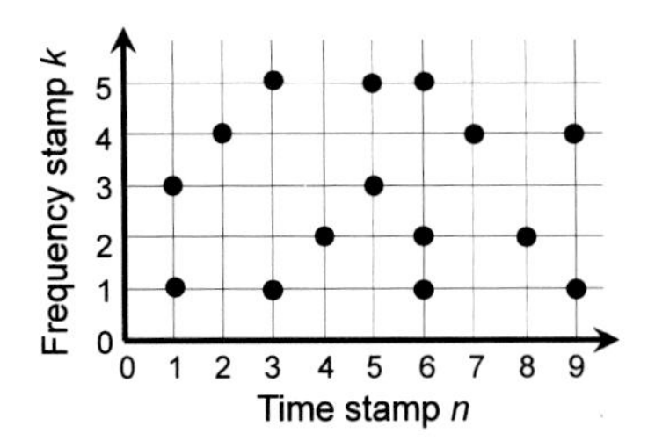

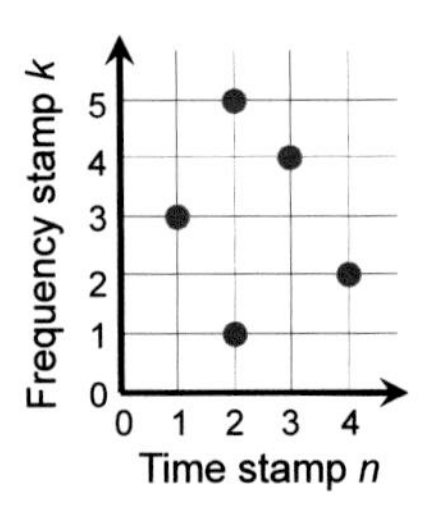

Exercise 7.2. Let $\mathcal{F}(\mathcal{D}) := \mathcal{C}(\mathcal{D})$ and $\mathcal{F}(\mathcal{Q}) := \mathcal{C}(\mathcal{Q})$ be specified as in the figure of Exercise 7.1. Determine the inverted lists and the indicator functions as in Figure 7.6. Then compute the matching function $\Delta_{\mathcal{F}}$ as in (7.8).

Exercise 7.3. In this exercise, we look at the survival probability of a hash that consists of two frequency stamps and a time stamp difference (see Section 7.1.4). Let $p \in [0,1]$ be the probability of a spectral peak surviving in the query audio fragment, and let $F \in \mathbb{N}$ denote the fan-out of the target zone. Assuming that the peak survival probability is independent and identically distributed, show that the joint probability of the anchor point and at least one target point in its target zone surviving is given by (7.16):

$$p \cdot (1-(1-p)^F).$$

Furthermore, compute the number $(1-(1-p)^F)$ for $p \in \{0.1, 0.2, 0.3, 0.4, 0.5\}$ in combination with different $F \in \{1, 5, 10, 20, 40\}$. Discuss the results and the kind of trade-offs involved.

Exercise 7.4. Let $\mathcal{F} = \mathbb{R}$ be a feature space and $c : \mathcal{F} \times \mathcal{F} \to \mathbb{R}_{\geq 0}$ be a local cost measure defined by $c(x,y) = |x-y|$ for $x, y \in \mathbb{R}$ (see also Exercise 3.10). Given the sequences $X = (x_1, \ldots, x_N) = (3,0,6)$ of length $N = 3$ and $Y = (y_1, \ldots, y_M) = (2,4,0,4,0,0,5,2)$ of length $M = 8$, compute the matching function $\Delta_{\mathrm{Diag}} : [0 : M-N] \to \mathbb{R}$ (see (7.20)) as well as the resulting best match (see (7.23)). Furthermore, compute the DTW-based matching function $\Delta_{\mathrm{DTW}} : [1 : M] \to \mathbb{R}$ using the step size set $\Sigma = \{(1,0), (0,1), (1,1)\}$ (see (7.29)) as well as the resulting optimal subsequence $Y(a^* : b^*)$ (see (7.25)).

Exercise 7.5. In this exercise, we show how the matching procedures of Section 7.2 can be applied to a concatenated feature sequence of different recordings, while avoiding matches across different recordings. As in Section 7.2.2, let $X = (x_1, \ldots, x_N)$ be a feature sequence of a query audio fragment. Furthermore, let $Y^i = (y_1^i, \ldots, y_{M_i}^i)$ be feature sequences of length $M_i \geq N$ of two database recordings indexed by $i \in \{1,2\}$. Let Δ_{Diag}^i be the two matching functions obtained by comparing X and Y^i for $i \in \{1,2\}$ (see (7.20)). Next, we concatenate both feature sequences by defining

$$Y := (y_1^1, \ldots, y_{M_1}^1, y^\infty, y_1^2, \ldots, y_{M_2}^2),$$

where y^∞ denotes a feature vector consisting of ∞ entries. Assume that $c(x, y^\infty) := \infty$ for any feature vector x. Furthermore, assume that the sum of the value ∞ with a finite value is defined to be ∞ and that the minimum over a set containing finite values as well as the value ∞ is defined to be the minimum over the finite values (see also Exercise 3.13). Using these calculation rules, let Δ_{Diag} be the matching function obtained by comparing X and Y. Describe the relation between Δ_{Diag}, Δ_{Diag}^1, and Δ_{Diag}^2. What happens in the case that Y is simply defined as the concatenation of Y^1 and Y^2 (without the additional y^∞ vector)?

Similarly, define the matching functions Δ_{DTW}, Δ^1_{DTW}, and Δ^2_{DTW} based on the step size set $\Sigma = \{(1,0),(0,1),(1,1)\}$ (see (7.29)) and discuss their relations. What happens when the step size condition $\Sigma = \{(2,1),(1,2),(1,1)\}$ is used? Describe a strategy to avoid matches across different recordings in this setting.

Exercise 7.6. Let $X = (x_1,x_2,\ldots,x_N)$ and $Y = (y_1,y_2,\ldots,y_M)$ be two feature sequences over the feature space $\mathcal{F}$, and let $c : \mathcal{F}\times\mathcal{F} \to \mathbb{R}$ be a local cost measure. The task of **subsequence DTW** is to determine the subsequence of Y that best matches the sequence X. This subsequence is given by

$$(a^*,b^*) := \underset{(a,b):1\leq a\leq b\leq M}{\operatorname{argmin}} \ \mathrm{DTW}\big(X\,,Y(a:b)\big)$$

(see (7.25)). Following Section 7.2.3, specify the subsequence DTW algorithm (using the step size set $\Sigma = \{(1,0),(0,1),(1,1)\}$) similar to Table 3.2. Given the cost matrix $\mathbf{C}$, the algorithm should output the accumulated cost matrix $\mathbf{D}$, the indices $a^*, b^* \in [1:M]$, as well as an optimal warping path between X and $Y(a^*:b^*)$.

Exercise 7.7. The goal of this exercise is to show how diagonal matching is related to DTW-based matching. Let $X = (x_1,x_2,\ldots,x_N)$ and $Y = (y_1,y_2,\ldots,y_M)$ be two sequences, and let Δ_{Diag} be the matching function based on diagonal matching (see Section 7.2.2). Furthermore, let Δ_{DTW} be the DTW-based matching function using the step size set $\Sigma = \{(1,1)\}$ (instead of using $\Sigma = \{(1,0),(0,1),(1,1)\}$ as in Section 7.2.3). First, describe how the DTW-based procedure needs to be modified when using $\Sigma = \{(1,1)\}$. Then, explain how Δ_{Diag} and Δ_{DTW} are related.

Exercise 7.8. For a sequence $S = (s_1,\ldots,s_L)$, let $\mathrm{Rev}(S) = (r_1,\ldots,r_L)$ with $r_\ell := s_{L-\ell+1}$, $\ell \in [1:L]$ denote the reversed sequence. Now, let $X = (x_1,x_2,\ldots,x_N)$ and $Y = (y_1,y_2,\ldots,y_M)$ be two feature sequences as in Section 7.2.3. Using the step size set $\Sigma = \{(1,0),(0,1),(1,1)\}$, let $\Delta_{\mathrm{DTW}}[X,Y]$ be the DTW-based matching function for X and Y and $\Delta_{\mathrm{DTW}}[\mathrm{Rev}(X),\mathrm{Rev}(Y)]$ be the one for $\mathrm{Rev}(X)$ and $\mathrm{Rev}(Y)$. Assume that the indices

$$(a^*,b^*) := \underset{(a,b):1\leq a\leq b\leq M}{\operatorname{argmin}} \ \mathrm{DTW}\big(X\,,Y(a:b)\big)$$

(see (7.25)) are uniquely determined. In Section 7.2.3, we showed that

$$b^* = \underset{m\in[1:M]}{\operatorname{argmin}} \ \Delta_{\mathrm{DTW}}[X,Y](m),$$

whereas a^* was obtained via backtracking. Show that a^* can also be computed without backtracking using the matching function $\Delta_{\mathrm{DTW}}[\mathrm{Rev}(X),\mathrm{Rev}(Y)]$.
[**Hint:** Study the relation between optimal paths that align X with subsequences of Y and optimal paths that align $\mathrm{Rev}(X)$ with subsequences of $\mathrm{Rev}(Y)$.]

Exercise 7.9. Let $\mathcal{F} = \mathbb{R}$ be a feature space and $s := s^a : \mathcal{F}\times\mathcal{F} \to \mathbb{R}$ a similarity measure defined by $s^a(x,y) := a - |x-y|$ for a constant $a \in \mathbb{R}$ and $x,y \in \mathbb{R}$ (see also Exercise 4.1). Given the sequences $X = (x_1,\ldots,x_N) = (1,0,4,2,1,3,0)$ of length $N = 7$ and $Y = (y_1,\ldots,y_M) = (2,3,1,3,6)$ of length $M = 5$, compute the optimal local alignment (best matching subsequences) of X and Y using the procedure described in Section 7.3.2. To this end, compute the similarity matrix $\mathbf{S}$ (see (7.31)) using $s = s^1$ (i.e., $a = 1$), the accumluated score matrix $\mathbf{D}$ (see (7.33)), the score-maximizing path P^* (see (7.32)), and the two induced segments $\pi_1(P^*)$ and $\pi_2(P^*)$ (see also Figure 7.20).

Then, in the same fashion, compute the optimal local alignment using the similarity measure $s = s^2$ (i.e., $a = 2$). What do you expect when further increasing the number a? Why is it problematic when all entries of $\mathbf{S}$ are positive?

Exercise 7.10. Let $X = (x_1,x_2,\ldots,x_N)$ and $Y = (y_1,y_2,\ldots,y_M)$ be two sequences over the feature space $\mathcal{F}$. A **partial match** of length $L \in \mathbb{N}_0$ between X and Y is defined to be a sequence

$P = ((n_1,m_1),\ldots,(n_L,m_L))$ of cells $(n_\ell,m_\ell) \in [1:N]\times[1:M]$, $\ell \in [1:L]$, which is strictly monotonically increasing:

$$n_1 < n_2 < \ldots < n_L \quad \text{and} \quad m_1 < m_2 < \ldots < m_L.$$

Given a similarity measure $s : \mathcal{F}\times\mathcal{F} \to \mathbb{R}$, define the similarity matrix $\mathbf{S}$ by $\mathbf{S}(n,m) := s(x_n,y_m)$ as in (7.31). Then, the total score $\sigma(P)$ of a partial match P is specified by

$$\sigma(P) := \sum_{\ell=1}^{L} \mathbf{S}(n_\ell,m_\ell).$$

Describe an algorithm based on dynamic programming as in Table 3.2 to compute an optimal (i.e., score-maximizing) partial match.

Exercise 7.11. Show that the definitions of the precision $\mathrm{P}_\mathcal{Q}(r)$ and recall $\mathrm{R}_\mathcal{Q}(r)$ at rank $r \in [1:K]$ in (7.45) and (7.46) agree with the definitions in (4.47) and (4.48), respectively. To this end, depending on r, define a suitable set $\mathcal{I}_+^{\mathrm{Est}}$.

Exercise 7.12. Let us consider a database $\{\mathcal{D}_1,\mathcal{D}_2,\ldots,\mathcal{D}_K\}$ consisting of $K = 8$ documents. Given a query document $\mathcal{Q}$, assume that we have a similarity measure that yields the following values $\gamma(\mathcal{Q},\mathcal{D}_k) \in \mathbb{R}$ for each $k \in [1:K]$:

k	1	2	3	4	5	6	7	8
$\gamma(\mathcal{Q},\mathcal{D}_k)$	0.7	2.6	3.6	3.5	3.2	3.7	1.5	3.1

Furthermore, let $\mathcal{I}_\mathcal{Q} = \{2,3,4,8\}$ be the set of the relevant items (see (7.43)). Calculate the precision $\mathrm{P}_\mathcal{Q}(r)$ and recall $\mathrm{R}_\mathcal{Q}(r)$ at rank $r \in [1:K]$. Furthermore, draw the corresponding precision–recall curve (as in Figure 7.21c). Finally, determine the break-even point, the maximal F-measure $\mathrm{F}_\mathcal{Q}^{\max}$ (see (7.47)), as well as the average precision $\overline{\mathrm{P}}_\mathcal{Q}$ (see (7.48)).

Exercise 7.13. Let us consider a PR curve $\{(\mathrm{P}_\mathcal{Q}(r),\mathrm{R}_\mathcal{Q}(r)) \mid r \in [1:K]\}$ for a ranked retrieval result over K database documents. Recall that the break-even point of the PR curve is the positive value where the precision equals the recall. Show that the break-even point exists if and only if there is at least one relevant document among the top $|\mathcal{I}_\mathcal{Q}|$ items of the ranked list. Furthermore, show that in this case $\mathrm{P}_\mathcal{Q}(r) = \mathrm{R}_\mathcal{Q}(r)$ if and only if $r = |\mathcal{I}_\mathcal{Q}|$.

Exercise 7.14. Show that the maximal F-measure of a PR curve is at least as large as the break-even point (if it exists; see Exercise 7.13). Give an example where the maximal F-measure and the break-even point do not coincide.

Exercise 7.15. Let us consider a database consisting of $K \in \mathbb{N}$ documents. Furthermore, let $\mathcal{Q}$ be a query document with $L := |\mathcal{I}_\mathcal{Q}| \in [1:K]$ relevant items. Assume that the relevant items are ranked by a retrieval system at the positions

$$r_1 < r_2 < \ldots < r_L,$$

where $r_\ell \in [1:K]$ for $\ell \in [1:L]$. (Recall from Section 7.3.3 that, the smaller the index r_ℓ, the higher the rank of the document.) Specify a formula for the average precision $\overline{\mathrm{P}}_\mathcal{Q}$ of this ranking (see (7.48)). Furthermore, assuming $K = 5$ and $L = 2$, calculate the average precision for all possible rankings.

Chapter 8
Musically Informed Audio Decomposition

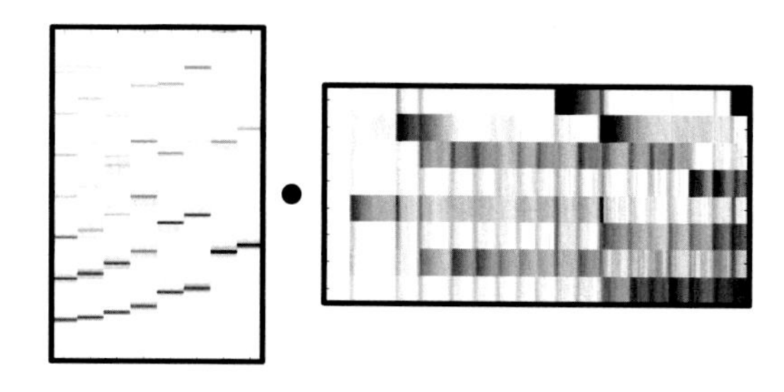

Audio signals are typically complex mixtures of different sound sources. The sound sources can be several people talking simultaneously in a room, different instruments playing together, or a speaker talking in the foreground with music being played in the background. The decomposition of a complex sound mixture into its constituent components is one of the central research topics in digital audio signal processing. Often, this task is referred to as **source separation**. A classical source separation scenario is the so-called **cocktail party problem**, where the objective is to separate the voice of a specific speaker from a mixture of conversations with multiple speakers and background noises.

In the field of music processing, there are many related issues commonly subsumed under the notion of source separation. In music, a source might correspond to a melody, a bass line, a drum track, or a general instrumental voice. For example, when producing a song, a singer, a guitarist, a keyboard player, and a drummer may be recorded separately. In the subsequent step, the resulting individual **tracks** are mixed down to create a cohesive whole. The task of source separation may be thought of as the inverse process. Given a recorded song, the objective is to recover the individual tracks as if they were played in an isolated fashion (see Figure 8.1).

Source separation methods often rely on specific assumptions such as the availability of multiple channels, where several microphones have been used to record the acoustic scene from different directions. Furthermore, the source signals to be identified are assumed to be independent in a statistical way. In music, however, such assumptions are not applicable in many cases. For example, musical sound sources may outnumber the available information channels, such as a string quartet recorded in two-channel stereo. Also, sound sources in music are typically highly correlated in time and frequency. Instruments follow the same rhythmic patterns and play notes which are harmonically related. This makes the separation of mu-

M. Müller, *Fundamentals of Music Processing*, https://doi.org/10.1007/978-3-030-69808-9_8

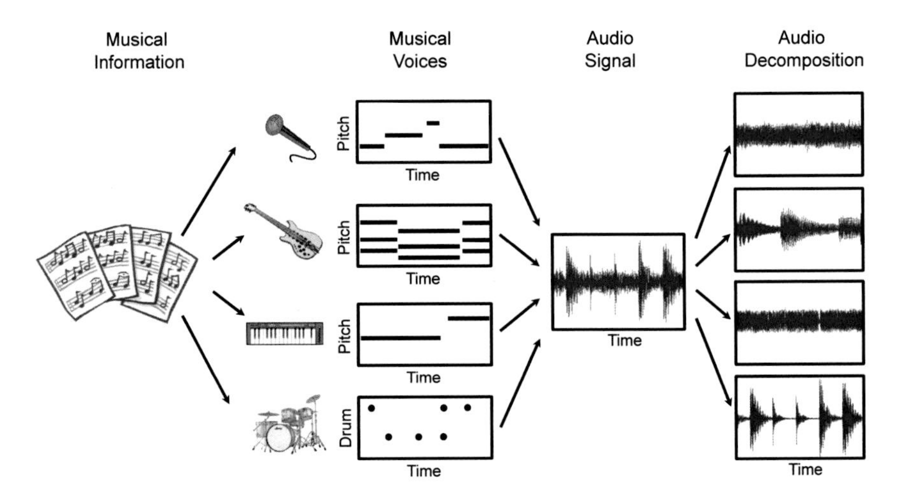

Fig. 8.1 Scenario of a musically informed decomposition of an audio signal into individual instrument tracks.

sical voices from a polyphonic sound mixture an extremely difficult and generally intractable problem [16].

When decomposing a music signal, one strategy is to exploit music-specific properties and additional musical knowledge. For example, the separation problem can be alleviated by exploiting the fact that the melody is often the leading voice, characterized by its dominant dynamics and by its temporal continuity. The track of a bass guitar may be extracted by specifically looking at the lower part of the frequency spectrum. A human singing voice can often be distinguished from other musical sources due to characteristic time–frequency patterns such as vibrato. Furthermore, the drum track may be isolated by exploiting the fact that most of its components are of percussive nature while the other sources are of more harmonic nature. Besides such acoustic cues, **score-informed source separation** strategies make use of the availability of score representations in order to support the separation process. The score provides valuable information in two respects. On the one hand, pitch and timing of note events provide rough guidance within the separation process. On the other hand, the score provides a natural way to specify the target sources to be separated. For example, as illustrated by Figure 8.1, the specification of musical voices in separate staves of the sheet music may be exploited to extract each individual voice from a given audio recording, where the score provides additional cues on the sources' spectral and temporal properties.

In this chapter, we discuss various approaches for decomposing a given music signal into sound sources or, more generally, into semantically meaningful components. In these approaches, we exploit additional information either in the form of specific acoustic properties of the components, or in the form of additional score information. Rather than giving a comprehensive overview of source separation and its many related tasks, we consider three specific scenarios along with some key techniques that are widely used in music processing and beyond.

In Section 8.1, we start with the task of decomposing an audio signal into **harmonic** and **percussive** components. The crucial observation is that harmonic sounds are typically reflected by horizontal structures in a spectrogram representation of the input signal, while percussive sounds form vertical structures. Based on this observation, we show how a spectrogram can be decomposed into two components that correspond to vertical and horizontal structures. We then discuss how the two components can be transformed back to the time domain by applying a kind of **inverse STFT**. This general reconstruction technique also plays an important role in the subsequent sections of this chapter.

In Section 8.2, we deal with the problem of extracting the main melody from a music recording or, to be more precise, with the problem of identifying the fundamental frequency trajectory corresponding to the melody's notes. The underlying assumption is that most notes of the main melody are manifested as **predominant frequency** trajectories over time. In particular, we introduce a generalized log-frequency spectrogram referred to as a **salience representation** along with a number of important techniques including **instantaneous frequency** estimation and **harmonic summation**. The extracted frequency trajectory can be used to decompose the recording into melody and accompaniment tracks—a problem referred to as melody separation.

Finally, in Section 8.3, we discuss a central technique referred to as **nonnegative matrix factorization** (NMF). In the music context, this technique can be used to approximate a magnitude spectrogram by a product of two nonnegative matrices. Intuitively, the first matrix represents the various spectral patterns (musical pitches) that occur in a recording, while the second matrix exhibits the time points where these spectral patterns are active. Within this scenario, we show how to successively incorporate score information to finally yield a **notewise decomposition** of the music signal.

In this final chapter on audio decomposition, our motivation is to present a challenging research direction with many as yet unsolved problems. We discuss a number of key techniques which are useful for a variety of music and multimedia processing tasks beyond source separation. Furthermore, we encounter a number of acoustic and musical properties of audio recordings that have been introduced and discussed in previous chapters, thus rounding off the book nicely.

8.1 Harmonic–Percussive Separation

Musical sounds can comprise a wide range of sound components with different acoustic qualities. As an illustration, let us reconsider the piano example from Figure 2.10 (see Section 2.1.4). In the spectrogram representation, one can observe horizontal lines that are stacked on top of each other. Recall that these horizontal structures correspond to the harmonics, the integer multiples of the fundamental

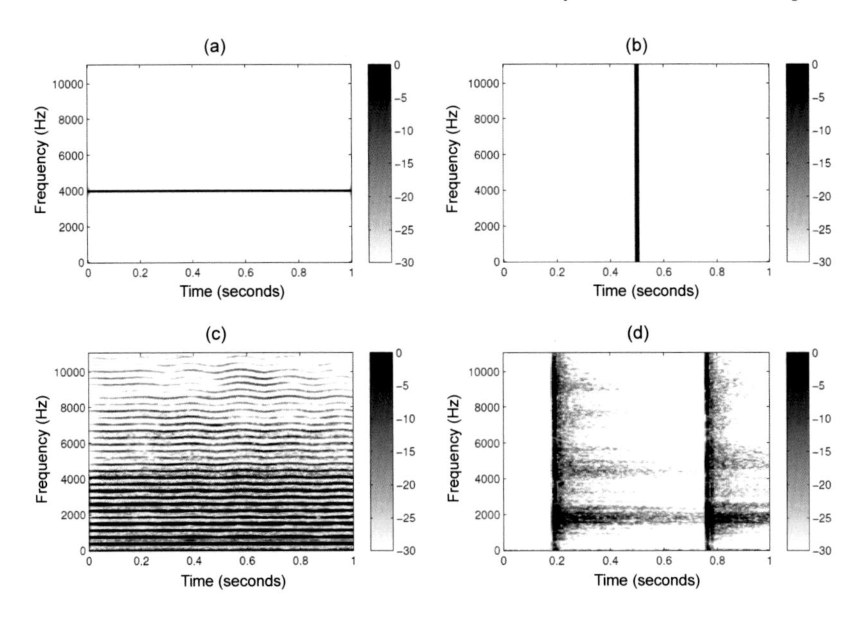

Fig. 8.2 Spectrogram of various signals (magnitudes are given in dB). **(a)** Ideal harmonic signal (sinusoid). **(b)** Ideal percussive signal (impulse). **(c)** Recording of a note played on a violin. **(d)** Recording of two click sounds generated by castanets.

frequency of a played note. Furthermore, one can observe vertical lines at the notes' start time positions. These vertical structures result from noise-like transients that occur in the attack phase of a piano sound.

Motivated by these observations, we consider two types of sound components: harmonic sounds and percussive sounds. Loosely speaking, a **harmonic** sound is what we perceive as pitched sound, what makes us hear melodies and chords. The prototype of a harmonic sound is the acoustic realization of a sinusoid, which corresponds to a horizontal line in a spectrogram representation (see Figure 8.2a). The sound of a violin played without vibrato is another typical example of what we consider a harmonic sound. Again, most of the observed structures in the spectrogram are of horizontal nature even though they are intermingled with noise-like components (see Figure 8.2c). On the other hand, a **percussive** sound is what we perceive as a clash, a knock, a clap, or a click. The prototype of a percussive sound is the acoustic realization of an impulse, which corresponds to a vertical line in a spectrogram representation (see Figure 8.2b). Recall from Section 2.3.3.2 that impulse-like sounds such as a drum stroke or a transient that occurs in the attack phase of a musical tone lead to many nonzero Fourier coefficients that are spread across the entire frequency spectrum. This is also demonstrated by Figure 8.2d, which shows the spectrogram of two click sounds generated by castanets. An important characteristic of such percussive sounds is that they do not have a pitch and are localized in time.

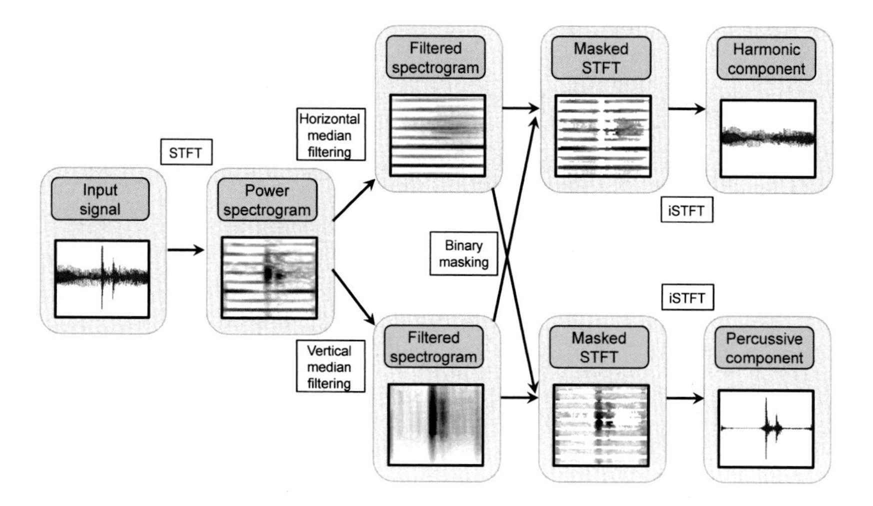

Fig. 8.3 Overview of a procedure for harmonic–percussive separation as suggested in [17].

The goal of **harmonic–percussive separation** (HPS) is to decompose a given audio signal into two parts—one consisting of the harmonic and another of the percussive events. This task is rather vague since it often remains unclear whether a sound event is actually of harmonic or percussive nature. Indeed, there are many sounds such as white noise or applause which are neither harmonic nor percussive.

We now approach this task from a more technical point of view, motivated by the observation that harmonic events tend to form horizontal structures and percussive events tend to form vertical structures in a spectrogram. Closely following the work by Fitzgerald [17], we introduce an HPS procedure as illustrated in Figure 8.3. The idea is to filter a spectrogram representation of the given signal in the horizontal direction (along time) to enhance harmonic events while suppressing percussive ones. Similarly, the spectrogram is filtered in the vertical direction (along frequency) to enhance percussive events while suppressing harmonic ones. The two resulting filtered spectrograms are used to generate time–frequency masks, which are then applied to the original spectrogram. From the masked spectrogram representations, the harmonic and percussive parts of the signal are obtained by applying an inverse STFT. In Section 8.1.1, we explain the mathematical details of this procedure. The issue of reconstructing a time-domain signal (waveform) from a modified STFT representation, which involves some unanticipated pitfalls, is discussed in Section 8.1.2. Finally, in Section 8.1.3, we discuss further examples and applications.

8.1.1 Horizontal–Vertical Spectrogram Decomposition

We now describe the HPS algorithm as proposed in [17] in detail. Let $x : \mathbb{Z} \to \mathbb{R}$ be a discrete-time representation of a sampled audio signal (see (2.19)). The objective is to decompose x into a harmonic component signal $x^{\mathrm{h}} : \mathbb{Z} \to \mathbb{R}$ and a percussive component signal $x^{\mathrm{p}} : \mathbb{Z} \to \mathbb{R}$ such that

$$x = x^{\mathrm{h}} + x^{\mathrm{p}}. \tag{8.1}$$

In the first step, we compute the discrete STFT $\mathcal{X}$ of the signal x (see (2.26) of Section 2.1.4). For convenience, we repeat its definition:

$$\mathcal{X}(n,k) := \sum_{r=0}^{N-1} x(r+nH)w(r)\exp(-2\pi ikr/N), \tag{8.2}$$

where $w : [0 : N-1] \to \mathbb{R}$ is a suitable window function of length N and H is the hop size parameter. To avoid boundary considerations at later stages, we may assume that $n \in \mathbb{Z}$ and $k \in \mathbb{Z}$ by applying a suitable **zero-padding** of the matrix $\mathcal{X}$ in the time as well as frequency direction. From $\mathcal{X}$, we derive the (power) spectrogram $\mathcal{Y}$ as in (2.29):

$$\mathcal{Y}(n,k) := |\mathcal{X}(n,k)|^2. \tag{8.3}$$

8.1.1.1 Median Filtering

In the separation process, harmonic events are regarded as horizontal structures in $\mathcal{Y}$ and percussive events as vertical structures. Let us fix a frequency index k_0 and define the function $\mathcal{Y}^{k_0} : \mathbb{Z} \to \mathbb{R}$ by setting $\mathcal{Y}^{k_0}(n) := \mathcal{Y}(n,k_0)$, $n \in \mathbb{Z}$. Then, a percussive event at some time instance $n_0 \in \mathbb{Z}$ results in a spike of the function $\mathcal{Y}^{k_0}$ at that instance (see Figure 8.4a). Similarly, let us fix a time index n_0 and define a function $\mathcal{Y}_{n_0} : \mathbb{Z} \to \mathbb{R}$ by setting $\mathcal{Y}_{n_0}(k) := \mathcal{Y}(n_0,k)$, $k \in \mathbb{Z}$. Then, a harmonic event of some frequency corresponding to the parameter $k_0 \in \mathbb{Z}$ results in a spike of the function $\mathcal{Y}_{n_0}$ at that position (see Figure 8.4b). In other words, percussive events can be regarded as outliers across time, while harmonic events can be regarded as outliers across frequency. This brings us to the concept of median filtering, which is used to reduce the effect of outliers in general sequences of real numbers. As suggested in [17], we apply these filters in the vertical and horizontal directions to reduce the effect of harmonic and percussive events, respectively.

The **median** of a finite list of numbers is the numerical value with the property that half the numbers fall below the value and half above it. The median can be computed by arranging all the numbers from lowest value to highest value and picking the middle one. If there is an even number of observations, then there is no single middle value; the median is then usually defined to be the mean of the two middle values. For example, the median of the list $(5,3,2,8,2)$ is 3, while the median of the

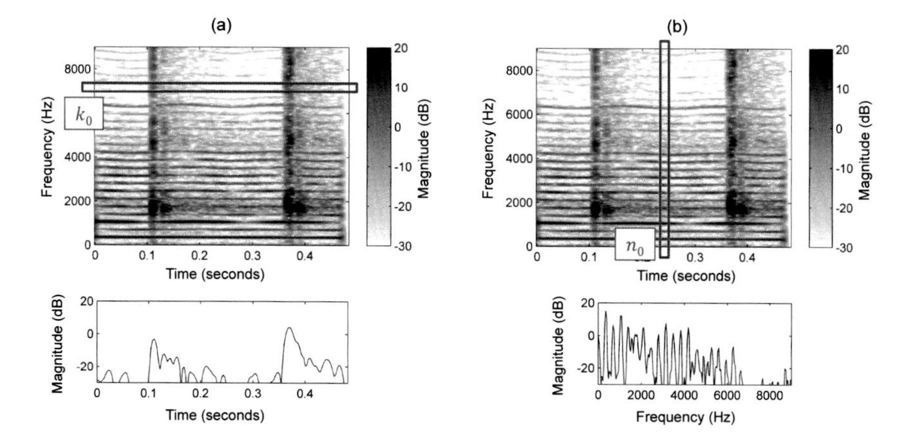

Fig. 8.4 Interpretation of harmonic and percussive components for an audio recording of a note played on a violin superimposed with two click sounds generated by castanets (similar to Figure 8.2). **(a)** Spectrogram $\mathcal{Y}$ and function $\mathcal{Y}^{k_0}$ for some fixed frequency parameter k_0. Percussive events lead to spikes in $\mathcal{Y}^{k_0}$. **(b)** Spectrogram $\mathcal{Y}$ and function $\mathcal{Y}_{n_0}$ for some fixed time parameter n_0. Harmonic events lead to spikes in $\mathcal{Y}_{n_0}$.

list $(5,3,2,8)$ is 4. More formally, let $A = (a_1,a_2,\ldots,a_L)$ be a list of length $L \in \mathbb{N}$ consisting of real numbers $a_\ell \in \mathbb{R}$, $\ell \in [1:L]$. First, the elements of A are sorted in ascending order. This results in a list $\tilde{A} = (\tilde{a}_1,\tilde{a}_2,\ldots,\tilde{a}_L)$ with $\tilde{a}_\ell \leq \tilde{a}_m$ for $\ell < m$, $\ell,m \in [1:L]$. Then, the median $\mu_{1/2}(A)$ of A is defined as

$$\mu_{1/2}(A) := \begin{cases} \tilde{a}_{(L+1)/2}, & \text{for } L \text{ being odd,} \\ (\tilde{a}_{L/2} + \tilde{a}_{L/2+1})/2, & \text{otherwise.} \end{cases} \tag{8.4}$$

The median can be applied in a local fashion to a sequence of real numbers. To this end, one replaces a given element in the sequence by the median defined by the elements that lie in a suitably defined neighborhood of the given element. This leads us to the concept of a **median filter** of length $L \in \mathbb{N}$. Let $A = (a_n \mid n \in \mathbb{Z})$ be a sequence of real numbers $a_n \in \mathbb{R}$ and assume that $L \in \mathbb{N}$ is odd. Then the sequence $\mu^L_{1/2}[A]$ is defined by

$$\mu^L_{1/2}[A](n) = \mu_{1/2}((a_{n-(L-1)/2},\ldots,a_{n+(L-1)/2})). \tag{8.5}$$

For example, consider the sequence $A = (\ldots,0,5,3,2,8,2,0,\ldots)$, where we assume that A is zero outside the shown values. Using $L = 3$, we obtain $\mu^L_{1/2}[A] = (\ldots,0,3,3,3,2,2,0,\ldots)$.

In our scenario, we apply the concept of median filtering to the spectrogram $\mathcal{Y}$ in two ways: once horizontally by considering rows of $\mathcal{Y}$ and once vertically by considering columns of $\mathcal{Y}$. This yields two filtered spectrograms which we denote by $\tilde{\mathcal{Y}}^{\mathrm{h}}$ and $\tilde{\mathcal{Y}}^{\mathrm{p}}$, respectively. More precisely, let L^{h} and L^{p} be odd length parameters,

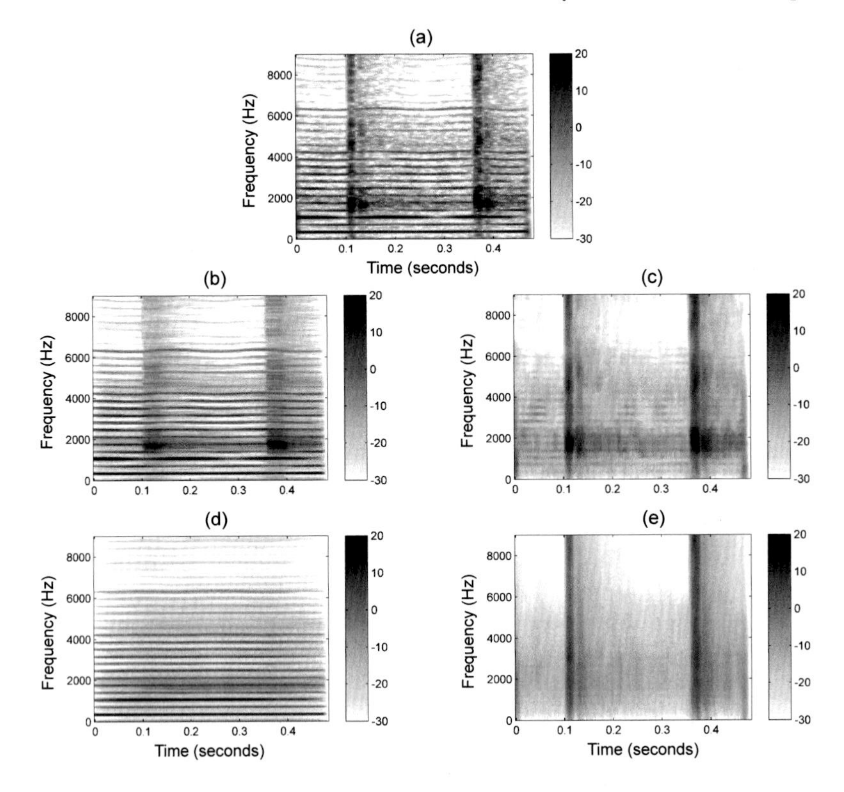

Fig. 8.5 Continuation of the example from Figure 8.4. **(a)** Original spectrogram $\mathcal{Y}$. **(b)** Filtered spectrogram $\tilde{\mathcal{Y}}^{\mathrm{h}}$ using a small length L^{h}. **(c)** Filtered spectrogram $\tilde{\mathcal{Y}}^{\mathrm{p}}$ using a small length L^{p}. **(d)** Filtered spectrogram $\tilde{\mathcal{Y}}^{\mathrm{h}}$ using a large length L^{h}. **(e)** Filtered spectrogram $\tilde{\mathcal{Y}}^{\mathrm{p}}$ using a large length L^{p}.

then we define

$$\tilde{\mathcal{Y}}^{\mathrm{h}}(n,k) := \mu_{1/2}((\mathcal{Y}(n-(L^{\mathrm{h}}-1)/2,k),\ldots,\mathcal{Y}(n+(L^{\mathrm{h}}-1)/2,k))), \quad (8.6)$$

$$\tilde{\mathcal{Y}}^{\mathrm{p}}(n,k) := \mu_{1/2}((\mathcal{Y}(n,k-(L^{\mathrm{p}}-1)/2),\ldots,\mathcal{Y}(n,k+(L^{\mathrm{p}}-1)/2))) \quad (8.7)$$

for $n,k \in \mathbb{Z}$ (assuming some suitable zero-padding of $\mathcal{Y}$).

As an example, let us come back to the audio recording used in Figure 8.4, which consists of a note played by a violin (harmonic component) and two castanet clicks (percussive component). Figure 8.5a shows the original spectrogram of the signal. When applying a median filter in the horizontal direction, the horizontal structures become more apparent, whereas the vertical structures vanish (see Figure 8.5b). Further increasing the median length parameter L^{h}, this effect becomes even stronger (see Figure 8.5d). When applying a median filter in the vertical direction, one obtains similar enhancement effects, this time for the percussive structures (see Figure 8.5c and Figure 8.5e).

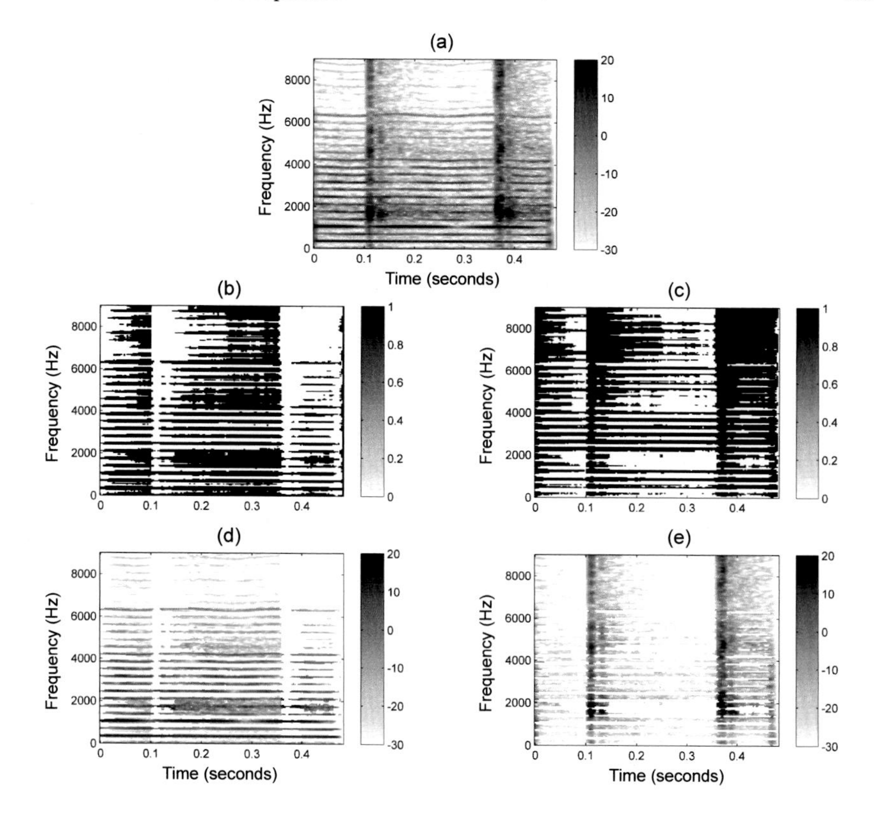

Fig. 8.6 Continuation of the example from Figure 8.5. For computing the masks, the filtered spectrograms from Figure 8.5d and Figure 8.5e are used. **(a)** Original spectrogram $\mathcal{Y}$. **(b)** Binary mask $\mathcal{M}^{\mathrm{h}}$. **(c)** Binary mask $\mathcal{M}^{\mathrm{p}}$. **(d)** Masked spectrogram $\mathcal{Y}^{\mathrm{h}}$. **(e)** Masked spectrogram $\mathcal{Y}^{\mathrm{p}}$.

8.1.1.2 Binary and Soft Masking

The two filtered spectrograms $\tilde{\mathcal{Y}}^{\mathrm{h}}$ and $\tilde{\mathcal{Y}}^{\mathrm{p}}$ are not directly applied for constructing the harmonic and percussive components of the signal. Instead, they are used to first generate two masks which, in turn, are then employed for "punching out" the desired components from the original spectrogram. There are various types of time–frequency masks one may derive from $\tilde{\mathcal{Y}}^{\mathrm{h}}$ and $\tilde{\mathcal{Y}}^{\mathrm{p}}$. The first type is referred to as a **binary mask**, where each time–frequency bin is assigned either the value one or the value zero. In the binary case, we define the two masks by setting

$$\mathcal{M}^{\mathrm{h}}(n,k) := \begin{cases} 1, & \text{if } \tilde{\mathcal{Y}}^{\mathrm{h}}(n,k) \geq \tilde{\mathcal{Y}}^{\mathrm{p}}(n,k), \\ 0, & \text{otherwise,} \end{cases} \tag{8.8}$$

$$\mathcal{M}^{\mathrm{p}}(n,k) := \begin{cases} 1, & \text{if } \tilde{\mathcal{Y}}^{\mathrm{h}}(n,k) < \tilde{\mathcal{Y}}^{\mathrm{p}}(n,k), \\ 0, & \text{otherwise} \end{cases} \tag{8.9}$$

for $n,k \in \mathbb{Z}$. For our running example, the two resulting binary masks are shown in Figure 8.6b and Figure 8.6c, respectively.

Instead of a binary (hard) decision, one can consider a relative weighting when comparing the magnitudes of spectral coefficients. This leads us to another type of mask also known as a **soft mask**. In this case, we define the two masks by setting

$$\mathcal{M}^{\mathrm{h}}(n,k) := \frac{\tilde{\mathcal{Y}}^{\mathrm{h}}(n,k)+\varepsilon/2}{\tilde{\mathcal{Y}}^{\mathrm{h}}(n,k)+\tilde{\mathcal{Y}}^{\mathrm{p}}(n,k)+\varepsilon}, \tag{8.10}$$

$$\mathcal{M}^{\mathrm{p}}(n,k) := \frac{\tilde{\mathcal{Y}}^{\mathrm{p}}(n,k)+\varepsilon/2}{\tilde{\mathcal{Y}}^{\mathrm{h}}(n,k)+\tilde{\mathcal{Y}}^{\mathrm{p}}(n,k)+\varepsilon} \tag{8.11}$$

for $n,k \in \mathbb{Z}$. The small positive value $\varepsilon > 0$ is added to avoid division by zero.

A (binary or soft) time–frequency mask expresses the extent to which each of the time–frequency bins belongs to the respective component. To obtain the component, the mask is applied to the original spectrogram by pointwise multiplication. In the case of the harmonic and percussive masks, this yields two masked versions $\mathcal{Y}^{\mathrm{h}}$ and $\mathcal{Y}^{\mathrm{p}}$ defined by

$$\mathcal{Y}^{\mathrm{h}}(n,k) := \mathcal{M}^{\mathrm{h}}(n,k) \cdot \mathcal{Y}(n,k), \tag{8.12}$$

$$\mathcal{Y}^{\mathrm{p}}(n,k) := \mathcal{M}^{\mathrm{p}}(n,k) \cdot \mathcal{Y}(n,k) \tag{8.13}$$

for $n,k \in \mathbb{Z}$. For an illustration, we refer to Figure 8.6d and Figure 8.6e. In the case of binary masks, a mask value of one preserves the value in the spectrogram, while a mask value of zero suppresses it. In other words, every time–frequency bin of $\mathcal{Y}$ is assigned either to $\mathcal{Y}^{\mathrm{h}}$ or to $\mathcal{Y}^{\mathrm{p}}$. In the case of soft masks, this assignment is not strict but proportionate as expressed by the masking weights. This kind of spectral manipulation is also known as **Wiener filtering**, which is an important concept in statistical digital signal processing. As it goes beyond the scope of the book, we refer to [21] for details on this topic.

We have not yet fully reached our goal. As indicated in (8.1), our objective is to decompose the signal x into a harmonic component signal x^{h} and a percussive component signal x^{p}. So far, we have decomposed the spectrogram $\mathcal{Y}$ of the signal into two components $\mathcal{Y}^{\mathrm{h}}$ and $\mathcal{Y}^{\mathrm{p}}$. The most convenient way to obtain two time-domain signals x^{h} and x^{p} is to apply the two masks directly to the original STFT $\mathcal{X}$, yielding two complex-valued masked STFTs $\mathcal{X}^{\mathrm{h}}$ and $\mathcal{X}^{\mathrm{p}}$:

$$\mathcal{X}^{\mathrm{h}}(n,k) := \mathcal{M}^{\mathrm{h}}(n,k) \cdot \mathcal{X}(n,k), \tag{8.14}$$

$$\mathcal{X}^{\mathrm{p}}(n,k) := \mathcal{M}^{\mathrm{p}}(n,k) \cdot \mathcal{X}(n,k) \tag{8.15}$$

for $n,k \in \mathbb{Z}$. Then, each of these masked STFTs is converted into a time-domain signal by applying an inverse STFT.

Note that this procedure is more problematic than it might seem at first glance. Simply using the same phase information of $\mathcal{X}$ for both components $\mathcal{X}^{\mathrm{h}}$ and $\mathcal{X}^{\mathrm{p}}$ does not account for a possible phase interference between different signal components (see Section 2.3.3.1). In general, the estimation of coherent phase information for different signal components is very hard or even intractable. For convenience, one often transfers the phase information of the mixture signal to the different components even though this is not correct. A second problem arises from the fact that, due to the windowing, the STFT may not be invertible. Only under certain restrictions imposed on the window function and the hop size used can one reconstruct a signal from its STFT. Finally, even in the case that the STFT is invertible, manipulating an STFT (e.g., by applying a mask) may cause problems in the reconstruction of a coherent time-domain signal. In the following section, we discuss the reconstruction problem in more detail.

8.1.2 Signal Reconstruction

In this section, we start with the problem of inverting the STFT and then discuss the case where the STFT has been modified. As usual, let $x:\mathbb{Z}\to\mathbb{R}$ be a discrete-time signal and $\mathcal{X}$ its STFT as computed in (8.2). Furthermore, let $w:[0:N-1]\to\mathbb{R}$ denote the underlying real-valued discrete window function of length $N\in\mathbb{N}$ and $H\in\mathbb{N}$ the hop size parameter. For notational convenience, we extend the window function to $w:\mathbb{Z}\to\mathbb{R}$ by setting $w(r)=0$ for $r\in\mathbb{Z}\setminus[0:N-1]$.

8.1.2.1 Signal Reconstruction from Original STFT

We first show that the signal x can be recovered from its STFT $\mathcal{X}$ under relatively mild conditions on the windowing process. Recall from Section 2.1.4 and Section 2.5.3 that each column of $\mathcal{X}$ is obtained by applying a DFT_N to a windowed section of the original signal x. More precisely, for a fixed frame parameter $n\in\mathbb{Z}$, let $x_n:\mathbb{Z}\to\mathbb{R}$ be the windowed signal defined by

$$x_n(r):=x(r+nH)w(r) \tag{8.16}$$

for $r\in\mathbb{Z}$ (see (2.146)). Then, the STFT coefficients $\mathcal{X}(n,k)$ for $k\in[0:N-1]$ are obtained via

$$(\mathcal{X}(n,0),\ldots,\mathcal{X}(n,N-1))^\top=\mathrm{DFT}_N\cdot(x_n(0),\ldots,x_n(N-1))^\top. \tag{8.17}$$

We have seen that the DFT_N is an invertible matrix with its inverse given by (2.118). Therefore, we can reconstruct the windowed signal x_n from the STFT by

$$(x_n(0),\ldots,x_n(N-1))^\top=\mathrm{DFT}_N^{-1}\cdot(\mathcal{X}(n,0),\ldots,\mathcal{X}(n,N-1))^\top \tag{8.18}$$

and $x_n(r)=0$ for $r\in\mathbb{Z}\setminus[0:N-1]$. To obtain the samples $x(r)$ of the original signal, we have to reverse the windowing process. How, if at all, can this be done? Let us consider the superposition over all suitably shifted versions of windowed sections of the signal:

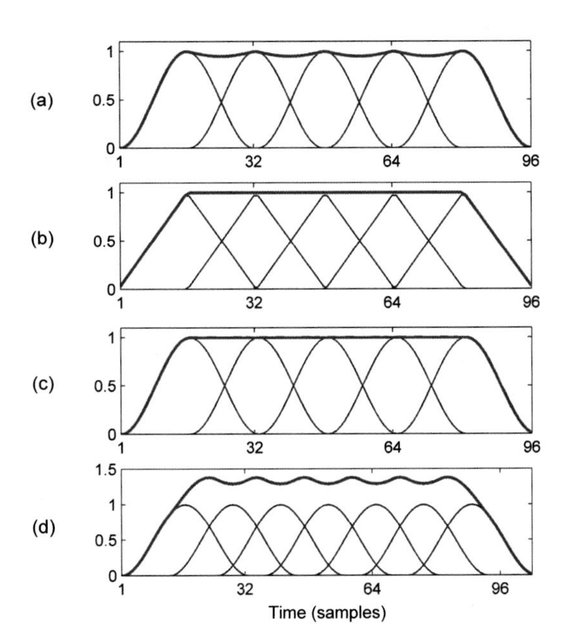

Fig. 8.7 Various window functions of length N and their time-shifted versions using a hop size H. The sum of the shown time-shifted versions is indicated by the thick red curve. **(a)** Hann window with hop size $H = N/2$ (see Figure 2.29b and (2.140)). **(b)** Triangular window with $H = N/2$ (see Figure 2.29b). **(c)** Squared sinusoidal window with $H = N/2$. **(d)** Squared sinusoidal window with $H = 3N/8$.

$$\sum_{n\in\mathbb{Z}} x_n(r-nH) = \sum_{n\in\mathbb{Z}} x(r-nH+nH)w(r-nH) \\ = x(r)\sum_{n\in\mathbb{Z}} w(r-nH). \tag{8.19}$$

Therefore, in the case that

$$\sum_{n\in\mathbb{Z}} w(r-nH) \neq 0 \tag{8.20}$$

for some $r \in \mathbb{Z}$, one can obtain the sample $x(r)$ via

$$x(r) = \frac{\sum_{n\in\mathbb{Z}} x_n(r-nH)}{\sum_{n\in\mathbb{Z}} w(r-nH)}. \tag{8.21}$$

In other words, if the condition (8.20) holds for all $r \in \mathbb{Z}$, one can reconstruct the original signal x from its STFT $\mathcal{X}$. For example, this is the case when the window function $w : [0 : N-1] \to \mathbb{R}$ is strictly positive and the hop size is smaller than or equal to the window length.

The good news is that it is not hard to find window functions along with hop sizes that satisfy the condition (8.20). For example, using the Hann window as defined in (2.140) and a hop size that is smaller than the window length, one can immediately see that the sum over the time-shifted windows is always positive (see also Figure 8.7a). Often, one chooses a window function and a hop size such that the stronger condition

$$\sum_{n\in\mathbb{Z}} w(r-nH) = 1 \tag{8.22}$$

for all $r \in \mathbb{Z}$ is fulfilled. In this case, one also says that the time-shifted window functions define a **partition of unity** of the discrete time axis $\mathbb{Z}$. For example, as illustrated by Figure 8.7b, one obtains such a partition when using the triangular window from Figure 2.29b and a hop size $H = N/2$ of half the window length. Similarly, one obtains a partition of unity when using the window $w : \mathbb{Z} \to \mathbb{R}$ defined by

$$w(r) := \begin{cases} \sin(\pi r/N)^2 & \text{if } r \in [0 : N-1], \\ 0 & \text{otherwise,} \end{cases} \tag{8.23}$$

and a hop size of $H = N/2$ (see Figure 8.7c and Exercise 8.4). As illustrated by Figure 8.7d, the property of being a partition of unity not only depends on the window function itself but also on the hop size parameter.

8.1.2.2 Signal Reconstruction from a Modified STFT

Let us assume that (8.20) holds for all $r \in \mathbb{Z}$, so that the original signal can be reconstructed from its STFT. In many applications, the STFT is manipulated by applying a time–frequency mask as in Section 8.1.1.2. This results in a **modified STFT** (MSTFT). One important question is whether there is a time-domain signal whose STFT coincides with the specified MSTFT. In this case, we say that the MSTFT is **valid**. In practice, however, it turns out that most of the modified STFTs are not valid.

This fact may be surprising at first sight. From the reconstruction described in Section 8.1.2.1, it seems straightforward to apply the following procedure: Assume that $\mathcal{X}^{\mathrm{Mod}}$ is the given MSTFT. In a first step, we apply the **inverse DFT** to each of the columns of $\mathcal{X}^{\mathrm{Mod}}$, yielding

$$(v_n(0), \ldots, v_n(N-1))^\top := \mathrm{DFT}_N^{-1}\left((\mathcal{X}^{\mathrm{Mod}}(n,0), \ldots, \mathcal{X}^{\mathrm{Mod}}(n,N-1))^\top\right) \tag{8.24}$$

for $n \in \mathbb{Z}$ (see (8.18)). Furthermore, we set $v_n(r) := 0$ for $r \in \mathbb{Z} \setminus [0 : N-1]$. Then, based on the overlap–add technique as specified in (8.21), we define a signal $x^{\mathrm{Rec}} : \mathbb{Z} \to \mathbb{R}$ by setting

$$x^{\mathrm{Rec}}(r) := \frac{\sum_{n\in\mathbb{Z}} v_n(r - nH)}{\sum_{n\in\mathbb{Z}} w(r - nH)} \tag{8.25}$$

for $r \in \mathbb{Z}$. Is there something wrong with the signal x^{Rec}? Yes, there is! In general, the STFT $\mathcal{X}^{\mathrm{Rec}}$ of the signal x^{Rec} is not the same as the modified STFT $\mathcal{X}^{\mathrm{Mod}}$. The reason is that, when applying the windowing to x^{Rec} as in (8.16), the resulting windowed sections x_n^{Rec} usually do not agree with the v_n obtained from (8.24).

To understand this better, let us have a look at the example shown in Figure 8.8. We start with a sampled signal $x : \mathbb{Z} \to \mathbb{R}$ which is a sinusoidal function with a spike close to the sample position $n = 32$, where we set $x(32) = 1$ and $x(33) = -1$. In Figure 8.8a, only the samples $x(n)$ for $n \in [1 : 64]$ are shown. Using the window from (8.23) with a length $N = 32$ and hop size $H = N/2 = 16$, we obtain windowed sections x_n for $n \in \mathbb{Z}$ and the resulting STFT $\mathcal{X}$ with coefficients $\mathcal{X}(n,k)$

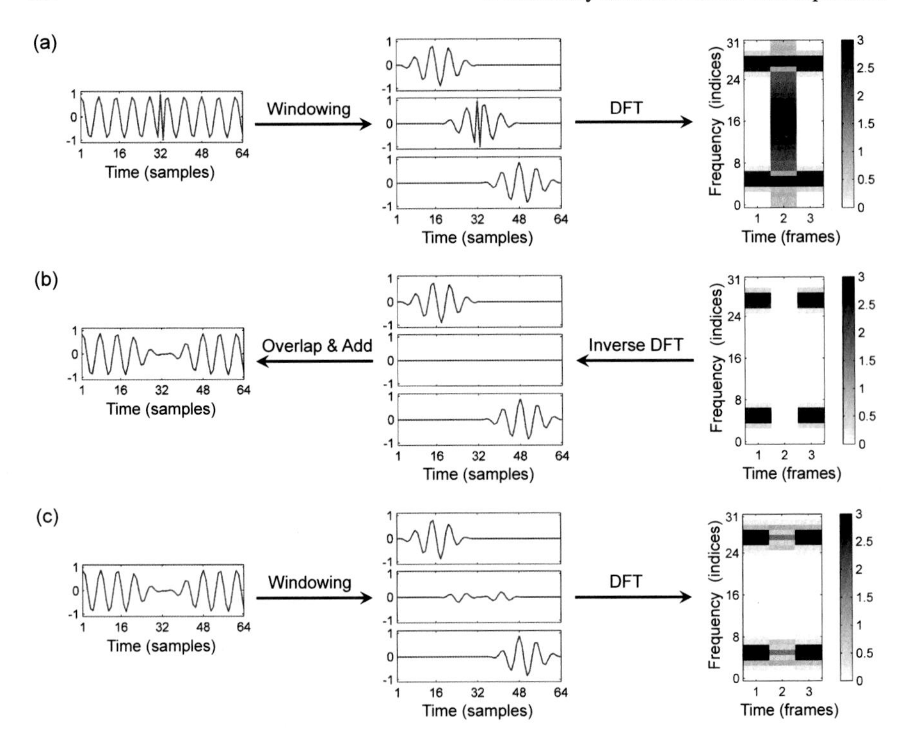

Fig. 8.8 **(a)** Signal x, windowed sections x_n, and magnitude of the STFT $\mathcal{X}$. In the visualization, only the frames for $n \in [1:3]$ are shown. **(b)** Magnitude of modified STFT $\mathcal{X}^{\mathrm{Mod}}$, signals v_n, and signal x^{Rec}. **(c)** Signal x^{Rec}, windowed sections x_n^{Rec}, and magnitude of STFT $\mathcal{X}^{\mathrm{Rec}}$.

for $n \in \mathbb{Z}$ and $k \in [0:N-1]$. Figure 8.8a shows the resulting windowed sections x_n as well as the magnitude STFT for the frames $n \in [1:3]$. The other frames are also computed, but not included in the visualization. The harmonic component of the signal is revealed in the form of horizontal lines across the frames, whereas the impulse-like component falls into the second frame.

Next, let us modify the STFT $\mathcal{X}$. For example, to remove the impulse-like component, we set all coefficients in the second frame $\mathcal{X}(2,k)$ to zero, yielding the modified STFT shown in Figure 8.8b. As described above, we apply the inverse DFT to the MSTFT, resulting in the signals v_n, which are suitably overlaid and added to yield x^{Rec} via (8.25). At this point, we want to make a remark regarding the symmetry properties of the DFT for real-valued signals (see Exercise 2.24). Recall that the signals v_n are real-valued if and only if $\mathcal{X}^{\mathrm{Mod}}(n,0) \in \mathbb{R}$ and $\mathcal{X}^{\mathrm{Mod}}(n,k) = \overline{\mathcal{X}^{\mathrm{Mod}}(n,N-k)}$ for all $k \in [1:N-1]$. In other words, in the case that one wants to obtain a real-valued signal in the reconstruction step, one needs to take care that the symmetry properties are preserved when modifying the STFT.

Using x^{Rec}, we compute the windowed sections x_n^{Rec} and the STFT $\mathcal{X}^{\mathrm{Rec}}$. As illustrated by Figure 8.8c, the STFT $\mathcal{X}^{\mathrm{Rec}}$ does not match the MSTFT $\mathcal{X}^{\mathrm{Mod}}$. This is due to the fact that the time-shifted analysis windows used for computing the STFT overlap with their adjacent windows. For example, computing the second frame of $\mathcal{X}^{\mathrm{Rec}}$ also includes information from the first and third windows. Intuitively speaking, by using the overlap–add procedure of (8.25), the information from the previous and subsequent frames is reintroduced into the current frame. Note that, even though the signals v_n and x_n^{Rec} may be different, the respective sums over these signals yield the same signal x^{Rec}.

Our example has shown that the STFT of the reconstructed signal x^{Rec} may not coincide with the specified MSTFT. One can show that, in general, there exists no time-domain signal that realizes a given MSTFT. Therefore, an important problem is to find ways for estimating a signal whose STFT is at least as close as possible to the MSTFT with regard to a suitably defined distance measure. This topic goes beyond the scope of this book. Following [20], we only outline one of these possible procedures. To measure the distance between a given MSTFT $\mathcal{X}^{\mathrm{Mod}}$ and an STFT $\mathcal{X}'$ of a signal x', we introduce the squared error $\Delta(\mathcal{X}^{\mathrm{Mod}},\mathcal{X}')$ defined by

$$\Delta(\mathcal{X}^{\mathrm{Mod}},\mathcal{X}') := \sum_{n\in\mathbb{Z}} \sum_{k\in[0:N-1]} |\mathcal{X}^{\mathrm{Mod}}(n,k) - \mathcal{X}'(n,k)|^2. \tag{8.26}$$

The objective is to find the signal x^* whose STFT $\mathcal{X}^*$ minimizes this error over all possible signals x':

$$x^* := \underset{x'}{\mathrm{argmin}}\,\Delta(\mathcal{X}^{\mathrm{Mod}},\mathcal{X}'). \tag{8.27}$$

It can be shown that this optimization problem has an explicit solution given by

$$x^*(r) = \frac{\sum_{n\in\mathbb{Z}} w(r-nH)v_n(r-nH)}{\sum_{n\in\mathbb{Z}} w(r-nH)^2}, \tag{8.28}$$

where the signals v_n are defined as in (8.24). Note that this procedure is similar in nature to the overlap–add techniques suggested in (8.25). The major difference is that in (8.28) the signals v_n are windowed with the analysis window before being overlaid and added. Furthermore, the additional windowing is compensated by normalizing with the sum of the squared windows. For further details and a discussion of alternative procedures, we refer to [20].

8.1.3 Applications

In many audio processing tasks, the relevant information lies in either the harmonic or the percussive component of an audio signal. For example, in chord recognition (see Chapter 5) one tries to capture and classify the harmonic properties of an audio signal, while the percussive properties are left unconsidered. Similarly, the presence of percussive components may become problematic when trying to determine the main melody within a polyphonic music recording (see Section 8.2). For other tasks, one may have the reverse situation. For example, when analyzing and classi-

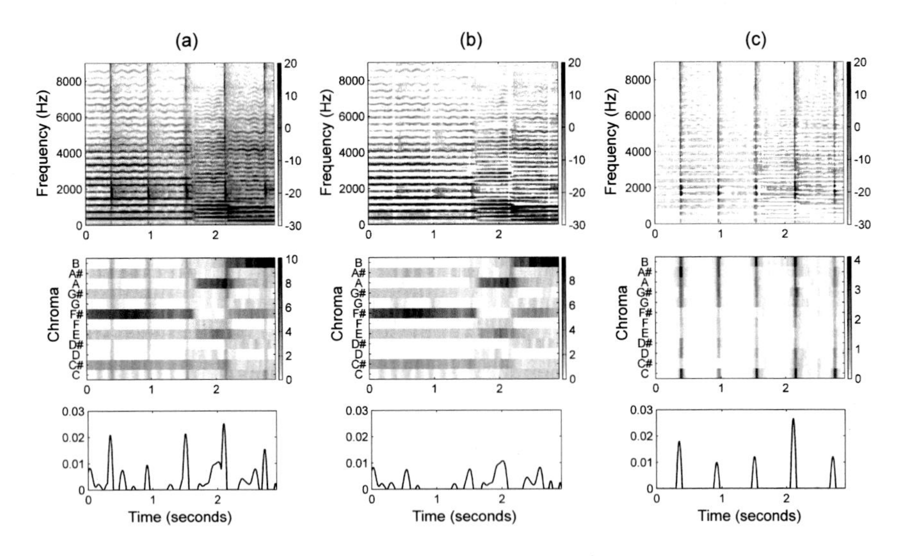

Fig. 8.9 Spectrogram (top), chroma (middle), and novelty representation (bottom) before and after applying an HPS decomposition. **(a)** Representations for the original signal x. **(b)** Representations for the harmonic component x^{h}. **(c)** Representations for the percussive component x^{p}.

fying drum sounds, most of the relevant information is contained in the percussive component of the audio signal. Also the measurement of transient-like phenomena, which is important in the context of onset detection (see Section 6.1), may be alleviated when removing tonal components beforehand. These scenarios indicate that harmonic–percussive separation can be a useful component for tackling a variety of music processing tasks. In the following, we further illustrate this by discussing two concrete examples (see Figure 8.9).

As a first example, let us consider how the output of an HPS procedure can be used to enhance chroma-based audio features. Recall from Section 3.1.2 that chroma features are designed for tasks where melodic and harmonic properties of music become important. Such properties are typically reflected by a concentration of the signal's energy in a small number of chroma bands. When computing chroma-based features, one starts with a decomposition of a spectrogram into pitchwise subbands (see (3.4)). In this decomposition, horizontal structures in the spectrogram lead to high concentration of the signal's energy in a few bands, whereas vertical structures lead to a flat energy distribution. Therefore, one way for improving the chroma representation is to first apply HPS to decompose the signal x into a harmonic component x^{h} and percussive component x^{p}. Then, the chroma features are computed only on the basis of the harmonic component x^{h}.

The effect of such an HPS-based preprocessing step is demonstrated by Figure 8.9, which shows the magnitude STFT (first row) and the resulting chroma representation (second row) for the original signal x, the harmonic component x^{h}, and the percussive component x^{p}. Of course, in this processing pipeline, one does not need to reconstruct the time-domain signal x^{h} prior to computing chroma features. Instead, one can directly use the masked magnitude STFT $\mathcal{Y}^{\mathrm{h}}$ (see (8.12)) for deriving the log-frequency and chroma representations. Also, the HPS step can be easily combined with further enhancement strategies such as logarithmic compression (see Section 3.1.2.1) or quantization and temporal smoothing (see Section 7.2.1). However, note that the various enhancement strategies may influence each other or may serve similar purposes. For example, quantization and temporal smoothing also aim at reducing the influence of percussive components, thus yielding an effect similar to the HPS-based preprocessing.

As a second example, let us consider the task of onset detection. As discussed in Section 6.1, a note onset often goes along with a transient-like sound component that is typically spread over the entire frequency spectrum. Therefore, HPS-based techniques may be used to enhance vertical time–frequency patterns before applying an onset detector. Figure 8.9 demonstrates the effect of such an approach, where a simple energy-based novelty function (see Section 6.1.1) is applied to the original signal as well as to the harmonic component x^{h} and the percussive component x^{p}. Note that other onset detectors based on spectral changes (see Section 6.1.2) may not benefit to the same degree from a prior harmonic–percussive decomposition. One reason is that the computation of spectral changes already involves some enhancement of percussive (vertical) structures by considering columnwise differences of adjacent spectral vectors.

Finally, we want to mention that harmonic–percussive decomposition is problematic for sounds that are neither of clearly harmonic nor of clearly percussive nature. For example, sounds such as applause or a heavily distorted electric guitar are often more or less randomly distributed among the two components. Also, depending on the parameter setting (length of the analysis window, length of the median filter), harmonic sounds may leak into the percussive component and vice versa. Finding suitable parameters often involves a delicate trade-off between leakage in one or the other direction. To cope with this problem, one possible strategy is to introduce a third residual component that captures all sounds that are neither harmonic nor percussive. For such an extension of HPS, we refer to [11] and Exercise 8.5.

8.2 Melody Extraction

In this section, we address a music processing task which is often referred to as **melody extraction**. Given a music recording, the objective is to estimate the sequence of frequency values that correspond to the main melody [38, 45]. Based on this estimation, the goal of **melody separation** is to decompose the music signal into a melody component that captures the main melodic voice and an accompaniment component that captures the remaining acoustic events.

Before we resume a technical viewpoint on melody extraction, let us first approach the concept of a melody from a more musical perspective. When asked to describe a specific song, we are often able to sing or hum the main melody. In

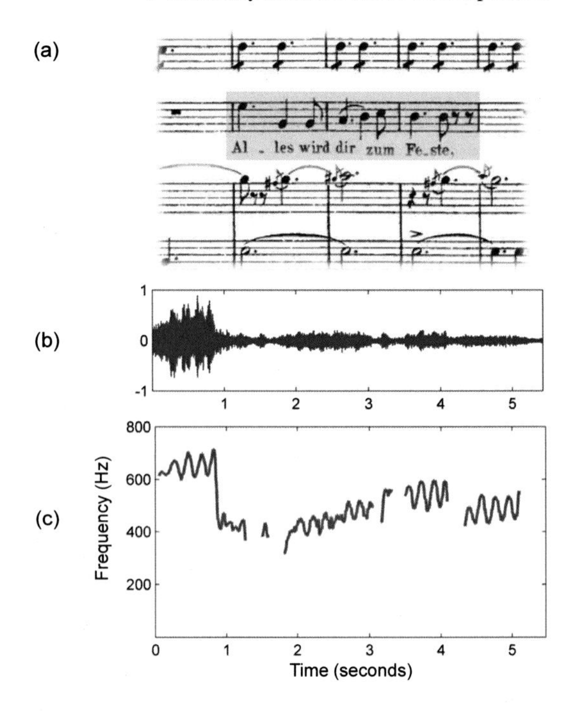

Fig. 8.10 Short excerpt of an aria from the opera "Der Freischütz" by Carl Maria von Weber. The main melody of this excerpt is performed by a soprano singer. **(a)** Sheet music representation. The melody is notated in a separate staff line underlaid with lyrics. **(b)** Waveform of a performance. **(c)** Sequence of fundamental frequency values corresponding to the melody.

general terms, a **melody** may be defined as a linear succession of musical tones expressing a particular musical idea. Because of the special arrangement of tones, a melody is perceived as a coherent entity, which gets stuck in a listener's head as the most memorable element of a song. As the original Greek term *melōidía* (meaning "singing" or "chanting") implies, a melody is often performed by a human voice. Of course, a melody may also be played by other instruments such as a violin in a concerto or a saxophone in a jazz piece. Oftentimes, the melody constitutes the leading element in a composition, appearing in the foreground, while the accompaniment is in the background. Sometimes melody and accompaniment may even be played on a single instrument such as a guitar or a piano. In any case, the melody typically stands out in one way or another. For example, the melody often comprises the higher notes in a musical composition, while the accompaniment consists of the lower notes. Or the melody is played by some instrument with a characteristic timbre (see Section 1.3.4). In some performances, the notes of a melody may feature easily discernible time–frequency patterns such as **vibrato**, **tremolo**, or **glissando** (i.e., a continuous glide from one pitch to another).

As an example, let us consider Figure 8.10, which shows a short excerpt of an opera aria. In the score representation (Figure 8.10a), the main melody is notated in a separate staff line underlaid with lyrics. In a performance by a soprano singer, the melody corresponds to a sequence of fundamental frequency values (Figure 8.10c). As opposed to the notated symbolic representation, some of the notes are smoothly connected. Furthermore, one can observe rather pronounced frequency modulations due to vibrato.

As with many other concepts in music processing, the notion of melody remains rather vague. In this book, we only consider a restricted scenario, making several simplifying assumptions (similar to [38, 45]). First, we consider the scenario where the music is given in the form of an audio recording (and not as a symbolic music representation). Therefore, rather than estimating a sequence of notes, our objective is to determine a sequence of frequency values that correspond to the notes' pitches. Such a frequency path over time, which may also capture continuous frequency glides and modulations, is referred to as a **frequency trajectory**. In particular, we are interested in the **fundamental frequency** values (also called **F0 values**) of the melody's notes (see Section 1.3.4). The resulting trajectory is also called an **F0-trajectory**. In the following, we restrict ourselves to music where the melody is predominantly being performed by a lead singer or a lead instrument. In particular, we assume that there is only one melody line at a time, which can be associated to a single sound source.

Based on these assumptions, our melody extraction problem can be regarded as the following signal processing task (see [45]): Given a recording, our objective is to automatically estimate the sequence of predominant F0-values that correspond to the notes played by the lead voice or instrument. Even this restricted task is more difficult than it may seem. First, in music with many instruments playing simultaneously, it is hard to attribute specific time–frequency patterns to notes of individual instruments. This task becomes even more difficult in the presence of resonance and reverberation effects, which further increase the overlap of different sound sources. Second, even after a successful estimation of fundamental frequencies, one still has to determine which of the F0-values belong to the predominant melody and which are part of the accompaniment.

Some of the challenges in automated melody extraction are illustrated by Figure 8.11, which shows two different spectrogram representations of a song's multitrack recording. In this example, the melody is performed by a male singer, who is accompanied by drums, piano, and guitar. When considering the recording only of the singing voice, the F0-trajectory of the melody is clearly visible (Figure 8.11a). However, note that some of the higher harmonics may contain more energy than the fundamental frequency. The task of finding the melody's F0-trajectory becomes much more challenging when several sound events occur at the same time (Figure 8.11b). In this case, even the automated estimation of the time intervals when the lead singer is active (see Figure 8.11c) becomes a nontrivial subtask [28].

In the following, we describe a typical procedure for predominant F0 estimation following [44]. As in most music processing tasks, the first step is to convert the audio signal into a time–frequency representation using an STFT. In Section 8.2.1, we introduce a technique referred to as **instantaneous frequency estimation**. This technique makes it possible to refine the frequency grid that is introduced by the discrete STFT. The discussion of the instantaneous frequency, which is derived by looking at the phase information, also puts the STFT and its properties in a different

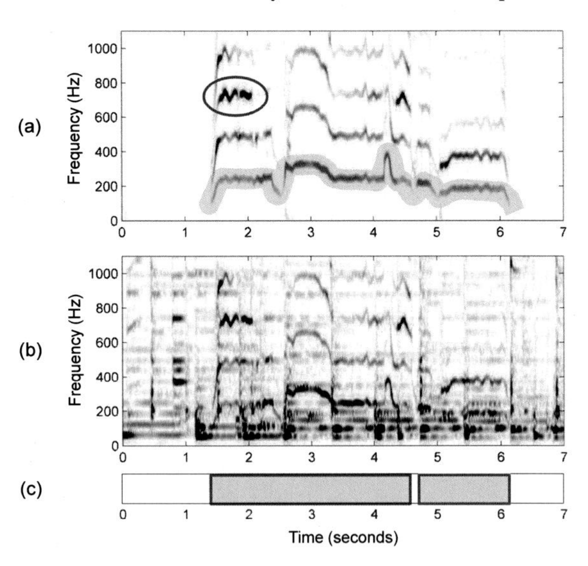

Fig. 8.11 Short excerpt of an audio recording. The song features a male singer (melody) and accompaniment (drums, piano, guitar). **(a)** Spectrogram of the isolated singing voice. The F0-trajectory of the melody is highlighted by the shaded region. The ellipsoid indicates that some higher harmonics may contain more energy than the fundamental frequency. **(b)** Spectrogram of the full recording including the singing voice and accompaniment. **(c)** Time positions where the singer is active.

light. Then, in Section 8.2.2, we show how the refined frequency estimates can be used to derive an improved log-frequency spectrogram. In a further processing step we use a technique called **harmonic summation** to obtain a **salience representation**. This representation is used as a basis for extracting the predominant melody. Simply looking at the maximal F0-value for each frame without additional assumptions may lead to numerous outliers and temporal discontinuities. In Section 8.2.3, we show how to stabilize the extraction process by introducing temporal continuity conditions and by incorporating additional knowledge as specified by a musical score. Finally, we indicate how to use the F0-trajectory in order to decompose a music recording into a melody and an accompaniment track.

8.2.1 Instantaneous Frequency Estimation

Let x denote the given music signal sampled at a rate of F_s Hertz. Furthermore, let $\mathcal{X}$ be its STFT as in (8.2) using a suitable window function of length $N \in \mathbb{N}$ and hop size $H \in \mathbb{N}$. In Section 2.5.3, we discussed the physical interpretation of the resulting Fourier coefficients $\mathcal{X}(n,k)$. The frame index $n \in \mathbb{Z}$ is associated to the physical time

$$T_\mathrm{coef}(n) := \frac{n \cdot H}{F_\mathrm{s}} \tag{8.29}$$

given in seconds (see (2.150)), and the frequency index $k \in [0:N/2]$ corresponds to the frequency

$$F_\mathrm{coef}(k) := \frac{k \cdot F_\mathrm{s}}{N} \tag{8.30}$$

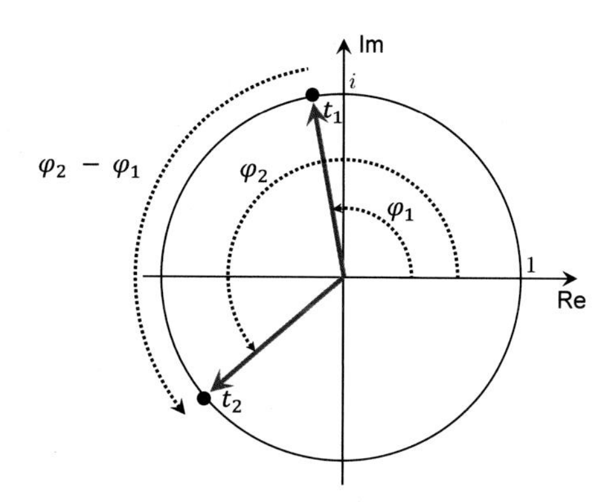

Fig. 8.12 Illustration for interpreting frequency as the quotient of the change $\varphi_2 - \varphi_1$ in angular position and the length $t_2 - t_1$ of the time interval, where φ_1 is the angular position at time t_1 and φ_2 the one at time t_2.

given in Hertz (see (2.149)). In particular, recall that the discrete STFT introduces a linear sampling of the frequency axis. In view of our F0 estimation task, the frequency resolution may not suffice to accurately capture continuous time–frequency patterns caused by vibrato or glissando. Furthermore, because of the logarithmic perception of frequency, the linear sampling of the frequency axis becomes particularly problematic for the low-frequency part of the spectrum. Increasing the frequency resolution by simply increasing the window length N is not a viable solution, since this process decreases the temporal resolution. In the following, we discuss a technique for obtaining an enhanced frequency estimation by exploiting the phase information encoded in the complex-valued STFT.

In order to explain this technique, let us start by recalling the main ideas of expressing and measuring frequency. As our prototypical oscillations, we considered sinusoidal functions each depending on a frequency parameter $\omega \in \mathbb{R}$ and a phase parameter $\varphi \in [0,1)$ (see (2.2)). For a fixed frequency parameter ω, we have seen that a sinusoid of arbitrary phase can be represented as a suitably weighted sum of a cosine and a sine function (see (2.54)). These two functions, in turn, can be regarded as the real and imaginary part of a complex-valued exponential function $\mathbf{exp}_\omega : \mathbb{R} \to \mathbb{C}$ (see (2.90)). Uniformly increasing the time parameter t, the exponential function $t \mapsto \mathbf{exp}_\omega(t)$ describes a circular motion around the unit circle. When projected onto the real and imaginary axes, this yields two sinusoidal motions (described by a cosine and a sine function). Thinking of the circular motion as a uniformly rotating wheel, the frequency parameter ω corresponds to the number of revolutions per unit time (in our case, the duration of one second). In other words, the frequency can be interpreted as the rate of rotation. Based on this interpretation, one can associate a frequency value with a rotating wheel for arbitrary time intervals $[t_1,t_2]$ with $t_1 < t_2$ (see Figure 8.12). To this end, one measures the angular position φ_1 at time t_1 and the angular position φ_2 at time t_2. The frequency is then defined as the change $\varphi_2 - \varphi_1$ in angular position divided by the length $t_2 - t_1$ of the time interval. In the limit case, when the time interval becomes arbitrarily small, one obtains the **instantaneous frequency** ω_{t_1} given by

$$\omega_{t_1} := \lim_{t_2 \to t_1} \frac{\varphi_2 - \varphi_1}{t_2 - t_1}. \tag{8.31}$$

When computing frequency using the change in angular position, one has to take care of **phase wrapping** as discussed in Section 6.1.3. Recall that we specify an angle in the form of the **principal value** of the normalized phase, which results in a number $\varphi \in [0,1)$. When computing phase differences, the choice of the principal value may produce discontinuities. One way to deal with such discontinuities is to use the concept of **phase unwrapping** as illustrated by Figure 6.9 of Section 6.1.3. Another way is to use the **principal argument function** $\Psi : \mathbb{R} \to [-0.5, 0.5]$, which we defined in (6.14). Recall that this function maps phase differences into the range $[-0.5, 0.5]$ by adding or subtracting a suitable integer value. We will apply this function later in (8.33).

What have we gained by looking at the frequency from a phase-based perspective? As said above, the discrete STFT $\mathcal{X}$ analyzes a given signal x at certain frequencies $\omega = F_{\mathrm{coef}}(k)$ which are spaced on a linear grid (see (8.30)). However, the frequencies occurring in the signal x may not exactly correspond to these frequencies. The main idea is to use phase differences across subsequent analysis frames to refine the frequency estimation beyond the frequency grid imposed by the STFT.

For the moment, let us assume a time-continuous perspective, fixing a frequency value $\omega \in \mathbb{R}$ and two time instances, say $t_1 \in \mathbb{R}$ and $t_2 \in \mathbb{R}$. Later, we will choose specific values that are related to the STFT parameters. Correlating the signal x with a windowed version of the analysis function $\mathbf{exp}_\omega$, one positioned at t_1 and one at t_2, we obtain two complex Fourier coefficients. Let φ_1 and φ_2 be the phases of these two coefficients, respectively. In the case that the signal x contains a strong frequency component of frequency ω, the two phases φ_1 and φ_2 should be consistent in the following way: A rotation of frequency ω that assumes the angular position φ_1 at time position t_1 should have the phase

$$\varphi^{\mathrm{Pred}} := \varphi_1 + \omega \cdot \Delta t, \tag{8.32}$$

where $\Delta t := t_2 - t_1$. Therefore, in the case that the signal x behaves similarly to the function $\mathbf{exp}_\omega$, one should have $\varphi_2 \approx \varphi^{\mathrm{Pred}}$. This case is illustrated by Figure 8.13a. However, what happens if the signal x oscillates, e.g., slightly slower than $\mathbf{exp}_\omega$ as illustrated by Figure 8.13b? In this case, the phase increment from time instance t_1 to instance t_2 for the signal x is less than the one for the prototype oscillation $\mathbf{exp}_\omega$. As a result, the phase φ_2 measured at t_2 is less than the predicted phase φ^{Pred}. Conversely, if x oscillates slightly faster than $\mathbf{exp}_\omega$, the phase φ_2 is larger than the predicted phase φ^{Pred} (see Figure 8.13c). To measure the difference between φ_2 and φ^{Pred}, we introduce the prediction error defined by

$$\varphi^{\mathrm{Err}} := \Psi(\varphi_2 - \varphi^{\mathrm{Pred}}). \tag{8.33}$$

The principal argument function Ψ (see (6.14)) ensures that the difference lies within the range $[-0.5, 0.5]$. The prediction error can be used to correct the frequency value ω to obtain a refined frequency estimate $\mathrm{IF}(\omega)$ for the signal x:

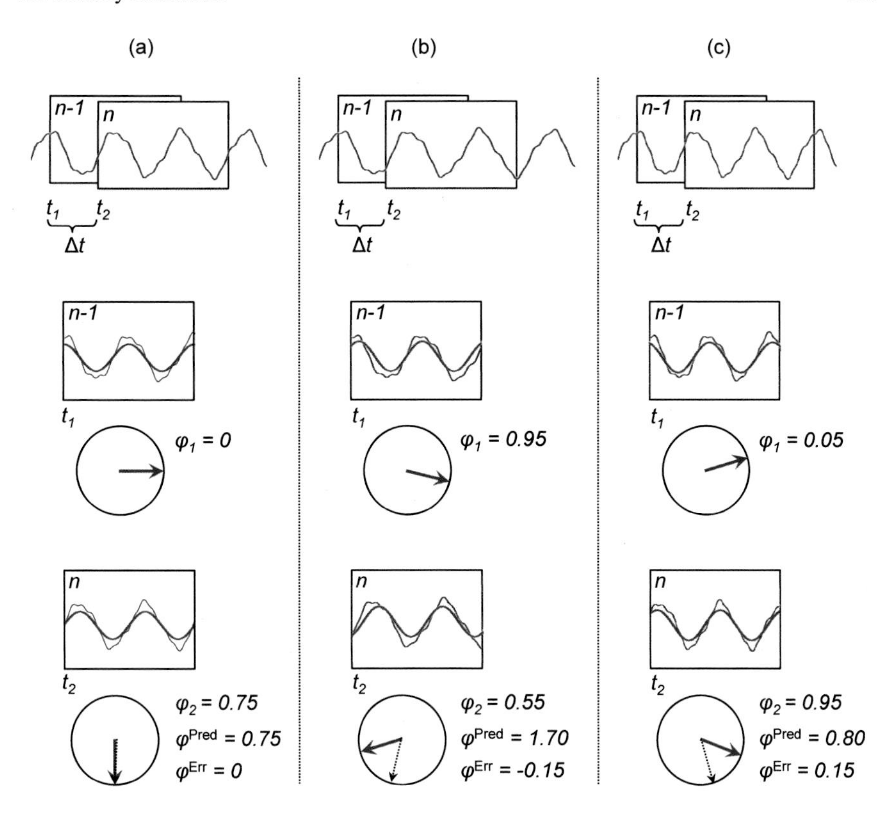

Fig. 8.13 Illustration for computing the instantaneous frequency for three different signals. **(a)** Signal x having the same frequency as the prototype sinusoid. **(b)** Signal x with a slightly lower frequency. **(c)** Signal x with a slightly higher frequency. From top to bottom, each of the subfigures shows the signal x (with an indication of two windowed sections corresponding to frames $n-1$ and n), the correlation between a prototype sinusoid and x at time instance t_1 and the measured phase φ_1, as well as the correlation between a prototype sinusoid and x at time instance t_2 and the measured phase φ_2. Furthermore, the predicted phase φ^{Pred} and the prediction error φ^{Err} are indicated.

$$\mathrm{IF}(\omega) := \omega + \frac{\varphi^{\mathrm{Err}}}{\Delta t}. \tag{8.34}$$

This value is also called the **instantaneous frequency** (IF) at ω. Strictly speaking, rather than referring to a single time instance, the instantaneous frequency refers—in this case—to an entire time interval $[t_1, t_2]$. In practice, however, this interval is typically chosen to be very small (on the order of a couple of milliseconds).

Let us come back to the examples shown in Figure 8.13 assuming that $t_1 = 0$ sec, $t_2 = 0.75$ sec, and $\omega = 1$ Hz. In Figure 8.13a, one measures $\varphi_1 = 0$ and $\varphi_2 = 0.75$. In this case, the predicted phase

$$\varphi^{\mathrm{Pred}} = \varphi_1 + \omega \cdot \Delta t = 0 + 1 \cdot 0.75 = 0.75 \tag{8.35}$$

coincides with φ_2 so that $\varphi^{\mathrm{Err}} = 0$ and $\mathrm{IF}(\omega) = \omega$. In the case of Figure 8.13b, one measures $\varphi_1 = 0.95$ and $\varphi_2 = 0.55$. On the other side, one obtains

$$\varphi^{\mathrm{Pred}} = \varphi_1 + \omega \cdot \Delta t = 0.95 + 1 \cdot 0.75 = 1.7. \tag{8.36}$$

This yields a prediction error of

$$\varphi^{\mathrm{Err}} := \Psi(0.55 - 1.7) = -0.15. \tag{8.37}$$

From this, we get the instantaneous frequency

$$\mathrm{IF}(\omega) = \omega + \frac{\varphi^{\mathrm{Err}}}{\Delta t} = 1 - \frac{0.15}{0.75} = 0.8, \tag{8.38}$$

which reflects the actual signal's frequency content well. Similarly, in the case of Figure 8.13c, one obtains the instantaneous frequency $\mathrm{IF}(\omega) = 1.2$ Hz. Note that the refinement of the frequency only works if the signal x contains a main frequency component close to ω. Furthermore, the time interval $[t_1 : t_2]$ should be small so that the difference between the unwrapped predicted phase and the unwrapped measured phase lies within the interval $[-0.5, 0.5]$.

We now apply the concept of instantaneous frequency for improving the frequency resolution of a discrete STFT. Using the polar coordinate representation (2.9), a Fourier coefficient $\mathcal{X}(n,k) \in \mathbb{C}$ can be written as

$$\mathcal{X}(n,k) = |\mathcal{X}(n,k)| \exp(2\pi i \varphi(n,k)) \tag{8.39}$$

with the phase $\varphi(n,k) \in [0,1)$. Recall that we have already used such a representation in the context of phase-based novelty detection (see (6.9) of Section 6.1.3). For the prototype oscillation, we use the frequency determined by the frequency parameter $k \in [0 : N/2]$ (see (8.30)):

$$\omega = F_{\mathrm{coef}}(k) = \frac{k \cdot F_{\mathrm{s}}}{N}. \tag{8.40}$$

Furthermore, the two time instances are determined by the positions of the previous frame and the current frame (see (8.29)):

$$t_1 = T_{\mathrm{coef}}(n-1) = \frac{(n-1) \cdot H}{F_{\mathrm{s}}} \quad \text{and} \quad t_2 = T_{\mathrm{coef}}(n) = \frac{n \cdot H}{F_{\mathrm{s}}}. \tag{8.41}$$

Finally, the measured phases at these time instances are the ones obtained by the STFT:

$$\varphi_1 = \varphi(n-1,k) \quad \text{and} \quad \varphi_2 = \varphi(n,k). \tag{8.42}$$

From this, we obtain the instantaneous frequency

$$F^{\mathrm{IF}}_{\mathrm{coef}}(k,n) := \mathrm{IF}(\omega) \tag{8.43}$$

as in (8.34). Using the above equations, one can easily derive the following formula for $F^{\mathrm{IF}}_{\mathrm{coef}}(k,n)$, which only depends on the measured phases as well as on the window length N, the hop size H, and the sampling rate F_{s} (see Exercise 8.6):

$$F^{\mathrm{IF}}_{\mathrm{coef}}(k,n) = (k+\kappa(k,n))\cdot\frac{F_{\mathrm{s}}}{N}, \tag{8.44}$$

where the **bin offset** $\kappa(k,n)$ is calculated as

$$\kappa(k,n) = \frac{N}{H}\cdot\Psi\left(\varphi(n,k)-\varphi(n-1,k)-\frac{k\cdot H}{N}\right). \tag{8.45}$$

Note that the quality of the estimated instantaneous frequency depends, among other parameters, on the length $\Delta t = t_2 - t_1 = H/F_{\mathrm{s}}$. Therefore, when applied to the discrete STFT, it is beneficial to use a small hop size H. On the downside, using a small hop size increases the computational cost for calculating the discrete STFT (see Exercise 8.7 for an alternative approach). In the following section, we will see some examples that illustrate the effect of using the instantaneous frequency instead of simply using the frequencies on the grid introduced by the discrete STFT (see Figure 8.14).

8.2.2 Salience Representation

For an accurate estimation of the predominant F0-values, we now derive a log-frequency spectrogram similar to the one introduced in Section 3.1.1. This time, however, we need a resolution that goes beyond the 128 pitch bands corresponding to the equal-tempered scale. Exploiting the instantaneous frequency, we show how the log-frequency spectrogram can be improved and refined, in particular in the low-frequency part of the spectrum.

8.2.2.1 Refined Log-Frequency Spectrogram

As preparation, let us recall the log-frequency spectrogram from Section 3.1.1. The idea was to pool the STFT coefficients by regarding the sets

$$P(p) = \{k : F_{\mathrm{pitch}}(p-0.5) \leq F_{\mathrm{coef}}(k) < F_{\mathrm{pitch}}(p+0.5)\} \tag{8.46}$$

for the pitch parameters $p\in[0:127]$ (see (3.3)). Instead of fixing a pitch and looking for all frequencies that lie in the resulting pitch band, one can also define a mapping $\mathrm{Bin}:\mathbb{R}\to\mathbb{Z}$ which assigns to a given frequency the corresponding pitch index. Indeed, it is not hard to prove (see Exercise 8.8) that

$$\mathrm{Bin}(\omega) := \left\lfloor 12\cdot\log_2\left(\frac{\omega}{440}\right)+69.5\right\rfloor \tag{8.47}$$

assigns to a given frequency $\omega \in \mathbb{R}$ the pitch $p := \mathrm{Bin}(\omega) \in \mathbb{Z}$ such that $\omega \in [F_{\mathrm{pitch}}(p-0.5), F_{\mathrm{pitch}}(p+0.5))$. From this, one obtains

$$P(p) := \{k : \mathrm{Bin}(F_{\mathrm{coef}}(k)) = p\}. \tag{8.48}$$

Each set $P(p)$ can be thought of as a bin, while $p = \mathrm{Bin}(\omega)$ yields the bin index associated to frequency ω. We now extend (8.47) by considering a more general bin assignment. To this end, let $\omega_{\mathrm{ref}} \in \mathbb{R}$ be a reference frequency which is to be assigned to the bin index 1. Furthermore, let $R \in \mathbb{R}$ (given in cents, see Section 1.3.2) be the desired resolution of the logarithmically spaced frequency axis. Then, for a frequency $\omega \in \mathbb{R}$ (given in Hertz), the bin index $\mathrm{Bin}(\omega)$ is defined as

$$\mathrm{Bin}(\omega) := \left\lfloor \frac{1200}{R} \cdot \log_2\left(\frac{\omega}{\omega_{\mathrm{ref}}}\right) + 1.5 \right\rfloor. \tag{8.49}$$

For example, $R = 100$ yields a subdivision of the frequency axis with a resolution of 100 cents (one semitone) per bin, which is the same resolution as in (8.47). Using $R = 10$ results in a finer subdivision of the frequency axis, where each bin corresponds to 10 cents (a tenth of a semitone).

Based on the bin mapping function (8.49), we now extend the definition (3.4) of the log-frequency spectrogram. Fixing a reference frequency ω_{ref} and a resolution R, let $B \in \mathbb{N}$ be the number of bins to be considered. For each bin index $b \in [1:B]$, we then define the set

$$P(b) := \{k : \mathrm{Bin}\left(F_{\mathrm{coef}}(k)\right) = b\}. \tag{8.50}$$

Furthermore, we set

$$\mathcal{Y}_{\mathrm{LF}}(n,b) := \sum_{k \in P(b)} |\mathcal{X}(n,k)|^2 \tag{8.51}$$

for each frame index $n \in \mathbb{Z}$ and bin index $b \in [1:B]$.

As an illustration, let us revisit the Weber excerpt from Figure 8.10, which also serves as our running example in the subsequent explanations. Figure 8.14a shows a spectrogram representation $\mathcal{Y}$ of the excerpt for frequencies up to 4000 Hz. In addition, to the right of the spectrogram, we provide a zoomed-in version which only covers the first second of audio between 250 Hz and 1000 Hz. Let us fix a reference frequency of $\omega_{\mathrm{ref}} = 100$ Hz and a resolution of $R = 10$ cents. In the following, we consider a maximal bin index B that corresponds to $\omega = 4000$ Hz, i.e., $B = \mathrm{Bin}(4000) = 640$. The resulting log-frequency spectrogram is shown in Figure 8.14b. Note that, in the visualizations, the labeling of the frequency axis is specified in Hertz rather than in bin indices. As demonstrated by the example, the linearly spaced frequency information in $\mathcal{Y}$ is expanded in a nonlinear, logarithmic fashion. This results in interpolation artifacts in the frequency direction, leading to a rather blurred representation in the lower part of $\mathcal{Y}_{\mathrm{LF}}$.

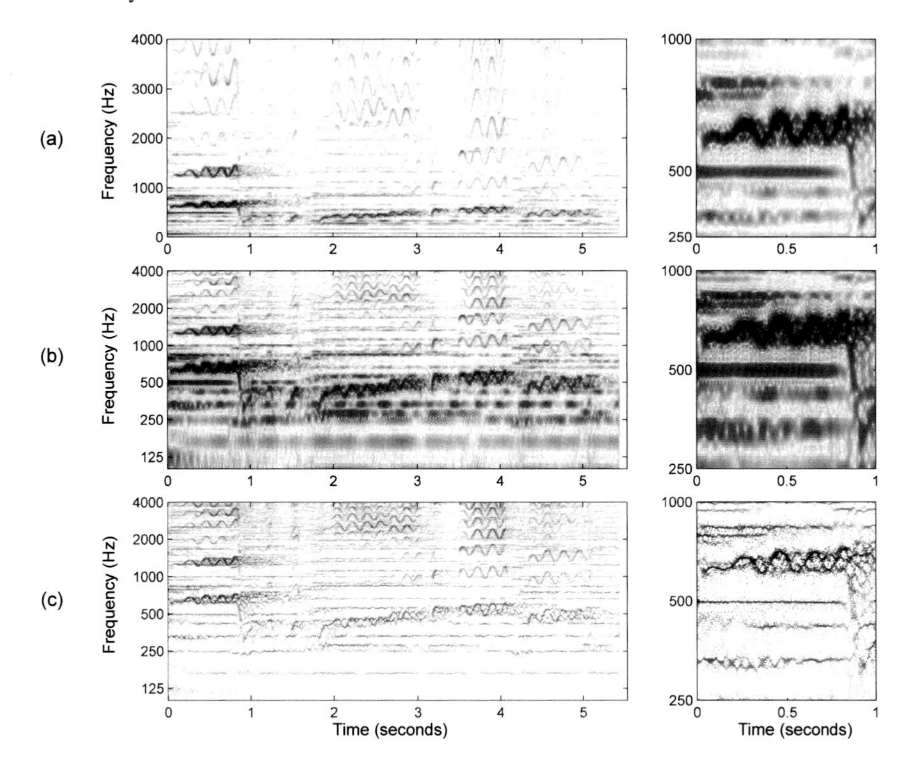

Fig. 8.14 Various spectrogram representations for a recording of a short excerpt of an aria from the opera "Der Freischütz" by Carl Maria von Weber (continuing the example from Figure 8.10). **(a)** Original spectrogram $\mathcal{Y}$ (left) and zoom into a section (right). **(b)** Log-frequency spectrogram $\mathcal{Y}_{\mathrm{LF}}$ using $R = 10$ and $\omega_{\mathrm{ref}} = 100$ Hz. **(c)** Enhanced log-frequency spectrogram $\mathcal{Y}_{\mathrm{LF}}^{\mathrm{IF}}$ employing the instantaneous frequency.

8.2.2.2 Using Instantaneous Frequency

We now discuss how this problem can be alleviated by using the instantaneous frequency as introduced in Section 8.2.1. Instead of taking the center frequencies $F_{\mathrm{coef}}(k)$, the idea is to employ the refined frequency estimates $F_{\mathrm{coef}}^{\mathrm{IF}}(k,n)$ (see (8.44)) for defining the sets

$$P^{\mathrm{IF}}(b,n) := \left\{k : \mathrm{Bin}\left(F_{\mathrm{coef}}^{\mathrm{IF}}(k,n)\right) = b\right\} \tag{8.52}$$

for $b \in [1:B]$ and $n \in \mathbb{Z}$. From this new bin assignment, we derive a refined log-frequency spectrogram $\mathcal{Y}_{\mathrm{LF}}^{\mathrm{IF}}$ by setting

$$\mathcal{Y}_{\mathrm{LF}}^{\mathrm{IF}}(n,b) := \sum_{k \in P^{\mathrm{IF}}(b,n)} |\mathcal{X}(n,k)|^2 \tag{8.53}$$

for each frame index $n \in \mathbb{Z}$ and bin index $b \in [1:B]$. The effect of this modification is illustrated by Figure 8.14c. By employing the instantaneous frequency, many of the characteristic time–frequency patterns in the lower part of the spectrum have become much sharper. Note that the estimation of the instantaneous frequency works particularly well for dominating spectral coefficients that can be clearly assigned to a single harmonic source. For example, the vertical blurring around 500 Hz (seen in the zoom visualizations of Figure 8.14) has been resolved nicely, resulting in a sharp horizontal line in $\mathcal{Y}_{\mathrm{LF}}^{\mathrm{IF}}$. However, the interference of several sound sources may lead to severe degradations in the IF estimates. Furthermore, the IF estimates may not be reliable at weaker time–frequency bins. Some of these problems can be alleviated by introducing spectral weighting and peak-picking strategies (see [44] for details). In the following, we discuss an enhancement strategy based on the fact that tonal time–frequency patterns are typically reinforced by the presence of harmonic partials.

8.2.2.3 Harmonic Summation

Recall that a sound event such as a musical tone is associated to a fundamental frequency along with its harmonic partials, which are (approximately) the integer multiples of the fundamental frequency (see Section 1.3.2). Therefore, a spectrogram representation of a recorded melody typically exhibits an entire family of frequency trajectories which are stacked on top of each other. This phenomenon is clearly visible in the examples of Figure 8.11 and Figure 8.14. In particular, one can observe several families of modulating frequency trajectories, which are the result of musical notes being sung with vibrato.

The multiple appearance of tonal time–frequency patterns can be exploited to improve a spectrogram representation. The idea is to jointly consider a frequency and its harmonics by forming suitably weighted sums—a technique also called **harmonic summation** (see [26, 44]). Let $H \in \mathbb{N}$ be the number of harmonics to be considered in the summation. Then, given a spectrogram representation $\mathcal{Y}$, we define a harmonic-sum spectrogram $\tilde{\mathcal{Y}}$ by setting

$$\tilde{\mathcal{Y}}(n,k) := \sum_{h=1}^{H} \mathcal{Y}(n,k \cdot h) \tag{8.54}$$

for $n,k \in \mathbb{Z}$ (assuming that $\mathcal{Y}$ is suitably zero-padded).

A similar construction can be applied for log-frequency spectrogram representations such as $\mathcal{Y}_{\mathrm{LF}}$ or $\mathcal{Y}_{\mathrm{LF}}^{\mathrm{IF}}$. In this case, however, one requires a small modification in the harmonic summation. In particular, working in the log-frequency domain, the relation between a frequency and its harmonics is not a multiplicative but an additive one. For the log-frequency spectrogram $\mathcal{Y}_{\mathrm{LF}}$, one obtains a harmonically enhanced version $\tilde{\mathcal{Y}}_{\mathrm{LF}}$ by setting

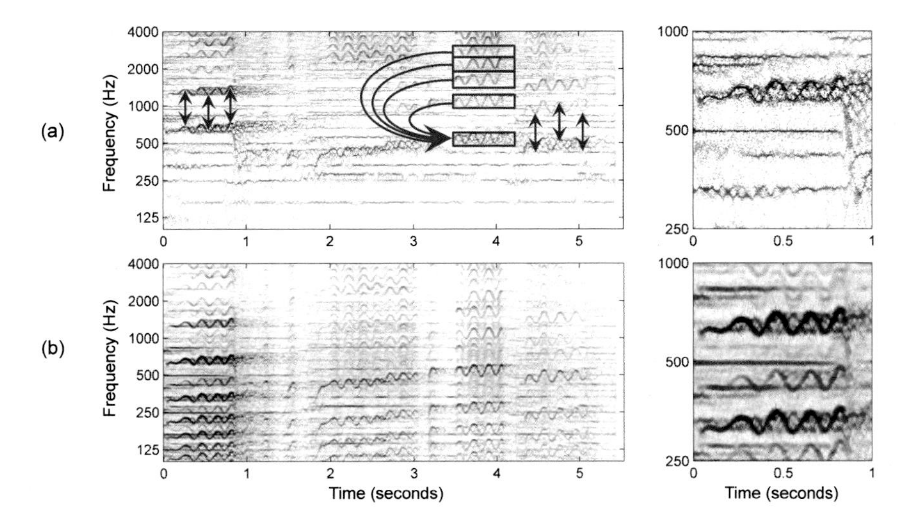

Fig. 8.15 **(a)** Illustration of harmonic summation for the enhanced log-frequency spectrogram $\mathcal{Y}_{\mathrm{LF}}^{\mathrm{IF}}$ from Figure 8.14c (along with a zoom into a section). **(b)** Resulting salience representation $\mathcal{Z} := \tilde{\mathcal{Y}}_{\mathrm{LF}}^{\mathrm{IF}}$.

$$\tilde{\mathcal{Y}}_{\mathrm{LF}}(n,b) := \sum_{h=1}^{H} \mathcal{Y}_{\mathrm{LF}}\left(n, b + \left\lfloor \frac{1200}{R}\log_2(h)\right\rfloor\right) \tag{8.55}$$

(see Exercise 8.9). Similarly, one obtains a harmonic-sum version

$$\mathcal{Z} := \tilde{\mathcal{Y}}_{\mathrm{LF}}^{\mathrm{IF}} \tag{8.56}$$

from $\mathcal{Y}_{\mathrm{LF}}^{\mathrm{IF}}$. Emphasizing the salience of tonal frequency components, $\mathcal{Z}$ is also referred to as a **salience representation** (see [44]). This representation will be used in Section 8.2.3 as a basis for extracting the predominant melody.

The process of harmonic summation, as well as the resulting salience representation, are illustrated by our Weber example in Figure 8.15. As shown in this example, the summation process amplifies the dominating harmonic or tonal components, while noise-like artifacts as introduced by the instantaneous frequency estimation are attenuated. On the downside, the process produces "ghost components" that appear particularly in the lower frequency regions of the spectrogram. This, however, is not a major problem when only looking for the predominant frequency trajectories, as we will see in the next section.

We want to make some concluding remarks on the summation process. When considering the spectrogram representation $\mathcal{Y}$, the frequency trajectories that correspond to the various harmonics are *scaled* versions of each other. This fact becomes particularly noticeable when looking at the vibrato patterns of Figure 8.14a, where the amplitudes of the frequency modulations are larger for the higher harmonics compared with the ones for the lower harmonics. In contrast, when considering a log-frequency spectrogram representation, the frequency trajectories are *translated* versions of each other (see Figure 8.15a). In this case, the amplitudes of the frequency modulations are the same for all harmonics. For a more formal treatment of this fact, we refer to Exercise 8.10.

8.2.3 Informed Fundamental Frequency Tracking

The construction of our salience representation $\mathcal{Z}$ from (8.56) is motivated by several assumptions and observations. First, the logarithmic frequency binning accounts for the logarithmic perception of frequency and the musical notion of a note's pitch. Using the instantaneous frequency improves the accuracy of the frequency estimates. Furthermore, the harmonic summation amplifies the regularly spaced frequency components. This accounts for the fact that a tone's energy is not only contained in the fundamental frequency, but spread over the entire harmonic spectrum.

Based on the assumption that the main melody corresponds to the predominant F0-value, a first strategy for melody extraction is to simply consider the frame-wise maximum of $\mathcal{Z}$. Let us formalize this procedure. In the following, a **frequency trajectory** is defined to be a function

$$\eta : \mathbb{Z} \rightarrow [1:B] \cup \{*\} \tag{8.57}$$

which assigns to each frame index $n \in \mathbb{Z}$ either a bin index $\eta(n) \in [1:B]$ or the symbol $\eta(n) = *$. The interpretation of $\eta(n) = *$ is that there is no melodic component at this time instance.

Taking the maximizing bin index of $\mathcal{Z}$ for each frame yields a frequency trajectory $\eta^{\max}$ defined by

$$\eta^{\max}(n) := \underset{b \in [1:B]}{\operatorname{argmax}}\ \mathcal{Z}(n,b). \tag{8.58}$$

As illustrated by our Weber example in Figure 8.16a, this procedure already yields an accurate estimation of the melody's F0-trajectory for most of the frames. However, at certain time positions, the calculated trajectory does not follow the singing voice, but some of the accompanying instruments. Also, the proposed method assigns an F0-value to each time frame, regardless of whether the melody is actually present or not. Finally, one can observe a number of outliers and temporal discontinuities that are due to confusions between the fundamental frequency and higher harmonics or lower ghost components introduced by the harmonic summation.

8.2.3.1 Continuity Constraints

We now discuss several ways for improving the melody extraction procedure by incorporating additional knowledge. As illustrated by Figure 8.16a, computing a frequency trajectory in a purely frame-wise fashion without considering any tem-

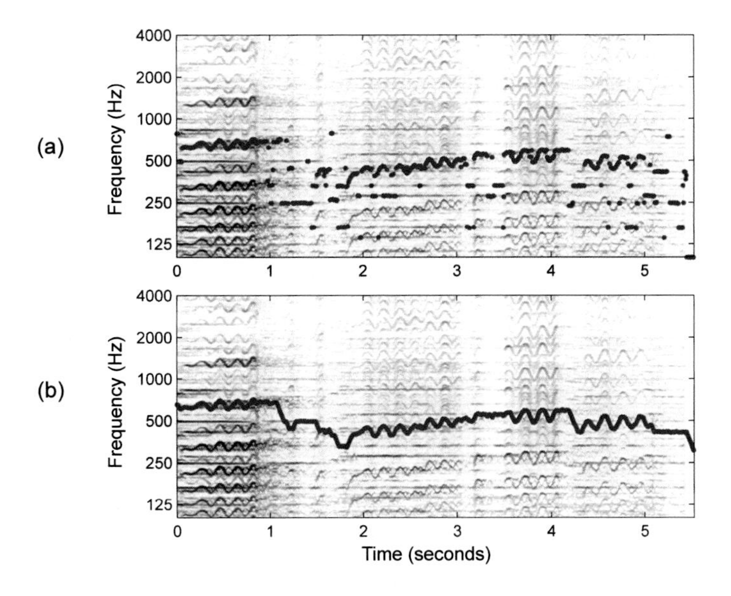

Fig. 8.16 Salience representation $\mathcal{Z}$ from Figure 8.15b with different frequency trajectories. **(a)** Frequency trajectory $\eta^{\max}$ obtained by frame-wise maximization. **(b)** Frequency trajectory η^{DP} using temporal continuity constraints.

poral context can lead to local discontinuities and random jumps. In practice, however, a melody's frequency trajectory is far more structured. Typically, it slowly changes over time with some occasional jumps between note transitions. Therefore, one needs a tracking procedure that can balance out the two conflicting conditions of temporal flexibility (to account for possible jumps) and temporal continuity (to account for smoothness properties).

We have already encountered a similar problem when discussing the task of chord recognition (see Chapter 5). In this context, we introduced temporal continuity into the frame-wise chord labeling process by using HMM-based models that involve transition probabilities (see Section 5.3). Inspired by the **Viterbi algorithm**, which solves the uncovering problem for HMMs (see Section 5.3.3.2), we now introduce a similar procedure for constructing a frequency trajectory based on **dynamic programming**.

Let $\mathcal{Z}$ be a salience representation that consists of N time frames and B frequency bins. The value $\mathcal{Z}(n,b)$ can be regarded as a score which expresses the likelihood that a frequency associated to bin $b \in [1:B]$ belongs to a dominating tonal component at time $n \in [1:N]$. To achieve temporal continuity, we slightly modify the concept of transition probabilities as used for Markov chains (see Section 5.3.1). To this end, we introduce a $(B \times B)$ matrix $\mathbf{T}$ that encodes the transition likelihood between bins. The value $\mathbf{T}(b,c) \in \mathbb{R}_{\geq 0}$ expresses the likelihood for moving from bin $b \in [1:B]$ at time frame $n \in [1:N-1]$ to bin $c \in [1:B]$ at the next frame $n+1$. For example, let us define $\mathbf{T}$ by setting

$$\mathbf{T}(b,c) := \frac{1}{1+|b-c|}. \tag{8.59}$$

Then the likelihood value $\mathbf{T}(b,c)$ becomes smaller when enlarging the bin distance $|b-c|$. This favors a trajectory that maintains (more or less) the same frequency value over time while punishing larger frequency jumps. Of course, there are many other ways for defining the matrix $\mathbf{T}$ in a meaningful way.

Given a salience representation $\mathcal{Z}$ and a transition likelihood matrix $\mathbf{T}$, we can associate to each trajectory $\eta : [1:N] \to [1:B]$ a **total score** $\sigma(\eta)$ by setting

$$\sigma(\eta) := \mathcal{Z}(1,\eta(1)) \cdot \prod_{n=2}^{N} \Big(\mathbf{T}(\eta(n-1),\eta(n)) \cdot \mathcal{Z}(n,\eta(n))\Big). \tag{8.60}$$

The desired trajectory η^{DP} is then defined to be the score-maximizing trajectory:

$$\eta^{\mathrm{DP}} := \underset{\eta}{\mathrm{argmax}}\, \sigma(\eta). \tag{8.61}$$

The maximal total score can be computed via dynamic programming similar to the Viterbi algorithm (see Table 5.2). Furthermore, the trajectory η^{DP} is obtained by applying a suitable backtracking procedure. The elaboration of the algorithmic details are left as an exercise (see Exercise 8.11).

In our Weber example, all outliers have been eliminated thanks to the continuity constraints introduced by the transition likelihoods (see Figure 8.16b). While the local frequency modulations are still captured well, the trajectory η^{DP} exhibits a continuous profile with smooth transitions between the different tones. Even though this may be a desirable property most of the time, discontinuities that are the result of abrupt note changes tend to be smoothed out. Furthermore, tracking errors still occur when the accompaniment exhibits tonal components that are stronger than the ones of the melody. This particularly holds for time positions where there is no melody at all. For this reason, melody extraction algorithms should include an estimation stage to determine where the melody is actually present and where it is not [28, 44].

8.2.3.2 Score-Informed Constraints

Besides incorporating continuity constraints, another strategy is to exploit additional musical knowledge about the melodic progression to support the F0-tracking process. For example, knowing the vocal range (soprano, alto, tenor, bass) of the singer, one may narrow down the search range of the expected F0-values. Or, having information about when the melody is actually present and when it is not, one can set all F0-values to the symbol '$*$' for the nonmelody frames. More generally, additional knowledge as described above can be used to define **constraint regions** within the time–frequency plane. The F0-tracking is then performed only in these specified regions. As an example of such an approach, we now discuss a **score-informed**

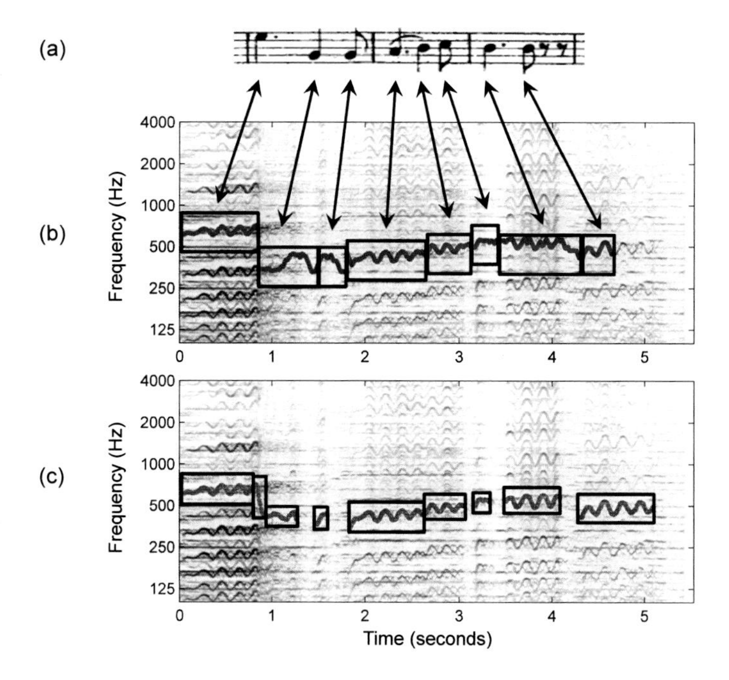

Fig. 8.17 **(a)** Musical score of the melody of our Weber example (see Figure 8.10). **(b)** Salience representation $\mathcal{Z}$ from Figure 8.15b with constraint regions obtained from synchronized note events and the resulting frequency trajectory η^{Score}. **(c)** Frequency trajectory η^{User} obtained from user-specified constraint regions.

procedure, where one assumes the availability of a score representation that underlies the given music recording. The additional instrumentation and note information provided by the score can then be used to guide the estimation of the melody's F0-trajectory.

The main idea of our score-informed tracking approach is illustrated by Figure 8.17. Recall from Section 1.1 that, rather than giving strict specifications, a musical score serves as a guide for performing a piece of music, leaving room for different interpretations. Reading the instructions in the score, a musician shapes the music by varying the tempo, dynamics, and articulation, thus creating a personal interpretation of the piece. In a first step, one needs to align the musical score and the given audio recording. In Chapter 3, we have studied **music synchronization** techniques for automating this process. In particular, based on **score–audio synchronization** (see Section 3.4), one can establish a temporal alignment between the score's note events and the physical time position, where they occur in the given music recording.

The aligned score information can be used to define a rectangular constraint region in the time–frequency plane for each note of the melody (see Figure 8.17b). The horizontal position and width of such a region correspond to the note's physical onset time and estimated duration, while the vertical position and height account for the note's pitch and the expected frequency deviations from its center frequency. For each of the constraint regions, one can compute a notewise frequency trajectory as described in Section 8.2.3.1. Assuming that there is no temporal overlap of the notes' constraint regions, one may assemble the notewise trajectories to form a single score-informed trajectory η^{Score}. In the case that a frame $n \in [1:N]$ is not covered by any of the constraint regions, one sets $\eta^{\mathrm{Score}}(n) := *$. Figure 8.17b shows the resulting score-informed trajectory η^{Score} for our Weber example.

As illustrated by the figure, score-informed constraint regions are an easy means to control the F0-extraction process. Suitably chosen regions may significantly reduce the confusion with harmonics and other musical voices. However, in the presence of synchronization inaccuracies and deviations between the score and audio recording, such a procedure can also become problematic. For example, in our Weber excerpt, the singer deviates from the specified score by singing a note C$^\sharp$5 instead of the notated B4 (second note to last in Figure 8.17a). As a result, the constraint region is misplaced, leading to a corrupted frequency trajectory. Similarly, incorrect estimations of the notes' durations introduce errors in the frequency estimation. For example, this happens for the last note of Figure 8.17a.

Many of these errors can be corrected by suitably adjusting the constraint regions. Figure 8.17c shows an example of user-optimized constraint regions along with the resulting F0-trajectory η^{User}.

8.2.3.3 Applications

The quality requirements of the extracted F0-trajectory very much depend on the applications in mind. As a first application, let us consider the problem of content-based audio retrieval (see Chapter 7). One retrieval scenario is known as **query-by-humming**, which can be regarded as a special case of version identification (see Section 7.3). In this scenario, the user specifies a query by singing or humming part of a melody. The objective is then to identify all audio recordings (or other music representations) that contain a melody similar to the specified query. One important step in solving this task is the generation of a melody database against which the sung query can be compared. To this end, one requires automated methods for extracting melodic information from audio recordings of polyphonic music. In this application scenario, however, one may not need a perfect F0-trajectory in order to identify a melody. The comparison of a query and a database document may be performed on the basis of a **mid-level representation** that tolerates local extraction errors, at least to a certain degree (see [46] for further details).

Other applications are less sensitive to inaccuracies in the extracted F0-values. One such application is the automated generation of a **Karaoke version** for a given song, where the main melodic voice is to be removed from the song's original recording. This leads us to the task of **melody separation**. Given a music signal x, the objective is to decompose the signal into a melody component x^{Mel} and an accompaniment component x^{Acc} such that

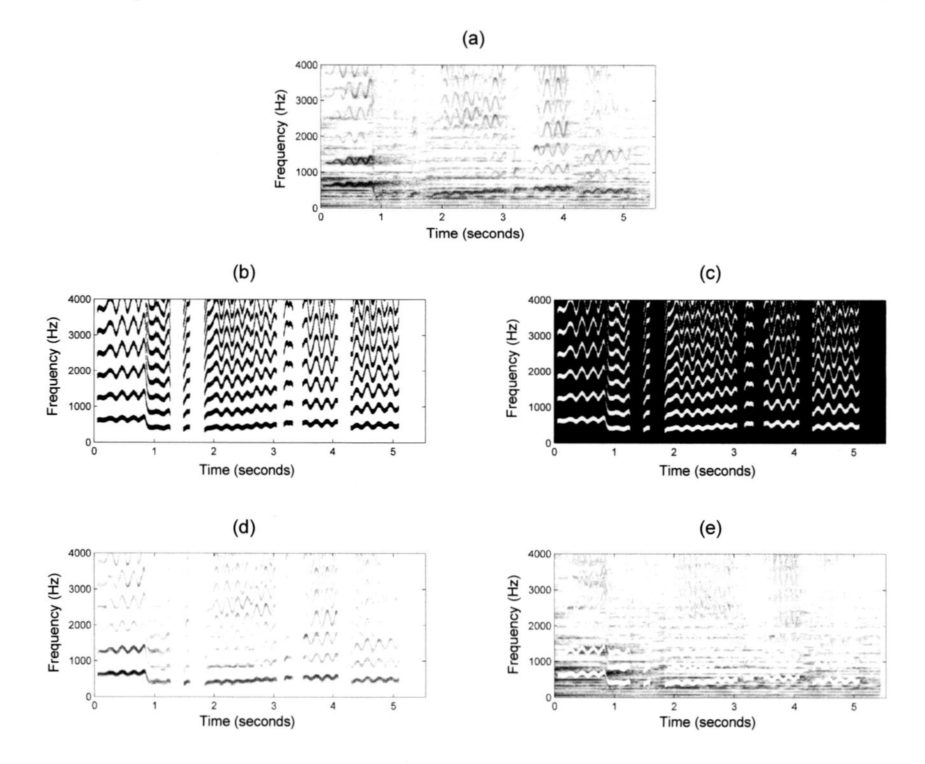

Fig. 8.18 **(a)** Magnitude of the STFT $\mathcal{X}$ of the original signal x. **(b)** Binary mask for the melodic component using the F0-trajectory from Figure 8.17c. **(c)** Complementary binary mask for the accompaniment component. **(d)** Magnitude of the masked STFT $\mathcal{X}^{\mathrm{Mel}}$. **(e)** Magnitude of the masked STFT $\mathcal{X}^{\mathrm{Acc}}$.

$$x = x^{\mathrm{Mel}} + x^{\mathrm{Acc}}. \tag{8.62}$$

One general approach is to first apply a melody extraction algorithm to derive the F0-trajectory of the main melody. Based on these F0-values and their harmonics, one can construct a binary mask for the melodic component as illustrated by Figure 8.18b. The binary mask of the accompaniment is defined to be the complement (see Figure 8.18c). Using masking techniques as described in Section 8.1.1.2, one then derives the two masked STFTs $\mathcal{X}^{\mathrm{Mel}}$ (see Figure 8.18d) and $\mathcal{X}^{\mathrm{Acc}}$ (see Figure 8.18e). From this, one obtains the signals x^{Mel} and x^{Acc} by applying signal reconstruction techniques as discussed in Section 8.1.2. Obviously, in this voice separation scenario, the requirements on the accuracy of the extracted F0-values are much higher than in the previously described retrieval scenario. Also note that, in this simplistic separation approach, nonharmonic properties of the singing voice such as fricative and plosive components (coming from consonants) have not been considered. Modeling and separating such nonharmonic components from sound mixtures generally constitutes a hard research question.

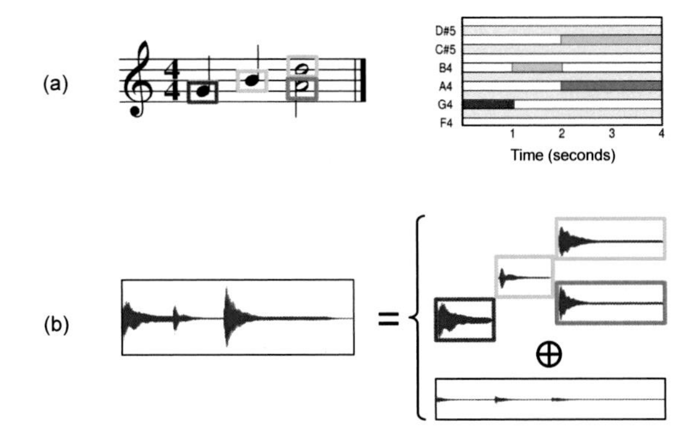

Fig. 8.19 (**a**) Score and piano-roll representation. (**b**) Score-informed decomposition of the corresponding music signal into notewise audio events and a residual signal.

8.3 NMF-Based Audio Decomposition

In the previous sections, we have studied source separation techniques for decomposing a music recording into a harmonic and percussive component as well as for extracting a melodic voice. In this section, we consider a scenario in which we decompose a music signal into a set of notewise audio events, where each audio event is directly associated with a note of a given musical score, see Figure 8.19. Having direct access to the audio recording in a notewise fashion opens up novel ways of editing and manipulating audio material (see Figure 8.27).

To obtain such a decomposition, we introduce in this section an important technique known as **nonnegative matrix factorization** (NMF). To build up some intuition, let us start with a simple example shown in Figure 8.20a. Suppose we are given a matrix V with nonnegative entries, which can be thought of as a time–frequency representation consisting of a sequence of spectral vectors. In our toy example, there are $N = 7$ time frames, and each of the vectors consists of $K = 11$ spectral coefficients. At first sight, the matrix does not seem to have a particular structure. However, it turns out that the matrix can be represented by only $R = 3$ different "prototype" spectral vectors (referred to as **templates**), which are suitably weighted and superimposed over time. Mathematically, this means that V can be factorized, i.e., V can be represented as a product of two matrices. To this end, the template vectors are stored in a $(K \times R)$ matrix W (called the **template matrix**), whereas the weights are stored in an $(R \times N)$ matrix H (called the **activation matrix**). Then, the original matrix V can be written as a matrix product $V = WH$. The advantage of such a factorization is that the factors W and H are often easier to understand and more accessible for further processing than the original matrix W.

Given a matrix V with nonnegative entries, the goal of NMF is to automatically find a matrix factorization similar to the one in our previous example. Mathematically, many different factorizations are possible. In NMF, one special requirement is that, besides the given matrix V, the factors W and H should also only contain nonnegative entries. As we will see, this additional requirement often allows for an immediate interpretation of W and H. On the downside, the computational process

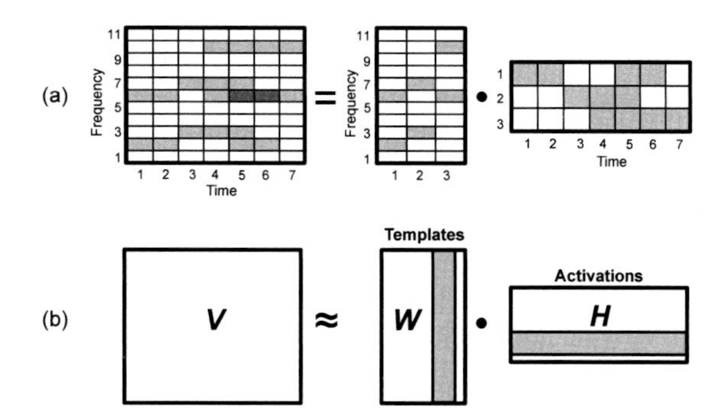

Fig. 8.20 **(a)** Decomposition of a nonnegative matrix V into a product of simpler nonnegative matrices W and H. (The shade of gray encodes the size of the matrix entries.) **(b)** Illustration of nonnegative matrix factorization (NMF).

becomes more complicated and the conditions are hard to satisfy. Therefore, to relax the problem, one looks for a factorization where the original matrix is represented at least in some approximate sense, i.e., $V \approx WH$ (see Figure 8.20b).

As in our example, the matrix V is regarded as a sequence of column vectors. The underlying assumption is that these vectors can be represented as a weighted superposition of a relatively small number of **template** vectors. The columns of W correspond to these templates. Furthermore, the rows of H—called **activations**—indicate where these templates occur in V (see Figure 8.20a).

In Section 8.3.1, we formally introduce NMF, which is one of the most successful machine learning techniques that has been applied to various problem areas, ranging from computer vision to text mining and audio processing. Following [27], we derive some update rules for learning the matrix factorization in an iterative fashion. This method is easy to implement as well as computationally efficient.

We then show in Section 8.3.2 how NMF-based techniques can be applied for audio decomposition by factoring the magnitude STFT into a product of two nonnegative matrices. In the ideal case, the first matrix represents the spectral patterns of the notes' pitches that occur in the piece of music, while the second matrix exhibits the time positions where these spectral patterns are active in the audio recording. Figure 8.21 illustrates such a factorization for a recording of the Prélude Op. 28, No. 4 by Frédéric Chopin, which also serves as our running example in the subsequent explanations. In this case, each template specified by the matrix W reflects how a note of a certain pitch is spectrally realized in V, and the activation matrix H looks similar to a corresponding piano-roll representation of the score (Figure 8.21b). In practice, however, it is often hard to predict which of the signal's properties are ultimately captured by the learned factors. To better control this factorization, we show how additional score information can be used to constrain NMF and to yield a more musically meaningful decomposition.

The score not only guides the factorization process, but also yields an intuitive and user-friendly representation for musically experienced users to specify the target sources or events to be separated. In Section 8.3.3, we discuss how the result of the score-informed NMF can be used for deriving various kinds of audio decompositions. In particular, we present an example application where a user can specify

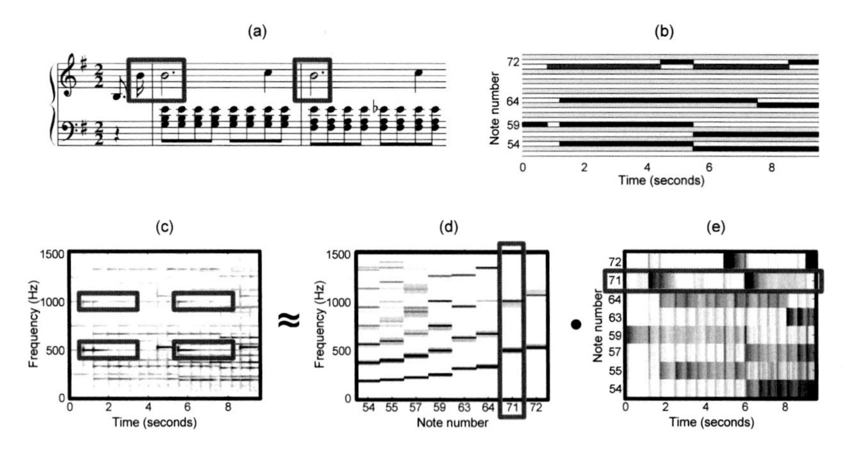

Fig. 8.21 Illustration of nonnegative matrix factorization $V \approx WH$. The example is based on an audio recording of the Prélude Op. 28, No. 4 by Frédéric Chopin. All information related to the note number $p = 71$ is highlighted by the red rectangular frames. **(a)** Musical score. **(b)** Piano-roll representation of the score synchronized to the audio recording. **(c)** Magnitude spectrogram of the audio recording used as matrix V. **(d)** Template matrix W. **(e)** Activation matrix H.

the desired audio manipulation within the score simply by editing some of the notes. These manipulations are then automatically transferred to a given audio recording.

8.3.1 Nonnegative Matrix Factorization

The general goal of **matrix decomposition** or **matrix factorization** is to represent a given matrix as a product of matrices that have specific properties. These properties can then be exploited for further processing, analysis, or storage. For example, in Section 2.4.3, we have seen how a factorization of the DFT matrix into a product of sparse matrices can be exploited to increase the computational efficiency (leading to the famous FFT algorithm). Besides computational purposes, matrix factorization techniques are also used for analytic purposes. For example, a given data matrix of some numerical observations is often hard to understand. In this context, a factorization of the matrix into a product of lower-rank matrices can reveal some inherent structures, which facilitates the interpretation of their meaning.

Nonnegative matrix factorization (NMF) is a technique where a matrix V with nonnegative entries is factored into two matrices W and H that also have only nonnegative entries. Typically, the matrices W and H are required to have a much lower rank than the original matrix V. The nonnegativity constraints often lead to a decomposition that allows for a semantically meaningful interpretation of the coefficients. However, in most cases, the resulting factorization problem has no exact solution, thus requiring optimization procedures for finding suitable numerical approximations. Closely following [27], we now give a formal definition of the NMF problem and introduce an iterative learning procedure for computing a factorization in practice.

8.3.1.1 Formal Definition of NMF

A matrix with real-valued coefficients is called **nonnegative** if all the coefficients are either zero or positive. Let $V \in \mathbb{R}_{\geq 0}^{K \times N}$ be such a nonnegative matrix having $K \in \mathbb{N}$ rows and $N \in \mathbb{N}$ columns. The dimensions K and N of the matrix V are usually thought to be large. Given a number $R \in \mathbb{N}$ smaller than both K and N, the goal of NMF is to find two nonnegative matrices $W \in \mathbb{R}_{\geq 0}^{K \times R}$ and $H \in \mathbb{R}_{\geq 0}^{R \times N}$ such that

$$V \approx W \cdot H. \tag{8.63}$$

This factorization is interpreted as follows (see [27]): The columns of V are regarded as K-dimensional data vectors (e.g., spectral vectors of a magnitude spectrogram), where N is the number of data vectors. This matrix is then approximately factorized into a $(K \times R)$ matrix W and an $(R \times N)$ matrix H. The parameter R, which is referred to as the **rank** of the factorization, is usually chosen to be smaller than K and N. Therefore, the number of coefficients in W and H is typically much smaller than the total number in V (i.e., $KR + RN \ll KN$), and the product WH can be thought of as a compressed version of the original matrix V.

Let us have a closer look at the meaning of the approximation by rewriting (8.63) column by column as

$$v \approx W \cdot h, \tag{8.64}$$

where v is the nth column of V and h the nth column of H, $n \in [1 : N]$. This shows that each data vector v is approximated by a linear combination of the columns of W, weighted by the entries of h. Therefore, W can be regarded as containing a kind of basis that is optimized for the linear approximation of the data in V. Since relatively few basis vectors are used to represent many data vectors, a good approximation can only be achieved if the basis vectors capture structures that are latent in the data. In this context, a good estimation of the parameter R, which defines the number of basis vectors used to approximate the data matrix V, constitutes a difficult problem by itself.

As already mentioned before, the basis vectors are also referred to as **template vectors**, whereas the weights specified by H are called **activations**. As opposed to arbitrary linear combinations as known from linear algebra, the linear combinations occurring in the NMF context only involve nonnegative weights of nonnegative template vectors. As a result, there are no effects such as destructive interferences, where a (positive) component can be canceled out by adding a kind of inverse (negative) component. Instead, the data vectors need to be explained in a purely constructive fashion only involving positive components.

To find an approximate factorization $V \approx W \cdot H$, we first need to specify a distance function that quantifies the quality of the approximation. There are many ways for defining such a distance function, leading to different NMF variants. In the following, we only consider one of these variants, which is based on the **Euclidean distance**. Let $A, B \in \mathbb{R}^{K \times N}$ be two matrices with coefficients A_{kn} and B_{kn} for $k \in [1:K]$ and $n \in [1:N]$. Then, the square of the Euclidean distance between A and B is defined by

$$\|A - B\|^2 := \sum_{k=1}^{K} \sum_{n=1}^{N} (A_{kn} - B_{kn})^2. \tag{8.65}$$

Based on this distance measure, we can formalize our NMF problem as follows: Given a nonnegative matrix $V \in \mathbb{R}_{\geq 0}^{K \times N}$ and a rank parameter R, minimize

$$\|V - WH\|^2 \tag{8.66}$$

with respect to $W \in \mathbb{R}_{\geq 0}^{K \times R}$ and $H \in \mathbb{R}_{\geq 0}^{R \times N}$. In other words, regarding $\|V - WH\|^2$ as a joint function of W and H, the objective is to find a minimum under the nonnegativity constraint for W and H.

For general matrices, this is a hard computational problem due to several reasons. First, it is in general difficult to enforce hard constraints such as nonnegativity on the solution of an optimization problem. Second, the joint optimization over both matrices W and H leads to computational challenges. In fact, when regarding $\|V - WH\|^2$ as a function of W only or H only, one can show that the resulting functions satisfy a strong property referred to as **convexity**. This property, which implies that any local minimum must be a global minimum, makes it possible to apply powerful tools from the field of convex analysis. However, $\|V - WH\|^2$ is not convex in both matrices together. Therefore, it is unrealistic to expect an algorithm that can solve this problem in the sense of finding a global minimum. However, there are many techniques from numerical optimization that can be applied to find—at least—local minima. In the following we discuss such a procedure which can be derived from the fundamental optimization strategy known as gradient descent.

8.3.1.2 Gradient Descent

Let us start by recalling some basic facts from differential calculus and numerical optimization (see, e.g., [34]). Suppose that we have a function

$$\varphi : \mathbb{R}^D \to \mathbb{R} \tag{8.67}$$

which depends on D variables, say $u_1, \ldots, u_D$. If the function φ is differentiable, one can compute the derivative with respect to one of those variables, with the others held constant. This yields a **partial derivative** denoted by $\frac{\partial \varphi}{\partial u_d}$ for each $d \in [1:D]$. Now, if φ is differentiable in a neighborhood of a point $\mathbf{p} \in \mathbb{R}^D$, then the **gradient** $\nabla \varphi(\mathbf{p})$ at $\mathbf{p}$ is defined as

$$\nabla \varphi(\mathbf{p}) := \left(\frac{\partial \varphi}{\partial u_1}(\mathbf{p}), \ldots, \frac{\partial \varphi}{\partial u_D}(\mathbf{p}) \right). \tag{8.68}$$

In this case, the function φ decreases fastest if one goes from $\mathbf{p}$ in the direction of the negative gradient of φ at $\mathbf{p}$. Motivated by this property, we define

$$\mathbf{p}' := \mathbf{p} - \gamma \cdot \nabla\varphi(\mathbf{p}) \tag{8.69}$$

for some parameter $\gamma \geq 0$, which is also referred to as the **step size**. Then, one can show that $\varphi(\mathbf{p}') \leq \varphi(\mathbf{p})$ in the case that the step size γ is small enough. Based on this fact, **gradient descent** tries to find a local minimum of φ by an iterative approach. Starting with a guess for a local minimum of φ at $\mathbf{p}^{(0)} \in \mathbb{R}^D$, one iteratively defines

$$\mathbf{p}^{(\ell+1)} := \mathbf{p}^{(\ell)} - \gamma^{(\ell)} \cdot \nabla\varphi(\mathbf{p}^{(\ell)}) \tag{8.70}$$

for $\ell = 0, 1, 2, \ldots$. Based on suitable step size parameters $\gamma^{(\ell)}$, which depend on the properties of φ and need to be adjusted at every iteration, one obtains

$$\varphi(\mathbf{p}^{(0)}) \geq \varphi(\mathbf{p}^{(1)}) \geq \varphi(\mathbf{p}^{(2)}) \geq \cdots \tag{8.71}$$

and the sequence $\mathbf{p}^{(0)}, \mathbf{p}^{(1)}, \mathbf{p}^{(2)}, \ldots$ hopefully converges to a local minimum of φ. With certain assumptions on $\gamma^{(\ell)}$ and φ, convergence to a local minimum can be guaranteed. When the function φ is convex, all local minima are also global minima, so in this case gradient descent converges to a global solution.

8.3.1.3 Learning the Factorization Using Gradient Descent

We now apply the concept of gradient descent to our problem of minimizing $\|V - WH\|^2$ as a function of W and H. Since the joint optimization is a very hard problem, one idea is to first fix the factor W and to optimize with regard to H, and then to fix the learned factor H and to optimize with regard to W. This process is then iterated, where the role of W and H is interchanged after each step.

For the moment, let us fix a matrix $W \in \mathbb{R}^{K \times R}$. We define a function $\varphi^W : \mathbb{R}^D \to \mathbb{R}$ with $D := RN$ by setting

$$\varphi^W(H) := \|V - WH\|^2 \tag{8.72}$$

for $H \in \mathbb{R}^{R \times N}$, where the matrix H is regarded as a D-dimensional vector. Furthermore, we denote the variables of the function φ^W by $H_{\rho\nu}$ for $\rho \in [1:R]$ and $\nu \in [1:N]$. Note that we use the parameters ρ and ν in order to distinguish them from the parameters r and n, which are later used as summation indices (see (8.73)). In order to apply gradient descent to φ^W, we need to compute the partial derivatives of φ^W with regard to all variables $H_{\rho\nu}$. After the following equations, we give more detailed explanations on the calculation steps:

$$\frac{\partial \varphi^W}{\partial H_{\rho\nu}} = \frac{\partial\left(\sum_{k=1}^{K}\sum_{n=1}^{N}\left(V_{kn} - \sum_{r=1}^{R} W_{kr}H_{rn}\right)^2\right)}{\partial H_{\rho\nu}} \tag{8.73}$$

$$= \frac{\partial\left(\sum_{k=1}^{K}\left(V_{k\nu} - \sum_{r=1}^{R} W_{kr}H_{r\nu}\right)^2\right)}{\partial H_{\rho\nu}} \tag{8.74}$$

$$= \sum_{k=1}^{K} 2\left(V_{k\nu} - \sum_{r=1}^{R} W_{kr}H_{r\nu}\right)\cdot(-W_{k\rho}) \tag{8.75}$$

$$= 2\left(\sum_{r=1}^{R}\sum_{k=1}^{K} W_{k\rho}W_{kr}H_{r\nu} - \sum_{k=1}^{K} W_{k\rho}V_{k\nu}\right) \tag{8.76}$$

$$= 2\left(\sum_{r=1}^{R}\left(\sum_{k=1}^{K} W^\top_{\rho k}W_{kr}\right)H_{r\nu} - \sum_{k=1}^{K} W^\top_{\rho k}V_{k\nu}\right) \tag{8.77}$$

$$= 2\left((W^\top WH)_{\rho\nu} - (W^\top V)_{\rho\nu}\right). \tag{8.78}$$

The sums in (8.73) are obtained by writing out the Euclidean distance and the matrix product in (8.72). To derive (8.74), we use the fact that a derivative of a summand that does not depend on ν (thus not depending on $H_{\rho\nu}$) must be zero. Therefore, only the summand with $n = \nu$ remains. Then, we apply the chain rule from calculus to obtain (8.75). In the next steps, we rearrange the sums and introduce the transposed matrix $W^\top$ with the property $W^\top_{\rho k} = W_{k\rho}$. This finally yields (8.78), which constitutes a compact matrix-based formulation of the partial derivatives.

Starting with an initial guess $H^{(0)} \in \mathbb{R}^{R\times N}$, we obtain from (8.78) the following additive update rules that reduce the squared Euclidean distance:

$$H_{rn}^{(\ell+1)} = H_{rn}^{(\ell)} - \gamma_{rn}^{(\ell)} \cdot \left(\left(W^\top WH^{(\ell)}\right)_{rn} - \left(W^\top V\right)_{rn}\right) \tag{8.79}$$

for $\ell = 0,1,2,\ldots$ and some suitable parameters $\gamma_{rn}^{(\ell)} \geq 0$. If the step sizes $\gamma_{rn}^{(\ell)}$ are set equal to some small positive number $\gamma^{(\ell)}$ (independent from r and n), we obtain the conventional gradient descent as described by (8.70). In (8.79), we use coordinate-dependent scaling factors $\gamma_{rn}^{(\ell)}$, thus yielding a generalized gradient descent.

Next, we fix a matrix $H \in \mathbb{R}^{R\times N}$ and define a function $\varphi^H : \mathbb{R}^D \to \mathbb{R}$ with $D := KR$ by setting

$$\varphi^H(W) := \|V - WH\|^2 \tag{8.80}$$

for $W \in \mathbb{R}^{K\times R}$. By a similar computation as in (8.78), one can derive the following additive update rules starting with an initial guess $W^{(0)} \in \mathbb{R}^{K\times R}$:

$$W_{kr}^{(\ell+1)} = W_{kr}^{(\ell)} - \gamma_{kr}^{(\ell)} \cdot \left(\left(W^{(\ell)}HH^\top\right)_{kr} - \left(VH^\top\right)_{kr}\right) \tag{8.81}$$

for $\ell = 0,1,2,\ldots$ and some suitable parameters $\gamma_{kr}^{(\ell)} \geq 0$.

At this point, let us reflect on the procedure we have obtained so far. We start with initialized matrices $W^{(0)} \in \mathbb{R}^{K\times R}$ and $H^{(0)} \in \mathbb{R}^{R\times N}$. Fixing $W^{(0)}$, we apply the update rules (8.79) to derive a matrix $H^{(1)}$. Fixing this matrix, we then use the update rules (8.81) to derive a matrix $W^{(1)}$. Alternately updating the activation and template matrices, this process is iterated, yielding matrices $H^{(\ell)}$ and $W^{(\ell)}$ for $\ell = 0, 1, 2, \ldots$. The hope is that the process converges to two matrices $H^{(\infty)}$ and $W^{(\infty)}$ which provide a local minimum for our optimization problem (8.66).

There are still several issues with this procedure. First, it is unclear how to choose the step size parameters to guarantee some kind of convergence. Second, applying the gradient descent for each of the factors separately, it is unclear if the alternation in the updates leads to a *joint* local minimum. Third, we have not yet accounted for the nonnegativity constraint. In other words, using the proposed additive update rules, some coefficients of the learned matrices may become negative even when starting from nonnegative matrices $W^{(0)}$ and $H^{(0)}$.

8.3.1.4 Multiplicative Update Rules

In the case of NMF, there is a simple yet powerful solution for enforcing the nonnegativity constraints. The crucial idea is to set the step size parameters to specific values with the result that the *additive* update rules become *multiplicative* update rules. More precisely, setting

$$\gamma_{rn}^{(\ell)} := \frac{H_{rn}^{(\ell)}}{\left(W^\top W H^{(\ell)}\right)_{rn}}, \tag{8.82}$$

the update rule (8.79) becomes

$$\begin{aligned} H_{rn}^{(\ell+1)} &= H_{rn}^{(\ell)} - \frac{H_{rn}^{(\ell)}}{\left(W^\top W H^{(\ell)}\right)_{rn}} \cdot \left(\left(W^\top W H^{(\ell)}\right)_{rn} - \left(W^\top V\right)_{rn}\right) \\ &= H_{rn}^{(\ell)} \cdot \frac{\left(W^\top V\right)_{rn}}{\left(W^\top W H^{(\ell)}\right)_{rn}}. \end{aligned} \tag{8.83}$$

Similarly, setting

$$\gamma_{kr}^{(\ell)} := \frac{W_{kr}^{(\ell)}}{\left(W^{(\ell)} H H^\top\right)_{kr}}, \tag{8.84}$$

the update rule (8.81) converts to

$$W_{kr}^{(\ell+1)} = W_{kr}^{(\ell)} \cdot \frac{\left(V H^\top\right)_{kr}}{\left(W^{(\ell)} H H^\top\right)_{kr}}. \tag{8.85}$$

The specified choices for the step size parameters in (8.82) and (8.84) are generally not small. Therefore, it may seem that there is no guarantee that the resulting update rules should cause the distance function (8.66) to decrease.

Surprisingly, this is indeed the case. Lee and Sung show in their seminal paper [27] that the Euclidean distance $\|V - W^{(\ell)} H^{(\ell)}\|$ is nonincreasing under the update rules (8.83) and (8.85). Furthermore, the Euclidean distance is invariant under

Algorithm: NMF ($V \approx WH$)

Input: Nonnegative matrix V of size $K \times N$
Rank parameter $R \in \mathbb{N}$
Threshold ε used as stop criterion

Output: Nonnegative template matrix W of size $K \times R$
Nonnegative activation matrix H of size $R \times N$

Procedure: Define nonnegative matrices $W^{(0)}$ and $H^{(0)}$ by some random or informed initialization. Furthermore set $\ell = 0$. Apply the following update rules (written in matrix notation):

(1) $H^{(\ell+1)} = H^{(\ell)} \odot (((W^{(\ell)})^\top V) \oslash ((W^{(\ell)})^\top W^{(\ell)} H^{(\ell)}))$

(2) $W^{(\ell+1)} = W^{(\ell)} \odot ((V(H^{(\ell+1)})^\top) \oslash (W^{(\ell)} H^{(\ell+1)} (H^{(\ell+1)})^\top))$

(3) Increase ℓ by one.

Repeat the steps (1) to (3) until $\|H^{(\ell)} - H^{(\ell-1)}\| \leq \varepsilon$ and $\|W^{(\ell)} - W^{(\ell-1)}\| \leq \varepsilon$ (or until some other stop criterion is fulfilled). Finally, set $H = H^{(\ell)}$ and $W = W^{(\ell)}$.

Table 8.1 Iterative algorithm for learning an NMF decomposition. The multiplicative update rules are given in matrix notation, where the operator $\odot$ denotes pointwise multiplication and the operator $\oslash$ pointwise division.

these updates if and only if $W^{(\ell)}$ and $H^{(\ell)}$ are at a stationary point of the distance (i.e., a point where the gradient is zero). The proof of these facts goes beyond the scope of this book, and we refer to [27] for details.

The multiplicative update rules and their properties have a number of remarkable implications. The first implication is that the matrix sequences $W^{(0)}, W^{(1)}, W^{(2)}, \ldots$ and $H^{(0)}, H^{(1)}, H^{(2)}, \ldots$ (as defined in Table 8.1) converge. Denoting the limit matrices by $W^{(\infty)}$ and $H^{(\infty)}$, the stationarity property implies that these matrices form a local minimum of the distance function (8.66). Another advantage of multiplicative update rules is that they are extremely easy to implement. Furthermore, in practice, the convergence turns out to be relatively fast in comparison with many other methods. Finally, one major benefit of using multiplicative update rules is that the nonnegativity constraints are enforced automatically. Indeed, starting with nonnegative matrices V, $W^{(0)}$, and $H^{(0)}$, the factors (8.82) and (8.84) are also nonnegative. As a result, the matrices $W^{(\ell)}$ and $H^{(\ell)}$ are nonnegative.

Table 8.1 summarizes the iterative learning procedure for computing an NMF decomposition based on multiplicative update rules. In practice, the iteration is performed until a specified stop criterion is fulfilled. For example, one may perform a certain number of iterations $\ell = 0, 1, 2, \ldots, L$ for some user-specified parameter $L \in \mathbb{N}$. As another stop criterion, one may look at the distances between two subsequently computed template matrices and activation matrices, respectively. Specifying a threshold $\varepsilon > 0$, the iteration may be stopped when $\|H^{(\ell+1)} - H^{(\ell)}\| \leq \varepsilon$ and $\|W^{(\ell+1)} - W^{(\ell)}\| \leq \varepsilon$. Note that this stop criterion does not say anything about the quality of the approximation $V \approx WH$ achieved by the procedure. Even in the limit case and even when converging to the global minimum (and not to a local one), the distance $\|V - WH\|$ may still be large. In particular, this may happen if the rank parameter R is chosen too small.

8.3.2 Spectrogram Factorization

We now show how our introduced NMF framework can be applied in practice for decomposing a given audio recording into musically meaningful sound events. The piano piece shown in Figure 8.21 will serve as our running example. Let us start with an individual note played on a piano, which corresponds to a time–frequency pattern as illustrated by Figure 1.23a. Intuitively, this pattern may be described by a spectral vector that has a pitch-dependent harmonic structure and some time-dependent weights that describe the sound's volume over time. Playing several notes, the resulting audio recording may be regarded as a superposition of the sound events that correspond to the individual notes.

Even though this model is too simplistic to correctly reflect all acoustic properties of the music recording, it justifies applying the NMF framework. In order to decompose the given music signal x, we apply NMF to the magnitude of the STFT $\mathcal{X}$ (see (8.2)). More precisely, we use the transposed version, yielding a nonnegative $(K \times N)$ matrix

$$V := |\mathcal{X}|^\top. \tag{8.86}$$

In the factorization $V \approx W \cdot H$, we expect that the template matrix W picks up the structure of the pitch-dependent spectral vectors, while the activation matrix H encodes when and how strongly the respective vectors are active. In this way, the matrix H yields a kind of weighted piano-roll representation for the notes' templates (see Figure 8.21d). For this kind of interpretation, the rank parameter R should correspond to the number of different pitches occurring in the piece of music. For example, in Figure 8.21a one can count eight different pitches, thus making $R = 8$ a good choice.

We now apply the NMF algorithm from Table 8.1 to V by iteratively updating the randomly initialized matrices $W^{(0)}$ and $H^{(0)}$. For our running example, both the initialized and learned template and activation matrices are shown in Figure 8.22. In this example, the rank parameter R was manually set to eight, allowing one template for each of the eight different musical pitches. However, as demonstrated by this example, there are various issues with this approach. Looking at the learned template matrix W, it is not clear to which sound or pitch a given template vector corresponds. While only a few of the template vectors reveal a clear harmonic structure, most of the templates seem to correspond to mixtures of notes rather than individual notes. This is also indicated by the activation patterns in H, which can hardly be associated to the piano-roll representation from Figure 8.21b. In summary, even though the original matrix V may be approximated well, simply applying the standard NMF approach based on a random initialization often produces a factorization that lacks clear musical semantics.

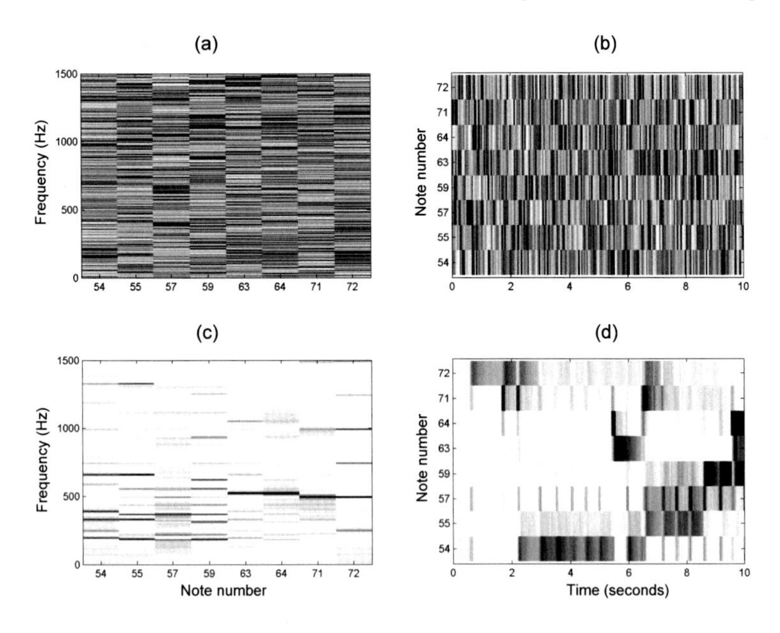

Fig. 8.22 NMF based on random initialization. The matrix V is the magnitude spectrogram from Figure 8.21c. (The matrix entries are encoded by different gray levels, which are lighter for small values and darker for large values.) **(a)** Initialization of $W^{(0)}$ using random values. **(b)** Initialization of $H^{(0)}$ using random values. **(c)** Learned W. **(d)** Learned H.

8.3.2.1 Template Constraints

To overcome these issues, one important idea is to guide the NMF processing by imposing additional suitable constraints on W and H. In this context, the multiplicative update rules again constitute a great advantage over additive rules. In the multiplicative case, zero-valued entries remain so during the entire learning process. Thanks to this property, one can enforce certain structures on W and H by setting suitable entries to zero in the initialization stage. Using this modified initialization, one can then apply the standard NMF approach from Table 8.1 in the subsequent learning process.

Let us apply this strategy to enforce a harmonic structure in the templates of W. The motivating observation is that many instruments such as a piano produce harmonic sounds, and that the templates should reflect this structure. Recall from Section 1.3.2 that a **harmonic sound** is one whose energy is concentrated around the **harmonics** of the **fundamental frequency**. To enforce such a structure in the templates, we can constrain the spectral energy between harmonics to be zero. More precisely, after assigning a musical pitch p to each template vector, we can use the center frequency $F_{\mathrm{pitch}}(p)$ (see (1.1)) associated with each pitch as an estimate of the fundamental frequency. From this, we can derive the approximate positions for the harmonics. As the exact frequencies are not known, a neighborhood around

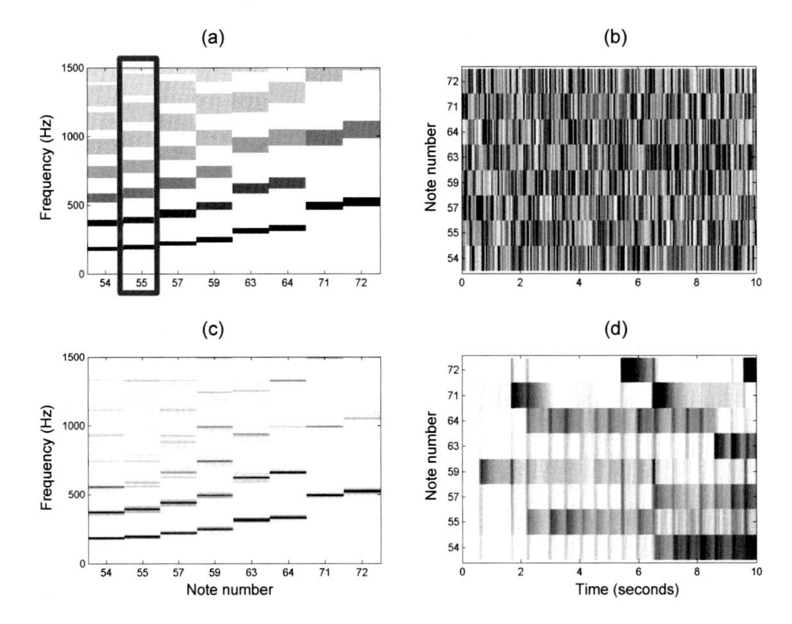

Fig. 8.23 NMF with a harmonically informed initialization of the template vectors. The matrix V is the magnitude spectrogram from Figure 8.21c. **(a)** Initialization of $W^{(0)}$ using harmonic templates. **(b)** Initialization of $H^{(0)}$ using random values. **(c)** Learned W. **(d)** Learned H.

these positions can then be initialized with nonzero values in the templates, while setting the remaining entries to zero. This kind of initialization is also illustrated by Figure 8.23a. For example, for a pitch corresponding to the note number $p = 55$, the frequency coefficients in a neighborhood of the harmonics $h \cdot F_{\mathrm{pitch}}(p)$ for $h = 1, 2, 3, \ldots$ are set to a nonzero value (e.g., the value of one or a value of $1/h$ as in our example), while the other coefficients are set to zero.

As illustrated by Figure 8.23c, the learning process based on multiplicative update rules not only retains this harmonic structure but further refines it. At the same time, the activation matrix reflects the presence and the intensity of the played notes much better (see Figure 8.23d). Furthermore, one can also observe vertical structures in the activation matrix. These structures are the result of short-time energy bursts that are related to the note onsets' transients.

8.3.2.2 Score-Informed Constraints

If additional prior knowledge is available, it can be exploited to further stabilize the factorization process. In this context, a musical score is particularly valuable. On a coarse level, we can extract *global* information from the score, such as which instruments are playing or which and how many pitches occur over the course of a piece of music. In our example, this information can be used to set the number of tem-

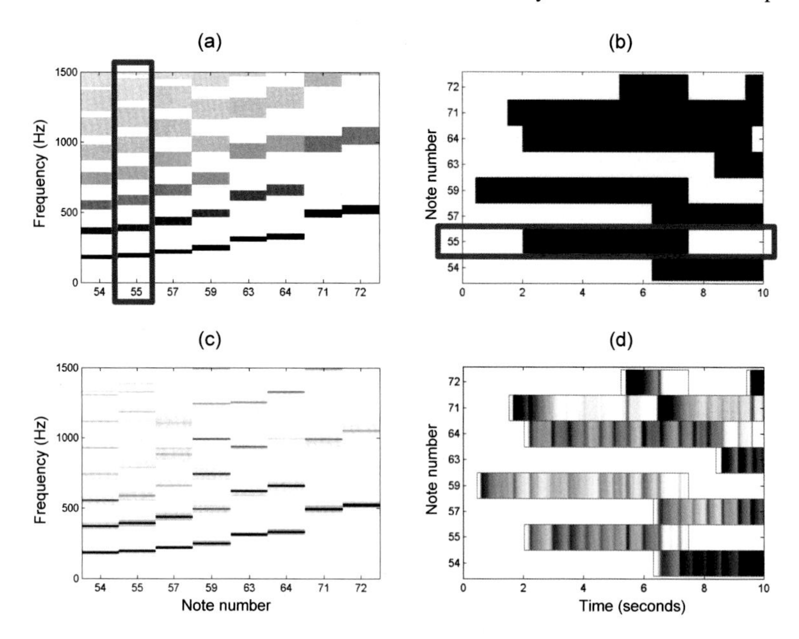

Fig. 8.24 NMF with a harmonically informed initialization of the template vectors and score-informed initialization of the activation matrix. The matrix V is the magnitude spectrogram from Figure 8.21c, and the score information comes from the synchronized piano-roll representation shown in Figure 8.21b. **(a)** Initialization of $W^{(0)}$ using harmonic templates. **(b)** Initialization of $H^{(0)}$ using score information. **(c)** Learned W. **(d)** Learned H.

plates automatically to $R = 8$. One can also refine the initialization of the templates using instrument-dependent harmonic models, which may be derived from example recordings of isolated notes.

On a finer level, one may also exploit *local* information about when notes are actually played. As in Section 8.2.3.2, we assume that a score aligned to a corresponding audio recording is available, i.e., that the note events specified by the score are aligned to the time positions where they occur in the audio recording. Such an alignment may be computed automatically by employing music synchronization techniques (see Chapter 3). Based on the synchronized score information, one can impose time constraints on where certain templates may become active. To this end, one initializes suitable regions in $H^{(0)}$ by setting the corresponding activation entries to one. The remaining entries are set to zero. Such a score-informed initialization is illustrated by Figure 8.24b. For example, the activation entries corresponding to pitch $p = 55$ are constrained to the time interval ranging from roughly $t = 2$ to $t = 7$ (given in seconds).

To account for possible alignment inaccuracies, the temporal boundaries for the constraint regions can be chosen rather generously. As a result, the activation matrix H can be interpreted as a coarse piano-roll representation of the synchronized score information. In a sense, the synchronization step can be seen as yielding a first

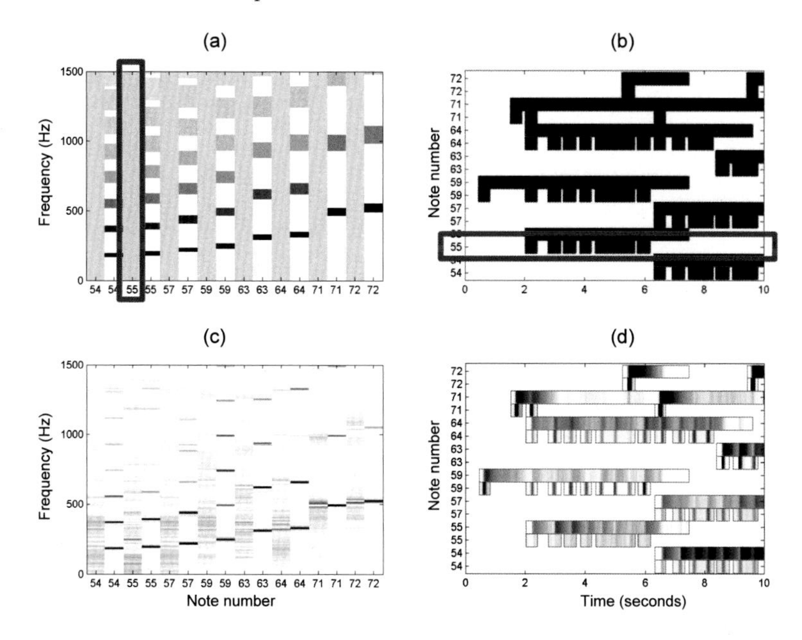

Fig. 8.25 Score-informed NMF using an extended NMF model with additional onset templates. The matrix V is the magnitude spectrogram from Figure 8.21c **(a)** Initialization of $W^{(0)}$ using harmonic templates. **(b)** Initialization of $H^{(0)}$ using score information. **(c)** Learned W. **(d)** Learned H.

approximate factorization, which is then refined by the NMF-based learning procedure. As expected, combining the activation constraints with those for the template vectors further stabilizes the factorization in the sense that the result better reflects the desired semantics. For example, most of the transients' short-time energy bursts, which are spread across activations of all pitches in Figure 8.23d, are suppressed in Figure 8.24d.

8.3.2.3 Onset Models

So far, our NMF-based decomposition only represents harmonic properties of the signal. To also account for percussive properties such as onsets, we now show how one may extend the NMF model by introducing additional templates. A first idea is to use one additional template that jointly explains the onsets for all pitches. However, since the spectral patterns for note onsets may not be completely independent from the respective pitch (as is the case with the piano), we use in the following one additional onset template for each pitch. In general, it is hard to predict the spectral shape of an onset template. Therefore, in contrast to the harmonic templates, we do not enforce any spectral constraints, but initialize the onset templates uniformly and let the learning process derive their respective shapes (see Figure 8.25a).

While it is hard to find meaningful spectral constraints for the onset templates, the short-time nature of note onsets makes it possible to introduce relatively strict temporal constraints on the activation side. Using the synchronized score information, one has a rough estimate for the onset positions. Within a small neighborhood around each of these positions (accounting for possible synchronization inaccuracies), we initialize the corresponding activation entries with one, while setting all remaining entries to zero. This strongly constrains the time points where onset templates are allowed to become active (see Figure 8.25b).

Let us have a look at how the additional onset templates affect the factorization. As illustrated by Figure 8.25c, the learned harmonic template vectors have a much cleaner harmonic structure compared with the ones in previous factorizations. The reason is that most of the transients' broadband energy is now captured by the onset templates. Furthermore, one can observe an impulse-like activation pattern at the beginning of note events for most of the onset templates. This indicates that these templates indeed represent onsets. Having a closer look at the onset templates, one can observe a spread of energy across many different spectral coefficients. Still, the spread is concentrated in regions around the fundamental frequency and the first harmonic.

In summary, we have seen that a combination of template and activation constraints may significantly stabilize and guide an NMF-based factorization process to yield a musically meaningful decomposition. In this process, constraints on the activation side can compensate for using relatively loose or even no constraints on the template side and vice versa. Note that the constraints introduced are hard in the sense that zero entries in $W^{(0)}$ and $H^{(0)}$ remain zero throughout the learning process. Therefore, one should use rather generous constraint regions to account for potential synchronization inaccuracies and to retain a certain degree of flexibility. Finally, we want to emphasize one of the most advantageous aspects of the proposed strategy. Using hard constraints by setting certain coefficients to zero allows for using exactly the same multiplicative update rules as in standard NMF. Thus the constrained procedure inherits the ease of implementation and computational efficiency of the original approach.

8.3.3 Audio Decomposition

The score-informed NMF procedure from Figure 8.25 yields a decomposition of the magnitude spectrum into musically meaningful template and activation matrices. In particular, the notewise activation constraints yield a mapping between nonzero coefficients in H and the score's note events (see Figure 8.25d). We now describe how this spectrogram decomposition can be employed to separate audio components that correspond to specific note groups. These note groups, which may encode a melody line, a certain motif, or the accompaniment, may be specified by means of a suitable labeling of the score representation.

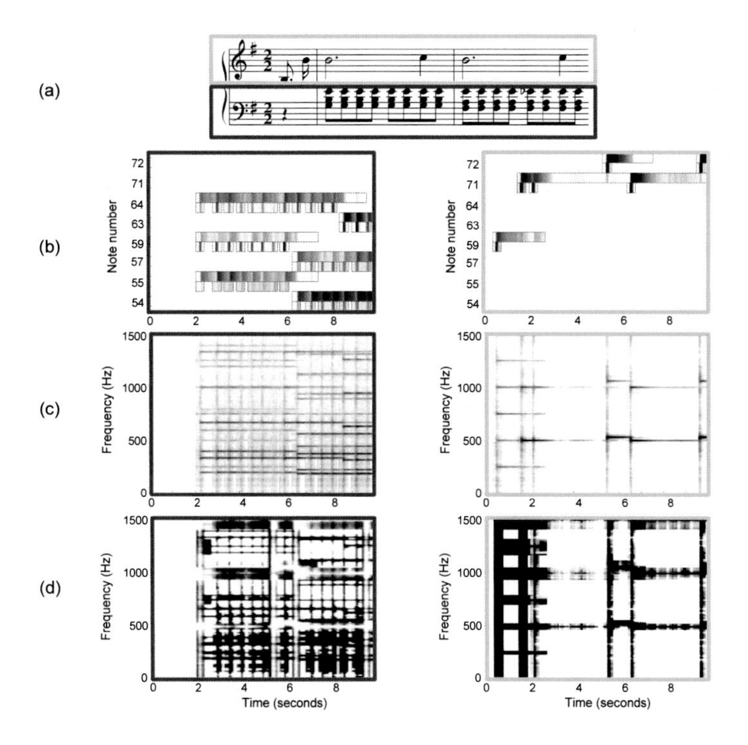

Fig. 8.26 Decomposition of a piano recording into signals corresponding to the notes of the left and the right hand. **(a)** Score representation (Prélude Op. 28, No. 4 by Frédéric Chopin) with a labeling of the notes for the left hand (indicated by the lower red box) and right hand (indicated by the upper yellow box). **(b)** Partition of the activation matrix H from Figure 8.25d into H^{L} and H^{R}. **(c)** Product matrices WH^{L} and WH^{R}. **(d)** Soft masks M^{L} and M^{R}.

8.3.3.1 Separation Process Using Spectral Masking

As an illustrative scenario, let us consider the task of decomposing the recording of our running example into two components. As indicated by Figure 8.26a, one component corresponds to the notes of the lower staff (the accompaniment played by the left hand of the pianist) and the other to the notes of the upper staff (the melody played by the right hand). Using the mapping between the activation constraints in H and the score's notes, we can split up H (see Figure 8.25d) into two matrices H^{L} and H^{R}, which contain the activations for the left and for the right hand, respectively (see Figure 8.26b). Multiplying these two matrices with the template matrix W, we obtain two matrices WH^{L} and WH^{R} as shown in Figure 8.26c. Intuitively, these two matrices can be regarded as the estimated magnitude STFTs for the desired components. To obtain time-domain signals, a first idea is to simply use the phase information of the original STFT $\mathcal{X}$ and to invert the resulting modified STFTs by applying the signal reconstruction method from Section 8.1.2.2. However, NMF-based models typically yield only a rough approximation of the original magnitude spectrogram, where spectral nuances may not be captured well. Therefore, the audio components reconstructed in this way may contain a number of audible artifacts.

Some of these artifacts may be removed or attenuated by using masking techniques as discussed in Section 8.1.1.2. Instead of directly using WH^{L} and WH^{R} for the reconstruction, the idea is to use these matrices to first define the matrices

$$M^{\mathrm{L}} := (WH^{\mathrm{L}}) \oslash (WH + \varepsilon) \qquad \text{and} \qquad M^{\mathrm{R}} := (WH^{\mathrm{R}}) \oslash (WH + \varepsilon) \tag{8.87}$$

(see Figure 8.26d), where the operator $\oslash$ denotes pointwise division and the small positive value $\varepsilon > 0$ is added to avoid division by zero. As in (8.10) and (8.11), these two matrices can then be used as **soft masks**. Applying the two masks to the original STFT $\mathcal{X}$ and employing the signal reconstruction method from Section 8.1.2.2, one obtains the two desired time-domain component signals for the left and the right hand notes, respectively. By using the masking-based approach, many of the spectral details of the original recording are preserved, even if they are not directly captured by the factors of the NMF decomposition. This often yields more acoustically appealing results. On the downside, by filtering the original audio data, masking may also retain more nontarget spectral components compared with a direct reconstruction from WH^{L} and WH^{R}.

8.3.3.2 Notewise Audio Processing

The described separation procedure can be applied for any group of notes or even individual notes specified by the score. Let us further formalize this process for the notewise case, which has already served as an example in Figure 8.19. The goal is to decompose a given audio recording x into notewise audio events x^m for $m \in [1:M]$, where M is the number of note events specified in the score, and a residual signal r such that

$$x = \sum_{m=1}^{M} x^m + r. \tag{8.88}$$

To obtain this decomposition, we apply the masking procedure described above for each note event separately. To this end, we introduce a binary constraint matrix $C^m \in \{0,1\}^{R\times N}$ that describes the activation constraints for note event $m \in [1:M]$. The union (OR-sum) of all C^m is used as initialization for the activation matrix. Based on the resulting score-informed decomposition $V \approx WH$, we define the notewise activation matrix $H^m := H \odot C^m$, where the operator $\odot$ denotes pointwise multiplication. Subsequently, we derive a spectral mask

$$M^m := (WH^m) \oslash (WH + \varepsilon) \tag{8.89}$$

as in (8.87). The mask M^m can be interpreted as a weighting matrix that reflects the contribution of the mth note event to the original spectrogram X. Then, as before, we define $\mathcal{X}^m := \mathcal{X} \odot M^m$ and apply an inverse STFT to obtain the audio event x^m (see Section 8.1.2). Finally, the residual signal is defined as

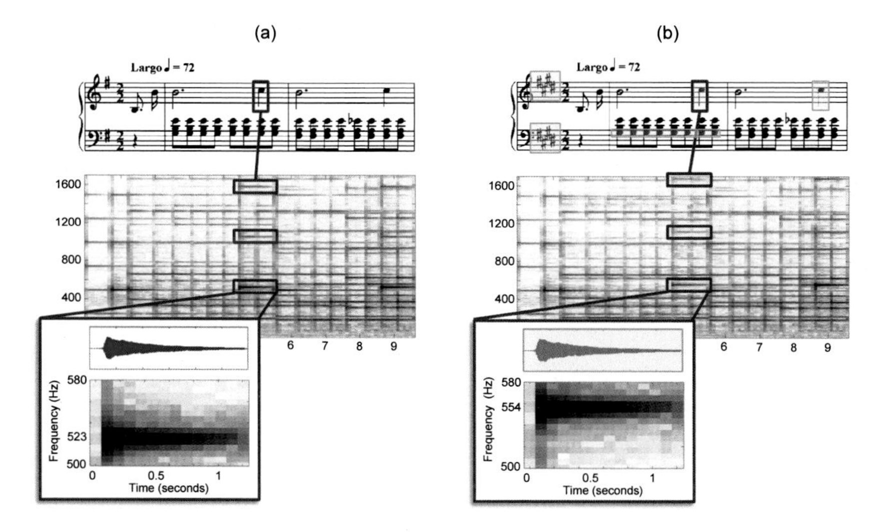

Fig. 8.27 Score-informed audio editing for Prélude Op. 28, No. 4 by Frédéric Chopin. **(a)** Original score and a recording's spectrogram. **(b)** Modified score and spectrogram.

$$r = x - \sum_{m=1}^{M} x^m. \tag{8.90}$$

The signal r holds a lot of valuable information since it may give deeper insights into the quality of the decomposition and separation process. The score-informed decomposition becomes problematic in the case that the recorded performance deviates from the musical score. More generally, synchronization inaccuracies and deviations in the alignment of the score events and their expected realization in the music recording typically lead to local errors in the decomposition. Further artifacts in the decomposition may be introduced by deviations in the expected tuning or by additional sound components caused by resonance or reverberation. Moreover, the model assumption that the harmonic partials' relative energy distribution (the harmonic structure of the templates) is independent of the volume is rather simplistic and may be violated for real sounds. Many of the resulting artifacts and errors are gathered in the residual component [9].

8.3.3.3 Audio Editing

As said before, the availability of score information can not only be exploited for the separation process, but also offers user-friendly access for interactive audio processing applications. Figure 8.27 gives an example of a score-informed audio editing application, where a user can easily specify the desired audio manipulation within the score simply by editing some of the notes. For example, a user may convert the piece originally written in E-minor (Figure 8.27a) into E-major (Figure 8.27b) by changing the **key signature** from one sharp to four sharps. This shifts the note C5 to C$^\sharp$5 and the note G3 to G$^\sharp$3. Based on the notewise audio decomposition, the same manipulations can be automatically transferred to a given audio recording. To this end, the original recording is first decomposed into the notewise audio events of the manipulated notes and a remainder component. Subsequently, the original pitch of each of the notewise audio events is suitably raised or lowered according to the specified score manipulations. Such sound modification techniques are also known as **pitch shifting** (see, e.g., [52]). Finally, the modified audio events are added back to the remainder component. By using similar strategies, it is also possible to change the duration or the volume of notes, to remove notes completely from the audio recording, or to add additional notes by copying and manipulating existing ones [9].

8.4 Summary and Further Readings

In this chapter, we studied various techniques for decomposing a music signal into its constituent components—a task closely related to what is generally referred to as source separation. Given a mixture signal containing different combinations of sources (e.g., vocals, drums, bass, and guitar), source separation aims to recover the individual source signals as if they were played in isolation [7]. Musical sources (within a mixture) often follow the same rhythmic patterns or play harmonically related notes. These strong correlations make music source separation a particularly challenging area of research. To make the decomposition of music recordings feasible, one general strategy is to use additional knowledge about the sources. For example, as noted in [7], musical sources often have a regular harmonic structure where groups of equally spaced frequencies are formed. Specific instruments may go along with characteristic frequency contours (e.g., vibrato patterns for singing). Other sources may correspond to temporal-spectral patterns that repeat over time [42].

This chapter has dealt with three audio decomposition scenarios where musical knowledge was exploited in one way or another. First, in Section 8.1, we exploited the property that percussive instruments typically exhibit structures in the frequency direction (short bursts of broadband energy) while harmonic instruments usually lead to structures in the time direction (slowly changing harmonics). This observation has motivated the method for decomposing a music signal into a harmonic and a percussive component [17]. Second, in Section 8.2, we studied techniques for automatically estimating frequency trajectories of harmonic sources [45]. Based on the assumption that the melody often correlates to the predominant fundamental frequency trajectory, we showed how this information can be used for decomposing a music signal into a melody component that captures the main melodic voice and an accompaniment component that captures the remaining acoustic events. Third, in Section 8.3, we discussed a score-informed approach for decomposing a music signal into notewise audio events [16]. As the underlying technique, we introduced the concept of nonnegative matrix factorization (NMF) [27], which is based on the assumption that characteristic events (in our case spectral vectors) repeatedly occur. While considering three selected decomposition scenarios, we studied various fundamental techniques, including signal reconstruction, STFT inversion, instantaneous frequency estimation, harmonic summation, and NMF. In the subsequent paragraphs, we give pointers to the literature that further discusses these topics.

As for the many tasks related to music source separation and signal decomposition, we only scratched the surface, and it is beyond the book's scope to give an overview of this extensive research area. For pointers to the literature, we refer to the overview article by Cano et al. [7]. In recent years, significant advances could be achieved using supervised learning methods based on deep neural networks (see, e.g., [22, 35, 41]). A good overview of current approaches is also provided by the Signal Separation Evaluation Campaign (SiSEC)—a community-based initiative to compare the performance of source separation systems based on standard datasets and evaluation metrics [50]. For open-source implementations of source separation systems based on deep learning, we refer to [31, 51].

Harmonic–Percussive Separation

The task of decomposing an audio signal into a harmonic and a percussive component has received much research interest, and various decomposition algorithms have been proposed. For example, in the earlier approach [14], the percussive part is separated by looking at signal components with an unstable phase behavior, while the remainder is defined to be the harmonic part of the signal. Subsequently, HPS approaches were proposed based on the observation that harmonic sounds reveal horizontal time–frequency structures, while percussive sounds reveal vertical ones. Based on this observation, Ono et al. [36] introduced an iterative diffusion technique applied to a spectrogram once in the horizontal and once in the vertical direction. Similar to the approach discussed in this chapter, spectral masking and signal reconstruction techniques are used to derive the harmonic and percussive signal components. Following the same lines, Fitzgerald [17] replaced the diffusion step with a simpler median filtering strategy, which turned out to yield similar results while having a much lower computational complexity. In Section 8.1.1, we closely followed Fitzgerald's approach. One problem with HPS is that certain signal components may be neither of harmonic nor of percussive nature (e.g., noise-like sounds such as applause). For an extension of HPS which introduces a third residual component, we refer to [11] (see also Exercise 8.5).

In Section 8.1.1.2, we discussed the topic of **time–frequency masking**. Applying a mask in the spectral domain in a pointwise fashion is equivalent to applying a suitable convolution filter in the time domain. In the case of soft time–frequency masking, this leads to an important concept in signal processing referred to as **Wiener filtering**. For details on this topic, we refer to [21].

Finally, in Section 8.1.2, we dealt with another important problem in digital signal processing, which is concerned with the **reconstruction** of a time-domain signal from a modified STFT. In their seminal paper [20], Griffin and Lim described several algorithms, including the overlap–add approach discussed in this chapter. Simply taking the original phase, this approach can lead to clicking and ringing artifacts in the reconstructed signal. As the main contribution of [20], the authors introduced a method that attenuates the inversion artifacts by iteratively modifying the original phase information. For variants and alternative phase estimation approaches, we refer to [37, 39, 43].

Melody Extraction

Main melody extraction, especially for the singing voice, is a central research topic in music information retrieval. For a comprehensive overview of related tasks and their applications, we refer to the overview articles [23, 32, 45]. In Section 8.2, we discussed a subtask of melody extraction, often referred to as **predominant** F0 **estimation** [4, 19, 38, 44]. As described in [38], the problem of main melody extraction is traditionally split into a preliminary analysis stage followed by a melody identification phase and concluded by a smoothing or tracking process. This is also the strategy presented in this chapter. Closely following [44], we introduced a salience representation used in the preliminary analysis stage. One important step was to refine the frequency grid introduced by a discrete STFT. To this end, we applied a technique for estimating the **instantaneous frequency**. Originally, this technique was used in the context of a **phase vocoder**, where one objective is to modify the time scale of an audio signal by exploiting the phase information [18]. An excellent tutorial on the phase vocoder and the underlying techniques can be found in [8]—this article also served as a source of inspiration for Section 8.2.1. As a second important technique, we discussed the concept of **harmonic summation**, which is a common technique used in fundamental frequency estimation [26, 44]. Finally, we want to note that data-driven methods based on deep neural networks have been successively applied for fundamental frequency tracking and for learning salience representations (see, e.g., [5, 4, 25]). For an example of a richly annotated multitrack dataset, as required for such supervised learning approaches, we refer to [6].

Numerous problems are conceptually similar or even equivalent to melody extraction [32]. For example, the melody is often performed by a solo or lead instrument. We refer to [41] for an overview of lead and accompaniment separation approaches. Another problem closely related to melody tracking is the extraction of the bass line from a music recording. The term **bass line** is typically used to refer to a linear succession of musical tones played with a bass guitar, a double bass, or a bass synthesizer. The bass line plays an important role in several music styles such as jazz or popular music. Therefore, explicit knowledge of the bass line is useful for a multitude of applications including chord extraction, downbeat estimation, genre classification, and so on [1]. Finally, we want to mention the problem of **drum extraction**, which requires techniques that are quite different from those used for melody extraction. Most research in drum extraction has targeted the Western drum kit composed of instrument classes including the bass drum, snare drum, cymbals, and hihat. Starting from solo drum signals, recent studies tackle the more realistic scenario of extracting and transcribing drum signals directly from polyphonic sound mixtures. For an overview and pointers to the literature, we refer to the comprehensive article by Wu et al. [56].

NMF-Based Audio Decomposition

In Section 8.3, we discussed **nonnegative matrix factorization** (NMF) as an example for an important machine learning and matrix decomposition technique. In their seminal paper, Lee and Seung [27] introduced and analyzed multiplicative update rules, laying the theoretical basis for a wide range of NMF applications. In music processing, NMF and its variants have been extensively used for tasks such as music transcription [3, 48], source separation [54], and structure analysis [55]. For an overview of NMF and its extensions, as well as applications to music processing and beyond, we refer to the overview articles [16, 49, 53, 57].

As discussed in Section 8.3.2, the application of standard NMF may lead to a decomposition that is hard to interpret from a musical perspective. Therefore, one often imposes certain constraints on the template and activation matrices. A typical approach is to enforce a harmonic structure in the templates, and temporal continuity in the activations (see, e.g., [3, 24, 40, 54]). In Section 8.3.2.2, we showed how synchronized score information can be exploited to guide the NMF decomposition process. Besides harmonic constraints [40] on the template side, the aligned note information can be used to introduce constraints on the activation side [15]. The application of dual constraints for the template as well as the activation matrix can significantly stabilize the factorization process, leading to musically meaningful audio decompositions. Of course, the availability of synchronized score information is a strong requirement, which only applies for limited classes of music.

The NMF-based procedures, as discussed in this chapter, rely on the assumption that the fundamental frequency associated with a musical pitch is approximately constant over time since the frequency position of harmonics in each template is fixed and cannot move up or down. While this assumption is valid for some instruments, such as a piano, which typically produces stable horizontal frequency trajectories, it is not true in general. For example, a violin may produce strong frequency modulations (vibrato) due to the way it is played. As a result, while a single note in the score is associated with a single musical pitch, its realization in the audio can be much more complex, involving a whole range of frequencies. To deal with such fluctuating fundamental frequencies, parametric signal models have been considered as extensions to NMF (see, e.g., [2, 13, 24]). In these approaches, the musical audio signal is modeled using a family of parameters that capture the fundamental frequency and its temporal fluctuations, the spectral envelope of instruments, or the amplitude progression. Such parameters often have an explicit acoustic or musical interpretation and allow for an integration of available score information.

We want to conclude this section by mentioning that significant progress has been achieved in many music processing tasks, including source separation, by apply-

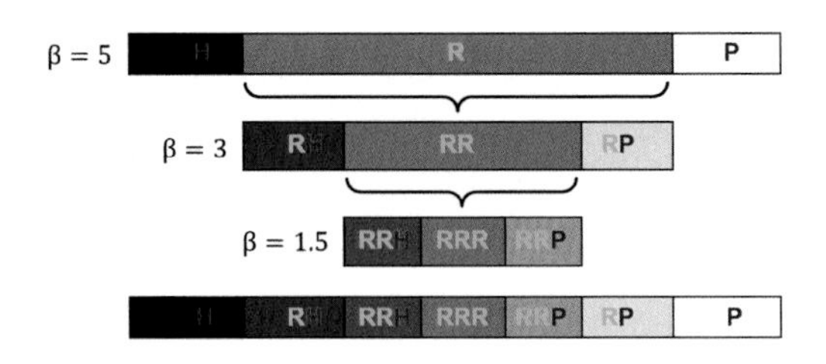

Fig. 8.28 Cascaded HRPS procedure with three cascades using different separation parameters β (see [29]).

ing deep learning (DL) techniques. Rather than requiring explicit modeling, these approaches take advantage of optimization techniques to train source models in a supervised manner, where both the mix and the isolated sources are required [7]. If a huge number of multi-track recordings are available, DL-based methods are capable of learning and separating even complex time–frequency patterns as occurring, e.g., in singing [23]. In other words, rather than exploiting a source's musical function (e.g., melody or bassline), such approaches perform the separation based on learned timbre- and instrument-specific cues [47]. For other scenarios, more traditional techniques such as NMF (while being explicit and instructive) may still be an alternative—in particular, when only little training material is available.

8.5 FMP Notebooks

Besides introducing basic source separation and audio decomposition tasks, the main purpose of this chapter was to study fundamental signal processing techniques such as signal reconstruction, STFT inversion, instantaneous frequency estimation, harmonic summation, and NMF. In **Part 8** of the FMP notebooks [33], we deepen the studies of these techniques from an implementation perspective while providing many instructive examples and illustrations.

We start in the **FMP Notebook** ***Harmonic–Percussive Separation (HPS)*** by providing an implementation of the simple yet beautiful decomposition approach originally suggested by Fitzgerald [17]. Closely following Section 8.1.1, we cover the required mathematical concepts such as median filtering, binary masking, and soft masking. Applied to spectrograms, the effects of these techniques immediately become clear when visualizing the processed matrices. The main parameters of the HPS implementation are the window length N and the hop size H of the STFT as well as the length parameters L^{h} and L^{p} of the median filters applied in the horizontal (time) and vertical (frequency) direction, respectively. To illustrate the role of and interplay between the different parameters, we conduct a systematic experiment where one can listen to the resulting harmonic and percussive sound components for various real-world music recordings. This way, one can understand acoustically how, e.g., the energy "flows" from the harmonic to the percussive component when increasing L^{h}. The opposite occurs when increasing L^{p}. In general, due to the interplay between the four parameters, it is not easy to predict the sound quality of

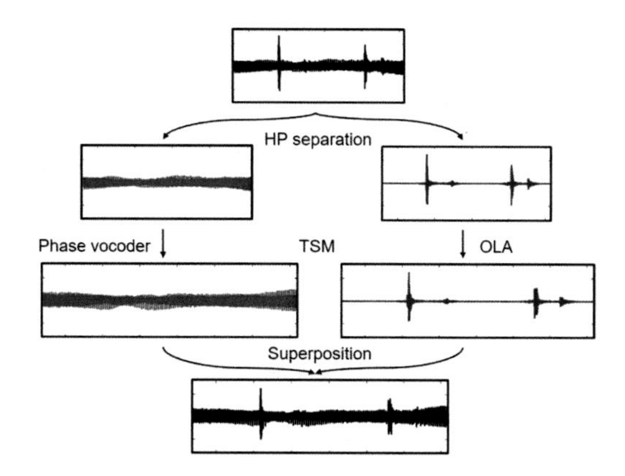

Fig. 8.29 Application of HPS for time-scale modification (TSM), where TSM based on the phase vocoder is used for the harmonic component and TSM based on OLA (overlap–add techniques) is used for the percussive component (see [12]).

the resulting components. In the **FMP Notebook *Harmonic–Residual–Percussive Separation (HRPS)***, we provide an implementation of the extended HPS approach by introducing a third, residual component (see [11] and Exercise 8.5). Again, students are encouraged to listen to the separated sound components computed for various music recordings. This HPRS approach has an additional separation parameter β, which can be used to adjust the size of the middle, residual component. We conclude the notebook by discussing a cascaded application of HRPS where the residual component is further decomposed using different separation factors β (see [29] and Figure 8.28).

In the HPS procedure, we need to reconstruct the time-domain signals for the harmonic and percussive sound components from modified STFTs. However, as discussed in Section 8.1.2.2, modified STFTs are typically not valid in the sense that there is no time-domain signal whose STFT coincides with the specified modified STFT. Intuitively, the problem arises from the STFT's overlapping windows, which reintroduce in the reconstruction some information from the previous and subsequent frames into the current frame. This fact, which is not easy to understand and often overlooked when employing black-box implementations for STFT inversion, may lead to unexpected signal artifacts. In the **FMP Notebook *Signal Reconstruction***, we provide Python code that yields such an example (similar to Figure 8.8). We strongly recommend that students experiment with such examples to gain a feeling on the intricacies of STFT inversion. As we already indicated in the **FMP Notebook *STFT: Inverse*** of **Part 2**, signal reconstruction needs to be regarded as an optimization problem, where the objective is to estimate a signal whose STFT is at least as close as possible to the modified STFT with regard to a suitably defined distance measure. Using a measure based on the mean square error leads to the famous approach originally introduced by Griffin and Lim [20]. This approach is often used as default in implementations of the inverse STFT, as, e.g., provided by the Python package `librosa`.

In the **FMP Notebook *Applications of HPS and HRPS***, we cover Python implementations for the applications sketched in Section 8.1.3. Similar to Figure 8.9, we show how to enhance a chroma representation by considering only a signal's

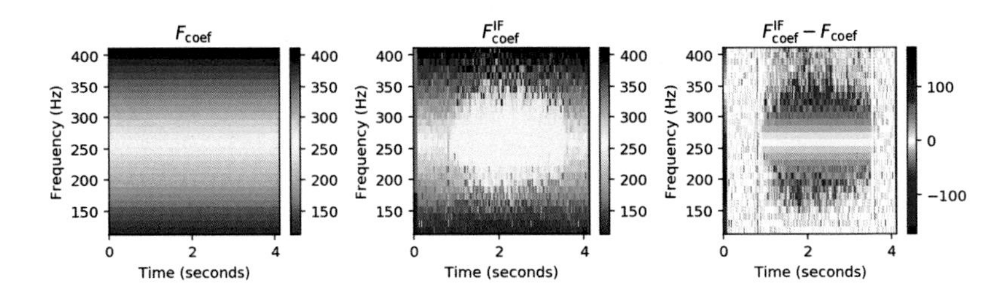

Fig. 8.30 Interpretation of time–frequency bins of an STFT as specified by the (frame-independent) frequency values $F_{\mathrm{coef}}(k)$ and the (frame-dependent) instantaneous frequency values $F^{\mathrm{IF}}_{\mathrm{coef}}(k,n)$. The bin offset (specified in Hertz) is given by $F^{\mathrm{IF}}_{\mathrm{coef}} - F_{\mathrm{coef}}$.

harmonic component and how to enhance a novelty representation by considering only a signal's percussive component. While these simple applications should by seen only as an illustration of the HPS decomposition's potential, we also sketch a more serious application in the context of time-scale modification (TSM), where the task is to speed up or slow down an audio signal's playback speed without changing its pitch. The main idea of the TSM approach by Driedger et al. [12] is to first split up the signal into a harmonic and percussive component using HPS. The two components are then processed separately using specialized TSM approaches—one that is specialized to stretch tonal elements of music and one that is specialized to preserve transients. The final TSM result is then obtained by superimposing the two modified components (see Figure 8.29). For an implementation of this procedure, which goes beyond the scope of the FMP notebooks, we refer to [10].

In the next notebooks, we turn to the topic of melody extraction, which serves as a motivating scenario for studying several important signal processing techniques. In the **FMP Notebook *Instantaneous Frequency Estimation***, we show how one can improve the frequency resolution of the discrete STFT by exploiting the information hidden in its phase. Looking at the phases of subsequent frames, as described in Section 8.2.1, allows for adjusting the STFT's frame-independent grid frequency $F_{\mathrm{coef}}(k)$ (see (8.30)) to obtain a frame-dependent instantaneous frequency $F^{\mathrm{IF}}_{\mathrm{coef}}(k,n)$ (see (8.44)). Besides providing an implementation, the notebook introduces visualizations that yield deeper insights into the IF estimation procedure. In particular, we look at a piano recording of the note C4 with fundamental frequency 261.5 Hz as an example (see Figure 8.30). The resulting visualization indicates that the IF estimation procedure assigns all frequency coefficients in a neighborhood of 261.5 Hz to exactly that frequency (see $F^{\mathrm{IF}}_{\mathrm{coef}}$ of Figure 8.30). Also the difference $F^{\mathrm{IF}}_{\mathrm{coef}} - F_{\mathrm{coef}}$, which corresponds to the bin offset computed in (8.45), is shown. The notebook closes with an experiment that indicates how the quality of the estimated instantaneous frequency depends on the hop-size parameter (with small hop sizes improving the IF estimate). In conclusion, an overall aim of the notebook is to emphasize the potential of the phase information—an aspect that is often neglected in a signal processing course.

As a second concept central to melody extraction, we introduce in the **FMP Notebook** ***Salience Representation*** a time–frequency representation that emphasizes the predominant frequency information. Closely following the explanations of Section 8.2.2, we provide a step-by-step implementation of the procedure. As our running example, we use a short excerpt of an aria from the opera "Der Freischütz" by Carl Maria von Weber. We first examine the shortcomings of logarithmic binning methods based on the STFT's linearly spaced frequency grid and then discuss the benefits when using the IF-based frequency refinement. Next, we introduce a Python function for harmonic summation following (8.55). Applied to our Weber example, the effect of harmonic summation does not seem to be huge—a disappointment that students often encounter when they put theory into practice. In our case, as we discuss in this notebook, a high frequency resolution in combination with the IF-based sharpening leads to small deviations across harmonically related frequency bins. To balance out these deviations, we introduce a simple method by introducing a smoothing step along the frequency axis. Continuing our Weber example, we show that this small modification increases the robustness of the harmonic summation, leading to significant improvements in the resulting salience representation. In general, when applying local operations to data that is sampled with high resolution, small deviations or outliers in the data may lead to considerable degradations. In such situations, additional filtering steps (e.g., convolution with a Gaussian kernel or median filtering) may help to alleviate some of the problems. Besides providing reference implementations, it is at the core of the FMP notebooks to also bring up practical issues and introduce small engineering tricks that may help in practice.

Assuming that the main melody corresponds to the strongest harmonic frequency component at each time point motivates the next topic covered by the **FMP Notebook** ***Fundamental Frequency Tracking***. Continuing our Weber example, we start by providing Python code for the visualization and sonification (using sinusoidal models) of frequency trajectories. In particular, listening to a trajectory's sonification superimposed with the original music recordings yields an excellent acoustic feedback on the trajectory's accuracy. Then, following Section 8.2.3, we provide implementations of different frequency tracking procedures, including a frame-wise approach, an approach using continuity constraints, and a score-informed approach. Again, the benefits and limitations of these approaches are made tangible through visualizations and sonifications of concrete examples. This again highlights the main purpose of the FMP notebooks: Instead of just passively following the concepts, the notebooks enable students to deepen their understanding by conducting experiments using their own examples.

Finally, in the **FMP Notebook** ***Melody Extraction and Separation***, we show how to integrate the algorithmic components learned in the previous notebooks to build a complete system. Based on the assumption that the melody correlates to the predominant fundamental frequency trajectory, we show how this information can be used for decomposing a music signal into a melody component that captures the main melodic voice and an accompaniment component that captures the remaining acoustic events. To this end, given a (previously estimated) predominant frequency trajectory, we construct a binary mask that takes harmonics into account

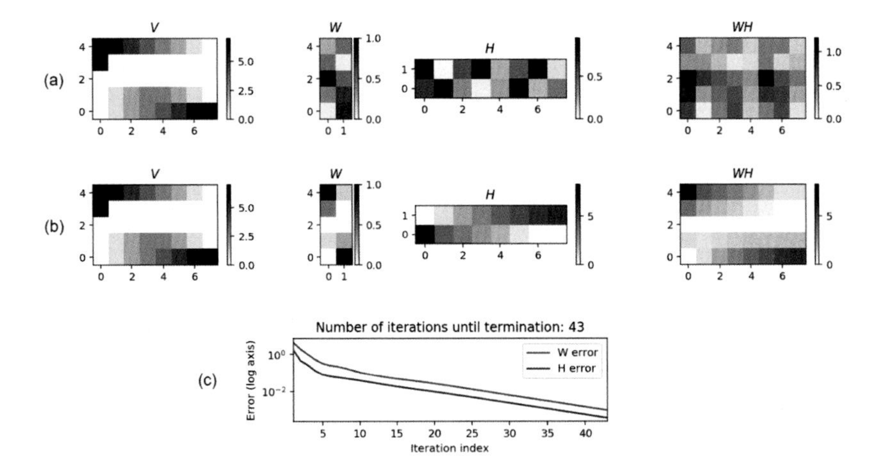

Fig. 8.31 NMF procedure applied to a toy example. **(a)** Matrix V and randomly initialized matrices W and H. **(b)** Matrix V and matrices W and H after training. **(c)** Error terms over iterations.

(see Section 8.2.3.3). In our implementation, we consider two variants, one based on a fixed size around each frequency bin and one based on a frequency-dependent size (where the neighborhood size increases linearly with the center frequency). Using such a binary mask and its complement, we apply the same signal reconstruction techniques as introduced in Section 8.1.2 to obtain component signals for the trajectory (i.e., the melody) and the rest (i.e., the accompaniment). Admittedly, the overall procedure is too simplistic to obtain state-of-the-art results in source separation. However, providing a full pipeline along with visual and acoustic analysis tools should invite students to explore the role of the various components and to start with their own research (e.g., using more advanced methods based on deep learning as provided by [31, 51]).

In the final three notebooks, we turn to a powerful and beautiful machine learning technique that is applicable for general data analysis far beyond the considered music scenario. Closely following the theory of Section 8.3.1, we provide in the **FMP Notebook *Nonnegative Matrix Factorization (NMF)*** a basic implementation of the NMF algorithm as specified in Table 8.1. There are several practical issues one needs to consider. First, for efficiency reasons, we use matrix-based operations for implementing the multiplicative update rules. Second, to avoid division by zero, a small value (machine epsilon) is added to the denominators in the multiplicative update rules. Third, we provide a parameter for controlling certain normalization constraints (e.g., enforcing that template vectors are normalized). Fourth, the implementation allows for specifying matrices used for initialization. Finally, different criteria may be used to terminate the iterative optimization procedure. Using explicit toy examples, we present some experiments that illustrate the functioning of the NMF procedure and discuss the role of the rank parameter (see also Figure 8.31).

For further extensions, implementations, and applications of NMF, we refer to the NMF toolbox [30].

The **FMP Notebook** ***NMF-Based Spectrogram Factorization*** yields an application of NMF to decompose a magnitude spectrogram into template and activation matrices that possess an explicit musical meaning. As in Section 8.3.2, we use the first measures of Chopin's Prélude Op. 28, No. 4 to demonstrate how one may integrate musical knowledge to guide the decomposition. In particular, we provide complete Python implementations for various initialization strategies using pitch-informed template constraints and score-informed activation constraints. Furthermore, to also account for percussive properties such as onsets, we implement the NMF model with additional onset templates as introduced in Section 8.3.2.3. This extended NMF model is then applied in the **FMP Notebook** ***NMF-Based Audio Decomposition*** for score-informed spectrogram factorization. In particular, we provide a full pipeline for decomposing a music recording into note-based sound events. Continuing our Chopin example, we decompose the recording into two components, where one component corresponds to the notes of the lower staff and the other to the notes of the upper staff (see also Figure 8.26). Providing the code and all the data required, this implementation reproduces the results originally introduced in [15]. Furthermore, we sketch the audio editing application based on notewise audio decomposition showing a video as presented in [9].

References

1. J. ABESSER, *Automatic Transcription of Bass Guitar Tracks Applied for Music Genre Classification and Sound Synthesis*, PhD thesis, Ilmenau University of Technology, Ilmenau, Germany, 2014.
2. E. BENETOS AND S. DIXON, *Multiple-instrument polyphonic music transcription using a temporally constrained shift-invariant model*, The Journal of the Acoustical Society of America (JASA), 133 (2013), pp. 1727–1741.
3. N. BERTIN, R. BADEAU, AND E. VINCENT, *Enforcing harmonicity and smoothness in bayesian non-negative matrix factorization applied to polyphonic music transcription*, IEEE Transactions on Audio, Speech, and Language Processing, 18 (2010), pp. 538–549.
4. R. M. BITTNER, *Data-Driven Fundamental Frequency Estimation*, PhD thesis, New York University, 2018.
5. R. M. BITTNER, B. MCFEE, J. SALAMON, P. LI, AND J. P. BELLO, *Deep salience representations for F0 tracking in polyphonic music*, in Proceedings of the International Society for Music Information Retrieval Conference (ISMIR), Suzhou, China, 2017, pp. 63–70.
6. R. M. BITTNER, J. SALAMON, M. TIERNEY, M. MAUCH, C. CANNAM, AND J. P. BELLO, *MedleyDB: A multitrack dataset for annotation-intensive MIR research*, in Proceedings of the International Society for Music Information Retrieval Conference (ISMIR), Taipei, Taiwan, 2014, pp. 155–160.
7. E. CANO, D. FITZGERALD, A. LIUTKUS, M. D. PLUMBLEY, AND F. STÖTER, *Musical source separation: An introduction*, IEEE Signal Processing Magazine, 36 (2019), pp. 31–40.
8. M. DOLSON, *The phase vocoder: A tutorial*, Computer Music Journal, 10 (1986), pp. 14–27.
9. J. DRIEDGER, H. GROHGANZ, T. PRÄTZLICH, S. EWERT, AND M. MÜLLER, *Score-informed audio decomposition and applications*, in Proceedings of the ACM International Conference on Multimedia (ACM-MM), Barcelona, Spain, 2013, pp. 541–544.
10. J. DRIEDGER AND M. MÜLLER, *TSM Toolbox: MATLAB implementations of time-scale modification algorithms*, in Proceedings of the International Conference on Digital Audio Effects (DAFx), Erlangen, Germany, 2014, pp. 249–256.
11. J. DRIEDGER, M. MÜLLER, AND S. DISCH, *Extending harmonic–percussive separation of audio signals*, in Proceedings of the International Society for Music Information Retrieval Conference (ISMIR), Taipei, Taiwan, October 2014, pp. 611–616.
12. J. DRIEDGER, M. MÜLLER, AND S. EWERT, *Improving time-scale modification of music signals using harmonic–percussive separation*, IEEE Signal Processing Letters, 21 (2014), pp. 105–109.
13. J.-L. DURRIEU, G. RICHARD, B. DAVID, AND C. FÉVOTTE, *Source/filter model for unsupervised main melody extraction from polyphonic audio signals*, IEEE Transactions on Audio, Speech, and Language Processing, 18 (2010), pp. 564–575.
14. C. DUXBURY, M. DAVIES, AND M. B. SANDLER, *Separation of transient information in audio using multiresolution analysis techniques*, in Proceedings of the International Conference on Digital Audio Effects (DAFx), Limerick, Ireland, 2001.
15. S. EWERT AND M. MÜLLER, *Using score-informed constraints for NMF-based source separation*, in Proceedings of the IEEE International Conference on Acoustics, Speech, and Signal Processing (ICASSP), Kyoto, Japan, March 2012, pp. 129–132.
16. S. EWERT, B. PARDO, M. MÜLLER, AND M. PLUMBLEY, *Score-informed source separation for musical audio recordings: An overview*, IEEE Signal Processing Magazine, 31 (2014), pp. 116–124.
17. D. FITZGERALD, *Harmonic/percussive separation using median filtering*, in Proceedings of the International Conference on Digital Audio Effects (DAFx), Graz, Austria, September 2010, pp. 246–253.
18. J. L. FLANAGAN AND R. M. GOLDEN, *Phase vocoder*, Bell System Technical Journal, 45 (1966), pp. 1493–1509.
19. M. GOTO, *A robust predominant-F0 estimation method for real-time detection of melody and bass lines in CD recordings*, in IEEE International Conference on Acoustics, Speech, and Signal Processing (ICASSP), vol. 2, 2000, pp. 757–760.

20. D. W. GRIFFIN AND J. S. LIM, *Signal estimation from modified short-time Fourier transform*, IEEE Transactions on Acoustics, Speech, and Signal Processing, 32 (1984), pp. 236–243.
21. M. H. HAYES, *Statistical Digital Signal Processing and Modeling*, Wiley, 1st ed., 1996.
22. P.-S. HUANG, M. KIM, M. HASEGAWA-JOHNSON, AND P. SMARAGDIS, *Joint optimization of masks and deep recurrent neural networks for monaural source separation*, IEEE/ACM Transactions on Audio, Speech, and Language Processing, 23 (2015), pp. 2136–2147.
23. E. J. HUMPHREY, S. REDDY, P. SEETHARAMAN, A. KUMAR, R. M. BITTNER, A. DEMETRIOU, S. GULATI, A. JANSSON, T. JEHAN, B. LEHNER, A. KRUPSE, AND L. YANG, *An introduction to signal processing for singing-voice analysis: High notes in the effort to automate the understanding of vocals in music*, IEEE Signal Processing Magazine, 36 (2019), pp. 82–94.
24. H. KAMEOKA, T. NISHIMOTO, AND S. SAGAYAMA, *A multipitch analyzer based on harmonic temporal structured clustering*, IEEE Transactions on Audio, Speech, and Language Processing, 15 (2007), pp. 982–994.
25. J. W. KIM, J. SALAMON, P. LI, AND J. P. BELLO, *CREPE: A convolutional representation for pitch estimation*, in Proceedings of the IEEE International Conference on Acoustics, Speech and Signal Processing (ICASSP), Calgary, Canada, 2018, pp. 161–165.
26. A. P. KLAPURI, *Multiple fundamental frequency estimation by summing harmonic amplitudes*, in Proceedings of the International Society for Music Information Retrieval Conference (ISMIR), 2006, pp. 216–221.
27. D. D. LEE AND H. S. SEUNG, *Algorithms for non-negative matrix factorization*, in Proceedings of the Neural Information Processing Systems (NIPS), Denver, Colorado, USA, November 2000, pp. 556–562.
28. B. LEHNER, R. SONNLEITNER, AND G. WIDMER, *Towards light-weight, real-time-capable singing voice detection*, in Proceedings of the International Society for Music Information Retrieval Conference (ISMIR), Curitiba, Brazil, 2013, pp. 53–58.
29. P. LÓPEZ-SERRANO, C. DITTMAR, AND M. MÜLLER, *Mid-level audio features based on cascaded harmonic–residual–percussive separation*, in Proceedings of the Audio Engineering Society Conference on Semantic Audio (AES), Erlangen, Germany, June 2017, pp. 32–44.
30. P. LÓPEZ-SERRANO, C. DITTMAR, Y. ÖZER, AND M. MÜLLER, *NMF toolbox: Music processing applications of nonnegative matrix factorization*, in Proceedings of the International Conference on Digital Audio Effects (DAFx), Birmingham, UK, 2019.
31. E. MANILOW, P. SEETHARMAN, AND J. SALAMON, *Open Source Tools & Data for Music Source Separation*, https://source-separation.github.io/tutorial, 2020.
32. M. MÜLLER, D. P. W. ELLIS, A. KLAPURI, AND G. RICHARD, *Signal processing for music analysis*, IEEE Journal on Selected Topics in Signal Processing, 5 (2011), pp. 1088–1110.
33. M. MÜLLER AND F. ZALKOW, *FMP Notebooks: Educational material for teaching and learning fundamentals of music processing*, in Proceedings of the International Society for Music Information Retrieval Conference (ISMIR), Delft, The Netherlands, 2019, pp. 573–580.
34. J. NOCEDAL AND S. J. WRIGHT, *Numerical Optimization*, Springer (Springer Series in Operations Research and Financial Engineering), 2006.
35. A. A. NUGRAHA, A. LIUTKUS, AND E. VINCENT, *Multichannel audio source separation with deep neural networks*, IEEE/ACM Transactions on Audio, Speech, and Language Processing, 24 (2016), pp. 1652–1664.
36. N. ONO, K. MIYAMOTO, J. LEROUX, H. KAMEOKA, AND S. SAGAYAMA, *Separation of a monaural audio signal into harmonic/percussive components by complementary diffusion on spectrogram*, in European Signal Processing Conference (EUSIPCO), Lausanne, Switzerland, 2008, pp. 240–244.
37. N. PERRAUDIN, P. BALAZS, AND P. L. SØNDERGAARD, *A fast Griffin–Lim algorithm*, in Proceedings of the IEEE Workshop on Applications of Signal Processing to Audio and Acoustics (WASPAA), New Paltz, NY, USA, October 2013, pp. 1–4.
38. G. E. POLINER, D. P. ELLIS, A. F. EHMANN, E. GÓMEZ, S. STREICH, AND B. ONG, *Melody transcription from music audio: Approaches and evaluation*, IEEE Transactions on Audio, Speech, and Language Processing, 15 (2007), pp. 1247–1256.

39. Z. Pruša, P. Balázs, and P. L. Søndergaard, *A noniterative method for reconstruction of phase from STFT magnitude*, IEEE/ACM Transactions on Audio, Speech, and Language Processing, 25 (2017), pp. 1154–1164.
40. S. A. Raczynski, N. Ono, and S. Sagayama, *Multipitch analysis with harmonic nonnegative matrix approximation*, in Proceedings of the International Society for Music Information Retrieval Conference (ISMIR), Vienna, Austria, September 2007, pp. 381–386.
41. Z. Rafii, A. Liutkus, F. Stöter, S. I. Mimilakis, D. FitzGerald, and B. Pardo, *An overview of lead and accompaniment separation in music*, IEEE/ACM Transactions on Audio, Speech, and Language Processing, 26 (2018), pp. 1307–1335.
42. Z. Rafii and B. Pardo, *Repeating pattern extraction technique (REPET): A simple method for music/voice separation.*, IEEE Transactions on Audio, Speech, and Language Processing, 21 (2013), pp. 71–82.
43. J. L. Roux and E. Vincent, *Consistent wiener filtering for audio source separation*, IEEE Signal Processing Letters, 20 (2013), pp. 217–220.
44. J. Salamon and E. Gómez, *Melody extraction from polyphonic music signals using pitch contour characteristics*, IEEE Transactions on Audio, Speech, and Language Processing, 20 (2012), pp. 1759–1770.
45. J. Salamon, E. Gómez, D. P. W. Ellis, and G. Richard, *Melody extraction from polyphonic music signals: Approaches, applications, and challenges*, IEEE Signal Processing Magazine, 31 (2014), pp. 118–134.
46. J. Salamon, J. Serrà, and E. Gómez, *Tonal representations for music retrieval: From version identification to query-by-humming*, International Journal of Multimedia Information Retrieval, 2 (2013), pp. 45–58.
47. O. Slizovskaia, L. Kim, G. Haro, and E. Gómez, *End-to-end sound source separation conditioned on instrument labels*, in Proceedings of the IEEE International Conference on Acoustics, Speech, and Signal Processing (ICASSP), 2019, pp. 306–310.
48. P. Smaragdis and J. C. Brown, *Non-negative matrix factorization for polyphonic music transcription*, in Proceedings of the IEEE Workshop on Applications of Signal Processing to Audio and Acoustics (WASPAA), 2003, pp. 177–180.
49. P. Smaragdis, C. Févotte, G. J. Mysore, N. Mohammadiha, and M. D. Hoffman, *Static and dynamic source separation using nonnegative factorizations: A unified view*, IEEE Signal Processing Magazine, 31 (2014), pp. 66–75.
50. F. Stöter, A. Liutkus, and N. Ito, *The 2018 Signal Separation Evaluation Campaign*, in Proceedings of the International Conference on Latent Variable Analysis and Signal Separation (LVA/ICA), vol. 10891 of Lecture Notes in Computer Science, Springer, 2018, pp. 293–305.
51. F. Stöter, S. Uhlich, A. Liutkus, and Y. Mitsufuji, *Open-Unmix – A reference implementation for music source separation*, Journal of Open Source Software, 4 (2019).
52. W. Verhelst and M. Roelands, *An overlap–add technique based on waveform similarity (WSOLA) for high quality time-scale modification of speech*, in Proceedings of the IEEE International Conference on Acoustics, Speech, and Signal Processing (ICASSP), Minneapolis, USA, 1993.
53. E. Vincent, N. Bertin, R. Gribonval, and F. Bimbot, *From blind to guided audio source separation: How models and side information can improve the separation of sound*, IEEE Signal Processing Magazine, 31 (2014), pp. 107–115.
54. T. Virtanen, *Monaural sound source separation by nonnegative matrix factorization with temporal continuity and sparseness criteria*, IEEE Transactions on Audio, Speech, and Language Processing, 15 (2007), pp. 1066–1074.
55. R. J. Weiss and J. P. Bello, *Unsupervised discovery of temporal structure in music*, IEEE Journal of Selected Topics in Signal Processing, 5 (2011), pp. 1240–1251.
56. C.-W. Wu, C. Dittmar, C. Southall, R. Vogl, G. Widmer, J. Hockman, M. Müller, and A. Lerch, *A review of automatic drum transcription*, IEEE/ACM Transactions on Audio, Speech, and Language Processing, 26 (2018), pp. 1457–1483.
57. G. Zhou, A. Cichocki, Q. Zhao, and S. Xie, *Nonnegative matrix and tensor factorizations: An algorithmic perspective*, IEEE Signal Processing Magazine, 31 (2014), pp. 54–65.

Exercises

Exercise 8.1. The arithmetic mean $\mu(A)$ of a list $A = (a_1, a_2, \ldots, a_L)$ that consists of real numbers $a_\ell \in \mathbb{R}$, $\ell \in [1:L]$ is defined by $\mu(A) := \left(\sum_{\ell=1}^{L} a_\ell\right)/L$. Let $A = (2,3,190,2,3)$. Compute the mean $\mu(A)$ as well as the median $\mu_{1/2}(A)$ (see (8.4)). Explain why the HPS algorithm described in Section 8.1 employs median filtering and not mean filtering.

Exercise 8.2. Let

$$\mathcal{Y} = \begin{bmatrix} 1 & 1 & 46 & 2 \\ 3 & 1 & 50 & 1 \\ 60 & 68 & 70 & 67 \\ 2 & 1 & 65 & 1 \end{bmatrix}$$

be a spectrogram. Assuming a suitable zero-padding, compute $\tilde{\mathcal{Y}}^{\mathrm{h}}$ as in (8.6) using $L^{\mathrm{h}} = 3$ and $\tilde{\mathcal{Y}}^{\mathrm{p}}$ as in (8.7) using $L^{\mathrm{p}} = 3$. Furthermore, compute the binary mask $\mathcal{M}^{\mathrm{h}}$ as in (8.8) and $\mathcal{M}^{\mathrm{p}}$ as in (8.9). Finally, apply the masks to the matrix $\mathcal{Y}$ using pointwise multiplication to derive the two matrices $\mathcal{Y}^{\mathrm{h}}$ as in (8.12) and $\mathcal{Y}^{\mathrm{p}}$ as in (8.13).

Exercise 8.3. Let F_{s} (given in Hz) be the sampling rate of a given signal x. Furthermore, let $N \in \mathbb{N}$ be the window length and $H \in \mathbb{N}$ the hop size of a discrete STFT. The filter lengths $L^{\mathrm{h}}, L^{\mathrm{p}} \in \mathbb{N}$ of the median filters used in the HPS approach are specified in terms of frame and frequency indices of the underlying STFT. In practice, it may be more convenient if a user can specify the filter length L^{h} in terms of seconds and L^{p} in terms of Hertz. Derive a formula that converts a time duration $\Delta_t \in \mathbb{R}$ given in seconds to a minimum filter length $L^{\mathrm{h}}(\Delta_t) \in \mathbb{N}$ given in frame indices covering this duration. Similarly, derive a formula that converts a frequency range $\Delta_\omega \in \mathbb{R}$ given in Hertz to a minimum filter length $L^{\mathrm{p}}(\Delta_\omega) \in \mathbb{N}$ given in frequency indices covering this range. Finally, assuming $F_{\mathrm{s}} = 22050$ Hz, $N = 1024$, and $H = 256$, determine $L^{\mathrm{h}}(\Delta_t)$ for $\Delta_t = 0.5$ sec and $L^{\mathrm{p}}(\Delta_\omega)$ for $\Delta_\omega = 600$ Hz.

Exercise 8.4. Show that one obtains a partition of unity (see (8.22)) when using the discrete window function $w : \mathbb{Z} \to \mathbb{R}$ defined by

$$w(r) := \begin{cases} \sin(\pi r/N)^2 & \text{if } r \in [0:N-1], \\ 0 & \text{otherwise} \end{cases}$$

(see (8.23)) and the hop size $H = N/2$ (assuming that N is even). What happens if the hop size $H = N/4$ (assuming that N is divisible by four) is used instead? Give a proof of your claim.

Exercise 8.5. One problem in harmonic–percussive separation (HPS) is that a sound may contain noise-like events (e.g., applause, distorted guitar) that are neither of harmonic nor of percussive nature. In this exercise, we study an extension to HPS by considering a third residual component which captures the sounds that lie "between" a clearly harmonic and a clearly percussive component. To this end, we introduce an additional parameter $\beta \in \mathbb{R}$ with $\beta \geq 1$ called the **separation factor**. Generalizing (8.8) and (8.9), we define the binary masks $\mathcal{M}^{\mathrm{h}}$, $\mathcal{M}^{\mathrm{p}}$, and $\mathcal{M}^{\mathrm{r}}$ for the clearly harmonic, the clearly percussive, and the residual components by setting

$$\mathcal{M}^{\mathrm{h}}(n,k) := \begin{cases} 1 & \text{if } \tilde{\mathcal{Y}}^{\mathrm{h}}(n,k) \geq \beta \cdot \tilde{\mathcal{Y}}^{\mathrm{p}}(n,k), \\ 0 & \text{otherwise}, \end{cases}$$

$$\mathcal{M}^{\mathrm{p}}(n,k) := \begin{cases} 1 & \text{if } \tilde{\mathcal{Y}}^{\mathrm{p}}(n,k) > \beta \cdot \tilde{\mathcal{Y}}^{\mathrm{h}}(n,k), \\ 0 & \text{otherwise}, \end{cases}$$

$$\mathcal{M}^{\mathrm{r}}(n,k) := 1 - \left(\mathcal{M}^{\mathrm{h}}(n,k) + \mathcal{M}^{\mathrm{p}}(n,k)\right).$$

Using these masks, derive a signal decomposition $x = x^{\mathrm{h}} + x^{\mathrm{p}} + x^{\mathrm{r}}$. Furthermore, discuss the role of the parameter β. How do the components change when successively increasing β?

Exercise 8.6. Derive the formula (8.44) for the instantaneous frequency $F_{\mathrm{coef}}^{\mathrm{IF}}(k,n)$ and the formula (8.45) for the bin offset $\kappa(k,n)$.

Exercise 8.7. We have seen in Section 8.2.1 that the quality of the estimated instantaneous frequency depends on the length $\Delta t = t_2 - t_1 = H/F_{\mathrm{s}}$. Therefore, it is beneficial to use a small hop size H. On the downside, using a small hop size increases the computational cost for calculating the discrete STFT. An alternative approach for obtaining good instantaneous frequency estimates is to keep the original hop size, but compute the STFT twice—the second time at a lag of just one sample. Discuss the benefits of this alternative approach over the strategy of simply reducing the hop size.

Exercise 8.8. Defining $\mathrm{Bin}(\omega) := \lfloor 12 \cdot \log_2(\omega/440) + 69.5 \rfloor$ for $\omega \in \mathbb{R}$ as in (8.47), show that $\omega \in [F_{\mathrm{pitch}}(p-0.5), F_{\mathrm{pitch}}(p+0.5))$ if and only if $\mathrm{Bin}(\omega) = p$ for $p \in \mathbb{Z}$.

Exercise 8.9. Let ω be a frequency and $h \cdot \omega$ its hth harmonic for some $h \in \mathbb{N}$. Considering the bin mapping function from (8.49), determine the relation between $\mathrm{Bin}(\omega)$ and $\mathrm{Bin}(h \cdot \omega)$. This relation explains the formula in (8.55) for the harmonic summation in the log-frequency domain.

Exercise 8.10. Let $\mathcal{Y}$ be a magnitude spectrogram with coefficients $\mathcal{Y}(n,k)$ for $n \in \mathbb{Z}$ and $k \in [0:K]$. Furthermore, for a given reference frequency ω_{ref} and a resolution R, let $\mathcal{Y}_{\mathrm{LF}}$ be the log-frequency magnitude spectrogram as defined in (8.51) with coefficients $\mathcal{Y}_{\mathrm{LF}}(n,b)$ for $n \in \mathbb{Z}$ and $b \in [1:B]$. Given a frequency trajectory $\eta : \mathbb{Z} \to [0:K]$ for $\mathcal{Y}$, describe how one can derive a corresponding trajectory $\eta_{\mathrm{LF}} : \mathbb{Z} \to [1:B]$ for $\mathcal{Y}_{\mathrm{LF}}$. Which problems may occur in this calculation?

Furthemore, let η^h and η_{LF}^h be the frequency trajectories of the first $H \in \mathbb{N}$ harmonics, $h \in [1:H]$. Note that $\eta^1 = \eta$ and $\eta_{\mathrm{LF}}^1 = \eta_{\mathrm{LF}}$. Describe the mathematical relations between these trajectories. Thinking of practical computations and real-world musical sounds, discuss some problems that may introduce inaccuracies in these relations.

Exercise 8.11. The goal of this exercise is to develop an efficient algorithm for computing a frequency trajectory with temporal continuity constraints (see Section 8.2.3.1). Given a salience representation $\mathcal{Z} \in \mathbb{R}_{\geq 0}^{N \times B}$ and a transition likelihood matrix $\mathbf{T} \in \mathbb{R}_{\geq 0}^{B \times B}$, let $\sigma(\eta)$ be the total score for a given trajectory $\eta : [1:N] \to [1:B]$ as defined in (8.60). Specify an algorithm based on dynamic programming and backtracking (similar to the Viterbi algorithm in Table 5.2) for determining the score-maximizing trajectory η^{DP} (see (8.61)).

Exercise 8.12. Fixing a matrix $H \in \mathbb{R}^{R \times N}$, let $\varphi^H : \mathbb{R}^D \to \mathbb{R}$ with $D := KR$ be defined by $\varphi^H(W) := \|V - WH\|^2$ for $W \in \mathbb{R}^{K \times R}$ (see (8.80)). Compute the gradient of φ^H (similar to the calculation of the gradient of φ^H in (8.73) to (8.78)). From this, derive the update rule as specified in (8.81).

Exercise 8.13. Show that, in the case of a "perfect" factorization $V = WH$, the matrices W and H are a fixed point of the multiplicative update rules (8.83) and (8.85).

Exercise 8.14. Let V be a $(K \times N)$ matrix with nonnegative entries. As in Figure 8.20a, we consider in this exercise an exact matrix factorization $V = WH$ with a nonnegative $(K \times R)$ matrix W and a nonnegative $(R \times N)$ matrix H. In the following examples, we have $N = 7$ and $K = 9$. Determine for each of the two matrices at least two decompositions $V = WH$ using $R = 3$:

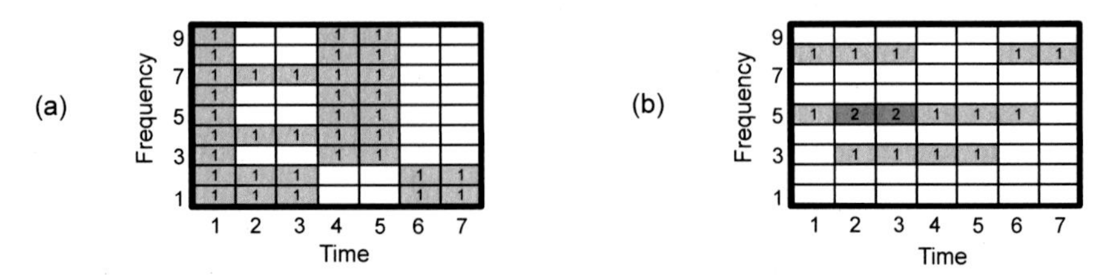

Explain why in these examples there are no exact factorizations when using $R = 2$.

Index

M. Müller, *Fundamentals of Music Processing*, https://doi.org/10.1007/978-3-030-69808-9

Printed in the United States
by Baker & Taylor Publisher Services